MW00444108

Pharmaceutical
Statistics

**For information on volumes 1–151 in the *Drugs and Pharmaceutical Science* Series,
Please visit www.informahealthcare.com**

FIFTH EDITION

Pharmaceutical Statistics

Practical and Clinical Applications

Sanford Bolton
Consultant
Tucson, Arizona, USA

Charles Bon
Biostudy Solutions, LLC
Wilmington, North Carolina, USA

informa
healthcare

New York   London

Informa Healthcare USA, Inc.
52 Vanderbilt Avenue
New York, NY 10017

© 2010 by Informa Healthcare USA, Inc.
Informa Healthcare is an Informa business

No claim to original U.S. Government works
Printed in the United States of America on acid-free paper
10 9 8 7 6 5 4 3 2 1

International Standard Book Number-10: 1-4200-7422-9
International Standard Book Number-13: 978-1-4200-7422-2 (Hardcover)

Library of Congress Cataloging-in-Publication Data

Bolton, Sanford, 1929–
    Pharmaceutical statistics : practical and clinical applications / Sanford Bolton,
Charles Bon. – 5th ed.
        p. ; cm. – (Drugs and the pharmaceutical sciences ; 203)
    Includes bibliographical references and index.
    ISBN-13: 978-1-4200-7422-2 (hardcover : alk. paper)
    ISBN-10: 1-4200-7422-9 (hardcover : alk. paper)   1. Pharmacy–Statistical methods.
I. Bon, Charles, 1949–   II. Title.   III. Series: Drugs and the pharmaceutical sciences ; 203.
    [DNLM: 1. Pharmacy–methods–Laboratory Manuals.   2. Statistics as Topic–Laboratory
Manuals.   W1 DR893B v.203 2009 / QV 25 B694p 2009]
    RS57.B65 2009
    615′.1072–dc22
                                                                    2009039659

For Corporate Sales and Reprint Permission call 212-520-2700 or write to: Sales Department, 52 Vanderbilt Avenue, 7th floor, New York, NY 10017.

Visit the Informa Web site at
www.informa.com

and the Informa Healthcare Web site at
www.informahealthcare.com

To my wife, Phyllis

always present,
always sensitive,
always inspirational

—S. B.

To Sanford Bolton
my mentor who kindled my love of statistics,
and to my wife, Marty,
who did the same for the other areas of my life

—C. B.

Preface

This is the fifth edition of *Pharmaceutical Statistics*. The first edition was published 25 years ago when there were no statistical texts, as far as I know, which were directed toward nonstatistician researchers in academia or the pharmaceutical industry. Although, such a book was not immediately recognized as being an important adjunct to pharmaceutical research, soon after its publication, the passage of time has clearly confirmed the need for a statistics book that is useful for the pharmaceutical scientist. The practical examples with a discussion of the pharmaceutical and clinical consequences have helped to give the pharmaceutical researcher another dimension.

When I first wrote this book in the early 1980s, using a typewriter and two fingers, one of my aims was to document my experience and have a book that could be my personal reference. In each new edition, I have added new material based on new experiences that I think will be useful to the pharmaceutical community as well as to enhance the book as my own reference.

This new edition has some new features. We have expanded some of the tables in the appendix to make them more complete. A more detailed explanation of one- and two-sided statistical tests and when they are applicable has been included. We have updated some of the material related to clinical trials. We have updated statistical applications to bioequivalence, as well as various designs used in bioequivalence studies. A program to calculate the number of subjects in bioequivalence trials under a number of assumptions has been added to the disk accompanying the book. We have also added some new material explaining in more detail the assumptions and applications of nonparametric methods, including application of the binomial distribution to put upper confidence limits on the proportion of successes and failures in a sample. We have included the application of confidence intervals for a ratio, using a method based on Fieller's Theorem. An interesting relationship between the mean and median of a sample is included, with a derivation.

Finally, we have done our best to remove typos and any errors that we have discovered from the fourth edition. Unfortunately, with so much material, it seems impossible to be perfect. However, we strive for perfection, to do our best, and we look forward to comments, criticisms, and ideas from our readers to improve the book, or include new material for the sixth edition.

Before leaving this introduction, again I give thanks to my teachers, my students, my colleagues, my readers, and my work with pharmaceutical problems from pharmaceutical firms of all sizes and shapes that continue to challenge and teach me.

I want to acknowledge those who have helped me both as a person and scientist, and helped me grow. In particular, I owe debts of gratitude to two mentors, now deceased, Dr. Takeru Higuchi and Dr. John Fertig. I acknowledge the institutions that encouraged me to write this book, and allowed me to apply the knowledge to apply statistical applications to pharmaceutical problems, that is, University of Wisconsin, Columbia University and St. John's University in Queens, NY. Finally, thanks to my family, friends, and students, all of whom have made my life more full and have been my family. Special thanks to my wife, Phyllis Bolton, Mohan Sondhi, Salah Ahmed, Spiro Spireas, Charles DiLiberti, Chuck Bon Jerry Reinstein, Robert and Maria Bell, Lama Pema, Mrs. Popoff, and The University of Arizona Guitar Department, to mention only a few.

Sanford Bolton

Contents

** A more advanced topic.

** A more advanced topic.

** A more advanced topic.

1 | Basic Definitions and Concepts

Statistics has its own vocabulary. Many of the terms that comprise statistical nomenclature are familiar: some commonly used in everyday language, with perhaps, somewhat different connotations. Precise definitions are given in this chapter so that no ambiguity will exist when the words are used in subsequent chapters. Specifically, such terms as *discrete* and *continuous variables, frequency distribution, population, sample, mean, median, standard deviation, variance, coefficient of variation (CV), range, accuracy,* and *precision* are introduced and defined. The methods of calculation of different kinds of means, the median, standard deviation, and range are also presented. When studying any discipline, the initial efforts are most important. The first chapters of this book are important in this regard. Although most of the early concepts are relatively simple, a firm grasp of this material is essential for understanding the more difficult material to follow.

1.1 VARIABLES AND VARIATION

Variables are the measurements, the values, which are characteristic of the data collected in experiments. These are the data that will usually be displayed, analyzed, and interpreted in a research report or publication. In statistical terms, these observations are more correctly known as *random variables*. Random variables take on values, or numbers, according to some corresponding probability function. Although we will wait until chapter 3 to discuss the concept of probability, for the present we can think of a random variable as the typical experimental observation that we, as scientists, deal with on a daily basis. Because these measurements may take on different values, repeat measurements observed under apparently identical conditions do not, in general, give the identical results (i.e., they are usually not exactly reproducible). Duplicate determinations of serum concentration of a drug one hour after an injection will not be identical no matter if the duplicates come from (a) the same blood sample or (b) from separate samples from two different persons or (c) from the same person on two different occasions. Variation is an inherent characteristic of experimental observations. To isolate and to identify particular causes of variability require special experimental designs and analysis. Variation in observations is due to a number of causes. For example, an assay will vary depending on

1. the instrument used for the analysis;
2. the analyst performing the assay;
3. the particular sample chosen;
4. unidentified, uncontrollable background error, commonly known as "noise."

This inherent variability in observation and measurement is a principal reason for the need of statistical methodology in experimental design and data analysis. In the absence of variability, scientific experiments would be short and simple: interpretation of experimental results from well-designed experiments would be unambiguous. In fact, without variability, single observations would often be sufficient to define the properties of an object or a system. Since few, if any, processes can be considered absolutely invariant, statistical treatment is often essential for summarizing and defining the nature of data, and for making decisions or inferences based on these variable experimental observations.

1.1.1 Continuous Variables

Experimental data come in many forms.* Probably the most commonly encountered variables are known as *continuous variables*. A continuous variable is one that can take on *any* value within some range or interval (i.e., within a specified lower and upper limit). The limiting factor for the total number of possible observations or results is the sensitivity of the measuring instrument. When weighing tablets or making blood pressure measurements, there are an infinite number of possible values that can be observed if the measurement could be made to an unlimited number of decimal places. However, if the balance, for example, is sensitive only to the nearest milligram, the data will appear as discrete values. For tablets targeted at 1 g and weighed to the nearest milligram, the tablet weights might range from 900 to 1100 mg, a total of 201 possible integral values (900, 901, 902, 903, . . ., 1098, 1099, 1100). For the same tablet weighed on a more sensitive balance, to the nearest 0.1 mg, values from 899.5 to 1100.4 might be possible, a total of 2010 possible values, and so on.

Often, continuous variables cannot be easily measured but can be ranked in order of magnitude. In the assessment of pain in a clinical study of analgesics, a patient can have a continuum of pain. To measure pain on a continuous numerical scale would be difficult. On the other hand, a patient may be able to differentiate slight pain from moderate pain, moderate pain from severe pain, and so on. In analgesic studies, scores are commonly assigned to pain severity, such as no pain = 0, slight pain = 1, moderate pain = 2, and severe pain = 3. Although the scores cannot be thought of as an exact characterization of pain, the value 3 does represent more intense pain than the values 0, 1, or 2. The scoring system above is a representation of a continuous variable by discrete "scores" that can be rationally ordered or ranked from low to high. This is commonly known as a rating scale, and the ranked data are on an ordinal scale. The rating scale is an effort to quantify a continuous, but subjective, variable.

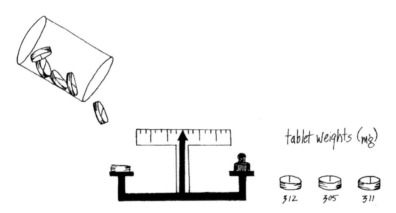

Tablet weights: an example of a variable measurement (a random variable).

1.1.2 Discrete Variables

In contrast to continuous variables, *discrete variables* can take on a countable number of values. These kinds of variables are commonly observed in biological and pharmaceutical experiments and are exemplified by measurements such as the number of anginal episodes in one week or the number of side effects of different kinds after drug treatment. Although not continuous, discrete data often have values associated with them that can be numerically ordered according to their magnitude, as in the examples given earlier of a rating scale for pain and the number of anginal episodes per week.

Discrete data that can be named (nominal), categorized into two or more classes, and counted are called categorical variables, or *attributes*; for example, the attributes may be different

---

* For a further discussion of different kinds of variables, see section 15.1.

side effects resulting from different drug treatments or the presence or absence of a defect in a finished product. These kinds of data are frequently observed in clinical and pharmaceutical experiments and processes. A finished tablet classified in quality control as "defective" or "not defective" is an example of a categorical or attribute type of variable. In clinical studies, the categorization of a patient by sex (male or female) or race is a classification according to attributes. When calculating ED_{50} or LD_{50}, animals are categorized as "responders" or "nonresponders" to various levels of a therapeutic agent, a categorical response. These examples describe variables that cannot be ordered. A male is not associated with a higher or lower numerical value than a female.

Continuous variables can always be classified into discrete classes where the classes are ordered. For example, patients can be categorized as "underweight," "normal weight," or "overweight" based on criteria such as those listed in Metropolitan Life Insurance tables of "Desirable Weights for Men and Women" [l]. In this example, "overweight" represents a condition that is greater than "underweight."

Thus we can roughly classify data as

1. continuous (blood pressure, weight);
2. discrete, associated with numbers and ordered (number of anginal episodes per week);
3. attributes: categorical, ordered (degree of overweight);
4. attributes: categorical, not ordered (male or female).

Classification by attributes: patients categorized by weight.

Underweight   Normal weight   Overweight

1.2 FREQUENCY DISTRIBUTIONS AND CUMULATIVE FREQUENCY DISTRIBUTIONS

1.2.1 Frequency Distributions

An important function of statistics is to facilitate the comprehension and meaning of large quantities of data by constructing simple data summaries. The *frequency distribution* is an example of such a data summary, a table or categorization of the frequency[†] of occurrence of variables in various class intervals. Sometimes a frequency distribution of a set of data is simply called a "distribution." For a sampling of continuous data, in general, a frequency distribution is constructed by classifying the observations (variables) into a number of discrete intervals. For categorical data, a frequency distribution is simply a listing of the number of observations in each class or category, such as 20 males and 30 females entered in a clinical study. This procedure results in a more manageable and meaningful presentation of the data.

[†] The frequency is the number of observations in a specified interval or class: for example, tablets weighing between 300 and 310 mg, or the number of patients who are female.

Table 1.1 Serum Cholesterol Changes (mg%) for 156 Patients After Administration of a Drug Tested for Cholesterol-Lowering Effect[a]

17	−12	25	−37	−29	−39
−22	0	−22	−63	34	−31
−64	−12	−49	5	−8	33
−50	−7	16	−11	−38	−17
0	−9	−21	1	2	−30
−32	−34	−14	−18	5	6
24	−6	−49	−8	−49	−37
−25	−12	14	10	−41	−66
−31	35	21	−19	−27	17
−6	−17	−6	1	−28	40
−31	17	−54	−27	−16	16
−44	10	−3	−3	5	6
−19	9	−10	−20	−9	−8
−10	−11	11	−39	19	−32
4	−15	−18	35	6	20
46	24	−27	−19	5	−60
27	23	−22	−1	12	−27
−13	−39	39	−34	−97	−26
38	14	−47	8	26	−15
−62	12	−53	11	21	−47
−54	−11	−5	0	55	34
−69	−11	−44	20	−50	19
0	−25	−24	−4	14	2
−34	16	−23	−71	−58	9
9	2	−2	−58	13	14
17	−13	−22	−3	−17	1

[a] A negative number means a decrease and a positive number means an increase.

Table 1.1 is a tabulation of serum cholesterol changes resulting from the administration of a cholesterol-lowering agent to a group of 156 patients. The data are presented in the order in which results were reported from the clinic.

A frequency distribution derived from the 156 cholesterol values is shown in Table 1.2. This table shows a tabulation of the frequency, or number, of occurrences of values that fall into the various class intervals of "serum cholesterol changes." Clearly, the condensation of the data as shown in the frequency distribution in Table 1.2 allows for a better "feeling" of the experimental results than do the raw data represented by the individual 156 results. For example, one can readily see that most of the patients had a lower cholesterol value in response to the drug (a negative change) and that most of the data lie between −60 and +19 mg%.

When constructing a frequency distribution, two problems must be addressed. The first problem is how many classes or intervals should be constructed, and the second problem is the specification of the width of each interval (i.e., specifying the upper and lower limit of each interval). There are no definitive answers to these questions. The choices depend on the nature

Table 1.2 Frequency Distribution of Serum Cholesterol Changes

Class interval		Frequency
−100 to −81	(−100.5 to −80.5)	1
−80 to −61	(−80.5 to −60.5)	6
−60 to −41	(−60.5 to −40.5)	16
−40 to −21	(−40.5 to −20.5)	31
−20 to −1	(−20.5 to −0.5)	40
+0 to +19	(−0.5 to + 19.5)	43
+20 to +39	(+19.5 to +39.5)	16
+40 to +59	(+39.5 to +59.5)	3

Data taken from Table 1.1.

Table 1.3 Frequency Distribution of Serum Cholesterol
Changes Using 16 Class Intervals

Class interval	Frequency
−100 to −91	1
−90 to −81	0
−80 to −71	1
−70 to −61	5
−60 to −51	6
−50 to −41	10
−40 to −31	14
−30 to −21	17
−20 to −11	22
−10 to −1	18
0 to +9	22
−10 to +19	21
+20 to +29	9
+30 to +39	7
+40 to +49	2
+50 to +59	1

of the data and good judgment. The number of intervals chosen should result in a table that considerably improves the readability of the data. The following rules of thumb are useful to help select the intervals for a frequency table:

1. Choose intervals that have significance in relation to the nature of the data. For example, for the cholesterol data, intervals such as 18 to 32 would be cumbersome and confusing. Intervals of width 10 or 20, such as those in Tables 1.2 and 1.3, are more easily comprehended and manipulated arithmetically.
2. Try not to have too many empty intervals (i.e., intervals with no observations). The half of the total number of intervals that contain the least number of observations should contain at least 10% of the data. The intervals with the least number of observations in Table 1.2 are the first two intervals (−100 to −81 and −80 to −61) and the last two intervals (+ 20 to +39 and +40 to +59) (one-half of the eight intervals), which contain 26% or 17% of the 156 observations.
3. Eight to twenty intervals are usually adequate.

Table 1.3 shows the same 156 serum cholesterol changes in a frequency table with 16 intervals. Which table gives you a better feeling for the results of this study, Table 1.2 or Table 1.3? (See also Exercise Problem 3.)

The width of all the intervals, in general, should be the same. This makes the table easy to read and allows for simple computations of statistics such as the mean and standard deviation. The intervals should be mutually exclusive so that no ambiguity exists when classifying values. In Tables 1.2 and 1.3, we have defined the intervals so that a value can be categorized only in one class interval. In this way, we avoid problems that can arise when observations are exactly equal to the boundaries of the class intervals. If the class intervals were defined so as to be continuous, such as −100 to −90, −90 to −80, −80 to −70, and so on, one must define the class to which a borderline value belongs, either the class below or the class above, a priori. For example, a value of −80 might be defined to be in the interval −80 to −70.

Another way to construct the intervals is to have the boundary values have one more "significant figure" than the actual measurements so that none of the values can fall on the boundaries. The extra figure is conveniently chosen as 0.5. In the cholesterol example, measurements were made to the nearest mg%; all values are whole numbers. Therefore, two adjacent values can be no less different than 1 mg%, +10, and +11, for example. The class intervals could then have a decimal of 0.5 at the boundaries, which means that no value can fall exactly on a boundary value. The intervals in parentheses in Table 1.2 were constructed in this manner. This categorization, using an extra figure that is halfway between the two closest possible values, makes sense from another point of view. After rounding off, a value of +20 can be considered to be between 19.5 and 20.5, and would naturally be placed in the interval 19.5 to 39.5, as shown in Table 1.2.

1.2.2 Stem-and-Leaf Plot

An expeditious and compact way of summarizing and tabulating large amounts of data, by hand, known as the *stem-and-leaf method* [2], is best illustrated with an example. We will use the data from Table 1.1 to demonstrate the procedure.

An ordered series of integers is conveniently chosen (see below) to cover the range of values. The integers consist of the first digit(s) of the data, as appropriate, and are arranged in a vertical column, the "stem." By adding another digit(s) to one of the integers in the stem column (the "leaves"), we can tabulate the data in class intervals as in a frequency table. For the data of Table 1.1, the numbers range from approximately -100 to $+60$. The stem is conveniently set up as follows:

−10	−7	−4	−1	+1	+4
−9	−6	−3	−0	+2	+5
−8	−5	−2	+0	+3	+6

In this example, the stem is the first digit(s) of the number and the leaf is the last digit. The first value in Table 1.1 is 17. Therefore, we place a 7 (leaf) next to the $+1$ in the stem column. The next value in Table 1.1 is -22. We place a 2 (leaf) next to -2 in the stem column, and so on. Continuing this process for each value in Table 1.1 results in the following stem-and-leaf diagram.

−10																							
−9	7																						
−8																							
−7	1																						
−6	4	2	9	3	6	0																	
−5	0	4	4	3	8	0	8																
−4	4	9	9	7	4	1	9	7															
−3	2	1	1	4	4	9	7	9	4	8	9	1	0	7	2								
−2	2	5	5	2	1	7	2	4	3	2	7	0	9	7	8	6	7						
−1	9	0	3	2	2	2	7	1	5	1	1	3	4	0	8	1	8	9	9	6	7	7	5
−0	6	7	9	6	6	3	5	2	8	3	1	4	3	8	9	8							
+0	0	4	0	9	0	9	2	5	1	1	8	0	2	5	5	6	5	6	6	2	9	1	
+1	7	7	0	7	4	2	6	6	4	1	0	1	9	2	6	4	3	7	6	9	4		
+2	4	7	4	3	5	1	0	1	0														
+3	8	9	9	5	4	3	4																
+4	6	0																					
+5	5																						
+6																							

This is a list of all the values in Table 1.1. The distribution of this data set is easily visualized with no further manipulation. However, if necessary, one can easily construct a frequency distribution from the configuration of data resulting from the stem-and-leaf tabulation. (Note that all categories in this particular example can contain as many as 10 different numbers except for the -0 category, which can contain only 9 numbers, -1 to -9 inclusive. This "anomaly" occurs because of the presence of both positive and negative values and the value 0. In this example, 0 is arbitrarily assigned a positive value.) In addition to the advantages of this tabulation noted above, the data are in the form of a histogram, which is a common way of graphically displaying data distributions (see chap. 2).

Table 1.4 Frequency Distribution of Tablet Potencies

Potency (mg)	W_i[a]	Frequency X_i[b]
89.5–90.5	1	90
90.5–91.5	0	91
91.5–92.5	2	92
92.5–93.5	1	93
93.5–94.5	5	94
94.5–95.5	1	95
95.5–96.5	2	96
96.5–97.5	7	97
97.5–98.5	10	98
98.5–99.5	8	99
99.5–100.5	13	100
100.5–101.5	17	101
101.5–102.5	13	102
102.5–103.5	9	103
103.5–104.5	0	104
104.5–105.5	0	105
105.5–106.5	5	106
106.5–107.5	4	107
107.5–108.5	0	108
108.5–109.5	0	109
109.5–110.5	2	110
	$\sum W_i = 100$	

[a] W_i is the frequency.
[b] X_i is the midpoint of the interval.

1.2.3 Cumulative Frequency Distributions

A large set of data can be conveniently displayed using a cumulative frequency table or plot. The data are first ordered and, with a large data set, may be arranged in a frequency table with n class intervals. The frequency, often expressed as a proportion (or percentage), of values equal to or less than a given value, X_i, is calculated for each specified value of X_i, where X_i is the upper point of the class interval ($i = 1$ to n). A plot of the cumulative proportion versus X can be used to determine the proportion of values that lie in some interval, that is, between some specified limits. The cumulative distribution for the tablet potencies in Table 1.4 is shown in Table 1.5 and

Table 1.5 Cumulative Frequency Distribution of Tablet Potencies

Potency, X_t(mg)[a]	Cumulative frequency ($\leq X$)	Cumulative proportion
90.5	1	0.01
92.5	3	0.03
93.5	4	0.04
94.5	9	0.09
95.5	10	0.10
96.5	12	0.12
97.5	19	0.19
98.5	29	0.29
99.5	37	0.37
100.5	50	0.50
101.5	67	0.67
102.5	80	0.80
103.5	89	0.89
106.5	94	0.94
107.5	98	0.98
110.5	100	1.00

Data taken from Table 1.4.
[a] X_t is the upper point of the class interval in Table 1.4, excluding null intervals.

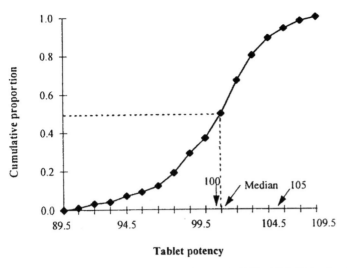

Tablet potency

Figure 1.1 Cumulative proportion plot for data in Table 1.5 (tablet potencies).

plotted in Figure 1.1. The cumulative proportion represents the proportion of values less than or equal to X_i (e.g., 29% of the values are less than or equal to 98.5). Also, for example, from an inspection of Figure 1.1, one can estimate the proportion of tablets with potencies between 100 and 105 mg inclusive, equal to approximately 0.48 (0.91 at 105 mg minus 0.43 at 100 mg). (See also Exercise Problem 5.)

The cumulative distribution is a very important concept in statistics. In particular, the application of the cumulative normal distribution, which is concerned with continuous data, will be discussed in chapter 3. A more detailed account of the construction and interpretation of frequency distributions is given in Refs. [3–5].

1.3 SAMPLE AND POPULATION

Understanding the concepts of samples and populations is important when discussing statistical procedures. *Samples* are usually a relatively small number of observations taken from a relatively large *population* or universe. The sample values are the observations, the data, obtained from the population. The population consists of data with some clearly defined characteristic(s). For example, a population may consist of all patients with a particular disease, or tablets from a production batch. The sample in these cases could consist of a selection of patients to participate in a clinical study, or tablets chosen for a weight determination. The sample is only part of the available data. In the usual experimental situation, we make observations on a relatively small sample in order to make inferences about the characteristics of the whole, the population. The totality of available data is the population or universe. When designing an experiment, the population should be clearly defined so that samples chosen are representative of the population. This is important in clinical trials, for example, where inferences to the treatment of disease states are crucial. The exact nature or character of the population is rarely known, and often impossible to ascertain, although we can make assumptions about its properties. Theoretically, a population can be finite or infinite in the number of its elements. For example, a finished package contains a finite number of tablets; all possible tablets made by a particular process, past, present, and future, can be considered infinite in concept. In most of our examples, the population will be considered to be infinite, or at least very large compared to the sample size. Table 1.6 shows some populations and samples, examples that should be familiar to the pharmaceutical scientist.

1.3.1 Population Parameters and Sample Statistics

"Any measurable characteristic of the universe is called a *parameter*" [6]. For example, the average weight of a batch of tablets or the average blood pressure of hypertensive persons in the United States are parameters of the respective populations. Parameters are generally

Table 1.6 Examples of Samples and Populations

Population	Sample
Tablet batch	Twenty tablets taken for content uniformity
Normal males between ages 18 and 65 years available to hospital	Twenty-four subjects selected for a phase I clinical study
Sprague–Dawley weaning rats	100 rats selected to test possible toxic effects of a new drug candidate
Analysts working for company X	Three analysts from a company to test a new assay method
Persons with diastolic blood pressure between 105 and 120 mm Hg in the United States	120 patients with diastolic pressure between 105 and 120 mm Hg to enter clinical study to compare two antihypertensive agents
Serum cholesterol levels of one patient	Blood samples drawn once a week for 3 months from a single patient

denoted by Greek letters; for example, the mean of the population is denoted as μ. Note that parameters are characteristic of the population, and are values that are usually unknown to us.

Quantities derived from the sample are called *sample statistics*. Corresponding to the true average weight of a batch of tablets is the average weight for the small sample taken from the population of tablets. We should be very clear about the nature of samples. Emphasis is placed here (and throughout this book) on the variable nature of such sample statistics. A parameter, for example, the mean weight of a batch of tablets, is a fixed value; it does not vary. Sample statistics are variable. Their values depend on the particular sample chosen and the variability of the measurement. The average weight of 10 tablets will differ from sample to sample because

1. we choose 10 different tablets at each sampling;
2. the balance (and our ability to read it) is not exactly reproducible from one weighing to another.

An important part of the statistical process is the characterization of a population by estimating its parameters. The parameters can be estimated by evaluating suitable sample statistics. The reader will probably have little trouble in understanding that the average weight of a sample of tablets (a sample statistic) estimates the true mean weight (a parameter) of the batch. This concept is elucidated and expanded in the remaining sections of this chapter.

1.4 MEASURES DESCRIBING THE CENTER OF DATA DISTRIBUTIONS

1.4.1 The Average
Probably the most familiar statistical term in popular use is the *average*, denoted by $\overline{X}$ (X bar). The average is also commonly known as the *mean* or *arithmetic average*. The average is a summarizing statistic and is a measure of the center of a distribution, particularly meaningful if the data are symmetrically distributed below and above the average. Symbolically, the mean is equal to

$$\frac{\sum_{i=1}^{N} X_i}{N} \tag{1.1}$$

the sum of the observations divided by the number of observations. $\sum_{i=1}^{N} X_i$ is the sum of the N values, each denoted by X_i, $(X_1, X_2, \ldots, X_n)$, where i can take on the values 1, 2, 3, 4, $\ldots$, n.[‡]

[‡] For the most part, when using summation notation in this book, we will not use the full notation, such as $\sum_{i=1}^{N} X_i$, but rather $\sum X$, the i notation being implied, unless otherwise stated.

The average of the values 7, 11, 6, 5, and 4 is

$$\frac{7 + 11 + 6 + 5 + 4}{5} = 6.6.$$

This is an unweighted average, each value contributing equally to the average.

1.4.2 Other Kinds of Averages

When averaging observations, we usually think of giving each observation equal weight. The usual formula for the average ($\sum X_i/N$) gives each value equal weight. If we believe that the values to be averaged do not carry the same weight, then we should use a weighted average. The average of three cholesterol readings 210, 180, and 270 is $(660)/3 = 220$. Suppose that the value of 210 is really the average of two values (200 and 220), we might want to consider giving this value twice as much weight as the other two values, resulting in an average

$$\frac{210 + 210 + 180 + 270}{4} = 217.5$$

or

$$\frac{2 \times 210 + 180 + 270}{2 + 1 + 1} = 217.5.$$

The formula for a weighted average, $\overline{X}_w$ is

$$\frac{\sum W_i X_i}{\sum W_i}, \tag{1.2}$$

where W_i is the weight assigned to the value X_i. The weights for the calculation of a weighted average are often the number of observations associated with the values X_i. This concept is illustrated for the calculation of the average for data categorized in the form of a frequency distribution. Table 1.4 shows a frequency distribution of 100 tablet potencies. The frequency is the number of observations of tablets in a given class interval, as defined previously. The frequency or number of tablets in a "potency" interval is the *weight* used in the computation of the weighted average. The value X associated with the weight is taken as the midpoint of the interval; for example, for the first interval, 89.5 to 90.5, $X_1 = 90$. Applying Eq. (1.2), the weighted average is $\sum W_i X_i / \sum W_i$:

$$\frac{1 \times 90 + 0 \times 91 + 2 \times 92 + 1 \times 93 + 5 \times 94 + \cdots + 4 \times 107 + 2 \times 110}{1 + 0 + 2 + 1 + 5 + \cdots + 4 + 2},$$

which equals $10{,}023/100 = 100.23$ mg.

It is not always obvious when to use a weighted average, and one should have a substantial knowledge of the circumstances and nature of the data in order to make this decision. In the previous example, if the 210 value (the average of two observations) came from one patient and the other values were single observations from two different patients, one may not want to use a weighted average. The reasoning in this example may be that this average is meant to represent the true average cholesterol of these three patients, each with different cholesterol levels. There does not seem to be a good reason to give twice as much weight to the "210" patient because that patient happened to have two readings. This may be more clearly seen if the patient had 100 readings and the other two patients only a single reading. The unweighted average would be very close to the average of the patient with the 100 readings and would not represent the average of the three patients. In this example, the average of three values (one value for each patient) would be a better representation of the average, $(210 + 180 + 270)/3 = 220$.

Table 1.7 Distribution of Particle Size of Powder

Midpoint Sieve size	Log sieve Size (Y)	Weight (W)	(WT) × (Y)
10[a]	2.3026	19.260	44.3478
30	3.4012	24.015	81.6797
50	3.1920	22.240	87.0034
70	4.2485	7.525	31.9699
90	4.4998	6.515	29.3163
150[b]	5.0106	20.445	102.4424
Sum		100.00	376.7595

[a] 10 is for sieve size less than 20, that is, between 0 and 20.
[b] 150 is substituted for >100.

If the four values were obtained from one patient where the 210 average came from one laboratory and the other two values from two different laboratories, the following reasoning might be useful to understand how to treat the data properly. If the different laboratories used the same analytical method that was expected to yield the same result, a weighted average would be appropriate (give twice the weight to the 210 value). If the laboratories have different methods that give different results for the same sample, an unweighted average may be more appropriate.

The distribution of particle size of a powdered blend is often based on the logarithm of the particle size (see sect. 10.1.1). The quantity (weight) of powder in a given interval of particle size may be considered a weighting factor when computing the average particle size. Table 1.7 shows the particle size distribution (frequency distribution) of a powder, where the class intervals are based on the logarithm of the sieve size fractions. The weighted average can be calculated as

$$\overline{X}_w = \frac{\sum \text{weight} \times (\log \text{ sieve size})}{\sum (\text{weights})}. \tag{1.3}$$

The weight is the percentage of powder found for a given particle size (or interval of sieve sizes). Note that for this example, the sieve size is taken as the midpoint of the untransformed class (sieve size) interval.

From Eq. (1.3), weighted average $= 376.7595/100.0 = 3.7676$. Since sieve size is in log terms, the antilog of $3.7676 = 43.3$ is an estimate of the average particle size. (For *more* advanced methods of estimating the parameters of particle size distributions, see Refs. [7,8].)

The calculation of the variance of a weighted average is dependent on the nature of the weighted average and an experienced statistician should be consulted if necessary (see SAS manual for options). This more advanced concept is discussed further in section 1.5.5.

Two other kinds of averages that are sometimes found in statistical procedures are the geometric and harmonic means. The *geometric mean* is defined as

$$\sqrt[n]{X_1 \cdot X_2 \cdot X_3 \cdots X_n}$$

or the nth root of the product of n observations.

The geometric mean of the numbers 50, 100, and 200 is

$$\sqrt[3]{50 \cdot 100 \cdot 200} = \sqrt[3]{1,000,000} = 100.$$

If a measurement of population growth shows 50 at time 0, 100 after one day, and 200 after two days, the geometric mean (100) is more meaningful than the arithmetic mean (116.7). The geometric mean is always less than or equal to the arithmetic mean, and is meaningful for data with logarithmic relationships. (See also sect. 15.1.1.) Note that the logarithm of $\sqrt[3]{50 \cdot 100 \cdot 200}$ is equal to $[\log 50 + \log 100 + \log 200]/3$, which is the average of the logarithms

Figure 1.2   Average illustrated as balancing forces.

of the observations. The geometric mean is the antilog of this average (the antilog of the average is 100).

The harmonic mean is the appropriate average following a reciprocal transformation (chap. 10). The harmonic mean is defined as

$$\frac{N}{\sum 1/X_i}.$$

For the three observations 2, 4, and 8 ($N = 3$), the harmonic mean is

$$\frac{3}{1/2 + 1/4 + 1/8} = 3.429.$$

1.4.3   The Median

Although the average is the most often used measure of centrality, the *median* is also a common measure of the center of a data set. When computing the average, very large or very small values can have a significant effect on the magnitude of the average. For example, the average of the numbers 0, 1, 2, 3, and 34 is 8. The arithmetic average acts as the fulcrum of a balanced beam, with weights placed at points corresponding to the individual values, as shown in Figure 1.2. The single value 34 needs four values, 0, 1, 2, and 3, as a counterbalance. Also, the median may be a more appropriate measure of central tendency for skewed distributions such as the log-normal distribution (see sect. 10.1.1).

The *median* represents the center of a data set, without regard for the distance of each point from the center. The median is the value that divides the data in half, half the values being less than and half the values greater than the median value. The median is easily obtained when the data are ranked in order of magnitude. The median of an odd number of *different*[§] observations is the middle value. For $2N + 1$ values, the median is the $(N + 1)$th ordered value. The median of the data 0, 1, 2, 3, and 34 is the third (middle) value, $2(N = 2, 2N +1 = 5$ values). By convention, the median for an even number of data points is considered to be the average of the two center points. For example, the median of the numbers, 0, 1, 2, and 3 is the average of the center points, 1 and 2, equal to $(1 + 2)/2 = 1.5$. The median is often used as a description of the center of a data set when the data have an asymmetrical distribution. In the presence of either extremely high or extremely low outlying values, the median appears to describe the distribution better than does the average. The median is more stable than the average in the presence of extreme observations. A very large or very small value has the same effect on the calculation of the median as any other value, larger or smaller than the median, respectively. On the other hand, as noted previously, very large and very small values have a significant effect on the magnitude of the mean.

The distribution of individual yearly incomes, which have relatively few very large values (the multimillionaires), serves as a good example of the use of the median as a descriptive statistic. Because of the large influence of these extreme values, the average income is higher than one might expect on an intuitive basis. The median income, which *is* less than the average income, represents a figure that is readily interpreted; that is, one-half of the population earns more (or less) than the median income.

The distribution of particle sizes for bulk powders used in pharmaceutical products *is* often skewed. In these cases, the median is a better descriptor of the centrality of the distribution than

---

§ If the median value is not unique, that is, two or more values are equal to the median, the median is calculated by interpolation (3).

is the mean [9]. The median is less efficient than the mean as an estimate of the center of a distribution; that is, the median is more variable [10]. For most of the problems discussed in this book, we will be concerned with the mean rather than the median as a measure of centrality.

An interesting, but not well documented, relationship between the mean and median shows that for positive numbers, the mean must be greater than half the median. This can be proven simply as follows:

Consider $2N + 1$ numbers whose median is "M" and mean is "m." We will choose an odd number of values so that the median is well defined. The mean, m, is the sum of all the numbers divided by $2N + 1$. Of the $2N + 1$ numbers, $N + 1$ is greater than or equal to the median, M. Therefore, m is greater than or equal to $(N + 1)M/(2N + 1)$. But $(N + 1)/(2N + 1) > \frac{1}{2}$. Therefore, $m > M/2$. Therefore the mean must be greater than half the median.

For example, consider the following extreme example. The data consist of the following values: 1, 1, 1, 999.5, 1000, 10,001,000. The median is 999.5. The mean is 571.8. 571.8 is greater than 999.5/2.

The median is also known as the *50th percentile* of a distribution. To compute percentiles, the data are ranked in order of magnitude, from smallest to largest. The nth percentile denotes a value below which $n\%$ of the data are found, and above which $(100 - n)\%$ of the data are found. The 10th, 25th, and 75th percentiles represent values below which 10%, 25%, and 75%, respectively, of the data occur. For the tablet potencies shown in Table 1.5, the 10th percentile is 95.5 mg; 10% of the tablets contain less than 95.5 mg and 90% of the tablets contain more than 95.5 mg of drug. The 25th, 50th, and 75th percentiles are also known as the first, second, and third quartiles, respectively.

The *mode* is less often used as the central, or typical, value of a distribution. The mode is the value that occurs with the greatest frequency. For a symmetrical distribution that peaks in the center, such as the normal distribution (see chap. 3), the mode, median, and mean are identical. For data skewed to the right (e.g., incomes), which contain a relatively few very large values, the mean is larger than the median, which is larger than the mode (Fig. 10.1).

1.5 MEASUREMENT OF THE SPREAD OF DATA

The mean (or median) alone gives no insight or information about the spread or range of values that comprise a data set. For example, a mean of five values equal to 10 may comprise the numbers

0, 5, 10, 15, and 20   or   5, 10, 10, 10, and 15.

The mean, coupled with the *standard deviation* or *range*, is a succinct and minimal description of a group of experimental observations or a data distribution. The standard deviation and the range are measures of the spread of the data; the larger the magnitude of the standard deviation or range, the more spread out the data are. A standard deviation of 10 implies a wider range of values than a standard deviation of 3, for example.

1.5.1 Range

The *range*, denoted as R, is the difference between the smallest and the largest values in the data set. For the data in Table 1.1, the range is 152, from -97 to $+55$ mg%. The range is based on only two values, the smallest and largest, and is more variable than the standard deviation (i.e., it is less stable).

1.5.2 Standard Deviation and Variance

The *standard deviation*, denoted as s.d. or S, is calculated as

$$\sqrt{\frac{\sum (X - \overline{X})^2}{N - 1}}, \tag{1.4}$$

where N is the number of data points (or sample size) and $\sum(X - \overline{X})^2$ is the *sum of squares* of the differences of each value from the mean, $\overline{X}$. The standard deviation is more difficult to calculate than is the range.

Table 1.8 Calculation of the Standard Deviation

X	$\overline{X}$	$X - \overline{X}$	$(X - \overline{X})^2$
101.8	103	−1.2	1.44
103.2	103	0.2	0.04
104.0	103	1.0	1.00
102.5	103	−0.5	0.25
103.5	103	0.5	0.25
$\sum X = 515$			$\sum (X - \overline{X})^2 = 2.98$
s.d. $= \sqrt{\frac{\sum (X - \overline{X})^2}{N - 1}} = \sqrt{\frac{2.98}{4}} = 0.86$			

Consider a group of data points: 101.8, 103.2, 104.0, 102.5, and 103.5. The mean is 103.0. Details of the calculation of the standard deviation are shown in Table 1.8. The difference between each value and the mean is calculated: $X - \overline{X}$. These differences are squared, $(X - \overline{X})^2$, and summed. The sum of the squared differences divided by $N - 1$ is calculated, and the square root of this result is the standard deviation.

With the accessibility of electronic calculators and computers, it is rare, nowadays, to hand compute a mean and standard deviation (or any other calculation, for that matter). Nevertheless, when computing the standard deviation by hand (or with the help of a calculator), a well-known shortcut computing formula is recommended. The shortcut is based on the identity

$$\sum (X - \overline{X})^2 = \sum X^2 - \frac{(\sum X)^2}{N}.$$

Therefore,

$$\text{s.d.} = \sqrt{\frac{\sum X^2 - (\sum X)^2 / N}{N - 1}}, \tag{1.5}$$

where $\sum X^2$ is the sum of each value squared and $(\sum X)^2$ is the square of the sum of all the values $[(\sum X)^2/N$ is also known as the *correction term*]. We will apply this important formula, Eq. (1.5), to the data above to illustrate the calculation of the standard deviation. This result will be compared to that obtained by the more time-consuming method of squaring each deviation from the mean (Table 1.8).

$$\sum (X - \overline{X})^2 = 101.8^2 + 103.2^2 + 104.0^2 + 102.5^2 + 103.5^2 - \frac{515^2}{5} = 2.98.$$

The standard deviation is $\sqrt{2.98/4} = 0.86$, as before.

The *variance* is the square of the standard deviation, often represented as S^2. The variance is calculated as

$$S^2 = \frac{\sum (X - \overline{X})^2}{N - 1}. \tag{1.6}$$

In the example of the data in Table 1.8, the variance, S^2, is

$$\frac{2.98}{4} = 0.745.$$

A question that often puzzles new students of statistics is: Why use $N - 1$ rather than N in the denominator in the expression for the standard deviation or variance [Eqs. (1.4) and (1.6)]?

The *variance of the population*, a parameter traditionally denoted as σ^2 (sigma squared), is calculated as[¶]:

$$\sigma^2 = \frac{\sum(X - \overline{X})^2}{N},$$ (1.7)

where N is the number of all possible values in the population. The use of $N - 1$ rather than N in the calculation of the variance of a *sample* (a sample statistic) makes the sample variance an *unbiased estimate* of the population variance. Because the sample variance is variable (a random variable), in any given experiment, S^2 will not be exactly equal to the true population variance, σ^2. However, in the long run, S^2 (calculated with $N - 1$ in the denominator) will equal σ^2, on the average. "On the average" means that if samples of size N were repeatedly randomly selected from the population, and the variance calculated for each sample, the averages of these calculated variance estimates would equal σ^2. Note that the sample variance is an estimate of the true population variance σ^2.

If S^2 estimates σ^2 on the average, the sample variance is an unbiased estimate of the population variance. It can be proven that the sample variance calculated with $N - 1$ in the denominator is an unbiased estimate of σ^2. To try to verify this fact by repeating exactly the same laboratory or clinical experiment (if the population variance were known) would be impractical. However, for explanatory purposes, it is often useful to illustrate certain theorems by showing what would happen upon repeated sampling from the same population. The concept of the unbiased nature of the sample variance can be demonstrated using a population that consists of three values: 0, 1, and 2. The population variance, $\sum(X - \overline{X})^2/3$, is equal to 2/3 [see Eq. (1.7)]. Using the repeated sample approach noted above, samples of size 2 are repeatedly selected at random from this population. The first choice is replaced before selection of the second choice so that each of the three values has an equal chance of being selected on both the first and second selection. (This is known as *sampling with replacement*.) The following possibilities of samples of size 2 are equally likely to be chosen:

0, 1; 1, 0; 0, 2; 2, 0; 1, 2; 2, 1; 1, 1; 2, 2; 0, 0

The *sample variance*[**] of these nine pairs are $[\sum(X - \overline{X})^2/(N - 1)]$ 0.5, 0.5, 2, 2, 0.5, 0.5, 0, 0, and 0, respectively. The average of the nine equally likely possible variances is

$$\frac{0.5 + 0.5 + 2 + 2 + 0.5 + 0.5 + 0 + 0 + 0}{9} = \frac{6}{9} = \frac{2}{3},$$

which is exactly equal to the population variance. This demonstrates the unbiased character of the sample variance. The sample standard deviation [Eq. (1.4)] is not an unbiased estimate of the *population standard deviation*, σ, which for a finite population is calculated as

$$\sqrt{\frac{\sum(X - \overline{X})^2}{N}}.$$ (1.8)

The observed variance is not dependent on the sample size. The sample variance will equal the true variance "on the average," but the variability of the estimated variance decreases as the sample size increases. The unbiased nature of a *sample estimate* of a *population parameter*, such as the variance or the mean, is a desirable characteristic. $\overline{X}$, the sample estimate of the true population mean, is also an unbiased estimate of the true mean. (The true mean is designated by the Greek letter μ. In general, population parameters are denoted by Greek letters as noted previously.)

[¶] Strictly speaking, this formula is for a population with a finite number of data points.
[**] For samples of size 2, the variance is simply calculated as the square of the difference of the values divided by 2, $d^2/2$. For example, the variance of 0 and 1 is $(1 - 0)^2/2 = 0.5$.

One should be aware that some calculators having a built-in function for calculating the standard deviation use N in the denominator of the formula for the standard deviation. As we have emphasized above, this is correct for the calculation of the population standard deviation (or variance), and will be close to the calculation of the sample standard deviation when N is large.

The value of $N - 1$ is also known as the *degrees of freedom* for the sample (later we will come across situations where degrees of freedom are less than $N - 1$). The concept of degrees of freedom (denoted as d.f.) is very important in statistics, and we will have to know the degrees of freedom for the variance estimates used in statistical tests to be described in subsequent chapters.

Another common misconception is that the standard deviation (or variance) of a sample becomes smaller as the sample size increases. The standard deviation of a sample is an estimate of the true standard deviation. The true standard deviation is a constant and does not change with a change in sample size. However, we can say that the estimate of the true standard deviation as observed in a sample is more reliable and less variable as the sample size increases. But, on the average, the standard deviation of a small or large sample will approximate the true standard deviation. As discussed later in this chapter (sect. 1.5.4), the standard deviation of a mean will decrease with larger sample sizes.

1.5.3 Coefficient of Variation

The variability of data may often be better described as a relative variation rather than as an absolute variation, such as that represented by the standard deviation or range. One common way of expressing the variability, which takes into account its relative magnitude, is the ratio of the standard deviation to the mean, s.d./$\overline{X}$. This ratio, often expressed as a percentage, is called the *coefficient of variation*, abbreviated as CV, or RSD, the relative standard deviation. A CV of 0.1 or 10% means that the s.d. is one-tenth of the mean. This way of expressing variability is useful in many situations. It puts the variability in perspective relative to the magnitude of the measurements and allows a comparison of the variability of different kinds of measurements. For example, a group of rats of average weight 100 g and s.d. of 10 g has the same relative variation (CV) as a group of animals with average weight 70 g and s.d. of 7 g. Many measurements have an almost constant CV, the magnitude of the s.d. being proportional to the mean. In biological data, the CV is often between 20% and 50%, and one would not be surprised to see an occasional CV as high as 100% or more. The relatively large CV observed in biological experiments is due mostly to "biological variation," the lack of reproducibility in living material. On the other hand, the variability in chemical and instrumental analyses of drugs is usually relatively small. Thus it is not unusual to find a CV of less than 1% for some analytical procedures.

1.5.4 Standard Deviation of the Mean (Standard Error of the Mean)

The s.d. is a measure of the spread of a group of individual observations, a measure of their variability. In statistical procedures to be discussed in this book, we are more concerned with making inferences about the mean of a distribution rather than with individual values. In these cases, the variability of the mean rather than the variability of individual values is of interest. The sample mean is a random variable, just as the individual values that comprise the mean are variable. Thus, repeated sampling of means from the same population will result in a distribution of means that has its own mean and s.d.

The *standard deviation of the mean*, commonly known as the *standard error of the mean*, is a measure of the variability of the mean. For example, the average potency of the 100 tablets shown in Table 1.4 may have been determined to estimate the average potency of the population, in this case, a production batch. An estimate of the variability of the mean value would be useful. The mean tablet potency is 100.23 mg and the s.d. is 3.687. To compute the s.d. of the mean (also designated as $S_{\overline{X}}$), we might assay several more sets of 100 tablets and calculate the mean potency of each sample. This repeated sampling would result in a group of means, each composed of 100 tablets, with different values, such as the five means shown in Table 1.9. The s.d. of this group of means can be calculated in the same manner as the individual values are calculated

Table 1.9 Means of Potencies of Five Sets of
100 Tablets Selected from a Production Batch

Sample	Mean potency
1	99.84
2	100.23
3	100.50
4	100.96
5	100.07

[Eq. (1.4)]. The s.d. of these five means is 0.431. We can anticipate that the s.d. of the means will be considerably smaller than the s.d. calculated from the 100 individual potencies. This fact is easily comprehended if one conceives of the mean as "averaging out" the extreme individual values that may occur among the individual data. The means of very large samples taken from the same population are very stable, tending to cluster closer together than the individual data, as illustrated in Table 1.9.

Fortunately, we do not have to perform real or simulated sampling experiments, such as weighing five sets of 100 tablets each, to obtain replicate data in order to estimate the s.d. of means. Statistical theory shows that the s.d. of mean values is equal to the s.d. calculated from the individual data divided by $\sqrt{N}$, where N is the sample size[††]:

$$S_{\overline{X}} = \frac{S}{\sqrt{N}}. \tag{1.9}$$

The s.d. of the numbers shown in Table 1.4 is 3.687. Therefore, the s.d. of the mean for the potencies of 100 tablets shown in Table 1.4 is estimated as $S/\sqrt{N} = 3.687/\sqrt{100} = 0.3687$. This theory verifies our intuition; the s.d. of means is smaller than the s.d. of the individual data points. The student should not be confused by the two estimates of the s.d. of the mean illustrated above. In the usual circumstance, the estimate is derived as $S/\sqrt{N}$ (0.3687 in this example). The data in Table 1.3 were used only to illustrate the concept of a s.d. of a mean. In any event, the two estimates are not expected to agree exactly; after all $S_{\overline{X}}$ is also a random variable and only estimates the true value, $\sigma/\sqrt{N}$.

As the sample size increases, the s.d. of the mean becomes smaller and smaller. We can reduce the s.d. of the mean, $S_{\overline{X}}$, to a very small value by increasing N. Thus means of very large samples hardly vary at all. The concept of the s.d. of the mean is important, and the student will find it well worth the extra effort made to understand the meaning and implications of $S_{\overline{X}}$.

1.5.5 Variance of a Weighted Average[‡‡]
The general formula for the variance of a weighted average is

$$S_w^2 = \frac{(\sum W_i^2 S_i^2)}{(\sum W_i)^2} \tag{1.10}$$

where S_i^2 is the variance of the ith observation. To compute the variance of the weighted mean, we would need to have an estimate of the variance of each observation.

If the weights of the observations are taken to be $1/S_i^2$ (the reciprocal of the variance, a common situation), then $S_w^2 = 1/\sum(1/S_i^2)$. This formula can be applied to the calculation of the variance of the grand average of a group of i means where the variance of the individual observations is constant, equal to S^2. (We know that the variance of the grand average is S^2/N, where $N = \sum n_i$.) The variance of each mean, S_i^2, is S^2/n_i, where n_i is the number of observations

---

[††] The variance of a mean, $S_{\overline{X}}^2$, is S^2/N.
[‡‡] This is a more advanced topic.

in group i. In this example, the weights are considered to be the reciprocal of the variance, and $S_w^2 = 1/\sum(n_i/S^2) = S^2/\sum n_i$. Of course, we need to know S^2 (or have an estimate) in order to calculate (or estimate) the variance of the average. An estimate of the variance, S^2, in this example is $\sum n_i(Y_i - \overline{Y}_w)^2/(N-1)$, where the n_i acts as the weights and N is the number of observations.

The following calculation can be used to estimate the variance where a specified number of observations is available as a measure of the weight (as in a set of means). The variance of a set of weighted data can be estimated as follows:

$$\text{estimated variance} = \frac{\sum W_i\left(Y_i - \overline{Y}_w\right)^2}{\sum W_i - 1}, \tag{1.11}$$

where W_i is the weight associated with Y_i, and $\overline{Y}_w = $ weighted average of Y.

A shortcut formula is

$$\frac{\left[\sum\left(W_iY_i^2\right) - \sum\left(W_iY_i\right)^2/\sum\left(W_i\right)\right]}{\sum W_i - 1}. \tag{1.12}$$

Example:

The diameters of 100 particles were measured with the results shown in Table 1.10.

From Eq. (1.12), the variance is estimated as $[89{,}375 - (2425)^2/100]/99 = 308.8.$ s.d. $= \sqrt{308.8} = 17.6$. The s.d. of the mean is $17.6/\sqrt{100} = 1.76$. Note: The weighted average is $2425/100 = 24.25$.

In this example, it makes sense to divide the corrected sum of squares by $(N-1)$, because this sum of squares is computed using data from 100 particles. In some cases, the computation of the variance is not so obvious.

1.6 CODING

From both a practical and a theoretical point of view, it is useful to understand how the mean and s.d. of a group of numbers are affected by certain arithmetic manipulations, particularly adding a constant to, or subtracting a constant from each value; and multiplying or dividing each value by a constant.

Consider the following data to exemplify the results described below:

$$\boxed{2, 3, 5, 10}$$

Mean $= \overline{X} = 5$

Variance $= S^2 = 12.67$

Standard deviation $= S = 3.56$

Table 1.10   Data for Calculation of Variance of a Weighted Mean

Diameter (m)	Midpoint	Number of particles = weight	Weight × midpoint	Weight × midpoint^2
Y_i	W_i	W_iY_i	$W_iY_i^2$	
0–10	5	25	125	625
10–20	15	35	525	7875
30–40	35	15	525	18,375
40–60	50	25	1250	62,500
Sum		100	2425	89,375

1. Addition or subtraction of a constant will cause the mean to be increased or decreased by the constant, but will not change the variance or s.d. For example, adding $+3$ to each value results in the following data:

> | 5, 6, 8, 13 |
>
> $\overline{X} = 8$
> $S = 3.56$

Subtracting 2 from each value results in

> | 0, 1, 3, 8 |
>
> $\overline{X} = 3$
> $S = 3.56$

This property may be used to advantage when hand calculating the mean and s.d. of very large or cumbersome numbers. Consider the following data:

> | 1251, 1257, 1253, 1255 |

Subtracting 1250 from each value we obtain

> | 1, 7, 3, 5 |
>
> $\overline{X} = 4$
> $S = 2.58$

To obtain the mean of the original values, add 1250 to the mean obtained above, 4. The s.d. is unchanged. For the original data

$\overline{X} = 1250 + 4 = 1254$
$S = 2.58$

This manipulation is expressed in Eq. (1.13) where X_i represents one of n observations from a population with variance σ^2. C is a constant and $\overline{X}$ is the average of the X_i's.

$$\text{Average}\,(X_i + C) = \sum \frac{X_i + C}{n} = \overline{X} + C$$

$$\text{Variance}\,(X_i + C) = \sigma^2 \tag{1.13}$$

2. If the mean of a set of data is $\overline{X}$ and the s.d. is S, multiplying or dividing each value by a constant k results in a new mean of $k\overline{X}$ or $\overline{X}/k$, respectively, and a new s.d. of kS or S/k, respectively. Multiplying each of the original values above by 3 results in

> | 6, 9, 15, 30 |
>
> $\overline{X} = 15\,(3 \times 5)$
> $S = 10.68\,(3 \times 3.56)$

Dividing each value by 2 results in

$$\boxed{1, 1.5, 2.5, 5}$$
$$\overline{X} = 2.5 \left(\frac{5}{2} \right)$$
$$S = 1.78 \left(\frac{3.56}{2} \right)$$

In general,

$$\text{Average} (C \cdot X_i) = C\overline{X}$$
$$\text{Variance} (C \cdot X_i) = C^2\sigma^2 \tag{1.14}$$

These results can be used to show that a set of data with mean $\overline{X}$ and s.d. equal to S can be converted to data with a mean of 0 and a s.d. of 1 (as in the "standardization" of normal curves, discussed in sect. 3.4.1). If the mean is subtracted from each value, and this result is divided by S, the resultant data have a mean of 0 and a s.d. of 1. The transformation is

$$\frac{X - \overline{X}}{S}. \tag{1.15}$$

Standard scores are values that have been transformed according to Eq. (1.15) [11]. For the original data, the first value 2 is changed to $(2 - 5)/3.56$ equal to -0.84. The interested reader may verify that transforming the values in this way results in a mean of 0 and a s.d. of 1.

1.7 PRECISION, ACCURACY, AND BIAS
When dealing with variable measurements, the definitions of *precision* and *accuracy*, often obscure and not distinguished in ordinary usage, should be clearly defined from a statistical point of view.

1.7.1 Precision
In the vocabulary of statistics, precision refers to the extent of variability of a group of measurements observed under similar experimental conditions. A precise set of measurements is compact. Observations, relatively close in magnitude, are considered to be precise as reflected by a small s.d. (Note that means are more precisely measured than individual observations according to this definition.) An important, sometimes elusive concept is that a precise set of measurements may have the same mean as an imprecise set. In most experiments with which we will be concerned, the mean and s.d. of the data are independent (i.e., they are unrelated). Figure 1.3 shows the results of two assay methods, each performed in triplicate. Both methods have an average result of 100%, but method II is more precise.

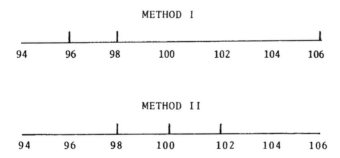

Figure 1.3   Representation of two analytical methods with the same accuracy but different precisions.

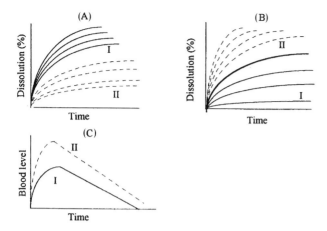

Figure 1.4 In vitro dissolution results for two formulations using two different methods and in vivo blood level versus time results. Methods A and B, in vitro; C, in vivo.

1.7.2 Accuracy

Accuracy refers to the closeness of an individual observation or mean to the true value. The "true" value is the result that would be observed in the absence of error (e.g., the true mean tablet potency or the true drug content of a preparation being assayed). In the example of the assay results shown in Figure 1.3, both methods are apparently equally accurate (or inaccurate).

Figure 1.4 shows the results of two dissolution methods for two formulations of the same drug, each formulation replicated four times by each method. The objective of the in vitro dissolution test is to simulate the in vivo oral absorption of the drug from the two dosage-form modifications. The first dissolution method, A, is very precise but does not give an accurate prediction of the in vivo results. According to the dissolution data for method A, we would expect that formulation I would be more rapidly and extensively absorbed in vivo. The actual in vivo results depicted in Figure 1.4 show the contrary result. The less precise method, method B in this example, is a more *accurate* predictor of the true in vivo results. This example is meant to show that a precise measurement need not be accurate, nor an accurate measurement precise.

Of course, the best circumstance is to have data that are both precise and accurate. If possible, we should make efforts to improve both the accuracy and precision of experimental observations. For example, in drug analysis, advanced electronic instrumentation can greatly increase the accuracy and precision of assay results.

1.7.3 Bias

Accuracy can also be associated with the term *bias*. The notion of bias has been discussed in section 1.4 in relation to the concept of unbiased estimates (e.g., the mean and variance). The meaning of bias in statistics is similar to the everyday definition in terms of "fairness." An accurate measurement, no matter what the precision, can be thought of as unbiased, because an accurate measurement is a "fair" estimate of the true result. A biased estimate is systematically either higher or lower than the true value. A biased estimate can be thought of as giving an "unfair" notion of the true value. For example, when estimating the average result of experimental data, the mean, $\overline{X}$, represents an estimate of the true population parameter, μ, and in this sense is considered accurate and unbiased. An average blood pressure reduction of 10 mm Hg due to an antihypertensive agent, derived from data from a clinical study of 200 patients, can be thought of as an unbiased estimate of the true blood pressure reduction due to the drug, provided that the patients are appropriately selected at "random." The true reduction in this case is the average reduction that would be observed if the antihypertensive effect of the drug were known for all members of the population (e.g., all hypertensive patients). The outcome of a single experiment, such as the 10 mm Hg reduction observed in the 200 patients above, will in all probability not be identical to the true mean reduction. But the mean reduction as observed

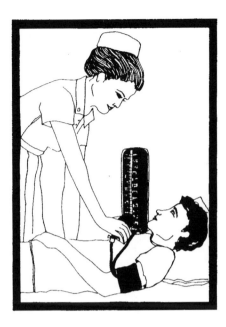 

Nurse 1 (before study)   Nurse 2 (during study)

Figure 1.5   Bias in determining the effect of an antihypertensive drug.

in the 200 patients is an accurate and unbiased assessment of the population average. A biased estimate is one which, on the average, does not equal the population parameter. In the example cited above for hypertensives, a biased estimate would result if for all patients one nurse took all the measurements before therapy and another nurse took all measurements during therapy, and each nurse had a different criterion or method for determining blood pressure. See Figure 1.5 for a clarification as to why this procedure leads to a biased estimate of the drug's effectiveness in reducing blood pressure. If the supine position results in higher blood pressure than the sitting position, the results of the study will tend to show a bias in the direction of too large a blood pressure reduction.

    The statistical estimates that we usually use, such as the mean and variance, are unbiased estimates. Bias often results from (a) the improper use of experimental design; (b) improper choice of samples; (c) unconscious bias, due to lack of blinding, for example; or (d) improper observation and recording of data, such as that illustrated in Figure 1.5.

1.8   THE QUESTION OF SIGNIFICANT FIGURES

The question of *significant figures* is an important consideration in statistical calculations and presentations. In general, the ordinary rules for retaining significant figures are not applicable to statistical computations. Contrary to the usual rules for retaining significant figures, one should retain as many figures as possible when performing statistical calculations, not rounding off until all computations are complete.

    The reason for not rounding off during statistical computations is that untenable answers may result when using computational procedures that involve taking differences between values very close in magnitude if values are rounded off prior to taking differences. This may occur when calculating "sums of squares" (the sum of squared differences from the mean) using the shortcut formula, Eq. (1.4), for the calculation of the variance or s.d. The shortcut formula for $\sum(X - \overline{X})^2$ is $\sum X^2 - (\sum X)^2/N$ that cannot be negative, and will be equal to zero only if all the data have the same value. If the two terms, $\sum X^2$ and $(\sum X)^2/N$, are very similar in magnitude, rounding off before taking their difference may result in a zero or negative difference. This problem is illustrated by calculating the s.d. of the three numbers 1.19, 1.20, and 1.21. If the squares of these numbers are first rounded off to two decimal places, the following calculation of the s.d. results:

$$S = \sqrt{\frac{\sum (X^2 - \sum X)^2 / N}{N-1}} = \sqrt{\frac{1.42 + 1.44 + 1.46 - 3.6^2/3}{2}}$$

$$= \sqrt{\frac{4.32 - 4.32}{2}} = 0.$$

The correct s.d. calculated without rounding off is 0.01.

Computers and calculators carry many digits when performing calculations and do not round off further unless instructed to do so. These instruments retain as many digits as their capacity permits through all arithmetic computations. The possibility of rounding off, even considering the large capacity of modern computers, can cause unexpected problems in sophisticated statistical calculations, and must be taken into account in preparing statistical software programs. These problems can usually be overcome by using special programming techniques.

At the completion of the calculations, as many figures as are appropriate to the situation can be presented. Common sense and the usual rules for reporting significant figures should be applied (see Ref. [9] for a detailed discussion of significant figures). Sokal and Rohlf [9] recommend that, if possible, observations should be measured with enough significant figures so that the range of data is between 30 and 300 possible values. This flexible rule results in a relative error of less than 3%. For example, when measuring diastolic blood pressure, the range of values for a particular group of patients might be limited to 60 to 130 mm Hg. Therefore, measurements to the nearest mm Hg would result in approximately 70 possible values, and would be measured with sufficient accuracy according to this rule. If the investigator can make the measurement only in intervals of 2 mm Hg (e.g., 70 and 72 mm Hg can be measured, but not 71 mm Hg), we would have 35 possible data points, which is still within the 30 to 300 suggested by this rule of thumb. Of course, rules should not be taken as "written in stone." All rules should be applied with judgment.

Common sense should be applied when reporting average results. For example, reporting an average blood pressure reduction of 7.42857 for 14 patients treated with an antihypertensive agent would not be appropriate. As noted above, most physicians would say that blood pressure is rarely measured to within 2 mm Hg. Why should one bother to report any decimals at all for the average result? When reporting average results, it is generally good practice to report the average with a precision that is "reasonable" according to the nature of the data. An average of 7.4 mm Hg would probably suffice for this example. If the average were reported as 7 mm Hg, for example, it would appear that too much information is suppressed.

KEY TERMS

Accuracy	Precision
Attributes	Random variable
Average ($\overline{X}$)	Range
Bias	Ranking
Coding	Rating scale
Coefficient of variation (CV)	Sample
Continuous variables	Significant figures
Correction term (CT)	Standard deviation (s.d., S)
Cumulative distribution	Standard error of the mean ($S_{\overline{X}}$)
Degrees of freedom (d.f.)	Standard score
Discrete variables	Treatment
Frequency distribution	Unbiased sample
Geometric mean	Universe
Harmonic mean	Variability
Mean ($\overline{X}$)	Variable
Median	Weighted average
Population	

EXERCISES

1. List three experiments whose outcomes will result in each of the following kinds of variables:
    (a) Continuous variables
    (b) Discrete variables
    (c) Ordered variables
    (d) Categorical (attribute) variables

2. What difference in experimental conclusions, if any, would result if the pain scale discussed in section 1.1 were revised as follows no pain $= 6$, slight pain $= 4$, moderate pain $= 2$, and severe pain $= 0$? (Hint: see sect. 1.6.)

3. (a) Construct a frequency distribution containing 10 class intervals from the data in Table 1.1.
    (b) Construct a cumulative frequency plot based on the frequency distribution from part (a).

4. What is the average result based on the frequency distribution in part (a) of problem 3? Use a weighted-average procedure.

5. From Figure 1.1, what proportion of tablets have potencies between 95 and 105 mg? What proportion of tablets have a potency greater than 105 mg?

6. Calculate the average and standard deviation of (a) the first 20 values in Table 1.1, and (b) the last 20 values in Table 1.1. If these data came from two different clinical investigators, would you think that the differences in these two sets of data can be attributed to differences in clinical sites? Which set, the first or last, is more precise? Explain your answer.

7. What are the median and range of the first 20 values in Table 1.1?

8. (a) If the first value in Table 1.1 were $+100$ instead of $+17$, what would be the values of the median and range for the first 20 values?
    (b) Using the first value as 100, calculate the mean, standard deviation, and variance. Compare the results for these first 20 values to the answers obtained in Problem 6.

§§**9. Given the following sample characteristics, describe the population from which the sample may have been derived. The mean is 100, the standard deviation is 50, the median is 75, and the range is 125.

**10. If the population average for the cholesterol reductions shown in Table 1.1 were somehow known to be 0 (the drug does not affect cholesterol levels on the average), would you believe that this sample of 156 patients gives an unbiased estimate of the true average? Describe possible situations in which these data might yield (a) biased results; (b) unbiased results.

**11. Calculate the average standard deviation using the sampling experiment shown in section 1.5.2 for samples of size 2 taken from a population with values of 0, 1, and 2 (with replacement). Compare this result with the population standard deviation. Is the sample standard deviation an unbiased estimate of the population standard deviation?

12. Describe another situation that would result in a biased estimate of blood pressure reduction as discussed in section 1.7.3 (Fig. 1.5).

13. Verify that the standard deviation of the values 1.19, 1.20, and 1.21 is 0.01 (see sect. 1.8). What is the standard deviation of the numbers 2.19, 2.20, and 2.21? Explain the result of the two calculations above.

---

§§ The double asterisk indicates optional, more difficult problems.

14. For the following blood pressure measurements: 100, 98, 101, 94, 104, 102, 108, 108, calculate (a) the mean, (b) the standard deviation, (c) the variance, (d) the coefficient of variation, (e) the range, and (f) the median.

**15. Calculate the standard deviation of the grouped data in Table 1.2. (Hint : $S^2 = \left[\sum N_i X_i^2 - (\sum N_i X_i)^2 / (\sum N_i) \right] / (\sum N_i - 1)$; see Ref. [3]. N_i = frequency per group with midpoint X_i)

16. Compute the arithmetic mean, geometric mean, and harmonic mean of the following set of data. 3, 5, 7, 11, 14, 57

   If these data were observations on the time needed to cure a disease, which mean would you think to be most appropriate?

17. If the weights are 2, 1, 1, 3, 1, and 2 for the numbers 3, 5, 7, 11, 14, and 57 (Exercise 16), compute the weighted average and variance.

REFERENCES

1. Berkow R. The Merck Manual, 14th ed. Rahway, NJ: Merck Sharp & Dohme Research Laboratories, 1982.
2. Tukey J. Exploratory Data Analysis. Reading, MA: Addison-Wesley, 1977.
3. Yule GU, Kendall MG. An Introduction to the Theory of Statistics, 14th ed. London: Charles Griffin, 1965.
4. Sokal RR, Rohlf FJ. Biometry. San Francisco, CA: W.H. Freeman, 1969.
5. Colton T. Statistics in Medicine. Boston, MA: Little, Brown, 1974.
6. Dixon WJ, Massey FJ Jr. Introduction to Statistical Analysis, 3rd ed. New York: McGraw-Hill, 1969.
7. United States Pharmacopeia, 9th Supplement. Rockville, MD: USP Convention, Inc., 1990:3584–3591.
8. Graham SJ, Lawrence RC, Ormsby ED, et al. Particle Size Distribution of Single and Multiple Sprays of Salbutamol Metered-Dose Inhalers (MDIs). Pharm Res 1995; 12:1380.
9. Lachman L, Lieberman HA, Kanig JL. The Theory and Practice of Industrial Pharmacy, 3rd ed. Philadelphia, PA: Lea & Febiger, 1986.
10. Snedecor GW, Cochran WG. Statistical Methods, 8th ed. Ames, IA: Iowa State University Press, 1989.
11. Rothman ED, Ericson WA. Statistics, Methods and Applications. Dubuque, IA: Kendall Hunt, 1983.

2 | DATA GRAPHICS

"The preliminary examination of most data is facilitated by the use of diagrams. Diagrams prove nothing, but bring outstanding features readily to the eye; they are therefore no substitute for such critical tests as may be applied to the data, but are valuable in suggesting such tests, and in explaining the conclusions founded upon them." This quote is from Ronald A. Fisher, the father of modern statistical methodology [1]. Tabulation of raw data can be thought of as the initial and least refined way of presenting experimental results. Summary tables, such as frequency distribution tables, are much easier to digest and can be considered a second stage of refinement of data presentation. Summary statistics such as the mean, median, variance, standard deviation, and the range are concise descriptions of the properties of data, but much information is lost in this processing of experimental results. Graphical methods of displaying data are to be encouraged and are important adjuncts to data analysis and presentation. Graphical presentations clarify and also reinforce conclusions based on formal statistical analyses. Finally, the researcher has the opportunity to design aesthetic graphical presentations that command attention. The popular cliché "A picture is worth a thousand words" is especially apropos to statistical presentations. We will discuss some key concepts of the various ways in which data are depicted graphically.

2.1 INTRODUCTION

The diagrams and plots that we will be concerned with in our discussion of statistical methods can be placed broadly into two categories:

1. Descriptive plots are those whose purpose is to transmit information. These include diagrams describing data distributions such as histograms and cumulative distribution plots (see sect. 1.2.3). Bar charts and pie charts are examples of popular modes of communicating survey data or product comparisons.
2. Plots that describe *relationships* between variables usually show an underlying, but unknown analytic relationship between the variables that we wish to describe and understand. These relationships can range from relatively simple to very complex, and may involve only two variables or many variables. One of the simplest relationships, but probably the one with greatest practical application, is the straight-line relationship between two variables, as shown in the Beer's law plot in Figure 2.1. Chapter 7 is devoted to the analysis of data involving variables that have a linear relationship.

When analyzing and depicting data that involve relationships, we are often presented with data in pairs (X, Y pairs). In Figure 2.1, the optical density Y and the concentration X are the data pairs. When considering the relationship of two variables, X and Y, one variable can often be considered the response variable, which is dependent on the selection of the second or causal variable. The response variable Y (optical density in our example) is known as the *dependent* variable. The value of Y depends on the value of the *independent* variable, X (drug concentration). Thus, in the example in Figure 2.1, we think of the value of optical density as being dependent on the concentration of drug.

2.2 THE HISTOGRAM

The *histogram*, sometimes known as a *bar graph*, is one of the most popular ways of presenting and summarizing data. All of us have seen bar graphs, not only in scientific reports but also in advertisements and other kinds of presentations illustrating the distribution of scientific data.

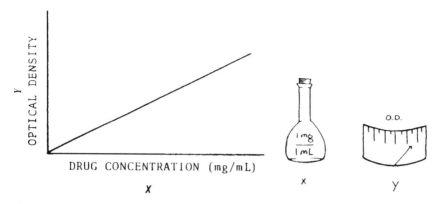

Figure 2.1 Beer's law plot illustrating a linear relationship between two variables.

The histogram can be considered as a visual presentation of a frequency table. The frequency, or proportion, of observations in each class interval is plotted as a bar, or rectangle, where the area of the bar is proportional to the frequency (or proportion) of observations in a given interval. An example of a histogram is shown in figure 2.2, where the data from the frequency table in Table 1.2 have been used as the data source. As is the case with frequency tables, class intervals for histograms should be of equal width. When the intervals are of equal width, the height of the bar is proportional to the frequency of observations in the interval. If the intervals are not of equal width, the histogram is not easily or obviously interpreted, as shown in Figure 2.2(B).

The choice of intervals for a histogram depends on the nature of the data, the distribution of the data, and the purpose of the presentation. In general, rules of thumb similar to that used

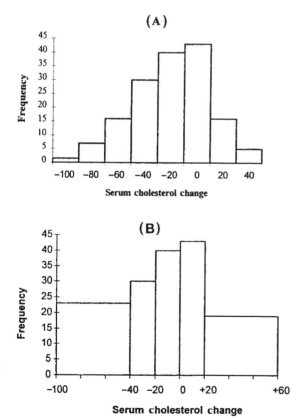

Figure 2.2 Histogram of data derived from Table 1.2.

for frequency distribution tables (sect. 1.2) can be used. Eight to twenty equally spaced intervals usually are sufficient to give a good picture of the data distribution.

2.3 CONSTRUCTION AND LABELING OF GRAPHS

Proper *construction and labeling* of graphs are crucial elements in graphical data representation. The design and actual construction of graphs are not in themselves difficult. The preparation of a *good* graph, however, requires careful thought and competent technical skills. One needs not only a knowledge of statistical principles, but also, in particular, computer and drafting competency. There are no firm rules for preparing good graphical presentations. Mostly, we rely on experience and a few guidelines. Both books and research papers have addressed the need for a more scientific guide to optimal graphics that, after all, is measured by how well the graph communicates the intended messages(s) to the individuals who are intended to read and interpret the graphs. Still, no rules will cover all situations. One must be clear that no matter how well a graph or chart is conceived, if the draftsmanship and execution is poor, the graph will fail to achieve its purpose.

A "good" graph or chart should be as *simple* as possible, yet clearly transmit its intended message. Superfluous notation, confusing lines or curves, and inappropriate draftsmanship (lettering, etc.) that can distract the reader are signs of a poorly constructed graph. The books *Statistical Graphics*, by Schmid [2], and *The Visual Display of Quantitative Information* by Tufte [3] are recommended for those who wish to study examples of good and poor renderings of graphic presentations. For example, Schmid notes that visual contrast should be intentionally used to emphasize important characteristics of the graph. Here, we will present a few examples to illustrate the recommendations for good graphic presentation as well as examples of graphs that are not prepared well or fail to illustrate the facts fairly.

Figure 2.3 shows the results of a clinical study that was designed to compare an active drug to a placebo for the treatment of hypertension. This graph was constructed from the *X, Y* pairs, *time* and *blood pressure*, respectively. Each point on the graph (+ , ■) is the average blood pressure for either drug or placebo at some point in time subsequent to the initiation of the study.

Proper construction and labeling of the typical rectilinear graph should include the following considerations:

1. A *title* should be given. The title should be brief and to the point, enabling the reader to understand the purpose of the graph without having to resort to reading the text. The title can be placed below or above the graph as in Figure 2.3.
2. The *axes* should be *clearly* delineated and *labeled.* In general, the zero (0) points of both axes should be clearly indicated. The ordinate (the *Y* axis) is usually labeled with the description parallel to the *Y* axis. Both the ordinate and abscissa (*X* axis) should be each appropriately

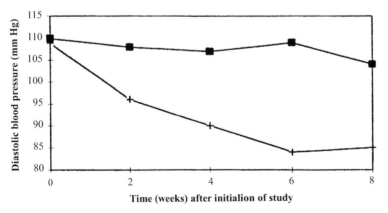

Figure 2.3 Blood pressure as a function of time in a clinical study comparing drug and placebo with a regimen of one tablet per day. ■, placebo (average of 45 patients); +, drug (average of 50 patients).

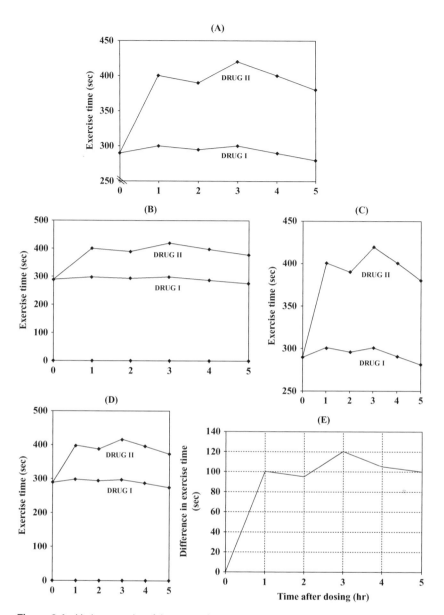

Figure 2.4 Various graphs of the same data presented in different ways. Exercise time at various time intervals after administration of single doses of two nitrate products. ◆ = Drug I, ■ = Drug II.

labeled and subdivided in units of equal width (of course, the X and Y axes almost always have different subdivisions). In the example in Figure 2.3, note the units of mm Hg and weeks for the ordinate and abscissa, respectively. Grid lines may be added [Fig. 2.4(E)] but, if used, should be kept to a minimum, not be prominent and should not interfere with the interpretation of the figure.

3. The numerical *values* assigned to the axes should be *appropriately spaced* so as to nicely cover the extent of the graph. This can easily be accomplished by trial and error and a little manipulation. The scales and proportions should be constructed to present a fair picture of the results and should not be exaggerated so to prejudice the interpretation. Sometimes, it may be necessary to skip or omit some of the data to achieve this objective. In these cases, the use of a "broken line" is recommended to clearly indicate the range of data not included in the graph (Fig. 2.4).

4. If appropriate, a *key* explaining the symbols used in the graph should be used. For example, at the bottom of Figure 2.3, the key defines ■ as the symbol for placebo and + for drug. In many cases, labeling the curves directly on the graph (Fig. 2.4) results in more clarity.
5. In situations where the graph is derived from laboratory data, inclusion of the *source* of the data (name, laboratory notebook number, and page number, for example) is recommended.

Usually graphs should stand on their own, independent of the main body of the text.

Examples of various ways of plotting data, derived from a study of exercise time at various time intervals after administration of a single dose of two long-acting nitrate products to anginal patients, are shown in Figures 2.4(A) to 2.4(E). All of these plots are accurate representations of the experimental results, but each gives the reader a different impression. It would be wrong to expand or contract the axes of the graph, or otherwise distort the graph, in order to convey an incorrect impression to the reader. Most scientists are well aware of how data can be manipulated to give different impressions. If obvious deception is intended, the experimental results will not be taken seriously.

When examining the various plots in Figure 2.4, one could not say which plot best represents the meaning of the experimental results without knowledge of the experimental details, in particular the objective of the experiment, the implications of the experimental outcome, and the message that is *meant* to be conveyed. For example, if an improvement of exercise time of 120 seconds for one drug compared to the other is considered to be significant from a medical point of view, the graphs labeled A, C, and E in Figure 2.4 would all seem appropriate in conveying this message. The graphs labeled B and D show this difference less clearly. On the other hand, if 120 seconds is considered to be of little medical significance, B and D might be a better representation of the data.

Note that in plot A of Figure 2.4, the ordinate (exercise time) is broken, indicating that some values have been skipped. This is not meant to be deceptive, but is intentionally done to better show the differences between the two drugs. As long as the zero point and the break in the axis are clearly indicated, and the message is not distorted, such a procedure is entirely acceptable.

Figures 2.4(B) and 2.5 are exaggerated examples of plots that may be considered not to reflect accurately the significance of the experimental results. In Figure 2.4(B), the clinically significant difference of approximately 120 seconds is made to look very small, tending to diminish drug differences in the viewer's mind. Also, fluctuations in the hourly results appear to be less than the data truly suggest. In Figure 2.5, a difference of 5 seconds in exercise time between the two drugs appears very large. Care should be taken when constructing (as well as reading) graphs so that experimental conclusions come through clear and true.

6. If more than one curve appears on the same graph, a convenient way to differentiate the curves is to use different symbols for the experimental points (e.g., ○, ×, Δ, □, +) and, if necessary, connecting the points in different ways (e.g., —.—.—., ..., –.–.–.–). A key or label is used, which is helpful in distinguishing the various curves, as shown in Figures 2.3 to 2.6. Other ways of differentiating curves include different kinds of crosshatching and use of different colors.

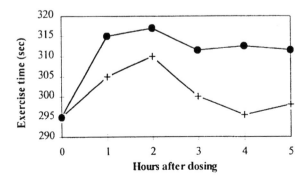

Figure 2.5 Exercise time at various time intervals after administration of two nitrate products. ●, product I; +, product II.

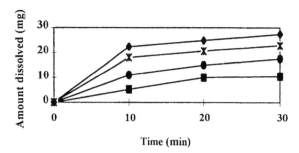

Figure 2.6 Plot of dissolution of four successive batches of a commercial tablet product. ♦ = batch I, ● = batch II, × = batch 3, ♦ = batch 4.

7. One should take care not to place too many curves on the same graph, as this can result in confusion. There are no specific rules in this regard. The decision depends on the nature of the data, and how the data look when they are plotted. The curves graphed in Figure 2.7 are cluttered and confusing. The curves should be presented differently or separated into two or more graphs. Figure 2.8 is a clearer depiction of the dissolution results of the five formulations shown in Figure 2.7.

8. The *standard deviation* may be indicated on graphs as shown in Figure 2.9. However, when the standard deviation is indicated on a graph (or in a table, for that matter), it should be made clear whether the variation described in the graph is an indication of the standard deviation (S) or the standard deviation of the mean ($S_{\bar{x}}$). The standard deviation of the mean, if appropriate, is often preferable to the standard deviation not only because the values on the graph are mean values, but also because $S_{\bar{x}}$ is smaller than the s.d., and therefore less cluttering. *Overlapping* standard deviations, as shown in Figure 2.10, should be avoided, as this representation of the experimental results is usually more confusing than clarifying.

9. The manner in which the points on a graph should be connected is not always obvious. Should the individual points be connected by straight lines, or should a smooth curve that approximates the points be drawn through the data? (See Fig. 2.11.) If the graphs represent functional relationships, the data should probably be connected by a smooth curve. For example, the blood level versus time data shown in Figure 2.11 are described most accurately by a smooth curve. Although, theoretically, the points should not be connected by straight lines as shown in Figure 2.11(A), such graphs are often depicted this way. Connecting the individual points with straight lines may be considered acceptable if one recognizes that this representation is meant to clarify the graphical presentation, or is done for some other appropriate reason. In the blood-level example, the area under the curve is proportional to the amount of drug absorbed. The area is often computed by the trapezoidal rule [4], and depiction of the data as shown in Figure 2.11(A) makes it easier to visualize and perform such calculations.

Figure 2.12 shows another example in which connecting points by straight lines is convenient but may *not* be a good representation of the experimental outcome. The straight line connecting the blood pressure at zero time (before drug administration) to the blood pressure after two weeks of drug administration suggests a gradual decrease (a linear decrease) in blood

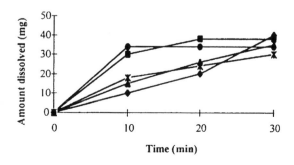

Figure 2.7 Plot of dissolution time of five different commercial formulations of the same drug. ● = product A, ■ = product B, × = product C, ▲ = product D, ♦ = product E.

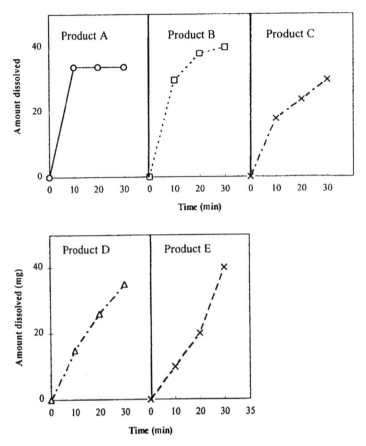

Figure 2.8 Individual plots of dissolution of the five formulations shown in Fig. 2.7.

pressure over the two-week period. In fact, no measurements were made during the initial two-week interval. The 10-mm Hg decrease observed after two weeks of therapy may have occurred before the two-week reading (e.g., in one week, as indicated by the dashed line in Fig. 2.12). One should be careful to ensure that graphs constructed in such a manner are not misinterpreted.

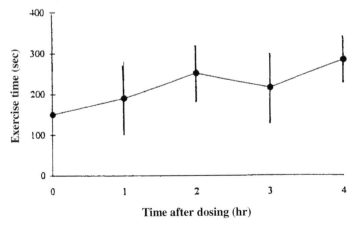

Figure 2.9 Plot of exercise time as a function of time for an antianginal drug showing mean values and standard error of the mean.

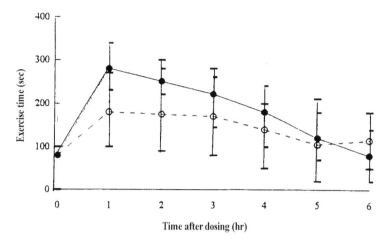

Figure 2.10   Graph comparing two antianginal drugs that is confusing and cluttered because of the overlapping standard deviations. •, drug A; o, drug B.

2.4   SCATTER PLOTS (CORRELATION DIAGRAMS)

Although the applications *of correlation* will be presented in some detail in chapter 7, we will introduce the notion of *scatter plots* (also called correlation diagrams or scatter diagrams) at this time. This type of plot or diagram is commonly used when presenting results of experiments. A typical scatter plot is illustrated in Figure 2.13. Data are collected in pairs (X and Y) with the objective of demonstrating a trend or relationship (or lack of relationship) between the X and Y variables. Usually, we are interested in showing a linear relationship between the variables (i.e., a straight line). For example, one may be interested in demonstrating a relationship (or correlation) between time to 80% *dissolution* of various tablet formulations of a particular drug

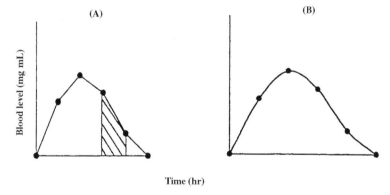

Figure 2.11   Plot of blood level versus time data illustrating two ways of drawing the curves.

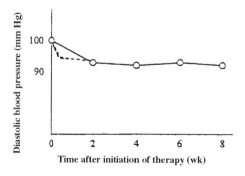

Figure 2.12   Graph of blood pressure reduction with time of antihypertensive drug illustrating possible misinterpretation that may occur when points are connected by straight lines.

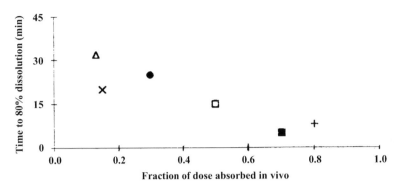

Figure 2.13 Scatter plot showing the correlation of dissolution time and in vivo absorption of six tablet formulations. △, formulation A; ×, formulation B; ●, formulation C; □, formulation D; ■, formulation E; +, formulation F.

and the *fraction of the dose absorbed* when human subjects take the various tablets. The data plotted in Figure 2.13 show pictorially that as dissolution increases (i.e., the time to 80% dissolution decreases) in vivo absorption increases. Scatter plots involve data pairs, X and Y, both of which are variable. In this example, *dissolution time* and *fraction absorbed* are both random variables.

2.5 SEMILOGARITHMIC PLOTS

Several important kinds of experiments in the pharmaceutical sciences result in data such that the *logarithm* of the response (Y) is linearly related to an independent variable, X. The semilogarithmic plot is useful when the response (Y) is best depicted as proportional changes relative to changes in X, or when the spread of Y is very large and cannot be easily depicted on a rectilinear scale. Semilog graph paper has the usual equal interval scale on the X axis and the logarithmic scale on the Y axis. In the logarithmic scale, equal intervals represent ratios. For example, the distance between 1 and 10 will exactly equal the distance between 10 and 100 on a logarithmic scale. In particular, first-order kinetic processes, often apparent in drug degradation and pharmacokinetic systems, show a linear relationship when log C is plotted versus time. First-order processes can be expressed by the following equation:

$$\log C = \log C_0 - \frac{kt}{2.3} \tag{2.1}$$

where C is the concentration at time t, C_0 the concentration at time 0, k the first-order rate constant, t the time, and log represents logarithm to the base 10.

Table 2.1 shows blood-level data obtained after an intravenous injection of a drug described by a one-compartment model [3].

Figure 2.14 shows two ways of plotting the data in Table 2.1 to demonstrate the linearity of the log C versus t relationship.

1. Figure 2.14(A) shows a plot of log C versus time. The resulting straight line is a consequence of the relationship of log concentration and time as shown in Eq. 2.1. This is an equation of a straight line with the Y intercept equal to log C_0 and a slope equal to $-k/2.3$. Straight-line relationships are discussed in more detail in chapter 8.

Table 2.1 Blood Levels After Intravenous Injection of Drug

Time after injection, t (hr)	Blood level, C (μg/mL)	Log blood level
0	20	1.301
1	10	1.000
2	5	0.699
3	2.5	0.398
4	1.25	0.097

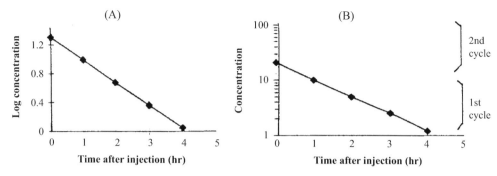

Figure 2.14   Linearizing plots of data from Table 2.1. (Plot A) log *C* versus time; (plot B) semilog plot.

2. Figure 2.14(B) shows a more convenient way of plotting the data of Table 2.1, making use of *semilog graph paper*. This paper has a logarithmic scale on the *Y* axis and the usual arithmetic, linear scale on the *X* axis. The logarithmic scale is constructed so that the spacing corresponds to the logarithms of the numbers on the *Y* axis. For example, the distance between 1 and 2 is the same as that between 2 and 4. (Log 2−log 1) is equal to (log 4−log 2). The semilog graph paper depicted in Figure 2.14(B) is two-cycle paper. The *Y* (log) axis has been repeated two times. The decimal point for the numbers on the *Y* axis is accommodated to the data. In our example, the data range from 1.25 to 20 and the *Y* axis is adjusted accordingly, as shown in Figure 2.14(B). The data may be plotted directly on this paper without the need to look up the logarithms of the concentration values.

2.6   OTHER DESCRIPTIVE FIGURES

Most of the discussion in this chapter has been concerned with plots that show relationships between variables such as blood pressure changes following two or more treatments, or drug decomposition as a function of time. Often occasions arise in which graphical presentations are better made using other more pictorial techniques. These approaches include the popular bar and pie charts. Schmid [2] differentiates bar charts into two categories: (a) *column charts* in which there is a vertical orientation and (b) *bar charts* in which the bars are horizontal. In general, the bar charts are more appropriate for comparison of categorical variables, whereas the column chart is used for data showing relationships such as comparisons of drug effect over time.

Bar charts are very simple but effective visual displays. They are usually used to compare some experimental outcome or other relevant data where the length of the bar represents the magnitude. There are many variations of the simple bar chart [2]; an example is shown in Figure 2.15. In Figure 2.15(A), patients are categorized as having a good, fair, or poor response. Forty percent of the patients had a good response, 35% had a fair response, and 25% had a poor response.

Figure 2.15(B) shows bars in pairs to emphasize the comparative nature of two treatments. It is clear from this diagram that Treatment *X* is superior to Treatment *Y*. Figure 2.15(C) is another way of displaying the results shown in Figure 2.15(B). Which chart do you think better sends the message of the results of this comparative study, Figure 2.15(B) or 2.15(C)? One should be aware that the results correspond only to the length of the bar. If the order in which the bars are presented is not obvious, displaying bars in order of magnitude is recommended. In the example in Figure 2.15, the order is based on the nature of the results, "Good," "Fair," and "Poor." Everything else in the design of these charts is superfluous and the otherwise principal objective is to prepare an aesthetic presentation that emphasizes but does not exaggerate the results. For example, the use of graphic techniques such as shading, crosshatching, and color, tastefully executed, can enhance the presentation.

Column charts are prepared in a similar way to bar charts. As noted above, whether or not a bar or column chart is best to display data is not always clear. Data trends over time usually are best shown using columns. Figure 2.16 shows the comparison of exercise time for two drugs using a column chart. This is the same data used to prepare Figure 2.4(A) (also, see Exercise Problem 8 at the end of this chapter).

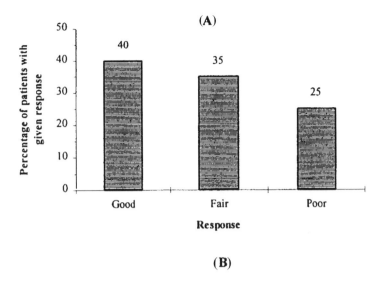

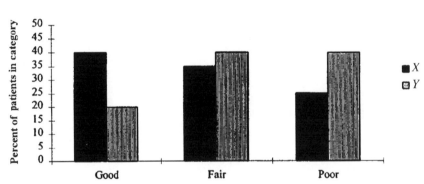

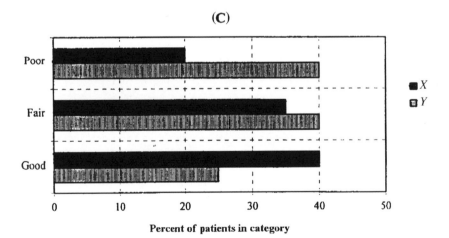

Figure 2.15 Graphical representation of patient responses to drug therapy.

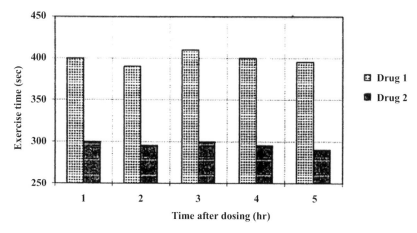

Figure 2.16   Exercise time for two drugs in the form of a column chart using data of Figure 2.4.

Pie charts are popular ways of presenting categorical data. Although the principles used in the construction of these charts are relatively simple, thought and care are necessary to convey the correct message. For example, dividing the circle into too many categories can be confusing and misleading. As a rule of thumb, no more than six sectors should be used. Another problem with pie charts is that it is not always easy to differentiate two segments that are reasonably close in size, whereas in the bar graph, values close in size are easily differentiated, since length is the critical feature.

The circle (or pie) represents 100%, or *all* of the results. Each segment (or slice of pie) has an area proportional to the area of the circle, representative of the contribution due to the particular segment. In the example shown in Figure 2.17(A), the pie represents the anti-inflammatory drug market. The slices are proportions of the market accounted for by major drugs in this therapeutic class. These charts are frequently used for business and economic descriptions, but can be applied to the presentation of scientific data in appropriate circumstances. Figure 2.17(B) shows the proportion of patients with good, fair, and poor responses to a drug in a clinical trial (see also Fig. 2.15).

Of course, we have not exhausted all possible ways of presenting data graphically. We have introduced the cumulative plot in section 1.2.3. Other kinds of plots are the stick diagram (analogous to the histogram) and frequency polygon [5]. The number of ways in which data can be presented is limited only by our own ingenuity. An elegant pictorial presentation of data can "make" a report or government submission. On the other hand, poor presentation of data can detract from an otherwise good report. The book *Statistical Graphics* by Calvin Schmid is recommended for those who wish detailed information on the presentation of graphs and charts.

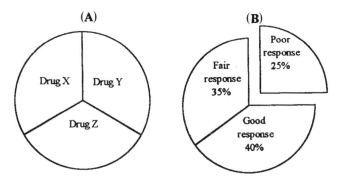

Figure 2.17   Examples of pie charts.

KEY TERMS

Bar charts   Independent variables
Bar graphs   Key
Column charts   Pie charts
Correlation   Scatter plots
Data pairs   Semilog plots
Dependent variables
Histogram

EXERCISES

1. Plot the following data, preparing and labeling the graph according to the guidelines outlined in this chapter. These data are the result of preparing various modifications of a
   formulation and observing the effect of the modifications on tablet hardness.

Formulation modification		
Starch (%)	Lactose (%)	Tablet hardness (kg)
10	5	8.3
10	10	9.1
10	15	9.6
10	20	10.2
5	5	9.1
5	10	9.4
5	15	9.8
5	20	10.4

   (Hint: Plot these data on a single graph where the Y axis is tablet hardness and the X axis
   is lactose concentration. There will be two curves, one at 10% starch and the other at 5%
   starch.)

2. Prepare a histogram from the data of Table 1.3. Compare this histogram to that shown in
   Figure 2.2(A). Which do you think is a better representation of the data distribution?

3. Plot the following data and label the graph appropriately.

Patient	X: response to product A	Y: response to product B
1	2.5	3.8
2	3.6	2.4
3	8.9	4.7
4	6.4	5.9
5	9.5	2.1
6	7.4	5.0
7	1.0	8.5
8	4.7	7.8

   What conclusion(s) can you draw from this plot if the responses are pain relief scores, where
   a high score means more relief?

4. A batch of tables was shown to have 70% with no defects, 15% slightly chipped, 10%
   discolored, and 5% dirty. Construct a pie chart from these data.

5. The following data from a dose–response experiment, a measure of physical activity, are the
   responses of five animals at each of three doses.

Dose (mg)	Responses
1	8, 12, 9, 14, 6
2	16, 20, 12, 15, 17
4	20, 17, 25, 27, 16

Plot the individual data points and the average at each dose versus (a) dose, (b) log dose.

6. The concentration of drug in solution was measured as a function of time.

Time (weeks)	Concentration
0	100
4	95
8	91
26	68
52	43

(a) Plot concentration versus time.
(b) Plot log concentration versus time.

7. Plot the following data and label the axes appropriately.

Patient	X: Cholesterol (mg%)	Y: Triglycerides (mg%)
1	180	80
2	240	180
3	200	70
4	300	200
5	360	240
6	240	200
Tablet	X: Tablet potency (mg)	Y: Tablet weight (mg)
1	5	300
2	6	300
3	4	280
4	5	295
5	6	320
6	4	290

8. Which figure do you think best represents the results of the exercise time study. Figure 2.16 or Figure 2.4(A)? If the presentation were to be used in a popular nontechnical journal read by laymen and physicians, which figure would you recommend?

REFERENCES
1. Fisher RA. Statistical Methods for Research Workers, 13th ed. New York, Hafner, 1963.
2. Schmid CF. Statistical Graphics. New York: Wiley, 1983.
3. Tufte ER. The Visual Display of Quantitative Data. Chelshire, CT: Graphics Press, 1983.
4. Gibaldi M, Perrier D. Pharmacokinetics, 2nd ed. New York: Marcel Dekker, 1982.
5. Dixon WJ, Massey FJ Jr. Introduction to Statistical Analysis, 3rd ed. New York: McGraw-Hill, 1969.

3 | Introduction to Probability: The Binomial and Normal Probability Distributions

The theory of statistics is based on probability. Some basic definitions and theorems are introduced in this chapter. This elementary discussion leads to the concept of a probability distribution, a mathematical function that assigns probabilities for outcomes in its domain. The properties of (a) the binomial distribution, a discrete distribution, and (b) the normal distribution, a continuous distribution, are presented. The normal distribution is the basis of modern statistical theory and methodology. One of the chief reasons for the pervasion of the normal distribution in statistics is the central limit theorem, which shows that means of samples from virtually all probability distributions tend to be normal for large sample sizes. Also, many of the probability distributions used in statistical analyses are based on the normal distribution. These include the t, F, and chi-square distributions. The binomial distribution is applicable to experimental results that have two possible outcomes, such as pass or fail in quality control, or cured or not cured in a clinical drug study. With a minimal understanding of probability, one can apply statistical methods intelligently to the simple but prevalent problems that crop up in the analysis of experimental data.

3.1 INTRODUCTION

Most of us have an intuitive idea of the meaning of probability. The meaning and use of probability in everyday life is a subconscious integration of experience and knowledge that allows us to say, for example: "If I purchase this car at my local dealer, the convenience and good service will *probably* make it worthwhile despite the greater initial cost of the car." From a statistical point of view, we will try to be more precise in the definition of probability. The *Random House Dictionary of the English Language* defines probability as "The likelihood of an occurrence expressed by the ratio of the actual occurrences to that of all possible occurrences; the relative frequency with which an event occurs, or is likely to occur." Therefore, the probability of observing an event can be defined as the proportion of such events that will occur in a large number of observations or experimental trials.

The approach to probability is often associated with odds in gambling or games of chance, and picturing probability in this context will help its understanding. When placing a bet on the outcome of a coin toss, the game of "heads and tails," one could reasonably *guess* that the probability of a head or tail is one-half (1/2) or 50%. One-half of the outcomes will be heads and one-half will be tails. Do you think that the probability of observing a head (or tail) on a single toss of the coin is exactly 0.5 (50%)? Probably not, a probability of 50% would result only if the coin is absolutely balanced. The only way to verify the probability is to carry out an extensive experiment, tossing a coin a million times or more and counting the proportion of heads or tails that result.

The gambler who knows that the odds in a game of craps favor the "house" will lose in the long run. Why should a knowledgeable person play a losing game? Other than for psychological reasons, the gambler may feel that a reasonably good chance of winning on any single bet is worth the chance, and maybe "Lady Luck" will be on his side. Probability is a measure of uncertainty. We may be able to predict accurately some average result in the long run, but the outcome of a single experiment cannot be anticipated with certainty.

3.2 SOME BASIC PROBABILITY

The concept of probability is "probably" best understood when discussing discontinuous or *discrete* variables. These variables have a countable number of outcomes. Consider an experiment in which only one of two possible outcomes can occur. For example, the result of treatment with an antibiotic is that an infection is either *cured* or *not cured* within five days. Although this situation is conceptually analogous to the coin-tossing example, it differs in the following respect. For the coin-tossing example, the probability can be determined by a rational examination of the nature of the experiment. If the coin is balanced, heads and tails are equally likely; the probability of a head is equal to the probability of a tail $= 0.5$. In the case of the antibiotic cure, however, the probability of a cure is not easily ascertained a priori, that is, prior to performing an experiment. If the antibiotic were widely used, based on his or her own experience, a physician prescriber of the product might be able to give a good estimate of the probability of a cure for patients treated with the drug. For example, in the physician's practice, he or she may have observed that approximately three of four patients treated with the antibiotic are cured. For this physician, the probability that a patient will be cured when treated with the antibiotic is approximately 75%.

A large multicenter clinical trial would give a better estimate of the probability of success after treatment. A study of 1000 patients might show 786 patients cured; the *probability of a cure* is estimated as 0.786 or 78.6%. This does not mean that the exact probability is 0.786. The exact probability can be determined only by treating the total population and observing the proportion cured, a practical impossibility in this case. In this context, it would be fair to say that exact probabilities are nearly always unknown.

3.2.1 Some Elementary Definitions and Theorems

1. $0 \leq P(A) \leq 1$ \hfill (3.1)

where $P(A)$ is the probability of observing event A. The probability of any event or experimental outcome, $P(A)$, cannot be less than 0 or greater than 1. An impossible event has a probability of 0. A certain event has a probability of 1.

2. If events $A, B, C, \ldots$ are *mutually exclusive*, the probability of observing A or B or $C \ldots$ is the sum of the probabilities of each event, $A, B, C, \ldots$. If two or more events are "mutually exclusive," the events cannot occur simultaneously, that is, if one event is observed, the other event(s) cannot occur. For example, we cannot observe both a head and a tail on a single toss of a coin.

$$P(A \text{ or } B \text{ or } C \ldots) = P(A) + P(B) + P(C) + \cdots \hfill (3.2)$$

An example frequently encountered in quality control illustrates this theorem. Among 1,000,000 tablets in a batch, 50,000 are known to be flawed, perhaps containing *specks* of grease. The probability of finding a *randomly* chosen tablet with specks is $50,000/1,000,000 = 0.05$ or 5%. The process of *randomly* choosing a tablet is akin to a lottery. The tablets are well mixed, ensuring that each tablet has an equal chance of being chosen. While blindfolded, one figuratively chooses a single tablet from a container containing the 1,000,000 tablets (see chapter 4 for a detailed discussion of random sampling). A gambler making an equitable bet would give odds of 19 to 1 against a specked tablet being chosen (1 of 20 tablets is specked). Odds are defined as

$$\frac{P(A)}{1 - P(A)}.$$

There are other defects among the 1,000,000 tablets. Thirty thousand, or 3%, have *chipped* edges and 40,000 (4%) are *discolored*. If these defects are mutually exclusive, the probability of observing any one of these events for a single tablet is 0.03 and 0.04, respectively [Fig. 3.1(A)]. According to Eq. (3.2), the probability of choosing an unacceptable tablet (specked, chipped, or discolored) at random is $0.05 + 0.03 + 0.04 = 0.12$, or 12%. (The probability of choosing an acceptable tablet is $1 - 0.12 = 0.88$.)

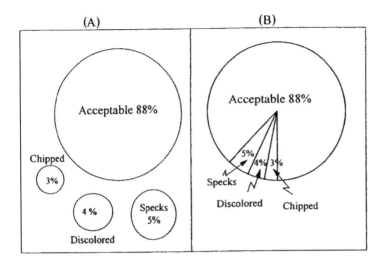

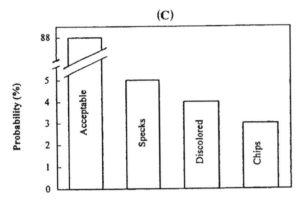

Figure 3.1   Probability distribution for tablet attributes.

3. $P(A) + P(B) + P(C) + \cdots = 1$ (3.3)

where $A, B, C, \ldots$ are mutually exclusive and exhaust all possible outcomes.

If the set of all possible experimental outcomes are mutually exclusive, the sum of the probabilities of all possible outcomes is equal to 1. This is equivalent to saying that we are certain that one of the mutually exclusive outcomes will occur.

All the four events in Figure 3.1 do not have to be mutually exclusive. In general:

4.   If two events are not mutually exclusive,

$$P(A \text{ or } B) = P(A) + P(B) - P(A \text{ and } B). \quad (3.4)$$

Note that if A and B are mutually exclusive, $P(A \text{ and } B) = 0$, and for two events, A and B, Eqs. (3.2) and (3.4) are identical. (A and B) means the simultaneous occurrence of A and B. (A or B) means that A or B or both A and B occur. For example, some tablets with chips may also be specked. If 20,000 tablets are both chipped *and* specked in the example above, one can verify that 60,000 tablets are specked *or* chipped.

$$P(\text{specked or chipped}) = P(\text{specked}) + P(\text{chipped}) - P(\text{specked or chipped})$$
$$= 0.05 + 0.03 - 0.02 = 0.06$$

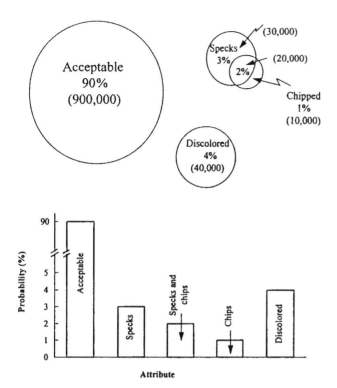

Figure 3.2 Distribution of tablet attributes where attributes are not all mutually exclusive.

The probability of finding a specked or chipped tablet is 0.06. Thirty thousand tablets are *only* specked, 10,000 tablets are *only* chipped, and 20,000 tablets are both specked and chipped; a total of 60,000 tablets specked or chipped. The distribution of tablet attributes under these conditions is shown in Figure 3.2. (Also, see Exercise Problem 23.)

With reference to this example of tablet attributes, we can enumerate all possible mutually exclusive events. In the former case, where each tablet was acceptable or had only a single defect, there are four possible outcomes (specked, chipped edges, discolored, and acceptable tablets). These four outcomes and their associated probabilities make up a *probability distribution*, which can be represented in several ways, as shown in Figure 3.1. The distribution of attributes where some tablets may be both specked and chipped is shown in Figure 3.2. The notion of a probability distribution is discussed further later in this chapter (sect. 3.3).

5. The multiplicative law of probability states that

$$P(A \text{ and } B) = P(A|B)\,P(B), \tag{3.5}$$

where $P(A|B)$ is known as the conditional probability of A given that B occurs. In the present example, the probability that a tablet will be specked given that the tablet is chipped is [from Eq. (3.5)]

$$P(\text{specked} \,|\, \text{chipped}) = \frac{P(\text{specked and chipped})}{P(\text{chipped})}$$

$$= \frac{0.02}{0.03} = \frac{2}{3}.$$

Referring to Figure 3.2, it is clear that 2/3 of the chipped tablets are also specked. Thus, the probability of a tablet being specked given that it is also chipped is 2/3.

3.2.2 Independent Events
In games of chance, such as roulette, the probability of winning (or losing) is theoretically the same on each turn of the wheel, irrespective of prior outcomes. Each turn of the wheel results in an independent outcome. The events, A and B, are said to be independent if a knowledge of B does not affect the probability of A. Mathematically, two events are independent if

$$P(A \mid B) = P(A). \tag{3.6}$$

Substituting Eq. (3.6) into Eq. (3.5), we can say that if

$$P(A \text{ and } B) = P(A)P(B), \tag{3.7}$$

then A and B are independent. When sampling tablets for defects, if each tablet is selected at random and the batch size is very large, the sample observations may be considered independent. Thus, in the example of tablet attributes shown in Figure 3.4, the probability of selecting an acceptable tablet (A) followed by a defective tablet (B) is

$$(0.88)(0.12) = 0.106.$$

The probability of selecting two tablets, both of which are acceptable, is $0.88 \times 0.88 = 0.7744$.

3.3 PROBABILITY DISTRIBUTIONS—THE BINOMIAL DISTRIBUTION
To understand probability further, one should have a notion of the concept of a probability distribution, introduced in section 3.2. A probability distribution is a mathematical representation (function) of the probabilities associated with the values of a random variable.

For discrete data, the concept can be illustrated by using the simple example of the outcome of antibiotic therapy introduced earlier in this chapter. In this example, the outcome of a patient following treatment can take on one of two possibilities: a cure with a probability of 0.75 or a failure with a probability of 0.25. Assigning the value 1 for a cure and 0 for a failure, the probability distribution is simply

$$f(1) = 0.75$$
$$f(0) = 0.25.$$

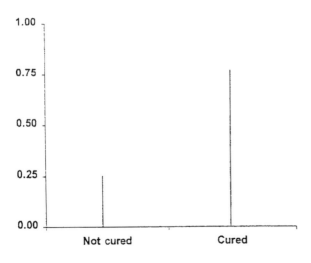

Figure 3.3 Probability distribution of a binomial outcome based on a single observation.

Table 3.1 Some Examples of Binomial Data in Pharmaceutical Research

Experiment or process	Dichotomous data
LD_{50} determination	Animals *live* or *die* after dosing. Determine dose that kills 50% of animals
ED_{50} determination	Drug is *effective* or *not effective.* Determine dose that is effective in 50% of animals
Sampling for defects	In quality control, product is sampled for defects. Tablets are *acceptable* or *unacceptable*
Clinical trials	Treatment is *successful* or *not successful*
Formulation modification	A. Palatability preference of *old* and *new* formulation B. New formulation is *more* or *less available* in crossover design

Figure 3.3 shows the probability distribution for this example, the random variable being the outcome of a patient treated with the antibiotic. This is an example of a binomial distribution. Another example of a binomial distribution is the coin-tossing game, heads or tails where the two outcomes have equal probability, 0.5. This binomial distribution ($p = 0.5$) has application in statistical methods, for example, the Sign test (sect. 15.2).

When a single observation can be dichotomized, that is, the observation can be placed into one of two possible categories, the binomial distribution can be used to define the probability characteristics of one or more such observations. The binomial distribution is a very important probability distribution in applications in pharmaceutical research. The few examples noted in Table 3.1 reveal its pervading presence in pharmaceutical processes.

3.3.1 Some Definitions

A *binomial trial* is a single binomial experiment or observation. The treatment of a single patient with the antibiotic is a binomial trial. The trial must result in only one of two outcomes, where the two outcomes are *mutually exclusive*. In the antibiotic example, the only possible outcomes are that a patient is either cured or not cured. In addition, only one of these outcomes is possible after treatment. A patient cannot be both cured and not cured after treatment. Each binomial trial must be *independent*. The result of a patient's treatment does not influence the outcome of the treatment for a different patient. In another example, when randomly sampling tablets for a

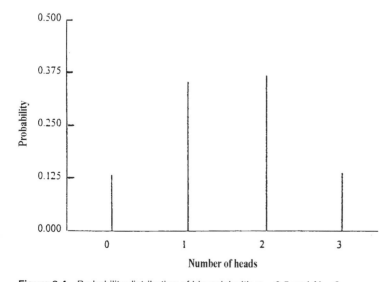

Figure 3.4 Probability distribution of binomial with $p = 0.5$ and $N = 3$.

binomial attribute, chipped or not chipped, the observation of a chipped tablet does not depend on or influence the outcome observed for any other tablet.

The binomial distribution is completely defined by two parameters: (a) the probability of one or the other outcome, and (b) the number of trials or observations, N. Given these two parameters, we can calculate the probability of any specified number of successes in N trials. For the antibiotic example, the probability of success is 0.75. With this information, we can calculate the probability that three of four patients will be cured ($N = 4$). We could also calculate this result, given the probability of failure (0.25). The probability of three of four patients being cured is exactly the same as the probability of one of four patients not being cured.

The probability of success (or failure) lies between 0 and 1. The probability of failure (the complement of a success) is 1 minus the probability of success $[1 - P(\text{success})]$.

Since the outcome of a binomial trial must be either success or failure, $P(\text{success}) + P(\text{failure}) = 1$ [see Eq. (3.3)].

The standard deviation of a binomial distribution with probability of success, p, and N trials is $\sqrt{pq/N}$, where $q = 1 - p$. The s.d. of the proportion of successes of antibiotic treatment in 16 trials is $\sqrt{0.75 \times 0.25/16} = 0.108$ (also see sect. 3.3.2).

The probability of the outcome of a binomial experiment consisting of N trials can be computed from the expansion of the expression

$$(p+q)^N, \tag{3.8}$$

where p is defined as the probability of success and q is the probability of failure. For example, consider the outcomes that are possible after three tosses of a coin. There are four ($N + 1$) possible results

1. three heads;
2. two heads and one tail;
3. two tails and one head;
4. three tails.

For the outcome of the treatment of three patients in the antibiotic example, the four possible results are

1. three cures;
2. two cures and one failure;
3. two failures and one cure;
4. three failures.

The probabilities of these events can be calculated from the individual terms from the expansion of $(p + q)^N$, where $N = 3$, the number of binomial trials.

$$(p+q)^3 = p^3 + 3p^2q + 3pq^2 + q^3$$

If $p = q = 1/2$, as is the case in coin tossing, then

$$p^3 = (1/2)^3 = 1/8 = P(\text{three heads})$$

$$3p^2q = 3/8 = P(\text{two heads and one tail})$$

$$3pq^2 = 3/8 = P(\text{two tails and one head})$$

$$q^3 = 1/8 = P(\text{three tails})$$

If $p = 0.75$ and $q = 0.25$, as is the case for the antibiotic example, then

$$p^3 = (0.75)^3 = 0.422 = P(3\,\text{cures})$$

$$3p^2q = 3(0.75)^2(0.25) = 0.422\ P(2\,\text{cures and 1 failure})$$

$$3pq^2 = 3(0.75)(0.25)^2 = 0.141\ P(1\,\text{cure and 2 failures})$$

$$q^3 = (0.25)^3 = 0.016 = P(3\,\text{failures})$$

The sum of the probabilities of all possible outcomes of three patients being treated or three sequential coin tosses is equal to 1 (e.g., $1/8 + 3/8 + 3/8 + 1/8 = 1$).

This is true of any binomial experiment because $(p + q)^N$ must equal 1 by definition (i.e., $p + q = 1$). The probability distribution of the coin-tossing experiment with $N = 3$ is shown in Figure 3.4. Note that this is a *discrete* distribution. The particular binomial distribution shown in the figure comprises only *four* possible outcomes (the four sticks).

A gambler looking for a fair game, one with equitable odds, would give odds of 7 to 1 on a bet that three heads would be observed in three tosses of a coin. The payoff would be eight dollars (including the dollar bet) for a one-dollar bet. A bet that either three heads or three tails would be observed would have odds of 3 to 1. (The probability of either *three heads* or *three tails* is $1/4 = 1/8 + 1/8$.)

To calculate exact probabilities in the binomial case, the expansion of the binomial, $(p + q)^N$ can be generalized by a single formula:

$$\text{Probability of } X \text{ successes in } N \text{ trials} = \binom{N}{X} p^x q^{N-X}. \tag{3.9}$$

$$\binom{N}{X} \text{ is defined as } \frac{N!}{X!(N - X)!}$$

(Remember that 0! is equal to 1.)

Consider the binomial distribution with $p = 0.75$ and $N = 4$ for the antibiotic example. This represents the distribution of outcomes after treating four patients. There are five possible outcomes

no patients are cured;
one patient is cured;
two patients are cured;
three patients are cured;
four patients are cured.

The probability that three of four patients are cured can be calculated from Eq. (3.9)

$$\binom{4}{3}(0.75)^3(0.25)^1 = \frac{4 \cdot 3 \cdot 2 \cdot 1}{1 \cdot 3 \cdot 2 \cdot 1}(0.42188)(0.25) = 0.42188.$$

The meaning of this particular calculation will be explained in detail in order to gain some insight into solving probability problems. There are four ways in which three patients can be cured and one patient not cured (Table 3.2). Denoting the four patients as A, B, C, and D, the probability that patients A, B, and C are cured and patient D is not cured is equal to

$$(0.75)(0.75)(0.75)(0.25) = 0.1055, \tag{3.10}$$

Table 3.2 Four Ways in Which Three of Four Patients Are Cured

	1	2	3	4
Patients cured	A, B, C	A, B, D	A, C, D	B, C, D
Patients not cured	D	C	B	A

where 0.25 is the probability that patient D will *not* be cured. There is no reason why any of the four possibilities shown in Table 3.2 should occur more or less frequently than any other (i.e., each possibility is equally likely). Therefore, the probability that the antibiotic will successfully cure exactly three patients is four times the probability calculated in Eq. (3.10)

$$4(0.1055) = 0.422.$$

The expression $\binom{4}{3}$ represents a combination, a selection of three objects, disregarding order, from four distinct objects. The combination, $\binom{4}{3}$, is equal to 4, and, as we have just demonstrated, there are four ways in which three cures can be obtained from four patients. Each one of these possible outcomes has a probability of $(0.75)^3(0.25)^1$. Thus, the probability of three cures in four patients is $4(0.75)^3 (0.25)^1$ as before.

The probability distribution based on the possible outcomes of an experiment in which four patients are treated with the antibiotic (the probability of a cure is 0.75) is shown in Table 3.3 and Figure 3.5. Note that the sum of the probabilities of the possible outcomes equals 1, as is also shown in the cumulative probability function plotted in Figure 3.5(B). The cumulative distribution is a nondecreasing function starting at a probability of zero and ending at a probability of 1. Figures 3.1 and 3.2, describing the distribution of tablet attributes in a batch of tablets, are examples of other discrete probability distributions.

Statistical hypothesis testing, a procedure for making decisions based on variable data is based on probability theory. In the following example, we use data observed in a coin-tossing game to decide whether or not we believe the coin to be loaded (biased).

You are an observer of a coin-tossing game and you are debating whether or not you should become an active participant. You note that only one head occurred among 10 tosses of the coin. You calculate the probability of such an event because it occurs to you that one head in 10 tosses of a coin is very unlikely; something is amiss (a "loaded" coin!). Thus, if the probability of a head is 0.5, the chances of observing one head in 10 tosses of a coin is less than 1 in 100 (Exercise Problem 18). This low probability suggests a coin that is not balanced. However, you properly note that the probability of any *single event* or outcome (such as one head in 10 trials) is apt to be small if N is sufficiently large. You decide to calculate the probability of this perhaps unusual result *plus* all other possible outcomes that are equally or less probable. In our example, this includes possibilities of no heads in 10 tosses, in addition to one or no tails in 10 tosses. These four probabilities (no heads, one head, no tails, and one tail) total approximately 2.2%. This is strong evidence in favor of a biased coin. Such a decision is based on the fact that the chance of obtaining an event as unlikely or less likely than one head in 10 tosses is about 1 in

Table 3.3 Probability Distribution for Outcomes of Treating Four Patients with an Antibiotic

Outcome	Probability
No cures	0.00391
One cure	0.04688
Two cures	0.21094
Three cures	0.42188
Four cures	0.31641

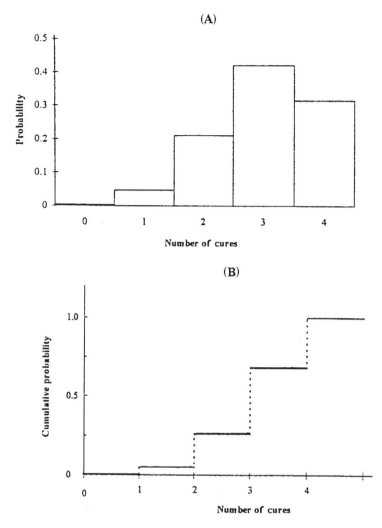

Figure 3.5   Probability distribution graph for outcomes of treating four patients with an antibiotic.

50 (2.2%) if the coin is *balanced.* You might wisely bet on tails on the next toss. You have made a decision: "The coin has a probability of less than 0.5 of showing heads on a single toss."

The probability distribution for the number of heads (or tails) in 10 tosses of a coin ($p = 0.5$ and $N = 10$) is shown in Figure 3.6. Note the symmetry of the distribution.

Although this is a discrete distribution, the "sticks" assume a symmetric shape similar to the normal curve. The two unlikely events in each "tail" (i.e., no heads or tails or one head or one tail) have a total probability of 0.022. The center and peak of the distribution is observed to be at $X = 5$, equal to NP, the number of trials times the probability of success. (See also Appendix Table IV.3, $p = 0.5$, $N = 10$.)

The application of binomial probabilities can be extended to more practical problems than gambling odds for the pharmaceutical scientist. When tablets are inspected for attributes or patients treated with a new antibiotic, we can apply a knowledge of the properties of the binomial distribution to estimate the true proportion or probability of success, and make appropriate decisions based on these estimates.

3.3.2   Summary of Properties of the Binomial Distribution

1. The binomial distribution is defined by N and p. With a knowledge of these parameters, the probability of any outcome of N binomial trials can be calculated from Eq. (3.9). We have

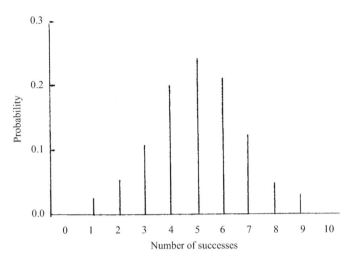

Figure 3.6 Probability distribution for $p = 0.5$ and $N = 10$.

noted that the sum of all possible outcomes of a binomial experiment with N trials is 1, which conforms to the notion of a probability distribution.

2. The results of a binomial experiment can be expressed either as the *number of successes* or as a *proportion*. Thus, if six heads are observed in 10 tosses of a coin, we can also say that 60% of the tosses are heads. If 16 defective tablets are observed in a random sample of 1000 tablets, we can say that 1.6% of the tablets sampled are defective. In terms of proportions, the *true mean* of the binomial population is equal to the probability of success, p. The sample proportion (0.6 in the coin-tossing example and 0.016 in the example of sampling for defective tablets) is an estimate of the true proportion.

3. The variability of the results of a binomial experiment is expressed as a standard deviation. For example, when inspecting tablets for the number of defectives, a different number of defective tablets will be observed depending on which 1000 tablets happen to be chosen. This variation, dependent on the particular sample inspected, is also known as *sampling error*. The s.d. of a binomial distribution can be expressed in two ways, depending on the manner in which the mean is presented (i.e., as a proportion or as the number of successes). The s.d. in terms of proportion of successes is

$$\sqrt{\frac{pq}{N}}. \tag{3.11}$$

In terms of number of successes, the s.d. is

$$\sqrt{Npq}, \tag{3.12}$$

where N is the sample size, the number of binomial trials. As shown in Eqs. (3.11) and (3.12), the s.d. is dependent on the value of P for binomial variables. The maximum s.d. occurs when $p = q = 0.5$, because Pq is maximized. The value of pq does not change very much with varying P and q until P or q reach low or high values, close to or more extreme than 0.2 and 0.8.

p	q	pq
0.5	0.5	0.25
0.4	0.6	0.24
0.3	0.7	0.21
0.2	0.8	0.16
0.1	0.9	0.09

4. When dealing with proportions, the variability of the observed proportion can be made as small as we wish by increasing the sample size [similar to the s.d. of the mean of samples of size N, Eq. (1.8)]. This means that we can estimate the proportion of "successes" in a population with very little error if we choose a sufficiently large sample. In the case of the tablet inspection example above, the variability (s.d.) of the proportion for samples of size 100 is

$$\sqrt{\frac{(0.016)(0.984)}{100}} = 0.0125.$$

By sampling 1000 tablets, we can reduce the variability by a factor of 3.16 ($\sqrt{100/1000} = 1/3.16$). The variability of the estimate of the true proportion (i.e., the sample estimate) is not dependent on the population size (the size of the entire batch of tablets in this example), but is dependent only on the size of the sample selected for observation. This interesting fact is true if the sample size is considerably smaller than the size of the population. Otherwise, a correction must be made in the calculation of the s.d. [4]. If the sample size is no more than 5% of the population size, the correction is negligible. In virtually all of the examples that concern us in pharmaceutical experimentation, the sample size is considerably less than the population size. Since binomial data are often easy to obtain, large sample sizes can often be accommodated to obtain very precise estimates of population parameters. An oft-quoted example is that a sample size of 6000 to 7000 randomly selected voters will be sufficient to estimate the outcome of a national election within 1% of the total popular vote. Similarly, when sampling tablets for defects, 6000 to 7000 tablets will estimate the proportion of a property of the tablets (e.g., defects) within, at most, 1% of the true value. (The least precise estimate occurs when $p = 0.5$.)

3.3.3 Confidence Limits with *N* Observations and Zero Successes or Failures
If one observes N independent binomial variables with zero successes (or failures), it is often of interest to place confidence limits on the true proportion of successes in the universe. As way of illustration, suppose we are testing an injectable product for sterility. There is no way of guaranteeing that all items in the batch will be sterile without 100% testing. Since the test may be destructive, a sample is taken. We expect to find all of the items tested to be sterile, that is, 100% sterile. One, then, may ask, what are the confidence limits for the true proportion of items in the batch that are sterile. The upper limit will be 100%. That is, if we see N items that are sterile, it is certainly possible that all of the items in the batch are sterile. The lower limit may be calculated as follows:

$$\text{Lower confidence limit} = p^N = P, \qquad\qquad (3.12A)$$

Where, p is the lower confidence limit, $P = 1 -$ probability of the confidence interval (e.g $(1 - 0.95$ for a 95% confidence interval) and N is the sample size. Note that the upper limit is 1.00.
Example:
One thousand (1000) items in a batch of 100,000 are tested for sterility with no failures (100% successes). What is the 95% confidence interval for the true proportion of sterile items in the batch.
Eq. (3.12A) can be written as $\ln(p) = \ln(P)/N$
$\ln(p) = \ln(1 - 0.95)/1000 = (-2.996/1000)$
$p = 0.997$.
That is, the 95% confidence interval for the proportion of sterile items is 0.997 to 1.00.
The 99% confidence interval is:
$\ln(p) = \ln(1 - 0.99)/1000 = (-4.61/1000)$
$p = 0.995$.
The 99% confidence interval for the proportion of sterile items is 0.9954 to 1.00.
(Note that $0.9954^{1000} = 0.01$.)

3.3.4 The Negative Binomial Distribution [5,6]

The negative binomial distribution does not have wide use in the pharmaceutical sciences, but can be useful in special situations. In a clinical trial, we might ask, for example, "How many successive cures can we expect to observe before seeing a failure, with a knowledge of the cure rate?" In quality control, we may be interested in the expected number of consecutive successes before a failure is observed, given the rate of failure. Another question might be, "What is the average number of consecutive good tablets observed before a failure is observed?" In general, the probability function is

$$= \binom{k+r-1}{k} \{P^r\}\{1 - P^r\}, \tag{3.12B}$$

where $0 < P < 1$, the probability of a success.

　　If $r = 1$, this is the probability distribution of failures before the first success. This can also be stated as the probability of success on the $(k+1)^{th}$ trial after k failures.

　　If $r = 1$, Eq. (3.11A) reduces to

$$\{P\}\{1 - P^k\}. \tag{3.12C}$$

　　Consider the following question: What is the probability that we will observe 50 good tablets before observing a split tablet on the 51st tablet. The probability of a split tablet is 0.01. Here, $p = 0.99$. (A good tablet is considered a failure in this context.) From Eq. (3.12C), the probability is

$$0.99 \left(1 - 0.99^{50}\right) = 0.599.$$

　　What is the average number of good tablets that a patient would take before he/she observes a split tablet. This can be calculated by using the negative binomial distribution.

　　The average number of good tablets before a split tablet is observed is P/q, where P is the probability of a good tablet (0.99) and q is $(1 - P)$ equal to 0.01.

$$\text{The average is } \frac{0.99}{0.01} = 99 \text{ tablets.}$$

　　Therefore, on the average, a patient would take 99 tablets before encountering a split tablet. If the tablet is taken once a day, it would take 99 days on the average before a split tablet was observed (5,6).

3.4 CONTINUOUS DATA DISTRIBUTIONS

Another view of probability concerns continuous data such as tablet dissolution time. The probability that any single tablet will have a particular specified dissolution result is 0, because the number of possible outcomes for continuous data is infinite. Probability can be conceived as the ratio of the number of times that an event occurs to the total number of possible outcomes. If the total number of outcomes is infinite, the probability of any single event is zero. This concept can be confusing. If one observes a large number of dissolution results, such as time to 90% dissolution, any particular observation might appear to have a finite probability of occurring. Analogous to the discussion for discrete data, could we not make an equitable bet that a result for dissolution of exactly 5 minutes 13 seconds, for example, would be observed? The apparent contradiction is due to the fact that data that are *continuous, in theory,* appear as *discrete* data *in practice* because of the limitations of measuring instruments, as discussed in chapter 1. For example, a sensitive clock could measure time to virtually any given precision (i.e., to small fractions of a second). It would be difficult to conceive of winning a bet that a 90% dissolution time would occur at a very specific time, where time can be measured to any specified degree of precision (e.g., 30 minutes 8.21683475... seconds).

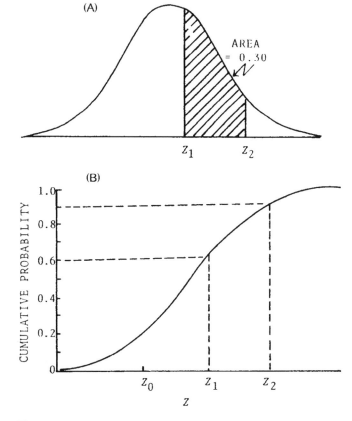

Figure 3.7 A normal distribution.

With *continuous* variables, we cannot express probabilities in as simple or intuitive a fashion as was done with discrete variables. Applications of calculus are necessary to describe concepts of probability with continuous distributions. Continuous cumulative probability distributions are represented by smooth curves (Fig. 3.7) rather than the step-like function shown in Figure 3.5(B). The area under the probability distribution curve (also known as the cumulative probability density) is equal to 1 for all probability functions. Thus the area under the normal distribution curve in Figure 3.7(A) is equal to 1.

3.4.1 The Normal Distribution

The normal distribution is an example of a continuous probability density function. The normal distribution is most familiar as the symmetrical, bell-shaped curve shown in Figure 3.8. A theoretical normal distribution is a continuous probability distribution and consists of an infinite number of values. In the theoretical normal distribution, the data points extend from positive infinity to negative infinity. It is clear that scientific data from pharmaceutical experiments cannot possibly fit this definition. Nevertheless, if real data conform reasonably well with the theoretical definition of the normal curve, adequate approximations, if not very accurate estimates of probability, can be computed based on normal curve theory.

The equation for the normal distribution (normal probability density) is

$$Y = \frac{1}{\sigma\sqrt{2\pi}} e^{-(1/2)(X-\mu)^2/\sigma^2}, \qquad (3.13)$$

where σ is the s.d., μ the mean, X the value of the observation, e the base of natural logarithms, $2.718\ldots$; and Y the ordinate of normal curve, a function of X.

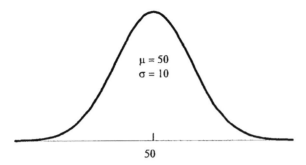

50

Figure 3.8 A typical normal curve.

The normal distribution is defined by its mean, μ, and its s.d., σ [see Eq. (3.13)]. This means that if these two parameters of the normal distribution are known, all the properties of the distribution are known. There are any number of different normal distributions. They all have the typical symmetrical, bell-shaped appearance. They are differentiated only by their means, a measure of location, and their s.d., a measure of spread. The normal curve shown in Figure 3.8 can be considered to define the distribution of the potencies of tablets in a batch of tablets. Most of the tablets have a potency close to the mean potency of 50 mg. The farther the assay values are from the mean, the fewer the number of tablets there will be with these more extreme values. As noted above, the spread or shape of the normal distribution is dependent on the s.d. A large s.d. means that the spread is large. In this example, a larger s.d. means that there are more tablets far removed from the mean, perhaps far enough to be out of specifications (Fig. 3.9).

In real-life situations, the distribution of a finite number of values often closely approximates a normal distribution. Weights of tablets taken from a single batch may be approximately normally distributed. For practical purposes, any continuous distribution can be visualized as being constructed by categorizing a large amount of data in small equilength intervals and constructing a histogram. Such a histogram can similarly be constructed for normally distributed variables.

Suppose that all the tablets from a large batch are weighed and categorized in small intervals or boxes (Fig. 3.10). The number of tablets in each box is counted and a histogram plotted as in Figure 3.11. As more boxes are added and the intervals made shorter, the intervals will eventually be so small that the distinction between the bars in the histogram is lost and a smooth curve results, as shown in Figure 3.12. In this example, the histogram of tablet weights looks like a normal curve.

Areas under the normal curve represent probabilities and are obtained by appropriate integration of Eq. (3.13). In Figure 3.7, the probability of observing a value between Z_1 and Z_2 is calculated by integrating the normal density function between Z_1 and Z_2.

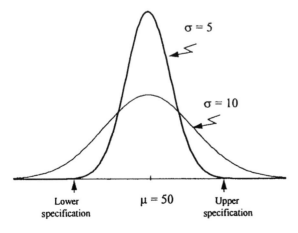

Lower
specification

$\mu = 50$

Upper
specification

Figure 3.9 Two normal curves with different standard deviations.

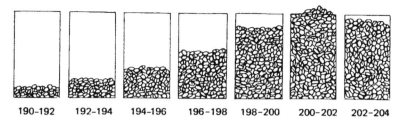

Figure 3.10 Categorization of tablets from a tablet batch by weight.

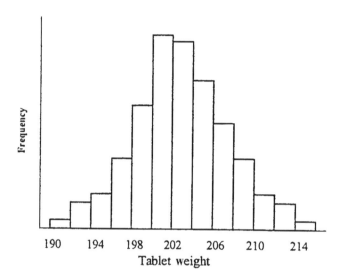

Figure 3.11 Histogram of tablet weights.

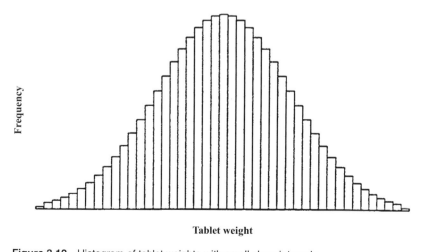

Figure 3.12 Histogram of tablet weights with small class intervals.

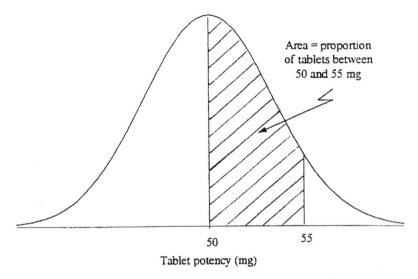

Figure 3.13 Area under normal curve as a representation of proportion of tablets in an interval.

This function is not easily integrated. However, tables are available that can be used to obtain the area between any two values of the variable, Z. Such an area is illustrated in Figure 3.7(A). If the area between Z_1 and Z_2 in Figure 3.7 is 0.3, the probability of observing a value between Z_1 and Z_2 is 3 in 10 or 0.3. In the case of the tablet potencies, the area in a specified interval can be thought of as the proportion of tablets in the batch contained in the interval. This concept is illustrated in Figure 3.13.

Probabilities can be determined directly from the cumulative distribution plot as shown in Figure 3.7(B) (see Exercise Problem 9). The probability of observing a value below Z_1 is 0.6. Therefore, the probability of observing a value between Z_1 and Z_2 is $0.9 - 0.6 = 0.3$.

There are an infinite number of normal curves depending on μ and σ. However, the area in any interval can be calculated from tables of cumulative areas under the *standard normal curve*. The standard normal curve has a mean of 0 and a s.d. of 1. Table IV.2 in App. IV is a table of cumulative areas under the standard normal curve, giving the area below Z (i.e., the area between $-\infty$ and Z). For example, for $Z = 1.96$, the area in Table IV.2 is 0.975. This means that 97.5% of the values comprising the standard normal curve are less than 1.96, lying between $-\infty$ and 1.96. The normal curve is symmetrical about its mean. Therefore, the area below -1.96 is 0.025 as depicted in Figure 3.14. The area between Z equal to -1.96 and $+1.96$ is 0.95. Referring to Table IV.2, the area below Z equal to $+2.58$ is 0.995, and the area below $Z = -2.58$ is 0.005. Thus the area between Z equal to -2.58 and $+2.58$ is 0.99. It would be very useful for the reader to memorize the Z values and the corresponding area between $\pm Z$ as shown in Table 3.4. These values of Z are commonly used in statistical analyses and tests.

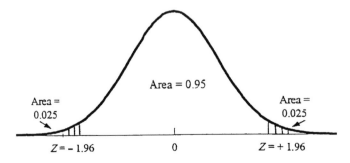

Figure 3.14 Symmetry of the normal curve.

Table 3.4 Area Between ±ʕ for Some
Commonly Used Values of Z

Z	Area between ±Z
0.84	0.60
1.00	0.68
1.28	0.80
1.65	0.90
1.96	0.95
2.32	0.98
2.58	0.99

The area in any interval of a normal curve with a mean and s.d. different from 0 and 1, respectively, can be computed from the standard normal curve table by using a transformation. The transformation changes a value from the normal curve with mean μ and s.d. σ, to the corresponding value, Z, in the standard normal curve. The transformation is

$$Z = \frac{X - \mu}{\sigma}. \tag{3.14}$$

The area (probability) between $-\infty$ and X (i.e., the area below X) corresponds to the value of the area below Z from the cumulative standard normal curve table. Note that if the normal curve that we are considering is the standard normal curve itself, transformation results in the identity

$$Z = \frac{X - 0}{1} = X.$$

Z is exactly equal to X, as expected. Effectively the transformation changes variables with a mean of μ and a s.d. of σ to variables with a mean of 0 and a s.d. of 1.

Suppose in the example of tablet potencies that the mean is 50 and the s.d. is 5 mg. Given these two parameters, what proportion of tablets in the batch would be expected to have more than 58.25 mg of drug? First we calculate the transformed value, Z. Then the desired proportion (equivalent to probability) can be obtained from Table IV.2. In this example, $X = 58.25$, $\mu = 50$, and $\sigma = 5$. Referring to Eq. (3.14), we have

$$Z = \frac{X - \mu}{\sigma}$$

$$= \frac{58.25 - 50}{5} = 1.65.$$

According to Table IV.2, the area between $-\infty$ and 1.65 is 0.95. This represents the probability of a tablet having 58.25 mg or less of drug. Since the question was, "What proportion of tablets in the batch have a potency greater than 58.25 mg?", the area above 58.25 mg is the correct answer. The area under the entire curve is 1; the area above 58.25 mg is $1 - 0.95$, equal to 0.05. This is equivalent to saying that 5% of the tablets have at least 58.25 mg (58.25 mg or more) of drug in this particular batch or distribution of tablets. This transformation is illustrated in Figure 3.15.

One should appreciate that since the normal distribution is a perfectly symmetrical continuous distribution that extends from $-\infty$ to $+\infty$, real data never exactly fit this model. However, data from distributions reasonably similar to the normal can be treated as being normal, with the understanding that probabilities will be approximately correct. As the data are closer to normal, the probabilities will be more exact. Methods exist to test if data can reasonably be expected to be derived from a normally distributed population [1]. In this book, when applying the normal distribution to data we will either (a) assume that the data are close to

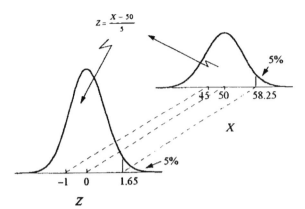

Figure 3.15 *Z* transformation for tablets with mean of 50 mg and s.d. of 5 mg.

normal according to previous experience or from an inspection of the data, or (b) that deviations from normality will not greatly distort the probabilities based on the normal distribution.

Several examples are presented below which further illustrate applications of the normal distribution.

Example 1: The U.S. Pharmacopia (USP) weight test for tablets states that for tablets weighing up to 100 mg, not more than 2 of 20 tablets may differ from the average weight by more than 10%, and no tablet may differ from the average weight by more than 20% [2]. To ensure that batches of a 100-mg tablet (labeled as 100 mg) will pass this test consistently, a statistician recommended that 98% of the tablets in the batch should weigh within 10% of the mean. One thousand tablets from a batch of 3,000,000 were weighed and the mean and s.d. were calculated as 101.2 ± 3.92 mg. Before performing the official USP test, the quality control supervisor wishes to know if this batch meets the statistician's recommendation. The calculation to answer this problem can be made by using areas under the standard normal curve if the tablet weights can be assumed to have a distribution that is approximately normal. For purposes of this example, the sample mean and s.d. will be considered equal to the true batch mean and s.d. Although not exactly true, the sample estimates will be close to the true values when a sample as large as 1000 is used. For this large sample size, the sample estimates are very close to the true parameters. However, one should clearly understand that to compute probabilities based on areas under the normal curve, both the mean and s.d. must be known. When these parameters are estimated from the sample statistics, other derived distributions can be used to calculate probabilities.

Figure 3.16 shows the region where tablet weights will be outside the limits, 10% from the mean ($\mu \pm 0.1\ \mu$), that is, 10.12 mg or more from the mean for an average tablet weight of 101.2 mg (101.2 ± 10.12 mg). The question to be answered is: What proportion of tablets are between

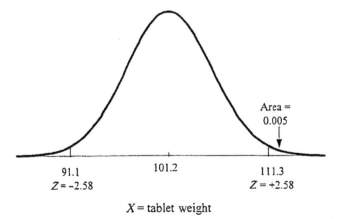

Figure 3.16 Distribution of tablets with mean weight 101.2 mg and s.d. equal to 3.92.

91.1 and 111.3 mg? If the answer is 98% or greater, the requirements are met. The proportion of tablets between 91.1 and 111.3 mg can be estimated by computing the area under the normal curve in the interval 91.1 to 111.3, the unshaded area in Figure 3.16. This can be accomplished by use of the Z transformation and the table of areas under the standard normal curve (Table IV.2). First, we calculate the areas below 111.3 by using the Z transformation

$$Z = \frac{X - \mu}{\sigma} = \frac{111.3 - 101.2}{3.92} = 2.58.$$

This corresponds to an area of 0.995 (see Table IV.2). The area above 111.3 is $(1 - 0.995) = 0.005$ or $1/200$. Referring to Figure 3.16, this area represents the probability of finding a tablet that weighs 111.3 mg or more. The probability of a tablet weighing 91.1 mg or less is calculated in a similar manner

$$Z = \frac{91.1 - 101.2}{3.92} = -2.58.$$

Table IV.2 shows that this area is 0.005; that is, the probability of a tablet weighing between $-\infty$ and 91.1 mg is 0.005. The probability that a tablet will weigh more than 111.3 mg or less than 91.1 mg is $0.005 + 0.005$, equal to 0.01. Therefore, 99% $(1.00 - 0.01)$ of the tablets weigh between 91.1 and 111.3 mg and the statistician's recommendation is more than satisfied. The batch should have no trouble passing the USP test.

The fact that the normal distribution is symmetric around the mean simplifies calculations of areas under the normal curve. In the example above, the probability of values exceeding Z equal to 2.58 is exactly the same as the probability of values being less than Z equal to -2.58. This is a consequence of the symmetry of the normal curve, 2.58 and -2.58 being equidistant from the mean. This is easily seen from an examination of Figure 3.16.

Although this batch of tablets should pass the USP weight uniformity test, if *some* tablets in the batch are out of the 10% or 20% range, there is a chance that a random sample of 20 will fail the USP test. In our example, about 1% or 30,000 tablets will be more than 10% different from the mean (less than 91.1 or more than 111.3 mg). It would be of interest to know the chances, albeit small, that of 20 randomly chosen tablets, more than 2 would be "aberrant." When 1% of the tablets in a batch deviate from the batch mean by 10% or more, the chances of finding more than 2 such tablets in a sample of 20 is approximately 0.001 (1/1000). This calculation makes use of the binomial probability distribution.

Example 2: During clinical trials, serum cholesterol, among other serum components, is frequently monitored to ensure that a patient's cholesterol is within the normal range, as well as to observe possible drug effects on serum cholesterol levels. A question of concern is: What is an abnormal serum cholesterol value? One way to define "abnormal" is to tabulate cholesterol values for apparently normal healthy persons, and to consider values very remote from the average as abnormal. The distribution of measurements such as serum cholesterol often has an approximately normal distribution.

The results of the analysis of a large number of "normal" cholesterol values showed a mean of 215 mg% and a s.d. of 35 mg%. This data can be depicted as a normal distribution as shown in Figure 3.17. "Abnormal" can be defined in terms of the proportion of "normal" values that fall in the extremes of the distribution. This may be thought of in terms of a gamble. By choosing to say that extreme values observed in a new patient are abnormal, we are saying that persons observed to have very low or high cholesterol levels could be "normal," but the likelihood or probability that they come from the population of normal healthy persons is small. By defining an abnormal cholesterol value as one that has a 1 in 1000 chance of coming from the distribution of values from normal healthy persons, cutoff points can be defined for abnormality based on the parameters of the normal distribution. According to the cumulative standard normal curve, Table IV.2, a value of Z equal to approximately 3.3 leaves 0.05% of the area in the upper tail. Because of the symmetry of the normal curve, 0.05% of the area is below $Z = -3.3$. Therefore, 0.1% (1/1000) of the values will lie outside the values of Z equal to ± 3.3

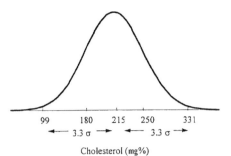

Cholesterol (mg%)

Figure 3.17   Distribution of "normal" cholesterol values.

in the standard normal curve. The values of X (cholesterol levels) corresponding to $Z = \pm 3.3$ can be calculated from the Z transformation.

$$Z = \frac{X - \mu}{\sigma} = \frac{X - 215}{35} = \pm 3.3$$

$$X = 215 \pm (3.3)(35) = 99 \text{ and } 331.$$

This is equivalent to saying that cholesterol levels that deviate from the average of "normal" persons by 3.3 s.d. units or more are deemed to be abnormal. For example, the lower limit is the mean of the "normals" minus 3.3 times the s.d. or $215 - (3.3)(35) = 99$. The cutoff points are illustrated in Figure 3.17.

Example 3: The standard normal distribution may be used to calculate the proportion of values in any interval from any normal distribution. As an example of this calculation, consider the data of cholesterol values in Example 2. We may wish to calculate the proportion of cholesterol values between 200 and 250 mg%.

Examination of Figure 3.18 shows that the area (probability) under the normal curve between 200 and 250 mg% is the probability of a value being less than 250 *minus* the probability of a value being less than 200. Referring to Table IV.2, we have

Probability of a value less than 250

$$\frac{250 - 215}{35} = 1 = Z \text{ probability} = 0.841.$$

Probability of a value less than 200

$$\frac{200 - 215}{35} = -0.429 = Z \text{ probability} = 0.334.$$

Therefore, the probability of a value falling between 250 and 200 is

$$0.841 - 0.334 = 0.507.$$

3.4.2   Central Limit Theorem

"Without doubt, the most important theorem in statistics is the central limit theorem" [3]. This theorem states that the distribution of sample means of size N taken from *any* distribution with a finite variance σ^2 and mean μ tends to be *normal* with variance σ^2/N and mean μ. We have previously discussed the fact that a sample mean of size N has a variance equal to σ^2/N. The new and important feature here is that if we are dealing with means of *sufficiently large sample size*, the means have a normal distribution, regardless of the form of the distribution from which the samples were selected.

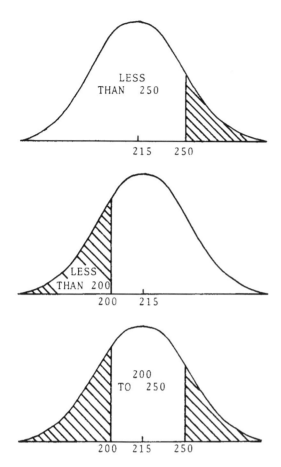

Figure 3.18 Illustration of the calculation of proportion of cholesterol values between 200 and 250 mg%.

How large is a "large" sample? The answer to this question depends on the form of the distribution from which the samples are taken. If the distribution is normal, any size sample will have a mean that is normally distributed. For distributions that deviate greatly from normality, larger samples will be needed to approximate normality than distributions that are more similar to the normal distributions (e.g., symmetrical distributions).

The power of this theorem is that the normal distribution can be used to describe most of the data with which we will be concerned, provided that the means come from samples of sufficient size. An example will be presented to illustrate how means of distributions far from normal tend to be normally distributed as the sample size increases. Later in this chapter, we will see that even the discrete binomial distribution, where only a very limited number of outcomes are possible, closely approximates the normal distribution with sample sizes as small as 10 in symmetrical cases (e.g. $P = q = 0.5$).

Consider a distribution that consists of outcomes 1, 2, and 3 with probabilities depicted in Figure 3.19. The probabilities of observing values of 1, 2, and 3 are 0.1, 0.3, and 0.6, respectively. This is an asymmetric distribution, with only three discrete outcomes. The mean is 2.5. Sampling from this population can be simulated by placing 600 tags marked with the number 3, 300 tags marked with the number 2, and 100 tags marked with the number 1 in a box. We will mix up the tags, select 10 (replacing each tag and mixing after each individual selection), and *compute the mean of the 10 samples.* A typical result might be five tags marked 3, four tags marked 2, and one tag marked 1, an average of 2.4. With a computer or programmable calculator, we can simulate this drawing of 10 tags. The distributions of 100 such means for samples of sizes 10 and 20 obtained from a computer simulation are shown in Figure 3.20. The distribution is closer to normal as the sample size is increased from 10 to 20. This is an empirical demonstration of

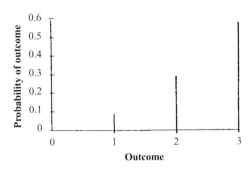

Figure 3.19 Probability distribution of outcomes 1, 2, and 3.

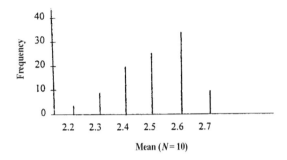

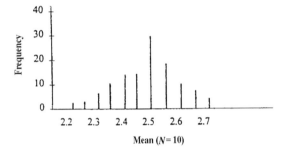

Figure 3.20 Distribution of means of sizes 10 and 20 from population shown in Figure 3.19.

the central limit theorem. Of course, under ordinary circumstances, we would not draw 100 samples each of size 10 (or 20) to demonstrate a result that can be proved mathematically.

3.4.3 Normal Approximation to the Binomial

A very important result in statistical theory is that the binomial probability distribution can be approximated by the normal distribution if the sample size is sufficiently large (see sect. 3.4.2). A conservative rule of thumb is that if NP (the product of the number of observations and the probability of success) and Nq are both greater than or equal to 5, we can use the normal distribution to approximate binomial probabilities. With symmetric binomial distributions, when $P = q = 0.5$, the approximation works well for NP less than 5.

To demonstrate the application of the normal approximation to the binomial, we will examine the binomial distribution described above, where $N = 10$ and $p = 0.5$. We can superimpose a normal curve over the binomial with $\mu = 5$ (number of successes) and standard deviation $\sqrt{NPq} = \sqrt{10(0.5)(0.5)} = 1.58$, as shown in Figure 3.21.

The probability of a discrete result can be calculated by using the binomial probability [Eq. (3.9)] or Table IV.3. The probability of seven successes, for example, is equal to 0.117. In a normal distribution, the probability of a single value cannot be calculated. We can only calculate the probability of a range of values within a specified interval. The area that approximately

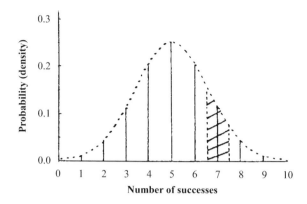

Figure 3.21 Normal approximation to binomial distribution: $NP = 5$ and s.d. $= 1.58$.

corresponds to the probability of observing seven successes in 10 trials is the area between 6.5 and 7.5, as illustrated in Figure 3.21. This area can be obtained by using the Z transformation discussed earlier in this chapter [Eq. (3.14)]. The area between 6.5 and 7.5 is equal to the area below 7.5 minus the area below 6.5.

Area below 6.5 $Z = \frac{6.5-5}{1.58} = 0.948$ from Table IV.2, area $= 0.828$.
Area below 7.5
$Z = \frac{7.5-5}{1.58} = 1.58$ from Table IV.2, area $= 0.943$.
Therefore, the area (probability) between 6.5 and 7.5 is

$$0.943 - 0.828 = 0.115.$$

This area is very close to the exact probability of 0.117.

The use of $X \pm 0.5$ to help estimate the probability of a discrete value, X, by using a continuous distribution (e.g., the normal distribution) is known as a continuity correction. We will see that the continuity correction is commonly used to improve the estimation of binomial probabilities by the normal approximation (chap. 5).

Most of our applications of the binomial distribution will involve data that allow for the use of the normal approximation to binomial probabilities. This is convenient because calculations using exact binomial probabilities are tedious and much more difficult than the calculations using the standard normal cumulative distribution (Table IV.2), particularly when the sample size is large.

3.5   OTHER COMMON PROBABILITY DISTRIBUTIONS

3.5.1   The Poisson Distribution

Although we will not discuss this distribution further in this book, the Poisson distribution deserves some mention. The Poisson distribution can be considered to be an approximation to the binomial distribution when the sample size is large and the probability of observing a specific event is small. In quality control, the probability of observing a defective item is often calculated by using the Poisson. The probability of observing X events of a given kind in N observations, where the probability of observing the event in a single observation is P, is

$$P(X) = \frac{\lambda^X e^{-\lambda}}{X!} \qquad (3.15)$$

where $\lambda = NP$, e the base of natural logarithms (2.718...), and N the number of observations.

We may use the Poisson distribution to compute the probability of finding one defective tablet in a sample of 100 taken from a batch with 1% defective tablets. Applying Eq. (3.15), we have

$$N = 100 \quad P = 0.01 \quad NP = \lambda = (100)(0.01) = 1$$

$$P(1) = \frac{(1)^1(e^{-1})}{1!} = e^{-1} = 0.368.$$

The exact probability calculated from the binomial distribution is 0.370. (See Exercise Problem 8.)

3.5.2 The *t* Distribution ("Student's *t*")

The *t* distribution is an extremely important probability distribution. This distribution can be constructed by repeatedly taking samples of size N from a normal distribution and computing the statistic

$$t = \frac{\bar{X} - \mu}{S/\sqrt{N}},$$

where $\bar{X}$ is the sample mean, μ the true mean of the normal distribution, and S the sample standard deviation. The distribution of the *t*'s thus obtained forms the *t* distribution. The exact shape of the *t* distribution depends on sample size (degrees of freedom), but the *t* distribution is symmetrically distributed about a mean of zero, as shown in Figure 3.22(A).

To elucidate further the concept of a sampling distribution obtained by repeated sampling, as discussed for the *t* distribution above, a simulated sampling of 100 samples each of size 4 ($N = 4$) was performed. These samples were selected from a normal distribution with mean 50 and standard deviation equal to 5, for this example. The mean and standard deviation of each sample of size 4 were calculated and a *t* ratio [Eq. (3.16)] constructed.

The distribution of the 100 *t* values thus obtained is shown in Table 3.5. The data are plotted (histogram) together with the theoretically derived *t* distribution with 3 degrees of freedom ($N - 1 = 4 - 1 = 3$) in Figure 3.23. Note that the distribution is symmetrically centered around a mean of 0, and that 5% of the *t* values are 3.18 or more units from the mean (theoretically).

3.5.3 The Chi-Square (χ^2) Distribution

Another important probability distribution in statistics is the chi-square distribution. The chi-square distribution may be derived from normally distributed variables, defined as the sum of squares of independent normal variables, each of which has mean 0 and standard deviation 1. Thus, if Z is normal with $\mu = 0$ and $\sigma = 1$,

$$\chi^2 = \sum Z_i^2. \qquad (3.17)$$

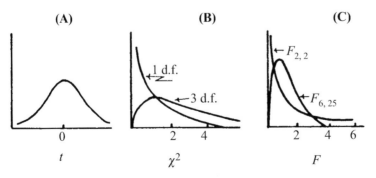

Figure 3.22 Examples of typical probability distributions.

Table 3.5 Frequency Distribution of 100 t Values Obtained by Simulated Repeat Sampling from a Normal Distribution with Mean 50 and Standard Deviation 5[a]

Class interval	Frequency
−5.5 to −4.5	1
−4.5 to −3.5	2
−3.5 to −2.5	2
−2.5 to −1.5	11
−1.5 to −0.5	18
−0.5 to +0.5	29
+0.5 to +1.5	21
+1.5 to +2.5	9
+2.5 to +3.5	4
+3.5 to +4.5	2
+4.5 to +5.5	1

[a] Sample size = 4.

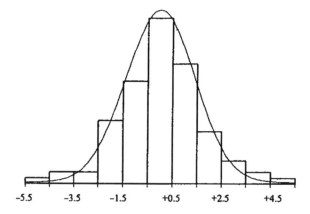

−5.5   −3.5   −1.5   +0.5   +2.5   +4.5

Figure 3.23 Simulated t distribution (d.f. $= 3$) compared to a theoretical t distribution.

Applications of the chi-square distribution are presented in chapters 5 and 15. The chi-square distribution is often used to assess probabilities when comparing discrete values from comparative groups, where the normal distribution can be used to approximate discrete probabilities.

As with the t distribution, the distribution of chi-square depends on degrees of freedom, equal to the number of independent normal variables as defined in Eq. (3.17). Figure 3.22(B) shows chi-square distributions with 1 and 3 degrees of freedom.

3.5.4 The F Distribution

After the normal distribution, the F distribution is probably the most important probability distribution used in statistics. This distribution results from the sampling distribution of the ratio of two independent variance estimates obtained from the same normal distribution. Thus, the first sample consists of N_1 observations and the second sample consists of N_2 observations

$$F = \frac{S_1^2}{S_2^2}. \tag{3.18}$$

The F distribution depends on two parameters, the degrees of freedom in the numerator ($N_1 - 1$) and the degrees of freedom in the denominator ($N_2 - 1$). This distribution is used to test for differences of means (analysis of variance) as well as to test for the equality of two variances. The F distribution is discussed in more detail in chapters 5 and 8 as applied to the comparison of two variances and testing of equality of means in the analysis of variance, respectively.

3.6 THE LOG-NORMAL DISTRIBUTION

The log-normal distribution results from a distribution, skewed to the right (see Fig. 10.1). Such data, when transformed into logs, exhibit the properties of a normal distribution. Thus, the logs of the original skewed data are normally distributed. The log-normal distribution is an important distribution in applications to the pharmaceutical sciences. Two important applications are in the analysis of bioequivalence data (see chap. 11) and particle size analysis. Some of the properties of a log-normal distribution are presented at this time as applied to the analysis of particle size of active drug substances and powders, such as excipients. This application is important, as the particle size of ingredients in common dosage forms may profoundly affect the therapeutic activity of the active ingredient.

3.6.1 Statistical Analysis of Particle Size

Typically, to characterize the particle size of a powdered substance, one defines the characteristics of the distribution of particles as previously described in this book (chap. 1), the mean, the standard deviation, and percentiles, particularly the 50th percentile or the median.

If the distribution of the particles followed an approximately normal distribution, the median and mean would be the "identical." However, experience shows that the particle size distribution is typically skewed to the right and follows an approximately log-normal distribution. We will not be concerned with the question of how to measure particle size, but for the present discussion, we will be measuring the diameter of the particles, assuming that the particles are perfect spheres. This measurement is also known as the spherical equivalent diameter. This is the diameter of a sphere that would have the equivalent volume of the irregular-shaped particle [7]. We will assume that the diameters have a log-normal distribution. Consider the data in Table 3.6 [7] that describe the distribution of diameters in the form of a frequency table.

The cumulative distribution is shown in Figure 3.24. Figure 3.24 is a special kind of graph that plots the cumulative distribution of the logs of the diameters on a probability scale, sometimes referred to as a probability plot. In this example, the X axis represents the area under a normal curve in terms of standard deviations (the standard normal curve). The Y axis represents the cumulative distribution of the logarithms of the diameters. Thus, the cumulative distribution of the logs of the diameters is conveniently shown on log-probability paper. If the distribution is log-normal, such a plot should show a straight line. That is, cumulative data plotted on probability paper will show a straight line if the data are normally distributed.

From Table 3.6 and Figure 3.24, the mean, median, and other percentiles, such as the 10th and 90th percentile, as well as the standard deviation can be ascertained for the log-transformed values. For example, referring to Figure 3.24, the median is represented by the 50% point on the probability scale, 10 μm for diameters (see below). Thus, the log-normal distribution can be conveniently characterized by these well-known parameters.

For example, inspection of Figure 3.24 shows that the 90th percentile is approximately 20 μm for the log-transformed diameters. Note, that for a log-normal distribution, the mean is larger than the median based on the original, untransformed numbers. The mean of the original, untransformed diameters is 12.25 mu. The median of the untransformed and transformed numbers does not change, 10 mu, when the median of the logs is back-transformed to the original numbers. The distribution of the logs of the diameters, however, will be normal, and shows a symmetric distribution where the mean and median are the same. Table 3.6 shows particle diameters in intervals (bins), and includes calculations based on the diameters and the mass or weight, as explained below.

As previously noted, typically, the distribution of diameters is considered to have a log-normal distribution. This distribution is based on the frequency of particles in the intervals describing the distribution (The first five columns in Table 3.6). The mean of the untransformed diameters is 12.25 mu as computed for data in the form of frequency tables (see sect. 1.2). Note that the mean of the log-transformed diameters is 2.300. The antilog of 2.300 is 9.97 or approximately 10, the same as the median.

Another way in which the particle distribution is described is by weight. If the density of the particles is considered constant, the weight will be proportional to the cube of the diameter. Note that the weight of a spherical particle is equal to volume x density. The volume of a

Table 3.6 Calculation of Some Average Diameters [7]

Interval	Diameter (D) (midsize)	Frequency (N)	Cumulative frequency (frequency)	Cum. % frequency (frequency)	Diameter × frequency (D × N)	ln(diameter)
1–1.4	1.2	2	2	0.2	2.4	0.182
1.4–2.0	1.7	5	7	0.7	8.5	0.531
2.0–2.8	2.4	14	21	2.1	33.6	0.875
2.8–3.6	3.2	60	81	8.1	192	1.163
3.6–6.0	4.8	100	181	18.1	480	1.569
6.0–7.6	6.8	190	371	37.1	1292	1.917
7.6–12.4	10	250	621	62.1	2500	2.303
12.4–15.6	14	160	781	78.1	2240	2.639
15.6–22.4	19	110	891	89.1	2090	2.944
22.4–29.6	26	70	961	96.1	1820	3.258
29.6–42.4	36	28	989	98.9	1008	3.584
42.4–59.6	51	10	999	99.9	510	3.932
59.6–84.4	72	1	1000	100	72	4.277
Total		1000			12248.5	
Average					12.2485	

Interval	Diameter (D) (midsize)	N × ln (diameter)	Cum. N × ln(D)	ND^3	Cum. ND^3	Cum. % (cum. ND^3)
1–1.4	1.2	0.365	0.365	3.456	3.5	6.01E-07
1.4–2.0	1.7	2.653	3.018	24.565	28.0	4.87E-06
2.0–2.8	2.4	12.257	15.274	193.536	221.6	3.85E-05
2.8–3.6	3.2	69.789	85.063	1966.08	2187.6	0.00038
3.6–6.0	4.8	156.862	241.925	11059.2	13246.8	0.002303
6.0–7.6	6.8	364.215	606.140	59742.08	72988.9	0.012687
7.6–12.4	10	575.646	1181.787	250000	322988.9	0.056143
12.4–15.6	14	422.249	1604.036	439040	762028.9	0.132458
15.6–22.4	19	323.888	1927.924	754490	1516518.9	0.263606
22.4–29.6	26	228.067	2155.991	1230320	2746838.9	0.477465
29.6–42.4	36	100.339	2256.329	1306368	4053206.9	0.704542
42.4–59.6	51	39.318	2295.648	1326510	5379716.9	0.935121
59.6–84.4	72	4.277	2299.924	373248	5752964.9	1
Total		2299.9242		5752964.917		
Average		2.2999242		5752.964917		

Average diameter = 12.2485; Standard deviation = 8.52766; CV = s.d./average diameter = 0.69622; GSD = Geometric s.d. from Figure 3.24 = ~1.9; CV = ~0.745; ln(GSD) = 0.642.

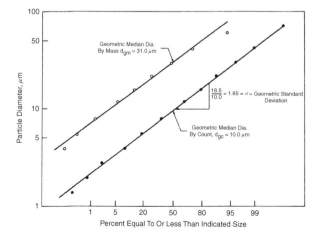

Figure 3.24 Log-probability plot of distribution from Table 3.6 [7].

sphere is 4/3 pi (radius)^3. The distribution can be described by the cube of the diameters (which are proportional to the weights of the particles) in each interval (see the last three columns in Table 3.6). The median weight is sometimes defined by particle size analysts as the particle *diameter* below which 50% of the weights lie. This is not the typical definition of the median. Using the usual definition, the median weight would be that weight that would have 50% of the weight below (and above) that value. For example, in Table 3.6, the usual definition would be the 50% cumulative point in the column labeled "Cum ND,^{3}" a value somewhat greater than 2,746,839, because that defines the weight that has 50% of the weight above and below this value. However, the median weight is often described in particle size analysis as *the diameter that corresponds to the median weight*, not the usual definition. From Figure 3.6 and Table 3.24, this value is 31 mu. Because this definition of the median weight, as often described by particle size analysts, is different from the usual definition, and there are several definitions of the median particle size [7,8], *one should clearly explain how the median is derived.*

Thus, there is some disconnect in these definitions. For example, the mean of the distribution would depend on whether we are talking about diameters or weights, and how the values are determined. For example, the determination of the mean of the particle distribution could be determined on a weight basis or diameter basis. Not only would the answers be different depending on the definition, but the distribution of particles will also depend on the definitions, whether we are talking about weight or diameters.

Other parameters can be determined from Figure 3.24 and Table 3.6. The coefficient of variation (CV = s.d./mean) for the diameters from Table 3.6 is 8.528/12.249 = 0.696. The geometric standard deviation (GSD) can be determined from the log-probability plot (Fig. 3.24). The standard deviation can be read from the plot as the distance of the cumulative diameters between the 50th percentile and 84th percentile. Note that one standard deviation encompasses 34% of the area above the mean or median. This GSD is approximately 1.9. It can be shown that the log (ln) of the GSD is equal to the CV of the untransformed data, equals approximately, 0.64. This is close to the observed CV of the untransformed diameters, 0.696. The CV based on the log-transformed diameters can be estimated by using the following equation:

$$CV = \sqrt{\{e^{(GSD2-1)}\}}$$
$$= \sqrt{\{e^{(0.642-1)}\}} = 0.74.$$

Note that these are theoretical concepts, so that one does not expect the observed and theoretical values to be identical. Of course, we do not expect data to exactly conform to a log-normal distribution, just as we do not expect real data to exactly conform to a normal distribution. (For more detailed discussion of particle size analysis, see Refs. [7,8].)

KEY TERMS

Binomial distribution
Binomial formula
Binomial trial
Central limit theorem
Chi-square distribution
Combinations
Conditional probability
Continuous distribution
Cumulative distribution
Density function
Discontinuous variable
Discrete distribution
Distribution
Equally likely
Event
Failure
F distribution
Independent events

Log-normal distribution
Multiplicative probability
Mutually exclusive
Negative binomial distribution
Normal distribution
Outcome
Poisson distribution
Population
Probability distribution
Proportion
Random
Randomly chosen
Standard normal distribution
Success
t distribution
Variability
Factorial
Z transformation

EXERCISES

1. Explain why do you think that a controlled multicenter clinical study better estimates the probability of a patient responding to treatment than the observations of a single physician in daily practice.

2. Describe the population that represents the multicenter antibiotic clinical study described in section 3.3.

3. Give three examples of probability distributions that describe the probability of outcomes in terms of attributes.

4. Explain why 30,000 tablets are only specked if 20,000 tablets are both chipped and specked as described in section 3.2. What is the probability, in the example described in section 3.2, of finding a specked tablet *or* a chipped tablet? (Hint: Count all the tablets that have either a speck or a chip.) See Eq. (3.4).

5. In a survey of hospital patients, it was shown that the probability that a patient has high blood pressure given that he or she is diabetic was 0.85. If 10% of the patients are diabetic and 25% have high blood pressure:
   (a) What is the probability that a patient has both diabetes and high blood pressure?
   (b) Are the conditions of diabetes and high blood pressure independent? [Hint: See Eqs. (3.5), (3.6), and (3.7).]

6. Show how the result 0.21094 is obtained for the probability of two of four patients being cured if the probability of a cure is 0.75 for each patient and the outcomes are independent (Table 3.2). (Enumerate all ways in which two of four patients can be cured, and compute the probability associated with each of these ways.)

7. What is the probability that three of six patients will be cured if the probability of a cure is 60%?

8. Calculate the probability of one success in 100 trials if $p = 0.01$.

9. From the cumulative plot in Figure 3.7(B), estimate the probability that a value, selected at random, will be (a) greater than Z_0; (b) less than Z_0.

10. What is the probability that a normal patient has a cholesterol value below 170 ($\mu = 215$, $\sigma = 35$)?

11. If the mean and standard deviation of the potency of a batch of tablets are 50 and 5 mg, respectively, what proportion of the tablets have a potency between 40 and 60 mg?

12. If a patient has a serum cholesterol value outside normal limits, does this mean that the patient is abnormal in the sense of having a disease or illness?

13. Serum sodium values for normal persons have a mean of 140 mEq/L and a s.d. of 2.5. What is the probability that a person's serum sodium will be between 137 and 142 mEq/L?

14. Data were collected over many years on cholesterol levels of normal persons in a New York hospital with the following results based on 100,000 readings. The mean is 205 mg%; the s.d. is 45. Assuming that the data have a normal distribution, what is the probability that a normal patient has a value greater than 280 mg%?

15. In the game of craps, two dice are thrown, each dice having an equal probability of showing one of the numbers 1 to 6 inclusive. Explain why the probability of observing a point of 2 (the sum of the numbers on the two dice) is 1/36.

16. Is the probability of observing two heads and one tail the same under the two following conditions: (a) simultaneously throwing three coins; (b) tossing one coin three consecutive times? Explain your answer.

17. What odds would you give of finding either none *or* one defective tablet in a sample of size 20 if the batch of tablets has 1% defective? Answer the same question if the sample size is 100.

18. What is the probability of observing exactly one head in 10 tosses of a coin?

§§19. The chance of obtaining a cure using conventional treatment for a particular form of cancer is 1%. A new treatment being tested cures two of the first four patients tested. Would you announce to the world that a major breakthrough in the treatment of this cancer is imminent? Explain your answer.

20. What is the s.d. for the binomial experiments described in Problems 17 and 19? (Answer in terms of NPq and Pq/N.)

§§21. In screening new compounds for pharmacological activity, the compound is administered to 20 animals. For a standard drug, 50% of the animals show a response on the average. Fifteen of the twenty animals show the response after administration of a new drug. Is the new drug a promising candidate? Why? [Hint: Compute the s.d. of the response based on $p = 0.5$. See if the observed response is more than 2 s.d.'s greater than 0.5.]

22. Using the binomial formula, calculate the probability that a sample of 30 tablets will show 0 or 1 defect if there are 1% defects in the batch. (What is the probability that there will be more than one defect in the sample of 30?)

23. The following expression can be used to calculate the probability of observing A or B or C (or any combination of A, B, C)

$$P(A \text{ or } B \text{ or } C) = P(A) + P(B) + P(C) - P(A \text{ and } B)$$
$$- P(A \text{ and } C) - P(B \text{ and } C) + P(A \text{ and } B \text{ and } C).$$

A survey shows that 85% of people with colds have cough, rhinitis, pain, or a combination of these symptoms. Thirty-five percent have at least cough, 50% have at least rhinitis, and 50% have at least pain. Twenty percent have (at least) cough and rhinitis, 15% have cough and pain, and 25% have rhinitis and pain. What percentage have all three symptoms?

REFERENCES

1. Hald A. Statistical Theory and Engineering Applications. New York: Wiley, 1965.
2. United States Pharmacopeia, 23rd Rev., and National Formulary, 18th ed. Rockville, MD: USP Pharmacopeial Convention, Inc., 1995.
3. Ostle B. Statistics in Research, 3rd ed. Ames, IA: Iowa State University Press, 1981.
4. Dixon WJ, Massey FJ Jr. Introduction to Statistical Analysis, 3rd ed. New York: McGraw-Hill, 1969.
5. Feller W. Probability Theory and Its Applications. Wiley and Sons, NY: Wiley, 1950:218.
6. http://en.wikipedia.org/wiki/Negative_binomial_distribution.
7. Stockham JD, Fochtman EG, eds. Particle Size Analysis. Ann Arbor, MI: Ann Arbor Science, 1977.
8. Jillavenkatesa A, Dapkunas S, Lum LH. Particle Size Characterization. Washington, D.C.: Material Science and Engineering Laboratory, U.S. Department of Commerce, National Institute of Standards and Technology, 2001:960–961.

§§ Optional, more difficult problems.

4 | Choosing Samples

The samples are the units that provide the experimental observations, such as tablets sampled for potency, patients sampled for plasma cholesterol levels, or tablets inspected for defects. The sampling procedure is an essential ingredient of a good experiment. An otherwise excellent experiment or investigation can be invalidated if proper attention is not given to choosing samples in a manner consistent with the experimental design or objectives. Statistical treatment of data and the inference based on experimental results depend on the sampling procedure. The way in which samples should be selected is not always obvious, and requires careful thought.

The implementation of the sampling procedure may be more or less difficult depending on the experimental situation, such as that which we may confront when choosing patients for a clinical trial, sampling blends, or choosing tablets for quality control tests. In this chapter, we discuss various ways of choosing samples and assigning treatments to experimental units (e.g., assigning different drug treatments to patients). We will briefly discuss various types of sampling schemes, such as simple random sampling, stratified sampling, systematic sampling, and cluster sampling. In addition, the use of random number tables to assign experimental units to treatments in designed experiments will be described.

4.1 INTRODUCTION

There are many different ways of selecting samples. We all take samples daily, although we usually do not think of this in a statistical sense. Cooks are always sampling their wares, tasting the soup to see if it needs a little more spice, or sampling a gravy or sauce to see if it needs more mixing. When buying a car, we take a test ride in a "sample" to determine if it meets our needs and desires.

The usual purpose of observing or measuring a property of a sample is to make some inference about the population from which the sample is drawn. In order to have reasonable assurance that we will not be deceived by the sample observations, we should take care that the samples are not biased. We would clearly be misled if the test car was not representative of the line, but had somehow been modified to entice us into a sale. We can never be sure that the sample we observe mirrors the entire population. If we could observe the entire population, we would then know its exact nature. However, 100% sampling is virtually never done. (One well-known exception is the U.S. census.) It is costly, time consuming, and may result in erroneous observations. For example, to inspect each and every one of 2 million tablets for specks, a tedious and time consuming task, would probably result in many errors due to fatigue of the inspectors.

Destructive testing precludes 100% sampling. To assay each tablet in a batch does not make sense. Under ordinary circumstances, no one would assay every last bit of bulk powder to ensure that it is not adulterated.

The sampling procedure used will probably depend on the experimental situation. Factors to be considered when devising a sampling scheme include

1. The nature of the population. For example, can we enumerate the individual units, such as packaged bottles of a product, or is the population less easily defined, as in the case of hypertensive patients?
2. The cost of sampling in terms of both time and money.
3. Convenience. Sometimes it may be virtually impossible to carry out a particular sampling procedure.
4. Desired precision. The accuracy and precision desired will be a function of the sampling procedure and sample size.

Sampling schemes may be roughly divided into *probability sampling* and *nonprobability sampling* (sometimes called authoritative sampling). Nonprobability sampling methods often are conceptually convenient and simple. These methods are considered as methods of *convenience* in many cases. Samples are chosen in a particular manner because alternatives are difficult. For example, when sampling powder from 10 drums of a shipment of 100 drums, those drums that are most easily accessible might be the ones chosen. Or, when sampling tablets from a large container, we may conveniently choose from those at the top. A "judgment" sample is chosen with possible knowledge that some samples are more "representative" than others, perhaps based on experience. A quality control inspector may decide to inspect a product during the middle of a run, feeling that the middle is more representative of the "average" product than samples obtained at the beginning or end of the run. The inspector may also choose particular containers for inspection based on knowledge of the manufacturing and bottling procedures. A "haphazard" sample is one taken without any predetermined plan, but one in which the sampler tries to avoid bias during the sampling procedure. Nonprobability samples often have a hidden bias, and it is not possible to apply typical statistical methods to estimate the population parameters (e.g., μ and σ) and the precision of the estimates. Nonprobability sampling methods should not be used unless probability sampling methods are too difficult or too expensive to implement.

We will discuss procedures and some properties of common *probability sampling methods.* Objects chosen to be included in probability samples have a known probability of being included in the sample and are chosen by some random device.

4.2 RANDOM SAMPLING

Simple *random sampling* is a common way of choosing samples. A random sample is one in which each individual (object) in the population to be sampled has an *equal chance of being selected.* The procedure of choosing a random sample can be likened to a bingo game or a lottery where the individuals (tags, balls, tablets, etc.) are thoroughly mixed, and the sample chosen at "random." This ensures that there is no bias; that is, *on the average*, the estimates of the population parameters (e.g., the mean) will be accurate. However, one should be aware, that in any single sample, random sampling does not ensure an accurate estimate of the mean and/or the standard deviation. An example of the lack of reliability of small samples can be shown based on the batting statistics of a 0.250 hitter in a given game. We would expect one hit in four times at bat, on the average. Suppose the batter comes up four times in the game. What is the probability that he will get exactly one hit? Applying the binomial theorem, the probability is 42%. See Problem number 12 at the end of this chapter.

Many statistical procedures are based on an assumption that samples are chosen at random. Simple random sampling is most effective when the variability is relatively small and uniform over the population [1].

In most situations, it is not possible to mix the objects that constitute the population and pick the samples out of a "box." But if all members of the population can be identified, a unique identification, such as a number, can be assigned to each individual unit. We can then choose the sample by picking numbers, randomly, from a box using a lottery-like technique. Usually, this procedure is more easily accomplished through the use of a table of random numbers. Random numbers have been tabulated extensively [2]. In addition to available tables, computer-generated random numbers may be used to select random samples or to assign experimental units randomly to treatments as described below.

4.2.1 Table of Random Numbers

Random numbers are frequently used as a device to choose samples to be included in a survey, a quality control inspection sample, or to assign experimental units to treatments such as assigning patients to drug treatments. The first step that is often necessary in the application of a table of random numbers is to assign a number to each of the experimental units in the population or to the units potentially available for inclusion in the sample. The numbers are assigned consecutively from 1 to N, where N is the number of units under consideration. The experimental units may be patients to be assigned to one of two treatments or bottles of tablets to be inspected for defects. We then choose a "starting point" in the table of random numbers, in some "random" manner. For example, we can close our eyes and point a finger on a page of the random number table, and this can be the starting point. Alternatively, the numbers thus

Choosing a random sample.

chosen can be thought of as the page, column, and row number of a new starting point. Using this random procedure, having observed the numbers 3674826, we would proceed to page 367, column 48, and row 26 in a book such as *A Million Random Digits* [2]. This would be the starting point for the random section. If the numbers designating the starting point do not correspond to an available page, row, or column, the next numbers in sequence (going down or across the page as is convenient) can be used, and so on.

Table IV.1 is a typical page from a table of random numbers. The exact use of the table will depend on the specific situation. Some examples should clarify applications of the random number table to randomization procedures.

1. A sample of 10 bottles is to be selected from a universe of 800 bottles. The bottles are numbered from 1 to 800 inclusive. A starting point is selected from the random number table and three-digit numbers are used to accommodate the 800 bottles. Suppose that the starting point is row 6 and column 21 in Table IV.1. (The first three-digit number is 177.) If a number greater than 800 appears or a number is chosen a second time (i.e., the same number appears twice or more in the table), skip the number and proceed to the next one. The first 10 numbers found in Table IV.1 with the starting point above and subject to the foregoing restraints are (reading down) 177, 703, 44, 127, 528, 43, 135, 104, 342, and 604 (Table 4.1). Note that we did not include 964 because there is no bottle with this number; only 800 bottles are available. These numbers correspond to the 10 bottles that will be chosen for inspection.
2. Random numbers may be used to assign patients randomly to treatments in clinical trials. Initially, the characteristics and source of the patients to be included in the trial should be carefully considered. If a drug for the treatment of asthma were to be compared to a placebo

Table 4.1 Excerpt from Table IV.1

	Column 21	
Row 6	17	7
	70	3
	04	4
	12	7
	52	8
	04	3
	13	5
	96	4
	10	4
	34	2
	60	4

treatment, the source (or population) of the samples to be chosen could be all asthmatics in this country. Clearly, even if we could identify all such persons, for obvious practical reasons it would not be possible to choose those to be included in the study using the simple random sampling procedure described previously.

In fact, in clinical studies of this kind, patients are usually recruited by an investigator (physician), and *all* patients who meet the protocol requirements and are willing to participate are included. Most of the time, patients in the study are randomly assigned to the two or more treatments by means of a table of random numbers or a similar "random" device. Consider a study with 20 patients designed to compare an active drug substance to an identically appearing placebo. As patients enter the study, they are assigned randomly to one of the treatment groups, 10 patients to be assigned to each group. One way to accomplish this is to "flip" a coin, assigning, for example, heads to the active drug product and tails to the placebo. After 10 patients have been assigned to one group, the remaining patients are assigned to the incomplete group.

A problem with a simple random assignment of this kind is that an undesirable allocation may result by chance. For example, although improbable, the first 10 patients could be assigned to the active treatment and the last 10 to the placebo, an assignment that the randomization procedure is intended to avoid. (Note that if the treatment outcome is associated with a time trend due to seasonal effects, physician learning, personnel changes, etc., such an assignment would bias the results.) In order to avoid this possibility, the randomization can be applied to subgroups of the sample, sometimes called a block randomization. For 20 patients, one possibility is to randomize in groups of 4, 2 actives and 2 placebos to be assigned to each group of 4. This procedure also ensures that if the study should be aborted at any time, approximately equal numbers of placebo and active treated patients will be included in the results. Another application of blocking is to adjust the randomization for baseline variables, such as sex, duration of disease, and so on.

If the randomization is performed in groups of 4 as recommended, the following patient allocation would result. (Use Table 4.1 for the random numbers as before, odd for placebo, even for active.)

Patient	Random no.	Drug	Comment
1	1	P	
2	7	P	
3	—	D	
4	—	D	Assign D to patient 3 and 4 to ensure equal allocation of D and P in the subgroup
5	0	D	
6	1	P	
7	5	P	
8	—	D	Assign D to patient 8 to ensure equal allocation of D and P in the subgroup
9	0	D	
10	1	P	
11	9	P	
12	—	D	Assign D to patient 12 to ensure equal allocation of D and P in the subgroup
13	1	P	
14	3	P	
15	—	D	
16	—	D	Assign D to patients 15 and 16 to ensure equal allocation of D and P in the subgroup
17	6	D	
18	7	P	
19	0	D	
20	—	P	Assign D to patient 20 to ensure equal allocation of D and P in the subgroup

Table 4.2 Excerpt from Table IV.1: Assignment of First 10 Numbers Between 1 and 20 to Placebo

	Column 11	
Row 11 -----	44	22 78 84 26 <u>04</u> 33 46 <u>09</u> 52
59 29 97 68 60	71	91 38 67 54 <u>13</u> 58 <u>18</u> 24 76
48 55 90 65 72	96	57 69 36 <u>10</u> 96 46 92 42 45
66 37 32 <u>20</u> 30	77	84 57 <u>03</u> 29 10 45 65 04 26
68 49 69 10 82	53	75 91 93 30 34 25 20 57 27
83 62 64 <u>11 12</u>	67	<u>19</u> ---------

The source and methods of randomization schemes for experiments or clinical studies should be documented for U.S. Food and Drug Administration submissions or for legal purposes. Therefore, it is a good idea to use a table of random numbers or a computer-generated randomization scheme for documentation rather than the coin-flipping technique. One should recognize, however that the latter procedure is perfectly fair, the choice of treatment being due to chance alone. Using a table of random numbers, a patient may be assigned to one treatment if an odd number appears and to the other treatment if an even number appears. We use single numbers for this allocation. If even numbers are assigned to drug treatment, the numbers in Table 4.1 would result in the following assignment to drug and placebo (read numbers down each column, one number at a time; the first number is 1, the second number is 7, the third number is 0, etc.).

Patient		Patient		Patient		Patient	
1	1 P	6	0 D	11	6 D	16	2 D
2	7 P	7	1 P	12	7 P	17	4 D
3	0 D	8	9 P	13	0 D	18	3 P
4	1 P	9	1 P	14	4 D		
5	5 P	10	3 P	15	2 D		

Since 10 patients have been assigned to placebo (P), the remaining two patients are assigned to drug (D). Again, the randomization can be performed in subgroups as described in the previous paragraph. If the randomization is performed in subgroups of size 4, for example, the first 4 patients would be assigned as follows: patients 1 and 2 to placebo (random numbers 1 and 7), and patients 3 and 4 to drug to attain equal allocation of treatments in this sample of 4.

Another approach is to number the patients from 1 to 20 inclusive as they enter the study. The patients corresponding to the first 10 numbers from the random number table are assigned to one of the two treatment groups. The remaining patients are assigned to the second treatment. In our example, the first 10 numbers will be assigned to placebo and the remaining numbers to drug. In this case, two-digit numbers are used from the random number table. (The numbers 1–20 have at most two digits.) Starting at row 11, column 11 in Table IV.1 and reading across, the numbers in Table 4.2 represent patients to be assigned to the first treatment group, placebo. Reading across, the first 10 numbers to appear that are between 1 and 20 (disregarding repeats), underlined in Table 4.2, are 4, 9, 13, 18, 10, 20, 3, 11, 12, and 19. These patients are assigned to placebo. The remaining patients, 1, 2, 5, 6, 7, 8, 14, 15, 16, and 17, are assigned to drug.

Randomization in clinical trials is discussed further in section 11.2.6.

4.3 OTHER SAMPLING PROCEDURES: STRATIFIED, SYSTEMATIC, AND CLUSTER SAMPLING

4.3.1 Stratified Sampling

Stratified sampling is a procedure in which the population is divided into subsets or strata, and random samples are selected from each strata. Stratified sampling is a recommended way of sampling when the strata are very different from each other, but objects within each stratum are alike. The precision of the estimated population mean from this sampling procedure is based on the variability within the strata. Stratified sampling will be particularly advantageous when this within-object variability is small compared to the variability between objects in different strata. In quality control procedures, items are frequently selected for inspection at random within specified time intervals (strata) rather than in a completely random fashion (simple random sampling). Thus we might sample 10 tablets during each hour of a tablet run. Often, the sample size chosen from each stratum is proportional to the size of the stratum, but in some circumstances, disproportionate sampling may be optimal. The computation of the mean and variance based on stratified sampling can be complicated, and the analysis of the data should take stratification into account [1]. In the example of the clinical study on asthmatics (see sect. 4.2.1), the stratification could be accomplished by dividing the asthmatic patients into subsets (strata) depending on age, duration of illness, or severity of illness, for example. The patients are assigned to treatments randomly within each subset. (See randomization in blocks above.) Note in this example that patients within each stratum are more alike than patients from different strata.

Consider an example of sampling tablets for drug content (assay) during a tablet run. If we believe that samples taken close in time are more alike than those taken at widely differing times, stratification would be desirable. If the tableting run takes 10 hours to complete, and a sample of 100 tablets is desired, we could take 10 tablets randomly during each hour, a stratified sample. This procedure would result in a more precise estimate of the average tablet potency than a sample of 100 tablets taken randomly over the entire 10-hour run.

Although stratified sampling often results in better precision of the estimate of the population mean, in some instances the details of its implementation may be more difficult than those of simple random sampling.

4.3.2 Systematic Sampling

Systematic sampling is often used in quality control. In this kind of sampling, every nth item is selected (e.g., every 100th item). The initial sample is selected in a random manner. Thus, a quality control procedure may specify that 10 samples be taken at a particular time each hour during a production run. The time during the hour for each sampling may be chosen in a random manner. Systematic sampling is usually much more convenient, and much easier to accomplish than simple random sampling and stratified sampling. It also results in a uniform sampling over the production run, which may result in a more precise estimate of the mean. Care should be taken that the process does not show a cyclic or periodic behavior, because systematic sampling will then not be representative of the process. The correct variance for the mean of a systematic sample is less than that of a simple random sample if the variability of the systematic sample is greater than the variability of the entire set of data.

To illustrate the properties of a systematic sample, consider a tableting process in which tablet weights tend to decrease during the run, perhaps due to a gradual decrease in tableting pressure. The press operator adjusts the tablet pressure every hour to maintain the desired weight. The tablet weights during the run are illustrated in Figure 4.1. If tablets are sampled 45 minutes after each hour, the average result will be approximately 385 mg, a biased result.

If the data appear in a random manner, systematic sampling may be desirable because it is simple and convenient to implement. As noted above, "systematic sampling is more precise than random sampling if the variance within the systematic sample is larger than the population variance as a whole." Another way of saying this is that systematic sampling is precise when units within the same sample are heterogeneous, and imprecise when they are homogeneous [3]. In the tableting example noted in the previous paragraph, the units in the sample tend to be similar (precise) and systematic sampling is a poor choice. (See Exercise Problem 11 for an example of construction of a systematic sample.)

4.3.3 Cluster Sampling

In cluster sampling, the population is divided into groups or clusters each of which contain "subunits." In single-stage cluster sampling, clusters are selected at random and all elements of the clusters chosen are included in the sample (4).

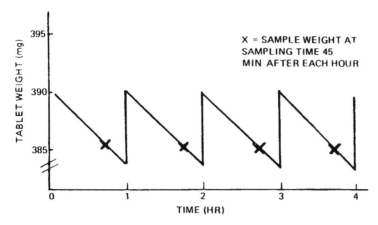

Figure 4.1 Illustration of problem with systematic sampling when process shows periodic behavior.

Two-stage cluster sampling may be used when there are many "primary" units, each of which can be "subsampled." For example, suppose that we wish to inspect tablets visually, packaged in the final labeled container. The batch consists of 10,000 bottles of 100 tablets each. The primary units are the bottles and the subsample units are the tablets within each bottle. Cluster sampling, in this example, might consist of randomly selecting a sample of 100 bottles, and then inspecting a random sample of 10 tablets from each of these bottles, thus the nomenclature, "two-stage" sampling. Often, cluster sampling is the most convenient way of choosing a sample. In the example above, it would be impractical to select 1000 tablets at random from the 1,000,000 packaged tablets (10,000 bottles × 100 tablets per bottle).

For a continuous variable such as tablet weights or potency, the estimate of the variance of the mean in two-stage cluster sampling is

$$(1 - f_1)S_1^2/n + [S_2^2/(\text{nm})](f_1(1 - f_2)) \tag{4.1}$$

where S_1^2 is the estimate of the variance among the primary unit means (the means of bottles). S_2^2 is the estimate of the variance of the subsample units, that is, units within the primary units (between tablets within bottles). f_1 and f_2 are the sampling fractions of the primary and subsample units, respectively. These are the ratios of units sampled to the total units available. In the present example of bottled tablets,

$f_1 = 100$ bottles/10,000 bottles $= 0.01$ (100 bottles are randomly selected from 10,000)
$f_2 = 10$ tablets/100 tablets $= 0.1$ (10 tablets are randomly selected from 100 for each of the 100 bottles)
$n =$ number of primary unit samples (100 in this example)
$m =$ number of units sampled from each primary unit (10 in this example).

If, in this example, S_1^2 and S_2^2 are 2 and 20, respectively, from Eq. (4.1), the estimated variance of the mean of 1000 tablets sampled from 100 bottles (10 tablets per bottle) is

$$\frac{(1 - 0.01)(2)}{100} + \left[\frac{20}{(100 \times 10)}\right](0.01)(0.9) = 0.01998.$$

If 1000 tablets are sampled by taking 2 tablets from each of 500 bottles, the estimated variance of the mean is

$$\frac{(1 - 0.05)(2)}{500} + \left[\frac{20}{(500 \times 2)}\right](0.05)(0.98) = 0.00478.$$

This example illustrates the increase in efficiency of sampling more primary units. The variance obtained by sampling 200 bottles is approximately one-half that of sampling 100 bottles.

If f_1 is small, the variance of the mean is related to the number of primary units sampled (n) equal to approximately S_1^2/n. Cost and time factors being equal, it is more efficient to sample more primary units and fewer subsample units given a fixed sample size. However, in many situations it is not practical or economical to sample a large number of primary units. The inspection of tablets in finished bottles is an example where inspection of many primary units (bottles) would be costly and inconvenient. See Exercise Problems 9 and 10 for further illustrations.

4.4 SAMPLING IN QUALITY CONTROL

Sampling of items for inspection, chemical, or physical analysis is a very important aspect of quality control procedures. For the moment, we will not discuss the important question: "What sample size should we take?" This will be discussed in chapter 6. What concerns us here is how to choose the samples. In this respect, the important points to keep in mind from a statistical point of view are as follows:

Tablet sampling

1. The sample should be "representative."
2. The sample should be chosen in a way that will be compatible with the objectives of the eventual data analysis.

For example, when sampling tablets, we may be interested in estimating the mean and standard deviation of the weight or potency of the tablet batch. If 20 tablets are chosen for a weight check during each hour for 10 hours from a tablet press (a stratified or systematic sample), the mean and standard deviation are computed in the usual manner if the production run is uniform resulting in random data. However, if warranted, the analysis should take into account the number of tablets produced each hour and the uniformity of production during the sampling scheme. For example, in a uniform process, an estimate of the average weight would be the average of the 200 tablets, or equivalently, the average of the averages of the 10 sets of 20 tablets sampled. However, if the rate of tablet production is doubled during the 9th and 10 hours, the averages obtained during these two hours should contribute twice the weight to the overall average as the average results obtained during the first eight hours. For further details of the statistical analysis of various sampling procedures, the reader is referred to Refs. [1,3].

Choosing a representative sample from a bulk powder, as an example, is often based on judgment and experience more than on scientific criteria (a "judgment" sample). Rules for sampling from containers and for preparing powdered material or granulations for assay are, strictly speaking, not "statistical" in nature. Bulk powder sampling schemes have been devised in an attempt to obtain a representative sample without having to sample an inordinately large amount of material. A common rule of thumb, taking samples from $\sqrt{N} + 1$ containers (N is the total number of containers), is a way to be reasonably sure that the material inspected is representative of the entire lot, based on tradition rather than on objective grounds. Using this rule, given a batch of 50 containers, we would sample ($\sqrt{50} + 1 = 8$) containers. The eight containers can be chosen using a random number table (see Exercise Problem 3).

Sampling plans for bulk powders and solid mixes such as granulations usually include the manner of sampling, the number of samples, and preparation for assay with an aim of obtaining a representative sample. One should bear in mind that a single assay will not yield information on variability. No matter what precautions we take to ensure that a single sample of a mix is representative of a batch, we can only estimate the degree of homogeneity by repeating the procedure one or more times on different portions of the mix. Repeat assays on the same sample gives an estimate of analytical error, not homogeneity of the mix. For a further discussion of this concept see chapter 13.

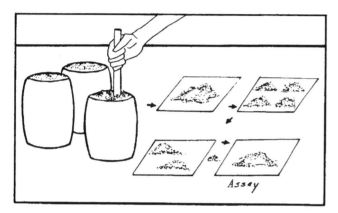

Sampling and assaying bulk powders.

An example of a procedure for sampling from large drums of a solid mixture is to insert a thief (a device for sampling bulk powders) and obtain a sample from the center of the container. A grain thief may be used to take samples from more than one part of the container. (If samples are to be taken for purposes of content uniformity, thieves that can sample small samples such as one or more tablets weights are recommended.) This procedure is repeated for an appropriate number of containers and the samples thoroughly mixed. The sample to be submitted for analysis is mixed further and quartered, rejecting two diagonal portions. The mixing and quartering is repeated until sufficient sample for analysis remains.

Other ideas on sampling for quality control and validation can be found in section 13.1.1.

KEY TERMS

Blocking
Cluster sample
Haphazard sample
Judgment sample
Multistage sample
Nonprobability sample
Probability sample
Representative sample

Sample
Sampling with replacement
Simple random sample
Stratified sample
Systematic sample
Table of random numbers
Two-stage cluster sample

EXERCISES

Use the table of random numbers (Tables IV.1) to answer the following questions.

1. Twenty-four patients are recruited for a clinical study, 12 patients to be randomly assigned to each of two groups, A and B. The patients come to the clinic and are entered into the

study chronologically, randomly assigned to treatment A or B. Devise a schedule showing to which treatment each of the 24 patients is assigned.

2. Devise a randomization scheme similar to that done in Problem 1 if 24 patients are to be assigned to three treatments.

3. Thirty drums of bulk material are to be sampled for analysis. How many drums would you sample? If the drums are numbered 1 to 30, explain how you chose drums and take the samples.

4. A batch of tablets is to be packaged in 5000 bottles each containing 1000 tablets. It takes four hours to complete the packaging operation. Ten bottles are to be chosen for quality control tests. Explain in detail how would you choose the 10 bottles.

5. Devise a randomization scheme to assign 20 patients to drug and placebo groups (10 patients in each group) using the numbers shown in Table 4.1 by using even numbers for assignment to drug and odd numbers for assignment to placebo.

6. Describe two different ways in which 20 tablets can be chosen during each hour of a tablet run.

7. One hundred bottles of a product, labeled 0 to 99 inclusive, are available to be analyzed. Analyze five bottles selected at random. Which five bottles would you choose to analyze?

8. A batch of tablets is produced over an eight-hour period. Each hour is divided into four 15-minute intervals for purposes of sampling. (Sampling can be done during 32 intervals, four per hour for eight hours.) Eight samples are to be taken during the run. Devise (a) a simple random sampling scheme, (b) a stratified sampling scheme, and (c) a systematic sampling scheme. Which sample would you expect to have the smallest variance? Explain.

9. The average potencies of tablets in 20 bottles labeled 1 to 20 are

Bottle number	Potency
1	312
2	311
3	309
4	309
5	310
6	308
7	307
8	305
9	306
10	307
11	305
12	301
13	303
14	300
15	299
16	300
17	300
18	297
19	296
20	294

(a) Choose a random sample of five bottles. Calculate the mean and standard deviation.
(b) Choose a systematic sample, choosing every 4th sample, starting randomly with one of the first four bottles. Calculate the mean and standard deviation of the sample.
(c) Compare the averages and standard deviations of the two samples and explain your results. Compare your results to those obtained by other class members.

10. Ten containers each contain four tablets. To estimate the mean potency, two tablets are to be randomly selected from three randomly chosen containers. Perform this sampling from the data shown below. Estimate the mean and variance of the mean. Repeat the sampling, taking three tablets from two containers. Explain your results. Compute the mean potency of all 40 tablets.

Container	Tablet potencies (mg)
1	290 289 305 313
2	317 300 285 327
3	288 322 306 299
4	281 305 309 289
5	292 295 327 283
6	286 327 297 314
7	311 286 281 288
8	306 282 282 285
9	313 301315 285
10	283 327 315 322

11. Twenty-four containers of a product are produced during eight minutes, three containers each minute. The drug content of each container is shown below

Minute		Container assay	
1	80	81	77
2	78	76	76
3	84	83	86
4	77	77	79
5	83	81	82
6	81	79	80
7	82	79	81
8	79	79	80

Eight containers are to be sampled and analyzed for quality control. Take a sample of eight as follows:

(a) Simple random sample.
(b) Stratified sample; take one sample at random each minute.
(c) Systematic sample; start with the first, second, or third container and then take every third sample thereafter.
Compute the mean and the variance of each of your three samples (a, b, and c). Discuss the results. Which sample gave the best estimate of the mean? Compare your results to those obtained from the other students in the class.

12. What is the probability that a batter with a 0.250 average will get exactly one hit in four times at bat? Answer: Probability of one hit in 4 times at bat is $4 \times (1/4) \times (3/4)^3 = 108/256 = 0.421875$, less than one-half of the time.

REFERENCES

1. Snedecor GW, Cochran WG. Statistical Methods, 8th ed. Ames, IA: Iowa State University Press, 1989.
2. The Rand Corporation. A Million Random Digits with 100,000 Normal Deviates. New York: The Free Press, 1966.
3. Cochran WG. Sampling Techniques, 3rd ed. New York: Wiley, 1967.
4. Stuart A. Basic Ideas of Scientific Sampling. London: Charles Griffith & Co., Ltd., 1976.

5 | Statistical Inference: Estimation and Hypothesis Testing

Parameter estimates obtained from samples are usually meant to be used to estimate the true population parameters. The sample mean and variance are typical estimators or predictors of the true mean and variance, and are often called "point" estimates. In addition, an interval that is apt to contain the true parameter often accompanies and complements the point estimate. These intervals, known as *confidence intervals*, can be constructed with a known a priori probability of bracketing the true parameters. Confidence intervals play an important role in the evaluation of drugs and drug products.

The question of *statistical significance* pervades much of the statistics commonly used in pharmaceutical and clinical studies. Advertising, competitive claims, and submissions of supporting data for drug efficacy to the FDA usually require evidence of superiority, effectiveness, and/or safety based on the traditional use of statistical hypothesis testing. This is the technique that leads to the familiar statement, "The difference is statistically significant" (at the 5% level or less, for example), words that open many regulatory doors. Many scientists and statisticians feel that too much is made of testing for statistical significance, and that decisions based on such statistical tests are often not appropriate. However, testing for statistical significance is one of the backbones of standard statistical methodology and the properties and applications of such tests are well understood and familiar in many experimental situations. This aspect of statistics is not only important to the pharmaceutical scientist in terms of applications to data analysis and interpretation, but is also critical to an understanding of the statistical process. Since much of the material following this chapter is based largely on a comprehension of the principles of hypothesis testing, the reader is urged to make special efforts to understand the material presented in this chapter.

5.1 STATISTICAL ESTIMATION (CONFIDENCE INTERVALS)

We will introduce the concept of statistical estimation and confidence intervals before beginning the discussion of hypothesis testing. Scientific experimentation may be divided into two classes: (a) experiments designed to estimate some parameter or property of a system, and (b) comparative experiments, where two or more treatments or experimental conditions are to be compared. The former type of experiment is concerned with estimation and the latter is concerned with hypothesis testing.

The term *estimation* in statistics has a meaning much like its meaning in ordinary usage. A population parameter is estimated based on the properties of a sample from the population. We have discussed the unbiased nature of the sample estimates of the true mean and variance, designated as $\overline{X}$ and S^2 (sects. 1.4 and 1.5). These sample statistics estimate the population parameters and are considered to be the best estimates of these parameters from several points of view.* However, the reader should understand that statistical conclusions are couched in terms of probability. Statistical conclusions are not invariant as may be the case with results of mathematical proofs. Without having observed the entire population, one can never be sure that the sample closely reflects the population. In fact, as we have previously emphasized, sample statistics such as the mean and variance are rarely equal to the population parameters.

---

* These "point" estimates are unbiased, consistent, minimum variance estimates. Among unbiased estimators, these have minimum variance, and approach the true value with high probability as the sample size gets very large.

Nevertheless, the sample statistics (e.g., the mean and variance) are the best estimates we have of the true parameters. Thus, having calculated $\overline{X}$ and S^2 for potencies of 20 tablets from a batch, one may very well inquire about the true average potency of the batch. If the mean potency of the *20 tablets is 49.8 mg*, the *best estimate* of the true batch mean is *49.8 mg*. This is known as the point estimate. Although we may be almost certain that the true batch mean is not exactly 49.8 mg, there is no reason, unless other information is available, to estimate the mean to be a value different from 49.8 mg.

The discussion above raises the question of the reliability of the sample statistic as an estimate of the true parameter. Perhaps one should hesitate in reporting that the true batch mean is 49.8 mg based on data from only 20 tablets. One might question the reliability of such an estimate. The director of quality control might inquire: "How close do you think the true mean is to 49.8 mg?" Thus, it is a good policy when reporting an estimate such as a mean to include some statement as to the reliability of the estimate. Does the 49.8-mg estimate mean that the true mean potency could be as high as 60 mg, or is there a high probability that the true mean is not more than 52 mg? This question can be answered by use of a *confidence interval*. A confidence interval is an interval within which we believe the true mean lies. We can say, for example, that the true batch mean potency is between 47.8 and 51.8 mg with 95% probability. The width of the interval depends on the properties of the population, the sample estimates of the parameters, and the degree of certainty desired (the probability statement).

Since most of the problems that we will encounter are concerned with the normal distribution, particularly sampling of means, we are most interested in confidence intervals for means. If the distribution of means is normal and σ is known, an interval with confidence coefficient, P (probability), can be computed using a table of the cumulative standard normal distribution, Table IV.2. A two-sided confidence interval, symmetric about the observed mean, is calculated as follows:

$$P\% \text{ confidence interval} = \overline{X} \pm \frac{Z_p\sigma}{\sqrt{N}} \tag{5.1}$$

where $\overline{X}$ is the observed sample mean, N the sample size, σ the population standard deviation, and Z_p the normal deviate corresponding to the $(P+1)/2$ percentile of the cumulative standard normal distribution (Table IV.2).

For the most commonly used 95% confidence interval, $Z = 1.96$, corresponding to $(0.95 + 1)/2 = 0.975$ of the area in the cumulative standard normal distribution. Other common confidence coefficients are 90% and 99%, having values of Z equal to 1.65 and 2.58, respectively. The probability statement, for example, 90%, 95%, 99%, depends on the context. Therefore, one cannot say that one probability is "better" than another. For example, in bioequivalence studies, a 90% confidence interval is most appropriate (see chap. 11). Inspection of Table IV.2 shows that the area in the tails of a normal curve between ± 1.65, ± 1.96, and ± 2.58 standard deviations from the mean is 90%, 95%, and 99%, respectively. This is illustrated in Figure 5.1 (see also Table 3.4).

Before presenting examples of the computation and use of confidence intervals, the reader should take time to understand the concept of a confidence interval. The confidence interval changes depending on the sample chosen because, although σ[†] and N remain the same, $\overline{X}$ varies from sample to sample. A confidence interval using the mean from any given sample may or may not contain the true mean. Without knowledge of the true mean, we cannot say whether or not any given interval contains the true mean. However, it can be proven that when intervals are constructed according to Eq. (5.1), $P\%$ (e.g., 95%) of such intervals will contain the true mean. Figure 5.2 shows how means of size N, taken from the same population, generate confidence intervals. Think of this as means of size 20, each mean generating a confidence interval [Eq. (5.1)]. For a 95% confidence interval, 19 of 20 such intervals will cover the true mean, μ, on the average. Any single interval has a 95% chance of covering the true mean, a

---

[†] σ is assumed to be known in this example.

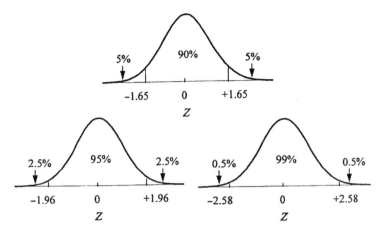

Figure 5.1 Areas in the tails of a standard normal curve.

priori. Of course, one would not usually take many means in an attempt to verify this concept, which can be proved theoretically. Under usual circumstances, only a single mean is observed and a confidence interval computed. This interval may not cover the true mean, but we know that 19 of 20 such intervals will cover the true mean.

Looking at the confidence interval from another point of view, suppose that a mean of 49.8 mg was observed for a sample size of 20 with $\sigma/\sqrt{N}$, $(\sigma_{\bar{x}})$ equal to 2. According to Eq. (5.1), the 95% confidence interval for the true mean is $49.8 \pm 1.96(2) = 45.9$ to 53.7 mg. Figure 5.3 shows that if the true mean were outside the range 45.9 to 53.7, the observation of the sample mean, 49.8 mg, would be very unlikely. The dashed curve in the figure represents the distribution of means of size 20 with a true mean of 54.7 and $\sigma_{\bar{x}} = 2$. In this example, the true mean is outside the 95% confidence interval, and the probability of observing a mean from this distribution as small as 49.8 mg or less is less than 1% (see Exercise Problem 1). Therefore, one could conclude that the true mean is probably not as great as 54.7 mg based on the observation of a mean of 49.8 mg from a sample of 20 tablets.

5.1.1 Confidence Intervals Using the *t* Distribution

In most situations in which confidence intervals are computed, σ, the true standard deviation, is unknown, but is estimated from the sample data. A confidence interval can still be computed based on the sample standard deviation, S. However, the interval based on the sample standard deviation will tend to be wider than that computed with a known standard deviation. This is reasonable because if the standard deviation is not known, one has less knowledge of the true distribution and consequently less assurance of the location of the mean.

The computation of the confidence interval in cases where the standard deviation is estimated from sample data is similar to that shown in Eq. (5.1) except that a value of t is substituted for the Z value

$$P\% \text{ confidence interval} = \overline{X} \pm \frac{tS}{\sqrt{N}}. \tag{5.2}$$

Values of t are obtained from the cumulative t table, Table IV.4, corresponding to a $P\%$ confidence interval.

The appropriate value of t depends on degrees of freedom (d.f.), a concept that we encountered in section 1.5.2. When constructing confidence intervals for means, the d.f. are equal to $N - 1$, where N is the sample size. For samples of size 20, d.f. $= 19$ and the appropriate values of t for 90%, 95%, or 99% confidence intervals are 1.73, 2.09, and 2.86, respectively. Examination of the t table shows that the values of t decrease with increasing d.f., and approach the corresponding Z values (from the standard normal curve) when the d.f. are large. This is expected, because when d.f. $= \infty$, the standard deviation is known and the t distribution coincides with

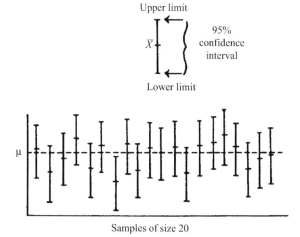

Figure 5.2   Concept of the confidence interval.

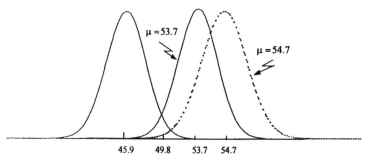

Figure 5.3   This figure shows that a mean of 49.8 is unlikely to be observed if the true mean is 54.7 (confidence interval = 45.9–53.7).

the standard normal distribution. We will talk more of the *t* distribution later in this chapter (see also sect. 3.5).

5.1.2   Examples of Construction of Confidence Intervals

Example 1: Confidence interval when σ is unknown and estimated from the sample. The labeled potency of a tablet dosage form is 100 mg. Ten individual tablets are assayed according to a quality control specification. The 10 assay results shown in Table 5.1 are assumed to be sampled from a normal distribution. The sample mean is 103.0 mg and the standard deviation is 2.22. A 95% confidence interval for the true batch mean [Eq. (5.1)] is

$$103 \pm 2.26 \left(\frac{2.22}{\sqrt{10}} \right) = 101.41 \text{ to } 104.59.$$

Table 5.1   Assay Results for 10 Randomly Selected Tablets (mg)

101.8	104.5
102.6	100.7
99.8	106.3
104.9	100.6
103.8	105.0
$\overline{X} = 103.0$	$S = 2.22$

Note that the t value is 2.26. This is the value of t with 9 d.f. ($N = 10$) for a 95% confidence interval taken from Table IV.4.

Example 2: Confidence interval when σ is known. Suppose that the standard deviation were known to be equal to 2.0. The 95% confidence interval for the mean is [Eq. (5.1)]

$$\overline{X} \pm \frac{1.96\sigma}{\sqrt{N}} = 103.0 \pm \frac{1.96(2.0)}{\sqrt{10}} = 101.76 \text{ to } 104.24.$$

The value 1.96 is obtained from Table IV.2 ($Z = 1.96$ for a two-sided symmetrical confidence interval) or from Table IV.4 for t with ∞ d.f.

Two questions arise from this example.

1. How can we know the s.d. of a batch of tablets without assaying every tablet?
2. Why is the s.d. used in Example 2 different from that in Example 1?

Although it would be foolhardy to assay each tablet in a batch (particularly if the assay were destructive, that is, the sample is destroyed during the assay process), the variance of a "stable" process can often be precisely estimated by averaging or pooling the variance over many batches (see also sect. 12.2 and App. I). The standard deviation obtained from this pooling is based on a large number of assays and will become very stable as long as the tableting process does not change. The pooled standard deviation can be assumed to be equal to or close to the true standard deviation (Fig. 5.4).

The answer to the second question has actually been answered in the previous paragraph. The variance of any single sample of 10 tablets will not be identical to the true variance, 2^2 or 4 in the example above. If the average variance over many batches can be considered equal to or very close to the true variance, the pooled variance is a better estimate of the variance than that obtained from 10 tablets. This presupposes that the variance does not change from batch to batch. Under these conditions, use of the pooled variance rather than the individual sample variance will result in a narrower confidence interval, on the average.

Example 3: Confidence Interval for a Proportion.(a) In a preclinical study, 100 untreated (control) animals were observed for the presence of liver disease. After six months, 25 of these animals were found to have the disease. We wish to compute a 95% confidence interval for

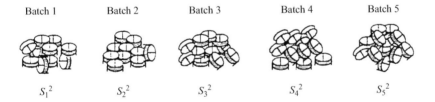

Batch 1	Batch 2	Batch 3	Batch 4	Batch 5
$S_1^{\,2}$	$S_2^{\,2}$	$S_3^{\,2}$	$S_4^{\,2}$	$S_5^{\,2}$

Pulled $S^2 = \overline{S}^2 = \sum S^2 / N$: N = number of batches

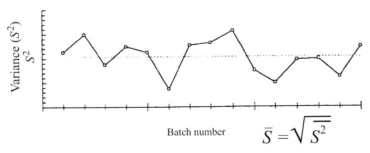

Figure 5.4 Pooling variances over batches, a good estimate of the true variance of a stable process (same sample size per batch).

the true proportion of animals who would have this disease if untreated (after six months). A confidence interval for a proportion has the same form as that for a mean. Assuming that the normal approximation to the binomial is appropriate, the confidence interval is approximately

$$\hat{p} \pm Z\sqrt{\frac{\hat{p}\hat{q}}{N}}, \tag{5.3}$$

where $\hat{p}$ is the observed proportion, $\hat{q} = 1 - \hat{p}$, Z the appropriate cutoff point from the normal distribution (Table IV.2), and N the sample size.

In the present example, a 95% confidence interval is

$$0.25 \pm 1.96\sqrt{\frac{(0.25)(0.75)}{100}} = 0.165 \text{ to } 0.335.$$

The true proportion is probably between 16.5% and 33.5%.[‡] Notice that the mean is equal to the observed proportion and that the normal approximation to the binomial distribution makes use of the Z value of 1.96 for the 95% confidence interval from the cumulative normal distribution. The standard deviation is computed from Eq. (3.11), $\sigma = \sqrt{\hat{p}\hat{q}/N}$.

A 99% confidence interval for the true proportion is

$$0.25 \pm 2.58\sqrt{\frac{(0.25)(0.75)}{100}} = 0.138 \text{ to } 0.362.$$

Note that the 99% confidence interval is wider than the 95% interval. The greater the confidence, the wider is the interval. To be 99% "sure" that the true mean is contained in the interval, the confidence interval must be wider than that which has a 95% probability of containing the true mean.

(b) To obtain a confidence interval for the true *number of animals* with liver disease when a sample of 100 shows 25 with liver disease, we use the standard deviation according to Eq. (3.12), $\sigma = \sqrt{N\hat{p}\hat{q}}$. A 95% confidence interval for the true *number* of diseased animals (where the observed number is $N\hat{p} = 25$) is

$$N\hat{p} \pm 1.96\sqrt{N\hat{p}\hat{q}} = 25 \pm 1.96\sqrt{(100)(0.25)(0.75)}$$
$$= 16.5 \text{ to } 33.5.$$

This answer is exactly equivalent to that obtained using proportions, in part (a) ($16.5/100 = 0.165$ and $33.5/100 = 0.335$). Further examples of symmetric confidence intervals are presented in conjunction with various statistical tests in the remaining sections of this chapter. In particular, confidence intervals for the true difference of two means or two proportions are given in sections 5.2.2, 5.2.3, and 5.2.6.

An interesting, special confidence interval that is useful for proportions is the case where 0 successes or failures are observed in N trials. For example, this situation arises in data from quality control and clinical trials. When inspecting individual items for sterility, we may observe zero defects in 1000 items inspected. In a clinical trial, we may observe no side effects of a particular kind in 200 patients. In these cases, it is of interest to put an upper bound on the proportion of failures, where failure in the above examples is the observation of a nonsterile item or a particular side effect. This can be calculated using a confidence interval. In these situations, we observe 100% successes and 0% failures, and the lower confidence interval is 0%. (We cannot have less than 0 failures.) We will put an upper limit on the true proportion of failures, equal to a lower limit on the proportion of successes. This may be thought of as a

---

[‡] Both $N\hat{p}$ and $N\hat{q}$ should be equal to or greater than 5 when using the normal approximation to the binomial (sect. 3.4.3).

one-sided confidence interval. To compute the probability of 0 failures in N trials, we can apply the binomial formula (from chap. 3)

$$\text{Probability of } X \text{ successes in } N \text{ trials} = \binom{N}{X} P^x q^{N-X}. \tag{3.9}$$

When $X = N$ (i.e., all N trials are successes), the probability of $X = N$ successes is P^N.

To obtain a lower limit for the proportion of successes based on a 95% confidence interval, we compute the value of p that results in a probability of 0.05, when we have all the observations successful. This computation uses the following formula:

$0.05 = p^N.$

A 95% confidence interval for the true proportion of failure is $1 - p$.

Suppose that inspection of 1000 ampoules in a batch of 30,000 shows that all items are sterile. With 95% confidence what is the upper limit of potential nonsterile ampoules in the batch.

$0.05 = p^{1000}$. The log of p can be calculated as $\log(0.05)/1000 = -0.002996$. $p =$ the antilog of $-0.002996 = 0.997$. The upper limit for the proportion of failures is $1 - 0.997 = 0.003$ or 3 in a thousand. We conclude that with 95% probability, there are no more than 3 failures in 1000 items. We see that it is impossible to guarantee 100% successes without inspecting each item (see also sampling, chap. 4). Certainly, this is not possible if the sampling is destructive. Of course, the intensity of sampling is dependent on cost and the potential risks to the consumer (patients) if failures exist in the batch.

Another useful application is the Negative Binomial, Time to Failure, described in chapter 3.

5.1.3 Asymmetric Confidence Intervals

5.1.3.1 One-Sided Confidence Intervals

In most situations, a two-sided confidence interval symmetric about the observed mean seems most appropriate. This is the shortest interval given a fixed probability. However, there are examples where a one-sided confidence interval can be more useful. Consider the case of a clinical study in which 18 of 500 patients treated with a marketed drug report headaches as a side effect. Suppose that we are only concerned with an "upper limit" on the proportion of drug-related headaches to be expected in the population of users of the drug. In this example, when constructing a 95% interval, we use a Z (or t) value that cuts off 5% of the area in the upper tail of the distribution, rather than the 2.5% in each tail excluded in a symmetric interval. Using the normal approximation to the binomial, the upper limit is

$$p + Z\sqrt{\frac{pq}{N}} = \frac{18}{500} + 1.65\sqrt{\frac{(0.036)(0.964)}{500}}$$

$$= 0.036 + 0.014 = 0.050.$$

Based on the one-sided 95% confidence interval, we conclude that the true proportion of headaches among drug users is probably not greater than 5%. Note that we make no statement about the lower limit, which must be greater than 0. Another application of a one-sided confidence interval is presented in section 7.5, as applied to the analysis of stability data. If a one-sided confidence interval is to be used for regulatory decisions or other "official" applications, the rationale for using a one-sided rather than a two-sided interval should be clearly explained prior to the experiment (see also sect. 5.1.3 explaining one-sided confidence intervals).

5.1.3.2 Other Asymmetric Confidence Intervals

In general, many P% confidence intervals can be constructed by suitably allocating $(1 - P)\%$ of the area to the lower and upper tails of the normal distribution. For example, a 95% confidence

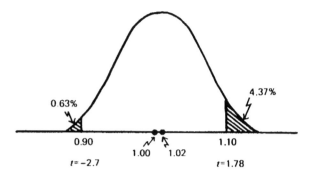

Figure 5.5 A 95% asymmetric confidence interval with $\overline{X} = 1.02$, s.d. $= 0.2$, and $N = 20$.

interval may be constructed by placing 1% of the area in the lower tail and 4% in the upper tail. This is not a common procedure and a good reason should exist before one decides to make such an allocation. Westlake [1,2] has proposed such an interval for the construction of confidence intervals in bioequivalence studies. In these studies, a ratio of some property (such as maximum serum concentration) of two products is compared. Westlake argues that an interval symmetric about the ratio 1.0 is more useful than one symmetric about the observed sample mean. The interval often has the great majority of the area in either the lower or upper tail, depending on the observed ratio. For a ratio greater than 1.0, most of the area will be in the upper tail and vice versa. Figure 5.5 illustrates this concept with a hypothetical example for products with an average ratio of 1.02. If the standard deviation is unknown and is estimated as 0.2 with 19 d.f. ($N = 20$), a 95% symmetric interval would be estimated as

$$1.02 \pm \frac{(2.1)(0.2)}{\sqrt{20}} = 1.02 \pm 0.094 = 0.926 \text{ to } 1.114.$$

To construct the Westlake interval, a symmetric interval about 1.0, detailed tables of the t distribution are needed [1]. In this example, t values of approximately 1.78 and -2.70 will cutoff 4.3% of the area in the upper tail and 0.7% in the lower tail, respectively. This results in an upper limit of $1.02 + 0.08 = 1.10$ and a lower limit of $1.02 - 0.12 = 0.90$, symmetric about 1.0 (1.0 ± 0.1).

Examples of confidence intervals for bioequivalence testing are given in chapters 11 and 15.

The remainder of this chapter will be concerned primarily with testing hypotheses, categorized as follows:

1. Comparison of the mean of a single sample (group) to some known or standard mean [single-sample (group) tests].
2. Comparison of means from two independent samples (groups) [two independent samples (groups) test, a form of the parallel-groups design in clinical trials].
3. Comparison of means from related samples (paired-sample tests).
4. One- and two-sample tests for proportions.
5. Tests to compare variances.

5.2 STATISTICAL HYPOTHESIS TESTING

To introduce the concept of hypothesis testing, we will use an example of the comparison of two treatment means (a two-sample test) that has many applications in pharmaceutical and clinical research. The details of the statistical test are presented in section 5.2.2. A clinical study is planned to compare the efficacy of a new antihypertensive agent to a placebo. Preliminary uncontrolled studies of the drug in humans suggest antihypertensive activity of the order of a drop of 10 to 15 mm Hg diastolic blood pressure. The proposed double-blind clinical trial is designed to study the effects of a once-a-day dose of tablets of the drug in a group of hypertensive patients. A second group of patients will receive an identical-appearing placebo.

Table 5.2 Average Results and Standard Deviation of a Clinical Study Comparing Drug and Placebo in the Treatment of Hypertension

	Drug	Placebo
Number of patients	11	10
Average blood pressure reduction (mm Hg)	10	1
Standard deviation	11.12	7.80

Blood pressure will be measured prior to the study and every two weeks after initiation of therapy for a total of eight weeks. For purposes of this presentation, we will be concerned only with the blood pressure at baseline (i.e., pretreatment) and after eight weeks of treatment. The variable that will be analyzed is the difference between the eight-week reading and the pretreatment reading. This difference, the change from baseline, will be called δ (delta). At the completion of the experiment, the average change from baseline will be compared for the active group and the placebo group in order to come to a decision concerning the efficacy of the drug in reducing blood pressure. The design is a typical *parallel-groups* design and the implementation of the study is straightforward. The problem, and question, that is of concern is: "What statistical techniques can be used to aid us in coming to a decision regarding the treatment (placebo and active drug) difference, and ultimately to a judgment of drug efficacy?"

From a qualitative and, indeed, practical point of view, a comparison of the *average* change in blood pressure for the active and placebo groups, integrated with previous experience, can give some idea of drug efficacy. Table 5.2 shows the average results of this study. (Only 21 patients completed the study.) Based on the results, our "internal computer" might reason as follows: "The new drug reduced the blood pressure by 10 mm Hg compared to a reduction of 1 mm Hg for patients on placebo. That is an impressive reduction for the drug"; or "The average reduction is quite impressive, but the sample size is small, less than 12 patients per group. If the raw data were available, it would be of interest to see how many patients showed an improvement when given the drug compared to the number who showed an improvement when given placebo." Particularly for small samples, one should examine the raw data. Such an examination of the clinical results may give an intuitive feeling of the effectiveness of a drug product. At one time, not very long ago, presentation of such experimental results accompanied by a subjective evaluation by the clinical investigator was important evidence in the support of efficacy of drugs. If the average results showed that the drug was no better than the placebo, the drug would probably be of little, if any, interest.

One obvious problem with such a subjective analysis is the potential lack of consistency in the evaluation and conclusions that may be drawn from the same results by different reviewers. Also, although some experimental results may appear to point unequivocally to either efficacy or lack of efficacy, the inherent variability of the experimental data may be sufficiently large to obscure the truth. In general, subjective perusal of data is not sufficient to separate drug-related effects from random variability. In particular, comparing average results from small samples without a proper statistical analysis can be problematic. Statistical hypothesis testing is an objective means of assessing whether or not observed differences between treatments can be attributed to experimental variation (error). Good experimental design and data analysis are essential if clinical studies are to be used as evidence for drug safety and efficacy. This is particularly critical when such evidence is part of a New Drug Application (NDA) for the FDA, or for use for advertising claims.

The statistical evaluation or test of treatment differences is based on the *ratio* of the *observed treatment difference* (drug minus placebo in this example) to the *variability* of the difference. A large observed difference between drug and placebo accompanied by small variability is the most impressive evidence of a real drug effect (Fig. 5.6).

The magnitude of the ratio can be translated into a probability or "statistical" statement relating to the true but unknown drug effect. This is the basis of the common statement "statistically significant," implying that the difference observed between treatments is real, not merely a result of random variation. Statistical significance addresses the question of whether or not the treatments truly differ, but does not necessarily apply to the *practical* magnitude of the drug

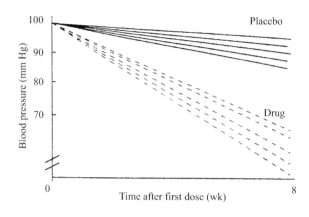

Figure 5.6 Mark of a real drug effect: a large difference between drug and placebo with small variation.

effect. The possibility exists that a small but real drug effect has no clinical meaning. Such judgments should be made by experts who can evaluate the magnitude of the drug effect in relation to the potential use of the drug vis-à-vis other therapeutic alternatives.

The preliminary discussion above suggests the procedure used in testing statistical hypotheses. Broadly speaking, data are first collected for comparative experiments according to an appropriate plan or design. For comparative experiments similar to that considered in our example, the *ratio* of the difference of the averages of the two treatments to its experimental error (standard deviation) is referred to an appropriate tabulated probability distribution. The treatment difference is deemed "statistically significant" if the ratio is sufficiently large relative to the tabulated probability values.

The testing procedure is based on the concept of a *null hypothesis*. The null hypothesis is a hypothetical statement about a parameter (such as the mean) that will subsequently be compared to the sample estimate of the parameter, to test for treatment differences. In the present example, the null hypothesis is

$$H_0 : \mu_1 = \mu_2 \quad \text{or} \quad \Delta = \mu_1 - \mu_2 = 0.$$

H_0 refers to the null hypothesis. μ_1 and μ_2 refer to the true blood pressure change from baseline for the two treatments. Δ is the hypothesized average difference of the change of blood pressure from baseline values for the new drug *compared* to placebo.

Δ = true average reduction in blood pressure due to drug minus true average
   reduction in blood pressure due to placebo

The sample estimate of Δ is designated as δ, and is assumed to have a normal distribution. The fact that H_0 is expressed as a specific difference (zero in this example), as opposed to a more general difference ($H_0 : \Delta \neq 0$), is an important concept. The test of "no difference" or some specific difference (e.g., $\Delta = 2$) is usually much more easily conceptualized and implemented than a test of some nonspecific difference.

The format of the null hypothesis statement is not always immediately apparent to those unfamiliar with statistical procedures. Table 5.3 shows some examples of how null hypothesis statements can be presented. The alternative hypothesis specifies alternative values of the parameter, which we accept as true if the statistical test leads to rejection of the null hypothesis. The alternative hypothesis includes values not specified in the null hypothesis. In our example, a reasonable alternative would include all values where the true values of the two means were not equal, typically stated as follows:

$$H_a : \mu_1 \neq \mu_2.$$

As noted above, the magnitude of the ratio of the (observed difference minus the hypothetical difference) to its variability, the s.d. of the observed difference, determines whether or

Table 5.3 Examples of the Null Hypothesis for Various Experimental Situations

Study	Null hypothesis	Comments
Effect of drug therapy on cholesterol level compared to placebo	$H_0 : \mu_1 = \mu_2$ or $H_0: \mu_1 - \mu_2 = 0$ or $H_0: \Delta = 0$	μ_1 refers to the true average cholesterol with drug and μ_1 refers to true average cholesterol with placebo
Effect of antibiotic on cure rate	$H_0 : p_0 = 0.8$	p_0 refers to the true proportion of patients cured; H_0 states that the hypothetical cure rate is 80%
Average tablet weight for quality control	$H_0 : w = 300$ mg	The target weight is a mean of 300 mg
Testing two mixing procedures with regard to homogeneity of the two mixes	$H_0 : \sigma_1^2 = \sigma_2^2$	The variance of the samples from the two procedures is hypothesized to be equal
Test to see if two treatments differ	$H_0 : \mu_1 \neq \mu_2$	This statement cannot be tested; H_0 must be specified as a specific difference or a limited range of differences

not H_0 should be accepted or rejected. A large ratio leads to rejection of H_0, and the difference is considered to be "statistically" significant. The specific details for testing simple hypotheses are presented below, beginning with the most elementary example, tests of a single mean.

5.2.1 Case I: Test of the Mean from a Single Population (One-Sample Tests), an Introduction to a Simple Example of Hypothesis Testing

The discussion above was concerned with a test to compare means from samples obtained from two groups, a drug group and a placebo group. The tests for a single mean are simpler in concept, and specific steps to construct this test are presented below. The process for other designs in which statistical hypotheses are tested is essentially the same as for the case described here. Other examples will be presented in the remainder of this chapter and, where applicable, in subsequent chapters of this book. The concept of hypothesis testing is important, and the student is well advised to make an extra effort to understand the procedures described below.

Data often come from a single population, and a comparison of the sample mean to some hypothetical or "standard" (known) value is desired. The examples shown in Table 5.4 are typical of those found in pharmaceutical research. The statistical test compares the observed value (a mean or a proportion, for example) to the hypothetical value.

To illustrate the procedure, we will consider an experiment to assess the effects of a change in manufacturing procedure on the average potency of a tablet product. A large amount of data was collected for the content of drug in the tablet formulation during a period of several years. The manufacturing process showed an average potency of 5.01 mg and a standard deviation of 0.11, both values considered to be equal to the true process parameters. A new batch was made with a modification of the usual manufacturing procedure. Twenty tablets were assayed

Table 5.4 Examples of Experiments Where a Single Population Mean Is Observed

Sample mean	Hypothetical or standard mean
Average tablet potency of N tablets	Label potency
Preference for product A in a paired preference test	50% are hypothesized to prefer product A
Average dissolution of N tablets	Quality control specifications
Proportion of patients cured by a new drug	Cure rate of P% based on previous therapy with a similar drug
Average cholesterol level of N patients under therapy	Hypothetical or standard value based on large amount of data collected by clinical laboratory
Average blood pressure reduction in N rats in preclinical study	Hypothetical average reduction considered to be of biological and clinical interest
Average difference of pain relief for two drugs taken by the same patients	Average difference (Δ) is hypothesized to be 0 if the drugs are identical

Table 5.5 Results of 20 Single-Tablet Assays from a Modification
of a Process with a Historical Mean of 5.01 mg

5.13	5.04	5.09	5.00
4.98	5.03	5.01	4.99
5.20	5.08	4.96	5.18
5.08	5.06	5.02	5.24
4.99	5.17	5.06	5.00

$\overline{X} = 5.0655$ mg $S = 0.0806$

σ (historical) $= 0.11$

and the results are shown in Table 5.5. The objective is to determine if the process modification results in a change of average potency from the process average of 5.01, the value of μ under the null hypothesis.

The steps for designing and analyzing this experiment are as follows:

1. *Careful planning* of the experiment ensures that the objectives of the experiment are addressed by an appropriate experimental design. The testing of a hypothesis where data are derived from a poorly implemented experiment can result in invalid conclusions. Proper design includes the choice and number of experimental units (patients, animals, tablets, etc.). Other considerations of experimental design and the manner in which observations are made are addressed in chapters 6, 8, and 11. Sample size may be determined on a scientific, statistical basis, but the choice is often limited by cost or time considerations, or the availability of experimental units. In the present example, the routine quality control content uniformity assay of 20 tablets was the determinant of sample size, a matter of convenience. The 20 tablets were chosen at random from the newly manufactured batch.

2. The *null hypothesis* and *alternative hypothesis* are defined prior to the implementation of the experiment or study. The usual test will be two sided

$$H_0 : \mu = \mu_0 \qquad H_a : \mu \neq \mu_0.$$

However, in the example below we will also discuss a one-sided test

$$H_0 : \mu \leq 5.01 \, \text{mg} \qquad H_a : \mu > 5.01 \, \text{mg}.$$

The objective of this experiment is to see if the average potency of the batch prepared with the modified procedure is different from that based on historical experience (5.01 mg). The null hypothesis takes the form of "no change," as discussed previously. To conclude that the new process has caused a change, we must demonstrate that the alternative hypothesis is true by rejecting the null hypothesis. The alternative hypothesis complements the null hypothesis. The two hypotheses are mutually exclusive and, together, in this example, cover all relevant possibilities that can result from the experiment. Either the average potency is 5.01 mg (H_0) or it is not (H_a). This is known as a *two-sided* (or *two-tailed*) test, suggesting that the average drug potency of the new batch can conceivably be smaller as well as greater than the historical process average of 5.01 mg. A *one-sided* test allows for the possibility of a difference in only one direction. Suppose that the process average of 5.01 mg suggested a preferential loss of drug during processing based on the theoretical amount added to the batch (e.g., 5.05 mg). The new procedure may have been designed to prevent this loss. Under these circumstances, one might hypothesize that the potency could only be greater (or, at least, not less) than the previous process average. Under this hypothesis, if the experiment reveals a lower potency than 5.01 mg, this result would be attributed to chance only; that is, although the average potency, in truth, is equal to or greater than 5.01 mg, chance variability may result in an experimental outcome where the observed average is "numerically" less than 5.01 mg. Such a result could occur, for example, as a result of a chance selection of

Table 5.6 Alpha and Beta Probabilities in Hypothesis Testing (Errors When Accepting or Rejecting H_0)

	H_0 is true	H_a (a specific alternative) is true
H_0 is rejected	Alpha (α)	$1 - $ beta
H_0 is accepted	$1 - $ alpha	Beta (β)

tablets of low potency for the assay sample. For a one-sided test, the null and alternative hypotheses may take the following form as noted above

$$H_0 : \mu \leq 5.01\,\text{mg} \qquad H_a : \mu > 5.01\,\text{mg}.$$

3. The *level of significance* is specified. This is the well-known p value associated with statements of statistical significance. The concept of the level of significance is crucial to an understanding of statistical methodology. The level of significance is defined as the probability that the statistical test results in a decision to reject H_0 (*a significant difference*) when, in fact, the treatments do not differ (H_0 is true). This concept will be clarified further when we describe the statistical test. By definition, the level of significance represents the chance of making a mistake when deciding to reject the null hypothesis. This mistake, or error, is also known as the *alpha* (α) *error* or error of the first kind (Table 5.6). Thus, if the statistical test results in rejection of the null hypothesis, we say that the difference is significant at the α level. If α is chosen to be 0.05, the difference is significant at the 5% level. This is often expressed, equivalently, as $p < 0.05$. Figure 5.7 shows values of $\overline{X}$ that lead to rejection of H_0 for a statistical test at the 5% level if σ is known.

The *beta* (β) *error* is the probability of accepting H_0 (no treatment difference) when, in fact, some specified difference included in H_a is the true difference. Although the evaluation of the β error and its involvement in sample-size determination is important, because of the complex nature of this concept, further discussion of this topic will be delayed until chapter 6.

The choice of magnitude of α, which should be established prior to the start of the experiment, rests on the experimenter or sponsoring organization. To make this choice, one should consider the risks or consequences that will result if an α error is made, that is, the error made when declaring that a significant difference exists when the treatments are indeed equivalent. Alpha should be defined prior to the experiment. It certainly would be unfair to choose an alpha after the results are obtained. Traditionally, α is chosen as 5% (0.05), although other levels such as 1% or 10% have been used. A justification for a level other than 5% should be forthcoming. An α error of 5% means that a decision that a significant difference exists (based on the rejection of H_0) has a probability of 5% (1 in 20) or less of being incorrect (*P less than or equal to* 0.05). Such a decision has credibility and is generally accepted as "proof" of a difference by regulatory agencies. When using the word "significant," one

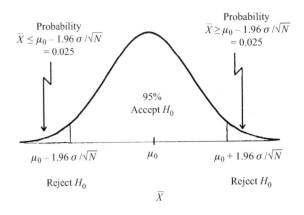

Figure 5.7 Region of rejection (critical region) in a statistical test (two-sided) at the 5% level with σ^2 known.

infers with a large degree of confidence that the experimental result does not support the null hypothesis.

An important concept is that if the statistical test results in a decision of *no significance*, the conclusion does *not* prove that H_0 is true or, in this case, that the average potency is 5.01 mg. Usually, "nonsignificance" is a weak statement, not carrying the clout or authority of the statement of "significance." Note that the chance of erroneously accepting H_0 is equal to β (Table 5.6). This means that β *percent* of the time, a nonsignificant result will be observed (H_0 is accepted as true), when a true difference specified by H_a or greater truly exists. Unfortunately, a good deal of the time when planning experiments, unlike α, β is not fixed in advance. The β level is often a result of circumstance. In most experiments, β is a consequence of the sample size, which is often based on considerations other than the size of β. However, the sample size is best computed with the aid of a predetermined value of β (see chap. 6). In our experiment, β was not fixed in advance. The sample of 20 tablets was chosen as a matter of tradition and convenience.

4. The *sample size*, in our example, has been fixed based on considerations that did not include β, as discussed above. However, the sample size can be calculated after α and β are specified, so that the experiment will be of sufficient size to have properties that will satisfy the choice of the α and β errors (see chap. 6 for further details).

5. After the experiment is completed, relevant statistics are computed. In this example and most situations with which we will be concerned, mean values are to be compared. It is at this point that the *statistical test of significance* is performed as follows. For a two-sided test, compute the ratio

$$Z = \frac{|\overline{X} - \mu_0|}{\sqrt{\sigma^2/N}} = \frac{|\overline{X} - \mu_0|}{\sigma/\sqrt{N}}. \tag{5.4}$$

The numerator of the ratio is the absolute value of the difference between the observed and hypothetical mean. (In a two-sided test, low or negative values as well as large positive values of the mean lead to significance.) The variance of $(\overline{X} - \mu_0)$[§] is equal to

$$\frac{\sigma^2}{N}.$$

The denominator of Eq. (5.4) is the standard deviation of the numerator. The Z ratio [Eq. (5.4)] consists of a difference, divided by its standard deviation. The ratio is exactly the Z transformation presented in chapter 3 [Eq. (3.14)], which transforms a normal distribution with mean μ and variance σ^2 to the standard normal distribution ($\mu = 0$, $\sigma^2 = 1$).

In general, σ^2 is unknown, but it can be estimated from the sample data, and the sample estimate, S^2, is then used in the denominator of Eq. (5.4). An important question is how to determine if the ratio

$$t = \frac{|\overline{X} - \mu_0|}{\sqrt{S^2/N}} \tag{5.5}$$

leads to a decision of "significant." This prevalent situation (σ^2 unknown); will be discussed below.

As discussed above, significance is based on a probability statement defined by α. More specifically, the difference is considered to be statistically significant (H_0 is rejected) if the observed difference between the sample mean and μ_0 is sufficiently large so that the observed or larger differences are improbable (probability of α or less, e.g., $p \leq 0.05$) if the null hypothesis is true ($\mu = 5.01$ mg). In order to calculate the relevant probability, the observations are assumed to be statistically independent and normally distributed.

[§] The variance of $(\overline{X} - \mu_0)$ is equal to the variance of $\overline{X}$ because μ_0 is constant and has a variance of 0.

With these assumptions, the ratio shown in Eq. (5.4) has a normal distribution with mean equal to 0 and variance equal to 1 (variance known, the standard normal distribution). The concept of the α error is illustrated in Figure 5.7. The values of $\overline{X}$ that lead to rejection of the null hypothesis define the "region of rejection," also known as the *critical region*. With a knowledge of the variance, the area corresponding to the critical region can be calculated using the standard normal distribution. The probability of observing a mean value in the critical region of the distribution defined by the null hypothesis is α. This region is usually taken as symmetrical areas in the tails of the distribution, with each tail containing $\alpha/2$ of the area ($2^1/_2\%$ in each tail at the 5% level) for a two-tailed test. Under the null hypothesis and the assumption of normality, $\overline{X}$ is normal with mean μ_0 and variance σ^2/N. The Z ratio [Eq. (5.4)] is a standard normal deviate, as noted above. Referring to Table IV.2, the values of $\overline{X}$ that satisfy

$$\frac{\overline{X} - \mu_0}{\sigma/\sqrt{N}} \leq -1.96 \quad \text{or} \quad \frac{\overline{X} - \mu_0}{\sigma/\sqrt{N}} \geq +1.96 \tag{5.6}$$

will result in rejection of H_0 at the 5% level. The values of $\overline{X}$ that lead to rejection of H_0 may be derived by rearranging Eq. (5.6).

$$\overline{X} \leq \mu_0 - \frac{1.96\sigma}{\sqrt{N}} \quad \text{or} \quad \overline{X} \geq \mu_0 + \frac{1.96\sigma}{\sqrt{N}} \tag{5.7}$$

or, equivalently,

$$\left|\overline{X} - \mu_0\right| \geq \frac{1.96\sigma}{\sqrt{N}}. \tag{5.8}$$

If the value of $\overline{X}$ falls in the critical region, as defined in Eqs. (5.7) and (5.8), the null hypothesis is rejected and the difference is said to be significant at the α (5%) level.

The statistical test of the mean assay result from Table 5.5 may be performed: (a) assuming that σ is known ($\sigma = 0.11$) or (b) assuming that σ is unknown, but estimated from the sample ($S = 0.0806$).

The following examples demonstrate the procedure for applying the test of significance for a *single mean*.

(a) *One-sample test, variance known.* In this case, we believe that the large quantity of historical data defines the standard deviation of the process precisely, and that this standard deviation represents the variation in the new batch. We assume, therefore, that σ^2 is known. In addition, as noted above, if the data from the sample are independent and normally distributed, the test of significance is based on the standard normal curve (Table IV.2). The ratio as described in Eq. (5.4) is computed using the known value of the variance. If the absolute value of the ratio is greater than that which cuts off $\alpha/2$ percent of the area (defining the two tails of the rejection region, Fig. 5.7), the difference between the observed and hypothetical means is said to be significant at the α level. *For a two-sided test, the absolute value of the difference is used because* both large positive and negative differences are considered evidence for rejecting the null hypothesis.

In this example, we will use a two-sided test, because the change in potency, if any, may occur in either direction, higher or lower. The level of significance is set at the traditional 5% level.

$$\alpha = 0.05$$

Compute the ratio [Eq. (5.4)]

$$Z = \frac{\left|\overline{X} - \mu_0\right|}{\sigma/\sqrt{N}} = \frac{|5.0655 - 5.01|}{0.11/\sqrt{20}} = 2.26.$$

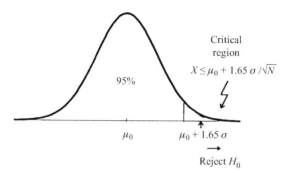

Critical
region

$X \leq \mu_0 + 1.65\, \sigma / \sqrt{N}$

95%

μ_0

$\mu_0 + 1.65\, \sigma$

Reject H_0

Figure 5.8 Rejection region for a one-sided test.

At the 5% level values of $|Z| \geq 1.96$ will lead to a declaration of significance for a two-sided test [Eq. (5.6)]. Therefore, the new batch can be said to have a potency different from previous batches (in this case, the mean is greater).

The level of significance is set *before* the actual experimental results are obtained. In the previous example, a one-sided test at the 5% level may be justified if convincing evidence were available to demonstrate that the new process would only result in mean results equal to or greater than the historical mean. If such a one-sided test had been deemed appropriate, the null hypothesis would be

$H_0 : \mu = 5.01$ mg.

The alternative hypothesis, $H_a : \mu > 5.01$ mg, eliminates the possibility that the new process can lower the mean potency. The concept is illustrated in Figure 5.8. Now the rejection region lies only in values of $\overline{X}$ greater than 5.01 mg, as described below. An observed value of $\overline{X}$ below 5.01 mg is considered to be due only to chance (or it may be of no interest to us in other situations).

The rejection region is defined for values of $\overline{X}$ equal to or greater than $\mu_0 + 1.65\sigma/\sqrt{N}$ [or, equivalently, $(\overline{X} - \mu_0)/(\sigma/N) \geq 1.65$] because 5% of the area of the normal curve is found above this value (Table IV.2). This is in keeping with the definition of α: If the null hypothesis is true, we will erroneously reject the null hypothesis 5% of the time. Thus, we can see that a *smaller* difference is needed for significance using a one-sided test; the Z ratio need only exceed 1.65 rather than 1.96 for significance at the 5% level. In the present example, values of $\overline{X} \geq [5.01 + 1.65(0.11)/\sqrt{20}] = 5.051$ will lead to significance for a one-sided test. Clearly, the observed mean of 5.0655 is significantly different from 5.01 ($p < 0.05$). Note that in a one-sided test, the sign of the numerator is important and the absolute value is not used.

Usually, statistical tests are two-sided tests. One-sided tests are warranted in certain circumstances. However, the choice of a one-sided test should be made a priori, and one must be prepared to defend its use. As mentioned above, in the present example, if evidence were available to show that the new process could not reduce the potency, a one-sided test would be acceptable. To have such evidence and convince others (particularly, regulatory agencies) of its validity is not always an easy task. Also, from a scientific point of view, two-sided tests are desirable because significant results in both positive and negative directions are usually of interest.

(b) *One-sample test, variance unknown.* In most experiments in pharmaceutical research, the variance is unknown. Usually, the only estimate of the variance comes from the experimental data itself. As has been emphasized in the example above, use of the cumulative standard normal distribution (Table IV.2) to determine probabilities for the comparison of a mean to a known value (μ_0) is valid only if the variance is known.

The procedure for testing the significance of the difference of an observed mean from a hypothetical value (one-sample test) when the variance is estimated from the sample data is the same as that with the variance known, with the following exceptions:

1. The variance is computed from the experimental data. In the present example, the variance is $(0.0806)^2$; the standard deviation is 0.0806 from Table 5.5.

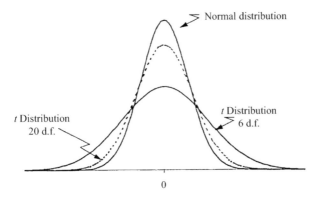

Figure 5.9 *t* distribution compared to the standard normal distribution.

2. The ratio is computed using S^2 instead of σ^2 as in Eq. (5.8a). This ratio

$$t = \frac{|\overline{X} - \mu_0|}{\sqrt{S^2/N}}$$

(5.8a)

is not distributed as a standard normal variable. If the mean is normally distributed, the ratio [Eq. (5.5)] has a *t* distribution. The *t* distribution looks like the standard normal distribution but has more area in the tails; the *t* distribution is more spread out. The shape of the *t* distribution depends on the d.f. As the d.f. increase the *t* distribution looks more and more like the standard normal distribution as shown in Figure 5.9. (Also, see sect. 3.5.2.) When the d.f. are equal to ∞ the *t* distribution is identical to the standard normal distribution (i.e., the variance is known).

The *t* distribution is a probability distribution that was introduced in section 5.1.1 and chapter 3. The area under the *t* distributions shown in Figure 5.9 is 1. Thus, as in the case of the normal distribution (or any continuous distribution), areas within specified intervals represent probabilities. However, unlike the normal distribution, there is *no* transformation that will change all *t* distributions (differing d.f.'s) to one "standard" *t* distribution. Clearly, a tabulation of all possible *t* distributions would be impossible. Table IV.4 shows commonly used probability points for representative *t* distributions. The values in the table are points in the *t* distribution representing cumulative areas (probabilities) of 80%, 90%, 95%, 97.5%, and 99.5%. For example, with d.f. = 10, 97.5% of the area of the *t* distribution is below a value of *t* equal to 2.23 (Fig. 5.10).

Note that when d.f. = ∞, the *t* value corresponding to a cumulative probability of 97.5% (0.975) is 1.96, exactly the same value as that for the standard normal distribution. Since the *t* distribution is symmetrical about zero, as is the standard normal distribution, a *t* value of −2.23 cuts off 1 − 0.975 = 0.025 of the area (d.f. = 10). This means that to obtain a significant

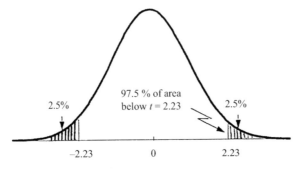

Figure 5.10 *t* distribution with 10 degrees of freedom.

difference of means at the 5% level for a two-sided test and d.f. equal to 10, the absolute value of the t ratio [Eq. (5.5)] must exceed 2.23. Thus the t values in the column headed "0.975" in Table IV.4 are values to be used for two-tailed significance tests at the 5% level (or for a two-sided 95% confidence interval). Similarly, the column headed "0.95" contains appropriate t values for significance tests at the 10% level for two-sided tests, or the 5% level for one-sided tests. The column headed "0.995" represents t values used for two-sided tests at the 1% level, or for 99% confidence intervals.

The number of d.f. used to obtain the appropriate value of t from Table IV.4 is the d.f. associated with the variance estimate in the denominator of the t ratio [Eq. (5.5)]. The d.f. for a mean are $N - 1$, or $19(20 - 1)$ in this example. The test is a two-sided test at the 5% level. The t ratio is

$$t = \frac{|\overline{X} - \mu_0|}{S/\sqrt{N}} = \frac{|5.0655 - 5.01|}{0.0806/\sqrt{20}} = 3.08.$$

The value of t needed for significance for a two-sided test at the 5% level is 2.09 (Table IV.4; 19 d.f.). Therefore, the new process results in a "significant" increase in potency ($p < 0.05$).

A 95% confidence interval for the true mean potency may be constructed as described in section 5.1.1 [Eq. (5.2)]

$$5.0655 \pm 2.09 \left(\frac{0.0806}{\sqrt{20}} \right) = 5.028 \text{ to } 5.103 \text{ mg}.$$

Note that the notion of the *confidence interval* is closely associated with the *statistical test*. If the confidence interval covers the hypothetical value, the difference is not significant at the indicated level, and vice versa. In our example, the difference was significant at the 5% level, and the 95% confidence interval does *not* cover the hypothetical mean value of 5.01.

Example 4: As part of the process of new drug research, a pharmaceutical company places all new compounds through an "antihypertensive" screen. A new compound is given to a group of animals and the reduction in blood pressure measured. Experience has shown that a blood pressure reduction of more than 15 mm Hg in these hypertensive animals is an indication for further testing as a new drug candidate. Since such testing is expensive, the researchers wish to be reasonably sure that the compound truly reduces the blood pressure by more than 15 mm Hg before testing is continued; that is, they will continue testing only if the experimental evidence suggests that the true blood pressure reduction is greater than 15 mm Hg with a high probability.

$$H_0 : \mu \leq 15 \text{ mm Hg reduction} \qquad H_a : \mu > 15 \text{ mm Hg reduction}$$

The null hypothesis is a statement that the new compound is unacceptable (blood pressure change is equal to or less than 15 mm Hg). This is typical of the concept of the null hypothesis. A rejection of the null hypothesis means that a difference probably exists. In our example, a true difference greater than 15 mm Hg means that the compound should be tested further. This is a *one-sided* test. Experimental results showing a difference of 15 mm Hg or less will result in a decision to accept H_0, and the compound will be put aside. If the blood pressure reduction exceeds 15 mm Hg the reduction will be tested for significance using a t test.

$$\alpha = 10\%(0.10)$$

The level of significance of 10% was chosen in lieu of the usual 5% level for the following reason. A 5% significance level means that 1 time in 20 a compound will be chosen as effective when the true reduction is less than 15 mm Hg. The company was willing to take a risk of 1 in 10 of following up an ineffective compound in order to reduce the risk of missing potentially effective compounds. One should understand that the choice of alpha and beta errors often is a compromise between reward and risk. We could increase the chances for reward, but we

Table 5.7 Blood Pressure Reduction Caused by a New Antihypertensive Compound in 10 Animals (mm Hg)

15	12
18	17
14	21
8	16
20	18
$\overline{X} = 15.9$	$S = 3.87$

could simultaneously increase the risk of failure, or, in this case, following up on an ineffective compound. Other things being equal, an increase in the α error decreases the β error; that is, there is a smaller chance of accepting H_0 when it is false. Note that the t value needed for significance is smaller at the 10% level than that at the 5% level. Therefore, a smaller reduction in blood pressure is needed for significance at the 10% level. The standard procedure in this company is to test the compound on 10 animals. The results shown in Table 5.7 were observed in a test of a newly synthesized potential antihypertensive agent.

The t test is [Eq. (5.5)]

$$t = \frac{15.9 - 15}{3.87/\sqrt{10}} = \frac{0.9}{1.22} = 0.74.$$

The value of t needed for significance is 1.38 (Table IV.4; one-sided test at the 10% level with 9 d.f.). Therefore, the compound is not sufficiently effective to be considered further. Although the average result was larger than 15 mm Hg, it was not sufficiently large to encourage further testing, according to the statistical criterion.

What difference (reduction) would have been needed to show a significant reduction, assuming that the sample variance does not change? Equation (5.5) may be rearranged as follows: $\overline{X} = t(S)/\sqrt{N} + \mu_0$. If $\overline{X}$ is greater than or equal to $t(S)/\sqrt{N} + \mu_0$, the average reduction will be significant, where t is the table value at the α level of significance with $(N - 1)$ d.f. In our example,

$$\frac{t(S)}{\sqrt{N}} + \mu_0 = \frac{(1.38)(3.87)}{\sqrt{10}} + 15 = 16.7.$$

A blood pressure reduction of 16.7 mm Hg or more (the critical region) would have resulted in a significant difference. (See Exercise Problem 10.)

5.2.2 Case II: Comparisons of Means from Two Independent Groups (Two Independent Groups Test)

A preliminary discussion of this test was presented in section 5.2. This most important test is commonly encountered in clinical studies (a parallel-groups design). Table 5.8 shows a few examples of research experiments that may be analyzed by the test described here. The data

Table 5.8 Some Examples of Experiments That May Be Analyzed by the Two- Independent- Groups Test

Clinical studies	Active drug compared to a standard drug or placebo; treatments given to different persons, one treatment per person
Preclinical studies	Comparison of drugs for efficacy and/or toxicity with treatments given to different animals
Comparison of product attributes from two batches	Tablet dissolution, potency, weight, etc., from two batches

of Table 5.2 will be used to illustrate this test. The experiment consisted of a comparison of an active drug and a placebo where each treatment is tested on different patients. The results of the study showed an average blood pressure reduction of 10 mm Hg for 11 patients receiving drug, and an average reduction of 1 mm Hg for 10 patients receiving placebo. The principal feature of this test (or design) is that treatments are given to two independent groups. The observations in one group are independent of those in the second group. In addition, we assume that the data within each group are normally and independently distributed.

The steps to be taken in performing the two independent groups test are similar to those described for the one-sample test (see sect. 5.2.1).

1. *Patients are randomly assigned to the two treatment groups.* (For a description of the method of random assignment, see chap. 4.) The number of patients chosen to participate in the study in this example was largely a consequence of cost and convenience. Without these restraints, a suitable sample size could be determined with a knowledge of β, as described in chapter 6. The drug and placebo were to be randomly assigned to each of 12 patients (12 patients for each treatment). There were several dropouts, resulting in 11 patients in the drug group and 10 patients in the placebo group.
2. *The null and alternative hypotheses are*

$$H_0: \mu_1 - \mu_2 = \Delta = 0 \qquad H_a : \Delta \neq 0.$$

We hypothesize no difference between treatments. A "significant" result means that treatments are considered different. This is a two-sided test. The drug treatment may be better or worse than placebo.
3. α *is set at 0.05.*
4. *The form of the statistical test depends on whether or not variances are known.* In the usual circumstances, the variances are unknown.

5.2.2.1 Two Independent -Groups Test, Variances Known
If the variances of both groups are known, the ratio

$$Z = \frac{\overline{X}_1 - \overline{X}_2 - (\mu_1 - \mu_2)}{\sqrt{\sigma_1^2/N_1 + \sigma_2^2/N_2}} \qquad (5.9)$$

has a normal distribution with mean 0 and standard deviation equal to 1 (the standard normal distribution). The numerator of the ratio is the difference between the observed difference of the means of the two groups $(\overline{X}_1 - \overline{X}_2)$ and the hypothetical difference $(\mu_1 - \mu_2$ according to $H_0)$. In the present case, and indeed in most of the examples of this test that we will consider, the hypothetical difference to be zero (i.e., $H_0 : \mu_1 - \mu_2 = 0$). The variability of $(\overline{X}_1 - \overline{X}_2)$[¶] (defined as the standard deviation) is equal to

$$\sqrt{\sigma_{\overline{X}_1}^2 + \sigma_{\overline{X}_2}^2}$$

[as described in App. I, if A and B are independent, $\sigma^2(A - B) = \sigma_A^2 + \sigma_B^2$]. Thus, as in the one-sample case, the test consists of forming a ratio whose distribution is defined by the standard normal curve. In the present example (test of an antihypertensive agent), suppose that the *variances* corresponding to drug and placebo are *known* to be 144 and 100, respectively. The rejection region is defined by α. For $\alpha = 0.05$, values of Z greater than 1.96 or less than -1.96 ($|Z| \geq 1.96$) will lead to rejection of the null hypothesis. Z is defined by Eq. (5.9).

[¶] The variance of $(\overline{X}_1 - \overline{X}_2) - (\mu_1 - \mu_2)$ is equal to the variance of $(\overline{X}_1 - \overline{X}_2)$ because μ_1 and μ_2 are constants and have a variance equal to zero.

For a two-sided test

$$Z = \frac{|\overline{X}_1 - \overline{X}_2|}{\sqrt{\sigma_1^2/N_1 + \sigma_2^2/N_2}},$$

$$\overline{X}_1 = 10, \quad \overline{X}_2 = 1, \quad N_1 = 11, \quad \text{and} \quad N_2 = 10.$$

Thus,

$$Z = \frac{|10 - 1|}{\sqrt{144/11 + 100/10}} = 1.87.$$

Since the absolute value of the ratio does not exceed 1.96, the difference is not significant at the 5% level. From Table IV.2, the probability of observing a value of Z greater than 1.87 is approximately 0.03. Therefore, the test can be considered significant at the 6% level [2(0.03) = 0.06 for a two-tailed test]. The probability of observing an absolute difference of 9 mm Hg or more between drug and placebo, if the two products are identical, is 0.06 or 6%.

We have set α equal to 5% as defining an unlikely event from a distribution with known mean (0) and variance ($144/11 + 100/10 = 23.1$). An event as far or farther from the mean (0) than 9 mm Hg can occur six times in a 100 if H_0 is true. Alternatively, the conclusion may be stated that the experimental results were not sufficient to reject H_0 because we set α at 5% a priori (i.e., before performing the experiment). In reality, there is nothing special about 5%. The use of 5% as the α level is based strongly on tradition and experience, as mentioned previously. Should significance at the 6% level result in a different decision than a level of 5%? To document efficacy, a significance level of 6% may not be adequate for acceptance by regulatory agencies. There has to be some cutoff point; otherwise, if 6% is acceptable, why not 7% and so on? However, for internal decisions or for leads in experiments used to obtain information for further work or to verify theories, 5% and 6% may be too close to "call." Rather than closing the door on experiments that show differences at $p = 0.06$, one might think of such results as being of "borderline" significance, worthy of a second look and/or further experimentation. In our example, had the difference between drug and placebo been approximately 9.4 mm Hg, we would have called the difference "significant," rejecting the hypothesis that the placebo treatment was equal to the drug.

P values are often presented with experimental results even though the statistical test shows nonsignificance at the predetermined α level. In this experiment, a statement that $p = 0.06$ ("The difference is significant at the 6% level") does not imply that the treatments are considered to be significantly different. We emphasize that if the α level is set at 5%, a decision that the treatments are different should be declared only if the experimental results show that $p \leq 0.05$. However, in practical situations, it is often useful for the experimenter and other interested parties to know the p value, particularly in the case of "borderline" significance.

5.2.2.2 Two- Independent -Groups Test, Variance Unknown

The procedure for comparing means of two independent groups when the variances are estimated from the sample data is the same as that with the variances known, with the following exceptions:

1. *The variance is computed from the sample data.* In order to perform the statistical test to be described below, in addition to the usual assumptions of normality and independence, *we assume that the variance is the same for each group.* (If the variances differ, a modified procedure can be used as described later in this chapter.) A rule of thumb for moderate-sized samples (N equal 10–20) is that the ratio of the two variances should not be greater than 3 to 4. Sometimes, in doubtful situations, a test for the equality of the two variances may be appropriate (see sect. 5.3) before performing the test of significance for means described here. To obtain an estimate of the common variance, first compute the variance of each group. The two

variances are *pooled* by calculating a weighted average of the variances, the best estimate of the true common variance. The weights are equal to the d.f., $N_1 - 1$ and $N_2 - 1$, for groups 1 and 2, respectively. N_1 and N_2 are the sample sizes for the two groups. The following formula may be used to calculate the pooled variance

$$S_p^2 = \frac{(N_1 - 1)S_1^2 + (N_2 - 1)S_2^2}{N_1 + N_2 - 2}. \tag{5.10}$$

Note that we do not calculate the pooled variance by first pooling together all of the data from the two groups. The pooled variance obtained by pooling the two *separate* variances will always be equal to or smaller than that computed from all of the data combined disregarding groups. In the latter case, the variance estimate includes the variability due to differences of means as well as that due to the variance within each group (see Exercise Problem 5). Appendix I has a further discussion of pooling variance.

2. *The ratio that is used for the statistical test is similar to Eq. (5.9).* Because the variance, S_p^2 (pooled variance), is estimated from the sample data, the ratio

$$t = \frac{(\overline{X}_1 - \overline{X}_2) - (\mu_1 - \mu_2)}{\sqrt{S_p^2/N_1 + S_p^2/N_2}} = \frac{(\overline{X}_1 - \overline{X}_2) - (\mu_1 - \mu_2)}{S_p\sqrt{1/N_1 + 1/N_2}} \tag{5.11}$$

is used instead of Z [Eq. (5.9)]. The d.f. for the distribution are determined from the variance estimate, S_p^2. This is equal to the d.f., pooled from the two groups, equal to $(N_1 - 1) + (N_2 - 1)$ or $N_1 + N_2 - 2$.

These concepts are explained and clarified, step by step, in the following examples.

Example 5: Two different formulations of a tablet of a new drug are to be compared with regard to rate of dissolution. Ten tablets of each formulation are tested, and the percent dissolution after 15 minutes in the dissolution apparatus is observed. The results are tabulated in Table 5.9. The object of this experiment is to determine if the dissolution rates of the two formulations differ. The test for the "significance" of the observed difference is described in detail as follows:

1. State the null and alternative hypotheses:

$$H_0 : \mu_1 = \mu_2 \qquad H_a : \mu_1 \neq \mu_2$$

Table 5.9 Percent Dissolution After 15 Minutes for Two Tablet Formulations

	Formulation A	Formulation B
	68	74
	84	71
	81	79
	85	63
	75	80
	69	61
	80	69
	76	72
	79	80
	74	65
Average	77.1	71.4
Variance	33.43	48.71
s.d.	5.78	6.98

μ_1 and μ_2 are the true mean 15-minute dissolution values for formulations A and B, respectively. This is a two-sided test. There is no reason to believe that one or the other formulation will have a faster or slower dissolution, a priori.

2. *State the significance level* $\alpha = 0.05$. The level of significance is chosen as the traditional 5% level.
3. *Select the samples.* Ten tablets taken at random from each of the two pilot batches will be tested.
4. *Compute the value of the t statistic [Eq. (5.11)]:*

$$\frac{\left|\overline{X}_1 - \overline{X}_2 - (\mu_1 - \mu_2)\right|}{S_p\sqrt{1/N_1 + 1/N_2}} = t = \frac{|77.1 - 71.4|}{S_p\sqrt{1/10 + 1/10}}$$

$\overline{X}_1 = 77.1$ and $\overline{X}_2 = 71.4$ (Table 5.9). $N_1 = N_2 = 10$ (d.f. $= 9$ for each group). S_p is calculated from Eq. (5.10)

$$S_p = \sqrt{\frac{9(33.43) + 9(48.71)}{18}} = 6.41.$$

Note that the *pooled standard deviation* is the *square root of the pooled variance*, where the pooled variance is a weighted average of the variances from each group. It is *not correct to average the standard deviations.* Although the sample variances of the two groups are not identical, they are "reasonably" close, close enough so that the assumption of equal variances can be considered to be acceptable. The assumption of equal variance and independence of the two groups is more critical than the assumption of normality of the data, because we are comparing means. Means tend to be normally distributed even when the individual data do not have a normal distribution, according to the central limit theorem. The observed value of t (18 d.f.) is

$$t = \frac{\left|\overline{X}_1 - \overline{X}_2\right|}{S_p\sqrt{1/N_1 + 1/N_2}} = \frac{|77.1 - 71.4|}{6.41\sqrt{2/10}} = 1.99.$$

Values of t equal to or greater than 2.10 (Table IV.4; d.f. $= 18$) lead to rejection of the null hypothesis. These values, which comprise the *critical region*, result in a declaration of "significance." In this experiment, the value of t is 1.99, and the difference is not significant at the 5% level ($p > 0.05$). This does not mean that the two formulations have the same rate of dissolution. The declaration of nonsignificance here probably means that the sample size was too small; that is, the same difference with a larger sample would be significant at the 5% level. Two different formulations are apt not to be identical with regard to dissolution. The question of statistical versus practical significance may be raised here. If the dissolutions are indeed different, will the difference of 5.7% (77.1–71.4%) affect drug absorption in vivo? A confidence interval on the difference of the means may be an appropriate way of presenting the results.

5.2.2.3 Confidence Interval for the Difference of Two Means

A confidence interval for the difference of two means can be constructed in a manner similar to that presented for a single mean as shown in section 5.1 [Eq. (5.2)]. For example, a confidence interval with a confidence coefficient of 95% is

$$(\overline{X}_1 - \overline{X}_2) \pm (t)S_p\sqrt{\frac{1}{N_1} + \frac{1}{N_2}}, \tag{5.12}$$

t is the value obtained from Table IV.4 with appropriate d.f., with the probability used for a two-sided test. (Use the column labeled "0.975" in Table IV.4 for a 95% interval.) For the example discussed above (tablet dissolution), a 95% confidence interval for the difference of the mean 15-minute dissolution values [Eq. (5.12)] is

$$(77.1 - 71.4) \pm 2.10(6.41)(0.447) = 5.7 \pm 6.02 = -0.32\% \text{ to } 11.72\%.$$

Thus the 95% confidence interval is from -0.32% to 11.72%.

5.2.2.4 Test of Significance If Variances of the Two Groups Are Unequal

If the two groups can be considered not to have equal variances and the variances are estimated from the samples, the usual t test procedure is not correct. This problem has been solved and is often denoted as the Behrens–Fisher procedure. Special tables are needed for the solution, but a good approximate test for the equality of two means can be performed using Eq. (5.13) [3].

$$t = \frac{(\overline{X}_1 - \overline{X}_2)}{\sqrt{S_1^2/N_1 + S_2^2/N_2}} \qquad (5.13)$$

If $N_1 = N_2 = N$, then the critical t is taken from Table IV.4 with $N - 1$ instead of the usual $2(N - 1)$ d.f. If N_1 and N_2 are not equal, then the t value needed for significance is a weighted average of the appropriate t values from Table IV.4 with $N_1 - 1$ and $N_2 - 1$ d.f.

$$\text{Weighted average of } t \text{ values} = \frac{w_1 t_1 + w_2 t_2}{w_1 + w_2},$$

where the weights are

$$w_1 = \frac{S_1^2}{N_1}, \qquad w_2 = \frac{S_2^2}{N_2}.$$

To make the calculation clear, assume that the means of two groups of patients treated with an antihypertensive agent showed the following reduction in blood pressure (mm Hg).

	Group A	Group B
Mean	10.7	7.2
Variance (S^2)	51.8	5.3
N	20	15

We have reason to believe that the variances differ, and for a two-sided test, we first calculate t' according to Eq. (5.13)

$$t' = \frac{|10.7 - 7.2|}{\sqrt{51.8/20 + 5.3/15}} = 2.04.$$

The critical value of t' is obtained using the weighting procedure. At the 5% level, t with 19 d.f. $= 2.09$ and t with 14 d.f. $= 2.14$. The weighted average t value is

$$\frac{(51.8/20)(2.09) + (5.3/15)(2.14)}{(51.8/20) + (5.3/15)} = 2.10.$$

Since t' is less than 2.10, the difference is considered to be not significant at the 5% level.

5.2.2.5 Overlapping Confidence Intervals and Statistical Significance

When comparing two independent treatments for statistical significance, sometimes people erroneously make conclusions based on the confidence intervals constructed from each treatment separately. In particular, if the confidence intervals overlap, the treatments are considered not to differ. This reasoning is not necessarily correct. The fallacy can be easily seen from the following example. Consider two independent treatments, A and B, representing two formulations of the same drug with the following dissolution results:

Treatment	N	Average	s.d.
A	6	37.5	6.2
B	6	47.4	7.4

For a two-sided test, the two-sample t test results in a t value of

$$t = \frac{|47.4 - 37.5|}{6.83\sqrt{1/6 + 1/6}} = 2.51.$$

Since 2.51 exceeds the critical t value with 10 d.f. (2.23), the results show significance at the 5% level.

Computation of the 95% confidence intervals for the two treatments results in the following:

Treatment A: $37.5 \pm (2.57)(6.2)\sqrt{1/6} = 30.99$ to 44.01.
Treatment B: $47.4 \pm (2.57)(7.4)\sqrt{1/6} = 39.64$ to 55.16.

Clearly, in this example, the individual confidence intervals overlap (the values between 39.64 and 44.01 are common to both intervals), yet the treatments are significantly different. The 95% confidence interval for the difference of the two treatments is

$$(47.4 - 37.5) \pm 8.79 = 1.1 \text{ to } 18.19.$$

As has been noted earlier in this section, if the 95% confidence interval does not cover 0, the difference between the treatments is significant at the 5% level.

5.2.2.6 Summary of t-Test Procedure and Design for Comparison of Two Independent Groups

The t-test procedure is essentially the same as the test using the normal distribution (Z test). The t test is used when the variance(s) are unknown and estimated from the sample data. The t distribution with ∞ d.f. is identical to the standard normal distribution. Therefore, the t distribution with ∞ d.f. can be used for normal distribution tests (e.g., comparison of means with variance known). When using the t test, it is necessary to compute a pooled variance. [With variances known, a pooled variance is not computed; see Eqs. (5.10) and (5.11).] An assumption underlying the use of this t test is that the variances of the comparative groups are the same. Other assumptions when using the t test are that the data from the two groups are independent and normally distributed. If the variances are considered to be unequal, use the approximate Behrens–Fisher method.

If H_0 is rejected (the difference is "significant"), one accepts the alternative, $H_a := \mu_1 \neq \mu_2$ or $\mu_1 - \mu_2 \neq 0$. The best estimate of the true difference between the means is the observed difference. A confidence interval gives a range for the true difference (see above). If the confidence interval covers 0, the statistical test is not significant at the corresponding alpha level.

Planning an experiment to compare the means of two independent groups usually requires the following considerations:

1. *Define the objective.* For example, in the example above, the objective was to determine if the two formulations differed with regard to rates of dissolution.

2. Determine the *number of samples* (experimental units) to be included in the experiment. We have noted that statistical methods may be used to determine the sample size (chap. 6). However, practical considerations such as cost and time constraints are often predominating factors. The *sample size* of the two groups *need not be equal* in this type of design, also known as a *parallel-groups* or *one-way analysis of variance* design. If the primary interest is the comparison of means of the two groups, equal sample sizes are optimal (assuming that the variances of the two groups are equal). That is, given the total number of experimental units available (patients, tablets, etc.), the most powerful comparison will be obtained by dividing the total number of experimental units into two equal groups. The reason for this is that $(1/N_1) + (1/N_2)$, which is in the denominator of the test ratio, is minimal when $N_1 = N_2 = N_t/2$ (N_t is the total sample size). In many circumstances (particularly in clinical studies), observations are lost due to errors, patient dropouts, and so on. The analysis described here is still valid, but some power will be lost. *Power* is the ability of the test to discriminate between the treatment groups. (Power is discussed in detail in chap. 6.) Sometimes, it is appropriate to use different sample sizes for the two groups. In a clinical study where a new drug treatment is to be compared to a standard or placebo treatment, one may wish to obtain data on adverse experiences due to the new drug entity in addition to comparisons of efficacy based on some relevant mean outcome. In this case, the design may include more patients on the new drug than the comparative treatment. Also, if the variances of two groups are known to be unequal, the optimal sample sizes will not be equal [4].

3. *Choose the samples.* It would seem best in many situations to be able to apply treatments to randomly chosen experimental units (e.g., patients). Often, practical considerations make this procedure impossible, and some compromise must be made. In clinical trials, it is usually not possible to select patients at random according to the strict definition of "random." We usually choose investigators who assign treatments to the patients available to the study in a random manner.

4. *Observations are made* on the samples. Every effort should be made to avoid bias. Blinding techniques and randomizing the order of observations (e.g., assays) are examples of ways to avoid bias. Given a choice, objective measurements, such as body weights, blood pressure, and blood assays, are usually preferable to subjective measurements, such as degree of improvement, psychological traits, and so on.

5. The *statistical analysis*, as described above, is then applied to the data. The statistical methods and probability levels (e.g., α) should be established prior to the experiment. However, one should not be immobilized because of prior commitments. If experimental conditions differ from that anticipated, and alternative analyses are warranted, a certain degree of flexibility is desirable. However, statistical theory (and common sense) shows that it is not fair to examine the data to look for all possible effects not included in the objectives. The more one looks, the more one will find. In a large data set, any number of unusual findings will be apparent if the data are examined with a "fine-tooth comb," sometimes called "data dredging." If such unexpected results are of interest, it is best to design a new experiment to explore and define these effects. Otherwise, large data sets can be incorrectly used to demonstrate a large number of unusual, but inadvertent, random, and inconsequential "statistically" significant differences.

5.2.3 Test for Comparison of Means of Related Samples (Paired-Sample *t* Test)

Experiments are often designed so that comparisons of two means are made on related samples. This design is usually more *sensitive* than the two independent groups *t* test. A test is more sensitive if the experimental variability is smaller. With smaller variability, smaller differences can be detected as statistically significant. In clinical studies, a paired design is often described as one in which each patient acts as his or her own "control." A bioequivalence study, in which each subject takes each of a test and reference drug product, is a form of paired design (see sect. 11.4).

In the paired-sample experiment, the two treatments are applied to experimental units that are closely related. If the same person takes both treatments, the relationship is obvious. Table 5.10 shows common examples of related samples used in paired tests.

Table 5.10 Examples of Related Samples

Clinical studies	Each patient takes each drug on different occasions (e.g., crossover study)
	Each patient takes each drug simultaneously, such as in skin testing; for example, an ointment is applied to different parts of the body
	Matched pairs: two patients are matched for relevant characteristics (age, sex, disease state, etc.) and two drugs randomly assigned, one to each patient
Preclinical studies	Drugs assigned randomly to littermates
Analytical development	Same analyst assays all samples
	Each laboratory assays all samples in collaborative test
	Each method is applied to a homogeneous sample
Stability studies	Assays over time from material from same container

The paired t test is identical in its implementation to the one-sample test described in section 5.2.1. In the paired test, the single sample is obtained by taking differences between the data for the two treatments for each experimental unit (patient or subject, for example). With N pairs of individuals, there are N data points (i.e., N differences). The N differences are designated as δ. Example 4, concerning the average reduction in blood pressure in a preclinical screen, was a paired-sample test in disguise. The paired data consisted of pre- and postdrug blood pressure readings for each animal. We were interested in the difference of pre- and postvalues (δ), the blood pressure reduction (see illustration below).

In paired tests, treatments should be assigned either in random order, or in some designed way, as in the crossover design. In the crossover design, usually one-half of the subjects receive the two treatments in the order A-B, and the remaining half of the subjects receive the treatments in the opposite order, where A and B are the two treatments. The crossover design is discussed in detail in chapter 11. With regard to blood pressure reduction, it is obvious that the order cannot be randomized. The pretreatment reading occurs before the post-treatment reading. The inflexibility of this ordering can create problems in interpretation of such data. The conclusions based on these data could be controversial because of the lack of a "control" group. If extraneous conditions that could influence the experimental outcome are different at the time of the initial and final observation (pre- and post-treatment), the treatment effect is "confounded" with the differences in conditions at the two points of observation. Therefore, randomization of the order of treatment given to each subject is important for the validity of this statistical test. For example, consider a study to compare two hypnotic drugs with regard to sleep-inducing effects. If the first drug were given to all patients before the second drug, and the initial period happened to be associated with hot and humid weather conditions, any observed differences between drugs (or lack of difference) would be "tainted" by the effect of the weather on the therapeutic response.

An important feature of the paired design is that the experimental units receiving the two treatments are, indeed, related. Sometimes, this is not as obvious as the example of the same

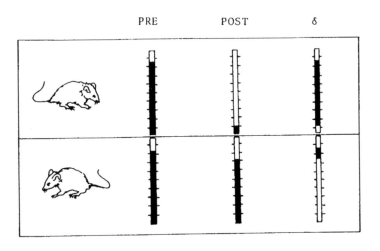

Table 5.11 Results of a Bioavailability Study Comparing a New Formulation (*A*) to a Marketed Form (*B*) with Regard to the Area Under the Blood-Level Curve

Animal	A	B	$\delta = B - A$	$A/B = R$
1	136	166	30	0.82
2	168	184	16	0.91
3	160	193	33	0.83
4	94	105	11	0.90
5	200	198	−2	1.01
6	174	197	23	0.88
			$\bar{\delta} = 18.5$	$\bar{R} = 0.89$
			$S_\delta = 13.0$	$S_R = 0.069$

patient taking both treatments. One can think of the concept of relatedness in terms of the paired samples being more alike than samples from members of different pairs. Pairs may be devised in clinical trials by pairing patients with similar characteristics, such as age, sex, severity of disease, and so on.

Example 6: A new formulation of a marketed drug is to be tested for bioavailability, comparing the extent of absorption to the marketed form on six laboratory animals. Each animal received both formulations in random order on two different occasions. The results, the area under the blood level versus time curve (AUC), are shown in Table 5.11.

$$H_0 : \Delta = 0^{**} \qquad H_a : \Delta \neq 0$$

This is a two-sided test, with the null hypothesis of equality of means of the paired samples. (The true difference is zero.) Before the experiment, it was not known which formulation would be more or less bioavailable if, indeed, the formulations are different. The significance level is set at 5%. From Table 5.11, the average difference is 18.5 and the standard deviation of the differences (δ values) is 13.0. The t test is

$$t = \frac{\bar{\delta} - \Delta}{S/\sqrt{N}}. \tag{5.14}$$

The form of the test is the same as the one-sample t test [Eq. (5.5)]. In our example, a two-sided test,

$$t = \frac{|18.5 - 0|}{13/\sqrt{6}} = 3.84.$$

For a two-sided test at the 5% level, a t value of 2.57 is needed for significance (d.f. = 5; there are six pairs). Therefore, the difference is significant at the 5% level. Formulation *B* appears to be more bioavailable.

In many kinds of experiments, *ratios* are more meaningful than differences as a practical expression of the results. In comparative bioavailability studies, the ratio of the AUCs of the two competing formulations is more easily interpreted than is their difference. The ratio expresses the *relative* absorption of the formulations. From a statistical point of view, if the AUCs for formulations *A* and *B* are normally distributed, the difference of the AUCs is also normally distributed. It can be proven that the ratio of the AUCs will *not* be normally distributed and the assumption of normality for the t test is violated. However, if the variability of the ratios is not great and the sample size is sufficiently "large," analyzing the ratios should give conclusions similar to that obtained from the analysis of the differences. Another alternative for the analysis of such data is the logarithmic transformation (see chap. 10), where the differences of the

---

** Δ is the hypothetical difference and δ the observed average difference.

logarithms of the AUCs are analyzed. (Also see chap. 11, Locke's approach, for an analysis of ratios.) For purposes of illustration, we will analyze the data in Table 5.11 using the ratio of the AUCs for formulations A and B. The ratios are calculated in the last column in Table 5.11.

The null and alternative hypotheses in this case are

$$H_0 : R_0 = 1 \qquad H_a : R_0 \neq 1,$$

where R_0 is the true ratio. If the products are identical, we would expect to observe an average ratio close to 1 from the experimental data. For the statistical test, we choose α equal to 0.05 for a two-sided test. Applying Eq. (5.5), where $\overline{X}$ is replaced by the average ratio $\overline{R}$

$$t = \frac{|\overline{R} - 1|}{S/\sqrt{6}} = \frac{|0.89 - 1|}{0.069/\sqrt{6}} = 3.85.$$

Note that this is a one-sample test. We are testing the mean of a single sample of ratios versus the hypothetical value of 1. Because this is a two-sided test, low or high ratios can lead to significant differences. As in the analysis of the differences, the value of t is significant at the 5% level. (According to Table IV.4, at the 5% level, t must exceed 2.57 for significance.)

A *confidence interval* for the average ratio (or difference) of the AUCs can be computed in a manner similar to that presented earlier in this chapter [Eq. (5.2)]. A 95% confidence interval for the true ratio A/B is given by

$$\frac{A}{B} = \overline{R} \pm \frac{t(S)}{\sqrt{N}} = 0.89 \pm \frac{2.57(0.069)}{\sqrt{6}} = 0.89 \pm 0.07 = 0.82 \text{ to } 0.96.$$

Again, the fact that the confidence interval does not cover the value specified by H_0 (1) means that the statistical test is significant at the 5% level.

A more complete discussion of the analysis of bioequivalence data as required by the FDA is given in chapter 11.

5.2.4   Normal Distribution Tests for Proportions (Binomial Tests)

The tests described thus far in this chapter (normal distribution and t tests as well as confidence intervals) can also be applied to data that are binomially distributed. To apply tests for binomial variables based on the normal distribution, a conservative rule is that the sample sizes should be sufficiently large so that both $N\hat{p}$ and $N\hat{q}$ are larger than or equal to 5. Where $\hat{p}$ is the observed proportion and $\hat{q} = 1 - \hat{p}$. For symmetric distributions ($p \cong 0.5$), this constraint may be relaxed somewhat. The binomial tests are based on the normal approximation to the binomial and, therefore, we use normal curve probabilities when making decisions in these tests. To obtain the probabilities for tests of significance, we can use the t table with ∞ d.f. or the standard normal distribution (Tables IV.4 and IV.2, respectively). We will also discuss the application of the χ^2 (chi-square) distribution to the problem of comparing the "means" of binomial populations.

5.2.4.1   Test to Compare the Proportion of a Sample to a Known or Hypothetical Proportion

This test is equivalent to the normal test of the mean of a single population. The test is

$$Z = \frac{\hat{p} - p_0}{\sqrt{p_0 q_0 / N}} \tag{5.15}$$

where $\hat{p}$ is the observed proportion and p_0 the hypothetical proportion under the null hypothesis $H_0 : p' = p_0$.

The test procedure is analogous to the one-sample tests described in section 5.2.1. Because of the discrete nature of binomial data, a correction factor is recommended to improve the normal approximation. The correction, often called the *Yates continuity correction*, consists of subtracting $1/(2N)$ from the absolute value of the numerator of the test statistic [Eq. (5.15)]

$$Z = \frac{|\hat{p} - p_0| - 1/(2N)}{\sqrt{p_0 q_0/N}}. \tag{5.16}$$

For a two-tailed test, the approximation can be improved as described by Snedecor and Cochran [5]. The correction is the same as the Yates correction if np is "a whole number or ends in 0.5." Otherwise, the correction is somewhat less than $1/(2N)$ (see Ref. [5] for details). In the examples presented here, we will use the Yates correction. This results in probabilities very close to those that would be obtained by using exact calculations based on the binomial theorem. Some examples should make the procedure clear.

Example 7: Two products are to be compared for preference with regard to some attribute. The attribute could be sensory (taste, smell, etc.) or therapeutic effect as examples. Suppose that an ointment is formulated for rectal itch and is to be compared to a marketed formulation. Twenty patients try each product under "blind" conditions and report their preference. The null hypothesis and alternative hypothesis are

$$H_0: p_a = p_b \quad \text{or} \quad H_0 : p_a = 0.5 \qquad H_a : p_a \neq 0.5,$$

where p_a and p_b are the hypothetical preferences for A and B, respectively. If the products are truly equivalent, we would expect one-half of the patients to prefer either product A or B. Note that is a *one-sample* test. There are two possible outcomes that can result from each observation: a patient prefers A or prefers B ($p_a + p_b = 1$).

We observe the proportion of preferences (successes) for A, where A is the new formulation. This is a two-sided test; very few or very many preferences for A would suggest a significant difference in preference for the two products. Final tabulation of results showed that 15 of 20 patients found product A superior (5 found B superior). Does this result represent a "significant" preference for product A? Applying Eq. (5.16), we have

$$Z = \frac{|15/20 - 0.5| - 1/40}{\sqrt{(0.5)(0.5)/20}} = 2.01.$$

Note the correction for continuity, $1/(2N)$. Also note that the denominator uses the value of pq based on the null hypothesis ($p_a = 0.5$), not the sample proportion ($0.75 = 15/20$). This procedure may be rationalized if one verbalizes the nature of the test. We assume that the preferences are equal for both products ($p_a = 0.5$). We then observe a sample of 20 patients to see if the results conform with the hypothetical preference. Thus, the test is based on a hypothetical binomial distribution with the expected number of preferences equal to 10 ($p_a \times 20$). See Figure 5.11, which illustrates the rejection region in this test. The value of $Z = 2.01$ (15 preferences in a sample of 20) is sufficiently large to reject the null hypothesis. A value of 1.96 or greater is

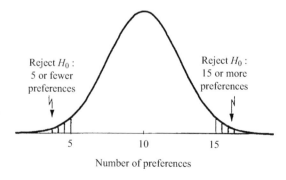

Reject H_0 :
5 or fewer
preferences

Reject H_0 :
15 or more
preferences

5   10   15

Number of preferences

Figure 5.11 Rejection region for the test of $p_a = 0.5$ for a sample of 20 patients ($\alpha = 0.05$, two-sided test).

significant at the 5% level (Table IV.2). The test of $p_0 = 0.5$ is common in statistical procedures. The sign test described in chapter 15 is a test of equal proportions (i.e., $p_0 = q_0 = 0.5$).

Example 8: A particularly lethal disease is known to result in 95% fatality if not treated. A new treatment is given to 100 patients and 10 survive. Does the treatment merit serious consideration as a new therapeutic regimen for the disease? We can use the normal approximation because the expected number of successes and failures both are ≥ 5, that is, $Np_0 = 5$ and $Nq_0 = 5$ and $Nq_0 = 95$ ($p_0 = 0.05$, $N = 100$). A one-sided test is performed because evidence supports the hypothesis that the treatment cannot worsen the chances of survival. The α level is set at 0.05. Applying Eq. (5.16), we have

$$H_0 : p_0 = 0.05 \qquad H_0 : p_0 > 0.05$$

$$Z = \frac{|0.10 - 0.05| - 1/200}{\sqrt{(0.05)(0.95)/100}} = 2.06.$$

Table IV.2 shows that a value of Z equal to 1.65 would result in significance at the 5% level (one-sided test). Therefore, the result of the experiment is strong evidence that the new treatment is effective ($p < 0.05$).

If either Np_0 or Nq_0 is less than 5, the normal approximation to the binomial may not be justified. Although this rule is conservative, if in doubt, in these cases, probabilities must be calculated by enumerating all possible results that are equally or less likely to occur than the observed result under the null hypothesis. This is a tedious procedure, but in some cases it is the only way to obtain the probability for significance testing [7]. Fortunately, most of the time, the sample sizes of binomial experiments are sufficiently large to use the normal approximation.

5.2.4.2 Tests for the Comparison of Proportions from Two Independent Groups

Experiments commonly occur in the pharmaceutical and biological sciences that involve the comparison of proportions from two independent groups. These experiments are analogous to the comparison of means in two independent groups using the t or normal distributions. For proportions, the form of the test is similar. With a sufficiently large sample size, the normal approximation to the binomial can be used, as in the single-sample test. For the hypothesis: $H_0 : p_a = p_b (p_a - p_b = 0)$, the test using the normal approximation is

$$Z = \frac{\hat{p}_a - \hat{p}_b}{\sqrt{\hat{p}_0 \hat{q}_0 (1/N_1 + 1/N_2)}}, \tag{5.17}$$

where $\hat{p}_a$ and $\hat{p}_b$, are the observed proportions in groups A and B, respectively, and N_1 and N_2 are the sample sizes for groups A and B, respectively, $\hat{p}_0$ and $\hat{q}_0$ are the "pooled" proportion of successes and failures. The pooled proportion, $\hat{p}_0$, is similar to the pooled standard deviation in the t test. For proportions, the results of the two comparative groups are pooled together and the "overall" observed proportion is equal to $\hat{p}_0$. Under the null hypothesis, the probability of success is the same for both groups, A and B. Therefore, the best estimate of the common probability for the two groups is the estimate based on the combination of data from the entire experiment. An example of this calculation is shown in Table 5.12. The pooled proportion, $\hat{p}_0$, is a weighted average of the two proportions. This is exactly the same as adding up the total number of "successes" and dividing this by the total number of observations. In the example in

Table 5.12 Sample Calculation for Pooling Proportions from Two Groups

Group I	Group II
$N = 20$	$N = 30$
$\hat{p}_1 = 0.8$	$\hat{p}_2 = 0.6$
$\hat{p}_0 = $ pooled $p = (20 \times 0.8 + 30 \times 0.6)/(20 + 30) = 0.68$	

Table 5.12, the total number of successes is 34, 16 in group I and 18 in group II. The total number of observations is 50, 30 + 20. The following examples illustrate the computations.

Example 9: In a clinical study designed to test the safety and efficacy of a new therapeutic agent, the incidence of side effects are compared for two groups of patients, one taking the new drug and the other group taking a marketed standard agent. Headache is a known side effect of such therapy. Of 212 patients on the new drug, 35 related that they had experienced severe headaches. Of 196 patients on the standard therapy, 46 suffered from severe headaches. Can the new drug be claimed to result in fewer headaches than the standard drug at the 5% level of significance? The null and alternative hypotheses are

$$H_0 : p_1 = p_2 (p_1 - p_2) = 0 \qquad H_a : p_1 \neq p_2.$$

This is a two-sided test. Before performing the statistical test, the following computations are necessary

$$\hat{p}_1 = \frac{35}{212} = 0.165$$

$$\hat{p}_2 = \frac{46}{196} = 0.235$$

$$\hat{p}_0 = \frac{81}{408} = 0.199 \,(\hat{q}_0 = 0.801).$$

Applying Eq. (5.17), we have

$$Z = \frac{|0.235 - 0.165|}{\sqrt{(0.199)(0.801)(1/212 + 1/196)}} = \frac{0.07}{0.0395} = 1.77.$$

Since a Z value of 1.96 is needed for significance at the 5% level, the observed difference between the two groups with regard to the side effect of "headache" is not significant ($p > 0.05$).

Example 10: In a preclinical test, the carcinogenicity potential of a new compound is determined by administering several doses to different groups of animals. A control group (placebo) is included in the study as a reference. One of the dosage groups showed an incidence of the carcinoma in 9 of 60 animals (15%). The control group exhibited 6 carcinomas in 65 animals (9.2%). Is there a difference in the proportion of animals with the carcinoma in the two groups ($\alpha = 5\%$)? Applying Eq. (5.17), we have

$$H_0 : p_1 = p_2 \qquad H_0 : p_1 \neq p_2$$

$$Z = \frac{|9/60 - 6/65|}{\sqrt{(15/125)(110/125)(1/60 + 1/65)}} = \frac{0.0577}{0.058} = 0.99.$$

Note that $\hat{p}_1 = 9/60 = 0.15$, $\hat{p}_2 = 6/65 = 0.092$, and $\hat{p}_0 = 15/125 = 0.12$.

Since Z does not exceed 1.96, the difference is not significant at the 5% level. This test could have been a one-sided test (a priori) if one were certain that the new compound could not lower the risk of carcinoma. However, the result is not significant at the 5% level for a one-sided test; a value of Z equal to 1.65 or greater is needed for significance for a one-sided test.

Example 11: A new operator is assigned to a tablet machine. A sample of 1000 tablets from this machine showed 8% defects. A random sample of 1000 tablets from the other tablet presses used during this run showed 5.7% defects. Is there reason to believe that the new operator produced more defective tablets than that produced by the more experienced personnel? We will perform a two-sided test at the 5% level, using Eq. (5.17).

$$Z = \frac{|0.08 - 0.057|}{\sqrt{(0.0685)(0.9315)(2/1000)}} = \frac{0.023}{0.0113} = 2.04$$

Since the value of Z (2.04) is greater than 1.96, the difference is significant at the 5% level. We can conclude that the new operator is responsible for the larger number of defective tablets produced at his station, assuming that there is no difference among tablet presses. (See also Exercise Problem 19.) If a continuity correction is used, the equivalent chi-square test with a correction as described below is recommended.**

There is some controversy about the appropriateness of a continuity correction in these tests. D'Agostino et al. [6] examined various alternatives and compared the results to exact probabilities. They concluded that for small sample sizes (N_1 and $N_2 < 15$), the use of the Yates continuity correction resulted in too conservative probabilities (i.e., probabilities were too high which may lead to a lack of rejection of H_0 in some cases).

They suggest that in these situations a correction should not be used. They also suggest an alternative analysis that is similar to the t test

$$t = \frac{|p_1 - p_2|}{\text{s.d.}\sqrt{1/N_1 + 1/N_2}},\tag{5.18}$$

where s.d. is the pooled standard deviation computed from the data considering a success equal to 1 and a failure equal to 0. The value of t is compared to the appropriate t value with $N_1 + N_2 - 2$ d.f. The computation for the example in Table 5.12 is as follows:

For group I, $S_1^2 = \frac{16 - (16^2/20)}{19} = 0.168.$

For group II, $S_2^2 = \frac{18 - (18^2/30)}{29} = 0.248.$

(Note that for group I the number of successes is 16 and the number of failures is 4. Thus, we have 16 values equal to 1 and 4 values equal to 0. The variance is calculated from these 20 values.) The pooled variance is

$$\frac{19 \times 0.168 + 29 \times 0.248}{48} = 0.216.$$

The pooled standard deviation is 0.465.
From Eq. (5.18),

$$t = \frac{|0.8 - 0.6|}{0.465\sqrt{1/20 + 1/30}} = 1.49.$$

The t value with 48 d.f. for significance at the 5% level for a two-sided test is 2.01. Therefore, the results fail to show a significant difference at the 5% level.

Fleiss [7] advocates the use of the Yates continuity correction. He states "Because the correction for continuity brings probabilities associated with χ^2 and Z into close agreement with the exact probabilities, the correction should always be used."

5.2.5   Chi-Square Tests for Proportions

An alternative method of comparing proportions is the chi-square (χ^2) test. This test results in identical conclusions as the binomial test in which the normal approximation is used as described above. The chi-square distribution is frequently used in statistical tests involving counts and proportions, as discussed in chapter 15. Here, we will show the application to fourfold tables (2×2 tables), the comparison of proportions in two independent groups.

The chi-square distribution is appropriate where the normal approximation to the distribution of discrete variables can be applied. In particular, when comparing two proportions, the chi-square distribution with 1 d.f. can be used to approximate probabilities. (The values for the

---

** The continuity correction can make a difference when making decisions based on the α level, when the statistical test is "just significant" (e.g., $p = 0.04$ for a test at the 5% level). The correction makes the test "less significant."

Table 5.13   Result of the Experiment Shown in Table 5.12 in the
Form of a Fourfold Table

	Group		
	I	II	Total
Number of successes	16	18	34
Number of failures	4	12	16
Total	20	30	50

x^2 distribution with one d.f. are exactly the square of the corresponding normal deviates. For example, the "95%" cutoff point for the chi-square distribution with 1 d.f. is 3.84, equal to 1.96^2.)

The use of the chi-square distribution to test for differences of proportions in two groups has two advantages: (a) the computations are easy and (b) a continuity correction can be easily applied. The reader may have noted that a continuity correction was not used in the examples for the comparison of two independent groups described above. The correction was not included because the computation of the correction is somewhat complicated. In the chi-square test, however, the continuity correction is relatively simple. The correction is most easily described in the context of an example. We will demonstrate the chi-square test using the data in Table 5.12. We can think of these data as resulting from a clinical trial where groups I and II represent two comparative drugs. The same results are presented in the *fourfold table* shown in Table 5.13.

The chi-square statistic is calculated as follows:

$$\chi^2 = \sum \frac{(O - E)^2}{E},$$
(5.19)

where O is the observed number in a cell (there are four cells in the experiment in Table 5.13; a cell is the intersection of a row and column; the upper left-hand cell, number of successes in group I, has the value 16 contained in it), and E the expected number in a cell.

The expected number is the number that would result if each group had the same proportion of successes and failures. The best estimate of the common p (proportion of successes) is the pooled value, as calculated in the test using the normal approximation above [Eq. (5.17)]. The pooled, p, $\hat{p}_0$, is 0.68 (34/50). With a probability of success of 0.68 (34/50), we would expect "13.6" successes for group I (20 × 0.68). The expected number of failures is 20 × 0.32 = 6.4. The expected number of failures can also be obtained by subtracting 13.6 from the total number of observations in group I, 20 − 13.6 = 6.4. Similarly, the expected number of successes in group II is 30 × 0.68 = 20.4. Again the number, 20.4, could have been obtained by subtracting 13.6 from 34.

This concept (and calculation) is illustrated in Table 5.14, which shows the expected values for Table 5.13. The marginal totals (34, 16, 20, and 30) in the "expected value" table are the same as in the original table, Table 5.13. In order to calculate the expected values, multiply the two marginal totals for a cell and divide this value by the grand total. This simple way of calculating

Table 5.14   Expected Values for the Experiment Shown in
Table 5.13

	Group		
	I	II	Total
Expected number of successes	13.6	20.4	34
Expected number of failures	6.4	9.6	16
Total	20.0	30.0	50

the expected values will be demonstrated for the upper left-hand cell, where the observed value is 16. The expected value is

$$\frac{(20)(34)}{50} = 13.6.$$

Once the expected value for one cell is calculated, the expected values for the remaining cells can be obtained by subtraction.

Expected successes in group II $= 34 - 13.6 = 20.4$.
Expected failures in group I $= 20 - 13.6 = 6.4$.
Expected failures in group II $= 16 - 6.4 = 9.6$.

Given the marginal totals and the value for any *one* cell, the values for the other three cells can be calculated. Once the expected values have been calculated, the chi-square statistic is evaluated according to Eq. (5.19).

$$\sum \frac{(O-E)^2}{E} = \frac{(16-13.6)^2}{13.6} + \frac{(18-20.4)^2}{20.4} + \frac{(4-6.4)^2}{6.4} + \frac{(12-9.6)^2}{9.6} = 2.206$$

The numerator of each term is $(\pm 2.4)^2 = 5.76$. Therefore, the computation of χ^2 can be simplified as follows:

$$\chi^2 = (O-E)^2 \left(\frac{1}{E_1} + \frac{1}{E_2} + \frac{1}{E_3} + \frac{1}{E_4} \right), \tag{5.20}$$

where E_1 through E_4 are the expected values for each of the four cells.

$$\chi^2 = (2.4)^2 \left(\frac{1}{13.6} + \frac{1}{20.4} + \frac{1}{6.4} + \frac{1}{9.6} \right) = 2.206$$

One can show that this computation is exactly equal to the square of the Z value using the normal approximation to the binomial. (See Exercise Problem 11.)

The d.f. for the test described above (the fourfold table) are equal to 1. In general, the d.f. for an $R \times C$ contingency table, where R is the number of rows and C is the number of columns, are equal to $(R-1)(C-1)$. The analysis of $R \times C$ tables is discussed in chapter 15.

Table IV.5, a table of points in the cumulative chi-square distribution, shows that a value of 3.84 is needed for significance at the 5% level (1 d.f.). Therefore, the test in this example is not significant; that is, the proportion of successes in group I is not significantly different from that in group II, 0.8 and 0.6, respectively.

To illustrate further the computations of the chi-square statistic and the application of the continuity correction, we will analyze the data in Example 10, where the normal approximation to the binomial was used for the statistical test. Table 5.15 shows the observed and expected values for the results of this preclinical study.

Table 5.15 Observed and Expected Values for Preclinical Carcinogenicity Study[a]

	Drug	Placebo	Total
Animals with carcinoma	9(7.2)	6 (7.8)	15
Animals without carcinoma	51(52.8)	59 (57.2)	110
Total	60	65	125

[a]Parenthetical values are expected values.

The uncorrected chi-square analysis results in a value of 0.98, $(0.99)^2$. (See Exercise Problem 18.) The continuity correction is applied using the following rule: If the fractional part of the difference $(O - E)$ is larger than 0 but ≤ 0.5, delete the fractional part. If the fractional part is greater than 0.5 or exactly 0, "reduce the fractional part to 0.5." Some examples should make the application of this rule clearer.

$O - E$	Corrected for continuity
3.0	2.5
3.2	3.0
3.5	3.0
3.9	3.5
3.99	3.5
4.0	3.5

In the example above, $O - E = \pm 1.8$. Therefore, correct this value to ± 1.5. The corrected chi-square statistic is [Eq. (5.20)]

$$(1.5)^2 \left(\frac{1}{7.2} + \frac{1}{7.8} + \frac{1}{52.8} + \frac{1}{57.2} \right) = 0.68.$$

In this example, the result is not significant using either the corrected or uncorrected values. However, when chi-square is close to significance at the α level, the continuity correction can make a difference. The continuity correction is more apparent in its effect on the computation of chi-square in small samples. With large samples, the correction makes less of a difference.

The chi-square test, like the normal approximation, is an approximate test, applying a continuous distribution to discrete data. The test is valid (close to correct probabilities) when the expected value in each cell is at least 5. This is an approximate rule. Because the rule is conservative, in some cases, an expected value in one or more cells of less than 5 can be tolerated. However, one should be cautious in applying this test if the expected values are too small.

5.2.6 Confidence Intervals for Proportions

Examples of the formation of a confidence interval for a proportion have been presented earlier in this chapter (Example 3). Although the confidence interval for the binomial is calculated using the standard deviation of the binomial based on the sample proportion, we should understand that in most cases, the s.d. is unknown. The sample standard deviation is an estimate of the true s.d., which for the binomial depends on the true value of the proportion or probability. However, when we use the sample estimate of the s.d. for the calculations, the confidence interval and statistical tests are valid using criteria based on the normal distribution (Table IV.2). We do not use the t distribution as in the procedures discussed previously.

The confidence interval for the true proportion or binomial probability, p_0 is

$$\hat{p} \pm Z \sqrt{\frac{\hat{p}\hat{q}}{N}}, \tag{5.3}$$

where $\hat{p}$ is the observed proportion in a sample of size N. The value of Z depends on the confidence coefficient (e.g., 1.96 for a 95% interval). Of 500 tablets inspected, 20 were found to be defective ($\hat{p} = 20/500 = 0.04$). A 95% confidence interval for the true proportion of defective tablets is

$$\hat{p} \pm 1.96 \sqrt{\frac{\hat{p}\hat{q}}{N}} = 0.04 \pm 1.96 \sqrt{\frac{(0.04)(0.96)}{500}}$$

$$= 0.04 \pm 0.017 = 0.023 \text{ to } 0.057.$$

To obtain a *confidence interval for the difference of two proportions* (two independent groups), when the "underlying proportions, p_1 and p_2 are not hypothesized to be equal (7), use the following formula:

$$(\hat{p}_1 - \hat{p}_2) \pm Z\sqrt{\frac{\hat{p}_1\hat{q}_1}{N_1} + \frac{\hat{p}_2\hat{q}_2}{N_2}} \tag{5.21}$$

where $\hat{p}_1$ and $\hat{p}_2$ are the observed proportions in groups I and II, respectively, and N_1, and N_2 are the respective sample sizes of the two groups. Z is the appropriate normal deviate (1.96 for a 95% confidence interval).

In the example of incidence of headaches in two groups of patients, the proportion of headaches observed in group I was $35/212 = 0.165$ and the proportion in group II was $46/196 = 0.235$. A 95% confidence interval for the difference of the two proportions, calculated from Eq. (5.21), is

$$(0.235 - 0.165) \pm 1.96\sqrt{\frac{(0.165)(0.835)}{212} + \frac{(0.235)(0.765)}{196}}$$

$$= 0.07 \pm 0.078 = -0.008 \text{ to } 0.148.$$

The difference between the two proportions was not significant at the 5% level in a two-sided test (see "Test for Comparison of Proportions from Two Independent Groups" in sect. 5.2.4). Note that 95% confidence interval covers 0, the difference specified in the null hypothesis ($H_0 : p_1 - p_2 = 0$).[††]

Fleiss [7] and Hauck and Anderson [8] recommend the use of a continuity correction for the construction of confidence intervals that gives better results than that obtained without a correction [Eq. (5.21)]. If a 90% or 95% interval is used, the Yates correction works well if N_1p_1, N_1q_1, N_2p_2, and N_2q_2 are all greater than or equal to 3. The 99% interval is good for N_1p_1, N_1q_1, N_2p_2, and N_2q_2 all greater than or equal to 5. The correction is $1/2N_1 + 1/2N_2$. Applying the correction to the previous example, a 95% confidence interval is

$$(0.235 - 0.165)$$
$$\pm \{1.96 [(0.165)(0.835)/212 + (0.235)(0.765)/196]^{1/2} + (1/424 + 1/392)\}$$
$$= 0.070 \pm 0.0825 = -0.0125 \text{ to } 0.1525.$$

We have noted previously that if the hypothesis test of equality of two proportions is statistically significant, the confidence interval for the difference of the proportions will not cover zero (and vice versa). Sometimes, this does not hold when comparing two proportions because the formulation for the hypothesis test of equal proportions is different from the confidence interval calculation. In this case, Fleiss [7] recommends changing the hypothesis test statistic by replacing the denominator in equation 5.17 with the square root of $(p_1q_1)/N_1 + (p_2q_2)/N_2$.

An approach to sample size requirements using confidence intervals for bioequivalence trials with a binomial variable is given in section 11.4.8.

5.3 COMPARISON OF VARIANCES IN INDEPENDENT SAMPLES

Most of the statistical tests presented in this book are concerned with means. However, situations arise where variability is important as a measure of a process or product performance. For example, when mixing powders for tablet granulations, one may be interested in measuring the homogeneity of the mix as may be indicated in validation procedures. The "degree" of homogeneity can be determined by assaying different portions of the mix, and calculating the

---

[††] The form of the confidence interval [Eq. (5.21)] differs from the form of the statistical test in that the latter uses the pooled variance [Eq. (5.17)]. Therefore, this relationship will not always hold for the comparison of two proportions.

standard deviation or variance. (Sample weights equal to that in the final dosage form are most convenient.) A small variance would be associated with a relatively homogeneous mix, and vice versa. Variability is also often of interest when assaying drug blood levels in a bioavailability study or when determining a clinical response to drug therapy. We will describe statistical tests appropriate for two situations: the comparison of two variances from independent samples, and the comparison of variances in related (paired) samples. The test for related samples will be presented in chapter 7 because methods of calculation involve material presented there. The test for the comparison of variances in independent samples described here assumes that the data in each sample are independent and normally distributed.

The notion of significance tests for two variances is similar to the tests for means (e.g., the t test). The null hypothesis is usually of the form

$$H_0 : \sigma_1^2 = \sigma_2^2.$$

For a two-sided test, the alternative hypothesis admits the possibility of either variance being larger or smaller than the other

$$H_0 : \sigma_1^2 \neq \sigma_2^2.$$

The statistical test consists of calculating the ratio of the two sample variances. The ratio has an F distribution with $(N_1 - 1)$ d.f. in the numerator and $(N_2 - 1)$ d.f. in the denominator. To determine if the ratio is "significant" (i.e., the variances differ), the observed ratio is compared to appropriate table values of F at the α level. The F distribution is not symmetrical and, in general, to make statistical decisions, we would need F tables with both upper and lower cutoff points.

Referring to Figure 5.12, if the F ratio falls between F_L and F_U the test is not significant. We do not reject the null hypothesis of equal variances. If the F ratio is below F_L or above F_U, we reject the null hypothesis and conclude that the variances differ (at the 5% level, the shaded area in the example of Fig. 5.12). The F table to test the equality of two variances is the same as that used to determine significance in analysis of variance tests to be presented in chapter 8 (Table IV.6). However, F tables for ANOVA usually give only the upper cutoff points ($F_{U,0.05}$ in Fig. 5.12, for example).

Nevertheless, it is possible to perform a two-sided test for two variances using the one-tailed F table (Table IV.6) by forming the ratio with the *larger variance in the numerator*. Thus, the ratio will always be equal to or greater than 1. The ratio is then referred to the usual ANOVA F table, but the level of significance is twice that stated in the table. For example, the values that must be exceeded for significance in Table IV.6 represent cutoff points at the 20%, 10% or 2% level if the larger variance is in the numerator. For significance at the 5% level, use Table 5.16, a brief table of the upper 0.025 cutoff points for some F distributions.

To summarize, for a two-sided test at the 5% level, calculate the ratio of the comparative variances with the larger variance in the numerator. (Clearly, if the variances in the two groups are identical, there is no need to perform a test of significance.) To be significant at the 5% level, the ratio must be equal to or greater than the tabulated upper 2.5% cutoff points (Table 5.16). For significance at the 10% level or 20% level, for a two-sided test, use the upper 5% or 10% points in Table IV.6A.

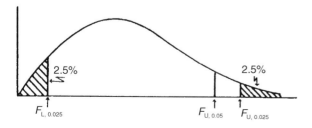

Figure 5.12 Example of two-sided cutoff points in an F distribution.

Table 5.16 Brief Table of Upper 0.025 Cutoff Points of the F Distribution

Degrees of in freedom denominator	Degrees of freedom in numerator											
	2	3	4	5	6	8	10	15	20	25	30	∞
2	39.0	39.2	39.3	39.3	39.3	39.4	39.4	39.4	39.5	39.5	39.5	39.5
3	16.0	15.4	15.1	14.9	14.7	14.5	14.4	14.3	14.2	14.1	14.1	13.9
4	10.6	10.0	9.6	9.4	9.2	9.0	8.8	8.7	8.6	8.5	8.5	8.3
5	8.4	7.8	7.4	7.2	7.0	6.8	6.6	6.4	6.3	6.3	6.2	6.0
6	7.3	6.6	6.2	6.0	5.8	5.6	5.5	5.3	5.2	5.1	5.1	4.9
7	6.5	5.9	5.5	5.3	5.1	4.9	4.8	4.6	4.5	4.4	4.4	4.1
8	6.1	5.4	5.1	4.8	4.7	4.4	4.3	4.1	4.0	3.9	3.9	3.7
9	5.7	5.1	4.7	4.5	4.3	4.1	4.0	3.8	3.7	3.6	3.6	3.3
10	5.5	4.8	4.5	4.2	4.1	3.9	3.7	3.5	3.4	3.4	3.3	3.1
15	4.8	4.2	3.8	3.6	3.4	3.2	3.1	2.9	2.8	2.7	2.6	2.4
20	4.5	3.9	3.5	3.3	3.1	2.9	2.8	2.6	2.5	2.4	2.4	2.1
24	4.3	3.7	3.4	3.2	3.0	2.8	2.6	2.4	2.3	2.3	2.2	1.9
30	4.2	3.6	3.3	3.0	2.9	2.7	2.5	2.3	2.2	2.1	2.1	1.8
40	4.1	3.5	3.1	2.9	2.7	2.5	2.4	2.2	2.1	2.0	1.9	1.6
∞	3.7	3.1	2.8	2.6	2.4	2.2	2.1	1.8	1.7	1.6	1.6	1.0

For a *one-sided test*, if the null hypothesis is

$$H_0: \sigma_A^2 \geq \sigma_B^2 \qquad H_0: \sigma_A^2 < \sigma_B^2.$$

Perform the test only if S_A^2 is smaller than S_B^2, with S_B^2 in the numerator. (If S_A^2 is equal to or greater than S_B^2, we cannot reject the null hypothesis.) Refer the ratio to Table IV.6 for significance at the 5% (or 1%) level. (The test is one -sided.)

One should appreciate that this statistical test is particularly sensitive to departures from the assumptions of normality and independence of the two comparative groups.

An example should clarify the procedure. Two granulations were prepared by different procedures. Seven random samples of powdered mix of equal weight (equal to the weight of the final dosage form) were collected from each batch and assayed for active material, with the results shown in Table 5.17. The test is to be performed at the 5% level: $H_0 : \sigma_1^2 = \sigma_2^2$; $H_a: \sigma_1^2 \neq \sigma_2^2$. For a two-sided test, we form the ratio of the variances with a σ_B^2, the larger variance in the numerator.

$$F = \frac{1.297}{0.156} = 8.3.$$

The tabulated F value with 6 d.f. in the numerator and denominator (Table 5.16) is 5.8. Therefore, the variances can be considered significantly different $(P < 0.05)$; granulation B is more variable than granulation A. If the test were performed at the 10% level, we would refer to the upper 5% points in Table IV.6, where a value greater than 4.28 would be significant.

Table 5.17 Assays from Samples from Two Granulations

Granulation *A*		Granulation *B*	
20.6	20.7	20.2	19.0
20.9	19.8	21.5	21.8
20.6	20.4	18.9	20.4
	21.0		21.0
$\overline{X} = 20.57$	$S^2 = 0.156$	$\overline{X} = 20.4$	$S^2 = 1.297$

If the test were *one sided, at the* 5% *level,* for example, with the null hypothesis

$$H_0 : \sigma_A^2 \geq \sigma_B^2 \qquad H_a : \sigma_A^2 < \sigma_B^2,$$

the ratio $1.297/0.156 = 8.3$ would be referred to Table IV.6 for significance. Now, a value greater than 4.28 would be significant at the 5% level.

If more than two variances are to be compared, the F test discussed above is not appropriate. Bartlett's test is the procedure commonly used to test the equality of more than two variances [1], as described in the following paragraph.

5.4   TEST OF EQUALITY OF MORE THAN TWO VARIANCES

The test statistic computation is shown in Eq. (5.22)

$$\chi^2 = \sum (N_i - 1) \ln S^2 - \sum [(N_i - 1) \ln S_1^2], \tag{5.22}$$

where S^2 is the pooled variance and S_i^2 is the variance of the ith sample.

The computations are demonstrated for the data of Table 5.18. In this example, samples of a granulation were taken at four different locations in a mixer. Three samples were analyzed in each of three of the locations, and five samples analyzed in the 4th location. The purpose of this experiment was to test the homogeneity of the mix in a validation experiment. Part of the statistical analysis requires an estimate of the variability within each location. The statistical test (analysis of variance, chap. 8) assumes homogeneity of variance within the different locations. Bartlett's test allows us to test for the homogeneity of variance (Table 5.8).

The pooled variance is calculated as the weighted average of the variances, where the weights are the d.f. $(N_i - 1)$.

$$\text{Pooled } S^2 = \frac{2 \times 3.6 + 2 \times 4.7 + 2 \times 2.9 + 4 \times 8.3}{2 + 2 + 2 + 4} = 5.56$$

$$\sum (N_i - 1) = 10$$

$$\sum [(N_i - 1) \ln S_i^2 = 2(1.2809) + 2(1.5476) + 2(1.0647)] + 4(2.1163) = 16.2516$$

$$\chi^2 = 10 \times \ln (5.56) - 16.2516 = 0.904.$$

To test χ^2 for significance, compare the result to the tabulated value of χ^2 (Table IV.5) with 3 d.f. (1 less than the number of variances being compared) at the appropriate significance level. A value of 7.81 is needed for significance at the 5% level. Therefore, we conclude that the variances do not differ. A significant value of χ^2 means that the variances are not all equal. This test is very sensitive to non-normality. That is, if the variances come from non-normal populations, the conclusions of the test may be erroneous.

See Exercise Problem 22 for another example where Bartlett's test can be used to test the homogeneity of variances.

Table 5.18   Results of Variability of Assays of Granulation at Six Locations in a Mixer

Location	N	N-1	Variance (S^2)	$\ln(S^2)$
A	3	2	3.6	1.2809
B	3	2	4.7	1.5476
C	3	2	2.9	1.0647
D	5	4	8.3	2.1163

Table 5.19 Short Table of Lower and Upper Cutoff Points for Chi-Square Distribution

Degrees of freedom	Lower 2.5%	Lower 5%	Upper 95%	Upper 97.5%
2	0.0506	0.1026	5.99	7.38
3	0.216	0.352	7.81	9.35
4	0.484	0.711	9.49	11.14
5	0.831	1.15	11.07	12.83
6	1.24	1.64	12.59	14.45
7	1.69	2.17	14.07	16.01
8	2.18	2.73	15.51	17.53
9	2.70	3.33	16.92	19.02
10	3.25	3.94	18.31	20.48
15	6.26	7.26	25.00	27.49
20	9.59	10.85	31.41	34.17
30	16.79	18.49	43.77	46.98
60	40.48	43.19	79.08	83.30
120	91.58	95.76	146.57	152.21

5.5 CONFIDENCE LIMITS FOR A VARIANCE

Given a sample variance, a confidence interval for the variance can be constructed in a manner similar to that for means. S^2/σ^2 is distributed as $\chi^2/$d.f. The confidence interval can be obtained from the chi-square distribution, using the relationship shown in Eq. (5.23).

$$\frac{S^2(n-1)}{\text{chi-square}_{\alpha/2}} \geq \sigma^2 \geq \frac{S^2(n-1)}{\text{chi-square}_{1-\alpha/2}} \tag{5.23}$$

For example, a variance estimate based on 10 observations is 4.2, with 9 d.f. For a 90% two-sided confidence interval, we put 5% of the probability in each of the lower and upper tails of the χ^2 distribution. From Table 5.19 and Eq. (5.23), the upper limit is

$$S^2(9/3.33) = 4.2(9/3.33) = 11.45.$$

The lower limit is

$$4.2(9/16.92) = 2.23.$$

The values, 3.33 and 16.92, are the cutoff points for 5% and 95% of the chi-square distribution with 9 d.f.. Thus, we can say that with 90% probability, the true variance is between 2.23 and 11.45. Exercise Problem 23 shows an example of a one-sided confidence interval for the variance from a content uniformity test.

5.5.1 Rationale for USP Content Uniformity Test

The USP content uniformity test was based on the desire for a plan that would limit acceptance to lots with sigma (RSD) less than 10% [9]. The main concern is to prevent the release of batches of product with excessive units outside of 75% to 125% of the labeled dose that may occur for lots with a large variability. If the observed RSD for 10 units is less than 6%, one can demonstrate that there is less than 0.05 probability that the true RSD of the lot is greater than 10%. A two-sided 90% confidence interval for an RSD of 6 for $N = 10$, can be calculated by taking the square root of the interval for the variance. In this example, the variance is 36 (RSD = 6). Following the logic of the previous example, the upper limit of the 90% confidence interval for the variance is $6^2(9/3.33) = 97.3$. Since the upper limit represents a one-sided 95% confidence limit, the upper limit for the standard deviation(s) is $\sqrt{97.3}$, approximately 10. See also Exercise Problem 24 at the end of this chapter.

5.6 TOLERANCE INTERVALS

Tolerance intervals have a wide variety of potential applications in pharmaceutical and clinical data analysis. A tolerance interval describes an interval in which a given percentage of the individual items lie, with a specified probability. This may be expressed as

Probability($L \leq$ % of population $\leq U$) where L is the lower limit and U is the upper limit.

For example, a tolerance interval might take the form of a statement such as, "There is 99% probability that 95% of the population is between 85 and 115." More specifically, we might say that there is 99% probability that 95% of the tablets in a batch have a potency between 85% and 115%. In order to be able to compute tolerance intervals, we must make an assumption about the data distribution. As is typical in statistical applications, the data will be assumed to have a normal distribution. In order to compute the tolerance interval, we need an estimate of the mean and standard deviation. These estimates are usually taken from a set of observed experimental data.

Given the d.f. for the estimated s.d., the limits can be computed from Table IV.19 in appendix IV. The factors in Table IV.19 represent multipliers of the standard deviation, similar to a confidence interval. Therefore, using these factors, the tolerance interval computation is identical to the calculation of a confidence interval.

P% tolerance interval containing X% of the population $= \overline{X} \pm t'$ (s.d.)

where t' is the appropriate factor found in Table IV.19.

The following examples are intended to make the calculation and interpretation clearer.

Table 5.20 Summary of Tests

Test		Section
Mean of single population	$t = \dfrac{\overline{X} - \mu}{S\sqrt{1/N}}$	5.2.1
Comparison of means from two independent populations (variances known)	$Z = \dfrac{\overline{X}_1 - \overline{X}_2}{\sqrt{\sigma_1^2/N_1 + \sigma_2^2/N_2}}$	5.2.2
Comparisons of means from two independent populations (variance unknown)	$t = \dfrac{\overline{X}_1 - \overline{X}_2}{S_p\sqrt{1/N_1 + 1/N_2}}$	5.2.2
Comparison of means from two related samples (variance unknown)[a]	$t = \dfrac{\hat{\delta}}{S\sqrt{1/N}}$	5.2.3
Proportion from a single population[b]	$Z = \dfrac{p - p_0}{\sqrt{p_0 q_0 N)}}$	
Comparison of two proportions from independent groups[b]	$Z = \dfrac{p - p_0}{\sqrt{p_0 q_0(1/N_1 + 1/N_2)}}$	5.2.4
Comparison of variances (two-sided test)	$F = \dfrac{S_1^2}{S_2^2}$ $(S_1^2 > S_2^2)$	5.2.4
Confidence limits for variance	$\dfrac{(n^2 - 1)S_{(n-1)}^2}{\chi_{\alpha/2}^2} \geq \sigma^2 \geq \dfrac{(n - 1)S_{(n-1)}^2}{\chi_{1-\alpha/2}^2}$	5.5

[a] If the variance is known, use the normal distribution.
[b] A continuity correction may be used (5.16 and 5.20).

Example 1. A batch of tablets was tested for content uniformity. The mean of the 10 tablets tested was 99.1% and the s.d. was 2.6%. Entering Table IV.19, for a 99% tolerance interval that contains 99.9% of the population with $N = 10$, the factor, $t\prime = 7.129$. Assuming a normal distribution of tablet potencies, we can say with 99% probability (99% "confidence") that 99.9% of the tablets are within 99.1 % $\pm 7.129 \times 2.6 = 99.1 \% \pm 18.5 = 80.6 \%$ to 117.6 %.

Example 2. In a bioequivalence study using a crossover design with 24 subjects, the ratio of test product to standard product was computed for each subject. One of the proposals for assessing individual bioequivalence is to compute a tolerance interval to estimate an interval that will encompass a substantial proportion of subjects who take the drug. The average of the 24 ratios was 1.05 with a s.d. of 0.3. A tolerance interval is calculated that has 95% probability of containing 75% of the population. The factor from Table IV.19 for $N = 24$ and 95% confidence is 1.557. The tolerance interval is $1.05 \pm 1.557 \times 0.3 = 1.05 \pm 0.47$. Thus, we can say that 75% of the patients will have a ratio between 0.58 and 1.52 with 95% probability. One of the problems with such an approach to individual equivalence is that the interval is dependent on the variability, and highly variable drugs will always show a wide variation of the ratio for different products. Therefore, using this interval as an acceptance criterion for individual equivalence may not be very meaningful. Also, this computation assumes a normal distribution, and individual ratios may deviate significantly from a normal distribution.

Table 5.20 summarizes some tests discussed in this chapter.

KEY TERMS

Alpha level	Nonsignificance
Alternative hypothesis	Normal curve test
Bartlett's test	Null hypothesis
Behrens–Fisher test	One-sample test
Beta error	One-sided test
Bias	One-way analysis of variance
Binomial trials	Paired-sample t test
Blinding	Parallel-groups design
Cells	Parameters
Chi-square test	Pooled proportion
Confidence interval	Pooled variance
Continuity correction	Power
Critical region	Preference tests
Crossover design	Randomization
Cumulative normal distribution	Region of rejection
Degrees of freedom	Sample size
Delta	Sensitive
Error	Significance
Error of first kind	t distribution
Estimation	t test
Expected values	Tolerance interval
Experimental error	Two-by-two table
Fourfold table	Two independent groups t test
F test	Two-tailed (sided) test
Hypothesis testing	Uncontrolled study
Independence	Variance
Independent groups	Yates correction
Level of significance	Z transformation
Marginal totals	

EXERCISES

1. Calculate the probability of finding a value of 49.8 or less if $\mu = 54.7$ and $\sigma = 2$.

2. If the variance of the population of tablets in Table 5.1 were known to be 4.84, compute a 99% confidence interval for the mean.

3. (a) Six analysts perform an assay on a portion of the same homogeneous material with the following results: 5.8, 6.0, 5.7, 6.1, 6.0, and 6.1. Place 95% confidence limits on the true mean.
   (b) A sample of 500 tablets shows 12 to be defective. Place a 95% confidence interval on the percent defective in the lot.
   (c) Place a 95% confidence interval on the difference between two products in which 50 of 60 patients responded to product A, and 25 of 50 patients responded to product B.

4. (a) Quality control records show the average tablet weight to be 502 mg with a standard deviation of 5.3. There are sufficient data so that these values may be considered known parameter values. A new batch shows the following weights from a random sample of six tablets: 500, 499, 504, 493, 497, and 495 mg. Do you believe that the new batch has a different mean from the process average?
   (b) Two batches of tablets were prepared by two different processes. The potency determinations made on five tablets from each batch were as follows: batch A: 5.1, 4.9, 4.6, 5.3, 5.5; batch B: 4.8, 4.8, 5.2, 5.0, 4.5. Test to see if the means of the two batches are equal.
   (c) Answer part (a) if the variance were unknown. Place a 95% confidence interval on the true average weight.

5. (a) In part (b) of Problem 4, calculate the variance and the standard deviation of the 10 values as if they were one sample. Are the values of the s.d. and S^2 smaller or larger than the values calculated from "pooling"?
   (b) Calculate the pooled s.d. above by "averaging" the s.d.'s from the two samples. Is the result different from the "pooled" s.d. as described in the text?

6.

Batch 1 (drug)	Pass/fail (improve, worsen)	Batch 2 (placebo)	Pass/fail (improve, worsen)
10.1	P	9.5	F
9.7	F	8.9	F
10.1	P	9.4	F
10.5	P	10.4	P
12.3	P	9.9	F
11.8	P	10.1	P
9.6	F	9.0	F
10.0	F	9.7	F
11.2	P	9.9	F
11.3	P	9.8	F

   (a) What are the mean and s.d. of each batch? Test for difference between the two batches using a t test.
   (b) What might be the "population" corresponding to this sample? Do you think that the sample size is large enough? Why? Ten objects were selected from each batch for this test. Is this a good design for comparing the average results from two batches?
   (c) Consider values above 10.0 a success and values 10.00 or less a failure. What is the proportion of successes for batch 1 and batch 2? Is the proportion of successes in batch 1 different from the proportion in batch 2 (5% level)?
   (d) Put 95% confidence limits on the proportion of successes with all data combined.

7. A new analytical method is to be compared to an old method. The experiment is performed by a single analyst. She selects four batches of product at random and obtains the following results.

Batch	Method 1	Method 2
1	4.81	4.93
2	5.44	5.43
3	4.25	4.30
4	4.35	4.47

(a) Do you think that the two methods give different results on the average?
(b) Place 95% confidence limits on the true difference of the methods.

8. The following data for blood protein (g/100 mL) were observed for the comparison of two drugs. Both drugs were tested on each person in random order.

Patient	Drug *A*	Drug *B*
1	8.1	9.0
2	9.4	9.9
3	7.2	8.0
4	6.3	6.0
5	6.6	7.9
6	9.3	9.0
7	7.6	7.9
8	8.1	8.3
9	8.6	8.2
10	8.3	8.9
11	7.0	8.3
12	7.7	8.8

(a) Perform a statistical test for drug differences at the 5% level.
(b) Place 95% confidence limits on the average differences between drugs A and B.

9. For examples 10 and 11, calculate the pooled p and q (p_0 and q_0).

10. In Example 4, perform a t test if the mean were 16.7 instead of 15.9.

11. Use the normal approximation and chi-square test (with and without continuity correction) to answer the following problem. A placebo treatment results in 8 patients out of 100 having elevated blood urea nitrogen (BUN) values. The drug treatment results in 16 of 100 patients having elevated values. Is this significantly different from the placebo?

12. Quality control records show that the average defect rate for a product is 2.8%. Two hundred items are inspected and 5% are found to be defective in a new batch. Should the batch be rejected? What would you do if you were the director of quality control? Place confidence limits on the *percent* defective and the *number* defective (out of 200).

¶¶**13. In a batch size of 1,000,000,5000 tablets are inspected and 50 are found defective.
(a) Put 95% confidence limits on the true number of defectives in the batch.
(b) At $\alpha = 0.05$, do you think that there could be more than 2% defective in the batch?
(**c) If you wanted to estimate the true proportion of defectives within $\pm\,0.1$% with 95% confidence, how many tablets would you inspect?

14. In a clinical test, 60 people received a new drug and 50 people received a placebo. Of the people on the new drug, 40 of the 60 showed a positive response and 25 of the 50 people on placebo showed a positive response. Perform a statistical test to determine if

---

¶¶ The double asterisk indicates optional, more difficult problems.

the new drug shows more of an effect than the placebo. Place a 95% confidence interval on the difference of proportion of positive response in the two test groups.

15. In a paired preference test, each of 100 subjects was asked to choose the preference between A and B. Of these 100, 60 showed no preference, 30 preferred A, and 10 preferred B. Is A significantly preferred to B?

16. Over a long period of time, a screening test has shown a response rate for a control of 20%. A new chemical shows 9 positive results in 20 observations (45%). Would you say that this candidate is better than the control? Place 99% confidence limits on the true response rate for the new chemical.

17. Use the chi-square test with the continuity correction to see if there is a significant difference in the following comparison. Two batches of tablets were made using different excipients. In batch A, 10 of 100 tablets sampled were chipped. In batch B, 17 of 95 tablets were chipped. Compare the two batches with respect to proportion chipped at the 5% level.

18. Show that the uncorrected value of chi-square for the data in Table 5.15 is 0.98.

19. Use the chi-square test, with continuity correction, to test for significance (5% level) for the data in Example 11.

20. Perform a statistical test to compare the variances in the two groups in Problem 6. $H_0 : \sigma_1^2 = \sigma_2^2; \quad H_a : \sigma_1^2 \neq \sigma_2^2$. Perform the test at the 10% level.

21. Compute the value of the corrected χ^2 statistic for data of Example 11 in 5.2.4. Compute the t value as recommended by D'Agostino et al. Compare the uncorrected value of Z with these results.

22. The homogeneity of a sample taken from a mixer was tested after 5, 10, and 15 minutes. The variances of six samples taken at each time were 16.21, 1.98, and 2.02. Based on the results of Bartlett's test for homogeneity of variances, what are your conclusions?

23. Six blend samples (unit dose size) show a variance of 9% (RSD = 3%). Compute a 95% one-sided upper confidence interval for the variance. Is this interval too large based on the official limit of 6% for RSD?

24. The USP content uniformity test for 30 units states that the RSD should not exceed 7.8%. Show that there is a 5% probability that the true RSD is less than 10%.

REFERENCES

1. Westlake WJ. Symmetrical confidence intervals for bioequivalence trials. Biometrics 1976; 32:741–744.
2. Westlake WJ. Bioavailability and bioequivalence of pharmaceutical formulations. In: Peace KE, ed. Statistical Issues in Drug Research and Development. New York: Marcel Dekker, 1990.
3. Snedecor GW, Cochran WG. Statistical Methods, 6th ed. Ames, IA: Iowa State University Press, 1967.
4. Cochran WG. Sampling Techniques, 3rd ed. New York: Wiley, 1967.
5. Snedecor GW, Cochran WG. Statistical Methods, 8th ed. Ames, IA: Iowa State University Press, 1989.
6. D'Agostino RB, Chase W, Belanger A. The appropriateness of some common procedures for testing the equality of two independent binomial populations. Am Stat 1988; 42:198
7. Fleiss J. Statistical Methods for Rates and Proportions, 2nd ed. New York: Wiley, 1981.
8. Hauck W, Anderson S. A comparison of large sample CI methods for the difference of two binomial probabilities. Am Stat 1986; 40:318.
9. Cowdery S, Michaels T. Pharmacopeial Forum. 1980:614.

6 | Sample Size and Power

The question of the size of the sample, the number of observations, to be used in scientific experiments is of extreme importance. Most experiments beg the question of sample size. Particularly when time and cost are critical factors, one wishes to use the minimum sample size to achieve the experimental objectives. Even when time and cost are less crucial, the scientist wishes to have some idea of the number of observations needed to yield sufficient data to answer the objectives. An elegant experiment will make the most of the resources available, resulting in a sufficient amount of information from a minimum sample size. For simple comparative experiments, where one or two groups are involved, the calculation of sample size is relatively simple. A knowledge of the α level (level of significance), β level (1 − power), the standard deviation, and a meaningful "practically significant" difference is necessary in order to calculate the sample size.

Power is defined as $1 - \beta$ (i.e., $\beta = 1 -$ power). Power is the ability of a statistical test to show significance if a specified difference truly exists. The magnitude of power depends on the level of significance, the standard deviation, and the sample size. Thus power and sample size are related.

In this chapter, we present methods for computing the sample size for relatively simple situations for normally distributed and binomial data. The concept and calculation of power are also introduced.

6.1 INTRODUCTION

The question of sample size is a major consideration in the planning of experiments, but may not be answered easily from a scientific point of view. In some situations, the choice of sample size is limited. Sample size may be dictated by official specifications, regulations, cost constraints, and/or the availability of sampling units such as patients, manufactured items, animals, and so on. The USP content uniformity test is an example of a test in which the sample size is fixed and specified [1].

The sample size is also specified in certain quality control sampling plans such as those described in MIL-STD-105E [2]. These sampling plans are used when sampling products for inspection for attributes such as product defects, missing labels, specks in tablets, or ampul leakage. The properties of these plans have been thoroughly investigated and defined as described in the document cited above. The properties of the plans include the chances (probability) of rejecting or accepting batches with a known proportion of rejects in the batch (sect. 12.3).

Sample-size determination in comparative clinical trials is a factor of major importance. Since very large experiments will detect very small, perhaps clinically insignificant, differences as being statistically significant, and small experiments will often find large, clinically significant differences as statistically insignificant, the choice of an appropriate sample size is critical in the design of a clinical program to demonstrate safety and efficacy. When cost is a major factor in implementing a clinical program, the number of patients to be included in the studies may be limited by lack of funds. With fewer patients, a study will be less sensitive. Decreased sensitivity means that the comparative treatments will be relatively more difficult to distinguish statistically if they are, in fact, different.

The problem of choosing a "correct" sample size is related to experimental objectives and the risk (or probability) of coming to an incorrect decision when the experiment and analysis are completed. For simple comparative experiments, certain prior information is required in order to compute a sample size that will satisfy the experimental objectives. The following considerations are essential when estimating sample size.

1. The α level must be specified that, in part, determines the difference needed to represent a statistically significant result. To review, the α level is defined as the risk of concluding that treatments differ when, in fact, they are the same. The level of significance is usually (but not always) set at the traditional value of 5%.

2. The β error must be specified for some specified treatment difference, Δ. Beta, β, is the risk (probability) of erroneously concluding that the treatments are not significantly different when, in fact, a difference of size Δ or greater exists. The assessment of β and Δ, the "practically significant" difference, *prior* to the initiation of the experiment, is not easy. Nevertheless, an educated guess is required. β is often chosen to be between 5% and 20%. Hence, one may be willing to accept a 20% (1 in 5) chance of not arriving at a statistically significant difference when the treatments are truly different by an amount equal to (or greater than) Δ. The consequences of committing a β error should be considered carefully. If a true difference of practical significance is missed and the consequence is costly, β should be made very small, perhaps as small as 1%. Costly consequences of missing an effective treatment should be evaluated not only in monetary terms, but should also include public health issues, such as the possible loss of an effective treatment in a serious disease.

3. The difference to be detected, Δ (that difference considered to have practical significance), should be specified as described in (2) above. This difference should not be arbitrarily or capriciously determined, but should be considered carefully with respect to meaningfulness from both a scientific and commercial marketing standpoint. For example, when comparing two formulas for time to 90% dissolution, a difference of one or two minutes might be considered meaningless. A difference of 10 or 20 minutes, however, may have practical consequences in terms of in vivo absorption characteristics.

4. A knowledge of the standard deviation (or an estimate) for the significance test is necessary. If no information on variability is available, an educated guess, or results of studies reported in the literature using related compounds, may be sufficient to give an estimate of the relevant variability. The assistance of a statistician is recommended when estimating the standard deviation for purposes of determining sample size.

To compute the sample size in a comparative experiment, (a) α, (b) β, (c) Δ, and (d) σ must be specified. The computations to determine sample size are described below (Fig. 6.1).

6.2 DETERMINATION OF SAMPLE SIZE FOR SIMPLE COMPARATIVE EXPERIMENTS FOR NORMALLY DISTRIBUTED VARIABLES

The calculation of sample size will be described with the aid of Figure 6.1. This explanation is based on normal distribution or t tests. The derivation of sample-size determination may appear complex. The reader not requiring a "proof" can proceed directly to the appropriate formulas below.

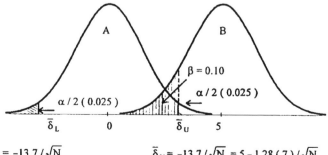

Figure 6.1   Scheme to demonstrate calculation of sample size based on α, β, Δ, and σ: $\alpha = 0.05$, $\beta = 0.10$, $\Delta = 5$, $\sigma = 7$; $H_0: \Delta = 0$, $H_a: \Delta = 5$.

6.2.1 Paired-Sample and Single-Sample Tests

We will first consider the case of a paired-sample test where the null hypothesis is that the two treatment means are equal: H_0: $\Delta = 0$. In the case of an experiment comparing a new antihypertensive drug candidate and a placebo, an average difference of 5 mm Hg in blood pressure reduction might be considered of sufficient magnitude to be interpreted as a difference of "practical significance" ($\Delta = 5$). The standard deviation for the comparison was known, equal to 7, based on a large amount of experience with this drug.

In Figure 6.1, the normal curve labeled A represents the distribution of differences with mean equal to 0 and σ equal to 7. This is the distribution under the null hypothesis (i.e., drug and placebo are identical). Curve B is the distribution of differences when the alternative, H_a: $\Delta = 5$,* is true (i.e., the difference between drug and placebo is equal to 5). Note that curve B is identical to curve A except that B is displaced 5 mm Hg to the right. Both curves have the same standard deviation, 7.

With the standard deviation, 7, known, the statistical test is performed at the 5% level as follows [Eq. (5.4)]:

$$Z = \frac{\bar{\delta} - \Delta}{\sigma/\sqrt{N}} = \frac{\bar{\delta} - 0}{7/\sqrt{N}}. \tag{6.1}$$

For a two-tailed test, if the absolute value of Z is 1.96 or greater, the difference is significant. According to Eq. (6.1), to obtain the significance

$$|\bar{\delta}| \geq \frac{\sigma Z}{\sqrt{N}} = \frac{7(1.96)}{\sqrt{N}} = \frac{13.7}{\sqrt{N}}. \tag{6.2}$$

Therefore, values of $\bar{\delta}$ equal to or greater than $13.7/\sqrt{N}$ (or equal to or less than $-13.7/\sqrt{N}$) will lead to a declaration of significance. These points are designated as $\bar{\delta}_L$ and $\bar{\delta}_U$ in Figure 6.1, and represent the cutoff points for statistical significance at the 5% level; that is, observed differences equal to or more remote from the mean than these values result in "statistically significant differences."

If curve B is the true distribution (i.e., $\Delta = 5$), an observed mean difference greater than $13.7/\sqrt{N}$ (or less than $-13.7/\sqrt{N}$) will result in the correct decision; H_0 will be rejected and we conclude that a difference exists. If $\Delta = 5$, observations of a mean difference between $13.7/\sqrt{N}$ and $-13.7/\sqrt{N}$ will lead to an *incorrect decision*, the acceptance of H_0 (no difference) (Fig. 6.1). By definition, the probability of making this *incorrect* decision is equal to β.

In the present example, β will be set at 10%. In Figure 6.1, β is represented by the area in curve B below $13.7/\sqrt{N}(\bar{\delta}_U)$, equal to 0.10. (This area, β, represents the probability of accepting H_0 if $\Delta = 5$.)

We will now compute the value of $\bar{\delta}$ that cuts off 10% of the area in the lower tail of the normal curve with a mean of 5 and a standard deviation of 7 (curve B in Figure 6.1). Table IV.2 shows that 10% of the area in the standard normal curve is below -1.28. The value of $\bar{\delta}$ (mean difference in blood pressure between the two groups) that corresponds to a given value of Z (-1.28, in this example) is obtained from the formula for the Z transformation [Eq. (3.14)] as follows:

$$\bar{\delta} = \Delta + Z_\beta \left(\frac{\sigma}{\sqrt{N}} \right)$$

$$Z_\beta = \frac{\bar{\delta} - \Delta}{\sigma/\sqrt{N}}. \tag{6.3}$$

Applying Eq. (6.3) to our present example, $\bar{\delta} = 5 - 1.28(7/\sqrt{N})$. The value of $\bar{\delta}$ in Eqs. (6.2) and (6.3) is identically the same, equal to $\bar{\delta}_U$. This is illustrated in Figure 6.1.

---

* Δ is considered to be the *true* mean difference, similar to μ. $\bar{\delta}$ will be used to denote the *observed* mean difference.

Table 6.1 Sample Size as a Function of Beta with $\Delta = 5$ and $\sigma = 7$: Paired Test ($\alpha = 0.05$)

Beta (%)	Sample size, N
1	36
5	26
10	21
20	16

From Eq. (6.2), $\bar{\delta}_U = 13.7/\sqrt{N}$, satisfying the definition of α. From Eq. (6.3), $\bar{\delta}_U = 5 - 1.28(7)/\sqrt{N}$, satisfying the definition of β. We have two equations in two unknowns ($\bar{\delta}_U$ and N), and N is evaluated as follows:

$$\frac{13.7}{\sqrt{N}} = 5 - \frac{1.28(7)}{\sqrt{N}}$$

$$N = \frac{(13.7 + 8.96)^2}{5^2} = 20.5 \cong 21.$$

In general, Eqs. (6.2) and (6.3) can be solved for N to yield the following equation:

$$N = \left(\frac{\sigma}{\Delta}\right)^2 (Z_\alpha + Z_\beta)^2, \tag{6.4}$$

where Z_α and Z_β[†] are the appropriate normal deviates obtained from Table IV.2. In our example, $N = (7/5)^2(1.96 + 1.28)^2 \cong 21$. A sample size of 21 will result in a statistical test with 90% power ($\beta = 10\%$) against an alternative of 5, at the 5% level of significance. Table 6.1 shows how the choice of β can affect the sample size for a test at the 5% level with $\Delta = 5$ and $\sigma = 7$.

The formula for computing the sample size if the standard deviation is known [Eq. (6.4)] is appropriate for a paired-sample test or for the test of a mean from a *single population*. For example, consider a test to compare the mean drug content of a sample of tablets to the labeled amount, 100 mg. The two-sided test is to be performed at the 5% level. Beta is designated as 10% for a difference of -5 mg (95 mg potency or less). That is, we wish to have a power of 90% to detect a difference from 100 mg if the true potency is 95 mg or less. If σ is equal to 3, how many tablets should be assayed? Applying Eq. (6.4), we have

$$N = \left(\frac{3}{5}\right)^2 (1.96 + 1.28)^2 = 3.8.$$

Assaying four tablets will satisfy the α and β probabilities. Note that $Z = 1.28$ cuts off 90% of the area under curve B (the "alternative" curve) in Figure 6.2, leaving 10% (β) of the area in the upper tail of the curve. Table 6.2 shows values of Z_α and Z_β for various levels of α and β to be used in Eq. (6.4). In this example, and most examples in practice, β is based on one tail of the normal curve. The other tail contains an insignificant area relating to β (the right side of the normal curve, B, in Fig. 6.1)

Equation (6.4) is correct for computing the sample size for a paired- or one-sample test if the standard deviation is known.

In most situations, the standard deviation is unknown and a prior estimate of the standard deviation is necessary in order to calculate sample size requirements. In this case, the estimate of the standard deviation replaces σ in Eq. (6.4), but the calculation results in an answer that is slightly too small. The underestimation occurs because the values of Z_α and Z_β are smaller than

---

[†] Z_β is taken as the positive value of Z in this formula.

Table 6.2   Values of Z_α and Z_β for Sample-Size Calculations

	Z_α		
	One sided	Two sided	Z_β^a
1%	2.32	2.58	2.32
5%	1.65	1.96	1.65
10%	1.28	1.65	1.28
20%	0.84	1.28	0.84

[a]The value of β is for a single specified alternative. For a two-sided test, the probability of rejection of the alternative, if true, (accept H_a) is virtually all contained in the tail nearest the alternative mean.

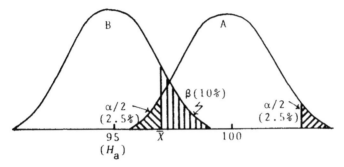

Figure 6.2   Illustration of the calculation of N for tablet assays. $\overline{X} = 95 + \sigma Z_\beta/\sqrt{N} = 100 - \sigma Z_\alpha/\sqrt{N}$.

the corresponding t values that should be used in the formula when the standard deviation is unknown. The situation is somewhat complicated by the fact that the value of t depends on the sample size (d.f.), which is yet unknown. The problem can be solved by an iterative method, but for practical purposes, one can use the appropriate values of Z to compute the sample size [as in Eq. (6.4)] and add on a few extra samples (patients, tablets, etc.) to compensate for the use of Z rather than t. Guenther has shown that the simple addition of $0.5Z_\alpha^2$, which is equal to approximately 2 for a two-sided test at the 5% level, results in a very close approximation to the correct answer [3]. In the problem illustrated above (tablet assays), if the standard deviation were *unknown* but *estimated* as being equal to 3 based on previous experience, a better estimate of the sample size would be $N + 0.5Z_\alpha^2 = 3.8 + 0.5(1.96)^2 \cong 6$ tablets.

6.2.2   Determination of Sample Size for Comparison of Means in Two Groups

For a two independent groups test (parallel design), with the standard deviation known and equal number of observations per group, the formula for N (where N is the sample size for each group) is

$$N = 2 \left(\frac{\sigma}{\Delta} \right)^2 (Z_\alpha + Z_\beta)^2. \tag{6.5}$$

If the standard deviation is unknown and a prior estimate is available (s.d.), substitute s.d. for σ in Eq. (6.5) and compute the sample size; but add on $0.25Z_\alpha^2$ to the sample size for each group.

Example 1: This example illustrates the determination of the sample size for a two independent groups (two-sided test) design. Two variations of a tablet formulation are to be compared with regard to dissolution time. All ingredients except for the lubricating agent were the same in these two formulations. In this case, a decision was made that if the formulations differed by 10 minutes or more to 80% dissolution, it would be extremely important that the experiment shows a statistically significant difference between the formulations. Therefore, the pharmaceutical scientist decided to fix the β error at 1% in a statistical test at the traditional 5% level. Data were available from dissolution tests run during the development of formulations of the drug and the standard deviation was *estimated* as 5 minutes. With the information presented above, the sample size can be determined from Eq. (6.5). We will add on $0.25Z_\alpha^2$ samples to the answer because the standard deviation is unknown.

$$N = 2\left(\frac{5}{10}\right)^2 (1.96 + 2.32)^2 + 0.25(1.96)^2 = 10.1.$$

The study was performed using 12 tablets from each formulation rather than the 10 or 11 suggested by the answer in the calculation above. Twelve tablets were used because the dissolution apparatus could accommodate six tablets per run.

Example 2: A bioequivalence study was being planned to compare the bioavailability of a final production batch to a previously manufactured pilot-sized batch of tablets that were made for clinical studies. Two parameters resulting from the blood-level data would be compared: area under the plasma level versus time curves (AUC) and peak plasma concentration (C_{max}). The study was to have 80% power ($\beta = 0.20$) to detect a difference of 20% or more between the formulations. The test is done at the usual 5% level of significance. Estimates of the standard deviations of the *ratios* of the values of each of the parameters [(final product)/(pilot batch)] were determined from a small pilot study. The standard deviations were different for the parameters. Since the researchers could not agree that one of the parameters was clearly critical in the comparison, they decided to use a "maximum" number of patients based on the variable with the largest relative variability. In this example, C_{max} was most variable, the ratio having a standard deviation of approximately 0.30. Since the design and analysis of the bioequivalence study is a variation of the paired t test, Eq. (6.4) was used to calculate the sample size, adding on $0.5Z_\alpha^2$, as recommended previously.

$$N = \left(\frac{\sigma}{\Delta}\right)^2 (Z_\alpha + Z_\beta)^2 + 0.5(Z_\alpha^2)$$

$$= \left(\frac{0.3}{0.2}\right)^2 (1.96 + 0.84)^2 + 0.5(1.96)^2 = 19.6. \tag{6.6}$$

Twenty subjects were used for the comparison of the bioavailabilities of the two formulations.

For sample-size determination for bioequivalence studies using FDA recommended designs, see Table 6.5 and section 11.4.4.

Sometimes the sample sizes computed to satisfy the desired α and β errors can be inordinately large when time and cost factors are taken into consideration. Under these circumstances, a compromise must be made—most easily accomplished by relaxing the α and β requirements[‡] (Table 6.1). The consequence of this compromise is that probabilities of making an incorrect decision based on the statistical test will be increased. Other ways of reducing the required sample size are (a) to increase the precision of the test by improving the assay methodology or carefully controlling extraneous conditions during the experiment, for example, or (b) to compromise by increasing Δ, that is, accepting a larger difference that one considers to be of practical importance.

Table 6.3 gives the sample size for some representative values of the ratio σ/Δ, α, and β, where the s.d. (s) is estimated.

6.3 DETERMINATION OF SAMPLE SIZE FOR BINOMIAL TESTS

The formulas for calculating the sample size for comparative binomial tests are similar to those described for normal curve or t tests. The major difference is that the value of σ^2, which is assumed to be the same under H_0 and H_a in the two-sample independent groups t or Z tests, is different for the distributions under H_0 and H_a in the binomial case. This difference occurs because σ^2 is dependent on P, the probability of success, in the binomial. The value of P will

[‡] In practice, α is often fixed by regulatory considerations and β is determined as a compromise.

Table 6.3 Sample Size Needed for Two-Sided *t* Test with Standard Deviation Estimated

	One-sample test								Two-sample test with *N* units per group							
	Alpha = 0.05				Alpha = 0.01				Alpha = 0.05				Alpha = 0.01			
	Beta =				Beta =				Beta =				Beta =			
Estimated S/Δ	0.01	0.05	0.10	0.20	0.01	0.05	0.10	0.20	0.01	0.05	0.10	0.20	0.01	0.05	0.10	0.20
4.0	296	211	170	128	388	289	242	191	588	417	337	252	770	572	478	376
2.0	76	54	44	34	100	75	63	51	148	106	86	64	194	145	121	96
1.5	44	32	26	20	58	54	37	30	84	60	49	37	110	82	69	55
1.0	21	16	13	10	28	22	19	16	38	27	23	17	50	38	32	26
0.8	14	11	9	8	19	15	13	11	25	18	15	12	33	25	21	17
0.67	11	8	7	6	15	12	11	9	18	13	11	9	24	18	15	13
0.5	7	6	5	4	10	8	8	7	11	8	7	6	14	11	10	8
0.4	6	5	4	4	8	7	6	6	8	6	5	4	10	8	7	6
0.33	5	4	4	3	7	6	6	5	6	5	4	4	8	6	6	5

be different depending on whether H_0 or H_a represents the true situation. The appropriate formulas for determining sample size for the one- and two-sample tests are

One-sample test

$$N = \frac{1}{2} \left[\frac{p_0 q_0 + p_1 q_1}{\Delta^2} \right] (Z_\alpha + Z_\beta)^2, \tag{6.7}$$

where $\Delta = p_1 - p_0$; p_1 is the proportion that would result in a meaningful difference, and p_0 is the hypothetical proportion under the null hypothesis.

Two-sample test

$$N = \left[\frac{p_1 q_1 + p_2 q_2}{\Delta^2} \right] (Z_\alpha + Z_\beta)^2, \tag{6.8}$$

where $\Delta = p_1 - p_2$; p_1 and p_2 are prior estimates of the proportions in the experimental groups. The values of Z_α and Z_β are the same as those used in the formulas for the normal curve or t tests. N is the sample size for each group. If it is not possible to estimate p_1 and p_2 prior to the experiment, one can make an educated guess of a meaningful value of Δ and set p_1 and p_2 both equal to 0.5 in the *numerator* of Eq. (6.8). This will maximize the sample size, resulting in a conservative estimate of sample size.

Fleiss [4] gives a fine discussion of an approach to estimating Δ, the practically significant difference, when computing the sample size. For example, one approach is first to estimate the proportion for the more well-studied treatment group. In the case of a comparative clinical study, this could very well be a standard treatment. Suppose this treatment has shown a success rate of 50%. One might argue that if the comparative treatment is additionally successful for 30% of the patients who do not respond to the standard treatment, then the experimental treatment would be valuable. Therefore, the success rate for the experimental treatment should be 50% + 0.3 (50%) = 65% to show a practically significant difference. Thus, p_1 would be equal to 0.5 and p_2 would be equal to 0.65.

Example 3: A reconciliation of quality control data over several years showed that the proportion of unacceptable capsules for a stable encapsulation process was 0.8% (p_0). A sample size for inspection is to be determined so that if the true proportion of unacceptable capsules is equal to or greater than 1.2% ($\Delta = 0.4\%$), the probability of detecting this change is 80% ($\beta = 0.2$). The comparison is to be made at the 5% level using a *one-sided* test. According to Eq. (6.7),

$$N = \frac{1}{2} \left[\frac{0.008 \cdot 0.992 + 0.012 \cdot 0.988}{(0.008 - 0.012)^2} \right] (1.65 + 0.84)^2$$

$$= \frac{7670}{2}$$

$$= 3835.$$

The large sample size resulting from this calculation is typical of that resulting from binomial data. If 3835 capsules are too many to inspect, α, β, and/or Δ must be increased. In the example above, management decided to increase α. This is a conservative decision in that more good batches would be "rejected" if α is increased; that is, the increase in α results in an increased probability of rejecting good batches, those with 0.8% unacceptable or less.

Example 4: Two antibiotics, a new product and a standard product, are to be compared with respect to the two-week cure rate of a urinary tract infection, where a cure is bacteriological evidence that the organism no longer appears in urine. From previous experience, the cure rate for the standard product is estimated at 80%. From a practical point of view, if the new product shows an 85% or better cure rate, the new product can be considered superior. The marketing division of the pharmaceutical company felt that this difference would support claims of better efficacy for the new product. This is an important claim. Therefore, β is chosen to be 1% (power = 99%). A two-sided test will be performed at the 5% level to satisfy FDA guidelines. The test is two-sided because, a priori, the new product is not known to be better or worse than the standard. The calculation of sample size to satisfy the conditions above makes use of Eq. (6.8); here $p_1 = 0.8$ and $p_2 = 0.85$.

$$N = \left[\frac{0.08 \cdot 0.2 + 0.85 \cdot 0.15}{(0.80 - 0.85)^2} \right] (1.96 + 2.32)^2 = 2107.$$

The trial would have to include 4214 patients, 2107 on each drug, to satisfy the α and β risks of 0.05 and 0.01, respectively. If this number of patients is greater than that can be accommodated, the β error can be increased to 5% or 10%, for example. A sample size of 1499 per group is obtained for a β of 5%, and 1207 patients per group for β equal to 10%.

Although Eq. (6.8) is adequate for computing the sample size for most situations, the calculation of N can be improved by considering the continuity correction [4]. This would be particularly important for small sample sizes

$$N' = \left[\frac{N}{4} \right] \left[1 + \sqrt{1 + \frac{8}{(N |p_2 - p_1|)}} \right]^2 ,$$

where N is the sample size computed from Eq. (6.8) and N' is the corrected sample size. In the example, for $\alpha = 0.05$ and $\beta = 0.01$, the corrected sample size is

$$N' = \left[\frac{2107}{4} \right] \left[1 + \sqrt{1 + \frac{8}{(2107 |0.80 - 0.85|)}} \right]^2 = 2186.$$

6.4   DETERMINATION OF SAMPLE SIZE TO OBTAIN A CONFIDENCE INTERVAL OF SPECIFIED WIDTH

The problem of estimating the number of samples needed to estimate the mean with a known precision by means of the confidence interval is easily solved by using the formula for the confidence interval (see sect. 5.1). This approach has been used as an aid in predicting election results based on preliminary polls where the samples are chosen by simple random sampling. For example, one may wish to estimate the proportion of voters who will vote for candidate A within 1% of the actual proportion.

We will consider the application of this problem to the estimation of proportions. In quality control, one can closely estimate the true proportion of percent defects to any given degree of precision. In a clinical study, a suitable sample size may be chosen to estimate the true proportion of successes within certain specified limits. According to Eq. (5.3), a two-sided confidence interval with confidence coefficient p for a proportion is

$$\hat{p} \pm Z \sqrt{\frac{\hat{p}\hat{q}}{N}}. \tag{6.3}$$

To obtain a 99% confidence interval with a width of 0.01 (i.e., construct an interval that is within ± 0.005 of the observed proportion, $\hat{p} \pm 0.005$),

$$Z_p \sqrt{\frac{\hat{p}\hat{q}}{N}} = 0.005$$

or

$$N = \frac{Z_p^2(\hat{p}\hat{q})}{(W/2)^2} \tag{6.9}$$

$$N = \frac{(2.58)^2(\hat{p}\hat{q})}{(0.005)^2}.$$

A more exact formula for the sample size for small values of N is given in Ref. [5].

Example 5: A quality control supervisor wishes to have an estimate of the proportion of tablets in a batch that weigh between 195 and 205 mg, where the proportion of tablets in this interval is to be estimated within ± 0.05 ($W = 0.10$). How many tablets should be weighed? Use a 95% confidence interval.

To compute N, we must have an estimate of $\hat{p}$ [see Eq. (6.9)]. If $\hat{p}$ and $\hat{q}$ are chosen to be equal to 0.5, N will be at a maximum. Thus, if one has no inkling as to the magnitude of the outcome, using $\hat{p} = 0.5$ in Eq. (6.9) will result in a sufficiently large sample size (probably, too large). Otherwise, estimate $\hat{p}$ and $\hat{q}$ based on previous experience and knowledge. In the present example from previous experience, approximately 80% of the tablets are expected to weigh between 195 and 205 mg ($\hat{p} = 0.8$). Applying Eq. (6.9),

$$N = \frac{(1.96)^2(0.8)(0.2)}{(0.10/2)^2} = 245.9.$$

A total of 246 tablets should be weighed. In the actual experiment, 250 tablets were weighed, and 195 of the tablets (78%) weighed between 195 and 205 mg. The 95% confidence interval for the true proportion, according to Eq. (5.3), is

$$p \pm 1.96\sqrt{\frac{\hat{p}\hat{q}}{N}} = 0.78 \pm 1.96\sqrt{\frac{(0.78)(0.22)}{250}} = 0.78 \pm 0.051.$$

The interval is slightly greater than $\pm 5\%$ because p is somewhat less than 0.8 (pq is larger for $p = 0.78$ than for $p = 0.8$). Although 5.1% is acceptable, to ensure a sufficient sample size, in general, one should estimate p closer to 0.5 in order to cover possible poor estimates of p.

If $\hat{p}$ had been chosen equal to 0.5, we would have calculated

$$N = \frac{(1.96)^2(0.5)(0.5)}{(0.10/2)^2} = 384.2.$$

Example 6: A new vaccine is to undergo a nationwide clinical trial. An estimate is desired of the proportion of the population that would be afflicted with the disease after vaccination. A good guess of the expected proportion of the population diseased without vaccination is 0.003. Pilot studies show that the incidence will be about 0.001 (0.1%) after vaccination. What size sample is needed so that the width of a 99% confidence interval for the proportion diseased in the vaccinated population should be no greater than 0.0002? To ensure that the sample size is sufficiently large, the value of p to be used in Eq. (6.9) is chosen to be 0.0012, rather than the expected 0.0010.

$$N = \frac{(2.58)^2(0.9988)(0.0012)}{(0.0002/2)^2} = 797,809.$$

The trial will have to include approximately 800,000 subjects in order to yield the desired precision.

6.5 POWER

Power is the probability that the statistical test results in rejection of H_0 when a specified alternative is true. The "stronger" the power, the better the chance that the null hypothesis will be rejected (i.e., the test results in a declaration of "significance") when, in fact, H_0 is false. The larger the power, the more sensitive is the test. *Power* is defined as $1 - \beta$. The larger the β error, the weaker is the power. Remember that β is an error resulting from *accepting* H_0 *when* H_0 *is false*. Therefore, $1 - \beta$ is the probability of *rejecting* H_0 *when* H_0 *is false*.

From an idealistic point of view, the power of a test should be calculated *before* an experiment is conducted. In addition to defining the properties of the test, power is used to help compute the sample size, as discussed above. Unfortunately, many experiments proceed without consideration of power (or β). This results from the difficulty of choosing an appropriate value of β. There is no traditional value of β to use, as is the case for α, where 5% is usually used. Thus, the power of the test is often computed after the experiment has been completed.

Power is best described by diagrams such as those shown previously in this chapter (Figs. 6.1 and 6.2). In these figures, β is the area of the curves represented by the alternative hypothesis that is included in the region of acceptance defined by the null hypothesis.

The concept of power is also illustrated in Figure 6.3. To illustrate the calculation of power, we will use data presented for the test of a new antihypertensive agent (sect. 6.2), a paired sample test, with $\sigma = 7$ and $H_0 : \Delta = 0$. The test is performed at the 5% level of significance. Let us suppose that the sample size is limited by cost. The sponsor of the test had sufficient funds to pay for a study that included only *12 subjects*. The design described earlier in this chapter (sect. 6.2) used 26 patients with β specified equal to 0.05 (power $= 0.95$). With 12 subjects, the power will be considerably less than 0.95. The following discussion shows how power is calculated.

The cutoff points for statistical significance (which specify the critical region) are defined by α, N, and σ. Thus, the values of $\bar{\delta}$ that will lead to a significant result for a two-sided test are as follows:

$$Z = \frac{|\bar{\delta}|}{\sigma/\sqrt{N}}$$

$$\bar{\delta} = \frac{\pm Z\sigma}{\sqrt{N}}.$$

In our example, $Z = 1.96$ ($\alpha = 0.05$), $\sigma = 7$, and $N = 12$.

$$\bar{\delta} = \frac{\pm(1.96)(7)}{\sqrt{12}} = \pm 3.96.$$

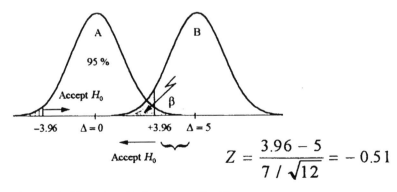

$$Z = \frac{3.96 - 5}{7 / \sqrt{12}} = -0.51$$

Figure 6.3   Illustration of beta or power $(1 - \beta)$.

Values of $\bar{\delta}$ greater than 3.96 or less than -3.96 will lead to the decision that the products differ at the 5% level. Having defined the values of $\bar{\delta}$ that will lead to rejection of H_0, we obtain the power for the alternative, H_a: $\Delta = 5$, by computing the probability that an average result, $\bar{\delta}$, will be greater than 3.96, if H_a is true (i.e., $\Delta = 5$).

This concept is illustrated in Figure 6.3. Curve B is the distribution with mean equal to 5 and $\sigma = 7$. If curve B is the true distribution, the probability of observing a value of $\bar{\delta}$ below 3.96 is the probability of accepting H_0 if the alternative hypothesis is true ($\Delta = 5$). This is the definition of β. This probability can be calculated using the Z transformation.

$$Z = \frac{3.96 - 5}{7/\sqrt{12}} = -0.51.$$

Referring to Table IV.2, the area below $+3.96$ ($Z = -0.51$) for curve B is approximately 0.31. The power is $1 - \beta = 1 - 0.31 = 0.69$. The use of 12 subjects results in a power of 0.69 to "detect" a difference of $+5$ compared to the 0.95 power to detect such a difference when 26 subjects were used. A power of 0.69 means that if the *true difference were 5 mm Hg*, the statistical test will result in *significance with a probability of 69%*; 31% of the time, such a test will result in acceptance of H_0.

A *power curve* is a plot of the power, $1 - \beta$, versus alternative values of Δ. Power curves can be constructed by computing β for several alternatives and drawing a smooth curve through these points. For a two-sided test, the power curve is symmetrical around the hypothetical mean, $\Delta = 0$, in our example. The power is equal to α when the alternative is equal to the hypothetical mean under H_0. Thus, the power is 0.05 when $\Delta = H_0$ (Fig. 6.4) in the power curve. The power curve for the present example is shown in Figure 6.4.

The following conclusions may be drawn concerning the *power* of a test if α is kept constant:

1. The larger the sample size, the larger the power.
2. The larger the difference to be detected (H_a), the larger the power. A large sample size will be needed in order to have strong power to detect a small difference.
3. The larger the variability (s.d.), the weaker the power.
4. If α is increased, power is increased (β is decreased) (Fig. 6.3). An increase in α (e.g., 10%) results in a smaller Z. The cutoff points are shorter, and the area of curve B below the cutoff point is smaller.

Power is a function of N, Δ, σ, and α.

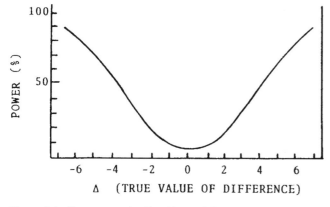

Figure 6.4 Power curve for $N = 12$, $\alpha = 0.05$, $\sigma = 7$, and H_0: $\Delta = 0$.

A simple way to compute the approximate power of a test is to use the formula for sample size [Eqs. (6.4) and (6.5). for example] and solve for Z_β. In the previous example, a single sample or a paired test, Eq. (6.4) is appropriate:

$$N = \left(\frac{\sigma}{\Delta}\right)^2 (Z_\alpha + Z_\beta)^2 \tag{6.4}$$

$$Z_\beta = \frac{\Delta}{\sigma}\sqrt{N} - Z_\alpha. \tag{6.10}$$

Once having calculated Z_β, the probability determined directly from Table IV.2 is equal to the power, $1 - \beta$. See the discussion and examples below.

In the problem discussed above, applying Eq. (6.10) with $\Delta = 5$, $\sigma = 7$, $N = 12$, and $Z_\alpha = 1.96$,

$$Z_\beta = \frac{5}{7}\sqrt{12} - 1.96 = 0.51.$$

According to the notation used for Z (Table 6.2), β is the area above Z_β. Power is the area below Z_β (power $= 1 - \beta$). In Table IV.2, the area above $Z = 0.51$ is approximately 31%. The power is $1 - \beta$. Therefore, the power is 69%.[§]

If N is small and the variance is unknown, appropriate values of t should be used in place of Z_α and Z_β. Alternatively, we can adjust N by subtracting $0.5Z_\alpha^2$ or $0.25Z_\alpha^2$ from the actual sample size for a one- or two-sample test, respectively. The following examples should make the calculations clearer.

Example 7: A bioavailability study has been completed in which the ratio of the AUCs for two comparative drugs was submitted as evidence of bioequivalence. The FDA asked for the power of the test as part of their review of the submission. (Note that this analysis is different from that presently required by FDA.) The null hypothesis for the comparison is H_0: $R = 1$, where R is the true average ratio. The test was two-sided with α equal to 5%. Eighteen subjects took each of the two comparative drugs in a paired-sample design. The standard deviation was calculated from the final results of the study, and was equal to 0.3. The power is to be determined for a difference of 20% for the comparison. This means that if the test product is truly more than 20% greater or smaller than the reference product, we wish to calculate the probability that the ratio will be judged to be significantly different from 1.0. The value of Δ to be used in Eq. (6.10) is 0.2.

$$Z_\beta = \frac{0.2\sqrt{16}}{0.3} - 1.96 = 0.707.$$

Note that the value of N is taken as 16. This is the inverse of the procedure for determining sample size, where $0.5Z_\alpha^2$ was added to N. Here we subtract $0.5Z_\alpha^2$ (approximately 2) from N; $18 - 2 = 16$. According to Table IV.2, the area corresponding to $Z = 0.707$ is approximately 0.76. Therefore, the power of this test is 76%. That is, if the true difference between the formulations is 20%, a significant difference will be found between the formulations 76% of the time. This is very close to the 80% power that was recommended before current FDA guidelines were implemented for bioavailability tests (where $\Delta = 0.2$).

Example 8: A drug product is prepared by two different methods. The average tablet weights of the two batches are to be compared, weighing 20 tablets from each batch. The average weights of the two 20-tablet samples were 507 and 511 mg. The pooled standard deviation was calculated to be 12 mg. The director of quality control wishes to be "sure" that if the average weights truly differ by 10 mg or more, the statistical test will show a significant difference, when

---

[§] The value corresponding to Z in Table IV.2 gives the power directly. In this example, the area in the table corresponding to a Z of 0.51 is approximately 0.69.

he was asked, "How sure?", he said 95% sure. This can be translated into a β of 5% or a power of 95%. This is a *two independent groups* test. Solving for Z_β from Eq. (6.5), we have

$$
\begin{aligned}
Z_\beta &= \frac{\Delta}{\sigma}\sqrt{\frac{N}{2}} - Z_\alpha \\
&= \frac{10}{12}\sqrt{\frac{19}{2}} - 1.96 = 0.609.
\end{aligned}
\tag{6.11}
$$

As discussed above, the value of N is taken as 19 rather than 20, by subtracting $0.25Z_\alpha^2$ from N for the two-sample case. Referring to Table IV.2, we note that the power is approximately 73%. The experiment does not have sufficient power according to the director's standards. To obtain the desired power, we can increase the sample size (i.e., weigh more tablets). (See Exercise Problem 10.)

6.6 SAMPLE SIZE AND POWER FOR MORE THAN TWO TREATMENTS (ALSO SEE CHAP. 8)

The problem of computing power or sample size for an experiment with more than two treatments is somewhat more complicated than the relatively simple case of designs with two treatments. The power will depend on the number of treatments and the form of the null and alternative hypotheses. Dixon and Massey [5] present a simple approach to determining power and sample size. The following notation will be used in presenting the solution to this problem.

Let $M_1, M_2, M_3 \ldots M_k$ be the hypothetical population means of the k treatments. The null hypothesis is $M_1 = M_2 = M_3 = M_k$. As for the two sample cases, we must specify the alternative values of M_i. The alternative means are expressed as a grand mean, $M_t \pm$ some deviation, D_i, where $\sum(D_i) = 0$. For example, if three treatments are compared for pain, Active A, Active B, and Placebo (P), the values for the alternative hypothesized means, based on a VAS scale for pain relief, could be $75 + 10$ (85), $75 + 10$ (85), and $75 - 20$ (55) for the two actives and placebo, respectively. The sum of the deviations from the grand mean, 75, is $10 + 10 - 20 = 0$. The power is computed based on the following equation:

$$
\psi^2 = \frac{\sum(M_i - M_t)^2/k}{S^2/n},
\tag{6.12}
$$

where n is the number of observations in each treatment group (n is the same for each treatment) and S^2 is the common variance. The value of ψ^2 is referred to Table 6.4 to estimate the required sample size.

Consider the following example of three treatments in a study measuring the analgesic properties of two actives and a placebo as described above. Fifteen subjects are in each treatment group and the variance is 1000. According to Eq. (6.12),

$$
\psi^2 = \frac{\{(85 - 75)^2 + (85 - 75)^2 + (55 - 75)^2\}/3}{1000/15} = 3.0.
$$

Table 6.4 gives the approximate power for various values of ψ, at the 5% level, as a function of the number of treatment groups and the d.f. for error for 3 and 4 treatments. (More detailed tables, in addition to graphs, are given in Dixon and Massey [5].) Here, we have 42 d.f. and three treatments with $\psi = \sqrt{3} = 1.73$. The power is approximately 0.72 by simple linear interpolation (42 d.f. for $\psi = 1.7$). The correct answer with more extensive tables is closer to 0.73.

Table 6.4   Factors for Computing Power for
Analysis of Variance

d.f. error	ψ	Power
Alpha $= 0.05$, $k = 3$		
10	1.6	0.42
	2.0	0.76
	2.4	0.80
	3.0	0.984
20	1.6	0.62
	1.92	0.80
	2.00	0.83
	3.0	>0.99
30	1.6	0.65
	1.9	0.80
	2.0	0.85
	3.0	>0.99
60	1.6	0.67
	1.82	0.80
	2.0	0.86
	3.0	>0.99
inf	1.6	0.70
	1.8	0.80
	2.0	0.88
	3.0	>0.99
alpha $= 0.05$, $k = 4$		
10	1.4	0.48
	2.0	0.80
	2.6	0.96
20	1.4	0.56
	2.0	0.88
	2.6	986
30	1.4	0.59
	2.0	0.90
	2.6	>0.99
60	1.4	0.61
	2.0	0.92
	2.6	>0.99
inf	1.4	0.65
	2.0	0.94
	2.6	>0.99

Table 6.4 can also be used to determine sample size. For example, how many patients per treatment group are needed to obtain a power of 0.80 in the above example? Applying Eq. (6.12),

$$\frac{\{(85 - 75)^2 + (85 - 75)^2 + (55 - 75)^2\}/3}{1000/n} = \psi^2.$$

Solve for ψ^2

$$\psi^2 = 0.2n.$$

We can calculate n by trial and error. For example, with $N = 20$,

$$0.2N = 4 = \psi^2 \quad \text{and} \quad \psi = 2.$$

For $\psi = 2$ and $N = 20$ (d.f. $= 57$), the power is approximately 0.86 (for d.f. $= 60$, power 0.86). For $N = 15$ (d.f. $= 42$, $\psi = \sqrt{3}$), we have calculated (above) that the power is approximately 0.72. A sample size of between 15 and 20 patients per treatment group would give a power of 0.80. In this example, we might guess that 17 patients per group would result in approximately 80% power. Indeed, more exact tables show that a sample size of 17($\psi = \sqrt{(0.2 \times 17)} = 1.85$) corresponds to a power of 0.79.

The same approach can be used for two-way designs, using the appropriate error term from the analysis of variance.

6.7   SAMPLE SIZE FOR BIOEQUIVALENCE STUDIES (ALSO SEE CHAP. 11)

In its early evolution, bioequivalence was based on the acceptance or rejection of a hypothesis test. Sample sizes could then be determined by conventional techniques as described in section 6.2. Because of inconsistencies in the decision process based on this approach, the criteria for acceptance was changed to a two-sided 90% confidence interval, or equivalently, two one-sided t test, where the hypotheses are $(\mu_1/\mu_2) < 0.8$ and $(\mu_1/\mu_2) > 1.25$ versus the alternative of $0.8 < (\mu_1/\mu_2) < 1.25$. This test is based on the antilog of the difference between the averages of the log-transformed parameters (the geometric mean). This test is equivalent to a two-sided 90% confidence interval for the ratio of means falling in the interval 0.80 to 1.25 in order to accept the hypothesis of equivalence. Again, for the currently accepted log-transformed data, the 90% confidence interval for the antilog of the difference between means must lie between 0.80 and 1.25, that is, $0.8 < \text{antilog}(\mu_1/\mu_2) < 1.25$. The sample-size determination in this case is not as simple as the conventional determination of sample size described earlier in this chapter. The method for sample-size determination for nontransformed data has been published by Phillips [6] along with plots of power as a function of sample size, relative standard deviation (computed from the ANOVA), and treatment differences. Although the theory behind this computation is beyond the scope of this book, Chow and Liu [7] give a simple way of approximating the power and sample size. The sample size for each sequence group is approximately

$$N = (t_{\alpha,\, 2N-2} + t_{\beta,\, 2N-2})^2 \left[\frac{CV}{(V - \delta)} \right]^2, \tag{6.13}$$

where N is the number of subjects per sequence, t the appropriate value from the t distribution, α the significance level (usually 0.10), $1 - \beta$ the power (usually 0.8), CV the coefficient of variation, V the bioequivalence limit, and δ the difference between products.

One would have to have an approximation of the magnitude of the required sample size in order to approximate the t values. For example, suppose that RSD $= 0.20$, $\delta = 0.10$, power is 0.8, and an initial approximation of the sample size is 20 per sequence (a total of 40 subjects). Applying Eq. (6.13)

$$n = (1.69 + 0.85)^2 [0.20/(0.20 - 0.10)]^2 = 25.8.$$

Use a total of 52 subjects. This agrees closely with Phillip's more exact computations. Dilletti et al. [8] have published a method for determining sample size based on the log-transformed variables, which is the currently preferred method. Table 6.5 showing sample sizes for various values of CV, power, and product differences is taken from their publication.

Based on these tables, using log-transformed estimates of the parameters would result in a sample size estimate of 38 for a power of 0.8, ratio of 0.9, and CV $= 0.20$. If the assumed ratio is 1.1, the sample size is estimated as 32.

Equation (6.13) can also be used to approximate these sample sizes using log values for V and δ: $n = (1.69 + 0.85)^2 [0.20/(0.223 - 0.105)]^2 = 19$ per sequence or 38 subjects in total, where 0.223 is the log of 1.25 and 0.105 is the absolute value of the log of 0.9.

Table 6.5 Sample Sizes for Given CV Power and Ratio (T/R) for Log-Transformed Parameters[a]

CV (%)	Power (%)	0.85	0.90	0.95	μ_r, μ_x 1.00	1.05	1.10	1.15	1.20
5.0	70	10	6	4	4	4	4	6	16
7.5		16	6	6	4	6	6	10	34
10.0		28	10	6	6	6	8	16	58
12.5		42	14	8	8	8	12	24	90
15.0		60	18	10	10	10	16	32	128
17.5		80	22	12	12	12	20	44	172
20.0		102	30	16	14	16	26	56	224
22.5		128	36	20	16	20	30	70	282
25.0		158	44	24	20	22	38	84	344
27.5		190	52	28	24	26	44	102	414
30.0		224	60	32	28	32	52	120	490
3.0	80	12	6	4	4	4	6	8	22
7.5		22	8	6	6	6	8	12	44
10.0		36	12	8	6	8	10	20	76
12.5		54	16	10	8	10	14	30	118
15.0		78	22	12	10	12	20	42	168
17.5		104	30	16	14	16	26	56	226
20.0		134	38	20	16	18	32	72	294
22.5		168	46	24	20	24	40	90	368
25.0		206	56	28	24	28	48	110	452
27.5		248	68	34	28	34	58	132	544
30.0		292	80	40	32	38	68	156	642
5.0	90	14	6	4	4	4	6	8	28
7.5		28	10	6	6	6	8	16	60
10.0		48	14	8	8	8	14	26	104
12.5		74	22	12	10	12	18	40	162
15.0		106	30	16	12	16	26	58	232
17.5		142	40	20	16	20	34	76	312
20.0		186	50	26	20	24	44	100	406
22.5		232	64	32	24	30	54	124	510
25.0		284	78	38	28	36	66	152	626
27.5		342	92	44	34	44	78	182	752
30.0		404	108	52	40	52	92	214	888

[a] *Source*: From Ref. [8].

For $\delta = 1.10$ (log $= 0.0953$), the sample size is: $n = (1.69 + 0.85)^2[0.20/(0.223 - 0.0953)]^2 = 16$ per sequence or 32 subjects in total.

If the difference between products is specified as zero (ratio $= 1.0$), the value for $t_{\beta, 2n-2}$ in Eq. (6.3) should be two sided (Table 6.2). For example, for 80% power (and a large sample size) use 1.28 rather than 0.84. In the example above with a ratio of 1.0 (0 difference between products), a power of 0.8, and a CV $= 0.2$, use a value of (approximately) 1.34 for $t_{\beta, 2n-2}$.

$$n = (1.75 + 1.34)^2[0.2/0.223]^2 = 7.7 \text{ per group or 16 total subjects.}$$

An Excel program to calculate the number of subjects required for a crossover study under various conditions of power and product differences, for both parametric and binary (binomial) data, is available on the disk accompanying this volume.

This approach to sample-size determination can also be used for studies where the outcome is dichotomous, often used as the criterion in clinical studies of bioequivalence (cured or not cured) for topically unabsorbed products or unabsorbed oral products such as sucralfate. This topic is presented in section 11.4.8.

KEY TERMS

Alpha level
Attribute
Beta error
Confidence interval
Delta
Power

Power curve
"Practical" significance
Sample size
Sampling plan
Sensitivity
Z transformation

EXERCISES

1. Two diets are to be compared with regard to weight gain of weanling rats. If the weight gain due to the diets differs by 10 g or more, we would like to be 80% sure that we obtain a significant result. How many rats should be in each group if the s.d. is estimated to be 5 and the test is performed at the 5% level?

2. How many rats per group would you use if the standard deviation were known to be equal to 5 in Problem 1?

3. In Example 3 where two antibiotics are being compared, how many patients would be needed for a study with $\alpha = 0.05$, $\beta = 0.10$, using a parallel design, and assuming that the new product must have a cure rate of 90% to be acceptable as a better product than the standard? (Cure rate for standard $= 80\%$).

4. It is hypothesized that the difference between two drugs with regard to success rate is 0 (i.e., the drugs are not different). What size sample is needed to show a difference of 20% significant at the 5% level with a β error of 10%? (Assume that the response rate is about 50% for both drugs, a *conservative* estimate.) The study is a two independent samples design (parallel groups).

5. How many observations would be needed to estimate a response rate of about 50% within $\pm 15\%$ (95% confidence limits)? How many observations would be needed to estimate a response rate of $20 \pm 15\%$?

6. Your boss tells you to make a new tablet formulation that should have a dissolution time (90% dissolution) of 30 minutes. The previous formulation took 40 minutes to 90% dissolution. She tells you that she wants an α level of 5% and that if the new formulation really has a dissolution time of 30 minutes or less, she wants to be 99% sure that the statistical comparison will show significance. (This means that the β error is 1%.) The s.d. is approximately 10. What size sample would you use to test the new formulation?

7. In a clinical study comparing the effect of two drugs on blood pressure, 20 patients were to be tested on each drug (two groups). The change in blood pressure from baseline mea-surements was to be determined. The s.d., measured as the difference among individuals' responses, is *estimated* from past experience to be 5.
   (a) If the statistical test is done at the 5% level, what is the power of the test against an alternative of 3 mm Hg difference between the drugs ($H_0 : \mu_1 = \mu_2$ or $\mu_1 - \mu_2 = 0$). This means: What is the probability that the test will show significance if the true difference between the drugs is 3 mm Hg or more ($H_a : \mu_1 - \mu_2 = 3$)?
   (b) What is the power if there are 50 people per group? α is 5%.

8. A tablet is produced with a labeled potency of 100 mg. The standard deviation is known to be 10. What size sample should be assayed if we want to have 90% power to detect a difference of 3 mg from the target? The test is done at the 5% level.

9. In a bioequivalence study, the ratio of AUCs is to be compared. A sample size of 12 subjects is used in a paired design. The standard deviation resulting from the statistical test is 0.25. What is the power of this test against a 20% difference if α is equal to 0.05?

10. How many samples would be needed to have 95% power for Example 8?

11. In a bioequivalence study, the maximum blood level is to be compared for two drugs. This is a crossover study (paired design) where each subject takes both drugs. Eighteen subjects entered the study with the following results. The observed difference is 10 µg/mL. The s.d. (from this experiment) is 40. A practical difference is considered to be 15 µg/mL. What is the power of the test for a 15-µg/mL difference for a two-sided test at the 5% level?

12. How many observations would you need to estimate a proportion within ±5% (95% confidence interval) if the expected proportion is 10%?

13. A parallel design is used to measure the effectiveness of a new antihypertensive drug. One group of patients receives the drug and the other group receives placebo. A difference of 6 mm Hg is considered to be of practical significance. The standard deviation (difference from baseline) is unknown but is estimated as 5 based on some preliminary data. Alpha is set at 5% and β at 10%. How many patients should be used in each group?

14. From Table 6.3, find the number of samples needed to determine the difference between the dissolution of two formulations for $\alpha = 0.05$, $\beta = 0.10$, $S = 25$, for a "practical" difference of 25 (minutes).

REFERENCES

1. United States Pharmacopeia, 23rd rev, and National Formulary, 18th ed. Rockville, MD: USP Pharmacopeial Convention, Inc., 1995.
2. U.S. Department of Defense Military Standard. Military Sampling Procedures and Tables for Inspection by Attributes (MIL-STD-105E). Washington, DC: U.S. Government Printing Office, 1989.
3. Guenther WC. Sample size formulas for normal theory tests. Am Stat 1981; 35:243.
4. Fleiss J. Statistical Methods for Rates and Proportions, 2nd ed. New York: Wiley, 1981.
5. Dixon WJ, Massey FJ Jr. Introduction to Statistical Analysis, 3rd ed. New York: McGraw-Hill, 1969.
6. Phillips KE. Power of the two one-sided tests procedure in bioequivalence. J Pharmacokinet Biopharm 1991; 18:137.
7. Chow S-C, Liu J-P. Design and Analysis of Bioavailability and Bioequivalence Studies. New York: Marcel Dekker, 1992.
8. Dilletti E, Hauschke D, Steinijans VW. Sample size determination: extended tables for the multiplicative model and bioequivalence ranges of 0.9 to 1.11 and 0.7 to 1.43. Int J Clin Pharmacol Toxicol 1991; 29:1.

7 | Linear Regression and Correlation

Simple linear regression analysis is a statistical technique that defines the functional relationship between two variables, X and Y, by the "best-fitting" straight line. A straight line is described by the equation, $Y = A + BX$, *where Y is the dependent variable* (ordinate), *X is the independent variable* (abscissa), *and A and B are the Y intercept* and *slope of the line*, respectively (Fig. 7.1).* Applications of regression analysis in pharmaceutical experimentation are numerous. This procedure is commonly used

1. to describe the relationship between variables where the functional relationship is known to be linear, such as in Beer's law plots, where optical density is plotted against drug concentration;
2. when the functional form of a response is unknown, but where we wish to represent a trend or rate as characterized by the slope (e.g., as may occur when following a pharmacological response over time);
3. when we wish to describe a process by a relatively simple equation that will relate the response, Y, to a fixed value of X, such as in stability prediction (concentration of drug versus time).

In addition to the specific applications noted above, regression analysis is used to define and characterize dose–response relationships, for fitting linear portions of pharmacokinetic data, and in obtaining the best fit to linear physical–chemical relationships.

Correlation is a procedure commonly used to characterize quantitatively the relationship between variables. Correlation is related to linear regression, but its application and interpretation are different. This topic is introduced at the end of this chapter.

7.1 INTRODUCTION

Straight lines are constructed from sets of data pairs, X and Y. Two such pairs (i.e., two points) uniquely define a straight line. As noted previously, a straight line is defined by the equation

$$Y = A + BX, \tag{7.1}$$

where A is the Y intercept (the value of Y when $X = 0$) and B is the slope ($\Delta Y / \Delta X$). $\Delta Y / \Delta X$ is $(Y_2 - Y_1)/(X_2 - X_1)$ for any two points on the line (Fig. 7.1). The slope and intercept define the line; once A and B are given, the line is specified. In the elementary example of only two points, a statistical approach to define the line is clearly unnecessary.

In general, with more than two X, y points,[†] a plot of y versus X will not *exactly* describe a straight line, even when the relationship is known to be linear. The failure of experimental data derived from truly linear relationships to lie exactly on a straight line is due to errors of observation (experimental variability). Figure 7.2 shows the results of four assays of drug samples of different, but known potency. The assay results are plotted against the known amount of drug. If the assays are performed without error, the plot results in a 45° line (slope $= 1$) which, if extended, passes through the origin; that is, the Y intercept, A, is 0 [Fig. 7.2(A)].

---

[*] The notation $Y = A + BX$ is standard in statistics. We apologize for any confusion that may result from the reader's familiarity with the equivalent, $Y = mX + b$, used frequently in analytical geometry.
[†] In the rest of this chapter, y denotes the experimentally observed point, and Y denotes the corresponding point on the least squares "fitted" line (or the true value of Y, according to context).

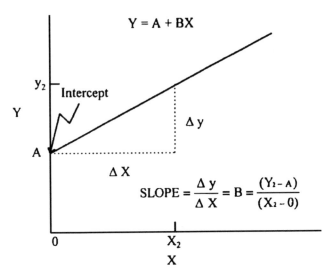

$$Y = A + BX$$

$$SLOPE = \frac{\Delta y}{\Delta X} = B = \frac{(Y_2 - A)}{(X_2 - 0)}$$

Figure 7.1 Straight-line plot.

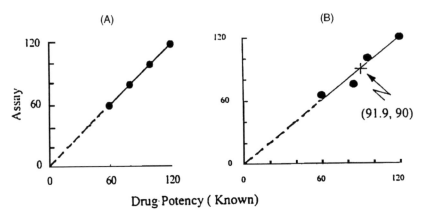

Drug Potency (Known)

Figure 7.2 Plot of assay recovery versus known amount: theoretical and actual data.

In this example, the equation of the line $Y = A + BX$ is $Y = 0 + 1(X)$, or $Y = X$. Since there is no error in this experiment, the line passes exactly through the four X, Y points.

Real experiments are not error free, and a plot of X, y data rarely exactly fits a straight line, as shown in Figure 7.2(B). We will examine the problem of obtaining a line to fit data that are not error free. In these cases, the line does not go exactly through all of the points. A "good" line, however, should come "close" to the experimental points. When the variability is small, a line drawn by eye will probably be very close to that constructed more exactly by a statistical approach [Fig. 7.3(A)]. With large variability, the "best" line is not obvious. What single line would you draw to best fit the data plotted in Figure 7.3(B)? Certainly, lines drawn through any two arbitrarily selected points will not give the best (or a unique) line to fit the totality of data.

Given N pairs of variables, X, Y, we can define the best straight line describing the relationship of X and y as the line that minimizes the sum of squares of the vertical distances of each point from the fitted line. The definition of "sum of squares of the vertical distances of each point from the fitted line" (Fig. 7.4) is written mathematically as $\sum(y - Y)^2$, where y represents the experimental points and Y represents the corresponding points on the fitted line. The line constructed according to this definition is called the *least squares* line. Applying techniques of

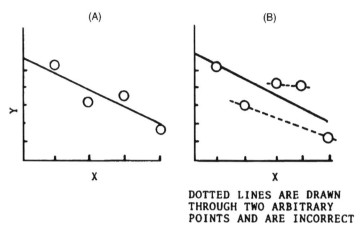

DOTTED LINES ARE DRAWN
THROUGH TWO ARBITRARY
POINTS AND ARE INCORRECT

Figure 7.3 Fit of line with variable data.

calculus, the slope and intercept of the least squares line can be calculated from the sample data as follows:

$$\text{Slope} = b = \frac{\sum(X - \overline{X})(y - \overline{y})}{\sum(X - \overline{X})^2} \tag{7.2}$$

$$\text{Intercept} = a = \overline{y} - b\overline{X} \tag{7.3}$$

Remember that the slope and intercept uniquely define the line.

There is a shortcut computing formula for the slope, similar to that described previously for the standard deviation

$$b = \frac{N\sum Xy - (\sum X)(\sum y)}{N\sum X^2 - (\sum X)^2}, \tag{7.4}$$

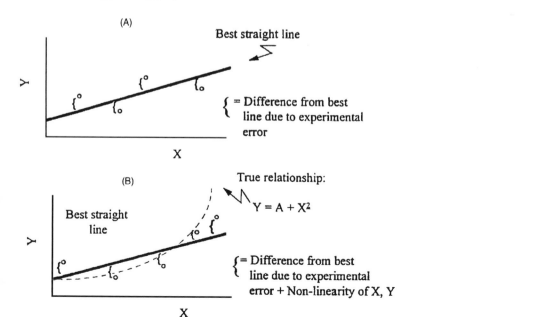

Figure 7.4 Lack of fit due to (**A**) experimental error and (**B**) nonlinearity.

Table 7.1   Raw Data from Figure 7.2(A) to Calculate the Least Squares Line

Drug potency, X	Assay, y	Xy
60	60	3600
80	80	6400
100	100	10,000
120	120	14,400
$\sum X = 360$	$\sum y = 360$	$\sum Xy = 34{,}400$
$\sum X^2 = 34{,}400$		

Table 7.2   Raw Data from Figure 7.2(B) Used to Calculate the Least Squares Line

Drug potency, X	Assay, y	Xy
60	63	3780
80	75	6000
100	99	9900
120	116	13,920
$\sum X = 360$	$\sum y = 353$	$\sum Xy = 33{,}600$
$\sum X^2 = 34{,}400$	$\sum y^2 = 32{,}851$	

where N is the number of X, y pairs. The calculation of the slope and intercept is relatively simple, and can usually be quickly computed using a computer (e.g., EXCEL) or with a hand calculator. Some calculators have a built-in program for calculating the regression parameter estimates, a and b.[‡]

For the example shown in Figure 7.2(A), the line that exactly passes through the four data points has a slope of 1 and an intercept of 0. The line, $Y = X$, is clearly the best line for these data, an exact fit. The least squares line, in this case, is exactly the same line, $Y = X$. The calculation of the intercept and slope using the least squares formulas, Eqs. (7.3) and (7.4), is illustrated below. Table 7.1 shows the raw data used to construct the line in Figure 7.2(A).

According to Eq. (7.4) ($N = 4$, $\sum X^2 = 34{,}400$, $\sum Xy = 34{,}400$, $\sum X = \sum y = 360$),

$$b = \frac{(4)(3600 + 6400 + 10{,}000 + 14{,}000) - (360)(360)}{4(34{,}400) - (360)^2} = 1$$

a is computed from Eq. (7.3); $a = \bar{y} - b\,\bar{X}(\bar{y} = \bar{X} = 90,\ b = 1)$. $a = 90 - 1(90) = 0$. This represents a situation where the assay results exactly equal the known drug potency (i.e., there is no error).

The actual experimental data depicted in Figure 7.2(B) are shown in Table 7.2. The slope b and the intercept a are calculated from Eqs. (7.4) and (7.3). According to Eq. (7.4),

$$b = \frac{(4)(33{,}600) - (360)(353)}{4(34{,}400) - (360)^2} = 0.915.$$

According to Eq. (7.3),

$$a = \frac{353}{4} - 0.915(90) = 5.9.$$

A perfect assay (no error) has a slope of 1 and an intercept of 0, as shown above. The actual data exhibit a slope close to 1, but the intercept appears to be too far from 0 to be attributed to random error. Exercise Problem 2 addresses the interpretation of these results as they relate to assay method characteristics.

[‡] a and b are the sample estimates of the true parameters, A and B.

This example suggests several questions and problems regarding linear regression analysis. The line that best fits the experimental data is an estimate of some true relationship between X and Y. In most circumstances, we will fit a straight line to such data only if we believe that the true relationship between X and Y is linear. The experimental observations will not fall exactly on a straight line because of variability (e.g., error associated with the assay). This situation (true linearity associated with experimental error) is different from the case where the underlying true relationship between X and Y is not linear. In the latter case, the lack of fit of the data to the least squares line is due to a combination of experimental error and the lack of linearity of the X, Y relationship (Fig. 7.4). Elementary techniques of simple linear regression will not differentiate these two situations: (a) experimental error with true linearity and (b) experimental error and nonlinearity. (A design to estimate variability due to both nonlinearity and experimental error is given in App. II.)

We will discuss some examples relevant to pharmaceutical research that make use of least squares linear regression procedures. The discussion will demonstrate how variability is estimated and used to construct estimates and tests of the line parameters A and B.

7.2 ANALYSIS OF STANDARD CURVES IN DRUG ANALYSIS: APPLICATION OF LINEAR REGRESSION

The assay data discussed previously can be considered as an example of the construction of a *standard curve* in drug analysis. Known amounts of drug are subjected to an assay procedure, and a plot of percentage recovered (or amount recovered) versus amount added is constructed. Theoretically, the relationship is usually a straight line. A knowledge of the line parameters A and B can be used to predict the amount of drug in an unknown sample based on the assay results. In most practical situations, A and B are unknown. The least squares estimates a and b of these parameters are used to compute drug potency (X) based on the assay response (y). For example, the least squares line for the data in Figure 7.2(B) and Table 7.2 is

$$Assay\ result = 5.9 + 0.915\ (potency). \tag{7.5}$$

Rearranging Eq. (7.5), an unknown sample that has an assay value of 90 can be predicted to have a true potency of

$$Potency = X = \frac{y - 5.9}{0.915}$$

$$Potency = \frac{90 - 5.9}{0.915} = 91.9.$$

This point (91.9, 90) is indicated in Figure 7.2 by a cross.

7.2.1 Line Through the Origin

Many calibration curves (lines) are known to pass through the origin; that is, the assay response must be zero if the concentration of drug is zero. The calculation of the slope is simplified if the line is forced to go through the point (0,0). In our example, if the intercept is *known* to be zero, the slope is (Table 7.2)

$$b = \frac{\sum Xy}{\sum X^2}$$

$$= \frac{33,600}{60^2 + 80^2 + 100^2 + 120^2} = 0.977. \tag{7.6}$$

The least squares line fitted with the zero intercept is shown in Figure 7.5. If this line were to be used to predict actual concentrations based on assay results, we would obtain answers that are different from those predicted from the line drawn in Figure 7.2(B). However, both lines have been constructed from the same raw data. "Is one of the lines correct?" or "Is one line better than the other?" Although one cannot say with certainty which is the better line, a thorough knowledge of the analytical method will be important in making a choice. For example, a nonzero intercept suggests either nonlinearity over the range of assays or the

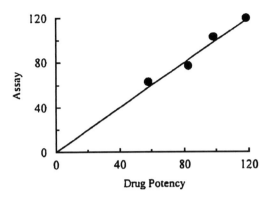

Figure 7.5 Plot of data in Table 7.2 with known (0, 0) intercept.

presence of an interfering substance in the sample being analyzed. The decision of which line to use can also be made on a statistical basis. A statistical test of the intercept can be performed under the null hypothesis that the intercept is 0 (H_0: $A = 0$, sect. 7.4.1). Rejection of the hypothesis would be strong evidence that the line with the positive intercept best represents the data.

7.3   ASSUMPTIONS IN TESTS OF HYPOTHESES IN LINEAR REGRESSION
Although there are no prerequisites for fitting a least squares line, the testing of statistical hypotheses in linear regression depends on the validity of several assumptions.

1.   *The X variable is measured without error.* Although not always *exactly* true, X is often measured with relatively little error and, under these conditions this assumption can be considered to be satisfied. In the present example, X is the potency of drug in the "known" sample. If the drug is weighed on a sensitive balance, the error in drug potency will be very small. Another example of an X variable that is often used, which can be precisely and accurately measured, is "time."

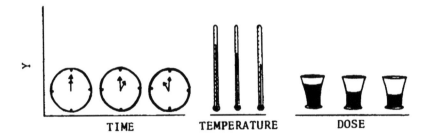

2.   *For each X, y is independent and normally distributed.* We will often use the notation $Y.x$ to show that the value of Y is a function of X.
3.   *The variance of y is assumed to be the same at each X.* If the variance of y is not constant, but is either known or related to X in some way, other methods (see sect. 7.7) are available to estimate the intercept and slope of the line [1].
4.   *A linear relationship exists between X and Y.* $Y = A + BX$, where A and B are the true parameters. Based on theory or experience, we have reason to believe that X and Y are linearly related.

These assumptions are depicted in Figure 7.6. Except for location (mean), the distribution of y is the same at every value of X; that is, y has the same variance at every value of X. In the example in Figure 7.6, the mean of the distribution of y's decreases as X increases (the slope is negative).

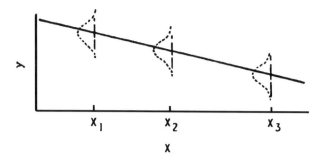

Figure 7.6 Normality and variance assumptions in linear regression.

7.4 ESTIMATE OF THE VARIANCE: VARIANCE OF SAMPLE ESTIMATES OF THE PARAMETERS

If the assumptions noted in section 7.3 hold, the distributions of *sample estimates* of the slope and intercept, b and a, are normal with means equal to B and A, respectively.§ Because of this important result, statistical tests of the parameters A and B can be performed using normal distribution theory. Also, one can show that the sample estimates are unbiased estimates of the true parameters (similar to the sample average, $\overline{X}$, being an unbiased estimate of the true mean, μ). The variances of the estimates, a and b, are calculated as follows:

$$\sigma_a^2 = \sigma_{Y,x}^2 \left[\frac{1}{N} + \frac{\overline{X}^2}{\sum (X - \overline{X})^2} \right] \qquad (7.7)$$

$$\sigma_b^2 = \frac{\sigma_{Y,x}^2}{\sum (X - \overline{X})^2}. \qquad (7.8)$$

$\sigma_{Y,x}^2$ is the variance of the response variable, y. An estimate of $\sigma_{Y,x}^2$ can be obtained from the closeness of the data to the least squares line. If the experimental points are far from the least squares line, the estimated variability is larger than that in the case where the experimental points are close to the least squares line. This concept is illustrated in Figure 7.7. If the data exactly fit a straight line, the experiment shows no variability. In real experiments the chance of an exact fit with more than two X, y pairs is very small. An unbiased estimate of $\sigma_{Y,x}^2$ is obtained from the sum of squares of deviations of the observed points from the fitted line as follows:

$$S_{Y,x}^2 = \frac{\sum (y - Y)^2}{N - 2} = \frac{\sum (y - \overline{y})^2 - b^2 [\sum (X - \overline{X})^2]}{N - 2}, \qquad (7.9)$$

where y is the observed value and Y is the predicted value of Y from the least squares line (Y = a + bX) (Fig. 7.7). The variance estimate, $S_{Y,x}^2$, has N − 2 rather than (N − 1) d.f. because two parameters are being estimated from the data (i.e., the slope and intercept).

When $\sigma_{Y,x}^2$ is unknown, the variances of a and b can be estimated, substituting $S_{Y,x}^2$ for $\sigma_{y,x}^2$ in the formulas for the variances [Eqs. (7.7) and (7.8)]. Equations (7.10) and (7.11) are used as the variance estimates, S_a^2 and S_b^2, when testing hypotheses concerning the parameters A and B. This procedure is analogous to using the sample estimate of the variance in the t test to compare sample means.

$$S_a^2 = S_{Y,x}^2 \times \left[\frac{1}{N} + \frac{\overline{X}^2}{\sum (X - \overline{X})^2} \right] \qquad (7.10)$$

$$S_b^2 = \frac{S_{Y,x}^2}{\sum (X - \overline{X})^2} \qquad (7.11)$$

§ a and b are calculated as linear combinations of the normally distributed response variable, y, and thus can be shown to be also normally distributed.

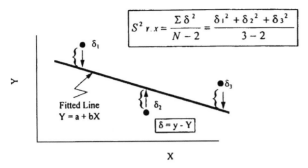

$$S^2{}_{Y.x} = \frac{\Sigma \delta^2}{N-2} = \frac{\delta_1{}^2 + \delta_2{}^2 + \delta_3{}^2}{3-2}$$

δ_1

Y

Fitted Line
Y = a + bX

δ_3

δ_2

$\delta = y - Y$

X

Figure 7.7   Variance calculation from least squares line.

7.4.1   Test of the Intercept, *A*

The background and formulas introduced previously are prerequisites for the construction of tests of hypotheses of the regression parameters A and B. We can now address the question of the "significance" of the Y intercept (a) for the line shown in Figure 7.2(B) and Table 7.2. The procedure is analogous to that of testing means with the t test. In this example, the null hypothesis is $H_0: A = 0$. The alternative hypothesis is $H_a: A \neq 0$. Here the test is two-sided; a priori, if the intercept is not equal to 0, it could be either positive or negative. A t test is performed as shown in Eq. (7.12). $S^2_{Y.x}$ and S^2_a are calculated from Eqs. (7.9) and (7.10), respectively.

$$t_{\text{d.f.}} = t_2 = \frac{|a - A|}{\sqrt{S^2_a}} \tag{7.12}$$

where $t_{\text{d.f.}}$ is the t statistic with $N - 2$ d.f., a is the observed value of the intercept, and A is the hypothetical value of the intercept. From Eq. (7.10)

$$S^2_a = S^2_{Y.x} \times \left[\frac{1}{N} + \frac{\overline{X}^2}{\sum (X - \overline{X})^2} \right]. \tag{7.10}$$

From Eq. (7.9)

$$S^2_{Y.x} = \frac{1698.75 - (0.915)^2(2000)}{2} = 12.15$$

$$S^2_a = 12.15 \left[\frac{1}{4} + \frac{(90)^2}{2000} \right] = 52.245.$$

From Eq. (7.12)

$$t_2 = \frac{|5.9 - 0|}{\sqrt{52.245}} = 0.82.$$

Note that this t test has 2 ($N - 2$) d.f. This is a weak test, and a large intercept must be observed to obtain statistical significance. To define the intercept more precisely, it would be necessary to perform a larger number of assays. If there is no reason to suspect a nonlinear relationship between X and Y, a nonzero intercept, in this example, could be interpreted as being due to some interfering substance(s) in the product (the "blank"). If the presence of a nonzero intercept is suspected, one would probably want to run a sufficient number of assays to establish its presence. A precise estimate of the intercept is necessary if this linear calibration curve is used to evaluate potency.

7.4.2 Test of the Slope, *B*

The test of the slope of the least squares line is usually of more interest than the test of the intercept. Sometimes, we may only wish to be assured that the fitted line has a slope other than zero. (A horizontal line has a slope of zero.) In our example, there seems to be little doubt that the slope is greater than zero [Fig. 7.2(B)]. However, the magnitude of this slope has a special physical meaning. A slope of 1 indicates that the amount recovered (assay) is equal to the amount in the sample, after correction for the blank (i.e., subtract the Y intercept from the observed reading of y). An observation of a slope other than 1 indicates that the amount recovered is some constant percentage of the sample potency. Thus we may be interested in a test of the slope versus 1.

$$H_0 : B = 1 \qquad H_a : B \neq 1$$

A t test is performed using the estimated variance of the slope, as follows:

$$t = \frac{b - B}{\sqrt{S_b^2}}. \tag{7.13}$$

In the present example, from Eq. (7.11),

$$S_b^2 = \frac{S_{y.x}^2}{\sum (X - \overline{X})^2} \tag{7.11}$$

$$= \frac{12.15}{2000} = 0.006075.$$

Applying Eq. (7.13), for a two-sided test, we have

$$t = \frac{|0.915 - 1|}{\sqrt{0.006075}} = 1.09.$$

This t test has 2 $(N - 2)$ d.f. (the variance estimate has 2 d.f.). There is insufficient evidence to indicate that the slope is significantly different from 1 at the 5% level. Table IV.4 shows that a t of 4.30 is needed for significance at $\alpha = 0.05$ and d.f. $= 2$. The test in this example has very weak power. A slope very different from 1 would be necessary to obtain statistical significance. This example again emphasizes the weakness of the statement "nonsignificant," particularly in small experiments such as this one. The reader interested in learning more details of the use and interpretation of regression in analytical methodology is encouraged to read chapter 5 in Ref. [2].

7.5 A DRUG STABILITY STUDY: A SECOND EXAMPLE OF THE APPLICATION OF LINEAR REGRESSION

The measurement of the rate of drug decomposition is an important problem in drug formulation studies. Because of the significance of establishing an expiration date defining the shelf life of a pharmaceutical product, stability data are routinely subjected to statistical analysis. Typically, the drug, alone and/or formulated, is stored under varying conditions of temperature, humidity, light intensity, and so on, and assayed for intact drug at specified time intervals. The pharmaceutical scientist is assigned the responsibility of recommending the expiration date based on scientifically derived stability data. The physical conditions of the stability test (e.g., temperature, humidity), the duration of testing, assay schedules, as well as the number of lots, bottles, and tablets that should be sampled must be defined for stability studies. Careful definition and implementation of these conditions are important because the validity and precision of the final recommended expiration date depends on how the experiment is conducted. Drug stability is discussed further in section 8.7.

The rate of decomposition can often be determined from plots of potency (or log potency) versus storage time, where the relationship of potency and time is either known or assumed to be linear. The current good manufacturing practices (CGMP) regulations [3] state that statistical criteria, including sample size and test (i.e., observation or measurement) intervals for each attribute examined, be used to assure statistically valid estimates of stability (211.166). The expiration date should be "statistically valid" (211.137, 201.17, 211.62).

The mechanics of determining shelf life may be quite complex, particularly if extreme conditions are used, such as those recommended for "accelerated" stability studies (e.g., high-temperature and high-humidity conditions). In these circumstances, the statistical techniques used to make predictions of shelf life at ambient conditions are quite advanced and beyond the scope of this book [4]. Although extreme conditions are commonly used in stability testing in order to save time and obtain a tentative expiration date, all products must eventually be tested for stability under the recommended commercial storage conditions. The FDA has suggested that at least three batches of product be tested to determine an expiration date. One should understand that different batches may show somewhat different stability characteristics, particularly in situations where additives affect stability to a significant extent. In these cases variation in the quality and quantity of the additives (excipients) between batches could affect stability. One of the purposes of using several batches for stability testing is to ensure that stability characteristics are similar from batch to batch.

The time intervals chosen for the assay of storage samples will depend to a great extent on the product characteristics and the anticipated stability. A "statistically" optimal design for a stability study would take into account the planned "storage" times when the drug product will be assayed. This problem has been addressed in the pharmaceutical literature [5]. However, the designs resulting from such considerations are usually cumbersome or impractical. For example, from a statistical point of view, the slope of the potency versus time plot (the rate of decomposition) is obtained most precisely if half of the total assay points are performed at time 0, and the other half at the final testing time. Note that $\sum(X - \overline{X})^2$ the denominator of the expression defining the variance of a slope [Eq. (7.8)], is maximized under this condition, resulting in a minimum variability of the slope. This "optimal" approach to designating assay sampling times is based on the assumption that the plot is linear during the time interval of the test. In a practical situation, one would want to see data at points between the initial and final assay in order to assess the magnitude of the decomposition as the stability study proceeds, as well as to verify the linearity of the decomposition. Also, management and regulatory requirements are better satisfied with multiple points during the course of the study. A reasonable schedule of assays at ambient conditions is 0, 3, 6, 9, 12, 18, and 24 months and at yearly intervals thereafter [6].

The example of the data analysis that will be presented here will be for a single batch. If the stability of different batches is not different, the techniques described here may be applied to data from more than one batch. A statistician should be consulted for the analysis of multibatch data that will require analysis of variance techniques [6,7]. The general approach is described in section 8.7.

Typically, stability or shelf life is determined from data from the first three production batches for each packaging configuration (container type and product strength) (see sect. 8.7). Because such testing may be onerous for multiple strengths and multiple packaging of the same drug product, matrixing and bracketing techniques have been suggested to minimize the number of tests needed to demonstrate suitable drug stability [8].

Assays are recommended to be performed at time 0 and 3, 6, 9, 12, 18 and 24 months, with subsequent assays at 12-month intervals as needed. Usually, three batches of a given strength and package configuration are tested to define the shelf life. Because many products have multiple strengths and package configurations, the concept of a "Matrix" design has been introduced to reduce the considerable amount of testing required. In this situation, a subset of all combinations of product strength, container type and size, and so on is tested at a given time point. Another subset is tested at a subsequent time point. The design should be balanced "such that each combinations of factors is tested to the same extent." All factor combinations should be tested at time 0 and at the last time point of the study. The simplest such design, called a "Basic Matrix 2/3 on Time Design," has two of the three batches tested at each time point, with all three batches tested at time 0 and at the final testing time, the time equal to the desired shelf life. Table 7.3 shows this design for a 36-month product. Tables of matrix designs show

Table 7.3 Matrix Design for Three Packages and Three Strengths

Batch strength		Package 1							Package 2							Package 3						
		3	6	9	12	18	24	36	3	6	9	12	18	24	36	3	6	9	12	18	24	36
1	5	X		X	X		X	X	X	X		X	X		X		X	X		X	X	X
1	10	X	X		X	X		X		X	X		X	X	X	X		X	X		X	X
1	15		X	X		X	X	X	X		X	X		X	X	X	X	X		X		X
2	5	X	X		X	X		X		X	X		X	X	X	X		X	X		X	X
2	10		X	X		X	X	X	X		X	X		X	X	X	X	X		X		X
2	15	X		X	X		X	X	X	X		X	X		X		X	X		X	X	X
3	5		X	X		X	X	X	X		X	X		X	X	X	X	X		X		X
3	10	X		X	X		X	X	X	X		X	X		X		X	X		X	X	X
3	15	X	X		X	X		X		X	X		X	X	X	X		X	X		X	X

Table 7.3A Matrix Design for Three Batches and Two Strengths

Time points for testing (mo)			0	3	6	9	12	18	24	36
S		Batch 1	T	T		T	T		T	T
T	S1	Batch 2	T	T		T	T	T		T
R		Batch 3	T		T		T		T	T
E										
N		Batch 1	T		T		T		T	T
G	S2	Batch 2	T	T		T	T	T		T
T		Batch 3	T		T		T		T	T
H										

designs for multiple packages (made from the same blend or batch) and for multiple packages and strengths. These designs are constructed to be symmetrical in the spirit of optimality for such designs. For example, this is illustrated in Table 7.3, looking only at the "5" strength for Package 1. Table 7.3 shows this design for a 36-month product with multiple packages and strengths (made from the same blend). For example, in Table 7.3, each batch is tested twice, each package from each batch is tested twice, and each package is tested six times at all time points between 0 and 36 months.

With multiple strengths and packages, other similar designs with less testing have been described [9].

The risks of applying such designs are outlined in the Guidance [8]. Because of the limited testing, there is a risk of less precision and shorter dating. If pooling is not allowed, individual lots will have short dating, and combinations not tested in the matrix will not have dating estimates. Read the guidance for further details. The FDA guidance gives examples of other designs.

The analysis of these designs can be complicated. The simplest approach is to analyze each strength and configuration separately, as one would do if there were a single strength and package. Another approach is to model all configurations including interactions. The assumptions, strengths, and limitations of these designs and analyses are explained in more detail in Ref. [9].

A Bracketing design [10] is a design of a stability program such that at any point in time only extreme samples are tested, such as extremes in container size and dosage. This is particularly amenable to products that have similar composition across dosage strengths and that intermediate size and strength products are represented by the extremes [10]. (See also FDA Guideline on Stability for further discussion as to when this is applicable.)

Suppose that we have a product in three strengths and three package sizes. Table 7.4 is an example of a Bracketing design [10].

Table 7.4 Example of Bracketing Design

Strength		Low			Medium			High		
Batch		1	2	3	4	5	6	7	8	9
Container	Small	T	T	T				T	T	T
	Medium									
	Large	T	T	T				T	T	T

Table 7.5 Tablet Assays from the Stability Study

Time, X (mo)	Assay,[a] y (mg)	Average
0	51, 51, 53	51.7
3	51, 50, 52	51.0
6	50, 52, 48	50.0
9	49, 51, 51	50.3
12	49, 48, 47	48.0
18	47, 45, 49	47.0

[a]Each assay represents a different tablet.

The testing designated by T should be the full testing as would be required for a single batch. Note that full testing would require nine combinations, or 27 batches. The matrix design uses four combinations, or 12 batches.

Consider an example of a tablet formulation that is the subject of a stability study. Three randomly chosen tablets are assayed at each of six time periods: 0, 3, 6, 9, 12, and 18 months after production, at ambient storage conditions. The data are shown in Table 7.5 and Figure 7.8.

Given these data, the problem is to establish an expiration date defined as that time when a tablet contains 90% of the labeled drug potency. The product in this example has a label of 50 mg potency and is prepared with a 4% overage (i.e., the product is manufactured with a target weight of 52 mg of drug). Note that FDA is currently discouraging the use of overages to compensate for poor stability.

Figure 7.8 shows that the data are variable. A careful examination of this plot suggests that a straight line would be a reasonable representation of these data. The application of least squares line fitting is best justified in situations where a theoretical model exists showing that the decrease in concentration is linear with time (a zero-order process in this example). The kinetics of drug loss in solid dosage forms is complex and a theoretical model is not easily derived. In the present case, we will assume that concentration and time *are* truly linearly related

$$C = C_0 - Kt, \tag{7.14}$$

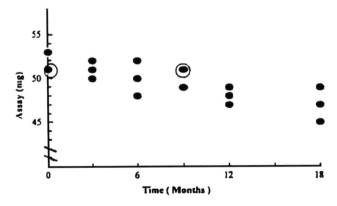

Figure 7.8 Plot of stability data from Table 7.3.

where C is the concentration at time t, C_0 the concentration at time 0 (Y intercept, A), K the rate constant ($-$ slope, $- B$), and t the time (storage time).

With the objective of estimating the shelf life, the simplest approach to the analysis of these data is to estimate the slope and intercept of the least squares line, using Eqs. (7.4) and (7.3). (An interesting exercise would be to first try and estimate the slope and intercept by eye from Fig. 7.8.) When performing the least squares calculation, note that each value of the time (X) is associated with three values of drug potency (y). When calculating C_0 and K, each "time" value is counted three times and N is equal to 18. From Table 7.3,

$$\sum X = 144 \qquad \sum y = 894 \qquad \sum Xy = 6984$$
$$\sum X^2 = 1782 \qquad \sum y^2 = 44,476 \qquad N = 18$$
$$\overline{X} = 8 \qquad \sum (X - \overline{X})^2 = 630 \qquad \sum (y - \overline{y})^2 = 74$$

From Eqs. (7.4) and (7.3), we have

$$b = \frac{N \sum Xy - \sum X \sum y}{N \sum X^2 - (\sum X)^2}$$
$$= \frac{18(6984) - 144(894)}{18(1782) - (144)^2} = \frac{-3024}{11,340} = -0.267 \, \text{mg/month} \tag{7.4}$$

$$a = \overline{y} - b\overline{X}$$
$$= \frac{894}{18} - (-0.267)\frac{144}{18} = 51.80. \tag{7.3}$$

The equation of the straight line best fitting the data in Figure 7.8 is

$$C = 51.8 - 0.267\,t. \tag{7.15}$$

The variance estimate, $S_{Y.x}^2$, represents the variability of tablet potency at a fixed time, and is calculated from Eq. (7.9)

$$S_{Y.x}^2 = \frac{\sum y^2 - (\sum y)^2 / N - b^2 \sum (X - \overline{X})^2}{N - 2}$$
$$= \frac{44,476 - (894)^2 / 18 - (-0.267)^2 (630)}{18 - 2} = 1.825.$$

To calculate the time at which the tablet potency is 90% of the labeled amount, 45 mg, solve Eq. (7.15) for t when C equals 45 mg.

$$45 = 51.80 - 0.267\,t$$
$$t = 25.5 \, \text{month}.$$

The best estimate of the time needed for these tablets to retain 45 mg of drug is 25.5 months (see the point marked with a cross in Fig. 7.9). The shelf life for the product will be less than 25.5 months if variability is taken into consideration. The next section, 7.6, presents a discussion of this topic. This is an average result based on the data from 18 tablets. For any single tablet, the time for decomposition to 90% of the labeled amount will vary, depending, for example, on the amount of drug present at time zero. Nevertheless, the shelf-life estimate is based on the average result.

7.6 CONFIDENCE INTERVALS IN REGRESSION ANALYSIS

A more detailed analysis of the stability data is warranted if one understands that 25.5 months is not the true shelf life, but only an estimate of the true value. A confidence interval for the estimate of time to 45 mg potency would give a range that probably includes the true value.

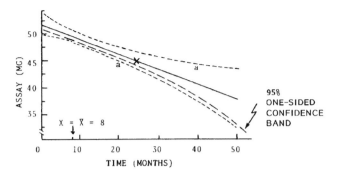

Figure 7.9 95% confidence band for "stability" line.

The concept of a confidence interval in regression is similar to that previously discussed for means. Thus the interval for the shelf life probably contains the true shelf life—that time when the tablets retain 90% of their labeled potency, on the average. The lower end of this confidence interval would be considered a conservative estimate of the true shelf life. Before giving the solution to this problem we will address the calculation of a confidence interval for Y (potency) at a given X (time). The width of the confidence interval for Y (potency) is not constant, but depends on the value of X, since Y is a function of X. In the present example, one might wish to obtain a range for the potency at 25.5 months' storage time.

7.6.1   Confidence Interval for *Y* at a Given *X*

We will construct a confidence interval for the true mean potency (Y) at a given time (X). The confidence interval can be shown to be equal to

$$Y \pm t(S_{Y.x})\sqrt{\frac{1}{N} + \frac{(X - \overline{X})^2}{\sum (X - \overline{X})^2}}. \tag{7.16}$$

t is the appropriate value ($N - 2$ d.f., Table IV.4) for a confidence interval with confidence coefficient P. For example, for a 95% confidence interval, use t values in the column headed 0.975 in Table IV.4.

In the linear regression model, y is assumed to have a normal distribution with variance $\sigma_{Y.x}^2$ at each X. As can be seen from Eq. (7.16), confidence limits for Y at a specified value of X depend on the *variance, degrees of freedom, number of data points* used to fit the line, and $X - \overline{X}$ the *distance of the specified X* (time, in this example) *from* $\overline{X}$, the average time used in the least squares line fitting. The confidence interval is smallest for the Y that corresponds to the value of X equal to $\overline{X}$, [the term, $X - \overline{X}$, in Eq. (7.16) will be zero]. As the value of X is farther from $\overline{X}$, the confidence interval for Y corresponding to the specified X is wider. Thus the estimate of Y is less precise, as the X corresponding to Y is farther away from $\overline{X}$. A plot of the confidence interval for every Y on the line results in a continuous confidence "band" as shown in Figure 7.9. The curved, hyperbolic shape of the confidence band illustrates the varying width of the confidence interval at different values of X, Y. For example, the 95% confidence interval for Y at $X = 25.5$ months [Eq. (7.16)] is

$$45 \pm 2.12(1.35)\sqrt{\frac{1}{18} + \frac{(25.5 - 8)^2}{630}} = 45 \pm 2.1.$$

Thus the result shows that the true value of the potency at 25.5 months is probably between 42.9 and 47.1 mg (45 ± 2.1).

7.6.2 A Confidence Interval for *X* at a Given Value of *Y*

Although the interval for the potency may be of interest, as noted above, this confidence interval does not directly answer the question about the possible variability of the shelf-life estimate. A careful examination of the two-sided confidence band for the line (Fig. 7.9) shows that 90% potency (45 mg) may occur between approximately 20 and 40 months, the points marked "*a*" in Figure 7.9. To obtain this range for *X* (time to 90% potency), using the approach of graphical estimation as described above requires the computation of the confidence band for a sufficient range of *X*. Also, the graphical estimate is relatively inaccurate. The confidence interval for the true *X* at a given *Y* can be directly calculated, although the formula is more complex than that used for the *Y* confidence interval [Eq. (7.16)].

This procedure of estimating *X* for a given value of *Y* is often called "inverse prediction." The complexity results from the fact that the solution for *X*, $X = (Y - a)/b$, is a quotient of variables. $(Y - a)$ and *b* are random variables; both have error associated with their measurement. The ratio has a more complicated distribution than a linear combination of variables such as is the case for $Y = a + bX$. The calculation of the confidence interval for the true *X* at a specified value of *Y* is

$$\frac{(X - g\overline{X}) \pm [t(S_{Y.x})/b]\left[\sqrt{(1 - g)/N + (X - \overline{X})^2/\sum(X - \overline{X})^2}\right]}{1 - g}, \tag{7.17}$$

where

$$g = \frac{t^2(S_{Y.x}^2)}{b^2 \sum(X - \overline{X})^2}$$

t is the appropriate value for a confidence interval with confidence coefficient equal to *P*; for example, for a two-sided 95% confidence interval, use values of *t* in the column headed 0.975 in Table IV.4.

A 95% confidence interval for *X* will be calculated for the time to 90% of labeled potency. The potency is 45 mg (*Y*) when 10% of the labeled amount decomposes. The corresponding time (*X*) has been calculated above as 25.5 months. For a two-sided confidence interval, applying Eq. (7.17), we have

$$g = \frac{(2.12)^2(1.825)}{(-0.267)^2(630)} = 0.183$$

$$X = 25.5 \qquad \overline{X} = 8 \qquad N = 18.$$

The confidence interval is

$$\frac{[25.5 - 0.183(8)] \pm [2.12(1.35)/(-0.267)][\sqrt{0.817/18 + (17.5)^2/630}]}{0.817}$$

$$= 19.8 \text{ to } 39.0 \text{ months.}$$

Thus, using a two-sided confidence interval, the true time to 90% of labeled potency is probably between 19.8 and 39.0 months. A conservative estimate of the shelf life would be the lower value, 19.8 months. If *g* is greater than 1, a confidence interval cannot be calculated because the slope is not significantly greater than 0.

The Food and Drug Administration has suggested that a one-sided confidence interval may be more appropriate than a two-sided interval to estimate the expiration date. For most drug products, drug potency can only decrease with time, and only the lower confidence band of the potency versus time curve may be considered relevant. (An exception may occur in the case of liquid products where evaporation of the solvent could result in an increased potency with time.) The 95% one-sided confidence limits for the time to reach a potency of 45 are computed using Eq. (7.17). Only the lower limit is computed using the appropriate t value that cuts off 5% of the area in a single tail. For 16 d.f., this value is 1.75 (Table IV.4), "g" = 0.1244. The calculation is

$$\frac{[25.5 - 0.1244(8)] + [1.75(1.35)/(-0.267)][\sqrt{0.8756/18 + (17.5)^2/630}}{0.8756}$$

$$= 20.6 \text{ months.}$$

The one-sided 95% interval for X can be interpreted to mean that the time to decompose to a potency of 45 is probably greater than 20.6 months. Note that the shelf life based on the one-sided interval is longer than that based on a two-sided interval (Fig. 7.9).

7.6.3 Prediction Intervals

The confidence limits for Y and X discussed above are limits for the *true values*, having specified a value of Y (potency or concentration, for example) corresponding to some value of X, or an X (time, for example) corresponding to a specified value of Y. An important application of confidence intervals in regression is to obtain confidence intervals for *actual future measurements* based on the least squares line.

1. We may wish to obtain a confidence interval for a value of Y to be actually measured at some value of X (some future time, for example).
2. In the example of the calibration (sect. 7.2), having observed a new value, y, after the calibration line has been established, we would want to use the information from the fitted calibration line to predict the concentration, or potency, X, and establish the confidence limits for the concentration at this newly observed value of y. This is an example of inverse prediction.

For the example of the stability study, we may wish to obtain a confidence interval for an actual assay (y) to be performed at some given future time, after having performed the experiment used to fit the least squares line (case 1 above).

The formulas for calculating a "prediction interval," a confidence interval for a future determination, are similar to those presented in Eqs. (7.16) and (7.17), with one modification. In Eq. (7.16), we add 1 to the sum under the square root portion of the expression. Similarly, for the inverse problem, Eq. (7.17) the expression $(1 - g)/N$ is replaced by $(N + 1)(1 - g)/N$. Thus the prediction interval for Y at a given X is

$$Y \pm t(S_{Y.x})\sqrt{1 + \frac{1}{N} + \frac{(X - \overline{X})^2}{\sum(X - \overline{X})^2}}. \tag{7.18}$$

The prediction interval for X at a specified Y is

$$\frac{(X - g\overline{X}) \pm [t(S)/b]\left[\sqrt{(N+1)(1 - g)/N + (X - \overline{X})^2/\sum(X - \overline{X})^2}\right]}{1 - g}. \tag{7.19}$$

The following examples should clarify the computations. In the stability study example, suppose that one wishes to construct a 95% confidence (prediction) interval for *an assay to be performed* at 25.5 months. (An actual measurement is obtained at 25.5 months.) This interval will be larger than that calculated based on Eq. (7.16), because the uncertainty now includes assay variability for the proposed assay in addition to the uncertainty of the least squares line. Applying Eq. (7.18) ($Y = 45$), we have

$$45 \pm 2.12(1.35)\sqrt{1 + \frac{1}{18} + \frac{17.5^2}{630}} = 45 \pm 3.55 \text{ mg.}$$

In the example of the calibration line, consider an unknown sample that is analyzed and shows a value (y) of 90. A prediction interval for X is calculated using Eq. (7.19). X is predicted to be 91.9 (see sect. 7.2).

$$g = \frac{(4.30)^2(12.15)}{(0.915)^2(2000)} = 0.134$$

$$\frac{[91.9 - 0.134(90)] \pm (4.3)(3.49)/0.915[\sqrt{5(0.866)/4 + (1.9)^2/2000}]}{0.866}$$

$$= 72.5 \text{ to } 111.9.$$

The relatively large uncertainty of the estimate of the true value is due to the small number of data points (four) and the relatively large variability of the points about the least squares line ($S_{Y.x}^2 = 12.15$).

7.6.4 Confidence Intervals for Slope (*B*) and Intercept (*A*)

A confidence interval can be constructed for the slope and intercept in a manner analogous to that for means [Eq. (6.2)]. The confidence interval for the slope is

$$b \pm t(S_b) = b \pm \frac{t(S_{Y.x})}{\sqrt{\sum(X - \overline{X})^2}}. \tag{7.20}$$

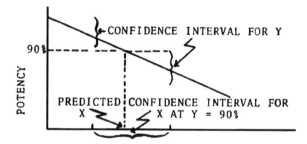

A confidence interval for the intercept is

$$a \pm t(S_a) = a \pm t(S_{Y.x}) \sqrt{\frac{1}{N} + \frac{\overline{X}^2}{\sum(X - \overline{X})^2}}. \tag{7.21}$$

A 95% confidence interval for the slope of the line in the stability example is [Eq. (7.20)]

$$(-0.267) \pm \frac{2.12(1.35)}{\sqrt{630}} = -0.267 \pm 0.114$$

$$= -0.381 \text{ to } -0.153.$$

A 90% confidence interval for the intercept in the calibration line example (sect. 7.2) is [Eq. (7.21)]

$$5.9 \pm 2.93(3.49)\sqrt{\frac{1}{4} + \frac{90^2}{2000}} = 5.9 \pm 21.2 = -15.3 \text{ to } 27.1.$$

(Note that the appropriate value of *t* with 2 d.f. for a 90% confidence interval is 2.93.)

7.7 WEIGHTED REGRESSION

One of the assumptions implicit in the applications of statistical inference to regression procedures is that the variance of *y* be the same at each value of *X*. Many situations occur in practice when this assumption is violated. One common occurrence is the variance of *y* being approximately proportional to X^2. This occurs in situations where *y* has a constant coefficient of variation (CV) and *y* is proportional to *X* ($y = BX$), commonly observed in instrumental methods of analysis in analytical chemistry. Two approaches to this problem are (*a*) a transformation of

Table 7.6 Analytical Data for a Spectrophotometric Analysis

Concentration (X)	Optical density (y)		CV	Weight (w)
5	0.105	0.098	0.049	0.04
10	0.201	0.194	0.025	0.01
25	0.495	0.508	0.018	0.0016
50	0.983	1.009	0.018	0.0004
100	1.964	2.013	0.017	0.0001

y to make the variance homogeneous, such as the log transformation (see chap. 10), and (b) a weighted regression analysis.

Below is an example of weighted regression analysis in which we assume a constant CV and the variance of y proportional to X^2 as noted above. This suggests a weighted regression, weighting each value of Y by a factor that is inversely proportional to the variance, $1/X^2$. Table 7.6 shows data for the spectrophotometric analysis of a drug performed at 5 concentrations in duplicate.

Equation (7.22) is used to compute the slope for the weighted regression procedure.

$$b = \frac{\sum w\,Xy - \sum w\,X\sum wy/\sum w}{\sum w\,X^2 - (\sum w\,X)^2/\sum w}. \tag{7.22}$$

The computations are as follows:

$\sum w = 0.04 + 0.04 + \ldots + 0.0001 + 0.0001 = 0.1042$

$\sum w\,Xy = (0.04)(5)(0.105) + (0.04)(5)(0.098) + \ldots + (0.0001)(100)(1.964) + (0.0001)(100)(2.013)$
$\qquad = 0.19983$

$\sum w\,X = 2(0.04)(5) + 2(0.01)(10) + \ldots + 2(0.0001)(100) = 0.74$

$\sum wy = (0.04)(0.105) + (0.04)(0.098) + \ldots + (0.0001)(1.964) + (0.0001)(2.013) = 0.0148693$

$\sum w\,X^2 = 2(0.04)(5)^2 + 2(0.01)(10)^2 + \ldots + 2(0.0001)(100)^2 = 10$

Therefore, the slope $b =$

$$\frac{0.19983 - (0.74)(0.0148693)/0.1042}{10 - (0.74)^2/0.1042} = 0.01986.$$

The intercept is

$$a = \overline{y}_w - b(\overline{X}_w), \tag{7.23}$$

where $\overline{y}_w = \sum wy/\sum w$ and $\overline{X}_w = \sum w\,X/\sum w$

$$a = 0.0148693/0.1042 - 0.01986(0.74/0.1042) = 0.00166. \tag{7.23a}$$

The weighted least squares line is shown in Figure 7.10.

7.8 ANALYSIS OF RESIDUALS

Emphasis is placed elsewhere in this book on the importance of carefully examining and graphing data prior to performing statistical analyses. The approach to examining data in this context is commonly known as Exploratory Data Analysis (EDA) [11]. One aspect of EDA is the examination of residuals. Residuals can be thought of as deviations of the observed data from

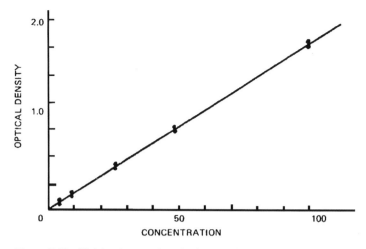

Figure 7.10 Weighted regression plot for data from Table 7.7.

the fit to the statistical model. Examination of residuals can reveal problems such as variance heterogeneity or nonlinearity. This brief introduction to the principle of residual analysis uses the data from the regression analysis in section 7.7.

The residuals from a regression analysis are obtained from the differences between the observed and predicted values. Table 7.7 shows the residuals from an unweighted least squares fit of the data of Table 7.6. Note that the fitted values are obtained from the least squares equation $y = 0.001789 + 0.019874(X)$.

If the linear model and the assumptions in the least squares analysis are valid, the residuals should be approximately normally distributed, and no trends should be apparent.

Figure 7.11 shows a plot of the residuals as a function of X. The fact that the residuals show a fan-like pattern, expanding as X increases, suggests the use of a log transformation or weighting procedure to reduce the variance heterogeneity. In general, the intelligent interpretation of residual plots requires knowledge and experience. In addition to the appearance of patterns in the residual plots that indicate relationships and character of data, outliers usually become obviously apparent [12].

Figure 7.12 shows the residual plot after a log (In) transformation of X and Y. Much of the variance heterogeneity has been removed.

For readers who desire more information on this subject, the book *Graphical Exploratory Data Analysis* [13] is recommended.

Table 7.7 Residuals from Least Squares Fit of Analytical Data (Table 7.6)

	Unweighted			Log transform	
Actual	Predicted value	Residual	Actual	Predicted value	Residual
0.105	0.101	+0.00384	−2.254	−2.298	+0.044
0.201	0.201	+0.00047	−1.604	−1.6073	+0.0033
0.495	0.499	−0.00364	−0.703	−0.695	−0.008
0.983	0.995	−0.0126	−0.017	−0.0004	−0.0166
1.964	1.989	−0.025	+0.675	+0.6863	−0.0113
0.098	0.101	−0.00316	−2.323	−2.298	−0.025
0.194	0.201	−0.00653	−1.640	−1.6073	−0.0033
0.508	0.499	+0.00936	−0.677	−0.6950	+0.018
1.009	0.995	+0.0135	+0.009	−0.0042	+0.0132
2.013	1.989	+0.00238	+0.700	0.6863	+0.0137

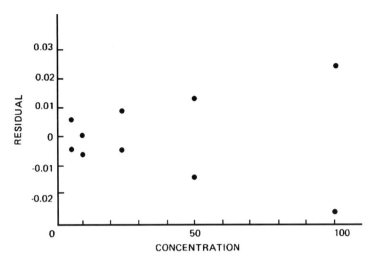

Figure 7.11   Residual plot for unweighted analysis of data of Table 7.6.

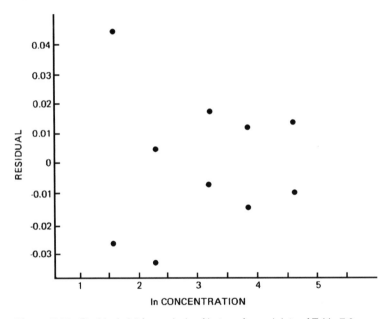

Figure 7.12   Residual plot for analysis of ln transformed data of Table 7.6.

7.9   NONLINEAR REGRESSION**

Linear regression applies to the solution of relationships where the function of Y is linear in the parameters. For example, the equation

$$Y = A + BX$$

is linear in A and B, the parameters. Similarly, the equation

$$Y = A + Be^{-x}$$

is also linear in the parameters. One should also appreciate that a linear equation can exist in more than two dimensions. The equation

$$Y = A + BX + CX^2,$$

an example of a quadratic equation, is linear in the parameters, A, B, and C. These parameters can be estimated by using methods of multiple regression (see App. III and Ref. [1]).

An example of a relationship that is nonlinear in this context is

$$Y = A + e^{BX}.$$

Here the parameter B is not in a linear form.

If a linearizing transformation can be made, then this approach to estimating the parameters would be easiest. For example, the simple first-order kinetic relationship

$$Y = Ae^{-BX}$$

is not linear in the parameters, A and B. However, a log transformation results in a linear equation

$$\ln Y = \ln A - BX.$$

Using the least squares approach, we can estimate $\ln A$ (A is the antilog) and B, where $\ln A$ is the intercept and B is the slope of the straight line when $\ln Y$ is plotted versus X. If statistical tests and other statistical estimates are to be made from the regression analysis, the assumptions of normality of Y (now $\ln Y$) and variance homogeneity of Y at each X are necessary. If Y is normal and the variances of Y at each X are homogeneous to start with, the $\ln$ transformation will invalidate the assumptions. (On the other hand, if Y is lognormal with constant CV, the log transformation will be just what is needed to validate the assumptions.)

Some relationships cannot be linearized. For example, in pharmacokinetics, the one-compartment model with first order absorption and excretion has the following form

$$C = D(e^{-ket} - e^{-kat})$$

where D, ke, and ka are constants (parameters). This equation cannot be linearized. The use of nonlinear regression methods can be used to estimate the parameters in these situations as well as the situations in which Y is normal with homogeneous variance prior to a transformation, as noted above.

The solutions to nonlinear regression problems require more advanced mathematics relative to most of the material in this book. A knowledge of elementary calculus is necessary, particularly the application of Taylor's theorem. Also, a knowledge of matrix algebra is useful in order to solve these kinds of problems. A simple example will be presented to demonstrate the principles. The general matrix solutions to linear and multiple regression will also be demonstrated.

In a stability study, the data in Table 7.8 were available for analysis. The equation representing the degradation process is

$$C = C_0 e^{-kt}. \tag{7.24}$$

Table 7.8 Data from a Stability Study

Time (t)	Concentration mg/L (C)
1 hr	63
2 hr	34
3 hr	22

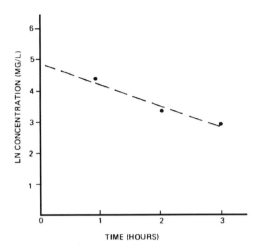

TIME (HOURS)

Figure 7.13   Plot of stability data from Table 7.8.

The concentration values are known to be normal with the variance constant at each value of time. Therefore, the usual least squares analysis will not be used to estimate the parameters C_0 and k after the simple linearizing transformation:

$$\ln C = \ln C_0 - kt.$$

The estimate of the parameters using nonlinear regression as demonstrated here uses the first terms of Taylor's expansion, which approximates the function and results in a linear equation. It is important to obtain good initial estimates of the parameters, which may be obtained graphically. In the present example, a plot of ln C versus time (Fig. 7.13) results in initial estimates of 104 for C_0 and +0.53 for k. The process then estimates a change in C_0 and a change in k that will improve the equation based on the comparison of the fitted data to the original data. Typical of least squares procedures, the fit is measured by the sum of the squares of the deviations of the observed values from the fitted values. The best fit results from an iterative procedure. The new estimates result in a better fit to the data. The procedure is repeated using the new estimates, which results in a better fit than that observed in the previous iteration. When the fit, as measured by the sum of the squares of deviations, is negligibly improved, the procedure is stopped. Computer programs are available to carry out these tedious calculations.

The Taylor expansion requires taking partial derivatives of the function with respect to C_0 and k. For the equation, $C = C_0 e^{-kt}$, the resulting expression is

$$dC = dC_0'(e^{-k't}) - dk'(C_0')(te^{-k't}). \tag{7.25}$$

In Eq. (7.25), dC is the change in C resulting from small changes in C_0 and k evaluated at the point, C_0' and k'. dC_0' is the change in the estimate of C_0, and dk' is the change in the estimate of k. $(e^{-k't})$ and $C_0'(te^{-k't})$ are the partial derivatives of Eq. (7.24) with respect to C_0 and k, respectively.

Equation (7.25) is linear in dC_0' and dk'. The coefficients of dC_0' and dk' are $(e^{-k't})$ and $-(C_0')(te^{-k't})$, respectively. In the computations below, the coefficients are referred to as $X1$ and $X2$, respectively, for convenience. Because of the linearity, we can obtain the least squares estimates of dC_0' and dk' by the usual regression procedures.

The computations for two iterations are shown below. The solution to the least squares equation is usually accomplished using matrix manipulations. The solution for the coefficients can be proven to have the following form:

$$B = (X' X)^{-1}(X'Y).$$

Table 7.9 Results of First Iteration

Time (t)	C	C'	dC	X1	X2
1	63	61.2	1.79	0.5886	−61.2149
2	34	36.0	−2.03	0.3465	−72.0628
3	22	21.2	0.79	0.2039	−63.6248
			$\sum dC^2 = 7.94$		

The matrix B will contain the estimates of the coefficients. With two coefficients, this will be a 2×1 (2 rows and 1 column) matrix.

In Table 7.9, the values of X1 and X2 are $(e^{-k't})$ and $(C_0')(te^{-k't})$, respectively, using the initial estimates of $C_0' = 104$ and $k' = +0.53$ (Fig. 7.13). Note that the fit is measured by the $\sum dC^2 = 7.94$.

The solution of $(X'X)^{-1}(X'Y)$ gives the estimates of the parameters, dC_0' and k'

$$
\begin{vmatrix} X'\,X \end{vmatrix}^{-1} \qquad\qquad \begin{vmatrix} X'\,Y \end{vmatrix}
$$

$$
\begin{vmatrix} 11.5236 & 0.06563 \\ 0.06563 & 0.00045079 \end{vmatrix} \begin{vmatrix} 0.5296 \\ -16.9611 \end{vmatrix} = \begin{vmatrix} 4.99 \\ 0.027 \end{vmatrix}
$$

The new estimates of C_0 and k are

$$C_0' = 104 + 4.99 = 108.99$$
$$k' = 0.53 + 0.027 = +0.557.$$

With these estimates, new values of C' are calculated in Table 7.10.

Note that the $\sum dC'^2$ is 5.85, which is reduced from 7.94, from the initial iteration. The solution of $(X'X)^{-1}(X'Y)$ is

$$
\begin{vmatrix} 12.587 & +0.06964 \\ +0.06964 & 0.0004635 \end{vmatrix} \begin{vmatrix} 0.0351 \\ -0.909 \end{vmatrix} = \begin{vmatrix} 0.378 \\ 0.002 \end{vmatrix}
$$

Therefore, the new estimates of C_0 and k are

$$C_0' = 108.99 + 0.38 = 109.37$$
$$k = 0.557 + 0.002 = 0.559.$$

The reader can verify that the new value of dC'^2 is now 5.74. The process is repeated until dC'^2 becomes stable. The final solution is $C_0 = 109.22$, k 0.558.

Another way of expressing the decomposition is

$$C = e^{\ln C_0 - kt}$$

Table 7.10 Results of Second Iteration

Time (t)	C	C'	dC'	X1	X2
1	63	62.4	0.6	0.5729	−62.4431
2	34	35.8	−1.8	0.3282	−71.5505
3	22	20.5	1.5	0.18806	−61.4896
			$\sum dC^2 = 5.85$		

or

$$\ln C = \ln C_0 - kt.$$

The ambitious reader may wish to try a few iterations using this approach. Note that the partial derivatives of C with respect to C_0 and k are $(1/C_0)(e^{\ln C_0 - kt})$ and $-t(e^{\ln C_0 - kt})$, respectively.

7.10 CORRELATION

Correlation methods are used to measure the "association" of two or more variables. Here, we will be concerned with two observations for each sampling unit. We are interested in determining if the two values are related, in the sense that one variable may be predicted from a knowledge of the other. The better the prediction, the better the correlation. For example, if we could predict the dissolution of a tablet based on tablet hardness, we say that dissolution and hardness are *correlated*. Correlation analysis assumes a linear or *straight-line relationship* between the two variables.

Correlation is usually applied to the relationship of continuous variables, and is best visualized as a *scatter plot* or correlation diagram. Figure 7.14(A) shows a scatter plot for two variables, tablet weight and tablet potency. Tablets were individually weighed and then assayed. Each point in Figure 7.14(A) represents a single tablet (X = weight, Y = potency). Inspection of this diagram suggests that weight and potency are *positively* correlated, as is indicated by the positive slope, or trend. Low-weight tablets are associated with low potencies, and vice versa. This positive relationship would probably be expected on intuitive grounds. If the tablet granulation is homogeneous, a larger weight of material in a tablet would contain larger amounts of drug. Figure 7.14(B) shows the correlation of tablet weights and dissolution rate. Smaller tablet weights are related to higher dissolution rates, a *negative* correlation (negative trend).

Inspection of Figure 7.14(A) and (B) reveals what appears to be an obvious relationship. Given a tablet weight, we can make a good "ballpark" estimate of the dissolution rate and

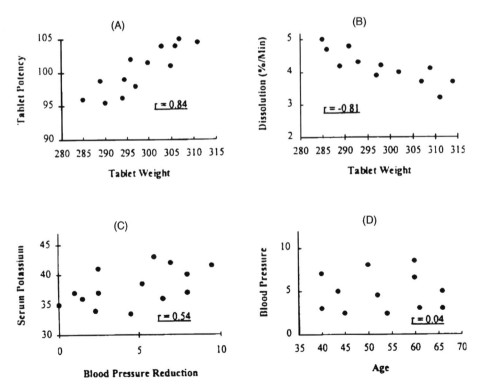

Figure 7.14 Examples of various correlation diagrams or scatter plots. The correlation coefficient, r, is defined in section 7.10.1.

potency. However, the relationship between variables is not always as apparent as in these examples. The relationship may be partially obscured by variability, or the variables may not be related at all. The relationship between a patient's blood pressure reduction after treatment with an antihypertensive agent and serum potassium levels is not as obvious [Fig. 7.14(C)]. There seems to be a trend toward higher blood pressure reductions associated with higher potassium levels—or is this just an illusion? The data plotted in Figure 7.14(D), illustrating the correlation of blood pressure reduction and age, show little or no correlation.

The various scatter diagrams illustrated in Figure 7.14 should give the reader an intuitive feeling for the concept of correlation. There are many experimental situations where a researcher would be interested in relationships among two or more variables. Similar to applications of regression analysis, correlation relationships may allow for prediction and interpretation of experimental mechanisms. Unfortunately, the concept of correlation is often misused, and more is made of it than is deserved. For example, the presence of a strong correlation between two variables does not necessarily imply a causal relationship. Consider data that show a positive relationship between cancer rate and consumption of fluoridated water. Regardless of the possible validity of such a relationship, such an observed correlation does not necessarily imply a causal effect. One would have to investigate further other factors in the environment occurring concurrently with the implementation of fluoridation, which may be responsible for the cancer rate increase. Have other industries appeared and grown during this period, exposing the population to potential carcinogens? Have the population characteristics (e.g., racial, age, sex, economic factors) changed during this period? Such questions may be resolved by examining the cancer rates in control areas where fluoridation was not enforced.

The correlation coefficient is a measure of the "degree" of correlation, which is often *erroneously* interpreted as a measure of "linearity." That is, a strong correlation is sometimes interpreted as meaning that the relationship between X and Y is a straight line. As we shall see further in this discussion, this interpretation of correlation is not necessarily correct.

7.10.1 Correlation Coefficient

The correlation coefficient is a quantitative measure of the relationship or correlation between two variables.

$$\text{Correlation coefficient} = r = \frac{\sum (X - \overline{X})(y - \overline{y})}{\sqrt{\sum (X - \overline{X})^2 \sum (y - \overline{y})^2}}. \tag{7.26}$$

A shortcut computing formula is

$$r = \frac{N \sum Xy - \sum X \sum y}{\sqrt{\left[N \sum X^2 - \left(\sum X \right)^2 \right] \left[N \sum y^2 - \left(\sum y \right)^2 \right]}}, \tag{7.27}$$

where N is the number of X, y pairs.

The correlation coefficient, r, may be better understood by its relationship to $S_{Y.x}^2$, the variance calculated from regression line fitting procedures. r^2 represents the relative reduction in the sum of squares of the variable y resulting from the fitting of the X, y line. For example, the sum of squares $\left[\sum (y - \overline{y})^2 \right]$ for the y values 0, 1, and 5 is equal to 14 [see Eq. (1.4)].

$$\sum (y - \overline{y})^2 = 0^2 + 1^2 + 5^2 - \frac{(0 + 1 + 5)^2}{3} = 14.$$

If these same y values were associated with X values, the sum of squares of y from the regression of y and X will be *equal to or less than* $\sum (y - \overline{y})^2$, or 14 in this example. Suppose that X and y values are as follows (Fig. 7.15):

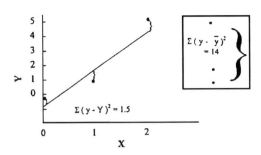

Figure 7.15 Reduction in sum of squares due to regression.

	X	y	Xy	
	0	0	0	$\sum (X - \overline{X})^2 = 2$
	1	1	1	
	2	5	10	$\sum (y - \overline{y})^2 = 14$
Sum	3	6	11	

According to Eq. (7.9), the sum of squares due to deviations of the y values from the regression line is

$$\sum (y - \overline{y})^2 - b^2 \sum (X - \overline{X})^2, \tag{7.28}$$

where b is the slope of the regression line (y on X). The term $b^2 \sum (X - \overline{X})^2$ is the reduction in the sum of squares due to the straight-line regression fit. Applying Eq. (7.28), the sum of squares is

$$14 - (2.5)^2(2) = 14 - 12.5 = 1.5 \quad \text{(the slope, } b, \text{ is 2.5).}$$

r^2 is the relative reduction of the sum of squares

$$\frac{14 - 1.5}{14} = 0.893 \qquad r = \sqrt{0.893} = 0.945.$$

The usual calculation of r, according to Eq. (7.27), is as follows:

$$\frac{3(11) - (3)(6)}{\sqrt{[3(5) - (3)^2][3(26) - (36)]}} = \frac{15}{\sqrt{6(42)}} = 0.945.$$

Thus, according to this notion, r can be interpreted as the relative degree of scatter about the regression line. If X and y values lie exactly on a straight line (a perfect fit), $S^2_{Y,x}$ is 0, and r is equal to ± 1; $+1$ for a line of positive slope and -1 for a line of negative slope. For a correlation coefficient equal to 0.5, $r^2 = 0.25$. The sum of squares for y is reduced 25%. A correlation coefficient of 0 means that the X, y pairs are not correlated [Fig. 7.14(D)].

Although there are no assumptions necessary to calculate the correlation coefficient, statistical analysis of r is based on the notion of a bivariate normal distribution of X and y. We will not delve into the details of this complex probability distribution here. However, there are two interesting aspects of this distribution that deserve some attention with regard to correlation analysis.

1. In typical correlation problems, *both X and y are variable*. This is in contrast to the linear regression case, where X is considered *fixed*, chosen, a priori, by the investigator.

2. In a bivariate normal distribution, X and y are linearly related. The regression of both X on y and y on X is a straight line.[¶] Thus, when statistically testing correlation coefficients, we are not testing for linearity. As described below, the statistical test of a correlation coefficient is a test of correlation or independence. According to Snedecor and Cochran, the correlation coefficient "estimates the degree of *closeness* of a linear relationship between two variables, Y and X, and the meaning of this concept is not easy to grasp" [11].

7.10.2 Test of Zero Correlation

The correlation coefficient is a rough measure of the degree of association of two variables. The degree of association may be measured by how well one variable can be predicted from another; the closer the correlation coefficient is to $+1$ or -1, the better the correlation, the better the predictive power of the relationship. A question of particular importance from a statistical point of view is whether or not an observed correlation coefficient is "real" or due to chance. If two variables from a bivariate normal distribution are uncorrelated (independent), the correlation coefficient is 0. Even in these cases, in actual experiments, random variation will result in a correlation coefficient different from zero. Thus, it is of interest to test an observed correlation coefficient, r, versus a hypothetical value of 0. This test is based on an assumption that y is a normal variable [11]. The test is a t test with $(N - 2)$ d.f., as follows:

$$H_0 : \rho = 0 \quad H_a : \rho \neq 0,$$

where ρ is the true correlation coefficient, estimated by r.

$$t_{N-2} = \frac{|r\sqrt{N - 2}|}{\sqrt{1 - r^2}}. \tag{7.29}$$

The value of t is referred to a t distribution with $(N - 2)$ d.f., where N is the sample size (i.e., the number of pairs). Interestingly, this test is identical to the test of the slope of the least squares fit, $Y = a + bX$ [Eq. (7.13)]. In this context, one can think of the test of the correlation coefficient as a test of the significance of the slope versus 0.

To illustrate the application of Eq. (7.29), Table 7.11 shows data of diastolic blood pressure and cholesterol levels of 10 randomly selected men. The data are plotted in Figure 7.16. r is calculated from Eq. (7.27)

$$r = \frac{N \sum Xy - \sum X \sum y}{\sqrt{\left[N \sum X^2 - (\sum X)^2\right]\left[N \sum y^2 - (\sum y)^2\right]}}$$

$$= \frac{10(260,653) - (3111)(825)}{\sqrt{[10(987,893) - 3111^2][10(69,279) - 825^2]}} = 0.809. \tag{7.30}$$

r is tested for significance using Eq. (7.29).

$$t_8 = \frac{\left|0.809\sqrt{8}\right|}{\sqrt{1 - (0.809)^2}} = 3.89.$$

A value of t equal to 2.31 is needed for significance at the 5% level (see Table IV.4). Therefore, the correlation between diastolic blood pressure and cholesterol is significant. The correlation is apparent from inspection of Figure 7.16.

[¶] The regression of y on X means that X is assumed to be the fixed variable when calculating the line. This line is different from that calculated when Y is considered the fixed variable (unless the correlation coefficient is 1, when both lines are identical). The slope of the line is $r\,S_y/S_x$ for the regression of y on X and $r\,S_x/S_y$ for x on Y.

Table 7.11   Diastolic Blood Pressure and Serum Cholesterol of 10 Persons

Person	Diastolic blood pressure (DBP), y	Cholesterol (C), X	Xy
1	80	307	24,560
2	75	259	19,425
3	90	341	30,690
4	74	317	23,458
5	75	274	20,550
6	110	416	45,760
7	70	267	18,690
8	85	320	27,200
9	88	274	24,112
10	78	336	26,208
	$\sum y = 825$	$\sum X = 3111$	$\sum Xy = 260,653$
	$\sum y^2 = 69,279$	$\sum X^2 = 987,893$	

Significance tests for the correlation coefficient versus values other than 0 are not very common. However, for these tests, the t test described above [Eq. (7.29)] should not be used. An approximate test is available to test for correlation coefficients other than 0 (e.g., H_0: $\rho = 0.5$). Since applications of this test occur infrequently in pharmaceutical experiments, the procedure will not be presented here. The statistical test is an approximation to the normal distribution, and the approximation can also be used to place confidence intervals on the correlation coefficient. A description of these applications is presented in Ref. [11].

7.10.3   Miscellaneous Comments

Before leaving the topic of correlation, the reader should once more be warned about the potential misuses of interpretations of correlation and the correlation coefficient. In particular, the association of high correlation coefficients with a "cause and effect" and "linearity" is not necessarily valid. Strong correlation *may* imply a direct causal relationship, but the nature of the measurements should be well understood before firm statements can be made about cause and effect. One should be keenly aware of the common occurrence of spurious correlations due to indirect causes or remote mechanisms.

The correlation coefficient does not test the linearity of two variables. If anything, it is more related to the slope of the line relating the variables. Linearity is assumed for the routine statistical test of the correlation coefficient. As has been noted above, the correlation coefficient measures the degree of correlation, a measure of the variability of a predictive relationship.

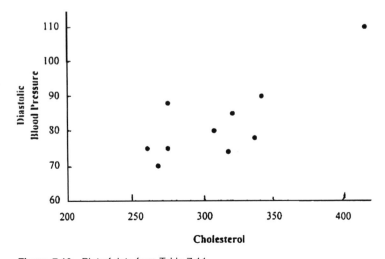

Figure 7.16   Plot of data from Table 7.11.

Table 7.12 Two Data Sets Illustrating Some Problems of Interpreting Correlation Coefficients

Set A		Set B	
X	y	X	y
−2	0	0	0
−1	3	2	4
0	4	4	16
+1	3	6	36
+2	0		

A proper test for linearity (i.e., do the data represent a straight-line relationship between X and Y?) is described in Appendix II and requires replicate measurements in the regression model. Usually, correlation problems deal with cases where both variables, X and y, are variable in contrast to the regression model where X is considered fixed. In correlation problems, the question of linearity is usually not of primary interest. We are more interested in the degree of association of the variables. Two examples will show that a high correlation coefficient does not necessarily imply "linearity" and that a small correlation coefficient does not necessarily imply lack of correlation (if the relationship is nonlinear).

Table 7.12 shows two sets of data that are plotted in Figure 7.17. Both data sets A and B show perfect (but nonlinear) relationships between X and y. Set A is defined by $Y = 4 - X^2$. Set B is defined by $Y = X^2$. Yet the correlation coefficient for set A is 0, an implication of no correlation, and set B has a correlation coefficient of 0.96, very strong correlation (*but not linearity!*). These examples should emphasize the care needed in the interpretation of the correlation coefficient, particularly in nonlinear systems.

Another example of data for which the correlation coefficient can be misleading is shown in Table 7.13 and Figure 7.18. In this example, drug stability is plotted versus pH. Five experiments were performed at low pH and one at high pH. The correlation coefficient is 0.994, a highly significant result ($p < 0.01$). Can this be interpreted that the data in Figure 7.18 are a good fit to a straight line? Without some other source of information, it would take a great deal of imagination to assume that the relationship between pH and $t_{1/2}$ is linear over the range of pH equal to 2.0 to 5.5. Even if the relationship were linear, had data been available for points in between pH 2.0 and 5.5, the fit may not be as good as that implied by the large value of r in this example. This situation can occur when one value is far from the cluster of the main body of data. One should be cautious in "over-interpreting" the correlation coefficient in these cases. When relationships between variables are to be quantified for predictive or theoretical reasons, regression procedures, if applicable, are recommended. Correlation, per se, is not as versatile or informative as regression analysis for describing the relationship between variables.

7.11 COMPARISON OF VARIANCES IN RELATED SAMPLES

In section 5.3, a test was presented to compare variances from two independent samples. If the samples are related, the simple F test for two independent samples is not valid [11]. Related, or

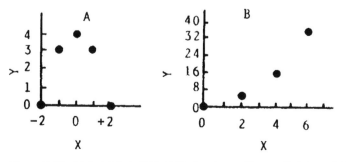

Figure 7.17 Plot of data in Table 7.12 showing problems with interpretation of the correlation coefficient.

Table 7.13   Data to Illustrate a Problem that Can Result
in Misinterpretation of the Correlation Coefficient

pH	Stability, $t_{1/2}$ (wk)
2.0	48
2.1	50
1.9	50
2.0	46
2.1	47
5.5	12

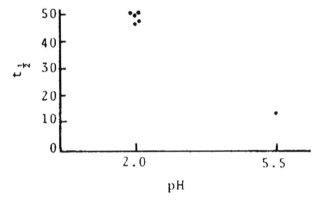

Figure 7.18   Plot of data from Table 7.10.

paired-sample tests arise, for example, in situations where the same subject tests two treatments, such as in clinical or bioavailability studies. To test for the equality of variances in related samples, we must first calculate the correlation coefficient and the F ratio of the variances. The test statistic is calculated as follows:

$$r_{ds} = \frac{F - 1}{\sqrt{(F + 1)^2 - 4r^2 F}}, \qquad (7.31)$$

where F is the ratio of the variances in the two samples and r is the correlation coefficient.

The ratio in Eq. (7.30), r_{ds}, can be tested for significance in the same manner as the test for the ordinary correlation coefficient, with $(N - 2)$ d.f., where N is the number of pairs [Eq. (7.29)]. As is the case for tests of the correlation coefficient, we assume a bivariate normal distribution for the related data. The following example demonstrates the calculations.

In a bioavailability study, 10 subjects were given each of two formulations of a drug substance on two occasions, with the results for AUC (area under the blood level versus time curve) given in Table 7.14.

The correlation coefficient is calculated according to Eq. (7.27).

$$r = \frac{(64,421)(10) - (781)(815)}{\sqrt{[(62,821)(10) - (781)^2][(67,087)(10) - (815)^2]}} = 0.699.$$

The ratio of the variances (Table 7.14), F, is

$$\frac{202.8}{73.8} = 2.75.$$

[Note: The ratio of the variances may also be calculated as $73.8/202.8 = 0.36$, with the same conclusions based on Eq. (7.31).]

Table 7.14 AUC Results of the Bioavailability Study (A vs. B)

Subject	Formulation	
	A	B
1	88	88
2	64	73
3	69	86
4	94	89
5	77	80
6	85	71
7	60	70
8	105	96
9	68	84
10	73	78
Mean	78.1	81.5
S^2	202.8	73.8

The test statistic, r_{ds}, is calculated from Eq. (7.31).

$$r_{ds} = \frac{2.75 - 1}{\sqrt{(2.75 + 1)^2 - 4(0.699)^2(2.75)}} = 0.593,$$

r_{ds} is tested for significance using Eq. (7.29).

$$t_8 = \frac{\left|0.593\sqrt{8}\right|}{\sqrt{1 - 0.593^2}} = 2.08.$$

Referring to the t table (Table IV.4, 8 d.f.), a value of 2.31 is needed for significance at the 5% level. Therefore, we cannot reject the null hypothesis of equal variances in this example. Formulation A appears to be more variable, but more data would be needed to substantiate such a claim.

A discussion of correlation of multiple outcomes and adjustment of the significance level is given in section 8.2.2.

KEY TERMS

Best-fitting line
Bivariate normal distribution
Confidence band for line
Confidence interval for X and Y
Correlation
Correlation coefficient
Correlation diagram
Dependent variable
Fixed value (X)
Independence
Independent variable
Intercept
Inverse prediction
Lack of fit
Linear regression
Line through the origin

Nonlinear regression
Nonlinearity
One-sided confidence interval
Prediction interval
Reduction of sum of squares
Regression
Regression analysis
Residuals
Scatter plot
Simple linear regression
Slope
$S^2_{Y,x}$
Trend
Variance of correlated samples
Weighted regression

EXERCISES

1. A drug seems to decompose in a manner such that appearance of degradation products is linear with time (i.e., $C_d = kt$).

t	C_d
1	3
2	9
3	12
4	17
5	19

   (a) Calculate the slope (k) and intercept from the least squares line.
   (b) Test the significance of the slope (test vs. 0) at the 5% level.
   (c) Test the slope versus 5 ($H_0: B = 5$) at the 5% level.
   (d) Put 95% confidence limits on C_d at $t = 3$ and $t = 5$.
   (e) Predict the value of C_d at $t = 20$. Place a 95% prediction interval on C_d at $t = 20$.
   (f) If it is known that $C_d = 0$ at $t = 0$, calculate the slope.

2. A Beer's law plot is constructed by plotting ultraviolet absorbance versus concentration, with the following results:

Concentration, X	Absorbance, y	Xy
1	0.10	0.10
2	0.36	0.72
3	0.57	1.71
5	1.09	5.45
10	2.05	20.50

   (a) Calculate the slope and intercept.
   (b) Test to see if the intercept is different from 0 (5% level). How would you interpret a significant intercept with regard to the actual physical nature of the analytical method?
   **(c) An unknown has an absorbance of 1.65. What is the concentration? Put confidence limits on the concentration (95%).

3. Five tablets were weighed and then assayed with the following results:

Weight (mg)	Potency (mg)
205	103
200	100
202	101
198	98
197	98

   (a) Plot potency versus weight (weight $= X$). Calculate the least squares line.
   (b) Predict the potency for a 200-mg tablet.
   (c) Put 95% confidence limits on the potency for a 200-mg tablet.

**This is a more advanced topic.

4. Tablets were weighed and assayed with the following results:

Weight	Assay	Weight	Assay
200	10.0	198	9.9
205	10.1	200	10.0
203	10.0	190	9.6
201	10.1	205	10.2
195	9.9	207	10.2
203	10.1	210	10.3

(a) Calculate the correlation coefficient.
(b) Test the correlation coefficient versus 0 (5% level).
(c) Plot the data in the table (scatter plot).

5. Tablet dissolution was measured in vitro for 10 generic formulations. These products were also tested in vivo. Results of these studies showed the following time to 80% dissolution and time to peak (in vivo).

Formulation	Time to 80% dissolution (min)	$T_{p(hr)}$
1	17	0.8
2	25	1.0
3	15	1.2
4	30	1.5
5	60	1.4
6	24	1.0
7	10	0.8
8	20	0.7
9	45	2.5
10	28	1.1

Calculate r and test for significance (versus 0) (5% level). Plot the data.

6. Shah et al. [14] measured the percent of product dissolved in vitro and the time to peak (in vivo) of nine phenytoin sodium products, with approximately the following results:

Product	Time to peak (hr)	Percentage dissolved in 30 min
1	6	20
2	4	60
3	2.5	100
4	4.5	80
5	5.1	35
6	5.7	35
7	3.5	80
8	5.7	38
9	3.8	85

Plot the data. Calculate the correlation coefficient and test to see if it is significantly different from 0 (5% level). (Why is the correlation coefficient negative?)

7. In a study to compare the effects of two pain-relieving drugs (A and B), 10 patients took each drug in a paired design with the following results (drug effectiveness based on a rating scale).

Patient	Drug A	Drug B
1	8	6
2	5	4
3	5	6
4	2	5
5	4	5
6	7	4
7	9	6
8	3	7
9	5	5
10	1	4

Are the drug effects equally variable?

8. Compute the intercept and slope of the least squares line for the data of Table 7.6 after a In transformation of both X and Y. Calculate the residuals and compare to the data in Table 7.7.

9. In a drug stability study, the following data were obtained:

Time (months)	Concentration (mg)
0	2.56
1	2.55
3	2.50
9	2.44
12	2.40
18	2.31
24	2.25
36	2.13

(a) Fit a least squares line to the data.
(b) Predict the time to decompose to 90% of label claim (2.25 mg).
(c) Based on a two-sided 95% confidence interval, what expiration date should be applied to this formulation?
(d) Based on a one-sided 95% confidence interval, what expiration date should be applied to this formulation?

††10. Fit the following data to the exponential $y = e^{ax}$. Use nonlinear least squares.

x	y
1	1.62
2	2.93
3	4.21
4	7.86

REFERENCES

1. Draper NR, Smith H. Applied Regression Analysis, 2nd ed. New York: Wiley, 1981.
2. Youden WJ. Statistical Methods for Chemists. New York: Wiley, 1964.
3. U.S. Food and Drug Administration. Current Good Manufacturing Practices (CGMP) 21 CFR. Washington, DC: Commissioner of the Food and Drug Administration, 2006:210–229.

††This is an optional, more difficult problem.

4. Davies OL, Hudson HE. Stability of drugs: accelerated storage tests. In: Buncher CR, Tsay J-Y, eds. Statistics in the Pharmaceutical Industry. New York: Marcel Dekker, 1994:445–479.

5. Tootill JPR. A critical appraisal of drug stability testing methods. J Pharm Pharmacol 1961; 13(suppl): 75T–86T.

6. Davis J. The Dating Game. Washington, DC: Food and Drug Administration, 1978.

7. Norwood TE. Statistical analysis of pharmaceutical stability data. Drug Dev Ind Pharm 1986; 12:553–560.

8. International Conference on Harmonization Bracketing and matrixing designs for stability testing of drug substances and drug products (FDA Draft Guidance) Step 2, Nov 9, 2000.

9. Nordbrock ET. Stability matrix designs. In: Chow S-C, ed. Encyclopedia of Pharmaceutical Statistics. New York: Marcel Dekker, 2000:487–492.

10. Murphy JR. Bracketing Design. In: Chow S-C, ed. Encyclopedia of Pharmaceutical Statistics. New York: Marcel Dekker, 2000:77.

11. Snedecor GW, Cochran WG. Statistical Methods, 8th ed. Ames, IA: Iowa State University Press, 1989.

12. Weisberg S. Applied Linear Regression. New York: Wiley, 1980.

13. duToit SHC, Steyn AGW, Stumpf RH. Graphical Exploratory Data Analysis. New York: Springer, 1986.

14. Shah VP, Prasad VK, Alston T, et al. In vitro in vivo correlation for 100 mg phenytoin sodium capsules. J Pharm Sci 1983; 72:306.

8 | Analysis of Variance

Analysis of variance, also known as *ANOVA*, is perhaps the most powerful statistical tool. ANOVA is a general method of analyzing data from designed experiments, whose objective is to *compare two or more group means*. The t test is a special case of ANOVA in which only two means are compared. By *designed experiments*, we mean experiments with a particular structure. Well-designed experiments are usually optimal with respect to meeting study objectives. The statistical analysis depends on the design, and the discussion of ANOVA therefore includes common statistical designs used in pharmaceutical research. ANOVA designs can be more or less complex. The designs can be very simple, as in the case of the t-test procedures presented in chapter 5. Other designs can be quite complex, sometimes depending on computers for their solution and analysis. As a rule of thumb, one should use the simplest design that will achieve the experimental objectives. This is particularly applicable to experiments otherwise difficult to implement, such as is the case in clinical trials.

8.1   ONE-WAY ANOVA

An elementary approach to ANOVA may be taken using the two independent groups t test as an example. This is an example of one-way ANOVA, also known as a "completely randomized" design. (Certain simple "parallel-groups" designs in clinical trials correspond to the one-way ANOVA design.) In the t test, the two treatments are assigned at random to different independent experimental units. In a clinical study, the t test is appropriate when two treatments are randomly assigned to different patients. This results in two groups, each group representing one of the two treatments. One-way ANOVA is used when we wish to test the equality of treatment means in experiments where two or more treatments are randomly assigned to different, independent experimental units. The typical null hypothesis is H_0: $\mu_1 = \mu_2 = \mu_3$, where μ_1 refers to treatment 1, and so on.

Suppose that 15 tablets are available for the comparison of three assay methods, 5 tablets for each assay. The one-way ANOVA design would result from a random assignment of the tablets to the three groups. In this example, five tablets are assigned to each group. Although this allocation (five tablets per group) is optimal with regard to the precision of the comparison of the three assay methods, it is not a necessary condition for this design. The number of tablets analyzed by each analytical procedure need not be equal for the purposes of comparing the mean results. However, one can say, in general, that symmetry is a desirable feature in the design of experiments. This will become more apparent as we discuss various designs. In the one-way ANOVA, symmetry can be defined as an equal number of experimental units in each treatment group.

We will pursue the example above to illustrate the ANOVA procedure. Five replicate tablets are analyzed in each of the three assay method groups, one assay per tablet. Thus we assay the 15 tablets, five tablets by each method, as shown in Table 8.1. If only two assay methods were to be compared, we could use a t test to compare the means statistically. If more than two assay methods are to be compared, the correct statistical procedure to compare the means is the one-way ANOVA.

ANOVA is a technique of separating the total variability in a set of data into component parts, represented by a statistical model. In the simple case of the one-way ANOVA, the model is represented as

$$Y_{ij} = \mu + G_i + \varepsilon_{ij}, \tag{8.1}$$

Table 8.1   Results of Assays Comparing Three
Analytical Methods

Method A	Method B	Method C
102	99	103
101	100	100
101	99	99
100	101	104
102	98	102
$\overline{X}$ 101.2	99.4	101.6
s.d. 0.84	1.14	2.07

where Y_{ij} is the jth response in treatment group i (e.g., $i = 3$, $j = 2$, second tablet in third group), G_i the deviation of the ith treatment (group) mean from the overall mean, μ; ε_{ij} the random error in the experiment (measurement error, biological variability, etc.) assumed to be normal with mean 0 and variance σ^2.

The model says that the response is a function of the true treatment mean ($\mu + G_i$) and a random error that is normally distributed, with mean zero and variance σ^2. In the case of a clinical study, $G_i + \mu$ is the true average of treatment i. If a patient is treated with an antihypertensive drug whose true mean effect is a 10-mm Hg reduction in blood pressure, then $Y_{ij} = 10 + \varepsilon_{ij}$, where Y_{ij} is the jth observation among patients taking the drug i. (Note that if treatments are identical, G_i is the same for all treatments.) The error, ε_{ij}, is a normally distributed variable, identically distributed for all observations. It is composed of many factors, including interindividual variation and measurement error. Thus the observed experimental values will be different for different people, a consequence of the nature of the assigned treatment and the random error, ε_{ij} (e.g., biological variation). Section 8.5 expands the concept of statistical models.

In addition to the assumption that the error is normal with mean 0 and variance σ^2, the errors must be independent. This is a very important assumption in the ANOVA model. The fact that the error has mean 0 means that some people will show positive deviations from the treatment mean, and others will show negative deviations; but on the average, the deviation is zero.

As in the t test, statistical analysis and interpretation of the ANOVA is based on the following assumptions:

1. The errors are normal with constant variance.
2. The errors (or observations) are independent.

As will be discussed below, ANOVA separates the variability of the data into parts, comparing that due to treatments to that due to error.

8.1.1   Computations and Procedure for One-Way ANOVA

ANOVA for a one-way design separates the variance into two parts, that due to *treatment differences* and that due to *error*. It can be proven that the *total sum of squares* (the squared deviations of each value from the overall mean)

$$\sum (Y_{ij} - \overline{Y})^2$$

is equal to

$$\sum (Y_{ij} - \overline{Y}_i)^2 + \sum N_i (\overline{Y}_i - \overline{Y})^2, \tag{8.2}$$

where $\overline{Y}$ is the overall mean and $\overline{Y}_i$ is the mean of the ith group. N_i is the number of observations in treatment group i. The first term in expression (8.2) is called the *within* sum of squares, and the second term is called the *between* sum of squares.

Table 8.2 Sample Data to Illustrate Eq. (8.2)

Group I (Y_{1j})	Group II (Y_{2j})	Group III (Y_{3j})
0	2	6
2	4	10
$\overline{Y}_t$ 1	3	8

$\overline{Y} = (1 + 3 + 8)/3 = (0 + 2 + 2 + 4 + 6 + 10)/6 = 4$

A simple example to demonstrate the equality in Eq. (8.2) is shown below, using the data of Table 8.2.

$$\sum (Y_{ij} - \overline{Y})^2 = \sum Y^2 - \frac{(\sum Y)^2}{N} = 160 - \frac{(24)^2}{6} = 64$$

$$\sum (Y_{ij} - \overline{Y}_i)^2 = (0 - 1)^2 + (2 - 1)^2 + (2 - 3)^2 + (4 - 3)^2 + (6 - 8)^2$$
$$= (10 - 8)^2 = 2 + 2 + 8 = 12$$

$$\sum N_i(\overline{Y}_i - \overline{Y})^2 = 2(1 - 4)^2 + 2(3 - 4)^2 + 2(8 - 4)^2 = 52.$$

Thus, according to Eq. (8.2), $64 = 12 + 52$.

The calculations for the analysis make use of simple arithmetic with shortcut formulas for the computations similar to that used in the t-test procedures. Computer programs are available for the analysis of all kinds of ANOVA designs from the most simple to the most complex. In the latter cases, the calculations can be very extensive and tedious, and use of computers may be almost mandatory. For the one-way design, the calculations pose no difficulty. In many cases, use of a pocket calculator will result in a quicker answer than can be obtained using a less accessible computer. A description of the calculations, with examples, is presented below.

The computational process consists first of obtaining the *sum of squares* (SS) for all of the data.

$$\text{Total sum of squares (SS)} = \sum (Y_{ij} - \overline{Y})^2. \tag{8.3}$$

The *total sum of squares* is divided into two parts: (a) the SS due to treatment differences (*between-treatment sum of squares*), and (b) the error term derived from the *within-treatment sum of squares*. The within-treatment sum of squares (within SS) divided by the appropriate degrees of freedom is the *pooled variance*, the same as that obtained in the t test for the comparison of two treatment groups. The ratio of the between-treatment mean square to the within-treatment mean square is a measure of treatment differences (see below).

To illustrate the computations, we will use the data from Table 8.1, a comparison of three analytical methods with five replicates per method. Remember that the objective of this experiment is to compare the average results of the three methods. We might think of method A as the standard, accepted method, and methods B and C as modifications of the method, meant to replace method A. As in the other tests of hypotheses described in chapter 5, we first state the null and alternative hypotheses as well as the significance level, prior to the experiment. For example, in the present case,*

$$H_0: \mu_A = \mu_B = \mu_C \qquad H_a: \mu_i \neq \mu_j \qquad \text{for any two means*}$$

---

* Alternatives to H_0 may also include more complicated comparisons than $\mu_i \neq \mu_j$; see example, section 8.2.1.

1. First, calculate the *total sum of squares* (total SS or TSS). Calculate $\sum(Y_{ij} - \overline{Y})^2$ [Eq. (8.3)] using all of the data, ignoring the treatment grouping. This is most easily calculated using the shortcut formula

$$\sum Y^2 - \frac{(\sum Y)^2}{N},$$ (8.4)

where $(\sum Y)^2$ is the grand total of all of the observations squared, divided by the total number of observations N, and is known as the *correction term, CT*. As mentioned in chapter 1, the correction term is commonly used in statistical calculations and is important in the calculation of the SS in the ANOVA.

$$\begin{aligned} TSS &= \sum Y^2 - \frac{(\sum Y)^2}{N} \\ &= (102^2 + 101^2 + \cdots + 103^2 + \cdots + 102^2) - \frac{(1511)^2}{15} \\ &= 152{,}247 - 152{,}208.07 = 38.93. \end{aligned}$$

2. The *between-treatment sum of squares* (between SS or BSS) is calculated as follows:

$$BSS = \sum \frac{T_i^2}{N_i} - CT,$$ (8.5)

where T_i is the sum of observations in treatment group i and N_i is the number of observations in treatment group i. N_i need not be the same for each group. In our example, the BSS is equal to

$$\left(\frac{506^2}{5} + \frac{497^2}{5} + \frac{508^2}{5}\right) - 152{,}208.07 = 13.73.$$

As previously noted, the *treatment* SS is a measure of treatment differences. A large SS means that the treatment differences are large. If the treatment means are identical, the treatment SS will be exactly equal to zero (0).

3. The *within-treatment sum of squares* (WSS) is equal to the difference between the TSS and BSS, that is, TSS = BSS + WSS. The WSS can also be calculated, as in the t test, by calculating $\sum(Y_{ij} - \overline{Y}_i)^2$ within each group, and pooling the results.

$$\begin{aligned} WSS &= TSS - BSS \\ &= 38.93 - 13.73 \\ &= 25.20. \end{aligned}$$ (8.6)

Having performed the calculations above, the SS for each "source" is set out in an "analysis of variance table," as shown in Table 8.3. The ANOVA table includes the *source, degrees of freedom, SS, mean square* (MS), and the *probability* based on the statistical test (*F* ratio).

Table 8.3 Analysis of Variance for the Data Shown in Table 8.1: Comparison of Three Analytical Methods

Source	d.f.	SS	Mean square	F
Between methods	2	13.73	6.87	$F = 3.27$[a]
Within methods	12	25.20	2.10	
Total	14	38.93		

[a] $0.05 < p < 0.10$.

The degrees of freedom, noted in Table 8.3, are calculated as $N - 1$ *for the total* (N is the total number of observations); *number of treatments minus one for the treatments; and for the within error, subtract d.f. for treatments from the total d.f.* In our example,

Total d.f. $= 15 - 1 = 14$
Between-treatment d.f. $= 3 - 1 = 2$
Within-treatment d.f. $= 14 - 2 = 12$

Note that for the within d.f., we have 4 d.f. from each of the three groups. Thus there are 12 d.f. for the within error term. The *mean squares* are equal to the SS divided by the d.f.

Before discussing the statistical test, the reader is reminded of the assumptions underlying the ANOVA model: *independence of errors, equality of variance,* and *normally distributed errors.*

8.1.1.1   Testing the Hypothesis of Equal Treatment Means

The *mean squares are variance estimates.* One can demonstrate that the variance estimated by the treatment mean square is a sum of the within variance plus a term that is dependent on treatment differences. If the treatments are identical, the term due to treatment differences is zero, and the between mean square (BMS) will be approximately equal to the within mean square (WMS) on the average. In any given experiment, the presence of random variation will result in nonequality of the BMS and WMS terms, even though the treatments may be identical. If the null hypothesis of equal treatment means is true, the distribution of the BMS/WMS ratio is described by the *F distribution.* Note that under the null hypothesis, both WMS and BMS are estimates of σ^2, the within-group variance.

The F distribution is defined by two parameters, d.f. in the numerator and denominator of the F ratio

$$F = \frac{\text{BMS (2 d.f.)}}{\text{WMS (12 d.f.)}} = \frac{6.87}{2.10} = 3.27.$$

In our example, we have an F with 2 d.f. in the numerator and 12 d.f. in the denominator. A test of significance is made by comparing the observed F ratio to a table of the F distribution with appropriate d.f. at the specified level of significance. The F distribution is an asymmetric distribution with a long tail at large values of F, as shown in Figure 8.1. (See also sects. 3.5 and 5.3.)

To tabulate all the probability points of all F distributions would not be possible. Tables of F, similar to the t table, usually tabulate points at commonly used α levels. The cutoff points ($\alpha = 0.01, 0.05, 0.10$) for F with n_1 and n_2 d.f. (numerator and denominator) are given in Table IV.6, the probabilities in this table (1%, 5% and 10%) are in the upper tail, usually reserved for one-sided tests. This table is used to determine statistical "significance" for the ANOVA. Although the alternative hypothesis in ANOVA (H_a: at least two treatment means not equal) is two sided, the ANOVA F test (BMS/WMS) uses the upper tail of the F distribution because,

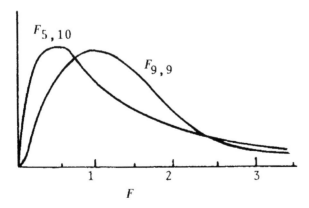

Figure 8.1   Some F distributions.

theoretically, the BMS cannot be smaller than the WMS.[†] (Thus, the F ratio will be less than 1 only due to chance variability.) The BMS is composed of the WMS plus a possible "treatment" term. Only large values of the F ratio are considered to be significant. In our example, Table 8.3 shows the F ratio to be equal to 3.27. Referring to Table IV.6, the value of F needed for significance at the 5% level is 3.89 (2 d.f. in the numerator and 12 d.f. in the denominator). Therefore, we cannot reject the hypothesis that all means are equal: method A = method B = method C ($\mu_A = \mu_B = \mu_C$).

8.1.2 Summary of Procedure for One-Way ANOVA
1. Choose experimental design and state the null hypothesis.
2. Define the α level.
3. Choose samples, perform the experiment, and obtain data.
4. Calculate the TSS and BSS.
5. Calculate the within SS as the difference between the TSS and the BSS.
6. Construct an ANOVA table with mean squares.
7. Calculate the F statistic (BMS/WMS).
8. Refer the F ratio statistic to Table IV.6 (n_1 and n_2 d.f., where n_1 is the d.f. for the BMS and n_2 is the d.f. for the WMS).
9. If the calculated F is equal to or greater than the table value for F at the specified α level of significance, at least two of the treatments can be said to differ.

8.1.3 A Common But Incorrect Analysis of the Comparison of Means from More Than Two Groups
In the example in section 8.1.1, if more than two assay methods are to be compared, the correct statistical procedure is a one-way ANOVA. A common error made by those persons not familiar with ANOVA is to perform three separate t tests on such data: comparing method A to method B, method A to method C, and method B to method C. This would require three analyses and "decisions," which can result in apparent contradictions. For example, decision statements based on three separate analyses could read

> Method A gives higher results than method $B (p < 0.05)$.
> Method A is not significantly different from method $C (p > 0.05)$.
> Method B is not significantly different from method $C (p > 0.05)$.

These are the conclusions one would arrive at if separate t tests were performed on the data in Table 8.1 (see Exercise Problem 1). One may correctly question: If A is larger than B, and C is slightly larger than A, how can C not be larger than B? The reasons for such apparent contradictions are (a) the use of different variances for the different comparisons, and (b) performing three tests of significance on the same set of data. ANOVA obviates such ambiguities by using a common variance for the single test of significance (the F test).[‡] The question of multiple comparisons (i.e., multiple tests of significance) is addressed in the following section.

8.2 PLANNED VERSUS A POSTERIORI (UNPLANNED) COMPARISONS IN ANOVA
Often, in an experiment involving more than two treatments, more specific hypotheses than the global hypotheses, $\mu_1 = \mu_2 = \mu_3 = \ldots$, are proposed in advance of the experiment. These are known as *a priori* or *planned comparisons*. For example, in our example of the three analytical methods, if method A is the standard method, we may have been interested in a comparison of each of the two new methods, B and C, with A (i.e., H_0: $\mu_A = \mu_C$ and $\mu_A = \mu_B$). We may proceed to make these comparisons at the conclusion of the experiment using the usual t-test

---

[†] This may be clearer if one thinks of the null and alternative hypotheses in ANOVA as H_a: $\sigma_B{}^2 = \sigma_w{}^2$; H_a: $\sigma_B{}^2 > \sigma_w{}^2$.

[‡] We have assumed in the previous discussion that the variances in the different treatment groups are the same. If the numbers of observations in each group are equal, the ANOVA will be close to correct in the case of moderate variance heterogeneity. If in doubt, a test to compare variances may be performed (see sect. 5.3).

procedure with the following proviso: *The estimate of the variance is obtained from the ANOVA, the pooled within mean square term.* This estimate comes from all the groups, not only the two groups being compared. ANOVA procedures, like the *t* test, assume that the variances are equal in the groups being tested.[‡] Therefore, the *within mean square* is the best estimate of the common variance. In addition, the increased d.f. resulting from this estimate results in increased precision and power (chap. 7) of the comparisons. A smaller value of *t* is needed to show "significance" compared to the *t* test, which uses only the data from a specific comparison, in general. Tests of only those comparisons planned a priori should be made using this procedure. This means that the α level (e.g., 5%) applies to each comparison.

Indiscriminate comparisons made after the data have been collected, such as looking for the largest differences as suggested by the data, will always result in more significant differences than those suggested by the stated level of significance. We shall see in section 8.2.1 that *a posteriori* tests (i.e., unplanned tests made after data have been collected) can be made. However, a "penalty" is imposed that makes it more difficult to find "significant" differences. This keeps the "experiment-wise" α level at the stated value (e.g., 5%). (For a further explanation, see sect. 8.2.1.) The statistical tests for the two planned comparisons as described above are performed as follows (a two independent groups *t* test with WMS equal to error, the pooled variance)

$$\text{Method } B \text{ versus method } A: \frac{|99.4 - 101.2|}{\sqrt{2.1(1/5 + 1/5)}} = 1.96.$$

$$\text{Method } C \text{ versus method } A: \frac{|101.6 - 101.2|}{\sqrt{2.1(1/5 + 1/5)}} = 0.44.$$

Since the *t* value needed for significance at the 5% level (d.f. = 12) is 2.18 (Table IV.4), neither of the comparisons noted previously is significant. However, when reporting such results, a researcher should be sure to include the actual averages. A confidence interval for the difference may also be appropriate. The confidence interval is calculated as described previously [Eq. (5.2)]; but remember to use the WMS for the variance estimate (12 d.f.). Also, the fact that methods *A* and *B* are not significantly different does not mean that they are the same. If one were looking to replace method *A*, other things being equal, method *C* would be the most likely choice.

If the comparison of methods *B* and *C* had been planned in advance, the *t* test would show a significant difference at the 5% level (see Exercise Problem 3). However, it would be unfair to decide to make such a comparison using the *t*-test procedure described above only after having seen the results. Now, it should be more clear why the ANOVA results in different conclusions from that resulting from the comparison of all pairs of treatments using separate *t* tests.

1. *The variance is pooled from all of the treatments.* Thus, it is the pooled variance from all treatments that is used as the error estimate. When performing separate *t* tests, the variance estimate differs depending on which pair of treatments is being compared. The pooled variance for the ordinary *t* test uses only the data from the specific two groups that are being compared. The estimates of the variance for each separate *t* test differ due to chance variability. That is, although an assumption in ANOVA procedures is that the variance is the same in all treatment groups, the *observed* sample variances will be different in different treatment groups because of the variable nature of the observations. This is what we have observed in our example. By chance, the variability for methods *A* and *B* was smaller than that for method *C*. Therefore, when performing individual *t* tests, a smaller difference of means is necessary to obtain significance when comparing methods *A* and *B* than that needed for the comparison of methods *A* and *C*, or methods *B* and *C*. Also, the d.f. for the *t* tests are 8 for the separate tests, compared to 12 when the pooled variance from the ANOVA is used. In

---

[‡] We have assumed in the previous discussion that the variances in the different treatment groups are the same. If the numbers of observations in each group are equal, the ANOVA will be close to correct in the case of moderate variance heterogeneity. If in doubt, a test to compare variances may be performed (see sect. 5.3).

conclusion, we obtain different results because we used different variance estimates for the different tests, which can result in ambiguous and conflicting conclusions.

2. The F test in the ANOVA takes into account the number of treatments being compared. An α level of 5% means that if all treatments are identical, 1 in 20 experiments (on the average) will show a significant F ratio. That is, the risk of erroneously observing a significant F is 1 in 20. If separate t tests are performed, each at the 5% level, for each pair of treatments (three in our example), the chances of finding at least one pair of treatments different in a given experiment will be greater than 5%, when the treatments are, in fact, identical. *We should differentiate between the two situations* (a) where we plan, a priori, specific comparisons of interest, and (b) where we make tests a posteriori suggested by the data. In case (a), each test is done at the α level, and each test has an α percent chance of being rejected if treatments are the same. In case (b), having seen the data we are apt to choose only those differences that are large. In this case, experiments will reveal differences where none truly exist much more than α percent of the time.

Multiple testing of data from the same experiment results in a higher significance level than the stated α level *on an experiment-wise basis*. This concept may be made more clear if we consider an experiment in which five assay methods are compared. If we perform a significance (t) test comparing each pair of treatments, there will be 10 tests, $(n)(n-1)/2$, where n is the number of treatments: $5(4)/2 = 10$ in this example. To construct and calculate 10 t tests is a rather tedious procedure. If treatments are identical and each t test is performed at the 5% level, the probability of finding at least one significant difference in an experiment will be much more than 5%. Thus the probability is very high that at the completion of such an experiment, this testing will lead to the conclusion that *at least* two methods are different. If we perform 10 separate t tests, the α level, on an experiment-wise basis, would be approximately 40%; that is, 40% of experiments analyzed in this way would show at least one significant difference, when none truly exists [1].

The Bonferroni method is often used to control the alpha level for multiple comparisons. For an overall level of alpha, the level is set at α/k for each test, where k is the number of comparisons planned. For the data of Table 8.1, for a test of two planned comparisons at an overall level of 0.05, each would be performed at the $0.05/2 = 0.025$ level. If the tests consisted of comparisons of the means (A vs. C) and (A vs. B), t tests could be performed. A more detailed t table than IV.4 would be needed to identify the critical value of t for a two-sided test at the 0.025 level with 12 d.f. This value lies between the tabled values for the 0.05 and 0.01 level and is equal to 2.56. The difference needed for significance at the 0.025 level is

$$2.56 \times \sqrt{2.1 \times \frac{2}{5}} = 2.35.$$

Since the absolute differences for the two comparisons (A vs. C) and (A vs. B) are 0.4 and 1.8, respectively, neither difference is statistically significant.

In the case of preplanned comparisons, significance may be found even if the F test in the ANOVA is not significant. This procedure is considered acceptable by many statisticians. Comparisons made after seeing the data that were not preplanned fall into the category of a *posteriori multiple* comparisons. Many such procedures have been recommended and are commonly used. Several frequently used methods are presented in the following section.

8.2.1 Multiple Comparisons in ANOVA

The discussion above presented compelling reasons to avoid the practice of using many separate t tests when analyzing data where more than two treatments are compared. On the other hand, for the null hypothesis of no treatment differences, a significant F in the ANOVA does not immediately reveal which of the multiple treatments tested differ. Sometimes, with a small number of treatments, inspection of the treatment means is sufficient to show obvious differences. Often, differences are not obvious. Table 8.4 shows the average results and ANOVA table for four drugs with regard to their effect on the reduction of pain, where the data are derived from subjective pain scores (see also Fig. 8.2). The null hypothesis is $H_0: \mu_A = \mu_B = \mu_C = \mu_D$. The alternative hypothesis here is that at least two treatment means differ. The α level is set at

Table 8.4 Average Results and ANOVA for Four Analgesic Drugs

	Reduction in pain with drugs			
	A	**B**	**C**	**D**
$\overline{X}$	4.5	5.7	7.1	6.3
S^2	3.0	4.0	4.5	3.8
S	1.73	2.0	2.12	1.95
N	10	10	10	10

	ANOVA			
Source	**d.f.**	**SS**	**Mean square**	**F**
Between drugs	3	36	12	$F_{3,36} = 3.14^a$
Within drugs	36	137.7	3.83	
Total	39	173.7		

$^a p < 0.05.$

5%. Ten patients were assigned to each of the four treatment groups. The *F* test with 3 and 36 d.f. is significant at the 5% level. An important question that we wish to address here is: Which treatments are different? Are all treatments different from one another, or are some treatments not significantly different? This problem may be solved using "multiple comparison" procedures. The many proposals that address this question result in similar but not identical solutions. Each method has its merits and deficiencies. We will present some approaches commonly used for performing a posteriori comparisons. Using these methods, we can test differences specified by the alternative hypothesis, as well as differences suggested by the final experimental data. These methods will be discussed with regard to comparing individual treatment means. Some of these methods can be used to compare any linear combination of the treatment means, such as the mean of drug *A* versus the average of the means for drugs *B*, *C*, and *D* [$\overline{A}$ vs. $(\overline{B} + \overline{C} + \overline{D})/3$].

For a further discussion of this problem, see the Scheffé method below.

8.2.1.1 Least Significant Difference

The method of "least significant difference" (LSD) proposed by R. A. Fisher, is the simplest approach to a posteriori comparisons. This test is a simple *t* test comparing all possible pairs of treatment means. (Note that this approach is not based on preplanned comparisons, discussed in the previous section.) However, the LSD method results in more significant differences than would be expected according to the α level. Because of this, many statisticians do not recommend

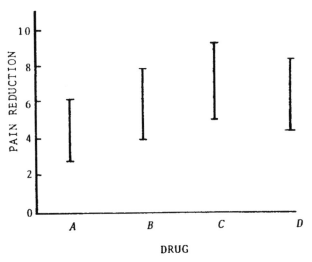

Figure 8.2 Result of pain reduction (± standard deviation) for four drugs with 10 patients per treatment group.

its use. The LSD test differs from the indiscriminate use of multiple t tests in that one proceeds (a) *only if the F test in the ANOVA is significant*, and (b) *the pooled (within mean square) variance is used as the variance estimate* in the t-test procedure. The LSD approach is illustrated using the data from Table 8.4.

$$\text{Since } t = \frac{\overline{X}_1 - \overline{X}_2}{\sqrt{S^2(1/N_1 + 1/N_2)}},$$

$$\text{LSD} = t\sqrt{S^2\left(\frac{1}{N_1} + \frac{1}{N_2}\right)}. \tag{8.7}$$

If the sample sizes are equal in each group ($N_1 = N_2 = N$),

$$\text{LSD} = t\sqrt{\frac{2s^2}{N}}, \tag{8.8}$$

where S^2 is the within mean square variance and t is the tabulated value of t at the α level, with appropriate degrees of freedom (d.f. = the number of degrees of freedom from the WMS of the ANOVA table). Any difference of two means that is equal to or exceeds the LSD is significant at the α level. From Table IV.4, the value of t at the 5% level with 36 d.f. is 2.03. The variance (from the ANOVA in Table 8.4) is 3.83. Therefore, the LSD is

$$\text{LSD} = 2.03\sqrt{\frac{2(3.83)}{10}} = 1.78.$$

The average pain reductions for drugs C and D are significantly greater than that for drug A ($\overline{C} - \overline{A} = 2.6$; $\overline{D} - \overline{A} = 1.8$).

Note that in the example shown in Table 8.1 (ANOVA table in Table 8.3), the F test is not significant. Therefore, one would not use the LSD procedure to compare the methods, after seeing the experimental results. If a comparison had been planned a priori, the LSD test could be correctly applied to the comparison.

8.2.1.2   Tukey's Multiple Range Test

Tukey's multiple range test is a commonly used multiple comparison test based on keeping the error rate at α (e.g., 5%) from an "experiment-wise" viewpoint. By "experiment-wise" we mean that if no treatment differences exist, the probability of finding at least one significant difference for a posteriori tests in a given experiment is α (e.g., 5%). This test is more conservative than the LSD test. This means that a larger difference between treatments is needed for significance in the Tukey test than in the LSD test. On the other hand, although the experiment-wise error is underestimated using the LSD test, the LSD test is apt to find real differences more often than will the Tukey multiple range test. (The LSD test has greater power.) Note that a trade-off exists. The easier it is to obtain significance, the greater the chance of mistakenly calling treatments different (α error), but the less chance of missing real differences (β error). The balance between these risks depends on the costs of errors in each individual situation. (See chap. 6 for a further discussion of these risks.)

In the multiple range test, treatments can be compared without the need for a prior significant F test. However, the ANOVA should always be carried out. The error term for the treatment comparisons comes from the ANOVA, the within mean square in the one-way ANOVA. Similar to the LSD procedure, a least significant difference can be calculated. Any difference of treatment means exceeding

$$Q\sqrt{\frac{S^2}{N}} \tag{8.9}$$

is significant. S^2 is the "error" variance from the ANOVA (within mean square for the one-way ANOVA) and N is the sample size. This test is based on equal sample sizes in each group. If the sample sizes are not equal in the two groups to be compared, an approximate method may be used with N replaced by $2N_1 N_2/(N_1 + N_2)$, where N_1 and N_2 are the sample sizes of the two groups. Q is the value of the "studentized range" found in Table IV.7A, a short table of Q at the 5% level. More extensive tables of Q may be found in Ref. [2] (Table A-18a). The value of Q depends on the *number of means being tested* (the number of treatments in the ANOVA design) and the *d.f. for error* (again, the within mean square d.f. in the one-way ANOVA). In the example of Table 8.4, the number of treatments is 4, and the d.f. for error are 36. From Table IV.7, the value of Q is approximately 3.81. Any difference of means greater than

$$3.81\sqrt{\frac{3.83}{10}} = 2.36$$

is significant at the 5% level. Therefore, this test finds only drugs A and C to be significantly different.

This test is more conservative than the LSD test. However, one must understand that the multiple range test tries to keep the error rate at α on an experiment-wise basis. In the LSD test, the error rate is greater than α for each experiment.

8.2.1.3 Scheffé Method

The Tukey method should be used if we are only interested in the comparison of treatment means after having seen the data. However, for more complicated comparisons (also known as *contrasts*) for a large number of treatments, the Scheffé method will often result in shorter intervals needed for significance. As in the Tukey method, the Scheffé method is meant to keep the α error rate at 5%, for example, on an experiment-wise basis. For the comparison of two means, the following statistic is computed:

$$\sqrt{S^2(k-1)F\left(\frac{1}{N_1}+\frac{1}{N_2}\right)}, \qquad\qquad (8.10)$$

where S^2 is the appropriate variance estimate (WMS for the one-way ANOVA), k is the number of treatments in the ANOVA design, and N_1 and N_2 are the sample sizes of the two groups being compared. F is the table value of F (at the appropriate level) with d.f. of $(k-1)$ in the numerator, and d.f. in the denominator equal to that of the error term in the ANOVA. Any difference of means equal to or greater than the value computed from expression (8.10) is significant at the α level. Applying this method to the data of Table 8.4 results in the following $[S^2 = 3.83, k = 4, F(3,36 \text{ d.f.}) = 2.86, N_1 = N_2 = 10]$:

$$\sqrt{3.83(3)(2.86)(1/10 + 1/10)} = 2.56.$$

Using this method, treatments A and C are significantly different. This conclusion is the same as that obtained using the Tukey method. However, treatments A and C barely make the 5% level; the difference needed for significance in the Scheffé method is greater than that needed for the Tukey method for this simple comparison of means. However, one should appreciate that the Scheffé method can be applied to more complicated contrasts with suitable modification of Eq. (8.10).

Suppose that drug A is a control or standard drug, and drugs B and C are homologous experimental drugs. Conceivably, one may be interested in comparing the results of the average of drugs B and C to drug A. From Table 8.4, the average of the means of drugs B and C is

$$\frac{5.7 + 7.1}{2} = 6.4.$$

For tests of significance of comparisons (contrasts) for the general case, Eq. (8.10) may be written as

$$\sqrt{(k-1)F \, V(\text{contrast})}, \tag{8.11}$$

where $(k-1)$ and (F) are the same as in Eq. (8.10), and $V(\text{contrast})$ is the variance estimate of the contrast. Here the contrast is

$$\frac{\overline{X}_B + \overline{X}_C}{2} - \overline{X}_A.$$

The variance of this contrast is (see also App. I)

$$\frac{S^2/N_B + S^2/N_C}{4} + \frac{S^2}{N_A} = S^2 \left(\frac{1}{20} + \frac{1}{10} \right) = \frac{3S^2}{20}.$$

(Note that $N_A = N_B = N_C = 10$ in this example.) From Eq. (8.11), a difference of $(\overline{X}_B + \overline{X}_C)/2 - \overline{X}_A$ exceeding

$$\sqrt{3(2.86)(3.83)\frac{3}{20}} = 2.22$$

will be significant at the 5% level. The observed difference is

$$6.4 - 4.5 = 1.9.$$

Since the observed difference does not exceed 2.22, the difference between the average results of drugs B and C versus drug A is not significant ($p > 0.05$). For a further discussion of this more advanced topic, the reader is referred to Ref. [3].

8.2.1.4 Newman–Keuls Test

The Newman–Keuls test uses the multiple range factor Q (see Tukey's Multiple Range Test) in a sequential fashion. In this test, the means to be compared are first arranged in order of magnitude. For the data of Table 8.4, the means are 4.5, 5.7, 6.3, and 7.1 for treatments A, B, D, and C, respectively.

To apply the test, compute the difference needed for significance for the comparison of 2, 3, ..., n means (where n is the total number of treatment means). In this example, the experiment consists of 4 treatments. Therefore, we will obtain differences needed for significance for 2, 3, and 4 means.

Initially, consider the first two means using the Q test

$$Q\sqrt{S^2/N}. \tag{8.12}$$

From Table IV.7, with 2 treatments and 36 d.f. for error, $Q = $ approximately 2.87. From Eq. (8.12),

$$Q\sqrt{S^2/N} = 2.87\sqrt{3.83/10} = 1.78.$$

For 3 means, find Q from Table IV.7 for $k = 3$

$$3.45\sqrt{3.83/10} = 2.14.$$

For 4 means, find Q from Table IV.7 for $k = 4$

$$3.81\sqrt{3.83/10} = 2.36.$$

Note that the last value, 2.36, is the same value as that obtained for the Tukey test.

Thus, the differences needed for 2, 3, and 4 means to be considered significantly different are 1.78, 2.14, and 2.36. This can be represented as follows

Number of treatments   2   3   4
Critical difference   1.78   2.14   2.36

The four ordered means are

$$
\begin{array}{cccc}
A & B & D & C \\
4.5 & 5.7 & 6.3 & 7.1
\end{array}
$$

The above notation is standard. Any two means connected by the same underscored line are not significantly different. Two means not connected by the underscored line are significantly different. Examination of the two underscored lines in this example shows that the only two means not connected are 4.5 and 7.1, corresponding to treatments A and C, respectively.

The determination of significant and nonsignificant differences follows. The difference between treatments A and C, covering 4 means, is equal to 2.6, which exceeds 2.36, resulting in a significant difference. The difference between treatments A and D is 1.8, which is less than the critical value of 2.14 for 3 means. This is described by the first underscore. (Note that we need not compare A and B or B and D since these will not be considered different based on the first underscore.) Treatments B, D, and C are considered to be not significantly different because the difference between B and C, encompassing 3 treatment means, is 1.4, which is less than 2.14. Therefore, a second underscore includes treatments B, D, and C.

8.2.1.5 Dunnett's Test

Sometimes experiments are designed to compare several treatments against a control but not among each other. For the data of Table 8.4, treatment A may have been a placebo treatment, whereas treatments B, C, and D are three different active treatments. The comparisons of interest are A versus B, A versus C, and A versus D. Dunnett [4,5] devised a multiple comparison procedure for treatments versus a control. The critical difference, D', for a two-sided test for any of the comparisons versus control is defined as

$$
D' = t'\sqrt{S^2\left(\frac{1}{N_1} + \frac{1}{N_2}\right)},
$$

where t' is obtained from Table IV.7B.

In the present example, p, the number of treatments to be compared to the control, is equal to 3, and d.f. $= 36$. For a two-sided test at the 0.05 level, the value of t' is 2.48 from Table IV.7B. Therefore the critical difference is

$$
2.48\sqrt{3.83\left(\frac{1}{10} + \frac{1}{10}\right)} = 2.17.
$$

Again, the only treatment with a difference from treatment A greater than 2.17 is treatment C. Therefore, only treatment C can be shown to be significantly different from treatment A, the control.

Those readers interested in further pursuing the topic of multiple comparisons are referred to Ref. [4].

8.2.2 Multiple Correlated Outcomes[§]

Many clinical studies have a multitude of endpoints that are evaluated to determine efficacy. Studies of antiarthritic drugs, antidepressants, and heart disease, for example, may consist of a measure of multiple outcomes. In a comparative study, if each measured outcome is evaluated

---

[§] A more advanced topic.

independently, the probability of finding a significant effect when the drugs are not different, for at least one outcome, is greater than the alpha level for the study. In addition, these outcomes are usually correlated. For example, relief of gastrointestinal distress and bloating may be highly correlated when evaluating treatment of gastrointestinal symptoms. If all the measures are independent, Bonferroni's inequality may be used to determine the significance level. For example, for five independent measures and a level of 0.01 for each measure, separate analyses of each measure will yield an overall alpha level of approximately 5% for the experiment as a whole (see sect. 8.2). However, if the measures are correlated, the Bonferroni adjustment is too conservative, making it more difficult to obtain significance. The other extreme is when all the outcome variables are perfectly correlated. In this case, one alpha level (e.g., 5%) will apply to all the variables. (One need test for only one of the variables; all other variables will share the same conclusion.) Dubey [6] has presented an approach to adjusting the alpha (α) level for multiple correlated outcomes. If we calculate the Bonferroni adjustment as $1 - (1 - \gamma)^k$ where k is the number of outcomes and γ is the level for testing each outcome, then the adjusted level for each outcome will lie between α (perfect correlation) and approximately α/k (no correlation). The problem can be formulated as

$$\alpha = \text{overall level of significance} = 1 - (1 - \gamma)^m, \tag{8.13}$$

where m lies between 1 and k, k being the number of outcome variables. If there is perfect correlation among all of the variables, $m = 1$, the level for each variable, γ is equal to α. For zero correlation, $m = k$, resulting in the Bonferroni adjustment. Dubey defines

$$m = k^{1-R_i}, \tag{8.14}$$

where R_i is an "average" correlation.

$$R_i = \sum_{i \neq j} \frac{R_{ij}}{k - 1}. \tag{8.15}$$

This calculation will be clarified in the example following this paragraph.

To obtain the alpha level for testing each outcome, γ, use Eq. (8.16) that is derived from Eq. (8.13) by solving for γ.

$$\gamma = 1 - (1 - \alpha)^{1/m} \tag{8.16}$$

The following example shows the calculation.

Suppose five variables are defined for the outcome of a study comparing an active and placebo for the treatment of heart disease: (1) trouble breathing, (2) pains in chest, (3) numbing/tingling, (4) rapid pulse, and (5) indigestion. The overall level for significance is set at 0.05. First, form the correlation matrix for the five variables. Table 8.5 is an example of such a matrix.

This matrix is interpreted for example, as the correlation between numbing/tingling and rapid pulse being 0.41 (variables 3 and 4), etc.

Table 8.5 Correlation Matrix for five Variables Measuring Heart "Disease"

	Variable				
	1	2	3	4	5
1	1.00	0.74	0.68	0.33	0.40
2	0.74	1.00	0.25	0.66	0.85
3	0.68	0.25	1.00	0.41	0.33
4	0.33	0.66	0.41	1.00	0.42
5	0.40	0.85	0.33	0.42	1.00

Calculate the "average" correlation, r_i from Eq. (8.15).

$$r_1 = \frac{\{0.74 + 0.68 + 0.33 + 0.40)}{4} = 0.538$$

$$r_2 = \frac{\{0.74 + 0.25 + 0.66 + 0.85)}{4} = 0.625$$

$$r_3 = \frac{\{0.68 + 0.25 + 0.41 + 0.33)}{4} = 0.425$$

$$r_4 = \frac{\{0.33 + 0.66 + 0.41 + 0.42)}{4} = 0.455$$

$$r_5 = \frac{\{0.40 + 0.85 + 0.33 + 0.42)}{4} = 0.500$$

The average correlation is

$$\frac{0.538 + 0.625 + 0.425 + 0.455 + 0.500}{5} = 0.509.$$

From Eq. (8.14),

$$m = k^{1-0.509} = 5^{0.491} = 2.203.$$

From Eq. (8.16), the level for each variable is adjusted to $\gamma = 1 - (1 - \alpha)^{1/m} = 1 - (1 - 0.05)^{1/2.203} = 0.023$.

Therefore, testing of the individual outcome variables should be performed at the 0.023 level.

Equation (8.13) can also be used to estimate the sample size of a study with multiple endpoints. Comelli [7] gives an example of a study with eight endpoints, and an estimated average correlation of 0.7. First, solve for γ, where alpha $= 0.05$ and R_i is 0.7.

$$\gamma = 1 - (1 - \alpha)^{1/m} = 0.05, \quad \text{where } m = 8^{(1-0.7)}.$$

γ is equal to 0.027. The sample size can then be computed by standard methods (see chap. 6). For the sample size calculation, use an alpha of 0.027 with desired power, and with the endpoint that is likely to show the smallest standardized treatment difference. For example, in a parallel design, suppose we wish to have a power of 0.8, and the endpoint with the smallest standardized difference is 0.5/1 (difference/standard deviation). Using Eq. (6.6), $N = 2(1/0.5)^2(2.21 + 0.84)^2 + 2 = 77$ per group.

8.3 ANOTHER EXAMPLE OF ONE-WAY ANOVA: UNEQUAL SAMPLE SIZES AND THE FIXED AND RANDOM MODELS

Before leaving the topic of one-way ANOVA, we will describe an example in which the sample sizes of the treatment groups are not equal. We will also introduce the notion of "fixed" and "random" models in ANOVA.

Table 8.6 shows the results of an experiment comparing tablet dissolution as performed by five laboratories. Each laboratory determined the dissolution of tablets from the same batch of a standard product. Because of a misunderstanding, one laboratory (D) tested 12 tablets, whereas the other four laboratories tested six tablets. The null hypothesis is

$$H_0: \mu_A = \mu_B = \mu_C = \mu_D = \mu_E,$$

and

$$H_a: \mu_i \neq \mu_j \quad \text{for at least two means.}$$

Table 8.6 Percent Dissolution After 15 Minutes for Tablets from a Single Batch Tested in Five Laboratories

	Laboratory				
	A	B	C	D	E
	68	55	78	75	65
	78	62	63	60	60
	63	67	78	66	66
	56	60	65	69	75
	61	67	70	58	75
	69	73	74	64	70
				71	
				71	
				65	
				77	
				60	
	—	—	—	63	—
Total	395	384	428	799	411
$\overline{X}$	65.8	64.0	71.3	66.6	68.5
s.d.	7.6	6.3	6.4	6.1	6.0

The ANOVA calculations are performed in an identical manner to that shown in the previous example (sect. 8.1.1). The ANOVA table is shown in Table 8.7. The F test for laboratories (4, 31 d.f.) is 1.15, which is *not* significant at the 5% level (Table IV.6). Therefore, the null hypothesis that the laboratories obtain the same average result for dissolution cannot be rejected.

$$\sum X = 2417 \quad \sum X^2 = 163,747 \quad N = 36$$

$$TSS = \sum X^2 - \frac{(\sum X)^2}{N} = 1472.306.$$

$$\text{Between Lab SS} = \frac{(395)^2}{6} + \frac{(384)^2}{6} + \frac{(428)^2}{6} + \frac{(799)^2}{12} + \frac{(411)^2}{6} - \frac{(2417)^2}{36} = 189.726.$$

Within lab SS = TSS − BSS = 1472.306 − 189.726 = 1282.58.

One should always question the validity of ANOVA assumptions. In particular, the assumption of independence may be suspect in this example.

Are tablets tested in sets of six, or is each tablet tested separately? If tablets are tested one at a time in separate runs, the results are probably independent. However, if six tablets are tested at one time, it is possible that the dissolution times may be related due to particular conditions that exist during the experiment. For example, variable temperature setting and mixing speed would affect all six tablets in the same (or similar) way. A knowledge of the particular experimental system and apparatus, and/or experimental investigation, is needed

Table 8.7 Analysis of Variance Table for the Data in Table 8.6 for Tablet Dissolution

Source	d.f.	SS	Mean square	F
Between labs	4	189.726	47.43	$F_{4,31} = 1.15$
Within labs	31	1282.58	41.37	
Total	35	1472.306		

to assess the possible dependence in such experiments. The assumption of equality of variance seems to be no problem in this experiment (see the standard deviations in Table 8.6).

8.3.1   Fixed and Random Models

In this example, the interpretation (and possible further analysis) of the experimental results depends on the nature of the laboratories participating in the experiment. The laboratories can be considered to be

1. the only laboratories of interest with respect to dissolution testing; for example, perhaps the laboratories include only those that have had trouble performing the procedure;
2. a random sampling of five laboratories, selected to determine the reproducibility (variability) of the method when performed at different locations.

The former situation is known as a *fixed model.* Inferences based on the results apply only to those laboratories included in the experiment. The latter situation is known as a *random model.* The random selection of laboratories suggests that the five laboratories are a sample chosen among many possible laboratories. Thus, inferences based on these results can be applied to all laboratories in the population of laboratories being sampled.

One way of differentiating a fixed and random model is to consider which treatment groups (laboratories) would be included if the experiment were to be run again. If the same groups would always be chosen in these perhaps hypothetical subsequent experiments, then the groups are fixed. If the new experiment includes different groups, the groups are random.

The statistical test of the hypothesis of equal means among the five laboratories is the same for both situations, *fixed* and *random.* However, in the random case, one may also be interested in estimating the variance. The estimates of the *within-laboratory* and *between-laboratory* variance are important in defining the reproducibility of the method. This concept is discussed further in section 12.4.1.

8.4   TWO-WAY ANOVA (RANDOMIZED BLOCKS)

As the one-way ANOVA is an extension of the two independent groups t test when an experiment contains more than two treatments, *two-way ANOVA* is an extension of the paired t test to more than two treatments. The two-way design, which we will consider here, is known as a randomized block design (the nomenclature in statistical designs is often a carryover based on the original application of statistical designs in agricultural experiments). In this design, treatments are assigned at random to each experimental unit or "block." (In clinical trials, where a patient represents a block, each patient receives each of the two or more treatments to be tested in random order.)

The randomized block design is advantageous when the levels of response of the different experimental units are very different. The statistical analysis separates these differences from the experimental error, resulting in a more precise (sensitive) experiment. For example, in the paired t test, taking differences of the two treatments should result in increased precision if the experimental units receiving the treatments are very different from each other, but they differentiate the treatments similarly. In Figure 8.3, the three patients are very different in their levels of response (blood pressure). However, each patient shows a similar difference between drugs A and B ($A > B$). In a two independent groups design (parallel groups), the experimental error is estimated from differences among experimental units within treatments. This is usually larger than the experimental error in a corresponding two-way design.

Another example of a two-way (randomized block) design is the comparison of analytical methods using product from different batches. The design is depicted in Table 8.8. If the batches have a variable potency, a rational approach is to run each assay method on material from each batch. The statistical analysis will separate the variation due to different batches from the other random error. The experimental error is free of batch differences, and will be smaller than that obtained from a one-way design using the same experimental material (product from different batches). In the latter case, material would be assigned to each analytical method at random.

A popular type of two-way design that deserves mention is that which includes pretreatment or baseline readings. This design, a *repeated measures* design, often consists of pretreatment readings followed by treatment and post-treatment readings observed over time. Repeated

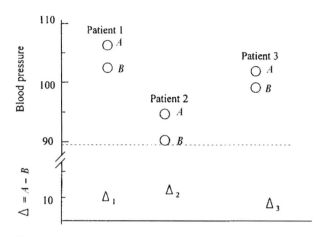

Figure 8.3 Increased precision in two-way designs.

measure designs are discussed further in chapter 11. In these designs, order (order is *time* in these examples) cannot be randomized. One should be careful to avoid bias in situations where a concomitant control is not part of these experiments. For example, suppose that it is of interest to determine if a drug causes a change in a clinical effect. One possible approach is to observe pretreatment (baseline) and post-treatment measurements, and to perform a statistical test (a paired t test) on the "change from baseline." Such an experiment lacks an adequate control group and interpretation of the results may be difficult. For example, any observed change or lack of change could be dependent on the time of observation, when different environmental conditions exist, in addition to any possible drug effect. A better experiment would include a *parallel* group taking a control product: a placebo or an active drug (positive control). The difference between change from baseline in the placebo group and test drug would be an unbiased estimate of the drug effect.

8.4.1 A Comparison of Dissolution of Various Tablet Formulations: Random and Fixed Models in Two-Way ANOVA

Eight laboratories were requested to participate in an experiment whose objective was to compare the dissolution rates of two generic products and a standard drug product. The purpose of the experiment was to determine (a) if the products had different rates of dissolution, and (b) to estimate the laboratory variability (differences) and/or test for significant differences among laboratories. If the laboratory differences are large, the residual or error SS will be substantially reduced compared to the corresponding error in the one-way design. If interaction is absent, we will be using the "within-laboratory" variability to test for differences among the products (see sect. 8.4.1.2). The laboratory SS and the product SS in the ANOVA are computed in a manner similar to the calculations in the one-way design. The residual SS is calculated as the total sum of squares (TSS) minus the laboratory and product SS. (The laboratory and product SS are also denoted as the row and column SS, respectively.) The *error* or *residual* SS, that part of the total

Table 8.8 Two-Way Layout for Analytical Procedures Applied to Different Batches of Material

| Batch | Analytical method | | | |
	A	B	C	...
1	← _____	_____	_____	_____ →
2	← _____	_____	_____	_____ →
3	← _____	_____	_____	_____ →
.				
.				
.				

Table 8.9 Tablet Dissolution After 30 Minutes for Three Products (Percent Dissolution)

Laboratory	Generic		Standard	Row total
	A	B		
1	89	83	94	266
2	93	75	78	246
3	87	75	89	251
4	80	76	85	241
5	80	77	84	241
6	87	73	84	244
7	82	80	75	237
8	68	77	75	220
Column total	666	616	664	1946
$\overline{X}$	83.25	77.0	83.0	
$\sum X^2 = 158{,}786$				

sum of squares remaining after subtracting out that due to rows and columns, is also often denoted as the *interaction* $(C \times R)$ SS.

The hypothesis of interest is

$$H_0: \mu_A = \mu_B = \mu_C.$$

That is, the average dissolution rates of the three products are equal. The level of significance is set at 5%. The experimental results are shown in Table 8.9.

The analysis proceeds as follows: Total sum of squares (TSS)

$$\text{Total sum of squares (TSS)} = \sum X^2 - CT = 89^2 + 93^2 + \cdots + 75^2 + 75^2 - \frac{(1946)^2}{24}$$
$$= 158{,}786 - 157{,}788.2 = 997.8.$$

Column sum of squares (CSS) or product SS

$$CSS = \frac{\sum C_j^2}{R} - CT = \frac{(666^2 + 616^2 + 664^2)}{8} - 157{,}788.2$$
$$= 200.3 \, (C_j \text{ is the total of column } j, \, R \text{ is the number of rows}).$$

Row sum of squares (RSS) or laboratory SS

$$RSS = \frac{\sum R_i^2}{C} - CT = \frac{(266^2 + 246^2 + \cdots + 220^2)}{3} - 157{,}788.2$$
$$= 391.8 \, (R_i \text{ is the total of row } i, \, C \text{ is the number of columns}).$$

Residual $(C \times R)$ sum of squares (ESS) $= TSS - CSS - RSS$

$$= 997.8 - 200.3 - 391.8 = 405.7.$$

The ANOVA table is shown in Table 8.10. The *d.f.* are calculated as follows:

Total $= N_t - 1$ $N_t =$ total number of observations
Column $= C - 1$ $C =$ number of columns
Row $= R - 1$ $R =$ number of rows
Residual $(C \times R) = (C - 1)(R - 1)$

Table 8.10 Analysis of Variance Table for the Data (Dissolution) from Table 8.8

Source	d.f.	SS	Mean square	F^{a}
Drug products	2	200.3	100.2	$F_{2,14} = 3.5$
Laboratories	7	391.8	56.0	$F_{7,14} = 1.9$
Residual ($C \times R$)	14	405.7	29.0	
Total	23	997.8		

[a] See the text for a discussion of proper F tests.

8.4.1.1 Tests of Significance
To test for differences among *products* (H_0: $\mu_A = \mu_B = \mu_C$), an F ratio is formed

$$\frac{\text{drug product mean square}}{\text{residual mean square}} = \frac{100.2}{29} = 3.5.$$

The F distribution has 2 and 14 d.f. According to Table IV.6, an F of 3.74 is needed for significance at the 5% level. Therefore, the products are not significantly different at the 5% level. However, had the a priori comparisons of each generic product versus the standard been planned, one could perform a t test for each of the two comparisons (using 29.0 as the error from the ANOVA), *generic A versus standard and generic B versus standard*. Generic A is clearly not different from the standard. The t test for generic B versus the standard is

$$t = \frac{|\overline{X}_B - \overline{X}_S|}{\sqrt{29(1/8 + 1/8)}} = \frac{6}{2.69} = 2.23.$$

This is significant at the 5% level (see Table IV.4; t with 14 d.f. $= 2.14$). Also, one could apply one of the multiple comparisons tests, such as the Tukey test described in section 8.2.1. According to Eq. (8.9), any difference exceeding $Q\sqrt{S^2(1/N)}$ will be significant. From Table IV.7, Q for 3 treatments and 14 d.f. for error is 3.70 at the 5% level. Therefore, the difference needed for significance for any pair of treatments for a posteriori tests is

$$3.70\sqrt{29\frac{1}{8}} = 7.04.$$

Since none of the means differ by more than 7.04, individual comparisons decided upon after seeing the data would show no significance in this experiment.

The test for laboratory differences is (laboratory mean square)/(residual mean square), which is an F test with 7 and 14 d.f. According to Table IV.6, this ratio is not significant at the 5% level (a value of 2.77 is needed for significance). As discussed further below, if *drug products* are a fixed effect, this test is valid only if interaction (drug product $\times$ laboratories) is absent. Under these conditions, the laboratories are not sufficiently different to show a significant F value at the 5% level.

8.4.1.2 Fixed and Random Effects in the Two-Way Model [§]
The proper test of significance in the two-way design depends on the model and the presence of *interaction*. The notion of interaction will be discussed further in the presentation of factorial designs (chap. 9). In the previous example, the presence of *interaction* means that the three products are ranked differently with regard to dissolution rate by at least some of the eight laboratories. For example, laboratory 2 shows that generic A dissolves fastest among the three products, with generic B and the standard being similar. On the other hand, laboratory 8 shows that generic A is the slowest-dissolving product. Interaction is conveniently shown graphically as in Figure 8.4. "Parallel curves" indicate no interaction.

[§] A more advanced topic.

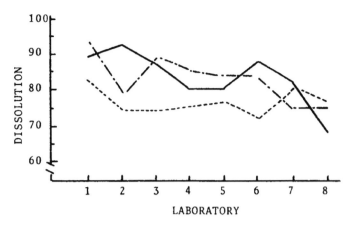

Figure 8.4 Average results of dissolution for eight laboratories. — · standard; — generic *A*; – – – generic *B*.

Of course, in the presence of error (variability), it is not obvious if the apparent lack of parallelism is real or is due to the inherent variability of the system. An experiment in which a lab makes a single observation on each product, such as is the case in the present experiment, usually contains insufficient information to make decisions concerning the presence or absence of interaction. To test for interaction, an additional error term is needed to test for the significance of the $C \times R$ residual term. In this case, the experiment should be designed to have replicates (at least duplicate determinations). In the absence of replication, it is best (usually) to assume that interaction is present. This is a conservative point of view. A knowledge of the presence or absence of interaction is important in order that one may choose the proper error term for statistical testing (the term in the denominator of the F test) as described below.

The concept of fixed and random effects was introduced under the topic of one-way ANOVA. A "fixed" category includes all the treatments of interest. In the present example, it is apparent that the columns, drug products, are fixed. We are only interested in comparing the two generic products with the standard. Otherwise, we would have included other products of interest in the experiment. On the other hand, the nature of the rows, laboratories, is not obvious. Depending on the context, laboratories may be either *random* or *fixed*. If the laboratories were selected as a random sample among many laboratories that perform such dissolution tests, then "laboratories" is a random factor. In the present situation, the laboratories are chosen as a means of replication in order to compare the dissolution of the three products. Then, inferences based on the result of the experiment are applied to the population of laboratories from which this sample of eight was drawn. We might also be interested in estimating the variance among laboratories in order to have some estimate of the difference to be expected when two or more laboratories perform the same test (see sect. 12.4.1). If the laboratories chosen were the only laboratories of interest, and inferences based on the experimental results apply only to these eight laboratories, then laboratories are considered to be fixed. Table 8.11 shows when the F tests in the two-way ANOVA are valid depending on the model and the presence of interaction.

Table 8.11   Tests in the Two-Way Analysis of Variance (One Observation Per Cell)

Columns	Rows	Interaction	Error term for the F test[a]
Fixed	Random	None	Residual ($C \times R$) or within
Random	Random	None	Residual ($C \times R$) or within
Random	Random	None	Residual ($C \times R$) or within
Fixed	Fixed	Present	Within
Fixed	Random	Present	Residual ($C \times R$) for fixed effect; use within for random effect
Random	Random	Present	Residual (CR)

[a] *Residual* is the usual residual mean square and includes ($C \times R$), column × row interaction. *Within* is the within mean square calculated from replicate determinations and will be called "error" in future discussions.

In the usual situation, columns are fixed (e.g., drug treatments, formulations) and rows are random (patients, batches, laboratories). In these cases, in the absence of replication, the proper test for columns is (column mean square)/(residual mean square).

Usually, the test for rows is not pertinent if rows are "random." For example, in a clinical study, in which two or more treatments are to be compared, the rows are "patients." The statistical test of interest in such situations is a comparison of the treatments; one does not usually test for patient differences. However, in many laboratory experiments, both column and row effects are of interest. In these cases, if significance testing is to be performed for *both row* and *column* effects (where either or both are fixed), it is a good idea to include proper replication (Table 8.11). *Duplicate assays* on the same sample such as may be performed in a dissolution experiment are not adequate to estimate the relevant variability. Replication in this example would consist of repeat runs, using different tablets for each run. An example of a two-way analysis in which replication is included is described in the following section.

8.4.2 Two-Way ANOVA with Replication

Before discussing an example of the analysis of two-way designs with replications, two points should be addressed regarding the implementation of such experiments.

1. It is best to have equal number of replications for each cell of the two-way design. In the dissolution example, this means that each lab replicates each formulation an equal number of times. If the number of replicates is very different for each cell, the analysis and interpretation of the experimental results can be very complicated and difficult.
2. The experimenter should be sure that the experiment is *properly* replicated. As noted above, merely replicating assays on the same tablet is not proper replication in the dissolution example. Replication is an independently run sample in most cases. Each particular experiment has its own problems and definitions regarding replication. If there is any doubt about what constitutes a proper replicate, a statistician should be consulted.

As an example of a replicated, two-way experiment, we will consider the dissolution data of Table 8.9. Suppose that the data presented in Table 8.9 are the average of two determinations (either *two* tablets or *two averages of six tablets each*—a total of 12 tablets). The actual duplicate determinations are shown in Table 8.12. We will consider "products" fixed and "laboratories" random.

The analysis of these data results in one new term in the ANOVA, that due to the *within-cell SS*. The *within-cell SS* represents the variability or error due to replicate determinations, and is the pooled SS from within the cells. In the example shown previously, the SS is calculated for

Table 8.12 Replicate Tablet Dissolution Data for Eight Laboratories Testing Three Products (Percent Distribution)

| Laboratory | Generic | | Standard | Row total |
	A	B		
1	87, 91	81, 85	93, 95	532
2	90, 96	74, 76	74, 82	492
3	84, 90	72, 78	84, 94	502
4	75, 85	73, 79	81, 89	482
5	77, 83	76, 78	80, 88	482
6	85, 89	70, 76	80, 88	488
7	79, 85	74, 86	71, 79	474
8	65, 71	73, 81	70, 80	440
Total	1332	1232	1328	3892
Average	83.25	77.0	83.0	

each cell, $\sum (X - \overline{X})^2$. For example, for the first cell (generic A in laboratory 1), $\sum (X - \overline{X})^2 = (87 - 89)^2 + (91 - 89)^2 = (87 - 91)^2/2 = 8$. The SS is equal to 8. The within SS is the total of the SS for the 24 (8×3) cells. The residual or interaction SS is calculated as the difference between the TSS and the sum of the column SS, row SS, and within-cell SS. The calculations for Table 8.12 are shown below.

$$\text{Total sum of squares} = \sum X^2 - \text{CT}$$

$$= 87^2 + 91^2 + 90^2 + \cdots + 71^2 + 79^2 + 70^2 + 80^2 - \frac{3892^2}{48}$$

$$= 318{,}160 - 315{,}576.3 = 2583.7.$$

$$\text{Product SS} = \frac{\sum C_j^2}{Rr} - \text{CT}$$
$$= \frac{1332^2 + 1232^2 + 1328^2}{16} - \frac{3892^2}{48} = 315{,}977 - 315{,}576.3$$
$$= 400.7$$

where C_j is the sum of observations in column j, R the number of rows, and r the number of replicates per cell.

$$\text{Laboratory SS} = \frac{\sum R_i^2}{Cr} - \text{CT}$$
$$= \frac{532^2 + 492^2 + \cdots + 440^2}{6} - \frac{3892^2}{48} = 316{,}360 - 315{,}576.3$$
$$= 783.7$$

where R_i is the sum of observations in row i, C the number of columns, and r the number of replicates per cell.

Within-cell SS**

$$\sum (X - \overline{X})^2, \text{ where the sum extends over all cells}$$

$$= \frac{(87 - 91)^2}{2} + \frac{(90 - 96)^2}{2} + \frac{(84 - 90)^2}{2} + \cdots + \frac{(70 - 80)^2}{2}$$
$$= 588.$$

$$\text{C} \times \text{R SS} = \text{TSS} - \text{PSS} - \text{LSS} - \text{WSS}$$
$$= 2583.7 - 400.7 - 783.7 - 588$$
$$= 811.3.$$

The ANOVA table is shown in Table 8.13. Note that the F test for drug products is identical to the previous test, where the averages of duplicate determinations were analyzed. However, the laboratory mean square is compared to the within mean square to test for laboratory differences. This test is correct if laboratories are considered either to be fixed (all FDA laboratories, for example), or random, when drug products are fixed (Table 8.13). For significance $F_{7,24}$ must exceed 2.43 at the 5% level (Table IV.6). The significant result for laboratories suggests that at least some of the laboratories may be considered to give different levels of response. For example, compare the results for laboratory 1 versus laboratory 8.

** For duplicate determinations, $\sum (X - \overline{X}) = (X_1 - X_2)^2/2$.

Table 8.13 ANOVA Table for the Replicated Dissolution Data Shown in Table 8.12

Source	d.f.	SS	Mean square	F[a]
Drug products	2	400.7	200.4	$F_{2,14} = 3.5$
Laboratories	7	783.7	112	$F_{7,24} = 4.6*$
$C \times R$ (residual)	14	811.3	58.0	$F_{14,24} = 2.37*$
Within cells (error)	24[b]	588	24.5	

[a] Assume drug products are fixed, laboratories random.
[b] d.f. for within cells is the pooled d.f., one d.f. for each of 24 cells; in general, d.f. $= R \times C (n - 1)$, where n is the number of replicates.
*$p < 0.05$.

Another statistical test, not previously discussed, is available in this analysis. The F test ($C \times R$ mean square/within mean square) is a test of *interaction*. In the absence of interaction (laboratory $\times$ drug product), the $C \times R$ mean square would equal the within mean square on the average. A value of the ratio sufficiently larger than 1 is an indication of interaction. In the present example, the F ratio is 2.37, 58.0/24.5. This is significant at the 5% level (see Table IV.6; $F_{14,24} = 2.13$ at the 5% level). The presence of a laboratory $\times$ drug product interaction in this experiment suggests that laboratories are not similar in their ability to distinguish the three products (Fig. 8.4).

8.4.3 Another Worked Example of Two-Way ANOVA[§]

Before leaving the subject of the basic ANOVA designs, we will present one further example of a two-way experiment. The design is a form of a factorial experiment, discussed further in chapter 9. In this experiment, three drug treatments are compared at three clinical sites. The treatments consist of two dosages of an experimental drug (low and high dose) and a control drug. Eight patients were observed for each treatment at each site. The data represent increased performance in an exercise test in asthmatic patients. The results are shown in Table 8.14. In order to follow the computations, the following table of totals (and definitions) should be useful.

$$CT = \frac{(371.5)^2}{72} = 1916.84$$

$R =$ number of rows $= 3$
$C =$ number of columns $= 3$
$r =$ number of replicates $= 8$
$R_i =$ total of row i (row 1 $= 108.9$, row 2 $= 140.7$, row 3 $= 121.9$)
$C_j =$ total of column j (column 1 $= 69.7$, column 2 $= 156.1$, column 3 $= 145.7$)

Table 8.14 Increase in Exercise Time for Three Treatments (Antiasthmatic) at Three Clinical Sites (Eight Patients Per Cell)

Site	Treatment			Cell means (standard deviation)		
	A (low dose)	B (high dose)	C (control)	A	B	C
I	4.0, 2.3, 2.1, 3.0 1.6, 6.4, 1.4, 7.0	3.6, 2.6, 5.5, 6.0 2.5, 6.0, 0.1, 3.1	5.1, 6.6, 5.1, 6.3 5.9, 6.2, 6.3, 10.2	3.475 (2.16)	3.675 (2.06)	6.463 (1.61)
II	2.4, 5.4, 3.7, 4.0 3.3, 0.8, 4.6, 0.8	6.6, 6.4, 6.8, 8.3 6.9, 9.0, 12.0,7.8	5.6, 6.4, 8.2, 6.5 4.2, 5.6, 6.4, 9.0	3.125 (1.68)	7.975 (1.86)	6.488 (1.52)
III	1.0, 1.3, 0.0, 5.1 0.2, 2.4, 4.5, 2.4	6.0, 8.1, 10.2, 6.6 7.3, 8.0, 6.8, 9.9	5.8, 4.1, 6.3, 7.4 4.5, 2.0, 6.8, 5.2	2.113 (1.88)	7.863 (1.52)	5.263 (1.73)

[§] A more advanced topic.

The cell totals are shown below

	A	B	C	Total
Site I	27.8	29.4	51.7	108.9
Site II	25	63.8	51.9	140.7
Site III	16.9	62.9	42.1	121.9
Total	69.7	156.1	145.7	371.5

The computations for the statistical analysis proceed as described in the previous example. The *within-cell mean square* is the pooled variance over the nine cells with 63 d.f. (7 d.f. from each cell). In this example (equal number of observations in each cell), the within-cell mean square is the average of the nine variances calculated from within-cell replication (eight values per cell). The computations are detailed below.

$$\text{Total sum of squares} = \sum X^2 - \text{CT}$$

$$= 4.0^2 + 2.3^2 + 2.1^2 + \cdots + 6.8^2 + 5.2^2 - \frac{(371.5)^2}{72}$$

$$= 2416.77 - 1916.84 = 499.93.$$

$$\text{CSS (treatment SS)} = \frac{\sum C_j^2}{Rr} - \text{CT} = \frac{69.7^2 + 156.1^2 + 145.7^2}{3 \times 8} - 1916.84 = 185.40.$$

$$\text{RSS (site SS)} = \frac{\sum R_i^2}{Cr} - \text{CT} = \frac{108.9^2 + 140.7^2 + 121.9^2}{3 \times 8} - 1916.84 = 21.30.$$

Within-cell mean square = pooled sum of squares from the nine cells

$$= \sum X^2 - \frac{\sum (\text{cell total})^2}{r} = 2416.7$$

$$- \frac{27.8^2 + 29.4^2 + 51.7^2 + \cdots + 42.1^2}{8} = 2416.77 - 2214.2 = 202.57.$$

$C \times R$ SS (treatment $\times$ site interaction SS)

$$= \text{total SS} - \text{treatment SS} - \text{site SS} - \text{within SS}$$
$$= 499.93 - 185.40 - 21.30 - 202.57$$
$$= 90.66.$$

Note the shortcut calculation for within SS using the squares of the cell totals. Also note that the $C \times R$ SS is a measure of *interaction* of sites and treatments. Before interpreting the results of the experiment from a statistical point of view, both the ANOVA table (Table 8.15) and a plot of the average results should be constructed (Fig. 8.5). The figure helps as a means of interpretation of the ANOVA as well as a means of presenting the experimental results to the "client" (e.g., management).

8.4.3.1 Conclusions of the Experiment Comparing Three Treatments at Three Sites: Interpretation of the ANOVA Table

The comparisons of most interest come from the treatment and treatment $\times$ site terms. The *treatment mean square* measures differences among the three treatments. The *treatment $\times$ site mean square* is a measure of how the three sites differentiate the three treatments. As is usually the case, interactions are most easily visualized by means of a plot (Fig. 8.5). The lack of "parallelism" is most easily seen as a difference between site I and the other two sites. Site I shows that treatment C has the greatest increase in exercise time, whereas the other two sites find treatment

Table 8.15 Analysis of Variance Table for the Data of Table 8.14 (Treatments and Sites Fixed)

Source	d.f.	SS	Mean square	F
Treatments	2	185.4	92.7	$F_{2,63} = 28.8$[a]
Sites	2	21.3	10.7	$F_{2,63} = 3.31$[b]
Treatment × site	4	90.66	22.7	$F_{4,63} = 7.05$[a]
Within	63	202.57	3.215	
Total	71	499.93		

[a]$p < 0.01$.
[b]$p < 0.05$.

B most efficacious. Of course, the apparent differences, as noted in Figure 8.5, may be due to experimental variability. However, the treatment × site interaction term (Table 8.15) is highly significant ($F_{4,63} = 7.05$). Therefore, this interaction can be considered to be real. The presence of interaction has important consequences on the interpretation of the results. The lack of consistency makes it difficult to decide if treatment B or treatment C is the better drug. Certainly, the decision would have been easier had all sites found the same drug best. The final statistical decision depends on whether one considers sites fixed or random. In this example treatments are fixed.

Case 1: Sites fixed. If both treatments and sites are fixed, the proper error term for treatments and sites is the within mean square. As shown in Table 8.15, both treatments and sites (as well as interaction) are significant. Inspection of the data suggests that treatments B and C are not significantly different, but that both of these treatments are significantly greater than treatment A (see Exercise Problem 11 for an a posteriori test). Although not of primary interest in such studies, the significant difference among sites may be attributed to the difference between site II and site I, site II showing greater average exercise times (due to higher results for treatment B). However, this difference is of less importance than the interaction of sites and treatments that exists in this study. Thus, although treatments B and C do not differ, on the average, in the fixed site case, site I is different from the other sites in the comparison of treatments B and C. One may wish to investigate further to determine the cause of such differences (e.g., different kinds of patients, different exercise equipment, etc.). If the difference between the results for treatments B and C were dependent on the type of patient treated, this would be an important parameter in drug therapy. In most multiclinic drug trials, clinical sites are selected at random, although it is impractical, if not impossible, to choose clinical sites in a truly random fashion (see also sect. 11.5). Nevertheless, the interpretation of the data is different if sites are considered to be a random effect.

Case 2: Sites random. If sites are random, and interaction exists, the correct error term for treatments is the treatment × site (interaction) mean square. In this case, the F test ($F_{2,4} = 4.09$) shows a lack of significance at the 5% level. The apparently "obvious" difference between treatment A and treatments B and C is not sufficiently large to result in significance because of the paucity of d.f. (4 d.f.). The disparity of the interpretation here compared to the fixed sites

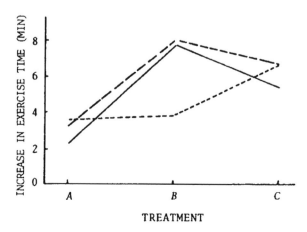

Figure 8.5 Plot of average results from data of Table 8.14. $--$ site II; $—$ site I; $———$ site III.

Table 8.16 Tests for Treatment Differences in Two-Way ANOVA with Replicate Observations (Treatments Fixed)

Rows	Interaction	Proper error term
Fixed	Present	Within mean square
Fixed	Absent	Within mean square or $C \times R$ mean square
Random	Present	$C \times R$ (interaction) mean square
Random	Absent	Within mean square (conservative test: use $C \times R$ mean square: pool $C \times R$ and within mean square—see the text)

case is due to the large interaction. The data suggest that differences among treatments are dependent on the site at which the drugs are tested. If the three sites are random selection from among many possible sites, this very small sample of sites does not give a reliable estimate of the population averages.

Table 8.16 abstracted from Table 8.11, shows the proper error terms for testing treatment differences, depending on whether sites (rows) are random or fixed. The testing also depends on whether or not there is interaction in the model. Ordinarily, it is not possible to predict the presence (or absence) of interaction in advance of the study. The conservative approach for statistical tests is to assume interaction exists. In this example, if sites are random, the $C \times R$ (interaction) mean square is the proper error term for treatments. Often, however, the interaction mean square has few d.f. This can considerably reduce the power of the test, as is the case in this example. In these situations, if the interaction mean square is not significant, the *interaction* and *within* mean squares may be pooled. This gives a pooled error term with more d.f. than either term alone. This is a controversial procedure, but can be considered acceptable if interaction is clearly not present.

8.4.4 Missing Data

Missing data can result from overt errors in measurements, patients not showing up for scheduled visits in a clinical trial, loss of samples, etc. In general, the problems of dealing with missing data are complex. Missing data can be considered to be caused by missing observations from a statistically valid, symmetric design. A common manifestation is when a "cell" is empty, that is, contains no values. A cell may be defined as the intersection of factor levels in a factorial or related design. For example, in a two-way crossover design, if a subject misses a visit, we have an empty cell. In a one-way design, missing values do not cause computational problems in general, because the analysis is valid when sample sizes in the independent groups are not equal.

For a two-way design with one missing value, the missing value may be estimated using the following formula:

$$Y_{ij} = \frac{rY_i + cY_j - y}{(r-1)(c-1)}, \tag{8.17}$$

where r is the number of rows, c the number of columns, Y_{ij} the observation in ith row and jth column, Y_i is the total of ith row, Y_j the total of jth column, and y the grand total of all observations.

For example, Table 8.17 shows data with a missing value in the second column and third row. From Eq. (8.17), $Y_{32} = (3 \times 10 + 3 \times 9 - 44)/[(3 - 1)(3 - 1)] = 3.25$.

An ANOVA is performed including the estimated observation (3.25), but the d.f. for error are reduced by 1 due to the missing observation. (See Exercise Problem 12.)

For more than one missing value and for further discussion, see Snedecor and Cochran [8]. For more complicated designs, computer software programs may be used to analyze data with missing values. One should be aware that in certain circumstances depending on the nature of the missing values and the design, a unique analysis may not be forthcoming. In some cases, some of the observed data may have to be removed in order to arrive at a viable analysis.

Another problem with missing data occurs in clinical studies with observations made over time where patients drop out prior to the anticipated completion of treatments (censored data). A common approach when analyzing such data where some patients start but do not complete

Table 8.17   Illustration of Estimation of a Single Missing Data Point

	Columns			
	1	2	3	Total
Rows				
1	3	5	6	14
2	7	4	9	20
3	4	—	6	10
Total	14	9	21	44

the study for various reasons, is to carry the last value forward. For example, in analgesic studies measuring pain, patients may give pain ratings over time. If pain is not relieved, patients may take a "rescue" medication and not complete the study as planned. The last pain rating on study medication would then be continued forward for the missed observation periods. For example, such a study might require pain ratings (1–5, where 5 is the worst pain and 1 is the least) every half-hour for six hours. Consider a patient who gives ratings of 5, 4, and 4 for hours 0 (baseline), half, and one hour, respectively. He then decides to take the rescue medication. The patient would be assigned a rating of 4 for all periods after one hour (1.5–6 hours, inclusive). Statistical methods are then used as usual. Other variations on the Last Value Carried forward (LVCF) concept is to carry forward either the best or worst reading prior to dropout as defined and justified in the study protocol. (See also sect. 11.2.7.) Other methods include the average of all observations for a given patient as the final result. These are still controversial and should be discussed with FDA prior to implementation. One problem with this approach occurs in disease states that are self-limiting. For example, in studies of single doses of analgesics in acute pain, if the study extends for a long enough period of time, pain will eventually be gone. To include patients who have dropped out prior to these extended time periods could bias the results at these latter times.

8.5   STATISTICAL MODELS[§]

Statistical analyses for estimating parameters and performing statistical tests are usually presented as linear models as introduced in section 8.1. (See also Apps. II and III.) The parameters to be included in the model are linearly related to the dependent variable in the form of

$$Y = B_0 X_0 + B_1 X_1 + B_2 X_2 + \cdots + \varepsilon, \tag{8.18}$$

where the Bs are the parameters to be estimated and the various X_i, represent the independent variables. Epsilon, ε, represents the random error associated with the experiment, and is usually assumed to be normal with mean 0 and variance, σ^2. This suggests that the estimate of Y is unbiased, with a variance, σ^2. For a simple model, where we wish to fit a straight line, the model would appear as

$$Y = B_0 X_0 + B_1 X_1,$$

where $X_0 = 1$ and X_1 (the independent variable) has a coefficient B_1.

In this example, we observe data pairs, X_i, Y_i, from which we estimate B_0 (intercept) and B_1 (slope). Again, this particular model represents the model of a straight line.

Although simple methods for analyzing such data have been presented in chapter 7, the data could also be analyzed using ANOVA based on the model. This analysis would first compute the TSS, which is the SS from a model with only a mean ($Y = \mu + \varepsilon$), with $N_t - 1$ d.f. This is the SS obtained as if all the data were in a single group. The $N_t - 1$ d.f. are based on the fact that, in the computation of SS, we are subtracting the mean from each observation before squaring. Having computed the SS from this simple model, a new SS would then be computed

[§] A more advanced topic.

from a model that looks like a straight line. Each observation is subtracted from the least squares line and squared (the residuals are subtracted from a model with two parameters, slope and intercept). The difference between the SS with one parameter (the mean) and the SS with two parameters (slope and intercept) has 1 d.f. and represents the SS due to the slope. The inclusion of a slope in the model reduces the SS. In general, as we include more parameters in the model, the SS is reduced. Eventually, if we have as many observations as terms in the model, we will have 0 residual SS, a perfect fit.

Typically, we include terms in the model that have meaning in terms of the experimental design. For example, for a one-way ANOVA (see sect. 8.1), we have separated the experimental material into k groups and assigned N_t subjects randomly to the k groups. The model consists of groups and a residual error, which represents the variability of observations within the groups

$$Y_{ik} = \mu + G_k + \varepsilon_{ik}.$$

μ represents the overall mean of the data, G_k represents the deviation from μ due to the kth group (i.e. kth group effect) (treatment), and ε_{ik} is the common variance (residual error). Note that the Xs are not written in the model statement, and are assumed to be equal to 1. A more detailed description of the model including three groups might look like this (see sect. 8.1)

$$Y_{i1} = \mu + G_1 + \varepsilon_{i1}, \quad Y_{i2} = \mu + G_2 + \varepsilon_{i2}, \quad Y_{i3} = \mu + G_3 + \varepsilon_{i3}$$

We then estimate μ, G_1, G_2, and G_3 from the model, and the residual is the error SS. Note that as before, ignoring groups, the total d.f. $= N_t - 1$. The fit of the model without groups, compared to the fit with groups (N_{k-1} d.f. for each group) has 2 d.f. [$(N_t - 1) - (N_t - 3)$])(that represent the SS for differences between groups. If groups have identical means, the residual SS will be approximately the same for the full model (three separate groups) and the reduced model (one group).

A somewhat more complex design is a two-way design, such as a randomized block design, where, for example, in a clinical study, each patient may be subjected to several treatments. This model includes both patient and group effects. The residual error is a combination of both group × patient interaction (GP) and within-individual variability. To separate these two sources of variability, patients would have to be replicated in each group (treatment). If such replication exists (see sect. 8.4.2), the model would appear as follows with g groups ($i = 1$ to g), and p patients ($j = 1$ to p) per group, each patient being replicated k times in each group

$$Y_{ijk} = \mu + G_i + P_j + GP_{ij} + \varepsilon_{ijk}.$$

Models may become complicated, but the procedure for their construction and analysis follows the simple approaches shown above. For experiments that are balanced (no missing data), the calculations are simple and give unambiguous results. For unbalanced experiments, the computations are more complex, and the interpretation is more difficult, sometimes impossible. Computer programs can analyze unbalanced data, but care must be taken to understand the data structure in order to make the proper interpretation (see also sect. 8.4.4).

8.6 ANALYSIS OF COVARIANCE[§]

The analysis of covariance (ANCOVA) combines ANOVA with regression. It is a way to increase precision and/or adjust for bias when comparing two treatments. ANCOVA uses observations (concomitant variables) that are taken independently of the test (outcome) variable. These concomitant observations are used to "adjust" the values of the test variable. This usually results in a statistical test that is more precise than the corresponding nonadjusted analysis. We look for covariates that are highly correlated with the experimental outcome, the greater the better (10). For example, the initial weight of a patient in a weight reduction study may be correlated with the weight reduction observed at the end of the study. Also, note that one may choose more than one covariate. One simple example is the use of baseline measurements when comparing

[§] A more advanced topic.

Table 8.18 Analytical Results for Eight Batches of Product Comparing Two Manufacturing Methods

Method I		Method II	
Raw material	Final product	Raw material	Final product
98.4	98.0	98.7	97.6
98.6	97.8	99.0	95.4
98.6	98.5	99.3	96.1
99.2	97.4	98.4	96.1
Average 98.70	97.925	98.85	96.30

the effect of two or more treatments. A common approach in such experiments is to examine the change from baseline (experimental observation–baseline) as discussed in sections 8.4 and 11.3.2. The analysis can also be approached using ANCOVA, where the baseline measurement is the covariate. The correction for baseline will then adjust the experimental observation based on the relationship of the two variables, baseline and outcome. Another example is the comparison of treatments where a patient characteristic, for example, weight, may be related to the clinical outcome; weight is the covariate. In these examples, assignment to treatment could have been stratified based on the covariate variable, for example, weight. ANCOVA substitutes for the lack of stratification by adjusting the results for the covariate, for example, weight. Refer to the chapter on regression (chap. 7) and to the section on one-way ANOVA (sect. 8.1) if necessary to follow this discussion. Ref. [9] is useful reading for more advanced approaches and discussion of ANCOVA.

In order to facilitate the presentation, Table 8.18 shows the results of an experiment comparing two manufacturing methods for finished drug product. In this example, the analysis of the raw material used in the product was also available.

If the two methods are to be compared using the four final (product) assays for each method, we would use a one-way ANOVA (independent sample t test in this example). The ANOVA comparing the two methods would be as shown in Table 8.19 and Table 8.21, columns 1 and 2.

The two methods yield different results at the 0.05 level ($p = 0.02$), with averages of 97.925 and 96.3, respectively. The question that one may ask is, "Are the raw material assays different for the products used in the test, accounting for the difference?" We can perform an ANOVA on the initial values to test this hypothesis. See Table 8.20 and Table 8.21, columns 1 and 3.

The average raw material assays for the lots used for the two methods are not significantly different (98.7 and 98.85). Thus, we may assume that the final assay results are not biased by possible differences in raw material. (Note that if the averages of the raw materials were different for the two methods, then one would want to take this into consideration when comparing the methods based on the final assay.) However, it is still possible that use of the initial values may reduce the variability of the comparison of methods due to the relationship between the raw material assay and the final. To account for this relationship, we can compute a linear fit of

Table 8.19 ANOVA Comparing Methods Based on Final Assay

Source	d.f.	Sum of squares	Mean square	F value	Pr > F
Between methods	1	5.28125	5.28125	9.88	0.0200
Within methods	6	3.20750	0.53458		
Total	7	8.48875			

Table 8.20 ANOVA Comparing Raw Material Assays

Source	d.f.	Sum of squares	Mean square	F value	Pr > F
Between methods	1	0.0450	0.0450	0.33	0.5847
Within methods	6	0.8100	0.1350		
Total	7	0.8550			

Table 8.21 Detailed Computations for ANCOVA for Data of Table 8.18

	(1)	(2)	(3)	(4)	(5)	(6)	(7)	(8)
Source	d.f.	Y Final assay	X Raw material	Final × raw	Slope	Reg. SS	d.f.	Res. SS
a. Within method A	3	0.6275	0.36	−0.33	−0.917	0.303	2	0.325
b. Within method B	3	2.58	0.45	−0.33	−0.733	0.242	2	2.338
c. Separate regressions	—	—	—	—	—	0.545	4	2.663
d. Within methods	6	3.2075	0.81	−0.66	−0.815	0.538	5	2.670
e. Between methods	1	5.28125	0.045	−0.4875	—	—	—	—
f. Total	7	8.48875	0.855	−1.1475	−1.343	1.541	6	6.948

Columns 1 and 2 are the simple ANOVA for the final assay.
Columns 1 and 3 are the simple ANOVA for the raw material assay.
Column 4 is the cross product SS $= \sum[(X - \overline{X})(Y - \overline{Y})]$.
Column 5 is computed as column 4/column 3 (final assay is the Y variable; raw material assay is the X variable).
Column 6 is column 3 × column 5 squared.
Column 7 is d.f. for residual (column 8).
Column 8 is column 2 − column 6.

the final assay result versus the raw material assay, and use the residual error from the fit as an estimate of the variance. The variance estimate should be smaller than that obtained when the relationship is ignored. The fitted lines for each method are assumed to be parallel, that is, the relationship between the covariate and the outcome variable (finished product assay) is the same for each method. With this assumption, the difference between methods, adjusted for the covariate, is the difference between the parallel lines at any value of the covariate, in particular the difference of the intercepts of the parallel lines. These concepts are illustrated in Figure 8.6.

Assumptions for covariance analysis include the following:

1. The covariate is not dependent on the experimental observation. That is, the covariate is not affected by the treatment (method). For example, an individual's weight measured prior to and during treatment by a cholesterol-reducing agent is not affected by his cholesterol reading(s).
2. The covariate is a fixed variable or the covariate and outcome variable have a bivariate normal distribution. The covariate is specified and measured before randomization to treatments.
3. Slopes for regression lines within each treatment group are equal, that is, the lines are parallel. If not, the analysis is still correct, but if interaction is suspected, we end up with an average effect. Interaction suggests that the comparison of treatments depends on the covariate level.

Covariance analysis is usually performed with the aid of a statistical software program as shown in Table 8.22. However, to interpret the output, it is useful to understand the nature of the

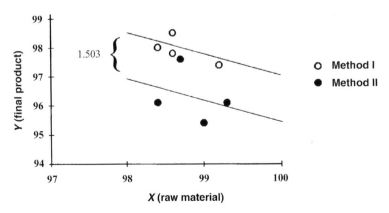

Figure 8.6 Illustration of adjusted difference of means in ANCOVA.

Table 8.22 ANCOVA Software Analysis of Data from Table 8.18

Source	d.f.	Sum of squares	Mean square	F ratio	Prob > F
X (Cov)	1	0.5377778	0.5377778	1.01	0.3616
A (Method)	1	4.278962	4.278962	8.01	0.0366
Error	5	2.669722	0.5339444		
Total (Adj)	7	8.488751			
Method	Means				
I	97.86389				
II	96.36112				

calculations. Table 8.21 is a complete table of the analysis for the example of the two analytical methods (Table 8.18). The following discussion refers to entries in Table 8.21.

The software (Table 8.22) computes the means, adjusted for the covariate, but does not perform a test for parallelism of the regression lines. A SAS program that includes Covariate × Method interaction in the model is a test for parallelism. To test for parallelism of the method versus covariate fitted lines, an analysis is performed to determine if the residual SS are significantly increased when all points are fitted to individual (two or more) parallel lines as compared to a fit to separate lines. (Note the similarity to stability analysis for pooling lots, sect. 8.7.) An F test comparing the variances is performed to determine significance

$$F_{\text{d.f.1,d.f.2}} = \frac{(\text{Residual SS parallel lines} - \text{residual SS separate lines})/(\text{groups} - 1)}{(\text{Residual SS separate lines})/\text{d.f.}}. \quad (8.19)$$

The residual SS from the parallel lines is obtained from a least squares fit (final product assay vs. raw material assay). These residual SS are calculated from line d in Table 8.21

$$\sum (y - \bar{y})^2 - b^2 \sum (X - \bar{X})^2 \quad \text{(see sect. 7.4)}.$$

This is equal to $(3.2075 - 0.815^2 \times 0.81) = 2.67$.

The residual SS when each method is fit separately is in line c, column (8) in Table 8.21. The analysis is in a form of the Gauss–Markov Theorem that describes an F test comparing two linear models, where one model has additional parameters. In this example, we fit a model with separate intercepts and slopes, a total of four parameters for the two methods, and compare the residual SS to a fit with common slope and separate intercepts, three parameters. The increase in the mean square residual due to the fit with less parameters is tested for significance using the F distribution as shown in Eq. (8.19). This is the same approach as that used to determine the pooling of stability lots as discussed in section 8.7.

$$F_{1,4} = \frac{(\text{Residual SS parallel lines} - \text{residual SS separate lines})/(2 - 1)}{(\text{Residual SS separate lines})/(8 - 4)}. \quad (8.20)$$

In this example, the F test with 1 and 4 d.f. is

$$\frac{(2.67 - 2.663)/1}{2.663/4} = 0.01,$$

which is not significant at $p < 0.05$ ($p > 0.9$). The lines can be considered to be parallel.

This computation may be explained from a different viewpoint. For a common slope, the residual SS is computed as $\sum (y - \bar{y})^2 - b^2 \sum (X - \bar{X})^2$. Here $\sum (y - \bar{y})^2$ and $\sum (X - \bar{X})^2$ are the sums of the sums of squares for each line, and b is the common slope (-0.815). From Table 8.21, line d, columns 2 to 4, the SS for the common line is

$$0.6275 + 2.58 - (-0.8148)^2(0.36 + 0.45) = 2.670.$$

For the separate lines (column 8 in Table 8.21), the sums of the SS is

$$SS = 0.325 + 2.338 = 2.663.$$

Another test of interest is the significance of the slope (vs. 0). If the test for the slope is not significant, the concomitant variable (raw material assay) is not very useful in differentiating the methods. The test for the slope is: within regression mean square/within residual mean square.

The residual mean square is that resulting from the fit of parallel lines (common slope).

In this example, from line d in Table 8.21, $F_{1,5} = 0.538/(2.67/5) = 1.01(p = 0.36)$. The common slope is -0.815 (line d, column 5) that is not significantly different from 0. Thus, we could conclude that use of the raw material assay as an aid in differentiating the methods is not useful. Nevertheless, the methods are significantly different both when we ignore the raw material assay ($p = 0.02$; Table 8.19) and when we use the covariance analysis (see below).

The test for difference of means adjusted for the covariate is a test for difference of intercepts of the parallel lines.

$$F_{1,5} = \frac{(\text{Residual SS total} - \text{residual SS within})/(\text{groups} - 1)}{(\text{Residual SS parallel lines})/(\text{d.f.})}.$$

In this example, $F_{1,5} = (6.948 - 2.670)/(2.670/5) = 8.01\ (p < 0.05)$ (see column 8, Table 8.21). This is a comparison of the fit with a common intercept (TSS) to the fit with separate intercepts (within SS) for the parallel lines.

The adjusted difference between methods can be calculated as the difference between intercepts or, equivalently, the distance between the parallel lines (Fig. 8.6). The adjusted means are calculated as follows [8].

The common slope is b. The intercept is $\overline{Y} - b\overline{X}$ (see Eq. 7.3, chap. 7). The difference of the intercepts is

$$(\overline{Y}_a - b\overline{X}_a) - (\overline{Y}_b - b\overline{X}_b) = \overline{Y}_a - \overline{Y}_b - b(\overline{X}_a - \overline{X}_b)$$
$$= 97.925 - 96.3 - (-0.815)(98.7 - 98.85)$$
$$= 1.503.$$

The difference between means adjusted for the raw material assay is 1.503.

8.6.1   Comparison of ANCOVA with Other Analyses

Two other common analyses use differences from baseline and ratios of the observed result to the baseline value when a concomitant variable, such as a baseline value, is available. For example, in clinical studies, baseline values are often measured in order to assess a treatment effect relative to the baseline value. Thus, once more, in addition to ANCOVA, two other ways of analyzing such data are analysis of differences from baseline or the ratio of the observed value and baseline value. The use of changes from baseline is a common approach that is statistically acceptable. If the covariance assumptions are correct, covariance should improve upon the difference analysis, that is, it should be more powerful in detecting treatment differences. The difference analysis and ANCOVA will be similar if the ANCOVA model approximates $Y = a + X$, that is, the slope of the X versus Y relationship is one (1). The use of ratios does disturb the normality assumption, but if the variance of the covariate is small, this analysis should be more or less correct. This model suggests that $Y/X = a$, where a is a constant. This is equivalent to $Y = aX$, a straight line that goes through the origin. [If the Y values, the experimentally observed results, are far from 0, and/or the X values are clustered close together, the statistical conclusions for ratios (observed/baseline) should be close to that from the ANCOVA.] See Exercise Problem 13 for further clarification.

A nonparametric ANCOVA is described in section 15.7.

8.7   ANOVA FOR POOLING REGRESSION LINES AS RELATED TO STABILITY DATA[§]

As discussed in chapter 7, an important application of regression and ANOVA is in the analysis of drug stability for purposes of establishing a shelf life. Accelerated stability studies are often used to establish a preliminary shelf life (usually 24 months), which is then verified by long-term studies under label conditions (e.g., room temperature). If more than one lot is to be used to establish a shelf life, then data from all lots should be used in the analysis. Typically, 3 production lots are put on stability at room temperature in order to establish an expiration date. The statistical analysis recommended by the FDA [10] consists of preliminary tests for pooling of data from the different lots. If both slopes and intercepts are considered similar for the multiple lots based on a statistical test, then data from all lots can be pooled. If not, the data may be analyzed as separate lots, or if slopes are not significantly different, a common slope with separate intercepts may be used to analyze the data. Pooling of all of the data gives the most powerful test (the longest shelf life) because of the increased d.f. and multiple data points. If lots are fitted separately, suggesting lot heterogeneity, expiration dating is based on the lot that gives the shortest shelf life. Separate fittings also result in poor precision because an individual lot will have fewer d.f. and less data points than that resulting from a pooled analysis. Degrees of freedom when fitting regression lines are $N - 2$, so that a stability study with 7 time points will have only 5 d.f. (0, 3, 6, 9, 12, 18, and 24 months). Fitting the data with a common slope will have intermediate precision compared to separate fits and a single combined fit.

The computations are complex and cannot be described in detail here, but the general principles will be discussed. The fitting is of the form of regression and covariance (see also sect. 8.6). The following model (Model 1) fits separate lines for each lot.

$$\text{Potency } (Y) = \sum a_i + \sum b_i X \quad \text{Model (1)}.$$

For three lots, the model contains six parameters, three intercepts, and three slopes. The residual error SS is computed with $N - 6$ d.f. for 3 lots, where N is the total number of data pairs. Thus, each of the three lines is fit separately, each with its own slope and intercept. Least squares theory, with the normality assumption (the dependent variable is distributed normally with the same variance at each value, X), can be applied to construct a test for equality of slopes. This is done by fitting the data with a reduced number of parameters, where there is a common slope for the lots tested. The fit is made to a model of the form

$$\text{Potency } (Y) = \sum a_i + b X \quad \text{Model (2)}.$$

For 3 lots, this fit has $N - 4$ d.f., where N is the total number of data pairs (X, Y) with the 3 intercepts and single slope accounting for the 4 d.f. Statistical theory shows that the following ratio, Eq. (8.21), has an F distribution

$$\frac{[\text{Residual SS from model (2)} - \text{Residual SS from model (1)}]/[P' - P]}{\text{Residual SS from model (1)}/[N - P']}. \tag{8.21}$$

If P' is the number of parameters to be fitted in Model (1), 6 for 3 lots, and P' is the number of parameters in Model (2), 4 for 3 lots, then the d.f. of this F statistic are $[P' - P]$ d.f. in the numerator (2 for 3 lots), and $N - P'$ d.f. in the denominator ($N - 6$ for 3 lots). If the F statistic shows significance, then the data cannot be pooled with a common slope, and separate fits for each line are used for predicting shelf life. A significant F ($p < 0.25$) suggests that a fit to individual lines is significantly better than a fit with a common slope, based on the increase in the sums of squares when the model with less parameters is fit. If slopes are not poolable, a 95% lower, if appropriate, one-sided (or 90% two-sided) confidence band about the fitted line for each lot is computed, and the expiration dates are determined for each batch separately.

If the F statistic is not significant, then a model with a common slope, but different intercepts, may be fit.

[§] A more advanced topic.

The most advantageous condition for estimating shelf life is when data from all lots can be combined. Before combining the data into a single line, a statistical test to determine if the lots are poolable is performed. In order to pool all of the data, a two-stage test is proposed by the FDA. First, the test for a common slope is performed as described in the preceding paragraph. If the test for a common slope is not significant ($p > 0.25$), a test is performed for a common intercept. This is accomplished by computing the residual SS for a fit to a single line (Model 3) minus the residual sums of squares for the reduced model with a common slope, Model 2, adjusted for d.f., and divided by the residual SS from the fit to the full model (separate slopes and intercepts), Model (1).

$$\text{Potency } (Y) = a + bX \quad \text{Model (3)}.$$

The F test for a common intercept is

$$\frac{\text{Residual SS from model (3)} - \text{residual SS from model (2)}/[P' - P]}{\text{Residual SS from model (1)}/[N - P']}. \tag{8.22}$$

For 3 lots, the F statistic has 2 d.f. in the numerator (2 parameter fit for a single line vs. a 4 parameter fit, 3 intercepts, and 1 slope for a fit with a common slope), and $N - 6$ d.f. in the denominator. Again, a significant F suggests that a fit using a common slope and intercept is not appropriate.

The FDA has developed a SAS program to analyze stability data using the above rules to determine the degree of pooling, that is, separate lines for each lot, a common slope for all lots, or a single fit with a common slope and intercept. A condensed version of the output of this program is described below.

The raw data is for three lots (A, B, and C), each with three assays at 0, 6, and 12 months.

	Lot		
Time (mo)	A	B	C
0	100	102	98
6	99	98	97
12	96	97	95

The output testing for pooling is derived from an ANCOVA with time as the covariate (Table 8.23). The ANOVA shows a common slope, indicated by line C with $p > 0.25$ ($p = 0.58566$). The test for a common intercept is significant, $p < 0.25$. Therefore, lines are fitted with a common slope but with separate intercepts.

Table 8.23 Modified and Annotated SAS Output from FDA Stability Program

Source	SS	d.f.	Mean square	F	p
A (pooled line)	9.67	4	2.42	3.10714	0.18935
B (intercept)	8.67	2	4.33	5.57143	0.09770
C (slope)	1.00	2	0.50	0.64286	0.58566
D (error)	2.33	3	0.78		

Key to sources of variation:
A = separate intercept, separate slope | common intercept, common slope. This is the residual SS from fit to a single line minus the residual SS from fits to separate lines. This is the SS attributed to model 3.
B = separate intercept, common slope | common intercept, common slope. This is the residual SS from a fit to a single line minus the residual SS from a fit with common slope and separate intercepts ($A - C$).
C = separate intercept, separate slope | separate intercept, common slope. This is the residual SS from a fit to a line with a common slope and separate intercepts line minus the residual SS from fits to separate lines. This is the SS attributed to model 2.
D = Residual. This is the residual SS from fits to separate lines ($9 - 6 = 3$ d.f.). This is the SS attribute to model 1.

The shelf life estimates vary from 20 to 25 months. The shortest time, 20 months, is used as the shelf life.

Stability analysis

Fitted line	Batch 1
	$Y = 100.33 - 0.333X$
Fitted line	Batch 2
	$Y = 101.00 - 0.333X$
Fitted line	Batch 3
	$Y = 98.67 - 0.333X$

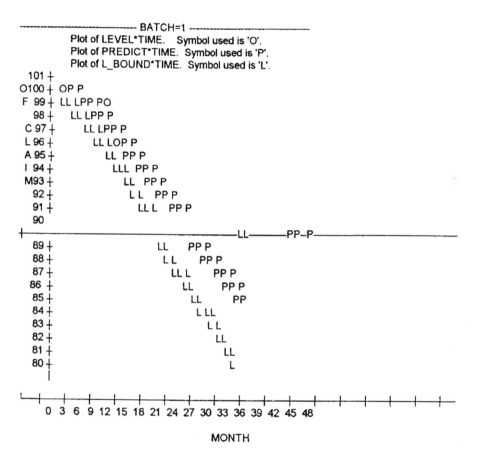

Stability Analysis: 95% one-sided lower confidence limits (separate intercepts and common slope)

Batch	Estimated dating period (mo/wk)
1	24
2	25
3	20

The data for each batch should be visually inspected to ensure that the average results based on these calculations have not hidden noncompliant or potentially noncompliant batches.

The FDA recommends using a significance level of 25% rather than the usual 5% level. The reason for this is the use of multilevel preliminary testing before coming to a decision. The use of a 25% level is somewhat arbitrary, and does not seem to have a clear theoretical rationale. This higher level of significance means that the criterion for pooling lots is more difficult to attain, thereby making it more difficult to establish the longer shelf life that results from pooling data from multiple lots. This may be considered to be a conservative rule from the point of view that shelf lives will not be overestimated. However, the analysis is open to interpretation, and it is the author's opinion that the 25% level of significance is too high.

Another problem with the FDA approach is that power is not considered in the evaluation. For example, if the model and assay precision is very good, lots that look similar with regard to degradation may not be poolable, whereas with very poor precision, lots that appear not to be similar may be judged poolable. Unfortunately, this problem is not easily solved. Finally, it is not clear why the FDA has not included a test for pooling based on a common intercept and separate slopes.

Nevertheless, the FDA approach has much to recommend it, as the problem is quite complex.

KEY TERMS

Alpha level
ANCOVA
ANOVA
ANOVA table
A posteriori comparisons
A priori comparisons
Assumptions
Between-treatment sum of squares or
  mean square (BSS or BMS)
Block
Bonferroni test
Completely randomized design
Components of variance
Contrasts
Control
Correction term
Degrees of freedom
Designed experiments
Dunnett's test
Error
Error sum of squares or mean square
  (ESS or EMS)
Experimental error
Experimental units
F distribution
Fixed model
Independence
Interaction
LSD procedure for multiple
  comparisons mean square
Missing data

Model
Multiple comparisons
Newman–Keuls' test
One-way analysis of variance
Parallel groups
Parallelism
Placebo
Pooled variance
Pooled regressions
Positive control
Power
Precision
Randomized block design
Random model
Repeated measures design
Replicates
Residual
Scheffé method for multiple comparisons
Shortcut computing formulas
Source
Stability
Sum of squares
Symmetry
Total sum of squares (TSS)
Treatments
Treatment sum of squares or mean square
T tests
Tukey's multiple range test
Two-way analysis of variance
Within sum of squares or mean square
  (WSS or WMS)

EXERCISES

1. Perform three separate t tests to compare method A to method B, method A to method C, and method B to method C in Table 8.1. Compare the results to that obtained from the ANOVA (Table 8.3).

2. Treatments A, B, and C are applied to six experiment subjects with the following results:

A	B	C
1	3	4
5	2	1

Perform an ANOVA and interpret the between-treatment mean square.

3. Repeat the t tests from Exercise Problem 1, but use the "pooled" error term for the tests. Explain why the results are different from those calculated in Problem 1. When is it appropriate to perform separate t tests?

4. It is suspected that four analysts in a laboratory are not performing accurately. A known sample is given to each analyst and replicate assays performed by each with the following results:

		Analyst	
I	II	III	IV
10	9	8	9
11	10	9	9
10	11	8	8

(a) State the null and alternative hypotheses.
(b) Is this a fixed or a random model?
(c) Perform an ANOVA. Use the LSD procedure to show which analysts differ if the "analyst" mean square is significant at the 5% level.
(d) Use Tukey's and Scheffé's multiple comparison procedures to test for treatment (analyst) differences. Compare the results to those in part (c).

5. Physicians from seven clinics in the United States were each asked to test a new drug on three patients. These physicians are considered to be among those who are expert in the disease being tested. The seventh physician tested the drug on only two patients. The physicians had a meeting prior to the experiment to standardize the procedure so that all measurements were uniform in the seven sites.
The results were as follows:

			Clinic			
1	2	3	4	5	6	7
9	11	6	10	5	7	12
8	9	9	10	3	7	10
7	13	9	7	4	7	—

(a) Perform an ANOVA.
(b) Are the results at the different clinics significantly different at the 5% level?
(c) If the answer to part (b) is yes, which clinics are different? Which multiple comparison test did you use?

§6. Are the following examples random or fixed? Explain.
(a) Blood pressure readings of rats are taken after the administration of four different drugs.

(b) A manufacturing plant contains five tablet machines. The same product is made on all machines, and a random sample of 100 tablets is chosen from each machine and weighed individually. The problem is to see if the machines differ with respect to the weight of tablets produced.

(c) Five formulations of the same product are compared. After six months, each formula is assayed in triplicate to compare stability.

(d) Same as part (b) except that the plant has 20 machines. Five machines are selected at random for the comparison.

(e) Ten bottles of 100 tablets are selected at random in clusters 10 times during the packaging of tablets (a total of 10,000 tablets). The number of defects in each bottle are counted. Thus we have 10 groups, each with 10 readings. We want to compare the average number of defects in each cluster.

7. Dissolution is compared for three experimental batches with the following results (each point is the time in minutes for 50% dissolution for a single tablet).
Batch 1: 15, 18, 19, 21, 23, 26
Batch 2: 17, 18, 24, 20
Batch 3: 13, 10, 16, 11, 9
(a) Is there a significant difference among batches?
(b) Which batch is different?
(c) Is this a fixed or a random model?

8. In a clinical trial, the following data were obtained comparing placebo and two drugs:

Patient	Placebo Predrug	Placebo Postdrug	Drug 1 Predrug	Drug 1 Postdrug	Drug 2 Predrug	Drug 2 Postdrug
1	180	176	170	161	172	165
2	140	142	143	140	140	141
3	175	174	180	176	182	175
4	120	128	115	120	122	122
5	165	165	176	170	171	166
6	190	183	200	195	192	185

(a) Test for treatment differences, using only postdrug values.
(b) Test for treatment differences by testing the change from baseline (predrug).
(c) For Problem 8(b), perform a posteriori multiple comparison tests (1) comparing all pairs of treatments using Tukey's multiple range rest and the Newman–Keuls' test and (2) comparing drug 1 and drug 2 to control using Dunnett's test.

9. Tablets were made on six different tablet presses during the course of a run (batch). Five tablets were assayed during the five-hour run, one tablet during each hour. The results are as follows:

Hour	Press 1	Press 2	Press 3	Press 4	Press 5	Press 6
1	47	49	46	49	47	50
2	48	48	48	47	50	50
3	52	50	51	53	51	52
4	50	47	50	48	51	50
5	49	46	50	49	47	49

(a) Are presses and hours fixed or random?
(b) Do the presses give different results (5% level)?

(c) Are the assay results different at the different hours (5% level)?

(d) What assumptions are made about the presence of interaction?

(e) If the assay results are significantly different at different hours, which hour(s) is different from the others?

§10. Duplicate tablets were assayed at hours 1, 3, and 5 for the data in Problem 9, using only presses 2, 4, and 6, with the following results:

Hour	Press		
	2	4	6
1	49, 52	49, 50	50, 53
3	50, 48	53, 51	52, 55
5	46, 47	49, 52	49, 53

If presses and hours are fixed, test for the significance of presses and hours at the 5% level. Is there significant interaction? Explain in words what is meant by interaction in this example.

11. Use Tukey's multiple range test to compare all three treatments (a posteriori test) for the data of Tables 8.13 and 8.14.

12. Compute the ANOVA for data of Table 8.17. Are treatments (columns) significantly different?

13. Perform an analysis of variance (one-way) comparing methods for the ratios (final assay/raw material assay) for data of Table 8.18. Compare probability level for methods to ANCOVA results.

REFERENCES

1. Snedecor GW, Cochran WG. Statistical Methods, 8th ed. Ames, IA: Iowa State University Press, 1989.
2. Dixon WJ, Massey FJ Jr. Introduction to Statistical Analysis, 3rd ed. New York: McGraw-Hill, 1969.
3. Scheffé H. The Analysis of Variance. New York: Wiley, 1964.
4. Dunnett C, Goldsmith C. When and How to do multiple comparisons. in: Buncher CR, Tsay J-Y, eds. Statistics in the Pharmaceutical Industry, 2nd ed. New York: Marcel Dekker, 1993:481–512.
5. Steel RGD, Torrie JH. Principles and Procedure of Statistics. New York: McGraw-Hill, 1960.
6. Dubey SD. Adjustment of P-values for multiplicities of Intercorrelating Symptoms. In: Buncher CR, Tsay J, eds. Statistics in the Pharmaceutical Industry, 2nd ed. New York: Marcel Dekker, 1993:513–528.
7. Comelli M. Multiple Endpoints. In: Chow S-C, ed. Encyclopedia of Pharmaceutical Statistics, Multiple Endpoints. New York: Marcel Dekker, 2000:333–344.
8. Snedecor GW, Cochran WG. Statistical Methods, 8th ed. Ames, IA: Iowa State University Press, 1989:273.
9. Permutt T. In: Chow S-C, ed. Encyclopedia of Pharmaceutical Statistics, Adjustments for Covariates. New York: Marcel Dekker, 2000:1–3.
10. FDA Stability Program, Moh-Jee Ng, Div. of Biometrics, Center for Drug Evaluation and Res., FDA 1992.

9 | Factorial Designs

Factorial designs are used in experiments where the effects of different factors, or conditions, on experimental results are to be elucidated. Some practical examples where factorial designs are optimal are experiments to determine the effect of pressure and lubricant on the hardness of a tablet formulation, to determine the effect of disintegrant and lubricant concentration on tablet dissolution, or to determine the efficacy of a combination of two active ingredients in an over-the-counter cough preparation. Factorial designs are the designs of choice for simultaneous determination of the effects of several factors and their interactions. This chapter introduces some elementary concepts of the design and analysis of factorial designs.

9.1 DEFINITIONS (VOCABULARY)

9.1.1 Factor

A *factor* is an *assigned variable* such as concentration, temperature, lubricating agent, drug treatment, or diet. The choice of factors to be included in an experiment depends on experimental objectives and is predetermined by the experimenter. A factor can be qualitative or quantitative. A *quantitative factor* has a numerical value assigned to it. For example, the factor "concentration" may be given the values 1%, 2%, and 3%. Some examples of *qualitative factors* are treatment, diets, batches of material, laboratories, analysts, and tablet diluent. Qualitative factors are assigned names rather than numbers. Although factorial designs may have one or many factors, only experiments with two factors will be considered in this chapter. Single-factor designs fit the category of one-way ANOVA designs. For example, an experiment designed to compare three drug substances using different patients in each drug group is a one-way design with the single-factor "drugs."

9.1.2 Levels

The levels of a factor are the values or designations assigned to the factor. Examples of levels are 30° and 50° for the factor 'temperature," 0.1 molar and 0.3 molar for the factor "concentration," and "drug" and "placebo" for the factor "drug treatment."

The *runs* or *trials* that comprise factorial experiments consist of all combinations of all levels of all factors. As an example, a two-factor experiment would be appropriate for the investigation of the effects of drug concentration and lubricant concentration on dissolution time of a tablet. If both factors were at two levels (two concentrations for each factor), four runs (dissolution determinations for four formulations) would be required, as follows:

Symbol	Formulation
(1)	Low drug and low lubricant concentration
a	Low drug and high lubricant concentration
b	High drug and low lubricant concentration
ab	High drug and high lubricant concentration

"Low" and "high" refer to the low and high concentrations preselected for the drug and lubricant. (Of course, the actual values selected for the low and high concentrations of drug will probably be different from those chosen for the lubricant.) The notation (symbol) for the various combinations of the factors, (1), a, b, ab, is standard. When both factors are at their low levels, we denote the combination as (1). When factor A is at its high level and factor B is at its low level, the combination is called a. b means that only factor B is at the high level, and ab means that both factors A and B are at their high levels.

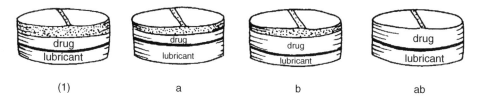

(1)   a   b   ab

9.1.3 Effects

The *effect* of a factor is the change in response caused by varying the level(s) of the factor. The *main effect* is the *effect* of a factor *averaged over all levels of the other factors.* In the previous example, a two-factor experiment with two levels each of drug and lubricant, the main effect due to drug would be the difference between the average response when drug is at the high level (runs b and ab) and the average response when drug is at the low level [runs (1) and a]. For this example the main effect can be characterized as a linear response, since the effect is the difference between the two points shown in Figure 9.1.

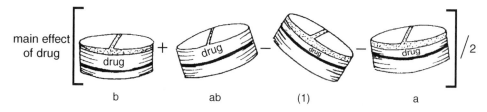

b   ab   (1)   a

More than two points would be needed to define more clearly the nature of the response as a function of the factor drug concentration. For example, if the response plotted against the levels of a quantitative factor is not linear, the definition of the main effect is less clear. Figure 9.2 shows an example of a curved (quadratic) response based on experimental results with a factor at three levels. In many cases, an important objective of a factorial experiment is to characterize the effect of changing levels of a factor or combinations of factors on the response variable.

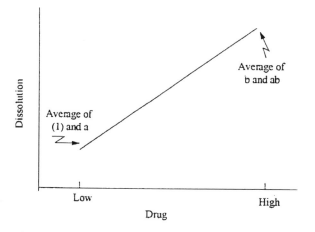

Figure 9.1 Linear effect of drug. a = lubricant, b = drug.

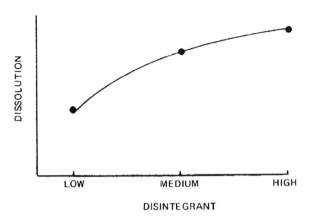

Figure 9.2   Nonlinear (quadratic) effect.

9.1.4   Interaction

Interaction may be thought of as a lack of "additivity of factor effects." For example, in a two-factor experiment, if factor *A* has an effect equal to 5 and factor *B* has an effect of 10, additivity would be evident if an effect of 15 (5 + 10) were observed when both *A* and *B* are at their high levels (in a two-level experiment). (It is well worth the extra effort to examine and understand this concept as illustrated in Fig. 9.3.)

  If the effect is greater than 15 when both factors are at their high levels, the result is *synergistic* (in biological notation) with respect to the two factors. If the effect is less than 15 when *A* and *B* are at their high levels, an *antagonistic* effect is said to exist. In statistical terminology, the lack of additivity is known as *interaction*. In the example above (two factors each at two levels), interaction can be described as the difference between the effects of drug concentration at the two lubricant levels. Equivalently, interaction is also the difference between the effects of lubricant at the two drug levels. More specifically, this means that the drug effect measured when the lubricant is at the low level [a − (1)] is *different* from the drug effect measured when the lubricant is at the high level (ab − b). If the drug effects are the same in the presence of both high and low levels of lubricant, the system is additive, and no interaction exists. Interaction is conveniently shown graphically as depicted in Figure 9.4. If the lines representing the effect of drug concentration at each level of lubricant are "parallel," there is no interaction. Lack of parallelism, as shown in Figure 9.4(B), suggests interaction. Examination of the lines in Figure 9.4(B) reveals that the effect of drug concentration on dissolution is dependent on the concentration of lubricant. The effects of drug and lubricant are not additive.

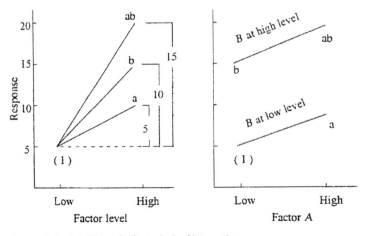

Figure 9.3   Additivity of effects: lack of interaction.

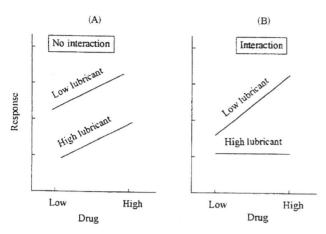

Figure 9.4 Illustration of interaction.

Factorial designs have many advantages [1]:

1. In the absence of interaction, factorial designs have maximum efficiency in estimating main effects.
2. If interactions exist, factorial designs are necessary to reveal and identify the interactions.
3. Since factor effects are measured over varying levels of other factors, conclusions apply to a wide range of conditions.
4. Maximum use is made of the data since all main effects and interactions are calculated from all of the data (as will be demonstrated below).
5. Factorial designs are orthogonal; all estimated effects and interactions are independent of effects of other factors. Independence, in this context, means that when we estimate a main effect, for example, the result we obtain is due only to the main effect of interest, and is not influenced by other factors in the experiment. In nonorthogonal designs (as is the case in many multiple-regression-type "fits"—see App. III), effects are not independent. *Confounding* is a result of lack of independence. When an effect is confounded, one cannot assess how much of the observed effect is due to the factor under consideration. The effect is influenced by other factors in a manner that often cannot be easily unraveled, if at all. Suppose, for example, that two drugs are to be compared, with patients from a New York clinic taking drug *A* and patients from a Los Angeles clinic taking drug *B*. Clearly, the difference observed between the two drugs is confounded with the different locations. The two locations reflect differences in patients, methods of treatment, and disease state, which can affect the observed difference in therapeutic effects of the two drugs. A simple factorial design where both drugs are tested in both locations will result in an "unconfounded," clear estimate of the drug effect if designed correctly, for example, equal or proportional number of patients in each treatment group at each treatment site.

9.2 TWO SIMPLE HYPOTHETICAL EXPERIMENTS TO ILLUSTRATE THE ADVANTAGES OF FACTORIAL DESIGNS

The following hypothetical experiment illustrates the advantage of the factorial approach to experimentation when the effects of multiple factors are to be assessed. The problem is to determine the effects of a special diet and a drug on serum cholesterol levels. To this end, an experiment was conducted in which cholesterol changes were measured in three groups of patients. Group *A* received the drug, group *B* received the diet, and group *C* received both the diet and drug. The results are shown below. The experimenter concluded that there was no interaction between drug and diet (i.e., their effects are additive).

Drug alone: decrease of 10 mg%
Diet alone: decrease of 20 mg%
Diet + drug: decrease of 30 mg%

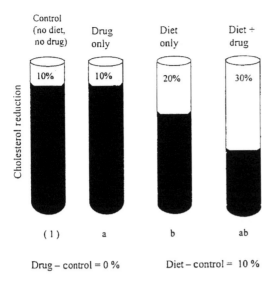

Drug – control = 0 %   Diet – control = 10 %

$$\text{Interaction} = 5\,\% = \frac{\{\,ab - b\,\} - \{\,a - (\,1\,)\,\}}{2}$$

Figure 9.5 Synergism in cholesterol lowering as a result of drug and diet.

However, suppose that patients given *neither* drug nor diet would have shown a decrease of serum cholesterol of 10 mg% had they been included in the experiment. (Such a result could occur because of "psychological effects" or seasonal changes, for example.) Under these circumstances, we would conclude that drug alone has no effect, that diet results in a cholesterol lowering of 10 mg%, and that the combination of drug and diet is synergistic. The combination of drug and diet results in a decrease of cholesterol equal to 20 mg%. This concept is shown in Figure 9.5.

Thus, without a fourth group, the control group (low level of diet and drug), we have no way of assessing the presence of interaction. This example illustrates how estimates of effects can be incorrect when pieces of the design are missing. Inclusion of a control group would have completed the factorial design, two factors at two levels. Drug and diet are the factors, each at two levels, either present or absent. The complete factorial design consists of the following four groups:

(1)   Group on normal diet without drug (drug and special diet at low level).
a   Group on drug only (high level of drug, low level of diet).
b   Group on diet only (high level of diet, low level of drug).
ab   Group on diet and drug (high level of drug and high level of diet).

The effects and interaction can be clearly calculated based on the results of these four groups (Fig. 9.5).

Incomplete factorial designs such as those described above are known as the *one-at-a-time* approach to experimentation. Such an approach is usually very *inefficient*. By performing the entire factorial, we usually have to do *less work*, and we get *more* information. This is a consequence of an important attribute of factorial designs: effects are measured with maximum precision. To demonstrate this property of factorial designs, consider the following hypothetical example. The objective of this experiment is to weigh two objects on an insensitive balance. Because of the lack of reproducibility, we will weigh the items in duplicate. The balance is in such poor condition that the zero point (balance reading with no weights) is in doubt. A typical one-at-a-time experiment is to weigh each object separately (in duplicate) in addition to a duplicate reading with no weights on the balance. The weight of item *A* is taken as the average of the readings with *A* on the balance minus the average of the readings with the pans empty. Under the assumption that the variance is the same for all weighings, regardless of the amount of material being weighed, the variance of the weight of A is the sum of the variances of the average weight of A and the average weight with the pans empty (see App. I)

$$\frac{\sigma^2}{2} + \frac{\sigma^2}{2} = \sigma. \tag{9.1}$$

Note that the variance of the *difference* of the average of two weighings is the *sum of the variances* of each weighing. (The variance of the average of *two* weighings is $\sigma^2/2$.)

Similarly, the variance of the weight of B is $\sigma^2 = \sigma^2/2 + \sigma^2/2$. Thus, based on six readings (two weighings each with the balance empty, with A and B on the balance), we have estimated the weights of A and B with variance equal to σ^2, where σ^2 is the variance of a single weighing.

In a factorial design, an extra reading(s) would be made, a reading with both A and B on the balance. In the following example, using a full factorial design, we can estimate the weight of A with the same precision as above using only 4 weighings (instead of 6). In this case the weighings are made without replication. That is, four weighings are made as follows:

(1)	Reading with balance empty	0.5 kg
a	Reading with item A on balance	38.6 kg
b	Reading with item B on balance	42.1 kg
ab	Reading with both items A and B on balance	80.5 kg

With a full factorial design, as illustrated above, the *weight of A* is estimated as (the main effect of A)

$$\frac{a - (1) + ab - b}{2}. \tag{9.2}$$

Expression (9.2) says that the estimate of the weight of A is the average of the weight of A alone minus the reading of the empty balance [a − (1)] and the weight of both items A and B minus the weight of B. According to the weights recorded above, the weight of A would be estimated as

$$\frac{38.6 - 0.5 + 80.5 - 42.1}{2} = 38.25 \, \text{kg}.$$

Similarly, the weight of B is estimated as

$$\frac{42.1 - 0.5 + 80.5 - 38.6}{2} = 41.75 \, \text{kg}.$$

Note how we use *all the data* to estimate the weights of A and B; the weight of B alone is used to help estimate the weight of A, and vice versa!

Interaction is measured as the average difference of the weights of A in the presence and absence of B as follows:

$$\frac{(ab - b) - [a - (1)]}{2}. \tag{9.3}$$

We can assume that there is no interaction, a very reasonable assumption in the present example. (The weights of the combined items should be the sum of the individual weights.) The estimate of interaction in this example is

$$\frac{(80.5 - 42.1) - (38.6 - 0.5)}{2} = 0.3.$$

The estimate of interaction is not zero because of the presence of random errors made on this insensitive balance.

Table 9.1   Eight Experiments for a 2^3 Factorial Design[a]

Combination	A	B	C
(1)	−	−	−
a	+	−	−
b	−	+	−
ab	+	+	−
c	−	−	+
ac	+	−	+
bc	−	+	+
abc	+	+	+

[a] −, factor at low level; +, factor at high level.

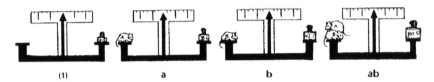

    (1)   a   b   ab

In this example, we have made *four* weighings. The variance of the main effects (i.e., the average weights of A and B) is σ^2, *exactly the same variance as was obtained using six weightings in the one-at-a-time experiment!*[*] We obtain the same precision with two-thirds of the work: four readings instead of six. In addition to the advantage of greater precision, if interaction were present, we would have had the opportunity to estimate the interaction effect in the full factorial design. *It is not possible to estimate the interaction in the one-at-a-time experiment.*

9.3 PERFORMING FACTORIAL EXPERIMENTS: RECOMMENDATIONS AND NOTATION

The simplest factorial experiment, as illustrated above, consists of four trials, two factors each at two levels. If three factors, A, B, and C, each at two levels, are to be investigated, eight trials are necessary for a full factorial design, as shown in Table 9.1. This is also called a 2^3 experiment, three factors each at two levels.

As shown in Table 9.1, in experiments with factors at two levels, the low and high levels of factors in a particular run are denoted by the absence or presence of the letter, respectively. For example, if all factors are at their low levels, the run is denoted as (1). If factor A is at its high level, and B and C are at their low levels, we use the notation a. If factors A and B are at their high levels, and C is at its low level, we use the notation ab, and so on.

Before implementing a factorial experiment, the researcher should carefully consider the experimental objectives vis-à-vis the appropriateness of the design. The results of a factorial experiment may be used (a) to help interpret the mechanism of an experimental system; (b) to recommend or implement a practical procedure or set of conditions in an industrial manufacturing situation; or (c) as guidance for further experimentation. In most situations where one is interested in the effect of various factors or conditions on some experimental outcome, factorial designs will be optimal.

The choice of factors to be included in the experimental design should be considered carefully. Those factors not relevant to the experiment, but which could influence the results, should be carefully controlled or kept constant. For example, if the use of different technicians, different pieces of equipment, or different excipients can affect experimental outcomes, but are not variables of interest, they should not be allowed to vary randomly, if possible. Consider an example of the comparison of two analytical methods. We may wish to have a single analyst

---

* The main effect of A, for example, is $[a - (1) + ab - b]/2$. The variance of the main effect is $(\sigma_a^2 + \sigma_{(1)}^2 + \sigma_{ab}^2 + \sigma_b^2)/4 = \sigma^2$. σ^2 is the same for all weighings (App. I).

perform both methods on the same spectrophotometer to reduce the variability that would be present if different analysts used different instruments. However, there will be circumstances where the effects due to different analysts and different spectrophotometers are of interest. In these cases, different analysts and instruments may be designed into the experiment as additional factors.

On the other hand, we may be interested in the effect of a particular factor, but because of time limitations, cost, or other problems, the factor is held constant, retaining the option of further investigation of the factor at some future time. In the example above, one may wish to look into possible differences among analysts with regard to the comparison of the two methods (an analyst × method interaction). However, time and cost limitations may restrict the extent of the experiment. One analyst may be used for the experiment, but testing may continue at some other time using more analysts to confirm the results.

The more extraneous variables that can be controlled, the smaller will be the residual variation. The residual variation is the random error remaining after the ANOVA removes the variability due to factors and their interactions. If factors known to influence the experimental results, but of no interest in the experiment, are allowed to vary "willy-nilly," the effects caused by the random variation of these factors will become part of the residual error. Suppose the temperature influences the analytical results in the example above. If the temperature is not controlled, the experimental error will be greater than if the experiment is carried out under constant-temperature conditions. The smaller the residual error, the more sensitive the experiment will be in detecting effects or changes in response due to the factors under investigation.

The choice of levels is usually well defined if factors are qualitative. For example, in an experiment where a product supplied by several manufacturers is under investigation, the levels of the factor "product" could be denoted by the name of the manufacturer: company X, company Y, and so on. If factors are quantitative, we can choose two or more levels, the choice being dependent on the size of the experiment (the number of trials and the amount of replication) and the nature of the anticipated response. If a response is known to be a linear function of a factor, two levels would be sufficient to define the response. If the response is "curved" (a quadratic response, for example[†]), at least three levels of the quantitative factor would be needed to characterize the response. Two levels are often used for the sake of economy, but a third level or more can be used to meet experimental objectives as noted above. A rule of thumb used for the choice of levels in two-level experiments is to divide extreme ranges of a factor into four equal parts and take the one-fourth ($^1/_4$) and three-fourths ($^3/_4$) values as the choice of levels [1]. For example, if the minimum and maximum concentrations for a factor are 1% and 5%, respectively, the choice of levels would be 2% and 4% according to this empirical rule.

The trials comprising the factorial experiment should be done in random order if at all possible. This helps ensure that the results will be unbiased (as is true for many statistical procedures). The fact that all effects are averaged over all runs in the analysis of factorial experiments is also a protection against bias.

9.4 A WORKED EXAMPLE OF A FACTORIAL EXPERIMENT

The data in Table 9.2 were obtained from an experiment with three factors each at two levels. There is no replication in this experiment. Replication would consist of repeating each of the eight runs one or more times. The results in Table 9.2 are presented in standard order. Recording the results in this order is useful when analyzing the data by hand (see below) or for input into computers where software packages require data to be entered in a specified or standard order. The standard order for a 2^2 experiment consists of the first four factor combinations in Table 9.2. For experiments with more than three factors, see Davies for tables and an explanation of the ordering [1].

The experiment that we will analyze is designed to investigate the effects of three components (factors)—stearate, drug, and starch—on the thickness of a tablet formulation. In this example, two levels were chosen for each factor. Because of budgetary constraints, use of more than two levels would result in too large an experiment. For example, if one of the three

---

[†] A quadratic response is of the form $Y = A + BX + CX^2$, where Y is the response and X is the factor level.

Table 9.2 Results of 2^3 Factorial Experiment: Effect of Stearate, Drug, and Starch Concentration on Tablet Thickness[a]

Factor combination	Stearate	Drug	Starch	Response (thickness) (cm × 10^3)
(1)	−	−	−	475
a	+	−	−	487
b	−	+	−	421
ab	+	+	−	426
c	−	−	+	525
ac	+	−	+	546
bc	−	+	+	472
abc	+	+	+	522

[a] −, factor at low level; +, factor at high level.

factors were to be studied at three levels, 12 formulations would have to be tested for a 2 × 2 × 3 factorial design. Because only two levels are being investigated, nonlinear responses cannot be elucidated. However, the pharmaceutical scientist felt that the information from this two-level experiment would be sufficient to identify effects that would be helpful in designing and formulating the final product. The levels of the factors in this experiment were as follows:

Factor	Low level (mg)	High level (mg)
A: Stearate	0.5	1.5
B: Drug	60.0	120.0
C: Starch	30.0	50.0

The computation of the main effects and interactions as well as the ANOVA may be done by hand in simple designs such as this one. Readily available computer programs are usually used for more complex analyses. (For *n* factors, an *n*-way analysis of variance is appropriate. In typical factorial designs, the factors are usually considered to be fixed.)

For two-level experiments, the effects can be calculated by applying the signs (+ or −) arithmetically for each of the eight responses as shown in Table 9.3. This table is constructed by placing a + or − in columns *A*, *B*, and *C* depending on whether or not the appropriate factor is at the high or low level in the particular run. If the letter appears in the factor combination, a + appears in the column corresponding to that letter. For example, for the product combination ab, a + appears in columns *A* and *B*, and a − appears in column *C*. Thus for column *A*, runs a, ab, ac, and abc have a + because in these runs, *A* is at the high level. Similarly, for runs (1), b, c, and bc, a − appears in column *A* since these runs have *A* at the low level.

Table 9.3 Signs to Calculate Effects in a 2^3 Factorial Experiment[a]

Factor combination	Level of factor in experiment			Interaction[b]			
	A	*B*	*C*	*AB*	*AC*	*BC*	*ABC*
(1)	−	−	−	+	+	+	−
a	+	−	−	−	−	+	+
b	−	+	−	−	+	−	+
ab	+	+	−	+	−	−	−
c	−	−	+	+	−	−	+
ac	+	−	+	−	+	−	−
bc	−	+	+	−	−	+	−
abc	+	+	+	+	+	+	+

[a] −, factor at low level; +, factor at high level.
[b] Multiply signs of factors to obtain signs for interaction terms in combination [e.g., *AB* at (1) = (−) × (−) = (+)].

Columns denoted by AB, AC, BC, and ABC in Table 9.3 represent the indicated interactions (i.e., AB is the interaction of factors A and B, etc.). The signs in these columns are obtained by multiplying the signs of the individual components. For example, to obtain the signs in column AB we refer to the signs in column A and column B. For run (1), the $+$ sign in column AB is obtained by multiplying the $-$ sign in column A times the $-$ sign in column B. For run a, the $-$ sign in column AB is obtained by multiplying the sign in column A $(+)$ times the sign in column B $(-)$. Similarly, for column ABC, we multiply the signs in columns A, B, and C to obtain the appropriate sign. Thus run ab has a $-$ sign in column ABC as a result of multiplying the three signs in columns A, B, and C: $(+) \times (+) \times (-)$.

The average effects can be calculated using these signs as follows. To obtain the average effect, multiply the response times the sign for each of the eight runs in a column, and divide the result by 2^{n-1}, where n is the number of factors (for three factors, 2^{n-1} is equal to 4). This will be illustrated for the calculation of the main effect of A (stearate). The main effect for factor A is

$$\frac{[-(1) + a - b + ab - c + ac - bc + abc]}{4}. \tag{9.4}$$

Note that the main effect of A is the average of all results at the high level of A minus the average of all results at the low level of A. This is more easily seen if formula (9.4) is rewritten as follows:

$$\text{Main effect of } A = \frac{a + ab + ac + abc}{4} - \frac{(1) + b + c + bc}{4}. \tag{9.5}$$

"Plugging in" the results of the experiment for each of the eight runs in Eq. (9.5), we obtain

$$\frac{[487 + 426 + 546 + 522 - (475 + 421 + 525 + 472)] \times 10^{-3}}{4} = 0.022 \text{ cm.}$$

The *main effect of* A *is interpreted* to mean that the net effect of increasing the stearate concentration from the low to the high level (averaged over all other factor levels) is to increase the tablet thickness by 0.022 cm. This result is illustrated in Figure 9.6.

The interaction effects are estimated in a manner similar to the estimation of the main effects. The signs in the column representing the interaction (e.g., AC) are applied to the eight responses, and as before the total divided by 2^{n-1}, where n is the number of factors. The interaction AC, for example, is defined as one-half the difference between the effect of A when

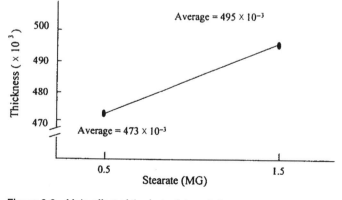

Figure 9.6 Main effect of the factor "stearate."

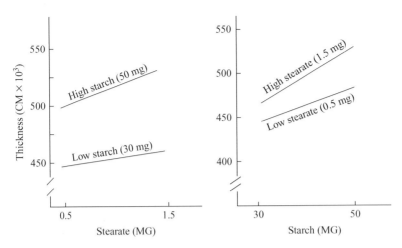

Figure 9.7 Starch × stearate interaction.

C is at the high level and the effect of A when C is at the low level (Fig. 9.7). Applying the signs as noted above, the AC interaction is estimated as

$$AC \text{ interaction} = \frac{1}{4}\{(abc + ac - bc - c) - [ab + a - b - (1)]\}. \tag{9.6}$$

The interaction is shown in Figure 9.7. With starch (factor C) at the high level, 50 mg, increasing the stearate concentration from the low to the high level (from 0.5 mg to 1.5 mg) results in an increased thickness of 0.0355 cm.[‡] At the low level of starch, 30 mg, increasing stearate concentration from 0.5 mg to 1.5 mg results in an increased thickness of 0.0085 cm. Thus stearate has a greater effect at the higher starch concentration, a possible starch × stearate interaction.

Lack of interaction would be evidenced by the same effect of stearate at both low and high starch concentrations. In a real experiment, the effect of stearate would not be identical at both levels of starch concentration in the absence of interaction because of the presence of experimental error. The statistical tests described below show how to determine the significance of observed nonzero effects.

The description of interaction is "symmetrical." The AC interaction can be described in two equivalent ways: (a) the effect of stearate is greater at high starch concentrations, or (b) the effect of starch concentration is greater at the high stearate concentration (1.5 mg) compared to its effect at low stearate concentration (0.5 mg). The effect of starch at low stearate concentration is 0.051. The effect of starch at high stearate concentration is 0.078. (Also see Fig. 9.7.)

The details of the analysis in this section is meant to give an insight into the interpretation of data resulting from a factorial experiment. In the usual circumstances, the analysis would be performed using a suitable computer program. To intelligently interpret the output from the program, it is essential that one understands the underlying principles and analysis.

9.4.1 Data Analysis

9.4.1.1 Method of Yates

Computers are usually used to analyze factorial experiments. However, hand analysis of simple experiments can give insight into the properties of this important class of experimental designs. A method devised by Yates for systematically analyzing data from 2^n factorial experiments (n factors each at two levels) is of historical interest and is demonstrated in Table 9.4. The data are first tabulated in standard order (see Ref. [1] for experiments with more than two levels).

[‡] $(1/2)(abc + ac - bc - c)$.

Table 9.4 Yates Analysis of the Factorial Tableting Experiment for Analysis of Variance

Combination	Thickness $(\times 10^3)$	(1)	(2)	(3)	Effect $(\times 10^3)(3)/4$	Mean square $(\times 10^6)(3)^2/8$
(1)	475	962	1809	3874	—	—
a	487	847	2065	88	22.0	968
b	421	1071	17	−192	−48.0	4608
ab	426	994	71	22	5.5	60.5
c	525	12	−115	256	64.0	8192
ac	546	5	−77	54	13.5	364.5
bc	472	21	−7	38	9.5	180.5
abc	522	50	29	36	9.0	162

The data are first added in pairs, followed by taking differences in pairs as shown in column (1) in Table 9.4.

$$475 + 487 = 962$$
$$421 + 426 = 847$$
$$525 + 546 = 1071$$
$$472 + 522 = 994$$
$$487 - 475 = 12$$
$$426 - 421 = 5$$
$$546 - 525 = 21$$
$$522 - 472 = 50$$

This addition and subtraction process is repeated sequentially on the n columns. (Remember that n is the number of factors, three columns for three factors.) Thus the process is repeated in column (2), operating on the results in column (1) of Table 9.4. Note, for example, that 1809 in column (2) is $962 + 847$ from column (1). Finally, the process is repeated, operating on column (2) to form column (3). Column (3) is divided by 2^{n-1} ($2^{n-1} = 4$ for three factors) to obtain the average effect. The mean squares for the ANOVA (described below) are obtained by dividing the square of column (n) by 2^n. For example, the mean square attributable to factor A is

$$\text{Mean square for } A = \frac{(88)^2}{8} = 968.$$

The mean squares are presented in an ANOVA table, as discussed below.

9.4.1.2 Analysis of Variance
The results of a factorial experiment are typically presented in an ANOVA table, as shown in Table 9.5. In a 2^n factorial, each effect and interaction has 1 degree of freedom. The error mean square for statistical tests and estimation can be estimated in several ways for a factorial experiment. Running the experiment with replicates is best. Duplicates are usually sufficient. However,

Table 9.5 Analysis of Variance for the Factorial Tableting Experiment

Factor	Source	d.f.	Mean square ($\times 10^6$)	F^a
A	Stearate	1	968	7.2^b
B	Drug	1	4608	34.3^c
C	Starch	1	8192	61.0^c
AB	Stearate × drug	1	60.5	
AC	Stearate × starch	1	364.5	2.7
BC	Drug × starch	1	180.5	
ABC	Stearate × drug × starch	1	162	

[a] Error mean square based on AB, BC, and ABC interactions, 3 d.f.
[b] $p < 0.1$.
[c] $p < 0.01$.

replication may result in an inordinately large number of runs. Remember that replicates do not usually consist of replicate analyses or observations on the same run. A true replicate usually is obtained by repeating the run, from "scratch." For example, in the 2^3 experiment described above, determining the thickness of several tablets from a single run [e.g., the run denoted by a (A at the high level)] would probably not be sufficient to estimate the experimental error in this system. The proper replicate would be obtained by preparing a new mix with the same ingredients, retableting, and measuring the thickness of tablets in this new batch.[§] In the absence of replication, experimental error may be estimated from prior experience in systems similar to that used in the factorial experiment. To obtain the error estimate from the experiment itself is always most desirable. Environmental conditions in prior experiments are apt to be different from those in the current experiment. In a large experiment, the experimental error can be estimated without replication by pooling the mean squares from higher order interactions (e.g., three-way and higher order interactions) as well as other interactions known to be absent, a priori. For example, in the tableting experiment, we might average the mean squares corresponding to the two-way interactions, AB and BC, and the three-way ABC interaction, if these interactions were known to be zero from prior considerations. The error estimated from the average of the AB, BC, and ABC interactions is

$$(60.5 + 180.5 + 162) \times \frac{10^{-6}}{3} = 134.2 \times 10^{-6}.$$

with 3 degrees of freedom (assuming that these interactions do not exist).

9.4.1.3   Interpretation

In the absence of interaction, the main effect of a factor describes the change in response when going from one level of a factor to another. If a large interaction exists, the main effects corresponding to the interaction do not have much meaning as such. Specifically, an AC interaction suggests that the effect of A depends on the level of C and a description of the results should specify the change due to A at each level of C. Based on the mean squares in Table 9.5, the effects that are of interest are A, B, C, and AC. Although not statistically significant, stearate and starch interact to a small extent, and examination of the data is necessary to describe this effect (Fig. 9.7). Since B does not interact with A or C, it is sufficient to calculate the effect of drug (B), averaged over all levels of A and C, to explain its effect. The effect of drug is to *decrease* the thickness by 0.048 mm when the drug concentration is raised from 60 to 120 mg [Table 9.4, column (3)/4].

9.5   FRACTIONAL FACTORIAL DESIGNS

In an experiment with a large number of factors and/or a large number of levels for the factors, the number of experiments needed to complete a factorial design may be inordinately large. For example, a factorial design with five factors each at two levels requires 32 experiments; a three-factor experiment each at three levels requires 27 experiments. If the cost and time considerations make the implementation of a full factorial design impractical, fractional factorial experiments can be used in which a fraction (e.g., $^1/_2$, $^1/_4$, etc.) of the original number of experiments can be run. Of course, something must be sacrificed for the reduced work. If the experiments are judiciously chosen, it may be possible to design an experiment so that effects that we believe are negligible are confounded with important effects. (The word "confounded" has been noted before in this chapter.) In fractional factorial designs, the negligible and important effects are indistinguishable, and thus confounded. This will become clearer in the first example.

To illustrate some of the principles of fractional factorial designs, we will discuss and present an example of a fractional design based on a factorial design where each of three factors is at two levels, a 2^3 design. Table 9.3 shows the eight experiments required for the full design. With the full factorial design, we can estimate seven effects from the eight experiments, the three main effects (A, B, and C), and the four interactions (AB, AC, BC, and ABC). In a $^1/_2$ replicate fractional design, we perform four experiments, but we can only estimate three effects. With

[§] If the tableting procedure in the different runs were identical in all respects (with the exception of tablet ingredients), replicates within each run would be a proper estimate of error.

Table 9.6 2^2 Factorial Design

Experiment	A level	B level	AB
(1)	−	−	+
a	+	−	−
b	−	+	−
ab	+	+	+

three factors, a $1/2$ replicate can be used to estimate the main effects, A, B, and C. The following procedure is used to choose the four experiments.

Table 9.6 shows the four experiments that define a 2^2 factorial design using the notation described in section 9.3.

To construct the $1/2$ replicate with three factors, we equate one of the effects to the third factor. In the 2^2 factorial the interaction, AB is equated to the third factor, C. Table 9.7 describes the $1/2$ replicate design for three factors. The four experiments consist of (1) c at the high level (a, b at the low level); (2) a at the high level (b, c at the low level); (3) b at the high level (a, c at the low level); and (4) a, b, c all at the high level.

From Table 9.7, we can define the confounded effects, also known as aliases. An effect is defined by the signs in the columns of Table 9.7. For example, the effect of A is

$$(a + abc) - (c + b).$$

Note that the effect of A is exactly equal to BC. Therefore, BC and A are confounded (they are aliases). Also note that $C = AB$ (by definition) and $B = AC$. Thus, in this design the main effects are confounded with the two factor interactions. This means that the main effects cannot be clearly interpreted if interactions are not absent or negligible. If interactions are negligible, this design will give fair estimates of the main effects. If interactions are significant, this design is not recommended.

Example 1. Davies [1] gives an excellent example of weighing three objects on a balance with a zero error in a $1/2$ replicate of a 2^3 design. This illustration is used because interactions are zero when weighing two or more objects together (i.e., the weight of two or more objects is the sum of the individual weights). The three objects are denoted as A, B, and C; the high level is the presence of the object to be weighed, and the low level is the absence of the object. From Table 9.7, we would perform four weighinings: A alone, B alone, C alone, and A, B, and C together (call this ABC).

1. The weight of A is the (weight of A + the weight of ABC − the weight of B − weight of C)/2.
2. The weight of B is the (weight of B + the weight of ABC − the weight of A − weight of C)/2.
3. The weight of C is the (weight of C + the weight of ABC − the weight of A − weight of B)/2.

As noted by Davies, this illustration is not meant as a recommendation of how to weigh objects, but rather to show how the design works in the absence of interaction. (See Exercise Problem 5 as another way to weigh these objects using a $1/2$ replicate fractional factorial design.)

Example 2. *A $1/2$ replicate of a 2^4 experiment:* Chariot et al. [2] reported the results of a factorial experiment studying the effect of processing variables on extrusion–spheronization of wet powder masses. They identified five factors each at two levels, the full factorial requiring 32 experiments. Initially, they performed a $1/4$ replicate, requiring eight experiments. One of the factors, extrusion speed, was not significant. To simplify this discussion, we will ignore this

Table 9.7 One-Half Replicate of 2^3 Factorial Design

Experiment	A level	B level	C = AB	AC	BC
c	−	−	+	−	−
a	+	−	−	−	+
b	−	+	−	+	−
abc	+	+	+	+	+

Table 9.8 One-Half Replicate of 2^4 Factorial Design (Extrusion–Spheronization of Wet Powder Masses)

		Parameter						
Experiment	A (min)	B (rpm)	C (kg)	D (mm)	$AB^a = CD$	$AC = BD$	$AD = BC$	Response
(1)	−	−	−	−	+	+	+	75.5
ab	+	+	−	−	+	−	−	55.5
ac	+	−	+	−	−	+	−	92.8
ad	+	−	−	+	−	−	+	45.4
bc	−	+	+	−	−	−	+	46.5
bd	−	+	−	+	−	+	−	19.7
cd	−	−	+	+	+	−	−	11.1
abcd	+	+	+	+	+	+	+	55.0

[a] Illustrates confounding.

factor for our example. The design and results are shown in Table 9.8. $A =$ spheronization time, $B =$ spheronization speed, $C =$ spheronization load, and $D =$ extrusion screen.

Note the confounding pattern shown in Table 9.8. The reader can verify these confounded effects (see Exercise Problem 6 at the end of this chapter). Table 9.8 was constructed by first setting up the standard 2^3 factorial (Table 9.3) and substituting D for the ABC interaction. For the estimated effects to have meaning, the confounded effects should be small. For example, if BC and AD were both significant, the interpretation of BC and/or AD would be fuzzy.

To estimate the effects, we add the responses multiplied by the signs in the appropriate column and divide by 4. For example, the effect of AB is

$$\frac{[75.5 + 55.5 - 92.8 - 45.4 - 46.5 - 19.7 + 11.14 + 55.0]}{4} = -1.825.$$

Estimates of the other effects are (see Exercise Problem 7)

$A = + 23.98$
$B = -12.03$
$C = + 2.33$
$D = - 34.78$
$AB = -1.83$
$AC = + 21.13$
$AD = + 10.83$

We cannot perform tests for the significance of these parameters without an estimate of the error (variance). The variance can be estimated from duplicate experiments, nonexistent interactions, or experiments from previous studies, for example. Based on the estimate above, factor A, D, and AC are the largest effects. To help clarify the possible confounding effects, eight more experiments can be performed. For example, the large effect observed for the interaction AC, spheronization time × spheronization load could be exaggerated due to the presence of a BD interaction. Without other insights, it is not possible to separate these two interactions (they are aliases in this design). Therefore, this design would not be desirable if the nature of these interactions is unknown. Data for the eight further experiments that complete the factorial design are given in Exercise Problem 8.

The conclusions given by Chariot et al. are as follows:

1. Spheronization time (factor A) has a positive effect on the production of spheres.
2. There is a strong interaction between factors A and C (spheronization time × spheronization load). Note that the BD interaction is considered to be small.
3. Spheronization speed (factor B) has a negative effect on yield.
4. The interaction between spheronization speed and spheronization load (BC) appears significant. The AD interaction is considered to be small.
5. The interaction between spheronization speed and spheronization time (AB) appears to be insignificant. The CD interaction is considered to be small.
6. Extrusion screen (D) has a very strong negative effect.

Table 9.9 Some Fractional Designs for Up to five Factors

Observations	Factors	Fraction of full factorial	Defining contrast	Confounding	Design
4	3	1/2	−ABC	Main effects confused with two-way interactions	(1), ab, ac, bc
8	4	1/2	ABCD	Main effects and three two-way interactions are not confused	(1), ab, ac, bc, ad, bd, cd, abcd
8	5	1/4	−BCE −ADE	Main effects confused with two-way interactions (see references note below)	(1), ad, bc, abcd, abe, bde, ace, cde
16	5	1/2	ABCDE	Main effects and two-factor interactions are not confused	(1), ab, ac, bc, ad, bd, cd, abcd, ae, be, ce, abce, de, abde, acde, bcde

See Refs. [1,3] for more detailed discussion and other designs.

Table 9.9 presents some fractional designs with up to eight observations. To find the aliases (confounded effects), multiply the defining contrast in the table by the effect under consideration. Any letter that appears twice is considered to be equal to 1. The result is the confounded effect. For example, if the defining contrast is $-ABC$ and we are interested in the alias of A, we multiply $-ABC$ by $A = -A^2BC = -BC$. Therefore, A is confounded with $-BC$. Similarly, B is confounded with $-AC$ and C is confounded with $-AB$.

9.6 SOME GENERAL COMMENTS

As noted previously, experiments need not be limited to factors at two levels, although the use of two levels is often necessary to keep the experiment at a manageable size. Where factors are quantitative, experiments at more than two levels may be desirable when curvature of the response is anticipated. As the number of levels increase, the size of the experiment increases rapidly and fractional designs are recommended.

The theory of factorial designs is quite fascinating from a mathematical viewpoint. Particularly, the algebra and arithmetic lead to very elegant concepts. For those readers interested in pursuing this topic further, the book *The Design and Analysis of Industrial Experiments*, edited by Davies, is indispensable [1]. This topic is also discussed in some detail in Ref. [4]. Applications of factorial designs in pharmaceutical systems have appeared in the recent pharmaceutical literature. Plaizier-Vercammen and De Neve investigated the interaction of povidone with low-molecular-weight organic molecules using a factorial design [5]. Bolton has shown the application of factorial designs to drug stability studies [6]. Ahmed and Bolton optimized a chromatographic assay procedure based on a factorial experiment [7].

KEY TERMS

Additivity
Aliases
Confounding
Main effect
One-at-a-time experiment
Replication
Effects
Factor
Fractional factorial designs

Half replicate
Interaction
Level
Residual variation
Runs
Standard order
2^n factorials
Yates analysis

EXERCISES

1. A 2^2 factorial design was used to investigate the effects of stearate concentration and mixing time on the hardness of a tablet formulation. The results below are the averages of the hardness of 10 tablets. The variance of an average of 10 determinations was estimated from replicate determinations as 0.3 (d.f. $= 36$). This is the error term for performing statistical tests of significance.

	Stearate	
Mixing time (min)	0.5%	1%
15	9.6 (1)	7.5 (a)
30	7.4 (b)	7.0 (ab)

(a) Calculate the ANOVA and present the ANOVA table.
(b) Test the main effects and interaction for significance.
(c) Graph the data showing the possible AB interaction.

2. Show how to calculate the effect of increasing stearate concentration at low starch level for the data in Table 9.2. The answer is an increased thickness of 0.085 cm. Also, compute the drug $\times$ starch interaction.

3. The end point of a titration procedure is known to be affected by (1) temperature, (2) pH, and (3) concentration of indicator. A factorial experiment was conducted to estimate the effects of the factors. Before the experiment was conducted, all interactions were thought to be negligible except for a pH $\times$ indicator concentration interaction. The other interactions are to be pooled to form the error term for statistical tests. Use the Yates method to calculate the ANOVA based on the following assay results:

Factor combination	Recovery (%)	Factor combination	Recovery (%)
(1)	100.7	c	99.9
a	100.1	ac	99.6
b	102.0	bc	98.5
ab	101.0	abc	98.1

(a) Which factors are significant?
(b) Plot the data to show main effects and interactions that are significant.
(c) Describe, in words, the BC interaction.

4. A clinical study was performed to assess the effects of a combination of ingredients to support the claim that the combination product showed a synergistic effect compared to the effects of the two individual components. The study was designed as a factorial with each component at two levels.
Ingredient A: low level, 0; high level, 5 mg
Ingredient B: low level, 0; high level, 50 mg
Following is the analysis of variance table:

Source	d.f.	MS	F
Ingredient A	1	150	12.5
Ingredient B	1	486	40.5
$A \times B$	1	6	0.5
Error	20	12	

The experiment consisted of observing six patients in each cell of the 2^2 experiment. One group took placebo with an average result of 21. A second group took ingredient A at a 5-mg dose with an average result of 25. The third group had ingredient B at a 50-mg dose with an average result of 29, and the fourth group took a combination of 5 mg of A and 50 mg of B with a result of 35. In view of the results and the ANOVA, discuss arguments for or against the claim of synergism.

5. The three objects in the weighing experiment described in section 9.5, Example 1, may also be weighed using the other four combinations from the 2^3 design not included in the example. Describe how you would weigh the three objects using these new four weighings. [Note that these combinations comprise a $1/2$ replicate of a fractional factorial with a different confounding pattern from that described in section 9.5. (Hint: See Table 9.9.)

6. Verify that the effects ($AB = CD$, $AC = BD$, and $AD = BC$) shown in Table 9.8 are confounded.

7. Compute the effects for the data in section 9.5, example 2 (Table 9.8).

8. ¶In example 2 in section 9.5 (Table 9.8), eight more experiments were performed with the following results:

Experiment	Response
a	78.7
b	56.9
c	46.7
ab	21.2
abc	67.0
abd	29.0
acd	34.9
bcd	1.2

Using the entire 16 experiments (the 8 given here plus the 8 in Table 9.7), analyze the data as a full 2^4 factorial design. Pool the three-factor and four-factor interactions (5 d.f.) to obtain an estimate of error. Test the other effects for significance at the 5% level. Explain and describe any significant interactions.

REFERENCES

1. Davies OL. The Design and Analysis of Industrial Experiments. New York: Hafner, 1963.
2. Chariot M, Frances GA, Lewis D, et al. A factorial approach to process variables of extrusion-spheronisation of wet powder masses. Drug Dev Ind Pharm 1987; 13(9–11):1639–1649.
3. Beyer WH, ed. Handbook of Tables for Probability and Statistics. Cleveland, OH: The Chemical Rubber Co., 1966.
4. Box GE, Hunter WG, Hunter JS. Statistics for Experimenters. New York: Wiley, 1978.
5. Plaizier-Vercammen JA, De Neve RE. Interaction of povidone with aromatic compounds II: Evaluation of ionic strength, buffer concentration, temperature, and pH by factorial analysis. J Pharm Sci 1981; 70:1252.
6. Bolton S. Factorial designs in pharmaceutical stability studies. J Pharm Sci 1983; 72:362.
7. Ahmed S, Bolton S. Factorial design in the study of the effects of selected liquid chromatographic conditions on resolution and capacity factors. J Liq Chromatogr 1990; 13:525.

¶ A more advanced topic.

10 | Transformations and Outliers

Critical examination of the data is an important step in statistical analyses. Often, we observe either what seem to be unusual observations (outliers) or observations that appear to violate the assumptions of the analysis. When such problems occur, several courses of action are available depending on the nature of the problem and statistical judgment. Most of the analyses described in previous chapters are appropriate for groups in which data are normally distributed with equal variance. As a result of the Central Limit theorem, these analyses perform well for data that are not normal provided the deviation from normality is not large and/or the data sets are not very small. (If necessary and appropriate, nonparametric analyses, chap. 15, can be used in these instances.) However, lack of equality of variance (heteroscedascity) in t tests, analysis of variance and regression, for example, is more problematic. The Fisher–Behrens test is an example of a modified analysis that is used in the comparison of means from two independent groups with unequal variances in the two groups (chap. 5). Often, variance heterogeneity and/or lack of normality can be corrected by a data transformation, such as the logarithmic or square-root transformation. Bioequivalence parameters such as AUC and C_{MAX} currently require a log transformation prior to statistical analysis. Transformations of data may also be appropriate to help linearize data. For example, a plot of log potency versus time is linear for stability data showing first-order kinetics.

Variance heterogeneity may also be corrected using an analysis in which each observation is weighted appropriately, that is, a weighted analysis. In regression analysis of kinetic data, if the variances at each time point differ, depending on the magnitude of drug concentration, for example, a weighted regression would be appropriate. For an example of the analysis of a regression problem requiring a weighted analysis for its solution, see chapter 7.

Data resulting from gross errors in observations or overt mistakes such as recording errors should clearly be omitted from the statistical treatment. However, upon examining experimental data, we often find unusual values that are not easily explained. The prudent experimenter will make every effort to find a cause for such aberrant data and modify the data or analysis appropriately. If no cause is found, one should use scientific judgment with regard to the disposition of these results. In such cases, a statistical test may be used to detect an outlying value. An outlier may be defined as an observation that is extreme and appears not to belong to the bulk of data. Many tests to identify outliers have been proposed and several of these are presented in this chapter.

10.1 TRANSFORMATIONS

A transformation applied to a variable changes each value of the variable as described by the transformation. In a *logarithmic (log) transformation*, each data point is changed to its logarithm prior to the statistical analysis. Thus the value 10 is transformed to 1 (i.e., log 10 = 1). The log transformation may be in terms of logs to the base 10 or logs to the base e ($e = 2.718 \ldots$), known as natural logs (In). For example, using natural logs, 10 would be transformed to 2.303 (ln 10 = 2.303). The *square-root* transformation would change the number 9 to 3.

Parametric analyses such as the t test and analysis of variance are the methods of choice in most situations where experimental data are continuous. For these methods to be valid, data are assumed to have a normal distribution with constant variance within treatment groups. Under appropriate circumstances, a transformation can change a data distribution that is not normal into a distribution that is approximately normal and/or can transform data with heterogeneous variance into a distribution with approximately homogeneous variance.

Table 10.1   Some Transformations Used to Linearize
Relationships Between Two Variables, X and Y

Function	Transformation	Linear form
$Y = Ae^{-BX}$	Logarithm of Y	$\ln Y = A - BX$
$Y = 1/(A + BX)$	Reciprocal of Y	$1/Y = A + BX$
$Y = X/(AX + B)$	Reciprocal of Y	$1/Y = A + B(1/X)$[a]

[a] A plot of $1/Y$ versus $1/X$ is linear.

Thus, data transformations can be used in cases where (1) the variance in regression and analysis of variance is not constant and/or (2) data are clearly not normally distributed (highly skewed to the left or right).

Another application of transformations is to linearize relationships such as may occur when fitting a least squares line (not all relationships can be linearized). Table 10.1 shows some examples of such linearizing transformations. When making linearizing transformations, if statistical tests are to be made on the transformed data, one should take care that the normality and variance homogeneity assumptions are not invalidated by the transformation.

10.1.1   The Logarithmic Transformation

Probably the most common transformation used in scientific research is the log transformation. Either logs to the base 10 ($\log_{10}$) or the base e, $\log_e$(ln) can be used. Data skewed to the right as shown in Figure 10.1 can often be shown to have an approximately log-normal distribution. A log-normal distribution is a distribution that would be normal following a log transformation, as illustrated in Figure 10.2. When statistically analyzing data with a distribution similar to that shown in Figure 10.1, a log transformation should be considered. One should understand that a reasonably large data set or prior knowledge is needed in order to know the form of the distribution. Table 10.2 shows examples of two data sets, listed in ascending order of magnitude. Data set A would be too small to conclude that the underlying distribution is not normal in the absence of prior information. Data set B, an approximately log-normal distribution, is strongly suggestive of non-normality. (See Exercise Problem 1.) One should understand that real data does not conform exactly to a normal or log-normal distribution. This does not mean that applying theoretical probabilities to data that approximate these distributions is not meaningful. If the distributions are reasonably close to a theoretical distribution, the statistical decisions will have alpha levels close to those chosen for the tests.

Two problems may arise as a consequence of using the log transformation.

1. Many people have trouble interpreting data reported in logarithmic form. Therefore, when reporting experimental results, such as means for example, a back transformation (the antilog) may be needed. For example, if the mean of the logarithms of a data set is 1.00, the antilog, 10, might be more meaningful in a formal report of the experimental results. The mean of a set of untransformed numbers is not, in general, equal to the antilog of the mean of the logs of these numbers. If the data are relatively nonvariable, the means calculated by these two methods will be close. The mean of the logs and the log of the mean will be identical only if each observation is the same, a highly unlikely circumstance. Table 10.3

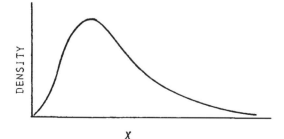

X

Figure 10.1   Log-normal distribution.

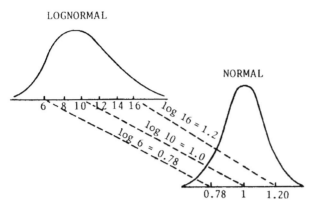

Figure 10.2 Transformation of a log-normal distribution to a normal distribution via the log transformation.

illustrates this concept. Note that the antilog of the mean of a set of log-transformed variables is the geometric mean (see chap. 1). This lack of "equivalence" can raise questions when someone reviewing the data is unaware of this divergence, "the nature of the beast," so to speak.

2. The second problem to be considered when making log transformations is that the log transformation that "normalizes" log-normal data also changes the variance. If the variance is not very large, the variance of the ln transformed values will have a variance approximately equal to $S^2/\overline{X}^2$. That is, the standard deviation of the data after the transformation will be approximately equal to the coefficient of variation (CV), $S/\overline{X}$. For example, consider the following data:

	X	**ln X**
	105	4.654
	102	4.625
	100	4.605
	110	4.700
	112	4.718
Mean	105.8	4.6606
s.d.	5.12	0.0483

Table 10.2 Two Data Sets That May Be Considered Lognormal

Data set A: 2, 17, 23, 33, 43, 55, 125, 135
Data set B: 10, 13, 40, 44, 55, 63, 115, 145, 199, 218, 231,
370, 501, 790, 795, 980, 1260, 1312, 1500, 4520

Table 10.3 Illustration of Why the Antilog of the Mean of the Logs Is Not Equal to the Mean of the Untransformed Values

	Case I			**Case II**	
	Original data	**Log transform**		**Original data**	**Log transform**
	5	0.699		4	0.603
	5	0.699		6	0.778
	5	0.699		8	0.903
	5	0.699		10	1.000
Mean	5	0.699	Mean	7	0.821
	Antilog (0.699) = 5			Antilog (0.821) = 6.62	

Table 10.4 Results of an Assay at Three Different Levels of Drug

| | At 40 mg | | At 60 mg | | At 80 mg | |
	Assay	Log assay	Assay	Log assay	Assay	Log assay
	37	1.568	63	1.799	82	1.914
	43	1.633	77	1.886	68	1.833
	42	1.623	56	1.748	75	1.875
	40	1.602	64	1.806	97	1.987
	30	1.477	66	1.820	71	1.851
	35	1.544	58	1.763	86	1.934
	38	1.580	67	1.826	71	1.851
	40	1.602	52	1.716	81	1.908
	39	1.591	55	1.740	91	1.959
	36	1.556	58	1.763	72	1.857
Average	38	1.578	61.6	1.787	79.4	1.897
s.d.	3.77	0.045	7.35	0.050	9.67	0.052
CV	0.10		0.12		0.12	

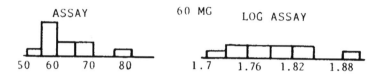

The CV of the original data is $5.12/105.8 = 0.0484$. The standard deviation of the ln transformed values is 0.0483, very close to the CV of the untransformed data. This property of the transformed variance can be advantageous when working with data groups that are both *lognormal* and have a *constant coefficient of variation*. If the standard deviation within treatment groups, for example, is not homogeneous but is proportional to the magnitude of the measurement, the CV will be constant. In analytical procedures, one often observes that the s.d. is proportional to the quantity of material being assayed. In these circumstances, the log (to the base e) transformation will result in data with homogeneous s.d. equal to CV. (The s.d. of the transformed data is approximately equal to CV*). This concept is illustrated in Example 1 that follows. Fortunately, in many situations, data that are approximately lognormal also have a constant CV. In these cases, the log transformation results in normal data with approximately homogeneous variance. The transformed data can be analyzed using techniques that depend on normality and homogeneous variance for their validity (e.g., ANOVA).

Example 1. Experimental data were collected at three different levels of drug to show that an assay procedure is linear over a range of drug concentrations. "Linear" means that a plot of the *assay results*, or a suitable transformation of the results, versus the *known concentration* of drug is a straight line. In particular, we wish to plot the results such that a linear relationship is obtained, and calculate the least squares regression line to relate the assay results to the known amount of drug. The results of the experiment are shown in Table 10.4. In this example, the assay results are unusually variable. This large variability is intentionally presented in this example to illustrate the properties of the log transformation. The skewed nature of the data in Table 10.4 suggests a log-normal distribution, although there are not sufficient data to verify the exact nature of the distribution. Also in this example, the s.d. increases with drug concentration. The s.d. is approximately proportional to the mean assay, an approximately constant CV (10–12%). Note that the log transformation results in variance homogeneity and a more symmetric data distribution (Table 10.4). Thus, there is a strong indication for a log transformation.

---

* The log transformation (log to the base 10) of data with constant CV results in data with s.d. approximately equal to CV/2.303.

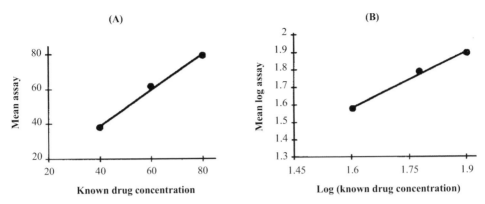

Figure 10.3 Plots of raw data means and log-transformed means for data of Table 10.4. (**A**) Means of untransformed data, (**B**) log transformation.

The properties of this relatively variable analytical method can be evaluated by plotting the known amount of drug versus the amount recovered in the assay procedure. Ideally, the relationship should be linear over the range of drug concentration being assayed. A plot of known drug concentration versus assay results is close to linear [Fig. 10.3(A)]. A plot of *log* drug concentration versus *log* assay is also approximately linear, as shown in Figure 10.3(B). From a statistical viewpoint, the log plot has better properties because the data are more "normal" and the variance is approximately constant in the three drug concentration groups as noted above. The line in Figure 10.3(B) is the least squares line. The details of the calculation are not shown here (see Exercise Problem 2 and chap. 7 for further details of the statistical line fitting).

When performing the usual statistical tests in regression problems, the assumptions include the following:

1. The data at each X should be normal (i.e., the amount of drug recovered at a given amount added should be normally distributed).
2. The assays should have the same variance at each concentration.

The log transformation of the assay results (Y) helps to satisfy these assumptions. In addition, in this example, the linear fit is improved as a result of the log transformation.

Example 2. In the pharmaceutical sciences, the logarithmic transformation has applications in kinetic studies, when ascertaining stability and pharmacokinetic parameters. First-order processes are usually expressed in logarithmic form (see also sect. 2.5)

$$\ln C = \ln C_0 - kt. \tag{10.1}$$

Least squares procedures are typically used to fit concentration versus time data in order to estimate the rate constant, k. A plot of concentration (C) versus time (t) is not linear for first-order reactions [Fig. 10.4(A)]. A plot of the log-transformed concentrations (the Y variable) versus time is linear for a first-order process [Eq. (10.1)]. The plot of log C versus time is shown in Figure 10.4(B), a semilog plot.

Thus, we may use linear regression procedures to fit a straight line to log C versus time data for first-order reactions. One should recognize, as before, that if statistical tests are performed to test the significance of the rate constant, for example, or when placing confidence limits on the rate constant, the implicit assumption is that log concentration is normal with constant variance at each value of X (time). These assumptions will hold, when linearizing such concentration versus time relationships if the *untransformed* values of "concentration" are *lognormal* with constant CV. In cases in which the assumptions necessary for statistical inference are invalidated by the transformation, one may question the validity of predictions based on

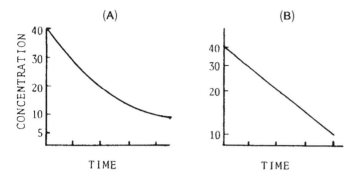

Figure 10.4   First-order plots. (**A**) Usual plot, (**B**) semilog plot.

least squares line fitting for first-order processes. For example, if the original, untransformed concentration values are normal with constant variance, the log transformation will distort the distribution and upset the constant variance condition. However, if the variance is small, and the concentrations measured are in a narrow range (as might occur in a short-term stability study to 10% decomposition), the log transformation will result in data that are close to normal with homogeneous variance. Predictions for stability during the short term based on the least squares fit will be approximately correct under these conditions.

Some properties of the log-normal distribution relevant to particle size analysis are also presented in section 3.6.1.

10.1.1.1   Analysis of Residuals

We have discussed the importance of carefully looking at and graphing data before performing transformations or statistical tests. The approach to examining data in this context is commonly known as exploratory data analysis, EDA, introduced in chapter 7. A significant aspect of EDA is the examination of residuals. Residuals are deviations of the observed data from the fit to the statistical model, the least squares line in this example. Figure 10.5 shows the residuals for the least squares fit of the data in Table 10.4, using the untransformed and transformed data analysis. Note that the residual plot versus dose shows the dependency of the variance on dose. The log response versus log dose shows a more uniform distribution of residuals.

Example 3. The log transformation may be used for data presented in the form of ratios. Ratios are sometimes used to express the comparative absorption of drug from two formulations based on the area under the plasma level versus time curve from a bioavailability study. Another way of comparing the absorptions from the two formulations is to test statistically the *difference* in absorption ($AUC_1 - AUC_2$), as illustrated in section 5.2.3. However, reporting results of relative absorption using a *ratio*, rather than a difference, has great appeal. The ratio can be interpreted in a pragmatic sense. Stating that formulation A is absorbed *twice* as much as formulation B has more meaning than stating that formulation A has an AUC 525 $\mu g \cdot hr/mL$ more than formulation B. (Note: The FDA Guidance for analysis of bioequivalence studies does not recommend this procedure.) A statistical problem that is evident when performing statistical tests on ratios is that the ratios of random variables will probably not be normally distributed. In particular, if both A and B are normally distributed, the ratio A/B does not have a normal distribution. On the other hand, the test of the differences of AUC has statistical appeal because the difference of two normally distributed variables is also normally distributed. The practical appeal of the *ratio* and the statistical appeal of *differences* suggest the use of a log transformation, when ratios seem most appropriate for data analysis.

The differences of logs is analogous to ratios; the difference of the logs is the log of the ratio: $\log A - \log B = \log(A/B)$. The antilog of the average difference of the logs will be close to the average of the ratios if the variability is not too large. The differences of the logs will

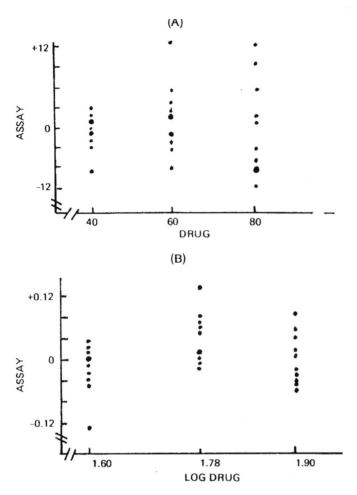

Figure 10.5 Residual plots from least squares line fitting of data from Table 10.4.

also tend to be normally distributed. But the normality assumption should not be a problem in these analyses because we are testing *mean* differences (again, the central limit theorem). After application of the log transformation, the data may be analyzed by the usual *t*-test (or ANOVA) techniques that assess treatment differences.

Table 10.5 shows AUC data for 10 subjects who participated in a bioavailability study. The analysis (a paired *t* test in this example) is performed on both the difference of the *logarithms* and the ratios. The *t* test for the ratios is a one-sample, two-sided test comparing the average ratio to 1 (H_0: $R = 1$) as shown in section 5.2.1.

t test for ratios:

$$H_0:\ R = 1$$
$$t = \frac{|1.025 - 1|}{0.378/\sqrt{10}} = 0.209.$$

95% confidence interval:

$$1.025 \pm \frac{2.26(0.378)}{\sqrt{10}} = 1.025 \pm 0.27.$$

Table 10.5   Results of the Bioavailability Study: Areas Under the Plasma Level Versus Time Curve

Subject	Product A		Product B		Ratio AUCs: A/B	Log A − Log B
	AUC	Log AUC	AUC	Log AUC		
1	533	2.727	651	2.814	0.819	−0.087
2	461	2.664	547	2.738	0.843	−0.074
3	470	2.672	535	2.728	0.879	−0.056
4	624	2.795	326	2.513	1.914	0.282
5	490	2.690	386	2.587	1.269	0.104
6	476	2.678	640	2.806	0.744	−0.129
7	465	2.667	582	2.765	0.799	−0.097
8	365	2.562	420	2.623	0.869	−0.061
9	412	2.615	545	2.736	0.756	−0.121
10	380	2.580	280	2.447	1.357	0.133
Average					1.025	−0.01077
s.d.					0.378	0.136

t test for difference of logs:

$$H_0: \log A - \log B = 0$$

$$t = \frac{|-0.01077|}{0.136/\sqrt{10}} = 0.250.$$

95% confidence interval:

$$-0.01077 \pm \frac{2.26(0.136)}{\sqrt{10}} = -0.01077 \pm 0.0972.$$

The confidence interval for the logs is −0.10797 to 0.08643. The antilogs of these values are 0.78 to 1.22. The confidence interval for the ratio is 0.75 to 1.30. Thus, the conclusions using both methods (ratio and difference of logs) are similar. Had the variability been smaller, the two methods would have been in better agreement.

	t test		Confidence interval	
Ratio	Difference of logs		Ratio	Difference of logs
0.209	0.250		0.75–1.30	0.78–1.22

Another interesting result that recommends the analysis of differences of logs rather than the use of ratios is a consequence of the *symmetry* that is apparent with the former analysis. With the log transformation, the conclusion regarding the equivalence of the products will be the same whether we consider the difference as (log A − log B) or (log B − log A). However, when analyzing ratios, the analysis of A/B will be different from the analysis of B/A. The product in the numerator has the advantage (see Exercise Problem 3). In the example in Table 10.5 the average ratio of B/A is 1.066. B appears slightly better than A. When the ratios are calculated as A/B, A appears somewhat better than B. The log transformation for bioavailability parameters, as has been recommended by others [1], is now routinely applied to analysis of bioequivalence data. This analysis is presented in detail in chapter 11.

For data containing zeros, very small numbers (close to zero) or negative numbers, using ratios or logarithms is either not possible or not recommended. Clearly, if we have a ratio with a zero in the denominator or a mixture of positive and negative ratios, the analysis and interpretation is difficult or impossible. Logarithms of negative numbers and zero are undefined. Therefore, unless special adjustments are made, such data are not candidates for a log transformation.

10.1.2   The Arcsin Transformation for Proportions

Another commonly used transformation is the arcsin transformation for proportions. The arcsin is the inverse sine function, also denoted as $\sin^{-1}$. Thus, if $\sin 45° = 0.7$, arcsin $0.7 = 45°$. Many calculators have a sine and inverse sine function available.

The problem that arises when analyzing proportions, where the data consist of proportions of widely different magnitudes, is the lack of homogeneity of variance. The variance homogeneity problem is a result of the definition of the variance for proportions, pq/N. If the proportions under consideration vary from one observation to another, the variance will also vary. If the proportions to be analyzed are approximately normally distributed (Np and $Nq \geq 5$; see chap. 5), the arcsin transformation will equalize the variances. The arcsin values can then be analyzed using standard parametric techniques such as ANOVA. When using the arcsin transformation, each proportion should be based on the same number of observations, N. If the number of observations is similar for each proportion, the analysis using arcsines will be close to correct. However, if the numbers of observations are very different for the different proportions, the use of the transformation is not appropriate. Also, for very small or very large proportions (less than 0.03 or greater than 0.97), a more accurate transformation is given by Mosteller and Youtz [2]. The following example should clarify the concept and calculations when applying the arcsin transformation.

Example 4. In preparation for a toxicological study for a new drug entity, an estimate of the incidence of a particular adverse reaction in untreated mice was desired. Data were available from previous studies, as shown in Table 10.6. The arcsin transformation is applied to the proportions as follows:

$$\text{Arcsin transformation} = \text{arcsin } \sqrt{p}. \qquad (10.2)$$

For example, in Table 10.6, the arcsin transformation of 10% (0.10) is arcsin $\sqrt{0.10}$, which is equal to $18.43°$.

The objective of this exercise is to estimate the incidence of the adverse reaction in normal, untreated animals. To this end, we will obtain the average proportion and construct a confidence interval using the arcsin-transformed data. The average arcsin is $26.197°$. The average proportions are not reported in terms of arcsines. As in the case of the log transformation, one should back transform the average transformed value to the original terms. In this example, we obtain the back transform as $\sin(\text{arcsin})^2$, or $\sin(26.197)^2 = 0.195$. This is very close to the average of the untransformed proportions, 20%. The *variance of a transformed proportion* can be shown to be equal to $820.7/N$, where N is the number of observations for each proportion [3]. Thus, in this example, the variance is $820.7/50 = 16.414$.

A confidence interval for the average proportion is obtained by finding the confidence interval for the average arcsin and back transforming to proportions. Ninety-five percent confidence interval: $\overline{X} \pm 1.96\sqrt{\sigma^2/N}$ [Eq. (5.1)]

$$26.197 \pm 1.96\sqrt{\frac{16.414}{6}} = 26.197 \pm 3.242.$$

Table 10.6   Incidence of an Adverse Reaction in Untreated Mice from Six Studies

	Proportion of mice showing adverse reaction	Arcsin P
	5/50 = 0.10	18.43
	12/50 = 0.24	29.33
	8/50 = 0.16	23.58
	15/50 = 0.30	33.21
	13/50 = 0.26	30.66
	7/50 = 0.14	21.97
Average	0.20	26.197°

Table 10.7 Summary of Some Common Transformations

Transformation	When used
Logarithm (log X)	s.d. $\propto \overline{X}$
Arcsin $(\sin^{-1})\sqrt{X}$	Proportions
Square root ($\sqrt{X}$ or $\sqrt{X} + \sqrt{X+1}$	(s.d.)$^2 \propto \overline{X}$
Reciprocal $(1/X)$	s.d. $\propto \overline{X}^2$

The 95% confidence interval for the average arcsin is 22.955° to 29.439°. This interval corresponds to an interval for the proportion of 15.2% to 24.2% (0.152–0.242).[†]

10.1.3 Other Transformations

Two other transformations that are used to correct deviations from assumptions for statistical testing are the *square-root* and *reciprocal transformations*. As their names imply, these transformations change the data as follows:

Square-root transformation: $X \rightarrow \sqrt{X}$
Reciprocal transformation: $X \rightarrow 1/X$

The square-root transformation is useful in cases where the variance is proportional to the mean. The situation occurs often where the data consist of counts, such as may occur in blood and urine analyses or microbiological data. If some values are 0 or very small, the transformation, $\sqrt{X} + \sqrt{X+1}$, has been recommended [4]. Different Poisson variables, whose variances equal their means, will have approximately equal variance after the square-root transformation (see Exercise Problem 6).

The reciprocal transformation may be used when the s.d. is proportional to the square of the mean [5]. The transformation is also useful where time to a given response is being measured. For some objects (persons) the time to the response may be very long and a skewed distribution results. The reciprocal transformation helps make the data more symmetrical.

Table 10.7 summarizes the common transformations discussed in this section.

10.2 OUTLIERS

Outliers, in statistics, refer to relatively small or large values that are considered to be different from, and not belong to, the main body of data. The problem of what to do with outliers is a constant dilemma facing research scientists. If the cause of an outlier is known, resulting from an obvious error, for example, the value can be omitted from the analysis and tabulation of the data. However, it is good practice to include the reason(s) for the omission of the aberrant value in the text of the report of the experimental results. For example, a container of urine, assayed for drug content in a pharmacokinetic study, results in too low a drug content because part of the urine was lost due to accidental spillage.

This is just cause to discard the data from that sample. In most cases, extreme values are observed without obvious causes, and we are confronted with the problem of how to handle the apparent outliers. Do the outlying data really represent the experimental process that is being investigated? Can we expect such extreme values to occur routinely in such experiments? Or was the outlier due to an error of observation? Perhaps the observation came from a population different from the one being studied. In general, aberrant observations *should not be arbitrarily discarded* only because they look too large or too small, perhaps only for the reason of making the experimental data look "better." In fact, the presence of such observations has sometimes been a clue to an important process inherent in the experimental system. Therefore, the question of what to do with outliers is not an easy one to answer. The error of either incorrectly including or excluding outlying observations will distort the validity of interpretation and conclusions of the experiment.

[†] $\sin(22.955°)^2 = 0.152$ and $\sin(29.439°)^2 = 0.242$.

Several statistical criteria for handling outlying observations will be presented here. These methods may be used if no obvious cause for the outlier can be found. If, for any reason, one or more outlying data are rejected, one has the option of (a) repeating the appropriate portion of the experiment to obtain a replacement value(s), (b) estimating the now "missing" value by statistical methods, or (c) analyzing the data without the discarded value(s). From a statistical point of view, the practice of looking at a set of data or replicates, and rejecting the value(s) that is most extreme (and possibly, rerunning the rejected point) is to be discouraged. Biases in the results are almost sure to occur. Certainly, the variance will be underestimated, since we are throwing out the extreme values, willy-nilly. For example, when performing assays, some persons recommend doing the assay in triplicate and selecting the two best results (those two closest together). In other cases, two assays are performed and if they "disagree," a third assay is performed to make a decision as to which of the original two assays should be discarded. Arbitrary rules such as these often result in incorrect decisions about the validity of results [6]. Experimental scientists usually have a very good intuitive "feel" for their data, and this should be taken into account before coming to a final decision regarding the disposition of outlying values. Every effort should be made to identify a cause for the outlying observation. However, in the absence of other information, the statistical criteria discussed below may be used to help make an objective decision. When in doubt, a useful approach is to analyze the data with and without the suspected value(s). If conclusions and decisions are the same with and without the extreme value(s), including the possible outlying observations would seem to be the most prudent action.

Statistical tests for the presence of outliers are usually based on an assumption that the data have a normal distribution. Thus, applying these tests to data that are known to be highly skewed, for example, would result too often in the rejection of legitimate data. If the national average income were to be estimated by an interview of 100 randomly selected persons, and 99 were found to have incomes of less than $100,000 while one person had an income of $1,000,000, it would be clearly incorrect to omit the latter figure, attributing it to a recording error or interviewer unreliability. The tests described below are based on statistics calculated from the observed data, which are then referred to tables to determine the level of significance. The significance level here has the same interpretation as that described for statistical tests of hypotheses (chap. 5). At the 5% level, an outlying observation may be incorrectly rejected 1 time in 20.

10.2.1 Dixon's Test for Extreme Values
The data in Table 10.8 represent cholesterol values (ordered according to magnitude) for a group of healthy, normal persons. This example is presented particularly, because the problem

Table 10.8 Ordered Values of Serum Cholesterol from 15 Normal Subjects

Subject	1	2	3	4	5	6	7	8	9	10	11	12	13	14	15
Cholesterol	165	188	194	197	200	202	205	210	214	215	227	231	239	249	297

that it represents has two facets. First, the possibility exists that the very low and very high values (165, 297) are the result of a recording or analytical error. Second, one may question the existence of such extreme values among *normal healthy* persons. Without the presence of an obvious error, one would probably be remiss if these two values (165, 297) were omitted from a report of "normal" cholesterol values in these normal subjects. However, with the knowledge that plasma cholesterol levels are approximately normally distributed, a statistical test can be applied to determine if the extreme values should be rejected.

Dixon has proposed a test for outlying values that can easily be calculated [7]. The set of observations are first ordered according to magnitude. A calculation is then performed of the *ratio* of the difference of the extreme value from one of its nearest neighboring values to the range of observations as defined below.

The formula for the *ratio*, r, depends on the sample size, as shown in Table IV.8. The calculated ratio is compared to appropriate tabulated values in Table IV.8. If the ratio is equal to or greater than the tabulated value, the observation is considered to be an outlier at the 5% level of significance.

The ordered observations are denoted as $X_1, X_2, X_3, \ldots, X_N$, for N observations, where X_1 is an extreme value and X_N is the opposite extreme. When $N = 3$ to 7, for example, the ratio $r = (X_2 - X_1)/(X_N - X_1)$ is calculated. For the five (5) values 1.5, 2.1, 2.2, 2.3, and 3.1, where 3.1 is the suspected outlier,

$$r = \frac{3.1 - 2.3}{3.1 - 1.5} = 0.5.$$

The ratio must be equal to or exceed 0.642 to be significant at the 5% level for $N = 5$ (Table IV.8). Therefore, 3.1 is not considered to be an outlier ($0.5 < 0.642$).

The cholesterol values in Table 10.8 contain two possible outliers, 165 and 297. According to Table IV.8, for a sample size of 15 ($N = 15$), the test ratio is

$$r = \frac{X_3 - X_1}{X_{N-2} - X_1}, \tag{10.3}$$

where X_3 is the third ordered value, X_1 is the smallest value, and X_{N-2} is the third largest value (two removed from the largest value).

$$r = \frac{194 - 165}{239 - 165} = \frac{29}{74} = 0.39.$$

The tabulated value for $N = 15$ (Table IV.8) is 0.525. Therefore, the value 165 cannot be rejected as an outlier.

The test for the largest value is similar, reversing the order (highest to lowest) to conform to Eq. (10.3). X_1 is 297, X_3 is 239, and X_{N-2} is 194.

$$r = \frac{239 - 297}{194 - 297} = \frac{-58}{-103} = 0.56.$$

Since 0.56 is greater than the tabulated value of 0.525, 297 can be considered to be an outlier, and rejected.

Consider an example of the results of an assay performed in triplicate.

94.5, 100.0, 100.4

Is the low value, 94.5, an outlier? As discussed earlier, triplicate assays have an intuitive appeal. If one observation is far from the others, it is often discarded, considered to be the result of some overt, but not obvious error. Applying Dixon's criterion ($N = 3$),

$$r = \frac{100 - 94.5}{100.4 - 94.5} = 0.932.$$

Surprisingly, the test does not find the "outlying" value small enough to reject the value at the 5% level. The ratio must be at least equal to 0.941 in order to reject the possible outlier for a sample of size 3. In the absence of other information, 94.5 is not obviously an outlier. The moral here is that what seems obvious is not always so. When one value of three appears to be "different" from the others, think twice before throwing it away.

After omitting a value as an outlier, the remaining data may be tested again for outliers, using the same procedure as described above with a sample size of $N - 1$.

10.2.2 The *T* Procedure

Another highly recommended test for outliers, the *T method* (Grubb's test), is also calculated as a ratio, designated T_n, as follows:

$$T_n = \frac{X_n - \overline{X}}{S}, \tag{10.4}$$

where X_n is either the smallest or largest value, $\overline{X}$ is the mean, and S is the s.d. If the extreme value is not anticipated to be high or low, prior to seeing the data, a test for the outlying value is based on the tabulation in Table IV.9. If the calculated value of T_n is equal to or exceeds the tabulated value, the outlier is rejected as an extreme value ($p \leq 0.05$). A more detailed table is given in Ref. [8].

For the cholesterol data in Table 10.7, T_n is calculated as follows:

$$T_n = \frac{297 - 215.5}{30.9} = 2.64,$$

where 297 is the suspected outlier, 215.5 is the average of the 15 cholesterol values, and 30.9 is the s.d. of the 15 values. According to Table IV.9, T_n is significant at the 5% level, agreeing with the conclusions of the Dixon test. The Dixon test and the T_n test may not exactly agree with regard to acceptance or rejection of the outlier, particularly in cases where the extreme value results in tests that are close to the 5% level. To maintain a degree of integrity in situations where more than one test is available, one should decide which test to use prior to seeing the data. On the other hand, for any statistical test, if alternative acceptable procedures are available, any difference in conclusions resulting from the use of the different procedures is usually of a small degree. If one test results in significance ($p < 0.05$) and the other just misses significance (e.g., $p = 0.06$), one can certainly consider the latter result close to being statistically significant at the very least.

10.2.3 Winsorizing

An interesting approach to the analysis of data to protect against distortion caused by extreme values is the process of Winsorization [7]. In this method, the extreme values, both low and high, are changed to the values of their closest neighbors. This procedure provides some protection against the presence of outlying values and, at the same time, very little information is lost. For the cholesterol data (Table 10.7), the extreme values are 165 and 297. These values are changed to that of their nearest neighbors, 188 and 249, respectively. This manipulation results in a data set with a mean of 213.9, compared to a mean of 215.5 for the untransformed data.

Winsorized estimates can be useful when *missing* values are known to be *extreme* values. For example, suppose that the two highest values of the cholesterol data from Table 10.7 were lost. Also, suppose that we know that these two missing values would have been the highest values in the data set, had we had the opportunity to observe them. Perhaps, in this example, the subjects whose values were missing had extremely high measurements in previous analyses;

or perhaps, a very rough assay was available from the spilled sample scraped off of the floor showing high levels of cholesterol. A reasonable estimate of the mean would be obtained by substituting 239 (the largest value after omitting 249 and 297) for the two missing values. Similarly, we could replace 165 and 188 by the third lowest value, 194. The new mean is now equal to 213.3, compared to a mean of 215.5 for the original data.

10.2.4 Overall View and Examples of Handling Outliers

The ultimate difficulty in dealing with outliers is expressed by Barnett and Lewis in the preface of their book on outliers [9]. "Even before the formal development of statistical methods, argument raged over whether, and on what basis, we should discard observations from a set of data on the grounds that they are 'unrepresentative,' 'spurious' or 'mavericks' or 'rogues.' The early emphasis stressed the contamination of the data by unanticipated and unwelcome errors or mistakes affecting some of the observations. Attitudes varied from one extreme to another: from the view that we should never sully the sanctity of the data by daring to adjudge its propriety, to an ultimate pragmatism expressing 'if in doubt, throw it out.'" They also quote Ferguson, "The experimenter is tempted to throw away the apparently erroneous values (the outliers) and not because he is certain that the values are spurious. On the contrary, he will undoubtedly admit that even if the population has a normal distribution, there is a positive although extremely small probability that such values will occur in an experiment. It is rather because he feels that other explanations are more plausible, and that the loss in accuracy of the experiment caused by throwing away a couple of good values is small compared to the loss caused by keeping even one bad value." Finally, in perspective, Barnett and Lewis state, "But, when all is said and done, the major problem in outlier study remains the one that faced the earliest workers in the subject—what is an outlier and how should we deal with it? We have taken the view that the stimulus lies in the subjective concept of surprise engendered by one, or a few, observations in a set of data. . . ."

Although most treatises on the use of statistics caution readers on the indiscriminate discarding of outlying results, and recommend that outlier tests be used with care, this does not mean that outlier tests and elimination of outlying results should never be applied to experimental data. The reason for omitting outliers from a data analysis is to improve the validity of statistical procedures and inferences. Certainly, if applied correctly for these reasons, outlier tests are to be commended. The dilemma is in the decision as to when such tests are appropriate. Most recommended outlier tests are very sensitive to the data distribution, and many tests assume an underlying normal distribution. Nonparametric outlier tests make less assumptions about the data distribution, but may be less discriminating.

Notwithstanding cautions about indiscriminately throwing out outliers, including outliers that are indeed due to causes that do not represent the process being studied, including outliers in the data analysis can severely bias the conclusions. When no obvious reason is apparent to explain an outlying value that has been identified by an appropriate statistical test, the question of whether or not to include the data is not easily answered. In the end, judgment is a very important ingredient in such decisions, since knowledge of the data distribution is usually limited. Part of "good judgment" is a thorough knowledge of the process being studied, in addition to the statistical consequences. If conclusions about the experimental outcome do not change with and without the outlier, both results can be presented. However, if conclusions are changed, then omission of the outlier should be justified based on the properties of the data.

Some examples should illustrate possible approaches to this situation.

Example 1. Analysis of a portion of a powdered mix comprised of 20 ground-up tablets (a composite) was done in triplicate with results of 75.1%, 96.9%, and 96.3%. The expected result was approximately 100%. The three assays represented three separate portions of the grind. A statistical test (see Table IV.8) suggested that the value of 75.1% is an outlier ($p < 0.05$), but there was no obvious reason for this low assay. Hypothetically, this result could have been caused by an erroneous assay, or more remotely, by the presence of one or more low potency tablets that were not well mixed with the other tablets in the grind. Certainly, the former is a more probable cause, but there is no way of proving this because the outlying sample is no longer available. It would seem foolhardy to reject the batch average of three results, 89.4%, without further investigation. There are two reasonable approaches to determining if, in fact, the 75.1%

value was a real result or an anomaly. One approach is to throw out the value of 75.1 based on the knowledge that the tablets were indeed ground thoroughly and uniformly and that the drug content should be close to 100%. Such a decision could have more credence if other tests on the product (e.g., content uniformity) supported the fact that 75.1 was an outlier. A second, more conservative approach would be to reassay the remaining portion of the mix to ensure that the 75.1 value could not be reproduced. How many more assays would be necessary to verify the anomaly? This question does not seem to have a definitive answer. This is a situation where scientific judgment is needed. For example, if three more assays were performed on the mix, and all assays were within limits, the average assay would be best represented by the five "good" assays (two from the first analysis and three from the second analysis). Scientifically, in this scenario, it would appear that including the outlier in the average would be an unfair representation of the drug content of this material. Of course, if an outlying result were found again during the reanalysis, the batch (or the 20 tablet grind) is suspect, and the need for a thorough investigation of the problem would be indicated.

Example 2. Consider the example above as having occurred during a content uniformity test, where one of 10 tablets gave an outlying result. For example, suppose 9 of 10 tablets were between 95% and 105%, and a single tablet gave a result of 71%. This would result in failure of the content uniformity test as defined in the USP. (No single tablet should be outside 75–125% of label claim.) The problem here (if no obvious cause can be identified) is that the tablet has been destroyed in the analytical process and we have no way of knowing if the result is indeed due to the tablet or some unidentified gross analytical error. This presents a more difficult problem than the previous one because we cannot assay the same homogenate from which the outlying observation originated. Other assays during the processing of this batch and the batch history would be useful in determining possible causes. If no similar problem had been observed in the history of the product, one might assume an analytical misfortune. As suggested in the previous example, if similar results had occurred in other batches of the product, a suggestion of the real possibility of the presence of outlying tablets in the production of this product is indicated. In any case, it would be prudent to perform extensive content uniformity testing, if no cause can be identified. Again, one may ask what is "extensive" testing? We want to feel "sure" that the outlier is an anomaly, not typical of tablets in the batch. Although it is difficult to assign the size of retesting on a scientific basis, one might use statistical procedures to justify the choice of a sample size. For example, using the concept of tolerance limits (sect. 5.6), we may want to be 99% certain that 99% of the tablets are between 85% and 115%, the usual limits for CU acceptance. In order to achieve this level of "certainty," we have to estimate the mean content (%) and the CV. (See App. V.)

Example 3. The results of a content uniformity test show 9 of 10 results between 91% and 105% of label, with one assay at 71%. This fails the USP content uniformity test, which allows a single assay between 75% and 125%, but none outside these limits. The batch records of the product in question and past history showed no persistent results of this kind. The "outlier" could not be attributed to an analytical error, but there was no way of detecting an error in sample handling or some other transient error that may have caused the anomaly. Thus, the 71% result could not be assigned a known cause with any certainty. Based on this evidence, rejecting the batch outright would seem to be rather a harsh decision. Rather, it would be prudent to perform further testing before coming to the ominous decision of rejection. One possible approach, as discussed in the previous paragraph, is to perform sufficient additional assays to ensure (with a high degree of probability) that the great majority of tablets are within 85% to 115% limits, a definition based on the USP content uniformity monograph. Using the tolerance limit concept (sect. 5.6), we could assay N new samples and create a tolerance interval that should lie within 85% to 115%. Suppose we estimate the CV as 3.5%, based on the nine good CU assays, other data accumulated during the batch production, and historical data from previous assays. Also, the average result is estimated as 98% based on all production data available. The value of t' for the tolerance interval for 99% probability that includes 99% of the tablets between 85% and 115% is 3.71. From Table IV.19, tolerance intervals, a sample of 35 tablets would give this confidence, provided that the mean and s.d. are as estimated, 98% and 3.5%, respectively. To protect against more variability and deviation from label, a larger sample would be more conservative. For

Table 10.9 SAS Output for Residuals for Data of Ryde et al. [13]

Obs	Subject	Seq	Period	Product	CO	AUC	YHAT	Resid	ERESID
1	1	1	1	1	0	106.3	93.518	12.7819	15.1863
2	1	1	2	2	1	36.4	75.638	−39.2375	15.1863
3	1	1	3	2	2	94.7	63.137	31.5625	15.1863
4	1	1	4	1	2	58.9	64.007	−5.1069	15.1863
5	2	1	1	1	0	149.2	139.518	9.6819	15.1863
6	2	1	2	2	1	107.1	121.638	−14.5375	15.1863
7	2	1	3	2	2	104.6	109.137	−4.5375	15.1863
8	2	1	4	1	2	119.4	110.007	9.3931	15.1863
9	3	1	1	1	0	134.8	155.543	−20.7431	15.1863
10	3	1	2	2	1	155.1	137.663	17.4375	15.1863
11	3	1	3	2	2	132.5	125.162	7.3375	15.1863
12	3	1	4	1	2	122.0	126.032	−4.0319	15.1863
13	4	1	1	1	0	108.1	82.193	25.9069	15.1863
14	4	1	2	2	1	84.9	64.312	20.5875	15.1863
15	4	1	3	2	2	33.2	51.812	−18.6125	15.1863
16	4	1	4	1	2	24.8	52.682	−27.8819	15.1863
17	6	2	1	2	0	85.0	88.081	−3.0806	15.3358
18	6	2	2	1	2	92.8	92.5250	0.2750	15.3358
19	6	2	3	1	1	81.9	80.0250	1.8750	15.3358
20	6	2	4	2	1	59.5	58.5694	0.9306	15.3358
21	7	2	1	2	0	64.1	83.9056	−19.8056	15.3358
22	7	2	2	1	2	112.8	88.3500	24.4500	15.3358
23	7	2	3	1	1	70.4	75.8500	−5.4500	15.3358
24	7	2	4	2	1	55.2	54.3944	0.8056	15.3358
25	8	2	1	2	0	15.3	29.5806	−14.2806	15.3358
26	8	2	2	1	2	30.1	34.0250	−3.9250	15.3358
27	8	2	3	1	1	22.3	21.5250	0.7750	15.3358
28	8	2	4	2	1	17.5	0.0694	17.4306	15.3358
29	9	2	1	2	0	77.4	74.9806	2.4194	15.3358
30	9	2	2	1	2	67.6	79.4250	−11.8250	15.3358
31	9	2	3	1	1	72.9	66.9250	5.9750	15.3358
32	9	2	4	2	1	48.9	45.4694	3.4306	15.3358
33	10	2	1	2	0	102.0	94.8806	7.1194	15.3358
34	10	2	2	1	2	106.1	99.3250	6.7750	15.3358
35	10	2	3	1	1	67.9	86.8250	−18.9250	15.3358
36	10	2	4	2	1	70.4	65.3694	5.0306	15.3358

example, suppose we decide to test 50 tablets, and the average is 97.5% with a s.d. of 3.7%. No tablet was outside 85% to 115%. The 99% tolerance interval is

$$97.5 \pm 3.385 \times 3.7 = 85.0 \text{ to } 110.0.$$

The lower limit just makes 85%. We can be 99% certain, however, that 99% of the tablets are between 85% and 110%. This analysis is evidence that the tablets are uniform. Note, that had we tested fewer tablets, say 45, the interval would have included values less than 85%. However, in this case, where the lower interval would be 84.8% (given the same mean and s.d.), it would appear that the batch can be considered satisfactory. For example, if we were interested in determining the probability of tablets having a drug content between 80% and 120%, application of the tolerance interval calculation results in a t' of $(97.5 - 80)/3.7 = 4.73$. Table IV.19 shows that this means that with a probability greater than 99%, 99.9% of the tablets are between 80% and 120%.

One should understand that the extra testing gives us confidence about the acceptability of the batch. We will never know if the original 71% result was real or caused by an error in the analytical process. However, if the 71% result was real, the additional testing gives us assurance that results as extreme as 71% are very unlikely to be detected in this batch. A publication [10] discussing the nature and possible handling of outliers is in Appendix V.

10.2.4.1 Lund's Method

The FDA has suggested the use of tables prepared by Lund [11] (Table IV.20) to identify outliers. This table compares the extreme residual to the standard error of the residuals (studentized residual), and gives critical values for the studentized residual at the 5% level of significance as a function of the number of observations and parameter d.f. in the model. For analysis of variance designs, these calculations may be complicated and use of a computer program is almost necessary. SAS [12] code is shown below to produce an output of the residuals and their standard errors, which should clarify the procedure and interpretation.

SAS Program to Generate Residuals and Standard Errors from a Two-Period Crossover Design for a Bioequivalence Study

```
Proc GLM;
Class subject product seq period;
model lcmax = seq subject(seq) product period;
lsmeans product/stderr;
estimate "test-ref" product −1 1;
output out = new p = yhat r = resid stdr = eresid;
proc print;
run;
```

The SAS output for the data of Ryde et al. [13] (without interaction and carryover) is shown in Table 10.9.

The largest residual is −39.2375 for Subject 1 in Period 2. The ratio of the residual to its standard error is $−39.2375/15.1863 = −2.584$. This model has 12 parameters and 36 observations. At the 5% level, from Table IV.20, the critical value is estimated at approximately 2.95. Therefore there are no "outliers" evident in this data at the 5% level.

KEY TERMS

Arcsin transformation	Parametric analyses
Back transformation	Ratios
Coefficient of variation	Reciprocal transformation
Dixon's test for outliers	Residuals
Exploratory data analysis	Skewed data
Fisher–Behrens test	Square-root transformation
Geometric mean	Studentized residual
Log transformation	T procedure
Nonparametric analyses	Tolerance interval
Ordered observations	Winsorizing
Outliers	

EXERCISES

1. Convert the data in Table 10.2, data set B, to logs and construct a histogram of the transformed data.

2. Fit the least squares line for the averages of log assay versus log drug concentration for the average data in Table 10.4.

Log *X*	Log *Y*
1.602	1.578
1.778	1.787
1.903	1.897

If an unknown sample has a reading of 47, what is the estimate of the drug concentration?

3. Perform a t test for the data of Table 10.5 using the ratio B/A (H_0: $R = 1$), and log B − log A (H_0: log B − log $A = 0$). Compare the values of t in these analyses to the similar analyses shown in the text for A/B and log A − log B.

4. Ten tablets were assayed with the following results: 51, 54, 46, 49, 53, 50, 49, 62, 47, 53. Is the value 62 an outlier? When averaging the tablets to estimate the batch average, would you exclude this value from the calculation? (Use both the Dixon method and the T method to test the value of 62 as an outlier.)

5. Consider 62 to be an outlier in Problem 4 and calculate the Winzorized average. Compare this to the average with 62 included.

6. A tablet product was manufactured using two different processes, and packaged in bottles of 1000 tablets. Five bottles were sampled from each batch (process) with the following results:

	Number of defective tablets per bottle									
	Process 1 bottle					Process 2 Bottle				
	1	2	3	4	5	1	2	3	4	5
No. of defects	0	6	1	3	4	0	1	1	0	1

Perform a t test to compare the average results for each process. Transform the data and repeat the t test. What transformation did you use? Explain why you used the transformation. [Hint: See transformations for Poisson variables.]

REFERENCES

1. Chow S-C, Liu J-P. Design and Analysis of Bioavailability and Bioequivalence Studies. New York: Marcel Dekker, 1992:18.
2. Mosteller F, Youtz C. Tables of the Freeman–Tukey transformations for the binomial and Poisson distributions. Biometrika 1961; 48:433–440.
3. Sokal RR, Rohlf FJ. Biometry. San Francisco, CA: W. H. Freeman, 1969.
4. Weisberg S. Applied Linear Regression. New York: Wiley, 1980.
5. Ostle B. Statistics in Research, 3rd ed. Ames, IA: Iowa State University Press, 1981.
6. Youden WJ. Statistical Methods for Chemists. New York: Wiley, 1964.
7. Dixon WJ, Massey FJ Jr. Introduction to Statistical Analysis, 3rd ed. New York: McGraw-Hill, 1969.
8. E-178–75, American National Standards Institute, Z1.14, 1975, p. 183.
9. Barnett V, Lewis T. Outliers in Statistical Data, 2nd ed. New York: Wiley, 1984.
10. Bolton S. Outlier tests and chemical assays. Clin Res Practices Drug Reg Affairs 1993; 10:221–232.
11. Lund RE. Tables for an approximate test for outliers in linear regression. Technometrics 1975; 17(4):473–476.
12. SAS Institute, Inc., Cary, N.C. 27513.
13. Chow S-C, Liu J-P. Design and Analysis of Bioavailability and Bioequivalence Studies. New York: Marcel Dekker, 1992:280.

11 | Experimental Design in Clinical Trials

The design and analysis of clinical trials is fertile soil for statistical applications. The use of sound statistical principles in this area is particularly important because of close FDA involvement, in addition to crucial public health issues that are consequences of actions based on the outcomes of clinical experiments. Principles and procedures of experimental design, particularly as applied to clinical studies, are presented. Relatively few different experimental designs are predominantly used in controlled clinical studies. In this chapter, we discuss several of these important designs and their applications.

11.1 INTRODUCTION

Both pharmaceutical manufacturers and FDA personnel have had considerable input in constructing guidelines and recommendations for good clinical protocol design and data analysis. In particular, the FDA has published a series of guidelines for the clinical evaluation of a variety of classes of drugs. Those persons involved in clinical studies have been exposed to the constant reminder of the importance of design in these studies. Clinical studies must be carefully devised and documented to meet the clinical objectives. Clinical studies are very expensive indeed, and before embarking, an all-out effort should be made to ensure that the study is on a sound footing. Clinical studies designed to "prove" or demonstrate efficacy and/or safety for FDA approval should be controlled studies, as far as is possible. A controlled study is one in which an adequate control group is present (placebo or active control), and in which measures are taken to avoid bias. The following excerpts from *General Considerations for the Clinical Evaluation of Drugs* show the FDA's concern for good experimental design and statistical procedures in clinical trials [1]:

1. Statistical expertise is helpful in the planning, design, execution, and analysis of clinical investigations and clinical pharmacology in order to ensure the validity of estimates of safety and efficacy obtained from these studies.
2. It is the objective of clinical studies to draw inferences about drug responses in well-defined target populations. Therefore, study protocols should specify the target population, how patients or volunteers are to be selected, their assignment to the treatment regimens, specific conditions under which the trial is to be conducted, and the procedures used to obtain estimates of the important clinical parameters.
3. Good planning usually results in questions being asked that permit direct inferences. Since studies are frequently designed to answer more than one question, it is useful in the planning phase to consider listing of the questions to be answered in order of priority.

The following are general principles that should be considered in the conduct of clinical trials:

1. Clearly state the objective(s).
2. Document the procedure used for randomization.
3. Include a suitable number of patients (subjects) according to statistical principles (see chap. 6).
4. Include concurrently studied comparison (control) groups.
5. Use appropriate blinding techniques to avoid patient and physician bias.
6. Use objective measurements when possible.
7. Define the response variable.
8. Describe and document the statistical methods used for data analysis.

11.2   SOME PRINCIPLES OF EXPERIMENTAL DESIGN AND ANALYSIS

Although many kinds of ingenious and complex statistical designs have been used in clinical studies, many experts feel that *simplicity* is the key in clinical study design. The implementation of clinical studies is extremely difficult. No matter how well designed or how well intentioned, clinical studies are particularly susceptible to Murphy's law: "If something can go wrong, it will!" Careful attention to protocol procedures and symmetry in design (e.g., equal number of patients per treatment group) often is negated as the study proceeds, due to patient dropouts, missed visits, carelessness, misunderstood directions, and so on. If severe, these deviations can result in extremely difficult analyses and interpretations. Although the experienced researcher anticipates the problems of human research, such problems can be minimized by careful planning.

We will discuss a few examples of designs commonly used in clinical studies. The basic principles of good design should always be kept in mind when considering the experimental pathway to the study objectives. In *Planning of Experiments*, Cox discusses the requirements for a good experiment [2]. When designing clinical studies, the following factors are important:

1. absence of bias;
2. absence of systematic error (use of controls);
3. adequate precision;
4. choice of patients;
5. simplicity and symmetry.

11.2.1   Absence of Bias

As far as possible, known sources of bias should be eliminated by blinding techniques. If a double-blind procedure is not possible, careful thought should be given to alternatives that will suppress, or at least account for possible bias. For example, if the physician can distinguish two comparative drugs, as in an open study, perhaps the evaluation of the response and the administration of the drug can be done by other members of the investigative team (e.g., a nurse) who are not aware of the nature of the drug being administered.

In a double-blind study, both the observer and patient (or subject) are unaware of the treatment being given during the course of the study. Human beings, the most complex of machines, can respond to drugs (or any stimulus, for that matter) in amazing ways as a result of their psychology. This is characterized in drug trials by the well-known "placebo effect." Also, a well-known fact is that the observer (nurse, doctor, etc.) can influence the outcome of an experiment if the nature of the different treatments is known. The subjects of the experiment can be influenced by words and/or actions, and unconscious bias may be manifested in the recording and interpretation of the experimental observations. For example, in analgesic studies, as much as 30% to 40% of patients may respond to a placebo treatment.

The double-blind method is accomplished by manufacturing alternative treatment dosage forms to be as alike as possible in terms of shape, size, color, odor, and taste. Even in the case of dosage forms that are quite disparate, ingenuity can always provide for double blinding. For example, in a study where an injectable dosage form is to be compared to an oral dosage form, the *double-dummy technique* may be used. Each subject is administered both an oral dose and an injection. In one group, the subject receives an active oral dose and a placebo injection, whereas in the other group, each subject receives a placebo oral dose and an active injection. There are occasions where blinding is so difficult to achieve or is so inconvenient to the patient that studies are best left "unblinded." In these cases, every effort should be made to reduce possible biases. For example, in some cases, it may be convenient for one person to administer the study drug, and a second person, unaware of the treatment given, to make and record the observation.

Examples of problems that occur when trials are not blinded are given by Rodda et al. [3]. In a study designed to compare an angiotensin converting enzyme (ACE) inhibitor with a beta-blocker, unblinded investigators tended to assign patients who had been previously unresponsive to beta-blockers to the ACE group. This allocation results in a treatment bias. The ACE group may contain the more seriously ill patients.

An important feature of clinical study design is randomization of patients to treatments. This topic has been discussed in chapter 4, but bears repetition. The randomization procedure as applied to various designs will be presented in the following discussion. Randomization is an integral and essential part of the implementation and design of clinical studies. Randomization will help to reduce potential bias in clinical experiments, and is the basis for valid calculations of probabilities for statistical testing.

11.2.2 Absence of Systematic Errors

Cox gives some excellent examples in which the presence of a systematic error leads to erroneous conclusions [2]. In the case of clinical trials, a systematic error would be present if one drug was studied by one investigator and the second drug was studied by a second investigator. Any observed differences between drugs could include "systematic" differences between the investigators. This ill-designed experiment can be likened to Cox's example of feeding two different rations to a group of animals, where each group of animals is kept together in separate pens. Differences in pens could confuse the ration differences. One or more pens may include animals with different characteristics that, by chance, may affect the experimental outcome. In the examples above, the experimental units (patients, animals, etc.) are not independent. Although the problems of interpretation resulting from the designs in the examples above may seem obvious, sometimes the shortcomings of experimental procedures are not obvious. We have discussed the deficiencies of a design in which a baseline measurement is compared to a post-treatment measurement in the absence of a control group. Any change in response from baseline to treatment could be due to changes in conditions during the intervening time period. To a great extent, systematic errors in clinical experiments can be avoided by the inclusion of an appropriate control group and random assignment of patients to the treatment groups.

11.2.3 Adequate Precision

Increased precision in a comparative experiment means less variable treatment effects and more efficient estimate of treatment differences. Precision can always be improved by increasing the number of patients in the study. Because of the expense and ethical questions raised by using large numbers of patients in drug trials, the sample size should be based on medical and statistical considerations that will achieve the experimental objectives described in chapter 6.

Often, an appropriate choice of experimental design can increase the precision. Use of baseline measurements or use of a crossover design rather than a parallel design, for example, will usually increase the precision of treatment comparisons. However, in statistics as in life, we do not get something for nothing. Experimental designs have their shortcomings as well as advantages. Properties of a particular design should be carefully considered before the final choice is made. For example, the presence of carryover effects will negate the advantage of a crossover design as presented in section 11.4.

Blocking is another way of increasing precision. This is the basis of the increased precision accomplished by use of the two-way design discussed in section 8.4. In these designs, the patients in a block have similar (and relevant) characteristics. For example, if age and sex are variables that affect the therapeutic response of two comparative drugs, patients may be "blocked" on these variables. Thus if a male of age 55 years is assigned to drug *A*, another male of age approximately 55 years will be assigned Treatment *B*. In practice, patients of similar characteristics are grouped together in a block and randomly assigned to treatments.

11.2.4 Choice of Patients

In most clinical studies, the choice of patients covers a wide range of possibilities (e.g., age, sex, severity of disease, concomitant diseases, etc.). In general, inferences made regarding drug effectiveness are directly related to the restrictions (or lack of restrictions) placed on patient eligibility as described in the study protocol. This is an important consideration in experimental design, and great care should be taken to describe that patients may be qualified or disqualified from entering the study.

11.2.5 Simplicity and Symmetry

Again we emphasize the importance of *simplicity*. More complex designs have more restrictions, and a resultant lack of flexibility. The gain resulting from a more complex design should be weighed against the expense and problems of implementation often associated with more sophisticated, complex designs.

Symmetry is an important design consideration. Often, the symmetry is obvious: In most (but not all) cases, experimental designs should be designed to have equal number of patients per treatment group, equal number of visits per patient, balanced order of administration, and an equal number of replicates per patient. Some designs, such as balanced incomplete block and partially balanced incomplete block designs, have a less obvious symmetry.

11.2.6   Randomization

Principles of randomization have been described in chapter 4. Randomization is particularly important when assigning patients to treatments in clinical trials, ensuring that the requirements of good experimental design are fulfilled and the pitfalls avoided [4]. Among other qualities, proper randomization avoids unknown biases, tends to balance patient characteristics, and is the basis for the theory that allows calculation of probabilities. Randomization ensures a balance in *the long run*. In any given experiment, two groups may not have similar characteristics due to chance. Therefore, it is important to carefully examine properties of the groups to assess if group differences could affect the experimental outcome. Use of covariance analysis can help overcome differences between groups as discussed in section 8.6.

In section 4.2, the advantages of randomization of patients in blocks is discussed. Table 11.1 is a short table of random permutations that gives random schemes for block sizes of 4, 5, 6, 8, and 10. This kind of randomization is also known as restricted randomization and allows for an approximate balance of treatment groups throughout the trial. As an example of the application of Table 11.1, consider a study comparing an active drug with placebo using a parallel design, with 24 patients per group (a total of 48 patients). In this case, a decision is made to group patients in blocks of 8, that is, for each group of eight consecutive patients, four will be on drug and four on placebo. In Table 11.1, we start in a random column in the section labeled "Blocks of 8," and select six sequential columns. Because this is a short table, we would continue into the first column if we had to proceed past the last column. (Note that this table is meant to illustrate the procedure and should not be used repeatedly in real examples or for sample sizes exceeding the total number of random assignments in the table. For example, there are 160 random assignments for blocks of size 8; therefore for a study consisting of more than 160 patients, this table would not be of sufficient size.) If the third column is selected to begin the random assignment, and we assign Treatment A to an odd number and Treatment B to an even number, the first eight patients will be assigned treatment as follows:

   B B A B B A A A.

11.2.7   Intent to Treat

In most clinical studies, there is a group of patients who have been administered drug who may not be included in the efficacy data analysis because of various reasons, such as protocol violations. This would include patients, for example, who (a) leave the study early for nondrug-related reasons, (b) take other medications that are excluded in the protocol, or (c) are noncompliant with regard to the scheduled dosing regimen, and so on. Certainly, these patients should be included in summaries of safety data, such as adverse reactions and clinical laboratory determinations. Under FDA guidelines, an analysis of efficacy data should be performed with these patients included as an "intent to treat" (ITT) analysis [5]. Thus, both an efficacy analysis including only those patients who followed the protocol, and an ITT analysis, which includes all patients randomized to treatments (with the possible exception of inclusion of ineligible patients, mistakenly included) are performed. In fact, the ITT analysis may take precedence over the analysis that excludes protocol violators. The protocol violators, or those patients who are not to be included in the primary analysis, should be identified, with reasons for exclusion, prior to breaking the treatment randomization code. The ITT analysis should probably not result in different conclusions from the primary analysis, particularly if the protocol violators and other "excluded" patients occur at random. In most circumstances, a different conclusion may occur for the two analyses only when the significance level is close to 0.05.

Table 11.1   Randomization in Blocks

BLOCKS OF 4																			
1	3	3	2	4	4	1	I	1	2	1	3	3	1	2	4	2	3	1	4
2	2	4	3	3	2	2	2	2	3	4	2	2	4	4	2	4	2	4	3
3	1	1	4	2	1	3	3	3	1	2	1	1	2	3	1	1	4	3	2
4	4	2	1	1	3	4	4	4	4	3	4	4	3	1	3	3	1	2	1
BLOCKS OF 5																			
4	4	1	3	5	5	4	2	5	5	3	5	4	3	2	2	3	2	5	4
2	5	3	5	2	3	5	5	1	1	2	2	2	4	3	5	4	3	1	2
3	3	5	4	1	2	1	3	4	3	5	4	1	5	4	3	2	4	4	3
1	2	4	2	3	1	3	4	2	4	4	3	5	2	5	1	1	1	2	1
5	1	2	1	4	4	2	1	3	2	1	1	3	1	1	4	5	5	3	5
BLOCKS OF 6																			
1	5	2	5	3	2	5	1	5	1	1	2	5	2	6	4	3	4	2	2
2	6	5	3	2	1	2	6	6	3	4	4	1	1	3	5	6	2	6	5
5	9	4	4	1	3	3	5	4	4	2	6	6	6	1	3	2	5	3	1
6	1	1	2	5	5	4	2	3	6	5	1	2	3	2	1	4	6	4	3
3	4	6	1	6	6	1	3	2	5	3	3	3	4	4	6	5	3	1	6
4	3	3	6	4	4	6	4	1	2	6	5	4	5	5	2	1	1	5	4
BLOCKS OF 8																			
7	4	2	4	1	2	1	5	3	4	4	8	5	3	5	2	2	5	1	6
8	2	4	5	8	5	5	2	4	5	6	6	4	5	4	7	8	3	7	7
4	3	1	6	3	6	3	4	5	2	7	5	1	1	3	6	6	6	8	5
1	5	6	3	2	7	8	8	2	1	3	1	3	8	6	3	3	8	5	1
2	8	8	1	7	8'	4	3	8	7	5	7	7	6	1	4	4	2	3	3
3	1	5	8	6	1	2	7	7	6	2	3	2	2	2	5	5	1	6	2
6	7	3	7	5	4	7	1	6	8	8	2	8	4	7	8	7	4	2	4
5	6	7	2	4	3	6	6	1	3	1	4	6	7	8	1	1	7	4	8
BLOCKS OF 10																			
1	9	4	1	3	4	1	4	6	8	9	9	10	9	5	5	6	6	4	3
4	6	5	8	2	7	4	5	3	9	7	6	6	1	1	4	3	2	9	2
5	2	3	4	7	8	5	9	9	2	10	8	10	7	4	3	9	7	10	9
9	8	6	10	8	9	8	10	5	7	2	4	4	4	10	10	4	1	2	7
2	10	8	9	1	6	6	8	4	10	5	2	9	2	6	1	1	9	7	5
10	3	9	5	6	2	9	1	8	1	1	3	5	8	8	8	7	3	3	10
8	4	7	7	9	3	10	7	1	4	3	7	3	3	2	9	2	5	1	8
3	5	2	2	5	1	7	6	7	5	8	1	7	5	3	6	5	8	5	1
6	7	10	3	10	5	3	3	2	6	4	10	8	6	9	2	8	4	6	6
7	1	1	6	4	10	2	2	10	3	6	5	2	10	7	7	10	10	8	4

If the conclusions differ for the two analyses, ITT results are sometimes considered to be more definitive. Certainly, an explanation should be given when conclusions are different for the two analyses. One should recognize that the issue of using an ITT analysis vis-à-vis an analysis including only "compliant" patients remains controversial.

11.3   PARALLEL DESIGN

In a parallel design, two or more drugs are studied, drugs being randomly assigned to different patients. Each patient is assigned a single drug. In the example presented here, a study was proposed to compare the response of patients to a new formulation of an antianginal agent and a placebo with regard to exercise time on a stationary bicycle at fixed impedance. An alternative approach would be to use an existing product rather than placebo as the comparative product. However, the decision to use placebo was based on the experimental objective: to demonstrate that the new formulation produces a measurable and significant increase in exercise time. A difference in exercise time between the drug and placebo is such a measure. A comparison of the new formulation with a positive control (an active drug) would not achieve the objective directly.

In this study, a difference in exercise time between drug and placebo of 60 seconds was considered to be of clinical significance. The standard deviation was estimated to be 65 based on change from baseline data observed in previous studies. The sample size for this study, for an alpha level of 0.05 and power of 0.90 (beta = 0.10), was estimated as 20 patients per group (see Exercise Problem 7). Therefore 40 patients were entered into the study, 20 each randomly assigned to placebo and active treatment. A randomization that obviates a long consecutive run of patients assigned to one of the treatments was used as described in section 11.2.6. Patients were randomly assigned to each treatment in groups of 10, with 5 patients to be randomly assigned to each treatment. This randomization was applied to each of the 4 subsets of 10 patients (40 patients total). From Table 11.1 starting in the fourth column, patients are randomized into the two groups as follows, placebo if an odd number appears and new formulation if an even number appears:

	Placebo	New formulation
Subset 1	1, 5, 6, 7, 9	2, 3, 4, 8, 10
Subset 2	11, 13, 15, 17, 18	12, 14, 16, 19, 20
Subset 3	22, 24, 27, 28, 29	21, 23, 25, 26, 30
Subset 4	31, 33, 36, 38, 39	32, 34, 35, 37, 40

The first subset is assigned as follows. The first number is 1; patient 1 is assigned to placebo. The second number (reading down) is 8; patient 2 is assigned to the new formulation (NF). The next two numbers (4, 10) are even. Patients 3 and 4 are assigned to NF. The next number is odd (9); patient 5 is assigned to Placebo. The next two numbers are odd and Patients 6 and 7 are assigned to Placebo. Patients 8, 9, and 10 are assigned to NF, placebo, and NF, respectively, to complete the first group of 10 patients. Entering column five, patient 11 is assigned to placebo, and so on.

An alternative randomization is to number patients consecutively from 1 to 40 as they enter the study. Using a table of random numbers, patients are assigned to placebo if an odd number appears, and assigned to the test product (NF) if an even number appears. Starting in the eleventh column of Table IV.1, the randomization scheme is as follows:

Placebo	New formulation
1, 6, 7, 8	2, 3, 4, 5
12, 13, 14	9, 10, 11
15, 18, 20	16, 17, 19
21, 22, 26	23, 24, 25
27, 28	29, 30, 31
32, 34, 35	33, 38, 39
36, 37	40

For example, the first number in column 11 is 7; patient number 1 is assigned to placebo. The next number in column 11 is 8; the second patient is assigned to the NF; and so on. A problem with this approach is that by chance we may observe a long string of consecutive of odd or even numbers, which would negate the purpose of the randomization as noted above.

Patients were first given a predrug exercise test to determine baseline values. The test statistic is the time of exercise to fatigue or an anginal episode. Tablets were prepared so that the placebo and active drug products were identical in appearance. Double-blind conditions prevailed. One hour after administration of the drug, the exercise test was repeated. The results of the experiment are shown in Table 11.2.

The key points in this design are as follows:

1. There are two independent groups (placebo and active, in this example). An equal number of patients are randomly assigned to each group.
2. A baseline measurement and a single post-treatment measurement are available.

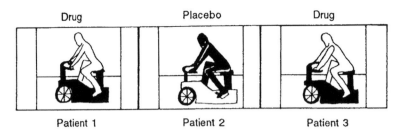

Drug	Placebo	Drug
Patient 1	Patient 2	Patient 3

This design corresponds to a one-way analysis of variance, or in the case of two treatments, a two independent groups t test. Since, in general, more than two treatments may be included in the experiment, the analysis will be illustrated using ANOVA.

When possible, pretreatment (baseline) measurements should be made in clinical studies. The baseline values can be used to help increase the precision of the measurements. For example, if the treatment groups are compared using differences from baseline, rather than the post-treatment exercise time, the variability of the measurements will usually be reduced. Using differences, we will probably have a better chance of detecting treatment differences, if they exist (increased power) [6]. "Subtracting out" the baseline helps to reduce the between-patient variability that is responsible for the variance (the "within mean square") in the statistical test. A more complex, but more efficient analysis is *analysis of covariance*. Analysis of covariance [6] takes baseline readings into account, and for an unambiguous conclusion, assumes that the slope of the response versus baseline is the same for all treatment groups. See "Analysis of Covariance" (sect. 8.6) for a more detailed discussion. Also, the interpretation may be more difficult than the simple "difference from baseline" approach.

To illustrate the results of the analysis with and without baseline readings, the data in Table 11.2 will be analyzed in two ways: (a) using only the post-treatment response,

Table 11.2 Results of the Exercise Test Comparing Placebo to Active Drug: Time (Seconds) to Fatigue or Angina

Placebo				Active drug (new formulation)			
	Exercise time				Exercise time		
Patient	Pre	Post	Post–Pre	Patient	Pre	Post	Post–Pre
1	377	345	−32	2	232	372	140
6	272	310	38	3	133	120	−13
7	348	347	−1	4	206	294	88
8	348	300	−48	5	140	258	118
12	133	150	17	9	240	340	100
13	102	129	27	10	246	393	147
14	156	110	−46	11	226	315	89
15	205	251	46	16	123	180	57
18	296	262	−34	17	166	334	168
20	328	297	−31	19	264	381	117
21	315	278	−37	23	241	376	135
22	133	124	−9	24	74	264	190
26	223	289	66	25	400	541	141
27	256	303	47	29	320	410	90
28	493	487	−6	30	216	301	85
32	336	309	−27	31	153	143	−10
34	299	281	−18	33	193	348	155
35	140	186	46	38	330	440	110
36	161	125	−36	39	258	365	107
37	259	236	−23	40	353	483	130
Mean	259	256	−3.05	Mean	226	333	107.2
s.d.	102	95	36.3	s.d.	83	106	51.5

post-treatment exercise time, and (b) comparing the difference from baseline for the two treatments. The reader is reminded of the assumptions underlying the t test and ANOVA: the variables should be independent, normally distributed with homogeneous variance. These assumptions are necessary for both post-treatment and difference analyses. Possible problems with lack of normality will be less severe in the difference analysis. The difference of independent non-normal variables will tend to be closer to normal than are the original individual data.

Before proceeding with the formal analysis, it is prudent to test the equivalence of the baseline averages for the two treatment groups. This test, if not significant, gives some assurance that the two groups are "comparable." We will use a two independent groups t test to compare baseline values (see sect. 5.2.2).

$$t = \frac{\overline{X}_1 - \overline{X}_2}{S_p\sqrt{1/N_1 + 1/N_2}}$$

$$= \frac{259 - 226}{S_p\sqrt{1/20 + 1/20}} = \frac{33}{93\sqrt{1/10}} = 1.12.$$

Note that the pooled standard deviation (93) is the pooled value from the baseline readings, $\sqrt{(102^2 + 83^2)/2}$. From Table IV.4, a t value of approximately 2.03 is needed for significance (38 d.f.) at the 5% level. Therefore, the baseline averages are not significantly different for the two treatment groups. If the baseline values are significantly different, one would want to investigate further the effects of baseline on response in order to decide on the best procedure for analysis of the data (e.g., covariance analysis, ratio of response to baseline, etc.).

11.3.1 ANOVA Using Only Post-Treatment Results

The average results for exercise time after treatment are 256 seconds for placebo and 333 seconds for the NF of active drug, a difference of 77 seconds (Table 11.2). Although the averages can be compared using a t test as in the case of baseline readings (above), the equivalent ANOVA is given in Table 11.3. The reader is directed to Exercise Problem 1 for the detailed calculations. According to Table IV.6A1, between groups (active and placebo) is significant at the 5% level.

11.3.2 ANOVA of Differences from the Baseline

When the baseline values are taken into consideration, the active drug shows an increase in exercise time over placebo of 110.25 seconds [107.2 − (−3.05)]. The ANOVA is shown in Table 11.4. The data analyzed here are the (post–pre) values given in Table 11.2. The F test for treatment differences is 61.3! There is no doubt about the difference between the active drug and placebo. The larger F value is due to the considerable reduction in variance as a result of including the baseline values in the analysis. The within-groups error term represents *within-* patient variation in this analysis. In the previous analysis for post-treatment results only, the within-groups error term represents the *between-*patient variation, which is considerably larger than the within-patient error. Although both tests are significant ($p < 0.05$) in this example, one can easily see that situations may arise in which treatments may not be *statistically* different based on a significance test if between-patient variance is used as the error term, but would be significant based on the smaller within-patient variance. Thus, designs that use the smaller within-patient variance as the error term for treatments are to be preferred, other things being equal.

Table 11.3 ANOVA Table for Post-Treatment Readings for the Data of Table 11.2

Source	d.f.	SS	MS	F
Between groups	1	59,213	59,213	$F_{1,38} = 5.86^*$
Within groups	38	383,787	10,099.7	
Total	39	443,000		

$^*p < 0.05$.

Table 11.4 Analysis of Variance for Differences from Baseline (Table 11.1)

Source	d.f.	SS	MS	F
Between groups	1	121,551	120,551	$F_{1,38} = 61.3^*$
Within groups	38	75,396	1984	
Total	39	196,947		

$^*p < 0.01$.

11.4 CROSSOVER DESIGNS AND BIOAVAILABILITY/BIOEQUIVALENCE STUDIES

In a typical crossover design, each subject takes each of the treatments under investigation on different occasions. Comparative bioavailability* or bioequivalence studies, in which two or more formulations of the same drug are compared, are usually designed as crossover studies. Perhaps the greatest appeal of the crossover design is that each patient acts as his or her own control. This feature allows for the direct comparison of treatments, and is particularly efficient in the presence of large interindividual variation. However, caution should be used when considering this design in studies where carryover effects or other interactions are anticipated. Under these circumstances, a parallel design may be more appropriate.

11.4.1 Description of Crossover Designs: Advantages and Disadvantages

The crossover (or changeover) design is a very popular, and often desirable, design in clinical experiments. In these designs, typically, two treatments are compared, with each patient or subject taking each treatment in turn. The treatments are typically taken on two occasions, often called *visits*, *periods*, or *legs*. The order of treatment is randomized; that is, either A is followed by B or B is followed by A, where A and B are the two treatments. Certain situations exist where the treatments are not separated by time, for example, in two visits or periods. For example, comparing the effect of topical products, locations of applications on the body may serve as the visits or periods. Product may be applied to each of two arms, left and right. Individuals will be separated into two groups, (1) those with Product A applied on the left arm and Product B on the right arm, and (2) those with Product B applied on the left arm and Product A on the right arm.

A————————→B B————————→A
First week   Second week   *or*   First week   Second week

This design may also be used for the comparison of more than two treatments. The present discussion will be limited to the comparison of two treatments, the most common situation in clinical studies. (The design and analysis of three or more treatment crossovers follows.) Crossover designs have great appeal when the experimental objective is the comparison of the performance, or effects, of two drugs or product formulations. Since each patient takes each product, the comparison of the products is based on *within*-patient variation. The *within*- or intrasubject variability will be smaller than the *between*- or intersubject variability used for the comparison of treatments in the one-way or parallel-groups design. Thus, crossover experiments usually result in greater precision than the parallel-groups design, where different patients comprise the two groups. Given an equal number of observations, the crossover design is more powerful than a parallel design in detecting product differences.

The crossover design is a type of Latin square. In a Latin square, the number of treatments equals the number of patients. In addition, another factor, such as order of treatment, is included in the experiment in a balanced way. The net result is an $N \times N$ array (where N is the number of treatments or patients) of N letters such that a given letter appears only once in a given row or column. This is most easily shown pictorially. A Latin square for four subjects taking four drugs is shown in Table 11.5. For randomizations of treatments in Latin squares, see Ref. [6].

---

* A bioavailability study, in our context, is defined as a comparative study of a drug formulation compared to an optimally absorbed (intravenous or oral solution) formulation.

Table 11.5   4 × 4 Latin Square: Four Subjects Take Four Drugs

Subject	Order in which drugs[a] are taken			
	First	**Second**	**Third**	**Fourth**
1	A	B	C	D
2	B	C	D	A
3	C	D	A	B
4	D	A	B	C

[a]Drugs are designated as A, B, C, D.

For the comparison of two formulations, a 2 × 2 Latin square ($N = 2$) consists of two patients each taking two formulations (A and B) on two different occasions in two "orders" as follows:

Patient	Occasion period	
	First	**Second**
1	A	B
2	B	A

The balancing of order ($A - B$ or $B - A$) takes care of time trends or other "period" effects, if present. (A period effect is a difference in response due to the occasion on which the treatment is given, independent of the effect due to the treatment.)

The 2 × 2 Latin square shown above is familiar to all who have been involved in bioavailability/bioequivalence studies. In these studies, the 2 × 2 Latin square is repeated several times to include a sufficient number of patients (see also Table 11.6). Thus, the crossover design can be thought of as a repetition of the 2 × 2 Latin square.

The crossover design has an advantage, previously noted, of increased precision relative to a parallel-groups design. Also, the crossover is usually more economical: one-half the number of patients or subjects have to be recruited to obtain the same number of observations as in a parallel design. (Note that each patient takes *two* drugs in the crossover.) Often, a significant part of the expense in terms of both time and money is spent recruiting and processing patients or volunteers. The advantage of the crossover design in terms of cost depends on the economics of patient recruiting, cost of experimental observations, as well as the relative within-patient/between-patient variation. The smaller the within-patient variation relative to the between-patient variation, the more efficient will be the crossover design. Hence, if a repeat observation on the same patient is very variable (nonreproducible), the crossover may not be very much better than a parallel design, cost factors being equal. This problem is presented and quantitatively analyzed in detail by Brown [7].

There are also some problems associated with crossover designs. A crossover study may take longer to complete than a parallel study because of the extra testing period. It should be noted, however, that if recruitment of patients is difficult, the crossover design may actually save time, because fewer patients are needed to obtain equal power compared to the parallel design. Another disadvantage of the crossover design is that missing data pose a more serious problem than in the parallel design. Since each subject must supply data on *two occasions* (compared to a single occasion in the parallel design), the chances of observations being lost to the analysis are greater in the crossover study. If an observation is lost in one of the legs of a two-period crossover, the data for that person carry very little information. When data are missing in the crossover design, the statistical analysis is more difficult and the design loses some efficiency. Finally, the administration of crossover designs in terms of management and patient compliance is somewhat more difficult than that of parallel studies.

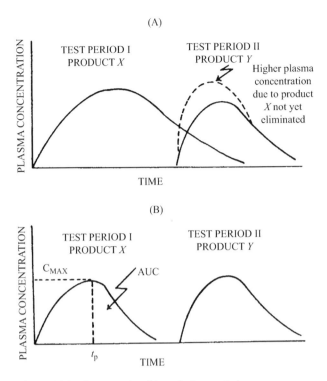

Figure 11.1   Carryover in a bioequivalence study.

Perhaps the most serious problem with the use of crossover designs is one common to all Latin square designs, the *possibility of interactions*. The most common interaction that may be present in crossover design is a *differential carryover* or residual effect. This effect occurs when the response on the second period (leg) is dependent on the response in the first period, and this dependency differs depending on which of the two treatments is given during the first period. Carryover is illustrated in Figure 11.1(A), where the short interval between administration of dosage forms X and Y is not sufficient to rid the body of drug when formulation X is given first. This results in an apparent larger blood level for formulation Y when it is given subsequent to formulation X. In the presence of differential carryover, the data cannot be properly analyzed except by the use of more complex designs (see replicate crossover designs in sect. 11.4.7). These special designs are not easily accommodated to clinical studies [8].

Figure 11.1(B) illustrates an example where a sufficiently long washout period ensures that carryover of blood concentration of drug is absent. The results depicted in Figure 11.1(A) show a carryover effect that could easily have been avoided if the study had been carefully planned. This example only illustrates the problem; often, carryover effects are not as obvious. These effects can be caused by such uncontrolled factors as psychological or physiological states of the patients, or by external factors such as the weather, clinical setting, assay techniques, and so on.

Grizzle has published an analysis to detect carryover (residual) effects [9]. When differential carryover effects are present, the usual interpretation and statistical analysis of crossover studies are invalid. Only the first period results can be used, resulting in a smaller, less sensitive experiment. An example of Grizzle's analysis is shown in this chapter in the discussion of bioavailability studies (sect. 11.4.2). Brown concludes that most of the time, in these cases, the parallel design is probably more efficient [7]. Therefore, if differential carryover effects are suspected prior to implementation of the study, an alternative to the crossover design should be considered (see below).

Because of the "built-in" individual-by-individual comparisons of products provided by the crossover design, the use of such designs in comparative clinical studies often seems very attractive. However, in many situations, where patients are being treated for a disease state, the design is either inappropriate or difficult to implement. In acute diseases, patients may be cured or improved so much after the first treatment that a "different" condition or state of illness is being treated during the second leg of the crossover. Also, psychological carryover has been observed, particularly in cases of testing psychotropic drugs.

The longer study time necessary to test two drugs in the crossover design can be critical if the testing period of each leg is of long duration. Including a possible *washout period* to avoid possible carryover effects, the crossover study will take at least *twice* as long as a parallel study to complete. In a study of long duration, there will be more difficulty in recruiting and maintaining patients in the study. One of the most frustrating (albeit challenging) facets of data analysis is data with "holes," missing data. Long-term crossover studies will inevitably have such problems.

11.4.2   Bioavailability/Bioequivalence Studies[†]

The assessment of "bioequivalence" (BE) refers to a procedure that compares the bioavailability of a drug from different formulations. Bioavailability is defined as the rate and extent to which the active ingredient or active moiety is absorbed from a drug product and becomes available at the site of action. For drug products that are not intended to be absorbed into the bloodstream, bioavailability may be assessed by measurements intended to reflect the rate and extent to which the active ingredient or active moiety becomes available at the site of action. In this chapter, we will not present methods for drugs that are not absorbed into the bloodstream (or absorbed so little as to be unmeasurable), but may act locally. Products containing such drugs are usually assessed using a clinical endpoint, using parallel designs discussed elsewhere in this chapter. Statistical methodology, in general, will be approached in a manner consistent with methods presented for drugs that are absorbed.

Thus, we are concerned with measures of the release of drug from a formulation and its availability to the body. BE can be simply defined by the relative bioavailability of two or more formulations of the same drug entity. According to 21 CFR 320.1, BE is defined as "the absence of a significant difference in the rate and extent to which the active ingredient or active moiety . . . becomes available at the site of drug action when administered . . . in an appropriately designed study."

BE is an important part of an NDA in which formulation changes have been made during and after pivotal clinical trials. BE studies, as part of Abbreviated New Drug Application (ANDA) submissions, in which a generic product is compared to a marketed, reference product, are critical parts of the submission. BE studies may also be necessary when formulations for approved marketed products are modified.

In general, most BE studies depend on accumulation of pharmacokinetic (PK) data that provide concentrations of drug in the bloodstream at specified time points following administration of the drug. These studies are typically performed, using oral dosage forms, on volunteers who are incarcerated (housed) during the study to ensure compliance with regard to dosing schedule as well as other protocol requirements. This does not mean that BE studies are limited to oral dosage forms. Any drug formulation that results in measurable blood concentrations after administration can be treated and analyzed in a manner similar to drugs taken orally. For drugs that act locally and are not appreciably absorbed, either a surrogate endpoint may be utilized in place of blood concentrations of drug (e.g., a pharmacodynamic response) or a clinical study using a therapeutic outcome may be necessary. Also, in some cases where assay methodology in blood is limited, or for other relevant reasons, measurements of drug in the urine over time may be used to assess equivalence.

To measure rate and extent of absorption for oral products, PK measures are used. In particular, model independent measures used are (a) area under the blood concentration versus

---

[†] Additional discussion of designs and analyses are given in Appendix X.

time curve (AUC) and the maximum concentration (C_{max}), which are measures of the amount of drug absorbed and the rate of absorption, respectively.

The time at which the maximum concentration occurs (t_{max}) is a more direct measure as an indicator of absorption rate, but is a very variable estimate.

Bioavailability/bioequivalence studies are particularly amenable to crossover designs. Virtually all such studies make use of this design. Most BE studies involve single doses of drugs given to normal volunteers, and are of short duration. Thus the disadvantages of the crossover design in long term, chronic dosing studies are not apparent in bioavailability studies. With an appropriate washout period between doses, the crossover is ideally suited for comparative bioavailability studies.

Statistical applications are essential for the evaluation of BE studies. Study designs are typically two-treatment, two-period (tttp) crossover studies with single or multiple (steady state) dosing, fasting or fed. Designs with more than two periods are now becoming more common, and are recommended in certain cases by the FDA. For long half-life drugs, where time is crucial, parallel designs may be desirable, but these studies use more subjects than would be used in the crossover design, and the implementation of parallel studies may be difficult and expensive. The final evaluation is based on parameter averages derived from the blood level curves, AUC, C_{max}, and t_{max}. Statistical analyses that have been recommended are varied, and the analyses presented here are typical of those recommended by regulatory agencies.

This section discusses some designs, their properties, and statistical evaluations.

Although crossover designs have clear advantages over corresponding parallel designs, their use is restricted, in general, as previously noted, because of potential differential carryover effects and confounded interactions. However, for BE studies, the advantages of these designs far outweigh the disadvantages. Because these studies are typically performed in healthy volunteers and are of short duration, the potential for carryover and interactions is minimal. In particular, the likelihood of differential carryover seems to be remote. Carryover may be observed if administration of a drug affects the blood levels of subsequent doses. Although possible, a carryover effect would be very unusual, particularly in single-dose studies with an adequate washout period. A washout period of at least seven half-lives is recommended. Even more unlikely, would be a differential carryover, which suggests that the carryover from one product is different from the carryover from the second product. A differential carryover effect can invalidate the second period results in a two-period crossover (see below). Because BE studies compare the same drug in different formulations, if a carryover exists at all, the carryover of two different formulations would not be expected to differ. This is not to say that differential carryover is impossible in these studies, but to this author's knowledge, differential carryover has not been verified in results of published BE studies, single or multiple dose. In the typical tttp design, differential carryover is confounded with other effects, and a test for carryover is not definitive. Thus, if such an effect is suspected, proof would require a more restrictive or higher order design, that is, a design with more than two periods. This problem will be discussed further as we describe the analysis and inferences resulting from these designs.

The features of the tttp design are as follows:

1. N subjects recruited for the study are separated into two groups, or two treatment *sequences*. N_1 subjects take the treatments in the order AB, and N_2 in the order BA, where $N_1 + N_2 = N$. For example, 24 (N) subjects are recruited and 12 (N_1) take the Generic followed by the Brand product, and 12 (N_2) take the Brand followed by the Generic. Note that the product may be taken as a single dose, in multiple doses, fasted or fed.
2. After administration of the product in the first period, blood levels of drug are determined at suitable intervals.
3. A washout period follows, which is of sufficient duration to ensure the "total" elimination of the drug given during the first period. An interval of at least nine drugs half-lives should be sufficient to ensure virtually total elimination of the drug. Often, a minimum of seven half-lives is recommended.
4. The alternate product is administered in the second period and blood levels determined as during Period 1.

FIRST PERIOD

SECOND PERIOD

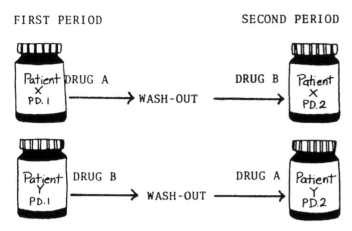

Crossover designs are planned so that each treatment is given an equal number of times in each period. This is most efficient and yields unbiased estimates of treatment differences if a period effect is present.

The blood is analyzed for each subject with both first and second periods analyzed concurrently (the same day). To detect possible analytical errors, the samples are usually analyzed chronologically (starting from the time 0 sample to the final sample), but with the identity of the product assayed unknown (sample blinding).

After the blood assays are complete, the blood level versus time curves are analyzed for the derived parameters, AUC_t (also noted as AUC_{0-t}), $AUC_{0-\infty}$, C_{max}, and t_{max} (t_p), for each analyte. AUC_t is the area to the last quantifiable concentration, and AUC_{inf} is AUC_t augmented by an estimate of the area from time t to infinity (C_t/K_e). This is shown and explained in Figure 11.2. A detailed analysis follows.

The analysis of the data consists of first determining the maximum blood drug concentration (C_{max}) and the area under the blood level versus time curve (AUC) for each subject, for each product. Often, more than one analyte is observed, for example, metabolites or multiple ingredients, all of which may need to be separately analyzed.

AUC is determined using the trapezoidal rule. The area between adjacent time points may be estimated as a trapezoid (Fig. 11.3). The area of each trapezoid, up to and including the final time point, where a measurable concentration is observed, is computed, and the sum of these areas is the AUC, designated as AUC_t. The area of a trapezoid is $1/2$ (base) (sum of two sides). For example, in Figure 11.3, the area of the trapezoid shown in the blood level versus time curve is 4. In this figure, C_{max} is 5 ng/mL and t_{max}, the time at which C_{max} occurs, is two hours. Having performed this calculation for each subject and product, the AUC and C_{max} values are transformed to their respective logarithms. Either natural logs (ln) or logs to the base

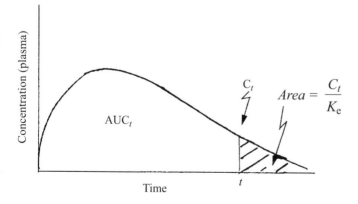

Figure 11.2 Derived parameters from bioequivalence study.

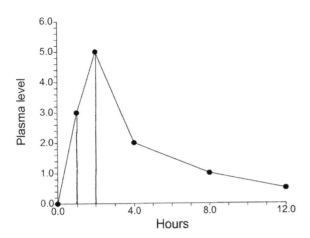

Figure 11.3 Illustration of trapezoidal rule.

10 (log) may be used. Typically, one uses the natural log, or ln. The details of the analysis are described later in this chapter. The analysis of AUC and C_{max} was not always performed on the logs of these values. Originally, the actual, observed (nontransformed) values of these derived parameters were used in the analysis. (This history will be discussed in more detail below.) However, examination of the theoretical derivations and mathematical expression of AUC and C_{max}, as well as the statistical properties, has led to the use of the logarithmic transformation. In particular, data appear to show that these values follow a log-normal distribution more closely than they do a normal distribution. The form of expression for AUC suggests a multiplicative model

$$AUC = FD/VK_e,$$

where F is fraction of drug absorbed, D is dose, V is volume of distribution, and K_e is elimination rate constant.

The distribution of AUC is complex because of the nonlinearity; it is a ratio. Ln(AUC) is equal to $\ln(F) + \ln(D) - \ln(V) - \ln(K_e)$. This is linear, and the statistical properties are more manageable. A similar argument can be made for C_{max}.

The present FDA requirement for equivalence is based on product ratios using a symmetric 90% confidence interval for the difference of the average parameters, after a log transformation. Earlier, according to FDA guidelines, the AUC and C_{max} were analyzed using the untransformed values of these derived parameters. Note that when using a clinical or pharmacodynamic endpoint (such as may be used in a parallel study when drug is not absorbed), the nontransformed data may be more appropriate and the "old" way of forming the confidence interval may still be used. This analysis is described below. However, at the present time, FDA is leaning toward an analysis based on Fieller's Theorem (Locke's Method). (These analyses, along with a log-transformed analysis, are described in the example following this discussion.)

11.4.2.1 Statistical Analysis
It is convenient to follow the statistical analysis and estimation of various effects by looking at the two sequences in the context of the model for this design:

Let
μ = overall mean
G_i = Effect of sequence group i ($i = 1, 2$)
S_{ik} = Effect of subject k in sequence i ($k = 1, 2, 3 \ldots N$)
P_j = Effect of period j ($j = 1, 2$)
$T_{t(ij)}$ = treatment effect t ($t = 1, 2$) in sequence i and period j
$Y_{ijk} = \mu + G_i + S_{ik} + P_j + T_t(ij) + e_{ijk}$

Table 11.6 Design for Two-Way Crossover Study

	Period I	Period II
Sequence I	A	B
Sequence II	B	A

The sequence × period interaction is the treatment effect (sequence × period is the comparison Period I–Period II for the two sequences; see Table 11.6).

e.g.,

$$\frac{\left[(A - B)_{\text{seq I}} - (B - A)_{\text{seq II}}\right]}{2} = A - B.$$

Suppose that carryover is present, but carryover is the same for both products. We can show that this would not bias the treatment comparisons. For the sake of simplicity, suppose that there is no period effect ($P_1 = P_2$). Also suppose that the direct treatment effects are $A = 3$ and $B = 2$. Both products have a carryover that adds 2 to the treatment (product) in the second period. (This would result in an additional value of 2 for the period effect.) Finally, assume that the effects for the sequences are equal; Sequence I = Sequence II. This means that the average results for subjects in Sequence I are the same as that for Sequence II. Based on this model, Product B in Period II would have a value of $2 + 2$ for carryover $= 4$. Product A in Period II has a value of $3 + 2 = 5$. Thus, the average difference between A and B is 1, as expected ($A = 3$ and $B = 2$). Table 11.7 shows these simulated data.

This same reasoning would show that equal carryover effects do not bias treatment comparisons in the presence of a period effect. (See Exercise Problem 11 at the end of this chapter.)

Differential carryover, where the two products have different carryover effects, is confounded with a sequence effect. This means that if the sequence groups have significantly different average results, one cannot distinguish this effect from a differential carryover effect in the absence of more definitive information. For example, one can show that if there is a sequence effect and no differential carryover, a differential carryover in the absence of a sequence effect could give the same result.

To help explain the confounding, assume that the difference between treatments is 0 (treatments are identical) and that Sequence I averages 2 units more than Sequence II (e.g., Sequence I = Sequence II + 2). Since subjects are assigned to the sequence groups at random, the differences should not be significant except by chance. With no carryover or period effects, the average results could be something like that shown in Table 11.8.

If Sequence I is the order A followed by B (AB) and Sequence II is the order BA, the treatment differences, $A - B$, would be $6 - 6 = 0$ in Sequence I, and $4 - 4 = 0$ in Sequence II. Treatment A is the same as Treatment B in Sequence I, and in Sequence II. However, this same result could occur as a result of differential carryover in the presence of treatment differences.

Table 11.9 shows the same results as Table 11.8 in a different format.

The data from Table 11.9 can be explained by assuming that A is 2 units higher than B (see Period I results), a carryover of $+2$ units when B follows A, and a carryover of -2 units when A follows B. The two explanations, a sequence effect or a differential carryover, cannot be separated in this two-way crossover design. The sequence effect is $G_1 - G_2$. The differential carryover is $\{[T_A(2) - T_A(1)] - [T_B(2) - T_B(1)]\}/2 = \{[T_A(2) + T_B(1)] - [T_B(2) + T_A(1)]\}/2$, which is exactly the sequence effect (average results in Sequence II − average results in Sequence I). The subscript $B(1)$ refers to average result for Product B in Period I.

Table 11.7 Simulated Data Illustrating Equal Carryover

	Period I	Period II
Sequence I	$A = 3$	$B = 4$
Sequence II	$B = 2$	$A = 5$

Table 11.8   Example of Sequence Effect

	Treatment *A*	Treatment *B*	Average
Sequence I	6	6	6
Sequence II	4	4	4
Average	5	5	

In practice, an ANOVA is performed, which results in significance tests for the sequence effect and an estimate of error for computing confidence intervals (see later in sect. 11.4.3).

The results of a typical single-dose BE study are shown in Table 11.10. These data were obtained from drug plasma level versus time determinations similar to those illustrated in Figure 11.1(B). Area under the plasma level versus time curve (AUC, a measure of absorption), time to peak plasma concentration (t_p), and the maximum concentration (C_{max}) are the parameters that are usually of most interest in the comparison of the bioavailability of different formulations of the same drug moiety.

The typical ANOVA for crossover studies will be applied to the AUC data to illustrate the procedure used to analyze the experimental results. In these analyses, the residual error term is used in statistical computations, for example, to construct confidence intervals. An ANOVA is computed for each parameter based on the model. The ANOVA table is not meant for the performance of statistical hypothesis tests, except perhaps to test the sequence effect, which uses the between subject within sequences mean square as the error term. Rather, the analysis removes some effects from the total variance to obtain a more "efficient" or pure estimate of the error term. It is the error term, or estimate of the within-subject variability (assumed to be equal for both products in this analysis), that is used to assess the equivalence of the parameter being analyzed. A critical assumption for the correct interpretation of the analysis is the *absence* of differential carryover effects, as discussed previously. Otherwise, the usual assumptions for ANOVA should hold. FDA statisticians encourage a careful statistical analysis of crossover designs. In particular, the use of a simple *t* test that ignores the possible presence of period and/or carryover effects is not acceptable.[‡] If period effects are present, and not accounted for in the statistical analysis, the analysis will be less sensitive. The error mean square in the ANOVA will be inflated due to inclusion of the period variance, and the width of the confidence interval will be increased. If differential carryover effects are present, the estimate of treatment differences will be biased (see sects. 11.4.1 and 11.4.2).

The usual ANOVA separates the total sum of squares into four components: subjects, periods, treatments, and error (residual). In the absence of differential carryover effects, the statistical test of interest is for treatment differences. The subject and period sum of squares are separated from the error term which then represents "intrasubject" variation. The subjects sum of squares (SS) can be separated into sequence SS and subject within sequence SS to test for the sequence effect. The sequence effect is confounded with carryover, and this test is described following the analysis without sequence effect.

Some history may be of interest with regard to the analysis recommended in the most recent FDA guidance [10]. In the early evolution of BE analysis, a hypothesis test was used at the 5% level of significance. The raw data were used in the analysis; that is, a logarithmic transformation was not recommended. The null hypothesis was simply that the products were equal, as opposed to the alternate hypothesis that the products were different. This had the obvious problem with regard to the power of the test. Products that showed nearly the same average

Table 11.9   Example of Differential Carryover Effect

	Period I	Period II	Average
Sequence I	$A = 6$	$B = 6$	6
Sequence II	$B = 4$	$A = 4$	4

[‡] In bioavailability studies, carryover effects are usually due to an inadequate washout period.

Table 11.10 Data for the Bioequivalence Study Comparing Drugs A and B

Subject	Order	AUC A	AUC B	Peak concentration A	Peak concentration B	Time to peak A	Time to peak B
1	AB	290	210	30	18	8	8
2	BA	201	163	22	19	10	4
3	AB	187	116	18	11	6	6
4	AB	168	77	20	14	10	3
5	BA	200	220	18	21	3	3
6	BA	151	133	25	16	4	6
7	AB	294	140	27	14	4	10
8	BA	97	190	16	23	6	6
9	BA	228	168	20	14	6	6
10	AB	250	161	28	19	6	4
11	AB	293	240	28	18	6	12
12	BA	154	188	16	20	8	8
	Mean	209.4	167.2	22.3	17.3	6.4	6.3
	Sum	2513	2006	268	207	77	76

results, but with very small variance, could show a significant difference, which may not be of clinical significance, and be rejected. Alternatively, products that showed large differences with large variance could show a nonsignificant difference, and be deemed equivalent. Similarly, products could be shown to be equivalent if a small sample size was used resulting in an undetected difference that could be clinically significant. Because of these problems, an additional caveat was added to the requirements. If the products showed a difference of less than 20%, was not statistically significant ($p > 0.05$), and the power of the study to detect a difference of 20% exceeded 80%, the products would be considered to be equivalent. This helped to avoid undersized studies and prevent products with observed large differences from passing the BE study. The following examples illustrate this problem.

Example 1. In a BE two-period, crossover study, with eight subjects, the test product showed an average AUC of 100, and the reference product showed an average AUC of 85. The observed difference between the products is (100–85)/85, or 17.6%.

The error term from the ANOVA (see below for description of the analysis) is 900, s.d. = 30. The test of significance (a t test with 6 d.f.) is

$$\frac{|100 - 85|}{\left[900\left(\frac{1}{8} + \frac{1}{8}\right)\right]^{1/2}} = 1.00.$$

This is not statistically significant at the 5% level (a t value of 2.45 for 6 d.f. is needed for significance). Therefore, the products may be deemed equivalent.

However, this test is underpowered based on the need for 80% power to show a 20% difference. A 20% difference from the reference is $0.2 \times 85 = 17$. The approximate power is (Eq. 6.11)

$$Z = [17/42.43][6]^{1/2} - 1.96 = -0.98.$$

Referring to a Table of the Cumulative Standard Normal Distribution, the approximate power is 16%. Although the test of significance did not reject the null hypothesis, the power of the test to detect a 20% difference is weak. Therefore, this product would not pass the BE requirements.

Example 2. In a BE two-period, crossover study, with 36 subjects, the test product showed an average AUC of 100, and the reference product showed an average AUC of 95. The products differ by approximately only 5%. The error term from the ANOVA is 100, s.d. = 10. The test of significance (a t test with 34 d.f.) is

$$\frac{|100 - 95|}{\left[100\left(\dfrac{1}{36} + \dfrac{1}{36}\right)\right]^{1/2}} = 2.12.$$

This is statistically significant at the 5% level (a t value of 2.03 for 34 d.f. is needed for significance). Therefore, the products may be deemed nonequivalent.

This test passes the criterion based on the need for 80% power to show a 20% difference. A 20% difference from the reference is $0.2 \times 95 = 19$. The approximate power is (see chap. 6)

$$Z = [19/14.14]\,[34]^{1/2} - 1.96 = 5.88.$$

The approximate power is almost 100%. Although the power of the test to detect a 20% difference is extremely high, the test of significance rejected the null hypothesis that the products were equal. Therefore, this Product would fail the BE requirements. In some cases, a Medical review would rule such a small difference as clinically non-significant and the product would be approved.

Other requirements at that time included the 75/75 rule [11]. This rule stated that 75% of the subjects in the study should have ratios of test/reference between 75% and 125%. This was an attempt to include a variability criterion in the assessment of study results. Unfortunately, this criterion has little statistical basis, and would almost always fail with highly variable drugs. In fact, if a highly variable drug (CV greater than 30–40%) is tested against itself, it would most likely fail this test. Eventually, this requirement was correctly phased out.

Soon after this phase in the evolution of BE regulations, the hypothesis test approach was replaced by the two one-sided t test or, equivalently, the 90% confidence interval approach [12]. This approach resolved the problems of hypothesis testing, and assumed that products that are within 20% of each other with regard to the major parameters, AUC and C_{max}, are therapeutically equivalent. For several years, this method was used without a logarithmic transformation. However, if the study data conformed better to a log-normal distribution than a normal distribution, a log transformation was allowed. An appropriate statistical test was applied to test the conformity of the data to these distributions.

The AUC data from Table 11.10 are analyzed below. To ease the explanation, the computations for the untransformed data are detailed. The log-transformed data are analyzed identically, and these results follow the untransformed data analysis. The sums of squares for treatments and subjects are computed exactly the same way as in the two-way ANOVA (see sect. 8.4). The new calculations are for the "period" (1 d.f.) and "sequence" (1 d.f.) sums of squares. We first show the analysis for periods. The analysis for sequence is shown when discussing the test for differential carryover. Two new columns are prepared for the "period" calculation. One column contains the data from the first period, and the second column contains data from the second period. For example, for the AUC data in Table 11.10, the data for the first period are obtained by noting the order of administration. Subject 1 took Product A during the first period (290); subject 2 took B during the first period (163); and so on. Therefore, the first period observations are

290, 163, 187, 168, 220, 133, 294, 190, 168, 250, 293, and 188 (sum = 2544).

The second period observations are

210, 201, 116, 77, 200, 151, 140, 97, 228, 161, 240, 154 (sum = 1975).

The "period" SS may be calculated as follows:

$$\frac{\left(\sum P_1\right)^2 + \left(\sum P_2\right)^2}{N} - CT, \tag{11.2}$$

where $\sum P_1$ and $\sum P_2$ are the sums of observations in the first and second periods, respectively, N is the number of subjects, and CT is the correction term. The following ANOVA and Table 11.10 will help clarify the calculations.

Calculations for ANOVA

$\sum X_t$ is the sum of all observations $= 4519$
$\sum X_A$ is the sum of observations for Product $A = 2513$
$\sum X_B$ is the sum of observations for Product $B = 2006$
$\sum P_1$ is the sum of observations for Period 1 $= 2544$
$\sum P_2$ is the sum of observations for Period 2 $= 1975$
$\sum X_t^2$ is the sum of the squared observations $= 929{,}321$
CT is the correction term $\frac{(\sum X_t)^2}{N_t} = \frac{(4519)^2}{24} = 850{,}890.04.$
Total SS $= \sum X_t^2 - \text{CT} = 78{,}430.96$
$\sum S_i$ is the sum of the observations for subject i (e.g., 500 for first subject)
Subject SS

$$= \frac{\sum (\sum S_i)^2}{2} - \text{CT} = \frac{500^2 + 364^2 + \ldots + 342^2}{2} - \text{CT} = 43{,}560.46$$

$$\text{Period sum of squares} = \frac{2544^2 + 1975^2}{12} - \text{CT} = 13{,}490.0$$

$$\text{Treatment sum of squares} = \frac{2513^2 + 2006^2}{12} - \text{CT} = 10{,}710.4$$

$$\begin{aligned}
\text{Error SS} &= \text{total SS} - \text{subject SS} - \text{period SS} - \text{treatment SS} \\
&= 78{,}430.96 - 43{,}560.46 - 13{,}490 - 10{,}710.38 \\
&= 10{,}670.1.
\end{aligned}$$

Note that the d.f. for error are equal to 10. The usual two-way ANOVA would have 11 d.f. for error (subjects $-$ 1) $\times$ (treatments $-$ 1). In this design, the error SS is diminished by the *period SS*, which has 1 d.f.

Again, the ANOVA is typically performed using appropriate computer programs. A General Linear Models (GLM) program is suitable with factors, sequence, subjects within sequence, treatment, and period.

11.4.2.2 Test for Carryover Effects
Dr. James Grizzle published a classic paper on analysis of crossover designs and presented a method for testing carryover effects (sequence effects in his notation) [9]. Some controversy exists regarding the usual analysis of crossover designs, particularly with regard to the assumptions underlying this analysis. Before using the Grizzle analysis, the reader should examine the original paper by Grizzle as well as the discussion by Brown, in which some of the problems of crossover designs are summarized [7].

One of the key assumptions necessary for a valid analysis and interpretation of crossover designs is the absence of differential carryover effects as has been previously noted. Data from Table 11.10 were previously analyzed using the typical crossover analysis, assuming that differential carryover was absent. Table 11.10 is reproduced as Table 11.11 (AUC only) to illustrate the computations needed for the Grizzle analysis.

The test for carryover, or sequence, effects is performed as follows:

1. Compute the SS due to carryover (or sequence) effects by comparing the results for group I to group II. (Note that these two groups, groups I and II, which differ in the order of treatment are designated as treatment "sequence" by Grizzle.) It can be demonstrated that in the absence of sequence effects, the average result for group I (A first, B second) is expected to be equal to the average result for group II (B first, A second). The SS is calculated as

$$\frac{(\sum \text{group I})^2}{N_1} + \frac{(\sum \text{group II})^2}{N_2} - \text{CT}.$$

Table 11.11   Data for AUC for the Bioequivalence Study Comparing Drugs *A* and *B*

	Group I (Treatment *A* first, *B* second)				Group II (Treatment *B* first, *A* second)		
Subject	***A***	***B***	**Total**	**Subject**	***A***	***B***	**Total**
1	290	210	500	2	201	163	364
3	187	116	303	5	200	220	420
4	168	77	245	6	151	133	284
7	294	140	434	8	97	190	287
10	250	161	411	9	228	168	396
11	293	240	533	12	154	188	342
Total	1482	944	2426	Total	1031	1062	2093

In our example the sequence SS is (1 d.f.)

$$\frac{(2426)^2}{12} + \frac{(2093)^2}{12} - \frac{(2426 + 2093)^2}{24} = 4620.375.$$

2. The proper error term to test the sequence effect is the within-group (sequence) mean square, represented by the SS between subjects within groups (sequence). This SS is calculated as follows:

$$\frac{1}{2}\sum (\text{subject total})^2 - (\text{CT})_\text{I} - (\text{CT})_\text{II},$$

where CT_I and CT_II are the correction terms for groups I and II, respectively. In our example, the within-group SS is

$$\frac{1}{2}(500^2 + 303^2 + 245^2 + \ldots + 364^2 + 420^2$$
$$+ \ldots 342^2) - \frac{(2426)^2}{12} - \frac{(2093)^2}{12} = 38{,}940.08.$$

This within-group (or subject within-sequence) SS has 10 d.f., 5 from each group. The mean square is $38{,}940/10 = 3894$.

3. Test the sequence effect by comparing the sequence mean square to the within-group mean square (*F* test).

$$F_{1,10} = \frac{4620.375}{3894} = 1.19$$

Referring to Table IV.6, the effect is not significant at the 5% level. (Note that in practice, this test is performed at the 10% level.) If the sequence (carryover) effect is not significant, one would proceed with the usual analysis and interpretation as shown in Table 11.12.

Table 11.12   Analysis of Variance Table for the Crossover Bioequivalence Study (AUC) Without Sequence Effect

Source	d.f.	SS	MS	*P*
Subjects	11	43,560.5	3960.0	
Period	1	13,490.0	13,490.0	
Treatment	1	10,710.4	10,710.4	$F_{1,10} = 12.6^*$
Error	10	10,670.1	1067.0	
Total	23	78,430.96		$F_{1,10} = 10.0^*$

$^*p < 0.05.$

If the sequence (carryover) effect is significant, the usual analysis is not valid. The recommended analysis uses only the first period results, deleting the data contaminated by the carryover, the second period results. Grizzle recommends that the preliminary test for carryover be done at the 10% level (see also the discussion by Brown [7]). For the sake of this discussion, we will compute the analysis as if the data revealed a significant sequence effect in order to show the calculations. Using only the first-period data, the analysis is appropriate for a one-way ANOVA design (sect. 8.1). We have two "parallel" groups, one on Product A and the other on Product B. The data for the first period are as follows:

Subject	A	Subject	B
1	290	2	163
3	187	5	220
4	168	6	133
7	294	8	190
10	250	9	168
11	293	12	188
Mean	247		177
S^2	3204.8		870.4

The ANOVA table is as follows:[§]

	d.f.	SS	MS	F
Between treatments	1	14,700	14,700	7.21
Within treatments	10	20,376	2037.6	

Referring to Table IV.6, an F value of 4.96 is needed for significance at the 5% level (1 and 10 d.f.). Therefore, in this example, the analysis leads to the conclusion of significant treatment differences.

The discussion and analysis above should make it clear that sequence or carryover effects are undesirable in crossover experiments. Although an alternative analysis is available, one-half of the data are lost (second period) and the error term for the comparison of treatments is usually larger than that which would have been available in the absence of carryover (within-subject versus between-subject variation). One should thoroughly understand the nature of treatments in a crossover experiment in order to avoid differential carryover effects if at all possible. (Note: Although at one time the presence of a sequence effect could cause rejection of a BE submission by FDA, at the present time if there are no circumstances that could cause carryover, the FDA review would take this into consideration as a spurious event.)

Since the test for carryover was set at 5% a priori, we will proceed with the interpretation, assuming that carryover effects are absent. (Again, note that this test is usually set at the 10% level in practice). Both period and treatment effects are significant ($F_{1,10} = 12.6$ and 10.0, respectively). The AUC values tend to be higher during the first period (on the average). This period (or order) effect does not interfere with the conclusion that Product A has a higher average AUC than that of Product B. The balanced order of administration of the two products in this design compensates equally for both products for systematic differences due to the period or order. Also, the ANOVA subtracts out the SS due to the period effect from the error term, which is used to test treatment differences.

If the design is not symmetrical, because of missing data, dropouts, or poor planning, a statistician should be consulted for the data analysis and interpretation. In an asymmetrical design, the number of observations in the two periods is different for the two treatment groups. This will always occur if there is an odd number of subjects. For example, the following scheme shows an asymmetrical design for seven subjects taking two drug products, A and B. In such situations, computer software programs can be used, which adjust the analysis and mean results for the lack of symmetry [13].

[§] This analysis is identical to a two-sample independent groups t test.

Subject	Period 1	Period 2
1	A	B
2	B	A
3	A	B
4	B	A
5	A	B
6	B	A
7	A	B

The complete ANOVA is shown in Table 11.13.

The statistical analysis in the example above was performed on AUC, which is a measure of relative absorption. The FDA recommends that plasma or urine concentrations be determined out to at least three half-lives, so that practically all the area under the curve will be included when calculating this parameter (by the trapezoidal rule, for example). Other measures of the rate and extent of absorption are time to peak and peak concentration. Often, more than one analyte is observed, for example, metabolites or multiple ingredients.

Much has been written and discussed about the expression and interpretation of bioequivalency/bioavailability data as a measure of rate and extent of absorption. When are the parameters AUC, t_p, and C_{max} important, and what part do they play in bioequivalency? The FDA has stated that products may be considered equivalent in the presence of different rates of absorption, particularly if these differences are designed into the product [14]. For example, for a drug that is used in chronic dosing, the extent of absorption is probably a much more important parameter than the rate of absorption. It is not the purpose of this presentation to discuss the merits of these parameters in evaluating equivalence, but only to alert the reader to the fact that BE interpretation need not be fixed and rigid.

The ANOVA for log AUC (AUC values are transformed to their natural logs) is shown in Table 11.14. Exercise Problem 9 at the end of this chapter requests the reader to construct this table. The procedure is identical to that shown for the untransformed data. *Analysis of the log-transformed parameters is currently required by the FDA. The critical parameters are AUC and C_{max}.*

11.4.3 Confidence Intervals in BE Studies

The scientific community is virtually unanimous in its opposition to the use of hypothesis testing for the evaluation of BE. Hypothesis tests are inappropriate in that products that are very close, but with small variance, may be deemed different, whereas products that are widely different, but with large variance, may be considered equivalent (not significantly different). (See previous discussion in sect. 11.4.2). The use of a confidence interval, the present criterion for equivalence, is more meaningful and has better statistical properties. (See chap. 5 for a discussion of confidence intervals.) Given the lower and upper limit of the ratio of the parameters, the user or prescriber of a drug can make an educated decision regarding the equivalence of alternative products. The confidence limits must lie between 0.8 and 1.25 based on the difference of the back-transformed averages of the log-transformed AUC and C_{max} results. This computation for AUC is shown below. For historical purposes and purposes of comparison, the confidence interval is computed using the nontransformed data (the old method) and the log-transformed

Table 11.13 ANOVA for Untransformed Data from Table 11.10 for AUC

Variable (source)	d.f.	SS	MS	F ratio	Prob > F
Sequence	1	4620.4	4620.4	1.19	0.3016
Subject (sequence)	10	38,940.1	3894.0	3.65	0.0265
Period	1	13,490.0	13,490.0	12.64	0.0052
Treat	1	10,710.4	10,710.4	10.04	0.0100
Residual	10	10,670.1	1067.0		
Total	23	78,430.96			

Table 11.14 ANOVA for Log-Transformed Data from Table 11.10 for AUC

Variable (source)	d.f.	SS	MS	F ratio	Prob > F
Sequence	1	0.0613	0.0613	0.46	0.5128
Subject (sequence)	10	1.332	0.1332	2.96	0.0507
Period	1	0.4502	0.4502	10.02	0.0101
Treat	1	0.2897	0.2897	6.44	0.0294
Residual	10	0.44955	0.04496		
Total	23	2.58307			

data (the current method). Note that a ratio based on the untransformed data may be used in certain special circumstances where a log transformation may be deemed inappropriate, such as data derived from a clinical study, where the data consist of a pharmacodynamic response or some similar outcome. (See also, the currently recommended analysis using Locke's Method based on Fieller's Theorem below.)

11.4.3.1 Locke's Method of Analysis (Confidence Interval for the Ratio of Two Normally Distributed Variables)

The confidence interval for the ratio of two variables is described in "Guidance for Industry, Center for Drug Evaluation and Research, Appendix V, Feb 1997 [15]." The computations assume normality of the variables. The example uses data supplied in the FDA document referenced above.

If two variables are both normally distributed, it is not statistically valid to place a confidence interval on ratios. Ratios of normally distributed variables are not normal. For example, the data in the FDA document are as follows:

Subject	Test	Reference	Ratio
2	−48.52	−22.2	2.19
3	−38.99	−18.65	2.09
4	−7.62	−22.42	0.34
7	0.98	−10.96	−0.09
9	−32.05	−37.4	0.86
11	−26.18	−26.73	0.98
12	−11.62	−12.56	0.93

The average ratio is 1.04 with a s.d. of 0.84 (the reader may verify these calculations). The 90% confidence interval is $1.04 \pm 1.94 \times 0.84 \times \sqrt{1/7}$ = approximately 0.42 to 1.66.

This is not correct. The correct calculations are as follows:

Calculate the mean and variance of the test and reference products

Mean of test = $AV_t = -23.43$
Mean of reference = $AV_r = -21.56$
Variance test = $\sigma^2_T = 323.13$
Variance reference = $\sigma^2_R = 80.10$.

Since the two variables are related or correlated (crossover design), calculate the covariance $= = \sigma_{TR} = \sum (t - AV_t)(r - AV_r)(N - 1)$, where N = sample size = 7 and covariance = 78.83.

A variable is defined as "G," where G must be greater than zero in order for the calculations to be valid.

$$G = \frac{(t^2 \sigma_R^2)}{(N \times AV_r^2)},$$

where N = sample size = 7

$$= \frac{(1.9432 \times 80.10)}{(7 \times 21.56^2)}$$
$$G = 0.093.$$

Then, apply the following formulas to calculate the confidence interval. (Note the similarity of these equations to the inverse equation to calculate the confidence interval for X, given Y in regression. This is a similar application of the calculation of a confidence interval for the ratio of two normal variables.)

$$K = \{AV_t^2/AV_r^2\} + \{\sigma_2^T/\sigma_R^2\}(1 - G) + \{\sigma_{TR}/\sigma_R^2\}[G(\sigma_{TR}/\sigma_R^2) - 2(AV_t/AV_r)]$$
$$= \{-23.43/ - 21.56\}^2 + \{323.13/80.1\}(1 - 0.093) +$$
$$\{78.83/80.1\}[0.093(78.3/80.1) - 2(-23.43/ - 21.56)]$$
$$K = 2.791.$$

Finally, calculate the 90% confidence interval ($t = 1.943$) as follows:

$$[(AV_t/AV_r) - G(\sigma_{TR}/\sigma_R^2)] \pm [(t/AV_r)\sqrt{\sigma_R^2 K/N}]/(1 - G).$$

$$= [(-23.43/21.56) - 0.0929(78.83/80.1)] \pm [(1.943/21.56) \text{ sqrt } (80.1 \times 2.791/7)]/(1 - 0.0929).$$

The 90% confidence interval is approximately 54% to 166%.

11.4.3.2 Nontransformed Data
The following discussion refers to the approach to the analysis of confidence intervals for BE prior to the present use of the logarithmic transformation. See also, above, the preferred method using Fieller's (Locke) Theorem.

90% confidence interval for AUC difference for data in Table 11.10

$$= \bar{\Delta} \pm t\sqrt{EMS\left(\frac{1}{N_1} + \frac{1}{N_2}\right)}$$

$$42.25 \pm 1.81\sqrt{\frac{1067}{6}} = 42.25 \pm 24.14 = 18.11 \text{ to } 66.39,$$

where 42.25 is the average difference of the AUCs, 1.81 the t value with 10 d.f., 1067 the variance estimate (Table 11.12), and $1/6 = 1/N_1 + 1/N_2$. The confidence interval can be expressed as an approximate percentage relative bioavailability by dividing the lower and upper limits for the AUC difference by the average AUC for Product B, the reference product as follows:

Average AUC for drug Product $B = 167.2$
Approximate 90% confidence interval for A/B
$= (167.2 + 18.11)/167.2$ to $(167.2 + 66.39)/167.2$
$= 1.108$ to 1.397.

Product A is between 11% and 40% more bioavailable than Product B. The ratio formed for the nontransformed data, as shown in the example above, has random variables in both the numerator and denominator. The denominator (the average value of the reference) is considered fixed in this calculation, when, indeed, it is a variable measurement. Also, the decision rule is not symmetrical with regard to the average results for the test and reference. That is, if the reference is 20% greater than the test, the ratio test/reference is not 0.8 but is $1/1.2 = 0.83$. Conversely, if the test is 20% greater than the reference, the ratio will be 1.2. Nevertheless, at one time this approximate calculation was considered satisfactory for the purposes of assessing BE. Note that the usual concept of power does not play a part in the approval process. It behooves the sponsor of the BE study to recruit sufficient number of subjects to help ensure approval based on this criterion. If the products are truly equivalent (the ratio of test/reference is truly between 0.8 and 1.2), the more subjects recruited, the greater the probability of passing the test. Note again that in this scenario the more subjects, the better the chance of passing. In practice, one chooses a sample size sufficiently large to make the probability of passing reasonably high. This probability may be defined as power in the context of proving equivalence. Sample size determination for various assumed differences between the test and reference products for various values of power (probability of passing the confidence interval criterion) has been published by Diletti et al. [20] (see Table 6.5).

The conclusions based on the confidence interval approach are identical to two one-sided t tests each performed at the 5% level [12,17]. The null hypotheses are

$$H_0 : \frac{A}{B} < 0.8 \quad \text{and} \quad H_0 : \frac{A}{B} > 1.25.$$

Note that with the log transformation, the upper limit is set at 1.25 instead of 1.2. This results from the properties of logarithms, where $\log (0.8) = -\log (1/0.8)$. If both tests are rejected, the products are considered to have a ratio of AUC and/or C_{max} between 0.8 and 1.25 and are taken to be equivalent. If either hypothesis (or both) is not rejected, the products are not considered to be equivalent.

The test product (A in Table 11.10) would not pass the FDA equivalency test because the upper confidence limit exceeds 1.25. For the two one-sided t tests, we test the observed difference versus the hypothetical difference needed to reach 80% and 125% of the standard product.

If the test product had an average AUC of 175 and the error were 1067, the product would pass the FDA criterion using the "old" method. The 90% confidence limits would be

$$175 - 167.2 \pm 1.81 \sqrt{\frac{1067}{6}} = -16.34 \text{ to } 31.94.$$

The 90% confidence limits for the ratio of the AUC of test product/standard product are calculated as

$$\frac{(167.2 - 16.34)}{167.2} = 0.902$$

$$\frac{(167.2 + 31.94)}{167.2} = 1.191.$$

The limits are within 0.8 and 1.25.
The two one-sided t tests are

$$H_0 : \frac{A}{B} < 0.8 \quad t = \frac{175 - 167.2 - [-33.4]}{\sqrt{1067/6}} = 3.09$$

$$H_0 : \frac{A}{B} > 1.25 \quad t = \frac{175 - 167.2 - [41.8]}{\sqrt{1067/6}} = 2.55,$$

where -33.4 represents 20% and 41.8 represents 25% of the reference.¶ Since both t values exceed 1.81, the table t for a one-sided test at the 5% level, the products are deemed to be equivalent.

¶ The former FDA criterion for the confidence interval was 0.8 to 1.20 based on nontransformed data. Therefore this presentation is hypothetical.

Westlake has discussed the application of a confidence interval that is symmetric about the ratio 1.0, the value that defines equivalent products. The construction of such an interval is described in section 5.1.

11.4.3.3   Log-Transformed Data (Current Procedure)

The log transform appears to be more natural when our interest is in the ratio of the product outcomes. The antilog of the difference of the average results gives the ratio directly [18].

Note that the *difference* of the logarithms is equivalent to the logarithm of the *ratio* [i.e., log $A - \log B = \log (A/B)$]. The antilog of the average difference of the logarithms is an estimate of the ratio of AUCs.

The ANOVA for the ln-transformed data is shown in Table 11.14.

The averages ln values for the test and standard products are

$$\overline{A} = 5.29751$$
$$\overline{B} = 5.07778$$
$$\overline{A} - \overline{B} = 5.29751 - 5.0778 = 0.21973.$$

The anti-ln of this difference, corresponding to the geometric mean of the individual ratios, is 1.246. This compares to the ratio of A/B for the untransformed values of 1.252.

$$0.21973 \pm 1.81\sqrt{0.045/6} = 0.06298 \text{ to } 0.37648.$$

The anti-ln of these limits are 1.065 to 1.457. The 90% confidence limits for the untransformed data are 1.108 to 1.397.

It is not surprising that both analyses give similar results and conclusions. However, in situations where the confidence interval is close to the lower and/or upper limits, the two analyses may result in different conclusions. A nonparametric approach has been recommended (but is not currently accepted by the FDA) if the data distribution is far from normal (see chap. 15). As discussed earlier, at one time, the FDA suggested an alternative criterion for proof of BE: at least 75% of the subjects should show the availability for a test product compared to the reference or standard formulation to be between 75% and 125%. This is called the *75/75 rule*. If 75% of the population *truly* shows at least 75% relative absorption of the test formulation compared to the standard, a sample of subjects in a clinical study will have a 50% chance of failing the test based on the FDA criterion. This criterion has little statistical basis and has fallen into disrepute. The concept of individual BE (sect. 11.4.6) is concerned with assessing the equivalence of products on an individual basis based on a more statistically based criterion.

11.4.4   Sample Size and Highly Variable Drug Products

Phillips [19] published sample sizes as a function of power, product differences, and variability. Diletti et al. [20] have published similar tables where the log transformation is used for the statistical analysis. These tables are more relevant to current practices. Table 6.4 shows sample sizes for the multiplicative (log-transformed) analysis, reproduced from the publication by Diletti. This table as well as more details on sample size estimation is given in section 6.5. (See also Excel program on the accompanying disk to calculate sample size under various assumptions.)

When the variation is large because of inherent biologic variability in the absorption and/or disposition of the drug (or due to the nature of the formulation), large sample sizes may be needed to meet the confidence interval criterion. Generally, using results of previous studies, one can estimate the within-subject variability from the residual error term in the ANOVA. This can be assumed to be the average of the within-subject variances of the two products. These variances cannot be separated in a two-period crossover design, nor can the variability be separately attributed to the drug itself or to the formulation effects. Thus, the variability estimate is some combination of both the drug and the formulation variances. A drug product is considered to be highly variable if the error variance shows a coefficient of variation (CV) of 30% or greater. There are many drug products that show such variability. CV's of 100% or more have been observed on occasion. To show equivalence for highly variable drug products, using the FDA criterion of a 90% confidence interval of parameter ratios of 0.8 to 1.25 requires a very large sample size.

For example, from Table 6.5, if the CV is 30% and the products differ by only 5%, a sample size of 40 is needed to have 80% power to show the products are equivalent. The FDA has been considering the problems of designing studies and interpreting results for variable drugs and/or drug products. This problem has been debated for some time, and a few recommendations have been proposed to deal with this problem. Although there is no single solution, possible alternatives include widening of the confidence interval criterion from 0.8 to 1.25 to 0.75 to 1.33 [21] and use of replicated or sequential designs. The European Agency for the Evaluation of Medicinal Products also makes provision for a wider interval provided it is prospectively defined and can be justified accordingly [22]. Another recommendation by Endrenyi [23] is to scale the ratio using the reference CV as the scaling factor. At the time of this writing, the FDA has published a guidance that includes a scaled analysis. This approach may be recommended for BE studies of highly variable products. This scaled analysis is described below. Individual BE in a replicate design to assess BE is also supposed to result in smaller sample sizes for highly variable drug products as compared to the corresponding two-period design. This solution to the problem is yet to be fully confirmed. Currently, products with large CVs require large studies, with an accompanying increased expense. Because these highly variable drugs have been established as safe and effective and have a history of efficacy and safety in the marketplace, increasing the confidence interval would be congruent with the drug's variability in practice. Scaled BE may provide an economical way of evaluating these drug products.

Note that for the determination of BE based on the final study results, power (computed a posteriori) plays no role in the determination of equivalence. However, to estimate the sample size needed before initiating the study, power is an important consideration. The greater the power one wishes to impose, where power is the probability of passing the 0.8 to 1.25 confidence interval, the more subjects will be needed. Usually, a power of 0.8 is used to estimate sample size. However, if cost is not important (or not excessive), a greater power (0.9, for example) can be used to gain more assurance of passing the study, assuming that the products are truly bioequivalent.

Equation (11.3) can be used to approximate the sample size needed for a specified power.

$$N = 2(t_{\alpha, 2N-2} + t_{\beta, 2N-2})^2 \left[\frac{CV}{(V - \delta)} \right]^2, \qquad (11.3)$$

where N is the total number of subjects required to be in the study; t the appropriate value from the t distribution (approximately 1.7); α the significance level (usually 0.1); $1 - \beta$ the power, usually 0.8; CV the coefficient of variation; V the BE limit (ln 1.25 = 0.223); and δ the difference between the products (for 5% difference, delta equals [ln(1.05) = 0.0488]).

If we assume a 5% difference between the products being compared, the number of subjects needed for a CV of 30% and power of 0.8 is: $N = 2 (1.7 + 0.86)^2 [0.3/(0.223 - .0488)]^2 =$ approximately 39 subjects, which is close to the 40 subjects from Table 6.5.

If the CV is 50%, we need approximately 108 subjects!

$$N = 2(1.7 + 0.86)^2 \left(\frac{0.5}{0.223 - 0.0488} \right)^2 = \text{approximately 108 subjects.}$$

It can be seen that with a large CV, studies become inordinately large.

BE studies are usually performed at a single site, where all subjects are recruited and studied as a single group. On occasion, more than one group is required to complete a study. For example, if a large number of subjects are to be recruited, the study site may not be large enough to accommodate the subjects. In these situations, the study subjects may be divided into two cohorts. Each cohort is used to assess the comparative products individually, as might be done in two separate studies. Typically, the two cohorts are of approximately equal size. The final assessment is based on a combination of both groups. The totality of data is analyzed with a new term in the ANOVA, a Treatment-by-Group interaction term.** This is a measure (on a log scale) of how the ratios of test to reference differ in the groups. For example, if the ratios are very much the same in each group, the interaction would be small or negligible. If interaction is large, as tested in the ANOVA, then the groups statistically should not be combined. However, if at least one of the groups individually passes the confidence interval criteria, then the test product might be acceptable. If interaction is not statistically significant ($p > 0.10$), then the confidence interval based on the pooled analysis, after dropping the interaction term, will determine acceptability. It is an advantage to pool the data, as the larger number of subjects increases power and there is a greater probability of passing the BE confidence interval, if the products are truly bioequivalent.

An interesting question arises if more than two groups are included in a BE study. As before, if there is no interaction, the data should be pooled. If interaction is evident, it is implied that at least one group is different from the others. Usually, it will be obvious which group is divergent from a visual inspection of the treatment differences in each group. The remaining groups may then be tested for interaction. Again, as before, if there is no interaction, the data should be pooled. If there is interaction, the aberrant group may be omitted, and the remaining groups tested, and so on. In rare cases, it may not be obvious which group or groups are responsible for the interaction. In that case, more statistical treatment may be necessary, and a statistician should be consulted. In any event, if any single group or pooled groups (with no interaction) passes the BE criteria, the test should pass. If a pooled study passes in the presence of interaction, but no single study passes, one may still argue that the product should pass, if there is no apparent reason for the interaction. For example, if the groups are studied at the same location under the identical protocol, and there is overlap in time among the treatments given to the different groups, as occurs often, there may be no obvious reason for a significant interaction. Perhaps, the result was merely due to chance. One may then present an argument for accepting the pooled results.

The following statistical models have been recommended for analysis of data in groups:

Model 1: GRP SEQ GRP*SEQ SUBJ(GRP*SEQ) PER(GRP) TRT GRP*TRT.
If the GRP*TRT term is not significant ($p > 0.10$), then reanalyze the data using Model 2.
Model 2: GRP SEQ GRP*SEQ SUBJ(GRP*SEQ) PER(GRP) TRT,

where GRP is the group, SEQ the sequence, GRP*SEQ the group-by-sequence, SUBJ(GRP*SEQ) the subject nested within group-by-sequence, PER(GRP) the period nested within group, TRT the treatment, and GRP*TRT the group-by-treatment interaction.

11.4.5   Outliers in BE Studies
An outlier is an observation far removed from the bulk of the observations.

The problems of dealing with outlying observations is discussed in some detail in section 10.2. These same problems exist in the analysis and interpretation of BE studies. Several kinds of outliers occur in BE studies. Analytical outliers may occur because of analytical errors, and these can usually be rectified by reanalyzing the retained blood samples. Another kind of outlier is a value that does not appear to fit the PK profile. If repeat analyses verify these values, one has little choice but to retain these values in the analysis. If such values appear rarely, they will usually not affect the overall conclusions since the individual results are a small part of the overall average results, such as in the calculation of AUC. An exception may occur if the aberrant value occurs at the time of the estimated C_{max}, where the outlier could be more influential. The biggest problem with outliers is when the outlier arises from a derived parameter (AUC or C_{max}) for an individual subject. The current FDA position is to disallow the exclusion of an outlier from the analysis solely on a statistical basis. However, if a clinical reason can be determined as a potential cause for the outlier and when the outlier appears to be due to the reference product, an outlier may be omitted from the analysis at the discretion of the FDA. The FDA also suggests

---

** Currently, FDA requires this only when groups are not from the same population or are dosed widely separated in time.

that the outlier be retested in a sample of 6 to 10 subjects from the original study to support the anomalous nature of the suspected outlier. Part of the reasoning for not excluding outliers is that one or two individual outliers suggest the possibility of a subpopulation that shows a difference between the products. Although theoretically possible, this author's opinion is that this is a highly unlikely event without definitive documentation. Also, using this reasoning, an outlying observation due to the reference product would suggest that the reference did not act uniformly among patients, suggesting a deficiency in the reference product. Another possible occasion for discarding an individual subject's result is the case where very little or no drug is absorbed. Explanations for this effect could be product-related or subject-related, but the true cause is unlikely to be known. Zero blood levels, in the absence of corroborating evidence for product failure, are most likely due to a failure of the subject. These problems remain controversial and should be dealt with on a case-by-case basis.

A more creative approach is possible in the case of replicate designs (see below). In these situations, the estimates of within-subject variability can be used to identify outliers. For example, if the within-subject variance for a given treatment is 0.61, but reduces to 0.04 when omitting the subject with the suspected outlier value, an F test can be performed comparing variances for the suspect data and the remaining data. The F ratio, in this example, is

$$F = \frac{0.61}{0.04} = 15.3.$$

The d.f. for the numerator are those for the variance estimate obtained using the results from all subjects and those for the denominator are those for the variance estimate obtained from the results omitting the suspected outlier. In the above example, if the numerator and denominator d.f. were 30 and 28, respectively, then an F value of 15.3 is highly significant ($p < 0.01$). An alternative analysis could be an ANOVA with and without the suspected outlier. An F test with 1 d.f. in the numerator and appropriate d.f. in the denominator would be:

[SS (all data) − SS (without outlier data)]/residual SS (all data)<

Another approach that has been used is to compare results for periods 1 and 2 versus periods 3 and 4 in a four-period fully replicated design.

Of course, if there is an obvious cause for the outlier, a statistical justification is not necessary. However, further evidence, even if only suspicious, is helpful.

If an outlier is detected, as noted above, the most conservative approach is to find a reason for the outlying observation, such as a transcription error, or an analytical error, or a subject that violated the protocol, and so on. In these cases, the data may be reanalyzed with the corrected data, or without the outlying data if due to analytical or protocol violation, for example.

If an obvious reason for the outlier is not forthcoming, one may wish to perform a new small study, replicating the original study, including the outlying subject along with a number of other subjects (at least five or six) from the original study. The results from the new study can be examined to determine if the data for the outlier from the original study are anomalous. It should be noted that the data from the small study are not used as a replacement for any of the original data, but serve only to confirm, or refute, that the suspected outlier subject is reproducibly an outlier. The procedure here is not fixed, but should be reasonable, and make sense. One can compare the test to reference ratios for the outlying subject in the two studies, and demonstrate that the data from the new study show that the outlying subject is congruent with the other subjects in the new study, for example.

11.4.6   Replicate Designs for BE Studies**

Replicate crossover designs may be defined as designs with more than two periods where products are given on more than one occasion. In the present context such replicate studies are studies in which individuals are administered one or both products on more than one occasion. FDA gives sponsors the option of using replicate design studies. Replicate studies can isolate

---

**A more advanced topic.

the within-subject variance of each product separately, as well as potential product-by-subject interactions.

The FDA recommends that submissions of studies with replicate designs be analyzed for average BE. The following (Table 11.15) is an example of the analysis of a two-treatment–four period replicate design to assess average BE. The design has each of two products, balanced in two sequences, *ABAB* and *BABA*, over four periods. Table 11.16 shows the results for C_{max} for a replicate study. Eighteen subjects were recruited for the study and 17 completed the study. An analysis using the usual approach for the two-treatment, two-period design, as discussed above, is not recommended. The FDA recommends use of a mixed model approach as in SAS PROC MIXED [9]. The recommended code is

PROC MIXED;

CLASSES SEQ SUBJ PER TRT;

MODEL LNCMAX = SEQ PER TRT/DDFM = SATTERTH;

RANDOM TRT/TYPE = FAO (2) SUB = SUBj G;

REPEATED/GRP = TRT SUB = SUBJ;

LSMEANS TRT;

ESTIMATE "T VS. R" TRT 1 − 1/CL ALPHA = 0.1;

RUN;

We will concentrate on the comparison of two products in three- or four-period designs. The FDA recommends using only two sequence designs because the interaction variability estimate, subject × formulation, will be otherwise confounded (see Ref. 24 for a comparison of the 2 and 4 sequence designs). The subject × formulation interaction is crucial because if this effect is substantial, the implication is that subjects do not differentiate formulations equally, that is, some subjects may give higher results for one formulation, and other subjects respond higher on the other formulation. Two sequence designs for three- and four-period studies are shown below. Although there are other designs available, these seem to have particularly good properties [16,24].

Three-period design	
Sequence	Period 1 2 3
1	A B B
2	B A A

Four-period design	
Sequence	Period 1 2 3 4
1	A B B A
2	B A A B

With replicate designs, carryover effects, within-subject variances and subject × formulation interactions can be estimated, unconfounded with other effects. Nevertheless, an unambiguous acceptable analysis is still not clear. Do we include all the effects in the model simultaneously or do we perform preliminary tests for inclusion in the model? What is the proper error term to construct a confidence interval on the average BE parameter (e.g., AUC)? Some estimates may not be available if all terms are included in the model. Therefore, preliminary testing may be necessary. These questions are not easy to answer and, despite their advantages, make the use of replicate designs problematic at the time of this writing.

The following is one way of proceeding with the analysis: Test for differential carryover. This term may be included in the model (along with the usual parameters) using a dummy variable, that is, 0 if treatment in Period 1, if Treatment B follows Treatment A, and 2 if Treatment A follows Treatment B. If differential carryover is not significant, remove it from the model. Include a term for subject × formulation interaction, and if this effect is large, the products may be considered bioequivalent (see sect. 11.4.6.1). Another problem that arises here is concerned with what error term should be used to construct the confidence interval for the average difference between formulations. The choices are among the within-subject variance (residual), the interaction term, or the residual with no interaction term in the model (pooled residual and interaction). The latter could be defended if the interaction term is small or not significant.

The analysis of studies with replicate designs would be very difficult without access to a computer program. Using SAS GLM, the following program can be used. (See below for FDA recommended approach.)

```
proc glm;
class sequence subject product period co;
model auc = period subject (sequence) product co;
lsmeans product/stderr;
estimate 'test-ref'product −11;
```

co is carryover

Using the data from Chow and Liu [16], a four-period design with nine subjects completing the study, the SAS output is as follows:

Dependent variable: AUC

Source	d.f.	Sum of squares	Mean square	F value	Pr > F
Model	13	40895.72505	3145.82500	8.25	0.0001
Error	22	8391.03801	381.41082		
Corrected total	35	49286.76306			

Dependent variable: AUC

Source	d.f.	Type I SS	Mean square	F value	Pr > F
SEQ	1	9242.13356	9242.13356	24.23	0.0001
SUBJECT (SEQ)	7	25838.61700	3691.23100	9.68	0.0001
PRODUCT	1	1161.67361	1161.67361	3.05	0.0949
PERIOD	3	4650.60194	1550.20065	4.06	0.0193
CO	1	2.69894	2.69894	0.01	0.9337

Source	d.f.	Type III SS	Mean square	F value	Pr > F
SEQ	1	8311.37782	8311.37782	21.79	0.0001
SUBJECT (SEQ)	7	25838.61700	3691.23100	9.68	0.0001
PRODUCT	1	975.69000	975.69000	2.56	0.1240
PERIOD	2	2304.85554	1152.42777	3.02	0.0693
CO	1	2.69894	2.69894	0.01	0.9337

Parameter	Estimate	T for HO: Pr > |T|		Std error of parameter estimate
test-ref	−10.98825000	−1.60	0.1240	6.87019569

Because carryover is not significant ($p > 0.9$), we can remove this term from the model and analyze the data with a subject × formulation (within sequence) term included in the model. The SAS output is as follows:

General linear models procedure
Dependent variable: AUC

Source	d.f.	Sum of squares	Mean squares	F value	Pr > F
Model	19	42490.87861	2236.36203	5.27	0.0008
Error	16	6795.88444	424.74278		
Corrected total	35	49286.76306			

Source	d.f.	Type I SS	Mean square	F value	Pr > F
SEQ	1	9242.13356	9242.13356	21.76	0.0003
SUBJECT (SEQ)	7	25838.61700	3691.23100	8.69	0.0002
PRODUCT	1	1161.67361	1161.67361	2.74	0.1177
PERIOD	3	4650.60194	1550.20065	3.65	0.0354
SUBJECT * PRODUCT(SEQ) (SEQ)	7	1597.85250	228.26464	0.54	0.7940

Source	d.f.	Type III SS	Mean square	F value	Pr > F
SEQ	1	9242.13356	9242.13356	21.76	0.0003
SUBJECT (SEQ)	7	25838.61700	3691.23100	8.69	0.0002
PRODUCT	1	1107.56806	1107.56806	2.61	0.1259
PERIOD	2	4622.20056	2311.10028	5.44	0.0157
SUBJECT * PRODUCT (SEQ)	7	1597.85250	228.26464	0.54	0.7940

The subject $\times$ product interaction is not significant ($p > 0.7$). Again the question of which error term to use for the confidence interval is unresolved. The choices are (a) interaction $= 228$, within-subject variance $= 425$, or pooled residual $= 365$. The d.f. will also differ depending on the choice. The simplest approach seems to be to use the pooled variance if the interaction term is not significant (the level must be defined). If interaction is significant, use the interaction term as the error. In the example given above, the analysis without interaction and carryover may be appropriate (also see sect. 11.4.6.1). The following analysis has an error term equal to 365.

Dependent variable: AUC

Source	d.f.	Sum of squares	Mean square	F value	Pr > F
Model	12	40893.02611	3407.75218	9.34	0.0001
Error	23	8393.73694	364.94508		
Corrected total	35	49286.76306			

Source	d.f.	Type III SS	Mean square	F value	Pr > F
SEQ	1	9242.13356	9242.13356	25.32	0.0001
SUBJECT (SEQ)	7	25838.61700	3691.23100	10.11	0.0001
PRODUCT	1	1107.56806	1107.56806	3.03	0.0949
PERIOD	3	4650.60194	1550.20065	4.25	0.0158

PRODUCT	AUC LSMEAN	Std err LSMEAN	Pr > |T| HO: LSMEAN = 0
1	87.7087500	4.5308014	0.0001
2	76.5462500	4.5308014	0.0001

Parameter	Estimate	T for HO: Parameter = 0	Pr > |T|	Std error of estimate
test-ref	−11.16250000	−1.74	0.0949	6.40752074

The complete analysis of replicate designs can be very complex and ambiguous, and is beyond the scope of this book. An example of the analysis as recommended by the FDA is shown later in this section. For an in-depth discussion of the analysis of replicate designs including estimation of sources of variability (see Refs. [16,24,25]).

The four-period design will be further discussed in the discussion of individual bioequivalence (IB), for which it is recommended. In a relatively recent guidance, the FDA [10] gives sponsors the option of using replicate design studies for all BE studies. However, at the time of this writing, the agency has ceased to recommend use of replicate studies although they may be useful in some circumstances. The purpose of these studies was to provide more information about the drug products than can be obtained from the typical, nonreplicated, two-period design. The FDA was interested in obtaining information from these studies to aid them in evaluation of the need for IB. In particular, replicate studies provide information on within-subject variance of each product separately, as well as potential product × subject interactions. As noted previously, the use of these designs and assessment of IB have been controversial, and its future in its present form is in doubt.

The FDA recommends that submissions of studies with replicate designs be analyzed for average BE [10]. Any analysis of IB will be the responsibility of the FDA, but will be only for internal use, not for evaluating BE for regulatory purposes.

The following is another example of the analysis of a two-treatment–four-period replicate design to assess average BE, as recommended by the FDA. This design has each of two products,

Table 11.15 Results of a Four-Period, Two-Sequence, Two-Treatment, Replicate Design (C_{max})

Subject	Product	Sequence	Period	C_{max}	$Ln(C_{max})$
1	Test	1	1	14	2.639
2	Test	1	1	16.7	2.815
3	Test	1	1	12.95	2.561
4	Test	2	2	13.9	2.632
5	Test	1	1	15.6	2.747
6	Test	2	2	12.65	2.538
7	Test	2	2	13.45	2.599
8	Test	2	2	13.85	2.628
9	Test	1	1	13.05	2.569
10	Test	2	2	17.55	2.865
11	Test	1	1	13.25	2.584
12	Test	2	2	19.8	2.986
13	Test	1	1	10.45	2.347
14	Test	2	2	19.55	2.973
15	Test	2	2	22.1	3.096
16	Test	1	1	22.1	3.096
17	Test	2	2	14.15	2.650
1	Test	1	3	14.35	2.664
2	Test	1	3	22.8	3.127
3	Test	1	3	13.25	2.584
4	Test	2	4	14.55	2.678
5	Test	1	3	13.7	2.617
6	Test	2	4	13.9	2.632
7	Test	2	4	13.75	2.621
8	Test	2	4	13.25	2.584
9	Test	1	3	13.95	2.635
10	Test	2	4	15.15	2.718
11	Test	1	3	13.15	2.576
12	Test	2	4	21	3.045
13	Test	1	3	8.75	2.169
14	Test	2	4	17.35	2.854
15	Test	2	4	18.25	2.904
16	Test	1	3	19.05	2.947
17	Test	2	4	15.1	2.715
1	Reference	1	2	13.5	2.603
2	Reference	1	2	15.45	2.738
3	Reference	1	2	11.85	2.472
4	Reference	2	1	13.3	2.588
5	Reference	1	2	13.55	2.606
6	Reference	2	1	14.15	2.650
7	Reference	2	1	10.45	2.347
8	Reference	2	1	11.5	2.442
9	Reference	1	2	13.5	2.603
10	Reference	2	1	15.25	2.725

(Continued)

Table 11.15   Results of a Four-Period, Two-Sequence, Two-Treatment, Replicate Design (C_{max}) *Continued*

11	Reference	1	2	11.75	2.464
12	Reference	2	1	23.2	3.144
13	Reference	1	2	7.95	2.073
14	Reference	2	1	17.45	2.859
15	Reference	2	1	15.5	2.741
16	Reference	1	2	20.2	3.006
17	Reference	2	1	12.95	2.561
1	Reference	1	4	13.5	2.603
2	Reference	1	4	15.45	2.738
3	Reference	1	4	11.85	2.472
4	Reference	2	3	13.3	2.588
5	Reference	1	4	13.55	2.606
6	Reference	2	3	14.15	2.650
7	Reference	2	3	10.45	2.347
8	Reference	2	3	11.5	2.442
9	Reference	1	4	13.5	2.603
10	Reference	2	3	15.25	2.725
11	Reference	1	4	11.75	2.464
12	Reference	2	3	23.2	3.144
13	Reference	1	4	7.95	2.073
14	Reference	2	3	17.45	2.859
15	Reference	2	3	15.5	2.741
16	Reference	1	4	20.2	3.006
17	Reference	2	3	12.95	2.561

balanced in two sequences, *ABAB* and *BABA*, over four periods. Table 11.15 shows the results for C_{max} for a replicate study. Eighteen subjects were recruited for the study and 17 completed the study. An analysis using the usual approach for the tttp design, as discussed above, is not recommended. The FDA [10] recommends use of a mixed model approach as in SAS PROC MIXED [13]. The recommended code is

```
PROC MIXED;
CLASSES SEQ SUBJ PER TRT;
MODEL LNCMAX = SEQ PER TRT/DDFM = SATTERTH;
RANDOM TRT/TYPE = FAO (2) SUB = SUBj G;
REPEATED/GRP = TRT SUB = SUBJ;
LSMEANS TRT;
ESTIMATE "T VS. R" TRT 1 − 1/CL ALPHA = 0.1;
RUN;
```

The abbreviated output is shown in Tables 11.16 and 11.17. Table 11.16 shows an analysis of the first two periods for ln (C_{max}) and untransformed C_{max}. Table 11.17 shows the output for the analysis of average BE using all four periods. Note that the confidence interval using the complete design (0.0592–0.1360) is not much different from that observed from the analysis of the first two periods (see Exercise at the end of the chapter), 0.0438, 0.1564. This should be expected because of the small variability exhibited by this product.

11.4.6.1   *Individual Bioequivalence*[††]
Another issue that has been introduced as a relevant measure of equivalence is "individual" bioequivalence (IB). This is in contrast to the present measure of "average" BE. Note that

---

[††] FDA has never accepted, nor currently endorses, this method, despite its having devoted resources to its development over a period >5 years. It is presented here due to its elegant statistical derivation from basic principles of drug interchangeability and its place in the history of bioequivalence testing in the U.S.

the evaluation of data from the tttp design results in a measure of average BE. Average BE addresses the comparison of average results derived from the tttp BE study, and does not consider differences of within-subject variance and interactions in the evaluation.

The IB approach is an attempt to evaluate the effect of changing products (brand to generic, for example) for an individual patient, considering the potential for a change of therapeutic effect

Table 11.16 ANOVA for Data from First Two Periods of Table 11.15

(A) LN TRANSFORMATION
Dependent variable: LNCMAX

Source	d.f.	Sum of squares	Mean square	F value	Pr > F
Model	18	1.65791040	0.09210613	10.34	0.0001
Error	15	0.13359312	0.00890621		
Corrected Total	33	1.79150352			

R square	CV	Root MSE	LNCMAX mean
0.925430	3.528167	0.09437271	2.67483698

Source	d.f.	Type I SS	Mean square	F value	Pr > F
SEQ	1	0.09042411	0.09042411	10.15	0.0061
SUBJ(SEQ)	15	1.48220203	0.09881347	11.09	0.000
PER	1	0.00039571	0.00039571	0.04	0.8359
TRT	1	0.08488855	0.08488855	9.53	0.0075

Least squares means

TRT	LNCMAX LSMEAN
Reference	2.62174427
Test	2.72185203

T for HO: Parameter	Pr > |T| Estimate	Std error of Parameter=0		Estimate	
T VS.R	0.10010777	3.09	0.0075	0.03242572	

(B) Dependent variable: CMAX

Source	d.f	Sum of squares	Mean square	F value	Pr > F
Model	18	381.26362847	21.18131269	9.07	0.0001
Error	15	35.01637153	2.33442477		
Corrected total	33	416.28000000			

R square	CV	Root MSE	CMAX mean
0.915883	10.25424	1.52788245	14.90000000

Source	d.f.	Type I SS	Mean square	F value	Pr > F
SEQ	1	18.59404514	18.59404514	7.97	0.0129
SUBJ(SEQ)	15	346.22095486	23.08139699	9.89	0.0001
PER	1	0.24735294	0.24735294	0.11	0.7493
TRT	1	16.20127553	16.20127553	6.94	0.0188

Least squares means

TRT	CMAX LSMEAN
Reference	14.1649306
Test	15.5479167

Dependent variable: CMAX

T for HO: Parameter	Pr > |T| Estimate	Std Error of Parameter=0		Estimate	
T VS. R	1.38298611	2.63	0.0188	0.52496839	

Table 11.17   Analysis of Data from Table 11.15 for Average Bioequivalence

ANALYSIS FOR LN-TRANSFORMED CMAX
The MIXED procedure
Class level information

Class	Concentrations values
SEQ	2 1 2
SUBJ	17 1 2 3 4 5 6 7 8 9 10 11 12 13
	14 15 16 17
PER	4 1 2 3 4
TRT	2 1 2

Covariance parameter estimates (REML)

Cov Parm	Subject	Group	Estimate
FA(1,1)	SUBJ		0.20078553
FA(2,1)	SUBJ		0.22257742
FA(2,2)	SUBJ		−0.00000000
DIAG	SUBJ	TRT 1	0.00702204
DIAG	SUBJ	TRT 2	0.00982420

Tests of Fixed Effects

Source	NDF	DDF	Type III F	Pr > F
SEQ	1	13.9	1.02	0.3294
PER	3	48.2	0.30	0.8277
TRT	1	51.1	18.12	0.0001

ESTIMATE statement results
Parameter T VS. R

Alpha = 0.1	Estimate		Std error	d.f.	t	Pr > |t|
	0.09755781	0.02291789		51.1	4.26	0.0001
	Lower		0.0592	Upper	0.1360	

Least squares means

Effect	TRT	LSMEAN	Std Error	d.f.	t	Pr > |t|
TRT	1	2.71465972	0.05086200	15	53.37	0.0001
TRT	2	2.61710191	0.05669416	15.3	46.16	0.0001

or increased toxicity when switching products [38]. This is a very difficult subject from both a conceptual and statistical point of view. Statistical methods and meaningful differences must be established to show differences in variability between products before this criterion can be implemented. Whether or not a practical approach can be developed, and whether or not such approaches will have meaning in the context of BE remains to be seen. Some of the statistical problems to be contemplated when implementing this concept include recommendations of specific replicated crossover designs to measure both within- and between-variance components as well as subject × product interactions, and definitions of limits that have clinical meaning. The issue is related to variability. Assuming that the average bioavailability is the same for both products as measured in a typical BE study, the question of IB appears to be an evaluation of formulation differences. Since the therapeutic agents are the same in the products to be compared, it is formulation differences that could result in excessive variability or differences in bioavailability that are under scrutiny. Some of the dilemmas are related to the inherent biologic variability of a drug substance. If a drug is very variable, we would expect large variability in studies of interchangeability of products. In particular, taking the same product on multiple occasions would show a lack of "reproducibility." The question that needs to be addressed is whether the new (generic) product would cause efficacy failure or toxicity when exchanged with the reference or brand product due to excessive variability. The onus is on the generic product. Product failure could be due to a change in the rate and extent of drug absorption as well as an increase in inter- and intrapatient variability. The FDA has spent some energy in addressing the problem of how to define and evaluate any changes incurred by the generic product. This is a difficult problem, not only in identifying the parameters to measure the variability, but also to define the degree of variability that would be considered excessive. For example, drugs that are very variable may be allowed more leniency in the criteria for "interchangeability" than less variable, narrow-therapeutic-range drugs.

The FDA has proposed an expression to define IB

$$\theta = \frac{[\delta^2 + \sigma_I^2 + (\sigma_T^2 - \sigma_R^2)]}{\sigma_R^2} \tag{11.4}$$

where δ is the difference between means of test and reference, σ_I^2 the subject $\times$ treatment interaction variance, σ_T^2 the within-subject test variance, and σ_R^2 the within-subject reference variance.

Equation 11.4 makes sense in that the comparison between test and reference products is scaled by the within-reference variance, thereby not penalizing very variable drug products when testing for BE. In addition, the expression contains a term for testing the mean difference, the interaction, and the difference between the within-subject variances. If the test product has a smaller within-subject variance than the reference, this favors the test product.

Before IB was to be considered a requirement from a regulatory point of view, data were accumulated from replicate crossover studies (three or more periods) to compile a database to assess the magnitude and kinds of intrasubject and formulation $\times$ subject variability that exist in various drug and product classes. The design and submission of such studies were more or less voluntary, and were analyzed for average BE. However, this gave the regulatory agency the opportunity to evaluate the data according to IB, and to evaluate the need for this new kind of criterion for equivalence. At the time of this writing, the FDA has rejected further development of this approach. The details of the design and analysis of these studies are presented below.

In summary, IB is an assessment that accounts for product differences in the variability of the PK parameters, as well as differences in their averages. IB evaluation is based on the statistical evaluation of the metric [Eq. (11.4)], which represents a "distance" between the products. In average BE, this distance can be considered the square of the difference in average results. In IB, in addition to the difference in averages, the difference between the within-subject variances for the two products, and the formulation $\times$ subject interaction (FS) are evaluated. In this section, we will not discuss the evaluation of population BE. The interested reader may refer to the FDA guidance [10].

The evaluation of IB is based on a 95% upper confidence limit on the metric, where the upper limit for approval, theta (θ), is defined as 2.4948. Note that we only look at the upper limit because the test is one-sided; that is, we are only interested in evaluating the upper value of the confidence limit, upon which a decision of passing or failing depends. A large value of the metric results in a decision of inequivalence. Referring to Eq. (11.4), a decision of inequivalence results when the numerator is large and the denominator is small in value. Large differences in the average results, combined with a large subject $\times$ formulation interaction, a large within-subject variance for the test product and a small within-subject variance for the reference product, will increase the value of theta (and vice versa).

Using the within-subject variance of the reference product in the denominator as a scaling device allows for a less stringent decision for BE in cases of large reference variances. That is, if the reference and test products appear to be very different based on average results, they still may be deemed equivalent if the reference within-subject variance is large. This can be a problem in interpretation of BE, because if the within-subject variance of the test product is sufficiently smaller than the reference, an unreasonably large difference between their averages could still result in BE [see Eq. (11.4)]. This could be described as a compensation feature or trade-off; that is, a small within-subject variance for the test product can compensate for a large difference in averages. To ensure that such apparently unreasonable conclusions will not be decisive, the FDA guidance has a proviso that the observed T/R ratio must be not more than 1.25 or less than 0.8.

11.4.6.2 Constant Scaling

The FDA guidance [10] also allows for a constant scaling factor in the denominator of Eq. (11.4). If the variance of the reference is very small, the IB metric may appear very large, even though the products are reasonably close. If the within-subject variance for the reference product is less than 0.04, a value of 0.04 may be used in the denominator, rather than the observed variance. This prevents an artificial inflation of the metric for cases of a small within-subject reference variance. This case will not be discussed further, but is a simple extension of the following discussion. The reader may refer to the FDA guidance for further discussion of this topic [10].

11.4.6.3 Statistical Analysis for IB

For average BE, the distribution of the difference in average results (log transformed) is known based on the assumption of a log-normal distribution of the parameters. One of the problems with the definition of BE based on the metric, Eq. (11.4), is that the distribution of the metric is complex, and cannot be easily evaluated. At an earlier evolution in the analysis of the metric, a bootstrap technique, a kind of simulation, was applied to the data to estimate its distribution. The nature of the distribution is needed to construct a confidence interval so that a decision rule of acceptance or rejection can be determined. This bootstrap approach was time consuming, and not exactly reproducible. An approximate "parametric" approach was recommended [26], which results in a hypothesis test that determines the acceptance rule. We refer to this approach as the "Hyslop" evaluation. This will be presented in more detail below.

To illustrate the use of the Hyslop approach, the data of Table 11.18 will be used. This data set has been studied by several authors during the development of methods to evaluate IB [27].

The details of the derivation and assumptions can be found in the FDA guidance [28] and the paper by Hyslop et al. [26].

The following describes the calculations involved and the definitions of some terms that are used in the calculations. The various estimates are obtained from the data of Table 11.18, using SAS [13], with the following code:

proc mixed data = Drug;

class seq subj per trt;

model ln $C_{\max}$ = seq per trt;

random int subject/subject = trt;

repeated/grp = trt sub = subj;

estimate "t vs.r" trt 1 − 1/cl alpha = 0.1;

run;

Table 11.19 shows the estimates of the variance components and average results for each product from the data of Table 11.18.

Basically, the Hyslop procedure obtains an approximate upper confidence interval on the sum of independent terms (variables) in the IB metric equation [Eq. (11.4)]. However, the statistical approach is expressed as a test of a hypothesis. If the upper limit of the CI is less than 0, the products are deemed equivalent, and vice versa. The following discussion relates to the scaled metric, where the observed reference within-subject variance is used in the denominator. An analogous approach is used for the case where the reference variance is small and the denominator is fixed at 0.04 (see Ref. [28]).

The IB criterion is expressed as

$$\theta = \frac{[\delta^2 + \sigma_d^2 + (\sigma_T^2 - \sigma_R^2)]}{\sigma_R^2}. \tag{11.5}$$

Table 11.18 Data from a Two-Treatment, Two-Sequence, Four-Period Replicated Design [20]

Subject	Sequence	Period	Product	Ln C_{max}
1	1	1	1	5.105339
1	1	3	1	5.090062
2	1	1	1	5.594340
2	1	3	1	5.459160
3	2	2	1	4.991792
3	2	4	1	4.693181
4	1	1	1	4.553877
4	1	3	1	4.682131
5	2	2	1	5.168778
5	2	4	1	5.213304
6	2	2	1	5.081404
6	2	4	1	5.333202
7	2	2	1	5.128715
7	2	4	1	5.488524
8	1	1	1	4.131961
8	1	3	1	4.849684
1	1	2	2	4.922168
1	1	4	2	4.708629
2	1	2	2	5.116196
2	1	4	2	5.344246
3	2	1	2	5.216565
3	2	3	2	4.513055
4	1	2	2	4.680278
4	1	4	2	5.155601
5	2	1	2	5.156178
5	2	3	2	4.987025
6	2	1	2	5.271460
6	2	3	2	5.035003
7	2	1	2	5.019265
7	2	3	2	5.246498
8	1	2	2	5.249127
8	1	4	2	5.245971

It can be shown that

$$\sigma_I^2 = \sigma_d^2 + 0.5(\sigma_T^2 + \sigma_R^2), \tag{11.6}$$

where σ_d^2 is the pure estimate of the subject $\times$ formulation interaction component. We can express this in the form of hypothesis test, where the IB metric is linearized as follows:
Substituting Eq. (11.6) into Eq. (11.5), and linearizing

$$\text{Let } \eta = (\delta)^2 + \sigma_I^2 + 0.5\,\sigma_T^2 - \sigma_R^2(-1.5 - \theta). \tag{11.7}$$

Table 11.19 Parameter Estimates from Analysis of Data of Table 4 with Some Definitions

μ'_T = mean of test; estimate = 5.0353
μ'_R = mean of reference; estimate = 5.0542
δ = difference between observed mean of test and reference = −0.0189
μ'^2_t = interaction variance; estimate = M_I = 0.1325
μ'^2_T = within-subject variance for the test product; estimate = M_T = 0.0568
μ'^2_R = within-subject variance for the reference product; estimate = M_R = 0.0584
n = degrees of freedom
s = number of sequences

Table 11.20 Computations for Evaluation of Individual Bioequivalence

Hq = (1 − alpha) level upper Confidence limit	Eq = point estimate	Uq = (Hq − Eq)²
$H_D = \left[\lvert\delta\rvert + t(1-\alpha,\ n-s)(1/s^2 \sum n_i^{-1}\ M_I)^{1/2}\right]^2$	$E_D = \delta^2$	U_D
$H_I = [(n-s) \cdot M_I]/\chi^2(\alpha,\ n-s)$	$E_I = M_I$	U_I
$H_T = [0.5 \cdot (n-s) \cdot M_T]/\chi^2(\alpha,\ n-s)$	$E_T = 0.5 \cdot M_T$	U_T
$H_R = [-(1.5+\theta_1) \cdot (n-s) \cdot M_R]/\chi^2(1-\alpha,\ n-s)$ [a]	$E_R = -(1.5+\theta_1) \cdot M_R$	U_R

[a]Note that we use the $1-\alpha$ percentile here because of the negative nature of this expression. $n = \sum n_i$; $s =$ number of sequences; $n_i =$ the number of subjects in sequence i.

We then form a hypothesis test with the hypotheses

$$H_0: \eta > 0 \qquad H_a: \eta > 0.$$

Howe's Method (Hyslop) effectively forms a CI for η by first finding an upper or lower limit for each component in η. Then, a simple computation allows us to accept or reject the null hypothesis at the 5% level (one-sided test). This is equivalent to seeing if an upper CI is less than the FDA-specified criterion, θ. Using Hyslop's Method, if the upper confidence limit is less than θ, the test will show a value less than 0, and the products are considered to be equivalent.

The computation for the method is detailed below.

We substitute the observed values for the theoretical values in Eq. (11.7). The observed values are shown in Table 11.19.

The next step is to compute the upper 95% confidence limits for the components in Eq. (11.7). Note that δ is normal with mean, true delta, and variance $2\sigma_d^2/N$. The variances are distributed as $(\sigma^2) \cdot \chi^2_{(n)}/n$ (where $n =$ d.f.). For example, $M_T \sim \sigma_T(n)^2 \chi^2_{(n)}/n$.

The equations for calculations are given in Table 11.20 [26].

$$H = \sum (E_i) + \sum (U_i)^{0.5} = -0.0720 + 0.3885 = 0.3165.$$

Table 11.21 shows the results of these calculations.
Examples of calculations

$$H_D = \left[\lvert -0.0189\rvert + 1.94 \cdot ((1/4) \cdot 0.1325/2)^{1/2}\right]^2 = 0.07213$$

$$H_I = ((6) \cdot 0.1325)/1.635 = 0.4862$$

$$H_T = (0.5 \cdot (6) \cdot 0.0568)/1.635 = 0.1042$$

$$H_R = (-(1.5+2.4948) \cdot (6) \cdot 0.0584)/12.59 = -0.1112$$

Table 11.21 Results of Calculations for Data of Table 11.20

H_i = confidence limit	E_i = point estimate	$U_i = (H - E)^2$
$H_d = 0.07213$	$E_d = 0.00357$	0.0052
$H_i = 0.4862$	$E_i = 0.1325$	0.1251
$H_t = 0.1042$	$E_t = 0.0284$	0.0057
$H_r = -0.1112$	$E_r = -0.2333$	0.0149
SUM	−0.0720	0.1509

If the upper CI exceeds zero, the hypothesis is rejected, and the products are bioinequivalent. This takes the form of a one-sided test of hypothesis at the 5% level.

Since this value (0.3165) exceeds 0, the products are considered to be inequivalent.

An alternative method to construct a decision criterion for IB based on the metric is given in Appendix IX.

11.4.6.4 *The Future*

At the present time, the design and analysis of BE studies use tttp designs with a log transformation of the estimated parameters. The 90% CI of the back-transformed difference of the average results for the comparative products must lie between 0.8 and 1.25 for the products to be deemed equivalent. Four-period replicate designs have been recommended on occasion for controlled-release products and, in some cases, very variable products. However, FDA recommends that these designs be analyzed for average BE. The results of these studies were analyzed for IB by the FDA to assess the need for IB; that is, is there a problem with formulation × subject interactions and differences between within-subject variance for the two products? The result of this venture showed that replicate designs were not needed, that is, the data does not show significant interaction or within-subject variance differences. IB may be reserved for occasions where these designs will be advantageous in terms of cost and time. In fact, recent communication with FDA suggests that IB requirements are not likely to continue in the present form. Some form of IB analysis may be optimal for very variable drugs, requiring less subjects than would be required using a tttp design for average BE. On the other hand, in the future if IB analysis shows the existence of problems with interaction and within-subject variances, it is possible that the four-period replicate design and IB analysis will be considered for at least some subset of drugs or drug products that exhibit problems. For very variable drug products, a scaled analysis has been proposed that would reduce the sample size relative to the usual crossover analysis (see below, sect. 11.4.9). Also, FDA is investigating the use of sequential designs, or add-on designs in the implementation of BE studies.

See Appendix X for a discussion of designs used in BE studies.

11.4.7 Sample Size for Test for Equivalence for a Dichotomous (Pass–Fail) Outcome

Tests for BE are usually based on an analysis of drug in body fluids (e.g., plasma). However, for drugs that are not absorbed, such as topicals and certain local acting gastrointestinal products (e.g., sucralfate), a clinical study is necessary. Often the outcome is based on a binomial outcome such as cured/not cured. See section 5.2.6 for confidence intervals for a proportion. A continuity correction is recommended. Makuch and Simon [29] have published a method for determining sample size for these studies, as well as other kinds of clinical studies where the objective is to determine equivalence. This reference is concerned particularly with cancer treatments where a less intensive treatment is considered to replace a more toxic treatment if the two treatments can be shown to be therapeutically equivalent. As for the case of BE studies with a continuous outcome, one needs to specify both alpha and beta errors in addition to a difference between the treatments that is considered important to estimate the required sample size.

In this approach, we assume a parallel-groups design (two independent groups), typical of these studies. To estimate the number of subjects required in the two groups, we will assume an equal number to be assigned to each group. An estimate of (1) the value of the proportion of subjects who will be "cured" or have a positive outcome for each treatment (P_1 and P_2), and (2) the difference between the treatments that are not clinically meaningful is needed. Makuch and Simon have shown that the number of subjects per group can be calculated from Eq. (11.4):

$$N = [P_1(1 - P_1) + P_2(1 - P_2)] \times \left\{ \frac{[Z_\alpha + Z_\beta]}{[\Delta - [P_1 - P_2]]} \right\}^2, \tag{11.8}$$

where delta (Δ) is the maximum difference between treatments considered to be of no clinical significance.

If we assume that the products are not different a priori, $P_1 = P_2 = P$, Eq. (11.4) reduces to

$$N = 2P(1 - P)\left\{\frac{[Z_\alpha + Z_\beta]}{\Delta}\right\}^2. \tag{11.9}$$

In a practical example, a clinical study is designed to compare the efficacy of a generic sucralfate to the brand product. The outcome is the healing of gastrointestinal ulcers. How many subjects should be entered in a parallel study with a dichotomous endpoint (healed/ not healed) if the expected proportion healed is 0.80 and the CI of the difference of the proportions should not exceed ±0.2? We wish to construct a two-sided 90% CI with a beta of 0.2 (power = 0.8). This means that with the required number of patients, we will be able to determine, with 90% confidence, if the healing rates of the products are within ±0.2. If indeed the products are equivalent, with a beta of 0.2, there is 80% probability that the CI for the difference between the products will fall within ±20%.

The values of Z for beta can be obtained from Table 6.2.

Note that if the products are not considered to be different with regard to proportion or probability of success, the values for beta will be based on a two-sided criterion. For example, for 80% power, use 1.28 (not 0.84). From Eq. (11.5),

$$N = 2(0.8)(1 - 0.8)\left\{\frac{[1.65 + 1.28]}{0.2}\right\}^2 = 69.$$

Sixty-nine subjects per group are required to satisfy the statistical requirements for the study.

If the criterion is made more similar to the typical BE criterion, we might consider the difference (delta) to be 20% of 0.8 or 16%, rather than the absolute 20%. If delta is 16%, the number of subjects per group will be approximately 108. (See Exercise Problem 12 at the end of this chapter.) The BE subject number calculator on the CD included with this book provides for the calculation of these subject numbers with the inclusion of a continuity correction often requested by FDA.

11.4.8   SCALED CRITERION FOR BE

The scaled criterion is currently endorsed by FDA for highly variable drug products [30]. A within-subject CV of 30% or greater is considered "highly variable." The recommended design is a three-period crossover with three sequences, TRR, RTR and RRT, where R is the reference and T is the test product. Thus, only the reference is replicated, and the within-subject variance can be estimated for the reference product. Although a minimum sample size of 24 is recommended, the appropriate sample size is determined by the sponsor. After a log transformation, the parameters (AUC and C_{max}) are calculated in addition to the within-subject variance of the reference product.

The statistical null hypothesis is

$$H_o : (X_T - X_R)^2/S_R^2 > \theta$$

The alternative hypothesis is

$$H_1 : (X_T - X_R)^2/S_R^2 \le \theta,$$

where θ is the scaled average BE limit, $X_T - X_R$ is the difference between the average parameter (AUC or C_{max}) after a log transformation, and S_R^2 is the calculated within-subject variance for the reference product.

θ is defined as $(\ln \Delta)^2/\sigma^2_{wo}$, where $\Delta = 1.25$ and $\sigma_{wo} = 0.25$.

Therefore, $\theta = 0.7967$.

BE is accepted if the null hypothesis is rejected and *the ratio of test to reference is between 0.8 and 1.25. Both criteria must be satisfied to declare* BE.

A 95% upper bound for $(X_T - X_R)^2/S_R^2$ from the BE study must be $\leq \theta$ in addition to the restriction of the ratio of test to reference parameters (0.8–1.25). As of this writing, a method for computing the upper bound is not forthcoming. Use of the "Hyslop" Method for individual BE, previously discussed and modified for this application, has been proposed.

11.4.9 NONINFERIORITY TRIALS

Noninferiority trials are related to BE studies in that in both cases we are not testing for differences. For noninferiority trials, we are testing that a test product is not worse than a reference product based on results of a clinical study. Again, we must define a value such that if the lower confidence bound (usually 95%) of the test treatment compared to the reference exceeds that value, the test treatment will be considered noninferior. This value should be defined in the protocol prior to seeing the study results, and is a value such that any value lower than the specified value would result in a conclusion of inferiority.

For example, comparing Test Drug X to Reference Drug Y, it was determined that a difference in average response of 2 units would be acceptable for purposes of noninferiority. That is, if study results showed that Drug X was no more than 2 units less than Drug Y, Drug X would be considered to be noninferior to Drug Y. The study showed that Drug X was 1 unit less than Drug Y. The 95% lower bound of this difference was 2.1, that is, based on the lower bound, Drug X, could be as much as 2.1 units less than Drug Y. Therefore, Drug X failed the noninferiority test. The lower confidence bound showed more than a 2 unit difference, and we can not conclude that Drug X is noninferior to Drug Y.

11.5 REPEATED MEASURES (SPLIT-PLOT) DESIGNS

Many clinical studies take the form of a baseline measurement followed by observations at more than one point in time. For example, a new antihypertensive drug is to be compared to a standard, marketed drug with respect to diastolic blood pressure reduction. In this case, after a baseline blood pressure is established, the patients are examined every other week for eight weeks, a total of four observations (visits) after treatment is initiated.

11.5.1 Experimental Design

Although this antihypertensive drug study was designed as a multiclinic study, the data presented here represent a single clinic. Twenty patients were randomly assigned to the two treatment groups, 10 to each group (see sect. 11.2.6 for the randomization procedure). Prior to drug treatment, each patient was treated with placebo, and blood pressure determined on three occasions. The average of these three measurements was the baseline reading.

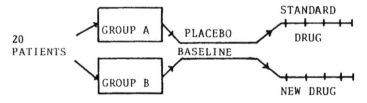

The baseline data were examined to ensure that the three baseline readings did not show a time trend. For example, a placebo effect could have resulted in decreased blood pressure with time during this preliminary phase.

Treatment was initiated after the baseline blood pressure was established. Diastolic blood pressure was measured every two weeks for eight weeks following initiation of treatment. (The dose was one tablet each day for the standard and new drug.) Two patients dropped out in the "standard drug" group, and one patient was lost to the "new drug" group, resulting in eight and nine patients in each treatment group. The results of the study are shown in Table 11.22 and Figure 11.4.

The design described above is commonly known in the pharmaceutical industry as a *repeated measures* or *split-plot* design. (This design is also denoted as an incomplete three-way or a partially hierarchical design.) This design is common in clinical or preclinical studies, where

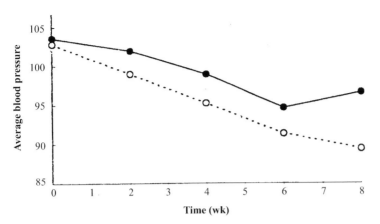

Figure 11.4 Plot of mean results from antihypertensive drug study. ●—standard drug; O—new drug.

two or more products are to be compared with multiple observations over time. The design can be considered as an extension of the one-way or parallel-groups design. In the present design (repeated measures), data are obtained at more than one time point. The result is two or more two-way designs, as can be seen in Table 11.22, where we have two two-way designs. The two-way designs are related in that observations are made at the same time periods. *The chief features of the repeated measures design* as presented here are as follows:

1. Different patients are randomly assigned to the different treatment groups, that is, a patient is assigned to only one treatment group.
2. The number of patients in each group need not be equal. Equal numbers of patients per group, however, result in optimum precision when comparing treatment means. Usually, these studies are designed to have the same number of patients in each group, but dropouts usually occur during the course of the study.
3. Two or more treatment groups may be included in the study.
4. Each patient provides more than one measurement over time.
5. The observation times (visits) are the same for all patients.
6. Baseline measurements are usually available.
7. The usual precautions regarding blinding and randomization are followed.

Although the analysis tolerates lack of symmetry with regard to the number of patients per group (see feature 2), the statistical analysis can be difficult if patients included in the study

Table 11.22 Results of a Comparison of Two Antihypertensive Drugs

| | | Standard drug | | | | | | New drug | | | |
| | | Week | | | | | | Week | | | |
Patient	Baseline	2	4	6	8	Patient	Baseline	2	4	6	8
1	102	106	97	86	93	3	98	96	97	82	91
2	105	103	102	99	101	4	106	100	98	96	93
5	99	95	96	88	88	6	102	99	95	93	93
9	105	102	102	98	98	8	102	94	97	98	85
13	108	108	101	91	102	10	98	93	84	87	83
15	104	101	97	99	97	11	108	110	95	92	88
17	106	103	100	97	101	12	103	96	99	88	86
18	100	97	96	99	93	14	101	96	96	93	89
						16	107	107	96	93	97
Mean	103.6	101.9	98.9	94.6	96.6	Mean	102.8	99.0	95.2	91.3	89.4

have missing data for one or more visits. In these cases, a statistician should be consulted regarding data analysis [31].

The usual assumptions of normality, independence, and homogeneity of variance for each observation hold for the split-plot analysis. In addition, there is another important assumption with regard to the analysis and interpretation of the data in these designs. The assumption is that the data at the various time periods (visits) are not correlated, or that the correlation is of a special form [32]. Although this is an important assumption, often ignored in practice, moderate departures from the assumption can be tolerated. Correlation of data during successive time periods often occurs such that data from periods close together are highly correlated compared to the correlation of data far apart in time. For example, if a person has a high blood pressure reading at the first visit of a clinical study, we might expect a similar reading at the subsequent visit if the visits are close in time. The reading at the end of the study is apt to be less related to the initial reading. The present analysis assumes that the correlation of the data is the same for all pairs of time periods, and that the pattern of the correlation is the same in the different groups (e.g., drug groups) [32]. If these assumptions are substantially violated, the conclusions based on the usual statistical analysis will not be valid. The following discussion assumes that this problem has been considered and is negligible [31].

11.5.2 ANOVA

The data of Table 11.22 will be subjected to the typical repeated measures (split-plot) ANOVA. As in the previous examples in this chapter, the data will be analyzed, corrected for baseline, by subtracting the baseline measurement from each observation. The measurements will then represent *changes from baseline*. The more complicated analysis of covariance is an alternative method of treating such data [31, 32]. More expert statistical help will usually be needed when applying this technique, and the use of a computer is almost mandatory. Subtracting out the baseline reading is easy to interpret and, generally, results in conclusions very similar to that obtained by covariance analysis. Table 11.23 shows the "changes from baseline" data derived from Table 11.22. For example, the first entry in this table, two weeks for the standard drug, is $106 - 102 = 4$.

When computing the ANOVA by hand (use a calculator), the simplest approach is to first compute the two-way ANOVA for each treatment group, "standard drug" and "new drug." The calculations are described in section 8.4. The results of the ANOVA are shown in Table 11.24. Only the sums of squares need to be calculated for this preliminary computation.

The final analysis combines the separate two-way ANOVAs and has two new terms, "weeks × drugs" interaction and "drugs," the variance represented by the difference between the drugs. The calculations are described below, and the final ANOVA table is shown in Table 11.25.

Table 11.23 Changes from Baseline of Diastolic Pressure for the Comparison of Two Antihypertensive Drugs

	Standard drug					New drug			
	Week					Week			
Patient	2	4	6	8	Patient	2	4	6	8
1	4	−5	−16	−9	3	−2	−1	−16	−7
2	−2	−3	−6	−4	4	−6	−8	−10	−13
5	−4	−3	−11	−11	6	−3	−7	−9	−9
9	−3	−3	−7	−7	8	−8	−5	−4	−17
13	0	−7	−17	−6	10	−5	−14	−11	−15
15	−3	−7	−5	−7	11	2	−13	−16	−20
17	−3	−6	−9	−5	12	−7	−4	−15	−17
18	−3	−4	−1	−7	14	−5	−5	−8	−12
					16	0	−11	−14	−10
Mean	−1.75	−4.75	−9	−7	Mean	−3.8	−7.6	−11.4	−13.3
Sum	−14	−38	−72	−56	Sum	−34	−68	−103	−120

Table 11.24 ANOVA for Changes from Baseline for Standard Drug and New Drug

	Standard drug		New drug	
Source	d.f.	Sum of squares	d.f.	Sum of squares
Patients	7	57.5	8	114.22
Weeks	3	232.5	3	486.97
Error	21	255.5	24	407.78
Total	31	545.5	35	1008.97

Patients' SS: Pool the SS from the separate ANOVAs ($57.5 + 114.22 = 171.72$ with $7 + 8 = 15$ d.f.).

Weeks' SS: This term is calculated by combining all the data, resulting in four columns (weeks), with 17 observations per column, 8 from the standard drug and 9 from the new drug. The calculation is

$$\frac{\sum C^2}{R_1 + R_2} - CT,$$

where C is the column sums of combined data and $R_1 + R_2$ is the sum of the number of rows,

$$= \frac{(-48)^2 + (-106)^2 + (-175)^2 + (-176)^2}{17} - \frac{(-505)^2}{68}$$

$$= 4420.1 - 3750.4 = 669.7.$$

Drug SS:

$$\text{Drug SS} = (CT_{SP}) + (CT_{NP}) - (CT_T),$$

where CT_{sp} is the correction term for the standard drug, CT_{NP} the correction term for the new product, and CT_T the correction term for the combined data.

$$\text{Drug SS} = \frac{(-180)^2}{32} + \frac{(-325)^2}{36} - \frac{(-505)^2}{68}$$

$$= 196.2.$$

Table 11.25 Repeated Measures (Split-Plot) ANOVA for the Antihypertensive Drug Study

Source	d.f.[a]	SS	MS	
Patients	15	171.7	11.45	
Weeks	3	669.7	223.23	
Drugs	1	196.2	196.2	$F_{1,15} = \dfrac{196.2}{11.45} = 17.1^*$
Weeks × drugs	3	49.8	16.6	
Error (within treatments)	45	663.3	14.74	$F_{1,15} = \dfrac{16.6}{14.74} = 1.1$
	67	1750.6		

[a] Degrees of freedom for "patients" and "error" are the d.f. pooled from the two-way ANOVAs. For "weeks" and "drugs," the d.f. are (weeks − 1) and (drugs − 1), respectively. For "weeks × drugs," d.f. are (weeks − 1) × (drugs − 1).
$^*p < 0.01$.

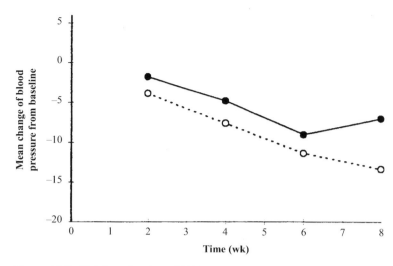

Figure 11.5 Plot from the data of Table 11.23 showing lack of significant interaction of weeks and drugs in experiment comparing standard and new antihypertensive drugs. ●—standard drug; o—new drug.

Weeks × drugs SS: This interaction term (see below for interpretation) is calculated as the pooled SS from the "week" terms in the separate two-way ANOVAs above, minus the week term for the final combined analysis, 669.7.

Weeks × drug SS = 232.5 + 486.97 − 669.7 = 49.8.

Error SS: The error SS is the pooled error from the two-way ANOVAs, 255.5 + 407.8 = 663.3.

11.5.2.1 Interpretation and Discussion

The terms of most interest are the "drugs" and "weeks × drugs" components of the ANOVA. "Drugs" measures the difference between the overall averages of the two treatment groups. The average reduction of blood pressure was (180/32) = 5.625 mm Hg for standard drug, and (325/36) = 9.027 mm Hg for the new drug. The F test for "drug" differences is (drug MS)/(patients MS) equal to 17.1 (1 and 15 d.f.; see Table 11.25). This difference is highly significant ($p < 0.01$). The significant result indicates that on the average, the new drug is superior to the standard drug with regard to lowering diastolic blood pressure.

The significant difference between the standard and new drugs is particularly meaningful if the difference is constant over time. Otherwise, the difference is more difficult to interpret. "Weeks × drugs" is a measure of interaction (see also chap. 9). This test compares the parallelism of the two "change from baseline" curves as shown in Figure 11.5. The F test for "weeks × drugs" uses a different error term than the test for "drugs." The F test with 3 and 45 d.f. is 16.6/14.74 = 1.1, as shown in Table 11.25. This nonsignificant result suggests that the pattern of response is not very different for the two drugs. A reasonable conclusion based on this analysis is that the new drug is effective (superior to the standard drug), and that its advantage beyond the standard drug is approximately maintained during the course of the experiment.

A significant nonparallelism of the two "curves" in Figure 11.5 would be evidence for a "weeks × drugs" interaction. For example, if the new drug showed a lower change in blood pressure than the standard drug at two weeks, and a higher change in blood pressure at eight weeks (the curves cross one another), interaction of weeks and drugs would more likely be significant. Interaction, in this example, would suggest that drug differences are dependent on the time of observation.

If interaction is present or the assumptions underlying the analysis are violated (particularly concerning the form of the covariance matrix) [31], a follow-up or an alternative is to perform p one-way ANOVAs at each of the p points in time. In the previous example, analyses

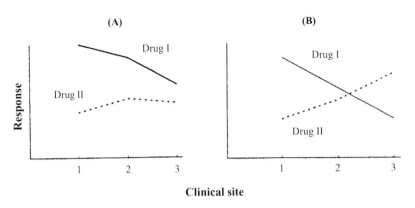

Figure 11.6 Two kinds of interaction: (**A**) one drug always better than another, but the difference changes for different clinical sites; (**B**) one drug better than another at sites 1 and 2 and worse at site 3.

would be performed at each of the four post-treatment weeks. A conclusion is then made on the results of these individual analyses (see Exercise Problem 8).

11.6 MULTICLINIC STUDIES

Most clinical studies carried out during late phase 2 or phase 3 periods of drug testing involve multiclinic studies. In these investigations, a common protocol is implemented at more than one study site. This procedure, recommended by the FDA, serves several purposes. It may not be possible to recruit sufficient patients in a study carried out by a single investigator. Thus multiclinic studies are used to "beef up" the sample size. Another very important consideration is that multiclinic studies, if performed at various geographic locations with patients representing a wide variety of attributes, such as age, race, socioeconomic status, and so on, yield data that can be considered representative under a wide variety of conditions. Multiclinic studies, in this way, guard against the possibility of a result peculiar to a particular single clinical site. For example, a study carried out at a single Veterans' Administration hospital would probably involve older males of a particular economic class. Also, a single investigator may implement the study in a unique way that may not be typical, and the results would be peculiar to his or her methods. Thus, if a drug is tested at many locations and the results show a similar measure of efficacy at all locations, one has some assurance of the general applicability of the drug therapy. In general, one should attempt to have more or less equal numbers of patients at each site, and to avoid having too few patients at sites.

However, there are instances where a drug has been found to be efficacious in the hands of some investigators and not for others. When this occurs, the drug effect is in some doubt unless one can discover the cause of such results. This problem is statistically apparent in the form of a treatment × site interaction. The comparative treatments (drug and placebo, for example) are not differentiated equally at different sites. A treatment × site interaction may be considered very serious when one treatment is favored at some clinical sites and the other favored at different sites. Less serious is the case of interaction where all clinics favor the same treatment, but some favor it more than others. These two examples of interaction are illustrated in Figure 11.6.

When interaction occurs, the design, patient population, clinical methods, protocol, and other possible problems should be carefully investigated and dissected, to help find the cause. The cause will not always be readily apparent, if at all. See section 8.4.3 for a further example and discussion of interactions in clinical studies. An important feature of multiclinic studies, as noted above, is that the same protocol and design should be followed at all sites.

Since one can anticipate missing values due to dropouts, missed visits, recording errors, and so on, an important consideration is that the design should not be so complicated that missing data will cause problems with the statistical interpretation or that the clinicians will have difficulty following the protocol. A simple design that will achieve the objective is to be preferred. Since parallel-groups designs are the most simple in concept, these should be preferred to some more esoteric design. Nevertheless, there are occasions where a more complex design would be appropriate providing that the study is closely monitored and the clinical investigators thoroughly educated.

11.7 INTERIM ANALYSES

Under certain conditions, it is convenient (and sometimes prudent) to look at data resulting from a study prior to its completion in order to make a decision to change the protocol procedure or requirements, or to abort the study early or to increase the sample size, for example. This is particularly compelling for a clinical study involving a disease that is life-threatening, is expensive, and/or is expected to take a long time to complete. A study may be stopped, for example, if the test treatment can be shown to be superior early on in the study. However, if the data are analyzed prior to study completion, a penalty is imposed in the form of a lower significance level to compensate for the multiple looks at the data (i.e., to maintain the overall significance level at alpha). The more occasions that one looks at and analyzes the data for significance, the greater the penalty, that is, the more difficult it will be to obtain significance at each analysis. The penalty takes the form of an adjustment of the alpha level to compensate for the multiple looks at the data. The usual aim is to keep the alpha level at a nominal level, for example 5%, considering the multiple analyses; this fixes the probability of declaring the treatments different when they are truly the same at, at most, 5%, taking into account the fact that at each look we have a chance of incorrectly declaring a significant difference. For example, if the significance level is 0.05 for a single look, two looks will have an overall significance level of approximately 0.08.

In addition to the advantage (time and money) of stopping a study early when efficacy is clearly demonstrated, there may be other reasons to shorten the duration of a study, such as stopping because of a drug failure, modifying the number of patients to be included, modifying the dose, and so on. If interim analyses are made for these purposes in phase 3 pivotal studies, an adjusted p level will probably be needed for regulatory purposes. Davis and Huang discuss this in more detail [33]. In any event, the approach to interim analyses should be clearly described in the study protocol, a priori; or, if planned after the study has started, the plan of the interim analysis should be communicated to the regulatory authorities (e.g., FDA). Even if interim looks do not affect the study procedure or outcome, such procedures should be clearly documented either in the study protocol or as an amendment to the protocol. One of the popular approaches to interim analyses was devised by O'Brien and Fleming [34], an analysis known as a group sequential method. The statistical analyses are performed after a group of observations have been accumulated rather than after each individual observation. The analyses should be clearly documented and should be performed by persons who cannot influence the continuation and conduct of the study.

The procedure and performance of these analyses must be described in great detail in the study protocol, including the penalties in the form of reduced "significance" levels. A very important feature of interim analyses is the procedure of breaking the randomization code. One should clearly specify who has access to the code and how the blinding of the study is maintained. It is crucial that the persons involved in conducting the study, clinical personnel and monitors alike, not be biased as a result of the analysis. This is of great concern to the FDA. Interim analyses should not be done willy-nilly, but should be planned and discussed with regulatory authorities. Associated penalties should be fixed in the protocol. As noted previously, this does not mean that interim analyses cannot and should not be performed as an afterthought if circumstances dictate their use during the course of the study. A Pharmaceutical Manufacturer's Association (PMA) committee [35] suggested the following to minimize potential bias resulting from an interim analysis. (1) "A Data Monitoring Committee (DMC) should be established to review interim results." The persons on this committee should not be involved in decisions regarding the progress of the study. (2) If the interim analysis is meant to terminate a study, the details should be presented in the protocol, a priori. (3) The results of the interim analysis should be confidential, known only to the DMC.

Sankoh [25] discusses situations where interim analyses have been used incorrectly from a regulatory point of view. In particular, he is concerned with unplanned interim analyses.

Table 11.26 Significance Levels for Two-Sided Group
Sequential Studies with an Overall Significance Level of 0.05
(According to O'Brien/Fleming)

Number of analysis (stages)	Analysis	Significance level
2	First	0.005
	Final	0.048
3	First	0.0005
	Second	0.014
	Final	0.045
4	First	0.0005
	Second	0.004
	Third	0.019
	Final	0.043
5	First	0.00001
	Second	0.001
	Third	0.008
	Fourth	0.023
	Final	0.041

These include (a) the lack of reporting these analyses and the consequent lack of adjustment of the significance level, (b) inappropriate adjustment of the level and inappropriate stopping rules, (c) interim analyses inappropriately labeled "administrative analyses," where actual data analyses have been carried out and results disseminated, (d) lack of documentation for the unplanned interim analysis, (e) and the importance of blinding and other protocol requirements.

An interim analysis may also be planned to adjust sample size. In this case, a full analysis should not be done. The analysis should be performed when the study is not more than half done, and only the variability should be estimated (not the treatment differences). Under these conditions, no penalty need be assessed. However, if the analysis is done near the end of the trial or if the treatment differences are computed, a penalty is required [25].

Table 11.26 shows the probability levels needed for significance for k looks (k analyses) at the data according to O'Brien and Fleming [34], where the data are analyzed at equal intervals during patient enrollment. For example, if the data are to be analyzed three times ($k = 3$, where k is the number of analyses or stages, including the final analysis), the analysis should be done after $\frac{1}{3}$, $\frac{2}{3}$ and all of the patients have been completed [36]. There are other schemes for group sequential interim analyses, including those that do not require analyses at equal intervals of patient completion [37].

For example, a study with 150 patients in each of two groups is considered for two interim analyses. This corresponds to three stages, two interim and one final analysis. The first analysis is performed after 100 patients are completed (50 per group) at the 0.0005 level. To show statistically significant differences, the product differences must be very large or obvious at this low level. If not significant, analyze the data after 200 patients are completed. A significance level of 0.014 must be reached to terminate the study. If this analysis does not show significance, complete the study. The final analysis must meet the 0.045 level for the products to be considered significantly different.

One can conjure up reasons as to why stopping a study early based on interim analysis is undesirable (less information on adverse effects or less information for subgroup analyses, for example). One possible solution to this particular problem in the case where the principle objective is to establish efficacy, is to use the results of the interim analysis for regulatory submission, if the study data meet the interim analysis p level, but to continue the study after the interim analysis, and then analyze the results for purposes of obtaining further information on adverse effects, and so on. However, in this procedure, one may face a dilemma if the study fails to show significance with regard to efficacy after including the remaining patients.

KEY TERMS

Analysis of covariance
AUC (area under curve)
Balance
Baseline measurements
Between-patient variation (error)
Bias
Bioavailability
Bioequivalence
Blinding
Carryover
Changeover design
C_{max}
Controlled study
Crossover design
Differential carryover
Double blind
Double dummy
75–75 rule
Experimental design
Grizzle analysis
Incomplete three-way ANOVA
Individual bioequivalence
Intent to treat

Interaction
Interim analyses
Latin square
Locke's Method
Log transformation
Multiclinic
Objective measurements
Parallel design
Period (visit)
Placebo effect
Positive control
Randomization
Repeated measures
Replicate designs
Scaled bioequivalence analysis
Sequences
Split plot
Symmetry
Systematic error
t_p
Washout period
Within-patient variation (error)
80% power to detect 20% difference

EXERCISES

1. (a) Perform the calculations for the ANOVA table (Table 11.3) from the data in Table 11.2.
    (b) Perform a t test comparing the differences from baseline for the two groups in Table 11.2. Compare the t value to the F value in Table 11.3.

2. Using the data in Table 11.10, test to see if the values of t_p are different for formulations A and B (5% level).

3. (a) Using the data in Table 11.10, compare the values of C_{max} for the two formulations (5% level). Calculate a confidence interval for the difference in C_{max}.
    (**b) Analyze the data for C_{max} using the Grizzle Method. Is a differential carryover effect present?

4. Analyze the AUC data in Table 11.10 using ratios of AUC (A/B). Find the average ratio and test the average for significance. (Note that H_0 is $AUC_A/AUC_B = 1.0$.) Assume no period effect.

5. Analyze the AUC data in Table 11.10 using logarithms of AUC. Compare the antilog of the average difference of the logs to the average ratio determined in Problem 4. Put a 95% confidence interval on the average difference of the logs. Take the antilogs of the lower and upper limit and express the interval as a ratio of the AUCs for the two formulations.

6. ** In a pilot study, two acne preparations were compared by measuring subjective improvement from baseline (10-point scale). Six patients were given a placebo cream and six different patients were given a cream with an active ingredient. Observations were made once a week for four weeks. Following are the results of this experiment:

** This is an optional, more difficult problem.

	Placebo					Active			
	Week					Week			
Patient	1	2	3	4	Patient	1	2	3	4
1	2	2	4	3	1	2	2	3	3
2	3	2	3	3	2	4	4	5	4
3	1	4	3	2	3	1	3	4	5
4	3	2	1	0	4	3	4	4	7
5	2	1	3	2	5	2	2	3	6
6	4	4	5	3	6	3	4	6	5

A score of 10 is complete improvement. A score of 0 is no improvement (negative scores mean a worsening of the condition). Perform an ANOVA (split plot). Plot the data as in Figure 11.4. Are the two treatments different? If so, how are they different?

7. For the exercise study described in section 11.3, the difference considered to be significant is 60 minutes with an estimated standard deviation of 55 minutes. Compute the sample size if the Type I (alpha) and Type II (beta) error rates are set at 0.05 and 0.10, respectively.

8. From the data in Table 11.23, test for a difference ($\alpha = 0.05$) between the two drugs at week 4.

9. Perform the ANOVA on the ln transformed bioavailability data (sect. 11.4.2, Table 11.10).

10. A clinical study is designed to compare three treatments in a parallel design. Thirty patients are entered into the study, 10 in each treatment group. The randomization is to be performed in groups of six. Show how you would randomize the 30 patients.

11. In the example in Table 11.7, suppose that a period effect of 3 existed in this study. This means that the observations in Period 2 are augmented by 3 units. Show that the difference between treatments is not biased, that is, the difference between A and B is 1.

12. Exercise: Compute the sample size for the example in section 11.4.8, assuming that a difference of 0.16 (16%) is a meaningful difference.

13. Compute the confidence interval using Locke's Method as described in section 11.4.3.

REFERENCES

1. Department of Health, Education and Welfare. General Considerations for the Clinical Evaluation of Drugs (GPO 017–012–00245–5) (Pub. HEW (FDA) 77–3040). Washington, D.C.: FDA Bureau of Drugs Clinical Guidelines, U.S. Government Printing Office, 1977.
2. Cox DR. Planning Experiments. New York: Wiley, 1958.
3. Rodda BE, Tsianco MC, Bolognese JA, et al. Clinical Development. In: Peace KE, ed. Statistical Issues in Drug Research and Development. New York: Marcel Dekker, 1990.
4. Buncher CR, Tsay J-Y. Statistics in the Pharmaceutical Industry, 2nd ed. New York: Marcel Dekker, 1994.
5. Fisher LD, et al. Intention to treat in clinical trials. In: Peace KE, ed. Statistical Issues in Drug Research and Development. New York: Marcel Dekker, 1990: 331–349.
6. Snedecor GW, Cochran WG. Statistical Methods, 7th ed. Ames, IA: Iowa University Press, 1980.
7. Brown BW Jr. The crossover experiment for clinical trials. Biometrics 1980; 36:69.
8. Cochran WG, Cox GM. Experimental Designs, 2nd ed. New York: Wiley, 1957.
9. Grizzle JE. The two-period change-over design and its use in clinical trials. Biometrics 1965; 21:467, 1974; 30:727.
10. Guidance for Industry (Draft). Bioavailability and Bioequivalence Studies for Orally Administered Drug Products, General Considerations. Rockville, MD: FDA, CDER, 2003.
11. Haynes JD. Change-over design and its use in clinical trials. J Pharm Sci 1981; 70:673.
12. Schuirman DL. On hypothesis testing to determine if the mean of a normal distribution is contained in a known interval. Biometrics 1981, 37:617.

13. The SAS System for Windows, Release 6.12. Cary, NC: SAS Institute Inc.
14. FDA bioavailability/bioequivalence regulations. Fed Regist 1977; 42:1624.
15. FDA. Guidance for Industry, Appendix V. New York: Center for Drug Evaluation and Research, 1997.
16. Chow S-C, Liu J-P. Design and Analysis of Bioavailability and Bioequivalence Studies. New York: Marcel Dekker, 1992:280.
17. Schuirman DL. A comparison of the two one-sided tests procedure and the power approach for assessing the equivalence of average bioavailability. J Pharmacokinet Biopharm 1987; 15:657–680.
18. Westlake WJ. Bioavailability and bioequivalence of pharmaceutical formulations. In: Peace KE, ed. Biopharmaceutical Statistics for Drug Development. New York: Marcel Dekker, 1988:329–351.
19. Phillips KE. Power of the two one-sided tests procedure in bioequivalence. J Pharmacokinet Biopharm 1990; 18:137.
20. Diletti E, Hauschke D, Steinijans VW. Sample size determination for bioequivalence assessment by means of confidence intervals. Int J Clin Pharmacol Ther Toxicol 1991; 29(1):1.
21. Boddy AW, Snikeris FC, Kringle RO, et al. An approach for widening the bioequivalence acceptance limits in the case of highly variable drugs. Pharm Res 1995; 12:1865.
22. Committee For Proprietary Medicinal Products (CPMP). Note for Guidance on the Investigation of Bioavailability and Bioequivalence. London: The European Agency for the Evaluation of Medicinal Products, Evaluation of Medicines for Human Use, 2001.
23. Tothfalusi L, Endrenyi L, Midha KK, et al. Evaluation of the bioequivalence of highly variable drugs and drug products. Pharm Res 2001; 18:728.
24. Jones G, Kenward MG. Design and Analysis of Cross-over Trials. London: Chapman and Hill, 1989.
25. Sankoh AJ. Interim analyses: an update of an FDA reviewer's experience and perspective. Drug Inf J 1995; 29:729.
26. Hyslop T, Hsuan F, Hesney M. A small sample confidence interval approach to assess individual bioequivalence. Stat Med 2000; 19:2885–2897.
27. Eckbohm, Melander H. The subject-by-formulation interaction as a criterion of interchangeability of drugs. Biometrics 1989; 45:1249–1254.
28. US Department of Health and Human Services. Statistical Approaches to Establishing Bioequivalence. Rockville, MD: FDA CDER, 2001.
29. Makuch R, Simon R. Sample size requirements for evaluating a conservative therapy. Cancer Treat Rep 1978; 62(7):1037.
30. Haidar SH, et al. Pharm Res 2008; 25:237–240.
31. Chinchilli VM. Clinical efficacy trials with quantitative Data. In: Peace KE, ed. Biopharmaceutical Statistics for Drug Development. New York: Marcel Dekker, 1988:353–394.
32. Winer BJ. Statistical Principles in Experimental Design, 2nd ed. New York: McGraw-Hill, 1971.
33. Davis R, Huang I. Interim analysis. In: Buncher CR, Tsay J-Y, eds. Statistics in the Pharmaceutical Industry, 2nd ed. New York: Marcel Dekker, 1994:267–285.
34. O'Brien PC, Fleming TR. A multiple testing procedure for clinical trials. Biometrics 1979; 35:549.
35. Summary from "Issues with Interim Analysis in the Pharmaceutical Industry," The PMA Biostatistics and Medical Ad Hoc Committee on Interim Analysis, 1990.
36. Berry DA. Statistical Methodology in the Pharmaceutical Sciences. New York: Marcel Dekker, 1990:294.
37. Geller N, Pocock S. Design and analysis of clinical trials with group sequential stopping rules. In: Peace KE, ed. Biopharmaceutical Statistics for Drug Development. New York: Marcel Dekker, 1988, Chapter 11.
38. Hauck WH, Anderson S. Measuring switchability and prescribability: When is average bioequivalence sufficient? J Pharmacokinet Biopharm 1994; 22:551.

12 | Quality Control

The science of quality control is largely statistical in nature, and entire books have been devoted to the application of statistical techniques to quality control. Statistical quality control is a key factor in process validation and the manufacture of pharmaceutical products. In this chapter, we discuss some common applications of statistics to quality control. These applications include Shewhart control charts, sampling plans for attributes, operating characteristic curves, and some applications to assay development, including components of variance analysis. The applications to quality control make use of standard statistical techniques, many of which have been discussed in previous portions of this book.

12.1 INTRODUCTION

Starting from raw materials to the final packaged container, quality control departments have the responsibility of assuring the integrity of a drug product with regard to safety, potency, and biological availability. If each and every item produced could be tested (100% testing), there would be little need for statistical input in quality control. Those individual dosage units that are found to be unsatisfactory could be discarded, and only the good items would be released for distribution. Unfortunately, conditions exist that make 100% sampling difficult, if not impossible. For example, if every dosage unit could be tested, the expense would probably be prohibitive both to manufacturer and consumer. Also, it is well known that attempts to test individually every item from a large batch (several million tablets, for example), result in tester fatigue, which can cause misclassifications of items and other errors. If testing is destructive, such as would be the case for assay of individual tablets, 100% testing is, obviously, not a practical procedure. However, 100% testing is not necessary to determine product quality precisely. Quality can be accurately and precisely estimated by testing only part of the total material (a sample). In general, quality control procedures require relatively small samples for inspection or analysis. Data obtained from this sampling can then be treated statistically to estimate population parameters such as potency, tablet hardness, dissolution, weight, impurities, content uniformity (variability), as well as to ensure the quality of attributes such as color, appearance, and so on.

In various parts of this book, we discuss data from testing finished products of solid dosage forms. The details of some of these tests are explained at the end of this chapter, section 12.7.

Statistical techniques are also used to monitor processes. In particular, control charts are commonly used to ensure that the average potency and variability resulting from a pharmaceutical process are stable. Control charts can be applied during *in-process* manufacturing operations, for *finished* product characteristics, and in *research and development* for repetitive procedures. Control charts are one of the most important statistical applications to quality control.

12.2 CONTROL CHARTS

Probably the best-known application of statistics to quality control that has withstood the test of time is the Shewhart control chart. Important attributes of the control chart are its simplicity and the visual impression that it imparts. The control chart allows for judgments based on an easily comprehended graph. The basic principles underlying the use of the control chart are described below.

12.2.1 Statistical Control

A process under statistical control is one in which the process is susceptible to variability due only to inherent, but unknown and uncontrolled *chance* causes. According to Grant [1]:

"Measured quality of manufactured product is always subject to a certain amount of variation as a result of chance. Some stable system of chance causes is inherent in any particular scheme of production and inspection. Variation within this stable pattern is inevitable. The reasons for variation outside this stable pattern may be discovered and corrected."

Using tablet manufacture as an example, where tablet weights are being monitored, it is not reasonable to expect that each tablet should have an identical weight, precisely equal to some target value. A tablet machine is simply not capable of producing identical tablets. The variability is due, in part, to (a) the variation of compression force, (b) variation in filling the die, and (c) variation in granulation characteristics. In addition, the balance used to weigh the tablets cannot be expected to give exactly reproducible weighings, even if the tablets could be identically manufactured. Thus, the weight of any single tablet will be subject to the vagaries of chance from the foregoing uncontrollable sources of error, in addition to other identifiable sources that we have not mentioned.

12.2.2 Constructing Control Charts

The process of constructing a control chart depends, to a great extent, on the process characteristics and the objectives that one wishes to achieve. A control chart for tablet weights can serve as a typical example. In this example, we are interested in ensuring that tablet weights remain close to a target value, under "statistical control." To achieve this objective, we will periodically sample a group of tablets, measuring the mean weight and variability. The mean weight and variability of each sample (*subgroup*) are plotted sequentially as a function of time. The control chart is a graph that has time or order of submission of sequential lots on the X axis and the average test result on the Y axis. The process average together with upper and lower limits are specified as shown in Figure 12.1. The preservation of order with respect to the observations is an important feature of the control chart. Among other things, we are interested in attaining a

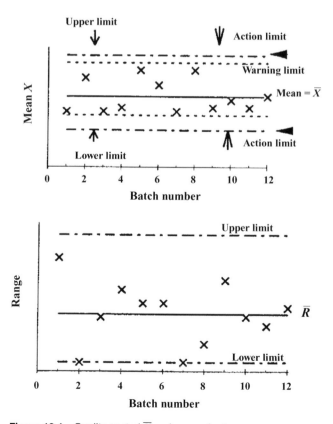

Figure 12.1   Quality control $\overline{X}$ and range charts.

state of statistical control and detecting *trends* or changes in the process average and variability. One can visualize such trends (mean and range) easily with the use of the control chart. The "consistency" of the data as reflected by the deviations from the average value is not only easily seen, but the chart provides a record of batch performance. This record is useful for regulatory purposes as well as for an in-house source of data.

As will be described subsequently, variability can be calculated on the basis of the standard deviation or the range. The range is easier to calculate than the standard deviation. Remember: The range is the difference between the lowest and highest value. If the sample size is not large (<10), the range is an efficient estimator of the standard deviation. Figure 12.1 shows an example of an "$\overline{X}$" (X bar or average) and "range" chart for tablet weights determined from consecutive tablet production batches.

12.2.2.1 Rational Subgroups

The question of how many tablets to choose at each sampling time (*rational subgroups*) and how often to sample is largely dependent on the nature of the process and the level of precision required. The larger the sample and the more frequent the sampling, the greater the precision, but also the greater will be the cost. If tablet samples are taken and weights averaged over relatively long periods of time, significant fluctuations that may have been observed with samples taken at shorter time intervals could be obscured. The subgroups should be as homogeneous as possible relative to the overall process. Subgroups are usually (but not always) taken as units manufactured close in time. For example, in the case of tablet production, consecutively manufactured tablets may be chosen for a subgroup. If possible, the subgroup sample size should be constant. Otherwise, the construction and interpretation of the control chart is more difficult. Four to five items per subgroup is usually an adequate sample size. Procedures for selecting samples should be specified under SOPs (standard operating procedures) in the quality control manual. In our example, 10 consecutive tablets are individually weighted at approximately one-hour intervals. Here the subgroup sample size is larger than the "usual" four or five, principally because of the simple and inexpensive measurement (weighing tablets). The average weight and range are calculated for each of the subgroup samples. One should understand that under ordinary circumstances the variation between individual items (tablets in this example) within a subgroup is due only to chance causes, as noted above. In the example, the 10 consecutive tablets are made almost at the same time. The granulation characteristics and tablet press effects are similar for these 10 tablets. Therefore, the variability observed can be attributed to causes that are not under our control (i.e., the inherent variability of the process).

12.2.2.2 Establishing Control Chart Limits

The principal use of the control chart is as a means of monitoring the manufacturing process. As long as the mean and range of the 10 tablet samples do not vary "too much" from subgroup to subgroup, the product is considered to be in control. To be "in control" means that the observed variation is due only to the random, uncontrolled variation inherent in the process, as discussed previously. We will define *upper and lower limits* for the mean and range of the subgroups. Values falling outside these limits are cause for alarm. The construction of these limits is based on normal distribution theory. We know, from chapter 3, that individual values from a normal distribution will be within 1.96 standard deviations of the mean 95% of the time, and within 3.0 (or 3.09) standard deviations of the mean 99.73% (or 99.8%) of the time (see Table IV.2). Therefore, the probability of observing a value outside these limits is small; only 1 in 20 in the former case and 2.7 in 1000 in the latter case. Two limits are often used in the construction of X (mean) charts as "warning" and "action" limits, respectively (Fig. 12.1). The warning limits are narrower than the action limits and do not require immediate action. If a process is subject only to random, chance variation, a value far from the mean is unlikely. In particular, a value more than 3.0 standard deviations from the mean is highly unlikely (2.7/1000), and can be considered to be probably due to some *systematic, assignable* cause. Such a "divergent" observation should signal the quality control unit to modify the process and/ or initiate an investigation into its cause. Of course, the "aberrant" value may be due only to chance. If so, subsequent means should fall close to the process average as expected. In some circumstances, one may wisely

Table 12.1 Tablet Weights and Ranges from a Tablet Manufacturing Process[a]

Date	Time	Mean, $\overline{X}$	Range
3/1	11 a.m.	302.4	16
	12 p.m.	298.4	13
	1 p.m.	300.2	10
	2 p.m.	299.0	9
3/5	11 a.m.	300.4	13
	12 p.m.	302.4	5
	1 p.m.	300.3	12
	2 p.m.	299.0	17
3/9	11 a.m.	300.8	18
	12 p.m.	301.5	6
	1 p.m.	301.6	7
	2 p.m.	301.3	8
3/11	11 a.m.	301.7	12
	12 p.m.	303.0	9
	1 p.m.	300.5	9
	2 p.m.	299.3	11
3/16	11 a.m.	300.0	13
	12 p.m.	299.1	8
	1 p.m.	300.1	8
	2 p.m.	303.5	10
3/22	11 a.m.	297.2	14
	12 p.m.	296.2	9
	1 p.m.	297.4	11
	2 p.m.	296.0	12

[a] Data are the average and range of 10 tablets.

make an observation on a new subgroup before the scheduled time, in order to verify the initial result. If two successive averages are outside the acceptable limits, chances are extremely high that a problem exists. An investigation to detect the cause and make a correction may then be initiated.

The procedure for constructing control charts will be illustrated using data on tablet weights as shown in Table 12.1 and Figure 12.2. Note that the $\overline{X}$ chart consists of an "average" or "standard" line along with upper and lower lines that represent the *action* lines. The *average* line may be determined from the history of the product, with regular updating, or may be determined from the product specifications. In this example, the average line is defined by the quality control specifications (standards) for this product, a target value of 300 mg. The *action* lines are constructed to represent ±3 standard deviations from the target value. This is also known as "3σ limits." Observations that lie outside these limits are a cause for action. Adjustments or other corrective action should *not* be implemented if the averages are within the action limits. Tampering with equipment and/or changing other established procedures while the process remains within limits should be avoided. Such interference will often result in increased variation.

In order to establish the *upper* and *lower* limits for the mean ($\overline{X}$), we need an estimate of the standard deviation, if it is not previously known. The standard deviation can be obtained from the replicates (10 tablets) of the subgroup samples that generate the means for the control chart. By pooling the variability from many subgroups ($N = 10$), a very good estimate of the true standard deviation, σ, can be obtained (see App. I). Note that an estimate of the standard deviation or range is needed before limits for the $\overline{X}$ chart can be established. If a "range" chart is used in conjunction with the $\overline{X}$ chart, the upper and lower limits for the $\overline{X}$ chart can be obtained from the range according to Table IV.10 (column A). These factors are derived from theoretical calculations relating the range and standard deviation. For example, in the long run, the range can be shown to be equal to 3.078 times the standard deviation for samples of size 10. If we wish

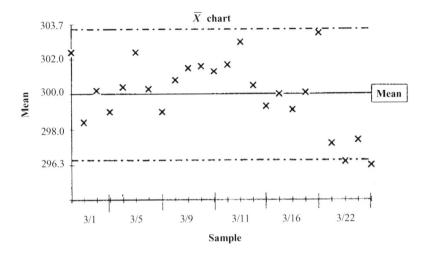

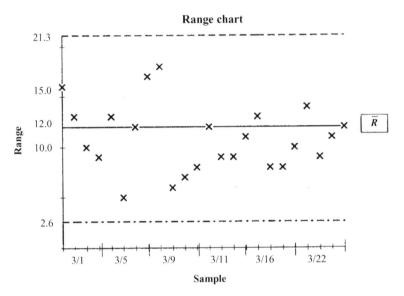

Figure 12.2 Control chart for tablet averages and range data from Table 12.1.

to establish 3σ limits about the *mean* of samples of size 10 (s.d. $= \sigma/\sqrt{10}$) using the range, the following relationship leads to the value 0.31 in Table IV.10 (see column A):

$$\overline{X} \pm \frac{3\sigma}{\sqrt{10}} = \overline{X} \pm \frac{3(\overline{R})}{(3.078)\sqrt{10}} = \overline{X} \pm 0.31\overline{R},$$

where $\overline{R}/3.078$ is the average range divided by 3.078, which on the average is equal to σ. Thus, if the average range is 12 for samples of size 10, the upper and lower control chart limits for $\overline{X}$ are

$$\overline{X} \pm 0.31\overline{R} = \overline{X} \pm 0.31(12) = \overline{X} \pm 3.72. \tag{12.1}$$

Note that the average range is simply the usual average of the range values, obtained in a manner similar to that for calculating the process average. Ranges obtained during the control charting process are averaged and updated as appropriate.

Table IV.10 also has factors for upper and lower limits for a *range chart*. The values in columns D_L and D_U are multiplied times the average range to obtain the lower and upper limits for the range. Usually, a range that exceeds the upper limit is a cause for action. A small value of the range shows good precision and may be disregarded in many situations. In the present example, the average range is set equal to 12 based on previous experience. For samples of size 10, D_L and D_U are 0.22 and 1.78, respectively. Therefore, the lower and upper limits for the range are

Lower limit: $0.22 \times 12 = 2.6$

Upper limit: $1.78 \times 12 = 21.3$. (12.2)

These limits are shown in the control chart for the range in Figure 12.2. See Figure 12.1 for another example of a range chart. Ordinarily, the sample size should be kept constant. If sample size varies from time to time, the limits for the control chart will change according to the sample size. If the sample sizes do not vary greatly, one solution to this problem is use an average sample size [2].

Having established the *mean* and the average *range*, the process is considered to be under control as long as the average and range of the subgroup samples fall within the lower and upper limits. If either the mean or range of a sample falls outside the limits, a possible "assignable" cause is suspected. The reason for the deviation should be investigated and identified, if possible. One should appreciate that a process can change in such a way that (a) only the average is affected, (b) only the variability is affected, or (c) both the average and variability are affected. These possibilities are illustrated in Figure 12.3.

In the example of tablet weights, one might consider the following as possible causes for the results shown in Figure 12.3. A change in average weight may be caused by a misadjustment of the tablet press. Increased variability may be due to some malfunction of one or more punches. Since 10 consecutive tablets are taken for measurement, if one punch gives very low weight tablets, for example, a large variability would result. A combination of lower weight and increased variability probably would be quickly detected if half of the punches were *sticking* in a random manner. Under these circumstances, the average ($\overline{X}$) would be substantially reduced and the range would be substantially increased relative to the values expected under *statistical control*.

The control charts shown in Figure 12.2 are typical. For the $\overline{X}$ chart, the mean was taken as 300 mg based on the target value as set out in the quality control standards. The upper and lower action limits were calculated on the basis of an average range of 12 and factor A in Table IV.10. The lower and upper action limits are 300 ± 3.72 mg or approximately 296.3 to 303.7 mg,

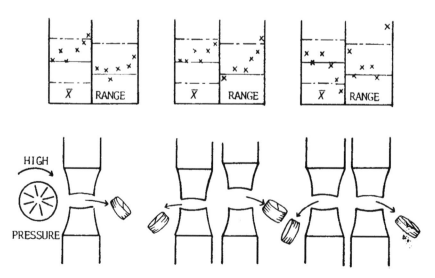

Figure 12.3 Representation of possible process changes as may be detected in a control chart procedure.

respectively. The process is out of control during the production of the batch produced on 3/22. This will be discussed further below. The range control chart shows that the process is in control with respect to this variable.

When the *standard deviation* rather than the range is computed for purposes of constructing control charts, the factors for calculating the limits for the $\overline{X}$ chart are different. The variability is monitored via a chart of the standard deviation of the subgroup rather than the range. Factors for setting limits for both $\overline{X}$ charts and "sigma" (standard deviation) charts may be found in Ref. [1].

If an outlying observation $(\overline{X}, R)$ is eliminated because an assignable cause has been found, that observation should be eliminated from future updating of the $\overline{X}$ and R charts.

12.2.3 Between-Batch Variation as a Measure of Variability (Moving Averages)

The discussion of control charts above dealt with a system that is represented by a regular schedule of production batches. The action limits for $\overline{X}$ were computed using the "within"-batch variation as measured by the variability between items in a "rational subgroup." The subgroup consists of a group of tablets manufactured under very similar conditions. For the manufacture of unit dosage forms with inherent heterogeneity, such as tablets, attempts to construct control charts that include different batches, based on within-subgroup variation, may lead to apparently excessive product failure and frustration. Sometimes, this unfortunate situation may result in the discontinuation of the use of control charts as an impractical statistical device. However, the nature of the manufacture of a heterogeneous mixture, such as the bulk granulations used for manufacturing tablets, lends itself to new sources of uncontrolled error. This error resides in the variability due to the different (uncontrolled) conditions under which different tablet batches are manufactured. One would be hard put to describe exactly why batch-to-batch differences should exist, or to identify the sources of these differences. Perhaps the dies and punches of the tablet press are subject to wear and erosion. Perhaps a new employee involved in the manufacturing process performs the job in a slightly different manner from his or her predecessor. Whatever the reason, such interbatch variation may exist.* In these cases, the within-subgroup variation underestimates the variation, and many readings will appear out of control. This is exemplified by the last batch in Table 12.1 and Figure 12.2.

Thus, when significant interbatch variation exists, the usual control chart will lead to many batches being out of control. If the cause of this variation cannot be identified or controlled, and the product consistently passes the official quality control specifications, other methods than the usual control chart may be used to monitor the process.

Use of the "Control Chart for Individuals" [1,2] seems to be one reasonable approach to monitoring such processes. The limits for the $\overline{X}$ chart are based on a moving range using two consecutive samples (Table 12.2). For example, the first value for the two-batch moving range is the range of batches 1 and 2 = 1.1(399.5 − 398.4). The second moving range is 399.5 − 398.8 = 0.7, and so on. The average moving range is 1.507. The average tablet weight of the 30 batches is 400.01. The average range is based on samples of 2. To estimate the standard deviation from the average range of samples of size 2, it can be shown that we should divide the average range by 1.128 (Table IV.10). The 3 sigma limits are $\overline{X} \pm 3(\overline{R}/1.128) = 400.01 \pm 3(1.507/1.128) = 400.01 \pm 4.01$. The range chart has an upper limit of 3.27(1.507) = 4.93. These charts are shown in Figure 12.4. Batch 13 is out of limits based on both the average and range charts.

The moving average method is another approach to construct control charts that can be useful in the presence of interbatch variation. In this method, we use only a single mean value for each batch, ignoring the individual values within the subgroup, if they are available. Thus, the data consist of a series of means over many batches as shown in Table 12.2. A three-batch moving average consists of averaging the present batch with the two immediately preceding batches. For example, starting with batch 3, the first value for the moving average chart is

$$\frac{398.4 + 399.5 + 398.8}{3} = 398.9.$$

---

* Process validation investigates and identifies such variation.

Table 12.2 Average Weight of 50 Tablets from 30 Batches of a Tablet Product: Example of the Moving Average

Batch	Batch average (mg)	Two-batch moving range	Three-batch moving average	Three-batch moving range
1	398.4	—	—	—
2	399.5	1.1	—	—
3	398.8	0.7	398.9	1.1
4	397.4	1.4	398.6	2.1
5	402.7	5.3	399.6	5.3
6	400.5	2.2	400.2	5.3
7	401.0	0.5	401.4	2.2
8	398.5	2.5	400.0	2.5
9	399.5	1.0	399.7	2.5
10	400.1	0.6	399.4	1.6
11	399.0	1.1	399.5	1.1
12	401.7	2.7	400.3	2.7
13	395.4	6.3	398.7	6.3
14	400.7	5.3	399.3	6.3
15	401.6	0.9	399.2	6.2
16	401.4	0.2	401.2	0.9
17	401.5	0.1	401.5	0.2
18	400.4	1.1	401.1	1.1
19	401.0	0.6	401.0	1.1
20	402.1	1.1	401.2	1.7
21	400.9	1.2	401.3	1.2
22	400.8	0.1	401.3	1.3
23	401.5	0.7	401.1	0.7
24	398.6	2.9	400.3	2.9
25	398.4	0.2	399.5	3.1
26	398.8	0.4	398.6	0.4
27	399.9	1.1	399.0	1.5
28	400.9	1.0	399.9	2.1
29	399.9	1.0	400.2	1.0
30	399.5	0.4	400.1	1.4

The second value is $(399.5 + 398.8 + 397.4)/3 = 398.6$. The calculation is similar to that used for the two-batch moving range in the example of the Control Chart for Individuals. The moving average values are plotted as in the ordinary control chart. Limits for the control chart are established from the moving range, which is calculated in a similar manner. The range of the present and the two immediately preceding batches is calculated for each batch. The average of these ranges is $\overline{R}$, the limits for the control chart are computed from Table IV.10. The computations of the moving average and range for samples of size 3 are shown in Table 12.2, and the data charted in Figure 12.5. The average weight was set at the targeted weight of 400 mg. The average moving range (from Table 12.2) is 2.35. The limits for the moving average chart are determined using the average range and the factor from Table IV.10 for samples of size 3.

$$400 \pm 1.02(2.35) = 400 \pm 2.4.$$

All of the moving average values fall within the limits based on the average moving range. In this analysis, the suspect batch number 13 is "smoothed" out when averaged with its neighboring batches. The upper limit for the range chart is $2.57(2.35) = 6.04$, which would be a cause to investigate the conditions under which batch number 13 was produced (Table 12.2). For further details of the construction and interpretation of moving average charts, see Refs. [1,3].

Another approach to the problem of between-batch variation is the difference chart. A good standard lot is set aside as the control. Each production lot is compared to the standard lot

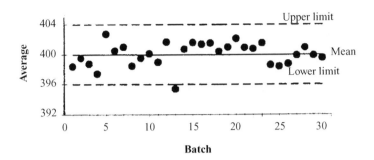

Average chart

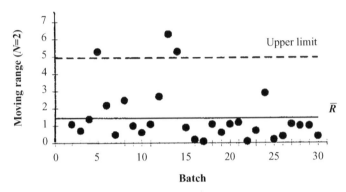

Range chart

Figure 12.4 Control charts for individuals from Table 12.2.

by taking samples of each. Both the control and production lots are measured and the difference of the means is plotted. The limits are computed as

$$0 \pm \frac{3}{\sqrt{n}}\sqrt{S_c^2 + S_p^2},$$

where S_c^2 and S_p^2 are the estimates of the variances of the control and production lots, respectively.

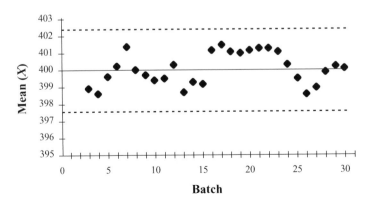

Figure 12.5 Moving average plot for tablet weight means from Table 12.2.

12.2.4 Quality Control Charts in Research and Development

Control charts may be advantageously conceived and used during assay development and validation, in preliminary research or formulation studies, and in routine pharmacological-screening procedures. During the development of assay methodology and validation, for example, by keeping records of assay results, an initial estimate of the assay standard deviation is available. The initial estimate can then be updated as data accumulate.

The following example shows the usefulness of control charts for control measurements in a drug-screening procedure. This test for screening potential anti-inflammatory drugs measures improvement of inflammation (guinea pig paw volume) by test compounds compared to a control treatment. A control chart was established to monitor the performance of the control drug (a) to establish the mean and variability of the control, and (b) to ensure that the results of the control for a given experiment are within reasonable limits (a validation of the assay procedure). The average paw volume difference (paw volume before treatment–paw volume after treatment) and the average range for a series of experiments are shown in Table 12.3. The control chart is shown in Figure 12.6.

As in the control charts for quality control, the mean and average range of the "process" were calculated from previous experiments. In this example, the screen had been run 20 times previous to the data of Table 12.3. These initial data showed a mean paw volume difference of 40 and a mean range ($\overline{R}$) of 9, which were used to construct the control charts shown in Figure 12.6. The subgroups consist of four animals each. Using Table IV.10, the action limits for the $\overline{X}$ and range charts were calculated as follows:

$$\overline{X} \pm 0.73\overline{R} = 40 \pm 0.73(9) = 33.4 \text{ to } 46.6 \,(\overline{X}\text{ chart})$$

$$\overline{R}(2.28) = 9(2.28) = 20.5 \text{ the upper limit for the range.}$$

Note that the lower limit for the range of subgroups consisting of four units is zero. Six of the 20 means are out of limits. Efforts to find a cause for the larger intertest variation failed. The procedures were standardized and followed carefully, and the animals appeared to be homogeneous. Because different shipments of animals were needed to proceed with these tests

Table 12.3 Average Paw Volume Difference and Range for a Screening Procedure (Four Guinea Pigs Per Test Group)

Test number	Mean	Range
1	38	4
2	43	3
3	34	3
4	48	6
5	38	24
6	45	4
7	49	5
8	32	9
9	48	5
10	34	8
11	28	12
12	41	10
13	40	22
14	34	5
15	37	4
16	43	14
17	37	6
18	45	8
19	32	7
20	42	13

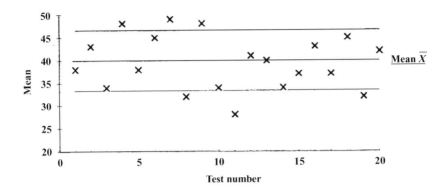

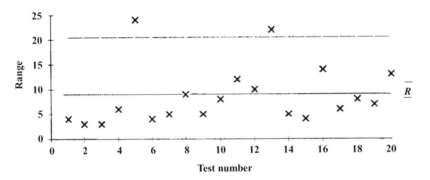

Figure 12.6 Control chart for means and range for control group in a pharmacological-screening procedure.

over time, the researchers felt that there was no way to "tighten up" the procedure. Therefore, as in the tablet weight example discussed in the preceding section, a new control chart was prepared based on the variability between test means. A moving average was recommended using *four* successive averages. Based on historical data, $\overline{X}$ was calculated as 39.7 with an average moving range of 12.5. The limits for the moving average graph are

$$39.7 \pm 0.73(12.5) = 30.6 \text{ to } 48.8.$$

The factor 0.73 is obtained from Table IV.10 for subgroup samples of size 4.

12.2.5 Control Charts for Proportions

Table 12.4 shows quality control data for the inspection of tablets where the measurement is an attribute, a binomial variable. Three hundred tablets are inspected each hour to detect various problems, such as specks, chips, color uniformity, logo, and so on. For this example, the defect

Table 12.4 Proportion of Chipped Tablets of 300 Inspected During Tablet Manufacture

Batch	Time			
	10 a.m.	11 a.m.	12 p.m.	1 p.m.
1	0.060	0.053	0.087	0.055
2	0.073	0.047	0.060	0.047
3	0.040	0.067	0.033	0.053
4	0.033	0.040	0.030	0.027
5	0.040	0.013	0.023	0.040
6	0.025	0.000	0.027	0.013

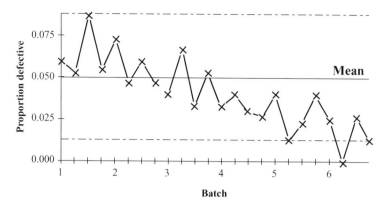

Figure 12.7 Control chart for proportion of tablets chipped.

under consideration is a chipped tablet. According to quality control specifications, this type of defect is considered of minor importance and an average of 5% chipped tablets is tolerable. This problem of chipped tablets was of recent origin, and the control chart was implemented as an aid to the manufacturing and research and development departments, who were looking into the cause of this defect. In fact, the 5% average had been written into the specifications as a result of the persistent appearance of the chipped tablets in recent batches. The data in Table 12.4 represent the first six batches where this attribute was monitored.

For the control chart, 5% defects was set as the average value. The action limits can be calculated from the standard deviation of a binomial. In this example, where 300 tablets were inspected, $N = 300$, $p = 0.05$, and $q = 0.95$ [$\sigma = \sqrt{pq/N}$, Eq. (3.11)].

$$\sigma = \sqrt{\frac{(0.05)(0.95)}{300}} = 0.0126.$$

The limits are $0.05 \pm 3\sigma = 0.05 \pm 3(0.0126) = 0.012$ to 0.088. Proportions below the lower limit indicate an improvement in the process in this example. Note that we can use the normal approximation to the binomial when calculating the 3σ limits, because both NP and Nq are greater than 5 (see sect. 3.4.3). The control chart is shown in Figure 12.7.

The chart clearly shows a trend with time toward less chipping. The problem seems to be lessening. Although no specific cause was found for this problem, increased awareness of the problem among manufacturing personnel may have resulted in more care during the tableting process.

12.2.6 Runs in Control Charts

The most important feature of the control chart is the monitoring of a process based on the average and control limits. In addition, control charts are useful as an aid in detecting trends that could be indicative of a lack of control. This is most easily seen as a long consecutive series of values that are within the control limits but (a) stay above (or below) the average or (b) show a steady increase (or decline). Statistically, such occurrences are described as "runs." For example, a run of 7 successive values that lie above the average constitutes a run of size 7. Such an event is probably not random because if the observed values are from a symmetric distribution and represent random variation about a common mean, the probability of 7 successive values being above the mean is $(1/2)^7 = 1/128$. In fact, the occurrence of such an event is considered to be suggestive of a trend and the process should be carefully watched or investigated.

In general, when looking for runs in a long series of data, the problem is that significant runs will be observed by chance when the process is under control. Nevertheless, with this understanding, it is useful to examine data to be forewarned of the possibility of trends and potential problems. The test for the number of runs above and below the median of a consecutive series of data is described in section 15.7. For the consecutive values 9.54, 9.63, 9.42, 9.86, 9.40, 9.31, 9.79, 9.56, 9.2, 9.8, and 10.1, the median is 9.56. The number of runs above and below the median is 8. According to Table IV.14, this is not an improbable event at the 5% level. If the consecutive values observed were 9.63, 9.86, 9.79, 9.8, 10.1, 9.56, 9.54, 9.42, 9.40, 9.31, and 9.2, the median is till 9.56, but the number of runs is 2. This shows a significant lack of randomness ($p < 0.05$). Also see Exercise Problem 12.

Duncan [2] describes a runs test that looks at the longest run occurring above or below the median. The longest run is compared to the values in Table IV.15. If the longest run is equal to or greater than the table value, the data are considered to be nonrandom. For the data of Table 12.1, starting with the data on the date 3/5 (ignore the data on 3/1 for this example), the median is 300.35. The longest run is 7. There are seven consecutive values above the median starting at 11 a.m. on 3/9. For $N = 20$, the table value in Table IV.15 is 7, and the data are considered to be significantly nonrandom ($p < 0.05$). Note that this test allows a decision of lack of control at the 5% level if a run of 7 is observed in a sequence of 20 observations.

For other examples of the application of the runs test, see Ref. [2]. Also see section 15.7 and Exercise Problem 11 in chapter 15.

In addition to the aforementioned criteria, that is, a point outside the control limits, a significant number of runs, or a single run of sufficient length, other rules of thumb have been suggested to detect lack of control. For example, a run of 2 or 3 outside the 2σ limits but within the 3σ limits, and runs of 4 or 5 between 1σ and 2σ limits can be considered cause for concern.

Cumulative sum control charts (cusum charts) are more sensitive to process changes. However, the implementation, construction, and theory of cusum charts are more complex than the usual Shewhart control chart. Ref. [4] gives a detailed explanation of the use of these control charts.

For more examples of the use of control charts, see chapter 13.

12.3 ACCEPTANCE SAMPLING AND OPERATING CHARACTERISTIC CURVES

Finished products or raw materials (including packaging components) that appear as separate units are inspected or analyzed before release for manufacturing purposes or commercial sale. The sampling and analytical procedures are specified in official standards or compendia (e.g., the USP), or in in-house quality control standards. The quality control procedure known as *acceptance sampling* specifies that a number of items be selected according to a scheduled sampling plan, and be inspected for attributes or quantitatively analyzed. The chief purpose of acceptance sampling is to make a decision regarding the acceptability of the material. Therefore, based on the inspection, a decision is made, such as "the material or lot is either accepted or rejected." Sampling plans for variables (quantitative measurements such as chemical analyses for potency) and attributes (qualitative inspection) are presented in detail in the U.S. government documents MIL-STD-414 and MIL-STD-105E, respectively [3,5].

A single *sampling plan for attributes* is one in which N items are selected at random from the population of such items. Each item is classified as defective or not defective with respect to the presence or absence of the attribute(s). If the sample size is small relative to the population size, this is a binomial process, and the properties of sampling plans for attributes can be derived using the binomial distribution. For example, consider the inspection of finished bottles of tablets for the presence of an intact seal. This is a binomial event; the seal is either intact or it is not intact. The sampling plan states the number of units to be inspected and the number of defects which, if found in the sample, leads to rejection of the lot. A typical plan may call for inspection of 100 items; if two or more are defective, reject the lot (batch). If one or less are defective, accept the lot. (The acceptance number is equal to one.) Theoretically, "100% inspection" will separate the good and defective items (seals in our example). In the absence of 100% inspection, there is no guarantee that the lot will have 0% (or any specified percentage) defects. Thus, underlying any sampling plan are two kinds of risks:

1. The *producer's or manufacturer's risk.* This is the risk or probability of rejecting (not releasing) the product, although it is really good. By "good" we mean that had we inspected every item, the batch would meet the criteria for release or acceptance. This risk reflects an unusually high number of defects appearing in the sample taken for inspection, by chance. The producer's risk can be likened to the α error, that is, rejecting the batch, even though it is good.

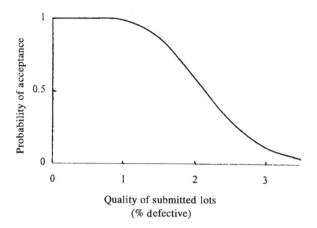

Figure 12.8 Operating characteristic curve for sampling plan N: sample 500 items—accept if 10 or less defective.

2. The *consumer's risk*. This is the probability that the product is considered acceptable (released), although, in truth, it would not be acceptable were it 100% inspected. The consumer's risk can be likened to the β error, that is, the batch is accepted even though it has a more than the acceptable number of defects.

There are any number of possible plans that, in addition to economic considerations, depend on

1. the number of items sampled;
2. the producer's risk;
3. the consumer's risk.

MIL-STD-105E is an excellent compilation of such plans [3]. Each plan gives the number of items to be inspected, and the number of defects in the sample needed to cause rejection of the lot. Each plan is accompanied by an *operating characteristic* (OC) *curve*. The OC curve shows the probability of accepting a lot based on the sampling plan specifications, given the true proportion of defects in the lot. A typical OC curve is shown in Figure 12.8.

The OC curve is a form of power curve (see sect. 6.5). The OC curve in Figure 12.8 is derived from a sampling plan (plan N from MIL-STD-105E) in which 500 items (bottles) are inspected from a lot that contains 30,000 items. If 11 or more items inspected are found to be defective, the lot is rejected. Inspection of Figure 12.8 shows that if the batch truly has 1% defect, the probability of accepting the lot is close to 99% when plan N is implemented. This plan is said to have an *acceptable quality level* (AQL) of 1%. An AQL of 1% means that the consumer will accept most of the product manufactured by the supplier if the level of defects is not greater than 1%, the specified AQL (i.e., 1%). In this example, with the AQL equal to approximately 1%, about 99% of the batches will pass this plan if the percent defects is 1% or less.

The plan actually chosen for a particular product and a particular attribute depends on the lot size and the nature of the attribute. If the presence (or absence) of an attribute (such as the integrity of a seal) is critical, then a stringent plan (a low AQL) should be adopted. If a defect is considered of minor importance, inspection for the presence of a defect can make use of a less stringent plan. MIL-STD-105E describes various plans for different lot (population) sizes, which range from less stringent for minor defects to more stringent for critical defects. These are known as levels of inspection, level I, II, or III. This document also includes criteria for contingencies for switching to more or less tight plans depending on results of prior inspection. A history of poor quality will result in a more stringent sampling plan and vice versa. If 2 of 2, 3, 4 or 5 consecutive lots are rejected, the normal plan is switched to the tightened plan. If five consecutive lots are accepted under the tightened plan, the normal plan is reinstated. If quality remains very good, reduced plans may be administered as described in MIL-STD-105E. The characteristics of the plan are defined by the AQL and the OC curve. For example,

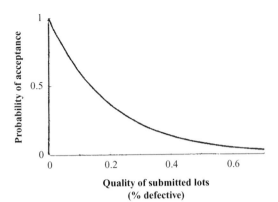

Quality of submitted lots
(% defective)

Figure 12.9 Operating characteristic curve for plan N: AQL = 0.025%.

for lot sizes of 10,001 to 35,000, the following are two of the possible plans recommended by MIL-STD-105E:

Plan	Sample size	Reject number[a] if AQL = 0.4%	1%
K	125	2	4
N	500	6	11

[a] Reject the lot if the number of defects (or more) are observed.

Plan N is a more "discriminating" plan than plan K. The larger sample size results in a greater probability of rejecting lots with more than AQL percentage of defects. For plan N, if there are 2% defects in the lot, the lot will be accepted approximately 57% of the time. For plan K, with 2% defects in the lot, the lot will be accepted 75% of the time. (See MIL-STD-105E [3] for OC curves. The OC curve for an AQL of 1% for plan N is shown in Fig. 12.8.)

In the present example, a defective seal is considered a critical defect and plan N will be implemented with an AQL of 0.025%. This means that lots with 0.025% (25 defects per 100,000 bottles) are considered acceptable. According to MIL-STD-105E, if one or more defects are found in a sample of 500 bottles, the lot is rejected.[†] This means that the lot is passed only if all 500 bottles are good. The OC curve for this plan is shown in Figure 12.9.

The calculations of the probabilities needed to construct the OC curve are not very difficult. These calculations have been presented in the discussion of the binomial distribution in chapter 3. As an illustration, we will calculate the probability of rejecting a lot using plan N with an AQL of 0.025%. As noted above, the lot will be rejected if one or more defects are observed in a sample of 500 items. Thus, the probability of accepting a lot with 0.025% defects is the probability of observing zero defects in a sample of 500. This probability can be calculated from Eq. (3.9)

$$\binom{N}{X} P^X q^{N-X} = \binom{500}{0} P^0 q^{500} = (0.00025)^0 (0.99975)^{500} = 0.88,$$

where 500 is the sample size, P the probability of a defect (0.00025), and q the probability of observing a bottle with an intact seal (0.99975). Thus, using this plan, lots with *0.025% defects will be passed 88% of the time*. A lot with 0.4% (4 defects per 1000 items) will be accepted with a probability of

$$\binom{500}{0} (0.004)^0 (0.996)^{500} = 0.13 \text{ (i.e., 13%)}.$$

[†] If the result of inspection calls for rejection, 100% inspection is a feasible alternative to rejection.

Copies of sampling plans K and N from MIL-STD-105E are shown in Tables 12.5 and 12.6.

In addition to the sampling plans discussed above, MIL-STD-105E also presents *multiple-sampling plans*. These plans use less inspection than single sampling plans, on the average. After the first sampling, one of three decisions may be made:

1. Reject the lot
2. Accept the lot
3. Take another sample

In a double-sampling plan, if a second sample is necessary, the final decision of acceptance or rejection is based on the outcome of the second sample inspection.

The theory underlying acceptance sampling for *variables* is considerably more complex than that for sampling for attributes. In these schemes, actual measurements are taken, such as assay results, dimensions of tablets, weights of tablets, measurement of containers, and so on. Measurements are usually more time consuming and more expensive than the observation of a binomial attribute. However, quantitative measurements are usually considerably less variable. Thus, there is a trade-off between expense and inconvenience, and precision. Many times, there is no choice. Official procedures may specify the type of measurement. Readers interested in plans for variable measurements are referred to MIL-STD-414 [5] and the book, "Quality Control and Industrial Statistics" [2] for details.

12.4 STATISTICAL PROCEDURES IN ASSAY DEVELOPMENT

Statistics can play an important role in assisting the analytical chemist in the development of assay procedures. A subcommittee of PMA (Pharmaceutical Manufacturers Association) statisticians developed a comprehensive scheme for documenting and verifying the equivalence of alternative assay procedures to a standard [6]. The procedure is called the *Greenbriar procedure* (named after the location where the scheme was developed). This approach includes a statistical design that identifies sources of variation such as that due to different days and different analysts. The design also includes a range of concentration of drug. The Greenbriar document emphasizes the importance of a thoughtful experimental design in assay development, a design that will yield data to answer questions raised in the study objectives. The procedure is too detailed to present here. However, for those who are interested, it would be a good exercise to review this document, a good learning experience in statistical application.

For those readers interested in pursuing statistical applications in assay and analytical development, two books, *Statistical Methods for Chemists* by Youden [7] and *The Statistical Analysis of Experimental Data*, by Mandel [8], are recommended. Both of these statisticians had long tenures with the National Bureau of Standards.

In this book, we have presented some applications of regression analysis in analytical methodology (see chaps. 7 and 13). Here, we will discuss the application of sample designs to identify and quantify factors that contribute to assay variability (components of variance).

12.4.1 Components of Variance‡

During the discussion of the one-way ANOVA design (sect. 8.1), we noted that the "between-treatment mean square" is a variance estimate that is composed of two different (and *independent*) variances: (a) that due to variability among units *within* a treatment group, and (b) that due to variability due to differences *between* treatment groups. If treatments are, indeed, identical, the ANOVA calculations are such that observed differences between treatment means will probably be accounted for by the within-treatment variation. In the ANOVA table, the ratio of the between-treatment mean square to the within-treatment mean square ($F = BMS/WMS$) will be approximately equal to 1 on the average when treatments are identical.

WITHIN TREATMENT VARIABILITY   BETWEEN TREATMENT VARIABILITY

Repeat Assays Using the
Same Method   METHOD A   METHOD B

‡ A more advanced topic [16].

Table 12.5   Sample Size Code Letter: *K* (Chart Shows Operating Characteristic Curves for Single Sampling Plans)[a]

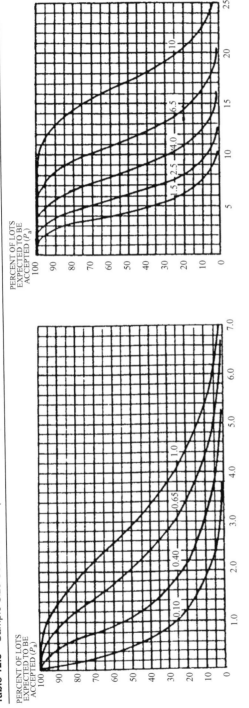

PERCENT OF LOTS EXPECTED TO BE ACCEPTED (P_a)

Tabulated values for operating characteristic curves for single sampling plans

P_a	Acceptable quality levels (normal inspection)											
	0.10	0.40	0.65	1.0	1.5	2.5	X	4.0	X	6.5	X	10
99.0	0.0081	0.119	0.349	0.658	1.43	2.33	2.81	3.82	4.88	5.98	8.28	10.1
95.0	0.0410	0.284	0.654	1.09	2.09	3.19	3.76	4.94	6.15	7.40	9.95	11.9
90.0	0.0840	0.426	0.882	1.40	2.52	3.73	4.35	5.62	6.92	8.24	10.9	13.0
75.0	0.230	0.769	0.382	2.03	3.38	4.77	5.47	6.90	8.34	9.79	12.7	14.9
50.0	0.554	1.34	2.14	2.94	4.54	6.14	6.94	8.53	10.1	11.7	14.9	17.3
25.0	1.11	2.15	3.14	4.09	5.94	7.75	8.64	10.4	12.2	13.9	17.4	20.0
10.0	1.84	3.11	4.26	5.35	7.42	9.42	10.4	12.3	14.2	16.1	19.8	22.5
5.0	2.40	3.80	5.04	6.20	8.41	10.5	11.5	13.6	15.6	17.5	21.4	24.2
1.0	3.68	5.31	6.73	8.04	10.5	12.8	18.3	16.1	18.3	20.4	24.5	27.5
	0.15	0.65	1.0	1.5	2.5	–	4.0	–	6.5	–	10	–
	Acceptable quality levels (tightend inspection)											

Sampling plans for sample size code letter K

Acceptable quality levels (normal inspection) (AQL values, each with Ac = Acceptance number, Re = Rejection number)

Type of sampling plan	Cumulative sample size	Less than 0.10 (▽)	0.10	0.15	0.25	0.40	0.65	1.0	1.5	2.5	—	4.0	—	6.5	—	10	Higher than 10 (△)
Single	125	▽	0 1	*(Use letter J)*	*(Use letter L)*	1 2	2 3	3 4	5 6	7 8	8 9	10 11	12 13	14 15	18 19	21 22	△
Double	80	▽	*			0 2	0 3	1 4	2 5	3 7	5 9	7 11	9 14	11 16			△
	160					1 2	3 4	4 5	6 7	8 9	12 13	18 19	23 24	26 27			
Multiple	32	▽				#	#	#	0 4	0 5	0 5	0 6	1 7	1 7	1 8	2 9	△
	64					0 2	0 3	0 4	0 5	1 7	2 7	3 8	3 9	4 10	6 12	7 14	
	96					0 2	0 3	1 4	2 6	3 9	4 9	6 10	7 12	8 13	11 17	13 19	
	128					0 3	1 4	2 5	3 7	5 11	6 11	8 13	10 15	12 17	16 22	19 25	
	160					1 3	2 5	3 6	5 8	7 12	8 12	11 15	14 17	17 20	22 25	25 29	
	192					1 3	3 5	4 7	7 9	10 13	11 14	14 17	18 20	21 23	27 29	31 33	
	224					2 3	4 5	6 7	9 10	13 14	14 15	18 19	21 22	25 26	32 33	37 38	

Acceptable quality levels (tightened inspection) (corresponding shifted AQL labels): Less than 0.15 | 0.15 | 0.25 | 0.40 | 0.65 | 1.0 | 1.5 | 2.5 | — | 4.0 | — | 6.5 | — | 10 | — | Higher than 10

[a] Quality of submitted lots (p, in percent defective for AQLs $\le$ 10; in defects per hundred units for AQLs > 10). *Note:* Figures on curves are acceptable quality levels (AQLs) for normal inspection. Curves for double and multiple sampling are matched as closely as practicable.

△ = Use next preceding sample size code letter for which acceptance and rejection numbers are available.

▽ = Use next subsequent sample size code letter for which acceptance and rejection numbers are available.

Ac = Acceptance number.

Re = Rejection number.

* = Use single sampling plan above (or alternatively use letter N).

\# = Acceptance not permitted at this sample size.

Table 12.6 Sample Size Code Letter: N (Chart N Shows Operating Characteristic Curves for Single Sampling Plans)[a]

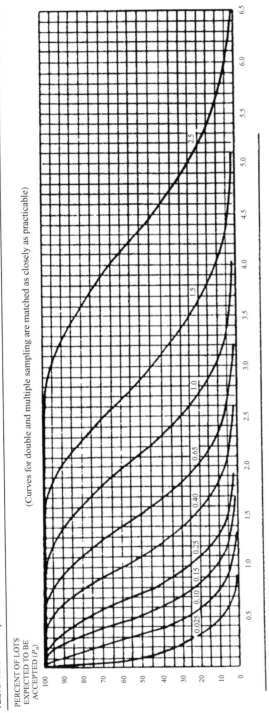

PERCENT OF LOTS EXPECTED TO BE ACCEPTED (P_a)

(Curves for double and multiple sampling are matched as closely as practicable)

Tabulated values for operation characteristic curves for single sampling plans

	Acceptable quality levels (normal inspection)											
P_a	0.025	0.10	0.15	0.25	0.40	0.65	X	1.0	X	1.5	X	2.5
99.0	0.0020	0.030	0.087	0.165	0.357	0.581	0.701	0.954	1.22	1.50	2.07	2.51
95.0	0.0103	0.071	0.164	0.273	0.523	0.796	0.939	1.23	1.54	1.85	2.49	2.98
90.0	0.0120	0.106	0.220	0.349	0.630	0.931	1.09	1.40	1.73	2.06	2.73	3.25
75.0	0.0576	0.192	0.345	0.507	0.844	1.19	1.37	1.72	2.08	2.45	3.18	3.74
50.0	0.139	0.336	0.535	0.734	1.13	1.53	1.73	2.13	2.53	2.93	3.73	4.33
25.0	0.277	0.539	0.784	1.02	1.48	1.94	2.16	2.60	3.04	3.48	4.35	4.99
10.0	0.461	0.778	1.06	1.34	1.86	2.35	2.60	3.08	3.56	4.03	4.95	5.64
5.0	0.599	0.949	1.26	1.55	2.10	2.63	2.89	3.39	3.89	4.38	5.34	6.05
1.0	0.921	1.328	1.68	2.01	2.62	3.20	3.48	4.03	4.56	5.09	6.12	6.87
	0.040	0.15	0.25	0.40	0.65	–	1.0	–	1.5	–	2.5	–
	Acceptable quality levels (tightened inspection)											

Sampling plans for sample size code letter N

Acceptable quality levels (normal inspection)

Values shown as **Ac Re** (acceptance number / rejection number).

Type of sampling plan	Cumulative sample size	Less than 0.025	0.025	0.040	0.065	—	0.10	0.15	0.25	0.40	0.65	—	1.0	—	1.5	—	2.5	Higher than 2.5
Single	500	▽	0 1	Use letter M	Use letter Q	Use letter P	1 2	2 3	3 4	5 6	7 8	8 9	10 11	12 13	14 15	18 19	21 22	△
Double	315	▽	*				0 2	0 3	1 4	2 5	3 7	4 8	5 9	6 10	7 11	9 14	11 16	△
Double	630						1 2	3 4	4 5	6 7	8 9	10 11	12 13	15 16	18 19	23 24	27 27	
Multiple	125	▽	*				# 2	# 2	# 3	# 4	0 4	0 4	0 5	1 6	1 7	2 8	2 9	△
Multiple	250						# 2	0 3	0 3	1 5	1 6	2 7	3 8	3 9	4 10	7 13	7 14	
Multiple	375						0 2	0 3	1 4	2 6	3 8	4 9	6 10	7 12	8 13	13 19	13 19	
Multiple	500						0 3	1 4	2 5	3 7	5 10	6 11	8 13	10 15	12 17	19 25	19 25	
Multiple	625						1 3	2 4	3 6	5 8	7 11	9 12	11 15	14 17	17 20	25 29	25 29	
Multiple	750						1 3	3 5	4 6	7 9	10 12	12 14	14 17	17 20	21 23	31 33	31 33	
Multiple	875						2 3	4 5	6 7	9 10	13 14	14 15	18 19	21 22	25 26	37 38	37 38	

Acceptable quality levels (tightened inspection)

		Less than 0.040	0.040	0.065	0.10	0.15	0.25	0.40	0.65	—	1.0	—	1.5	—	2.5	Higher than 2.5

ª Quality of submitted lots (p, in percent defective for AQLs ⩽ 10; in defects per hundred units for AQLs > 10).
Note: Figures on curves are acceptable quality levels (AQLs) for normal inspection. Curves for double and multiple sampling are matched as closely as practicable.

△ = Use next preceding sample size code letter for which acceptance and rejection numbers are available.

▽ = Use next subsequent sample size code letter for which acceptance and rejection numbers are available.

Ac = Acceptance number.

Re = Rejection number.

* = Use single sampling plan above (or alternatively use letter R).

= Acceptance not permitted at this sample size.

Table 12.7 Design to Analyze Components of Variance for the Tablet Assay

	Tablets (treatment groups)									
	1	2	3	4	5	6	7	8	9	10
Assay	48	49	49	55	48	54	45	47	53	50
Results	51	50	52	55	47	52	49	49	50	51
Mean	49.5	49.5	50.5	55	47.5	53	47	48	51.5	50.5
					Grand average = 50.2					

In certain situations (particularly when treatments are a random effect), one may be less interested in a statistical test of treatment differences, but more interested in separately estimating the variability due to different treatment groups *and* the variability within treatment groups. We will consider an example of a quality control procedure for the assay of finished tablets. Here, we wish to characterize the assay procedure by estimating the sources of variation that make up the variability of the analytical results performed on different, distinct tablets. This variability is composed of two parts: (a) that due to analytical error, and (b) that due to tablet heterogeneity. A oneway ANOVA design such as that shown in Table 12.7 will yield data to answer this objective. In the example shown in the table, 10 tablets are each analyzed in duplicate. Duplicate determinations were obtained by grinding each tablet separately, and then weighing two portions of the ground mixture for assay. The manner in which replicates (duplicates, in this example) are obtained is important, not only in the present situation, but also in most examples of statistical designs. Here we can readily appreciate that analytical error, the variability due to the analytical procedure only, is represented by differences in the analytical results of the two "identical" portions of a homogeneously ground tablet. This variability is represented by the "within" error in the ANOVA table shown in Table 12.8. The "within"-mean square is the pooled variance *within* treatment groups, where a group, in this example, is a single tablet.

The *between-tablet* mean square is an estimate of both *assay* (analytical *error*) and the *variability of drug content in different tablets* (tablet heterogeneity) as noted above. If tablets were identical, individual tablet assays would not be the same because of analytical error. In reality, in addition to analytical error, the drug assay is variable due to the inherent heterogeneity of such dosage forms. Variability between tablet assays is larger than that which can be accounted for by analytical error alone. This is the basis for the F test in the ANOVA [(between-mean square)/(within-mean square)]. Large differences in the drug content of different tablets result in a large value of the between-tablet mean square. This concept is illustrated in Figure 12.10, which shows an example of the distribution of *actual* drug content in a theoretical batch of tablets. The distribution of tablet assays is more spread out than the drug content distribution, because the variation based on the assay results of the different tablets include components due to *actual drug content variation plus assay error*.

Based on the theoretical model for the one-way ANOVA, section 8.1 (random model), it can be shown that the between-mean square is a combination of the assay error and tablet variability as follows:

$$BMS = n\sigma_T^2 + \sigma_w^2, \tag{12.3}$$

where n is the number of replicates in the design (based on equal replication in each group, two assays per tablet in our example), σ_T^2 the variance due to tablet drug content heterogeneity,

Table 12.8 Analysis of Variance for the Tablet Assay Data from Table 12.7

Source	d.f.	SS	MS
Between tablets	9	112.2	12.47
Within tablets	10	27.0	2.70
Total	19	139.2	

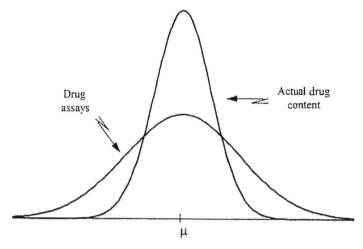

Figure 12.10 Distribution of actual drug content compared to distribution of analytical results of tablets (these are theoretical, hypothetical distributions).

and σ_W^2 is the within-treatment (assay) variance. In our example, $n = 2$, and the between-mean square is an estimate of $2\sigma_T^2 + \sigma_W^2$. The *within*-tablet mean square is an estimate of σ_W^2, equal to 2.70 (Table 12.8). The estimate of σ_T^2 from Eq. (12.3) is $(\text{BMS} - \sigma_W^2)/n$

$$\text{Estimate of } \sigma_T^2 = \frac{\text{between MS} - 2.70}{2} = \frac{12.47 - 2.70}{2} = 4.9.$$

In this manner we have estimated the *two components* of the between-treatment mean square term

$$\sigma_W^2 = 2.7 \text{ and } \sigma_T^2 = 4.9.$$

The purpose of the experiment above, in addition to estimating the components of variance, would often include an estimation of the overall average of drug content based on the 20 assays (Table 12.7). The average assay result is 50.2 mg. The estimates of the variance components can be used to estimate the variance of an average assay result, consisting of m tablets with n assay replicates per tablet. We use the fact that the variance of an average is equal to the variance divided by N, where N is equal to mn, the total number of observations. According to Eq. (12.3), the variance of the average result can be shown to be equal to

$$\frac{n\sigma_T^2 + \sigma_W^2}{mn}. \tag{12.4}$$

The variance estimate of the average assay result (50.2) for the data in Table 12.7, where $m = 10$ and $n = 2$, is

$$\frac{2(4.9) + 2.7}{10(2)} = 0.62.$$

Note that this result is exactly equal to the between-mean square divided by 20.

According to Eq. (12.4), the variance of *single* assays performed on *two* separate tablets, for example, is equal to ($m = 2, n = 1$)

$$\frac{4.9 + 2.7}{2} = 3.8.$$

Note that the variance of a *single assay* of a *single tablet* is $\sigma_T^2 + \sigma_W^2$. Similarly, the variance of the average of two assays performed on a single tablet ($m = 1$, $n = 2$) is $(2\sigma_T^2 + \sigma_W^2)/2$ (see Exercise Problem 11). The former method, where *two tablets* were each *assayed once*, has greater precision than duplicate assays on a single tablet. Given the same number of assays, the procedure that uses more tablets will always have better precision. The "best" combination of the number of tablets and replicate assays will depend on the particular circumstances, and includes time and cost factors. In some situations, it may be expensive or difficult to obtain the experimental material (e.g., obtaining patients in a clinical trial). Sometimes, the actual observation may be easily obtained, but the procedure to prepare the material for observation may be costly or time consuming. In the case of tablet assays, it is conceivable that the grinding of the tablets, dissolving, filtration, and other preliminary treatment of the sample for assay might be more expensive than the assay itself (perhaps automated). In such a case, replicate assays on ground material may be less costly than assaying separate tablets, where each tablet must be crushed and ground, dissolved, and filtered prior to assay. However, such situations are exceptions. Usually, in terms of precision, it is cost effective to average results obtained from different tablets.

The final choice of how many tablets to use and the total number of assays will probably be a compromise depending on the precision desired and cost constraints. The same precision can be obtained by assaying different combinations of numbers of tablets (m) with different numbers of replicate determinations (n) on each tablet. Time-cost considerations can help make the choice. Suppose that we have decided that a sufficient number of assays should be performed so that the variance of the average result is equal to approximately 1.5. In our example, where the variance estimates are $S_T^2 = 4.9$ and $S_W^2 = 2.7$, the average of five single-tablet assays would satisfy this requirement

$$S_T^2 = \frac{4.9 + 2.7}{5} = 1.52.$$

As noted above, the variance of a single-tablet assay is $S_T^2 + S_W^2$. An alternative scheme resulting in a similar variance of the mean result is to assay four tablets, each in duplicate

$$(m = 4, n = 2).$$

$$S_{\overline{X}}^2 = \frac{2(4.9) + 2.7}{8} = 1.56.$$

The latter alternative requires eight assays compared to five assays in the former scheme. However, the latter method uses only four tablets compared to the five tablets in the former procedure. The cost of a tablet would probably not be a major factor with regard to the choice of the alternative procedures. In some cases, the cost of the item being analyzed could be of major importance. In general, for tablet assays, in the presence of a *large assay variation*, if the analytical procedure is automated and the preparation of the tablet for assay is complex and costly, the procedure that uses less tablets with more replicate assays per tablet could be the best choice.

12.4.1.1 Nested Designs

Designs for the estimation of variance components often fall into a class called *nested* or completely hierarchical designs. The example presented above can be extended if we were also interested in ascertaining the variance due to differences in average drug content between *different batches* of tablets. We are now concerned with estimating (a) between-batch variability, (b) between-tablet (within batches) variability, and (c) assay variability. Between-batch variability exists because, despite the fact that the target potency is the same for all batches, the actual mean potency varies due to changing conditions during the manufacture of different batches. This concept has been discussed under the topic of control charts.

A design used to estimate the variance components, including batch variation, is shown in Table 12.9 and Figure 12.11. In this example, four batches are included in the experiment, with

Table 12.9 Nested Design for Determination of Variance Components

Batch	A			B			C			D		
Tablet	1	2	3	1	2	3	1	2	3	1	2	3
	50.6	49.1	51.1	50.1	51.0	50.2	51.4	52.1	51.1	49.0	47.2	48.9
	50.5	48.9	51.1	49.0	50.9	50.0	51.7	52.0	51.9	49.0	47.6	48.5
	50.8	48.5	51.4	49.4	51.6	49.8	51.8	51.4	51.6	48.5	47.6	49.2

ANOVA

Source	d.f.	SS	MS	Expected MS[a]
Between batches	3	48.6875	16.229	$\sigma_W^2 + 3\sigma_T^2 + 9\sigma_B^2$
Between tablets (within batches)	8	17.52	2.190	$\sigma_W^2 + 3\sigma_T^2$
Between assays (within tablets)	24	2.50	0.104	σ_W^2

[a]Coefficient for σ_T^2 = replicate assays; coefficient for σ_B^2 = replicate assays times the number of tablets per batch.

three tablets selected from each batch (tablets nested in batches), and three replicate assays of each tablet (replicate assays nested in tablets). This design allows the estimate of variability due to batch differences, tablet differences, and analytical error. The calculations for the ANOVA will not be detailed (see Ref. [9]) but the arithmetic is straightforward and is analogous to the analysis in the previous example.

The *mean squares* (MS) calculated from the *ANOVA estimate* the *true variances* indicated in the column "expected MS." The coefficients of the variances from the expected mean squares and the estimates of the three "sources" of variation can be used to estimate the components of variance. The variance components, σ_B^2, σ_T^2, and σ_W^2 may be estimated as follows from the mean square and expected mean square columns in Table 12.9.

$$S_W^2 = 0.104$$
$$S_W^2 + 3S_T^2 = 2.190 \quad S_T^2 = 0.695$$
$$S_W^2 + 3S_T^2 + 9S_B^2 = 16.229 \quad S_B^2 = 1.56$$

An estimate of the variance of single-tablet assays randomly performed within a single batch is $S_W^2 + S_T^2 = 0.799$. If tablets are randomly selected from different batches, the variance estimate of single-tablet assays is $S_W^2 + S_T^2 + S_B^2 = 2.36$.

Nested designs should be symmetrical to be easily analyzed and interpreted. The symmetry is reflected by the equal number of tablets from each batch, and the equal number of replicates per tablet. Missing or lost data result in difficulties in estimating the variance components [10].

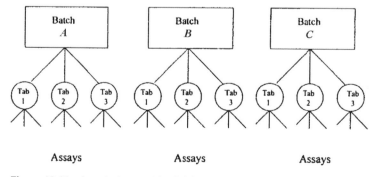

Figure 12.11 A nested or completely hierarchical design to estimate variance components (three of four batches are shown).

12.5 ESTABLISHING IN-HOUSE LIMITS

An important consideration in establishing standards is to evaluate limits for release of products. The two important kinds of release limits are "official" limits, such as stated in the USP or in regulatory submissions, and "in-house" limits that are narrower than the "official" limits. The purpose of in-house limits is to obtain a greater degree of assurance that the true attributes of the product are within official limits when the product is released. Thus, in-house limits decrease the consumer risk. If a product shows measurable decomposition during its shelf life, the in-house release specifications must be more narrow than the official limits to compensate for the product instability.

In the absence of instability, in-house limits should be sufficiently within the official limits to ensure the integrity of the product considering the variability of the measurement (assay). For the case of a *homogeneous sample* (e.g., solutions or a composite sample of a solid dosage form), the variability of the assay may be accounted for by analytical error. An important consideration is to use a proper estimate of the analytical variability. A distinction should be made between within-day variability and between-day variability. For this application, the variability of the analytical method should be estimated as between-day variability. The reason for this is that the variability of an assay on any given day will be dependent on assay conditions on that day, and is apt to be larger than the within-day variability (differences among replicate assays on the same day). For solid dosage forms, the variability of the final assay is a combination of analytical error and tablet heterogeneity (that is, in the absence of analytical error, two separate samples will differ in drug content due to the fact that perfect mixing is not possible in a powder mix). In this case, the estimate of assay variability should not ignore these components of variance. (See discussion of components of variance.)

The examples below show the calculation for a lower limit for in-house release specifications, but the same reasoning will apply for an upper in-house release specification.

$$
\begin{aligned}
\text{LRL} &= \text{Lower official limit} + t \times S \\
\text{LRL} &= \text{Lower release specification}
\end{aligned} \tag{12.5}
$$

For a 95% one-sided confidence interval, t is determined from a t table with d.f. based on the estimate of the assay standard deviation, S. The standard deviation is obtained from between-day replicates during assay development or from a standard product assayed on different days. For tablets, the proper standard deviation should include tablet heterogeneity, that is, replicate assays on different composites. A standard deviation estimated from replicates done on the same day (sometimes estimated from control charts) is not the correct standard deviation.

If, according to SOPs, the assay for release is done in duplicate, one might be tempted to divide the last term in Eq. (12.5) by $\sqrt{2}$. This is not strictly correct because the duplicates refer to within-day variability. If the duplicates were done on two separate days (an unlikely procedure) and on separate composites, then the division by $\sqrt{2}$ would be more correct. If replicates are used for the final assay, one could estimate the correct error if an estimate of the within- and between-day components of variance (based on assay of different composites) is available.

$$
S^2_{total} = \frac{S^2_{between} + S^2_{within}}{n},
$$

where n = number of replicates (separate sets of composites). In this case, the number of d.f. can be estimated using Saterthwaite's (see below) approximation. An alternative way of estimating the s.d., if product heterogeneity is not a factor, is to perform replicate determinations on a standard product over time and compute the s.d. of the average results. Some examples should clarify the procedure.

Example 1. Single assays on a portion of a cough syrup are performed as one of the tests for the release of the product. The assay has a s.d. of 2.1 based on the results of the assay performed on a single stable batch on 15 different occasions (days). From Table IV.4, the value of t with 14 d.f.

for a one-sided 95% confidence interval is 1.76. If the official limits are 90% to 110%, in-house limits of

$$90\% + 1.76 \times 2.1 = 93.7$$
$$110\% - 1.76 \times 2.1 = 106.3$$

mean that if the assay falls within 93.7% and 106.3%, the probability that the true batch mean is out of official specifications (90%–110%) is less than 5%.

Example 2. Single assays on a composite of 20 tablets are performed as one of the tests for the release of a product. During development of the product and the assay, an experimental batch of tablets was assayed on 20 different days (a different composite each day). This assay was identical to the composite assay, a 20 tablet composite. The drug in the dosage form is very stable. The s.d. (19 d.f.) is 2.1. From Table IV.4, the value of t with 19 d.f. for a one-sided 95% confidence interval is 1.73. If the official limits are 90% to 110%, the in-house limits are

$$90\% + 1.73 \times 2.1 \quad \text{and} \quad 110\% - 1.73 \times 2.1$$
$$93.63\% \text{ to } 106.37\%.$$

Example 3. Consider the situation in Example 2 where the assay is performed in duplicate and the average result is reported as a basis for releasing the batch. The duplicate determination is performed on two portions of the same 20 tablet composite on the same day. The variability of the result is a combination of tablet content heterogeneity, and within- and between-day assay variability. Since the same composite is assayed twice, the variance is

$$\frac{[S^2_{\text{tablet heterogeneity}}]}{20} + S^2_{\text{(assay) between}} + \frac{[S^2_{\text{(assay) within}}]}{2}. \tag{12.6}$$

If one considers the first term to be small relative to the last two terms, the s.d. can be computed with estimates of the within- and between-day variance components. These estimates could be obtained from historical data, including data garnered during the assay development. The important point to remember is that the computation is not straightforward because of the need to estimate variance components and the d.f. based on these estimates. Assuming that the between-day variance component of the assay is 0, we could calculate the limits as follows.

Assume that the first two terms in Eq. (12.6) are small and that the assay variability has been estimated based on 15 assays with s.d. = 2.1. The average of duplicate assays on the same composite would have in-house limits of

$$90\% + 1.76 \times 2.1/\sqrt{2} \text{ and } 110\% - 1.76 \times 2.1\sqrt{2}$$
$$92.6 \text{ to } 107.4\%.$$

If the tablet variability, $[S^2_{\text{tablet heterogeneity}}]/20$, is large compared to assay variability (probably a rare occurrence), performing duplicate assays on the same composite will not yield much useful information. In this case, to get more precision, one can assay separate 20 tablet composites (see Exercise Problem 13 at the end of this chapter).

Allen et al. [4] discuss the setting of in-house limits when a product is susceptible to degradation. This situation is complicated by the fact that the in-house limits must now take into consideration an estimate of the rate of degradation with its variability, as well as the variability due to the assay. Obviously, the in-house release limits should be within the official limits. In particular, for the typical case where the slope of the degradation plot is negative, we are concerned with the lower limit. If the official lower limit is 90%, the in-house release limit should be greater than 90% by an amount equal to the estimated amount of drug degraded during the shelf life plus another increment due to assay variability. The following notation is somewhat different from Allen et al., but the equations are otherwise identical. The lower release limit (LRL) can be calculated as shown in Eq. (12.7).

$$LRL = OL - DEGRAD + t \times \left(\frac{S_d^2 + S_a^2}{n} \right)^{1/2} \tag{12.7}$$

where OL is the official lower limit; DEGRAD the predicted amount of degradation during shelf life = average slope of stability regression lines × shelf life; S_d^2 the variance of total degradation = shelf life^2 × S_{slope}^2.

Note: Variance of slope = $S_{y \cdot x}^2 / \sum (X - \overline{X})^2$
Var ($k \times$ variable) = $k^2 \times S^2$ (variable) where k is a constant
S_a^2 = variance of assay
Note: S_a^2 is added because the assay performed at release is variable.

    Another problem in computing the LRL is computation of d.f. for the one-sided 95% t distribution. The problem results from the fact that d.f. are associated with two variance estimates. When combining independent variance estimates, Satterthwaite approximation can be used to estimate the d.f. associated with the combined variance estimate [Eq. (12.8)].
    For the linear combination, L, where

$$L = a_1 S_1^2 + a_2 S_2^2 + \dots$$

the d.f. for L are approximately

$$d.f. = \frac{(a_1 S_1^2 + a_2 S_2^2 + \dots)^2}{(a_1 S_1^2)^2 / v_1 + (a_2 S_2^2)^2 / v_2 + \dots} \tag{12.8}$$

where v_i is d.f. for variance i.
    The following example (from Allen) illustrates the calculation for the release limits.

OL = 90%
Average slope = −0.20%/month
shelf life = 24 months
DEGRAD = −0.20 × 24 = −4.8%
S_a = 1.1%
Standard error of the slope = 0.03%
S_d = 0.03 × 24 = 0.72%
d.f. = 58
t = 1.67
n = 2 (duplicate assays)

    If more than one lot is used for the computation, the lots should not be pooled without a preliminary test. Otherwise, an average slope may be used. In the case of multiple lots, the computations are not as straightforward as illustrated, and statistical assistance may be necessary.
    Note the precautions on the variance of duplicate assays as discussed above.

$$LRL = 90 + 4.8 + 1.67 \times \left(\frac{0.72^2 + 1.1^2}{2} \right)^{1/2}$$
$$= 96.6\%.$$

The lower release specification is set at 96.6%.

12.6 SOME STATISTICAL ASPECTS OF QUALITY AND THE "BARR DECISION"

The science of quality control is largely based on statistical principles, in part because we take small samples and make inferences about the large population (e.g., a batch). Following is a discussion of a few topics that illustrate some statistical ways of looking at data.

What is a good sample size? The FDA often seeks information on the rationale for sample sizes in SOPs. Are we taking enough samples? How many samples should we use for analysis? Actually, this is not an easy question to answer in many cases and that is why the question is asked so often. To answer this question from a statistical point of view, one has to answer a few questions, not all of them easy (chap. 6). For example, we need an estimate of the s.d. and definitions of alpha and beta levels for a given meaningful difference, if the data suggest some comparison.

Often the sample size is fixed based on other considerations such as official specifications. Cost is a major consideration. As an example, consider the composite assay for tablets as one of the QC release criteria. Twenty tablets are assayed to represent a million or more tablets in many cases.

Is this sample large enough? The sample size needed to make such an estimate depends on the precision (s.d.) of the data and the desired precision of the estimate in which we are interested, the mean of the 20 tablets in this case. For the composite assay test, we are required to assay at least 20 tablets. Suppose that tablet variability (RSD) as determined from CU tests is about 3% and the analytical error (RSD) is 1%. Based on this information, we can estimate the variability of the composite assay. The content uniformity variation is due to tablet heterogeneity, which includes weight variation and potency variation, in addition to analytical error.

$$S^2_{content\ uniformity} = S^2_{weight} + S^2_{potency} + S^2_{analytical}.$$

The tablet heterogeneity variance is the content uniformity variance minus the analytical variance.

$$S^2_{potency} + S^2_{weight} = S^2_{content\ uniformity} - S^2_{analytical} = (3)^2 - (1)^2 = 8.$$

We could even estimate the potency variation separately from weight variation if an estimate of weight variation is available (from QC tests for example).

The variability of the average of 20 tablets (without analytical error) is

$$S^2_{composite} = \frac{8}{20} = 0.4.$$

If we assay a mixture of 20 tablets, the variance including analytical error is

$$S^2 = 0.4 + 1 = 1.4$$
$$S = 1.18.$$

Do you think that the average of a randomly selected sample of 20 tablets gives an accurate representation of the batch? We might answer this question by looking at a confidence interval for the average content based on these data. Assume that the analytical error is well established and, for this calculation, 9 d.f. (based on CU data) are reasonable for the *t* value needed for the calculation of the confidence interval. If the observed composite assay is 99.3%, a 95% confidence interval for the true average is

$$99.3\% \pm 2.262 \times 1.18 = 96.6\ to\ 102.0.$$

If this is not satisfactory (too wide), we could reduce the interval width by performing replicate assays of the composite or, perhaps, by using more tablets in the composite. For example, duplicate assays from a single composite may be calculated as follows:

$$S^2 = 0.4 + \frac{1}{2} = 0.9$$
$$S = 0.95.$$

Note that the assay variance is reduced by half, but the variance due to tablet heterogeneity is not changed because we are using the same composite. The confidence interval for the duplicates is

$$99.3 \pm 2.262 \times 0.95 = 97.2\% \text{ to } 101.4\%.$$

Using more than 20 tablets would decrease the CI slightly. If we used 40 tablets with a single assay, the variance would be

$$S^2 = \frac{8}{40} + 1 = 1.2$$

and the CI would be 96.8 to 101.8.

When combining independent variance estimates, Satterthwaite approximation can be used to estimate the d.f. associated with the combined variance estimate. The formula [Eq. (12.8)] is presented in section 12.5

$$\text{d.f.} = \frac{(a_1 S_1^2 + a_2 S_2^2 + \ldots)^2}{(a_1 S_1^2)^2/v_1 + (a_2 S_2^2)^2/v_2 + \ldots}, \tag{12.8}$$

where v_i is d.f. for variance i.
For example, suppose the estimates of variance have the d.f. as follows:

$$S_{\text{analytical}}^2 = 2 \text{ with } 15 \text{ d.f.}$$
$$S_{\text{weight}}^2 = 9 \text{ with } 9 \text{ d.f.}$$
$$S_{\text{potency}}^2 = 1 \text{ with } 6 \text{ d.f.}$$

The d.f. for an estimate of content uniformity are based on the following linear combination:

$$1 \times S_{\text{analytical}}^2 + 1 \times S_{\text{weight}}^2 + 1 \times S_{\text{potency}}^2.$$

From Eq. (12.8),

$$\text{d.f.} = \frac{(9 + 2 + 1)^2}{(4/15 + 81/9 + 1/6)} = 15.3.$$

Estimating the d.f. using this approximation is less good for the differences of variances as compared to the sum of variances.

Example. Limits based on analytical variation are to be set for release of a product. The lower limit is 90%. In-house limits are to be sufficiently above 90% so the probability of an assay being below 90% is less than 0.05. Calculate the release limits where a single assay is done on a composite of 20 tablets. The assay RSD is 3% based on 25 d.f. Tablet heterogeneity (RSD) is estimated as 1% based on 9 d.f.

The estimated variance of the composite assay is ($a_1 = 1/20$, $s_1^2 = 1$, $a_2 = 1$, $S_2^2 = 3$)

$$\frac{1}{20} + 3^2 = 9.05\%$$

d.f. $\approx (3^2 + 1^2/20)^2/[3^2 \times (1/25) + (1/20)^2 \times (1/9)] = 25.3$

Assuming 26 d.f., $t = 1.71$
The lower limit is $90 + 1.71 \times \sqrt{9.05} = 95.1$.
Therefore, the lower in-house limit is 95.1%.

   Blend Samples. What are some properties of three dose weight samples for blend testing? This has been interpreted in different ways, such as (a) take three sample weights and assay the whole sample. (b) Take three sample weights and assay a single dose weight without mixing the sample. (Tread lightly when transferring the sample to the laboratory.) (c) Take three sample weights, mix thoroughly and assay a single sample. Based on the Barr decision [11], the latter (c) appears to be preferable. Some firms have been requested to sample the blend (3 dose weights) and to impose limits of 90% to 110% for each sample. One might ask if this standard is too restrictive, too liberal, or just right? To help evaluate this procedure, consider the case of a firm that assays three samples, each of single dosage weights. How might the above criterion for acceptance compare to that for 10 dosage units in which all must be between 85% and 115%?

   Some approximate calculations to see if the 90% to 110% limits are fair for the blend samples can shed some light on the nature of the specifications. Suppose that the assay is right on, at 100%. Suppose, also, that 99.9% of the tablets in the batch are between the 85% to 115% CU limits. The probability of 10 of 10 tablets passing if each has a probability of 0.999 of passing is (binomial theorem)

$$0.999^{10} = 0.99.$$

   If the tablets are distributed normally, the s.d. is about 4.6. This is based on the fact that a normal distribution with a mean of 100 and a s.d. of 4.6 will have 99.9% of the values between 85 and 115. This same distribution will have 97% of the tablets between 90 and 110. The probability of 3 of three units being between 90 and 110 is

$$0.97^3 = 0.91,$$

which is less than the probability of passing for the final tablet content uniformity test.

   The FDA has recommended that the limits for the blend samples be 90% to 110%. Since the probability of passing the final tablet CU test is 0.99 under these circumstances, the chances of failing the blend uniformity test may not seem fair, unless you believe that the blend should be considerably more uniform than the final tablets.

   What limits would be fair to make this acceptance criterion (3/3 must pass) equivalent to the USP test given the above estimates. Let the probability of passing the blend test $= 0.99$ to make the test equivalent to that for the finished tablets.

$$p^3 = 0.99,$$

where p is the probability of a single blend sample passing.

$$p = 0.9967$$

   That is, to make the probability of passing (3/3) the same as the final CU test, we would assume that 99.67% of the samples should be within limits. Assuming a normal distribution with a RSD of 4.6%, this corresponds to acceptance limits of about 87.5 to 112.5. This would seem fair. However, what are the consequences if the 3-dosage unit weight is composited? In this case, we are assaying the average of 3 tablet weights. These assays should be less variable with a s.d. less than 4.6, the s.d. of single unit weights. Although the variability of the average of 3 dosage weights will be smaller than a single dosage weight, the exact s.d. cannot be defined because the nature of tablet heterogeneity cannot be defined. For the sake of this example, let us assume that the s.d. is 2.66 ($4.6/\sqrt{3}$). Would 90% to 110% be fair limits for each of three blend samples, each consisting of 3 dosage weights? We can compute the probability of a single sample (3 tablet weights) passing using normal distribution theory.

$$Z = 10/2.66 = 3.76$$

probability $(90 < \text{assay} < 110) = 0.99983$.
　　　The probability of three samples passing is

$$0.99983^3 = 0.999.$$

　　　Although this test would be easier to pass than the final tablet content uniformity test, it is based on an assumption of the value of the s.d. for the three unit weight samples, an unknown!

12.7　IMPORTANT QC TESTS FOR FINISHED SOLID DOSAGE FORMS (TABLETS AND CAPSULES)

Important finished solid dosage form tests include

1. content uniformity;
2. assay;
3. dissolution.

　　　In this section, a description of these tests is presented. Included also is the f_2 test for comparing dissolution profiles of two different products with the same active ingredient (as is often done when comparing the dissolution of generic and brand products).

12.7.1　Content Uniformity

The content uniformity is a test to assess and control the variability of solid dosage forms. Although the sampling of the batch for these tests is not specified, good statistical practice recommends some kind of random or representative sampling [12]. This test consists of two stages. For tablets, 30 units are set aside to be tested. In the first stage, individually assay 10 tablets. If all tablets assay between 85% and 115% of label claim and the RSD is less than or equal to 6, the test passes. If the test does not pass, and no tablet is outside 75% or 125%, assay the remaining individual 20 tablets (Stage 2). The test passes if, of the 30 tablets, not more than one tablet is outside 85% to 115% of label claim, no tablet is outside 75% to 125%, and the RSD is less than or equal to 7.8%.

　　　For capsules, the first stage is the same as for tablets, except that one capsule may lie outside of 85% to 115%, but none outside 75% to 125%. The second stage assays 20 more capsules and of the total of 30 capsules, no more than three capsules can be outside 85% to 115%, none outside 75% to 125% and the RSD not more than 7.8%.

12.7.2　Assay

The potency of the final product is based on the average of (at least) 20 dosage units. Twenty random or representative units are ground into a "homogeneous" mix using a suitable method. A sample(s) of this mix is assayed. This assay must be within the limits specified in the USP or a specified regulatory document. Typically, but not always, the assay must be within 90% to 110% of label claim.

12.7.3　Dissolution

The FDA guidance for "Dissolution Testing of Immediate Release Oral Dosage Forms" succinctly describes methods for testing and evaluating dissolution data [13]. Dissolution testing evaluates the dissolution behavior of the drug from a dosage form as a function of time. Thus,

Table 12.10 USP Dissolution Test Acceptance Criteria

Stage	Number tested	Criteria
Stage 1 (S1)	6	Each unit not less than $Q + 5\%$
Stage 2 (S2)	6	Average of 12 units (S1 + S2) equal to or greater than and no unit less than $Q - 15\%$
Stage 3 (S3)	12	Average of 24 units (S1 + S2 + S3) equal to or greater than Q; and not more than 2 units are less than $Q - 15\%$, and no unit is less than $Q - 25\%$

Q is the dissolution specification in percent dissolved.

the typical dissolution-vs.-time curve shows the cumulative dissolution of drug over time. Provided a sufficient quantity of solvent is available, 100% of the drug should be dissolved, given enough time. The procedure for dissolution testing is described in the USP. Briefly, the procedure requires that individual units of the product (for solid dosage forms) be placed in a dissolution apparatus that typically accommodates six separate units. The volume and nature of the dissolution medium is specified (e.g., 900 mL of 0.1 N HCl), and the containers, rotating basket or paddle (USP), are then agitated at a prescribed rate in a water bath at 37°C. Portions of the solution are removed at specified times and analyzed for dissolved drug. Usually, dissolution specifications for immediate-release drugs are determined as a single point in time. Table 12.10 shows the USP Dissolution Test Acceptance Criteria [14], which may be superseded by specifications in individual drug monographs. For controlled-release products and during development, dissolution at multiple time points, resulting in a dissolution profile (Fig. 12.12) is necessary.

The principal purposes of dissolution testing are threefold: (1) for quality control, dissolution testing is one of several tests to ensure the uniformity of product from batch to batch. (2) Dissolution is used to help predict bioavailability for formulation development. For the latter purpose, it is well known that dissolution characteristics may predict the rate and extent of absorption of drugs in some cases, particularly if dissolution is the rate-determining step for drug absorption. Thus, although not always reliable, dissolution is probably the best predictor of bioavailability presently available. (3) Finally, dissolution may be used as a measure of change when formulation or manufacturing changes are made to an existing formulation.

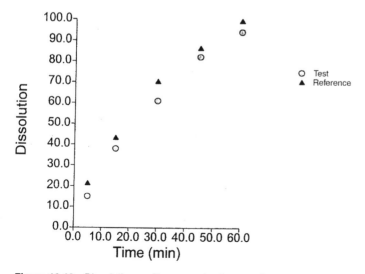

Figure 12.12 Dissolution profile comparing test to reference products.

Table 12.11 Comparison of Test and Reference Dissolution Profiles

| Time (min) | % Dissolution | | |
	Test	Reference	Difference ($R_i - T_i$)
5	15	21	6
15	38	43	5
30	61	70	9
45	82	86	4
60	94	99	5

The so-called f_2 method can be used to compare two dissolution profiles. The formula for the computation of f_2 is as follows:

$$f_2 = 50 \log \left\{ \left[1 + \left(\frac{1}{N} \right) \sum (R_i - T_i)^2 \right]^{-0.5} \times 100 \right\},$$

where N is the number of time points; R_i and T_i are the dissolution of reference and test products at time i.

Consider the following example (Table 12.11 and Fig. 12.12).

$$f_2 = 50 \log \left\{ \left[1 + \frac{1}{N} \sum (R_i - T_i)^2 \right]^{-0.5} \times 100 \right\}$$

$$= 50 \log \left\{ \left[1 + \frac{1}{5} \times (36 + 25 + 81 + 16 + 25) \right]^{-0.5} \times 100 \right\}$$

$$= 50 \log \left\{ \left[1 + \frac{1}{5} \times 183 \right]^{-0.5} \times 100 \right\} = 60.6$$

f_2 must be greater than 50 to show similarity.

f_2 should not be absolute. There are situations where the use of this test does not give results that give reasonable conclusions. For example, with rapidly dissolving drugs, large differences at early time points could result in an f_2 value less than 50 when the dissolution profiles seem to be similar. Also, the method should be used and interpreted with care when few data points are available.

Consider another example (Table 12.12 and Fig. 12.13).

$$f_2 = 50 \log \left\{ \left[1 + \left(\frac{1}{4} \right) (576 + 36 + 9 + 1) \right]^{-0.5} \times 100 \right\} = 45.$$

These products differ only at the very early, and probably variable, time point. Yet, they are not considered similar using this test. As noted, an interpretation of these kinds of data should be made with caution.

Table 12.12 A Second Comparison of Test and Reference Dissolution Profiles

| Time (min) | % Dissolution | |
	Test	Reference
5	51	75
10	89	95
15	93	96
30	97	98

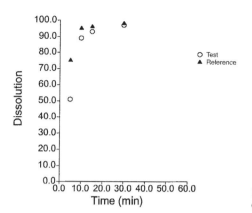

Figure 12.13 Dissolution profile comparing test to reference products for fast dissolving products.

12.8 OUT OF SPECIFICATION (OOS) RESULTS

A discussion of OOS results (failing assay) is presented in Appendices V and VI. These articles were prompted by the Barr decision and FDA's interpretation of Judge Wolin's decision [11]. Since these articles were published, the FDA has published a guidance for "Investigating Out of Specification (OOS) Test Results for Pharmaceutical Production," which addresses these problems and more clearly defines procedures to be followed if an OOS result is observed [15].

The following is a synopsis of the document and comments on topics relevant to this book. All OOS results should be investigated, whether or not the batch is rejected. It is important to find causes that would help maintain the integrity of the product in future batches. The laboratory data should first be inspected for accuracy before any test solutions are discarded. If no errors are apparent, a "complete failure investigation should follow." If an obvious error occurs, the analysis should be aborted, and immediately documented. The thrust of the investigation is to distinguish between a laboratory error and problems with the manufacture of the product. Of course, the optimal procedure would be to have the opportunity to retest the suspect sample if it is still available. In any event, if a laboratory error is verified, the OOS result will be invalidated.

In the laboratory phase of the investigation, various testing procedures are defined. Retesting is a first option if there is not an obvious laboratory error. This is a test of the same sample that yielded the OOS result. For example, for a solution, an aliquot of that same solution may be tested. For a powdered composite, a new weighing from the same composite may be tested. The analysis should be performed by a person other than the one who obtained the OOS result. This retesting could confirm a mishandling of the sample or an instrumental error, for example. The SOPs should define how many assays are necessary to confirm a retesting result. The number of retests should be based on sound scientific and statistical procedures. (See Appendix for an example of a basis for retesting.) However, an OOS result that cannot be documented as a laboratory error, in itself, may not be sufficient to reject the batch. All analytical and other QC results should "be reported and considered in batch release decisions."

Resampling is sampling not from the original sample, but from another portion of the batch. This may be necessary when the original sample is not available or was not prepared properly, for example. These results may further indicate manufacturing problems, or may help verify the OOS result as an anomaly.

The document also discusses averaging (see also App. VII). Averaging is useful when measuring several values from a homogeneous mixture. If the individual results are meant to measure variability, it is clear that averaging without showing individual values is not tolerable. In any event when reporting averages, the individual values should be documented. All of these procedures should be clearly spelled out in the appropriate SOPs. It is of interest that the document discusses the case where three assays yield values of 89, 89, and 92, with a lower limit of 90. Clearly, this should raise some questions, although the FDA document states that this by itself does not necessarily mean that the batch will be failed.

Finally, the FDA does allow the use of outlier tests as long as the procedure is clearly documented in the SOPs. As a final comment, common sense and good scientific judgment are required to make sensible decisions in this controversial area.

KEY TERMS

Acceptance sampling
Action limits
AQL
Batch variation
Between- and within-batch variation
Chance variation
Components of variance
Nested designs
Operating characteristic (OC)
OOS (out of specification)
Power curve
Producer's (manufacturer's) risk
Proportion (p) charts
Range chart
Rational subgroups
Release limits
Resampling

Consumer's risk
Control chart
Control chart for differences
Control chart for individuals
Expected mean square
f_2
Moving average chart
Runs
Sampling for attributes
Sampling for variables
Sampling plan
Statistical control
Upper and lower limits
Warning limits
$\overline{X}$ charts
100% inspection

EXERCISES

1. Duplicate assays are performed on a finished product as part of the quality control procedure. The average of assays over many batches is 9.95 and the average range of the duplicates is 0.10 mg. Calculate upper and lower limits for the $\overline{X}$ chart and the range chart.

2. Past experience has shown the percentage of defective tablets to be 2%. What are the lower and upper 3σ limits for samples of size 1000?

3. A raw material assay shows an average percentage of 47.6% active with an average range of 1.2 based on triplicate assay. Construct a control chart for the mean and range.

4. What is the probability of rejecting a batch of product that truly has 1.0% rejects (defects) if the sampling plan calls for sampling 100 items and rejecting the batch if two or more defects are found?

5. The initial data for the assay of tablets in production runs are as follows (10 tablets per batch):

Batch	Mean	Range
1	10.0	0.3
2	9.8	0.4
3	10.2	0.4
4	10.0	0.2
5	10.1	0.5
6	9.8	0.4
7	9.9	0.2
8	9.9	0.5
9	10.3	0.3
10	10.2	0.6

Construct an $\overline{X}$ and range chart based on this "initial" data. Comment on observations out of limits.

6. A sampling plan for testing sterility of a batch of 100,000 ampuls is as follows. Test 100 ampuls selected at random. If there are no rejects, pass the batch. If there are one or more rejects, reject the batch. If 50 of the 100,000 ampuls are not sterile, what is the probability that the batch will pass?

7. A new method was tried by four analysts in triplicate.[§]

1	2	3	4
115	105	131	129
120	130	152	121
112	106	141	130

Perform an analysis of variance (one-way). Estimate the components of variance (between-analyst and within-analyst variance). What is the variance of the mean assay result if three analysts each perform four assays (a total of 12 assays)? What is the variance if four analysts each perform duplicate assays (a total of eight assays)? If the first analysis by an analyst costs $5 and each subsequent assay by that analyst costs $1, which of the two alternatives is more economical?

8. Construct an $\overline{X}$ chart for the data of Table 12.2, using the moving average procedure. Use the moving average to obtain $\overline{X}$ and $\overline{R}$ for the graph, from the first 15 batches. Plot results for first 15 batches only.

9. Duplicate assays were run for quality control purposes for production batches. The first 10 days of production resulted in the following data: (a) 10.1, 9.8; (2) 9.6, 10.0; (3) 10.0, 10.1; (4) 10.3, 10.3; (5) 10.2, 10.8; (6) 9.3, 9.9; (7) 10.1, 10.1; (8) 10.4, 10.6; (9) 10.9, 11.0; (10) 10.3, 10.4.
   (a) Calculate the mean, average range, and average standard deviation.
   (b) Construct a control chart for the mean and range and plot the data on the chart.

10. What are the lower and upper limits for the range for the example of the moving average discussed at the end of section 12.2.3?

11. What is the variance of the average of duplicate assays performed on the same tablet where the between-tablet variance is 4.9 and the within tablet variance is 2.7? Compare this to the variance of the average of singles assays performed on two different tablets.

12. How did 8 runs arise from the data in the example discussed in section 12.2.5?

13. For an assay that is being used to determine in-house limits, the within- and between-day variances are estimated as 0.3% and 0.5%, respectively. Tablet heterogeneity is 4%. The assay is performed in duplicate on the same day from the same composite.
   (a) Compute the in-house limits if the official specifications are 90% and 110% and there are 25 d.f. for the assay.
   (b) Compute in-house limits if single assays are performed on two different composites on the same day.

REFERENCES

1. Grant EL. Statistical Quality Control, 4th ed. New York: McGraw-Hill, 1974.
2. Duncan AJ. Quality Control and Industrial Statistics, 5th ed. Homewood, IL: Irwin, 1986.
3. U.S. Department of Defense. MIL-STD-105E, Military Sampling Procedures and Tables for Inspections by Attributes. Washington, DC: Superintendent of Documents, U.S. Government Printing Office, 1989.
4. Allen PV, Dukes GR, Gerger ME. Determination of release limits: a general methodology. Pharm Res 1991; 8:1210.
5. U.S. Department of Defense. MIL-STD-414, Sampling Plans. Washington, DC: Superintendent of Documents, U.S. Government Printing Office, 1989.
6. Haynes JD, Pauls J, Platt R. Statistical Aspects of a Laboratory Study for Substantiation of the Validity of an Alternate Assay Procedure, "The Green-Briar Procedure." Final Report of the Standing Committee on Statistics to the PMA/QC Section, March 14, 1977.
7. Youden WJ. Statistical Methods for Chemists. New York: Wiley, 1964.

[§] Optional more difficult problem.

8.   Mandel J. The Statistical Analysis of Experimental Data. New York: Interscience, 1964.
9.   Bennet CA, Franklin NL. Statistical Analysis in Chemistry and the Chemical Industry. New York: Wiley, 1963.
10. Steel R, Torrie J. Principles and Procedures of Statistics. New York: McGraw-Hill Book Co., Inc., 1960.
11. United Stales v. Barr Labs, Inc., Consolidated Docket No. 92–1744 (AMW).
12. Bergum J, Utter M. Process validation. In: Chow S-C, ed. Encyclopedia of Pharmaceutical Statistics. New York: Marcel Dekker, 2000:422–440.
13. US Dept of Health and Human Services, FDA, CDER, 1997.
14. United States Pharmacopeia, 23, 1995 pp. 1791–1793.
15. Investigating Out of Specification (OOS) Test Results for Pharmaceutical Production US Dept of Health and Human Services, FDA, CDER, Sept 1998.
16. Box GE, Hunter WG, Hunter JS. Statistics for Experimenters. New York: Wiley, 1978.

13 | Validation

Although validation of analytical and manufacturing processes has always been important in pharmaceutical quality control, recent emphasis on their documentation by the FDA has resulted in a more careful look at the implementation of validation procedures. The FDA defines process validation as "... a documented program which provides a high degree of assurance that a specific process will consistently produce a product meeting its predetermined specification and quality attributes" [1]. Pharmaceutical process validation consists of well-documented, written procedures ensuring that a specific pharmaceutical technology is capable of and is attaining what is specified in official or in-house specifications, for example, a specified precision and accuracy of an assay procedure or the characteristics of a finished pharmaceutical product. Validation can be categorized as either *prospective* or *retrospective.* Prospective validation should be applied to new drug entities or formulations in anticipation of the product's requirements and expected performance. Berry [2,3] and Nash [4] have reviewed the physical–chemical and pharmaceutical aspects of process validation.

Retrospective validation may be the most convenient and effective way of validating processes for an existing product. Data concerning the key in-process and finished characteristics of an existing product are always available from previously manufactured batches. Usually, there is sufficient information available to demonstrate whether or not the product is being manufactured in a manner that meets the specifications expected of it.

13.1 PROCESS VALIDATION

In order to achieve a proper validation, an in-depth knowledge of the pharmaceutical process is essential. Since the end result of the process is variable (e.g., sterility, potency assay, tablet hardness, dissolution characteristics), statistical input is essential to validation procedures. For example, experimental design and data analysis are integral parts of assay and process validation.

For new products, prospective process validation studies are recommended based on GMP guidelines. Products already marketed may not have been validated for various reasons, for example, products marketed before the formal introduction of validation procedures. Retrospective or concurrent validation methods are used to validate processes for products that have not been validated previously. To recommend specific procedures for validation is difficult because of the variety of products and conditions used during their manufacture. However, there are some common procedures, including issues of sampling, assaying, and statistical analysis, that deserve some discussion.

13.1.1 Retrospective Validation

The GMP guidelines referring to validation [1] suggest that either retrospective or prospective validation may be used to validate a process. Retrospective validation would be applicable for a product that has been on the market for which adequate data are available for evaluation. Although there is no theoretical lower limit on the number of lots needed for such an evaluation, 20 lots have been suggested as an approximate lower limit [5]. In fact, judgment is needed when deciding what constitutes an adequate number of batches. For example, for a product that is made infrequently, or for a product that has an impeccable history of quality, a small number of batches (perhaps 5–10) may be sufficient. Retrospective validation consists of an evaluation of product characteristics over time. The characteristics should consist of attributes that reflect the consistency, accuracy, and safety of the product. For solid dosage forms, the chief characteristics to be evaluated are typically blend uniformity, content uniformity, final assay, dissolution, and hardness (for tablets). The most simple and direct way of evaluating and displaying these characteristics is via control charts (see sect. 12.2.2). Each attribute could be charted giving a visual display of the batch history. All batches that were released to the market should be included. However, if a rejected batch is clearly out of specifications, inclusion in the charting calculations could bias the true nature of the process. Certainly, if not included in the control charts, the absence of such batches should be clearly documented. Again, scientific judgment would be needed to make decisions regarding the inclusion of such batches in the control charts. The control chart not only allows an evaluation of the consistency of the process, but also can be helpful in identifying problems and as an aid in setting practical in-house release limits. Thus, retrospective validation is a useful evaluative procedure, and, representing a relatively large number of batches over a long period of time gives detailed information on the product performance.

13.1.2 Prospective Validation
On the other hand, prospective validation must always be performed for a new product during initial development and production. Usually, the first three production batches are evaluated in great detail in order to demonstrate consistency and accuracy. The important feature of prospective validation is that the attributes measured reflect the important or critical characteristics of the process. This requires a knowledge of the process. Having identified these features, an experimental design and sampling plan that captures the relevant measurements is needed. Each type of dosage form or product is different and may require different considerations.

One should be careful to distinguish process validation from formulation development and optimization. The validation process follows the formulation and processing conditions (such as mixing) "optimization," critical attributes having been evaluated and determined for the manufacturing process. At this point, the question of whether or not the process results in a consistent, reproducible product is the primary concern.

13.1.3 Sampling in Process Validation
Sampling is an important consideration during process validation. The answers to where, when, and how to take samples, as well as sample size and how many samples to be taken, are often not obvious. Judgment is important and no firm rules can be given. For example, during the validation procedure for solid dose forms, samples are taken (1) during the blend stage, (2) when core tablets are produced if applicable, and (3) from the finished tablets. We speak of random samples during these procedures, but, in fact, it is not possible, or practical, to take samples randomly during production. Rather, we try to take samples that are representative of the material being tested. For example, during the blend testing, we sample from the mixer or drums, ensuring that the samples are taken from locations that are representative. Samples taken from a blender or mixer, for purposes of validation, should include areas where good mixing may be suspect because of the geometry of the blender. The number and size of samples to be taken depend on the purpose of the study. For purposes of testing drug content or potency, one or more well selected large size, composited samples may suffice. For purposes of testing uniformity during a validation study, many smaller size samples are necessary. A sample size equal to no more than three dosage units has been suggested in a recent court decision [5], but sample sizes as small as one dosage unit are now becoming routine. Where electrostatic effects may cause the assay of small samples to be biased, single dose weights washed out of the thief, or larger sample sizes may give more reliable results. Although there are no rules for the number of samples to be taken, certainly 10 suitably selected samples should be sufficient when performing time-mix studies to determine the optimal mixing time. Having validated the process, for routine blend testing, assay of three to six samples, selected representatively, should be sufficient. The number of samples, if any, to be taken during production depends on the product and the process. For many products, blend testing may be eliminated for production lots after the process has been validated. A product that has a history of performing well will need less extensive sampling than one that has shown a propensity for problems during its history.

During routine production, if blend assays are indicated, samples are typically taken from drums rather than the mixer, as a matter of convenience. Also, sampling from the drums represents a further step in the process, so that if the blend is satisfactory at this stage, one has more confidence in the integrity of the batch than if samples are taken only from the blender. When drums are tested for blend uniformity and drug content during validation, each drum may be sampled. In addition, some or all of the drums may be sampled more than once; top, middle, and bottom, for example. Some companies sample the first and last drums extensively, from top, middle, and bottom, and the intervening drums only once. Again, as with sampling from the mixer, the size of the sample requires judgment, based on the nature of the product and the objective of the test.

The assays obtained from the drums can be analyzed for drug potency and uniformity. These assays should show a relatively small RSD, so that one has confidence that the RSD of the final product will be within limits, based on content uniformity assays. Although we cannot ensure that every portion of the mix is identical since the product is by nature not uniform, we would hope that the uniformity is good based on RSD requirements for the finished product (less than 6%). For example, if the RSD at an intermediate stage (such as a blend) were greater than 5%, some doubt would exist about the adequate uniformity of the mix.

In addition to blend testing for purposes of process validation or routine QC sampling, sampling for intermediate product testing (e.g., tablet cores) and final product testing is important. Some sampling procedures have been reviewed in chapter 4. Product is usually sampled by selecting units throughout the run from the production lines, by QC personnel. The sample is then submitted for assay as a composite of, for example, tablets over the entire run. This is a form of stratified sampling, tablets being selected every hour during the run, for example. Since final product analysis is usually specified in official documents (content uniformity, dissolution, and a potency composite assay), the number of samples to be analyzed is prespecified. The samples to be analyzed are taken, at random, from the units supplied by the QC department. For validation, additional assays are usually performed to ensure uniformity and compliance. For example, content uniformity and dissolution tests may be performed on dosage units selected from the beginning, middle, and end of the run. Coated products may be tested from different coating pans. For all of these tests, in validation studies, analysis of both average drug potency and uniformity is important. Statistical tests are less useful than statistical estimation in the form of confidence intervals or point estimates. For example, in a validation study, if content uniformity tests are run for tablets at the beginning, middle, and end of the run, we would look at the results from each of the three content uniformity tests to ensure that the RSD was similar and comfortably within the official limits without subjecting the data to a rigorous statistical test.

There are no specific rules. The GMPs and validation guidelines are only recommendations. If a standard procedure is implemented within a company, the procedure should be examined with regard to each product, to ensure that a particular product is not unique in some way that would require a variation in the testing procedure. Careful testing, based on good judgment and science, benefits both the consumer and the manufacturer.

Statistical analysis of the data is useful. However, statistical methods should be used to aid in an understanding of the data only. Hypothesis testing may not be useful, in part because of power considerations. Scientific judgment should prevail.

Several examples will be given with solutions to illustrate the "validation" train of thought. There is no unique statistical approach to any single problem in most practical situations. In validation procedures, in particular, there will be more than one way of attacking a problem. What is most needed is a clear idea of the problem and some common sense. In all of the following examples, statistical methods will be used that have been discussed elsewhere in this book.

Example 1: Retrospective validation. Quality control data are available for an ointment that has been manufactured during a period of approximately one year. The in-process (bulk) product is assayed in triplicate for each batch (top, middle, and bottom of the mixing tank). The finished product consists of either a 2-oz container or a 4-oz container, or both. A single assay is performed on each size of the finished product. The assay results for the eight batches manufactured are shown in Table 13.1.

Table 13.1 Results of Bulk and Finished Tablet Assays of Eight Batches

Batch	In-process bulk material (%)				Finished product (%)		
	Top	Middle	Bottom	Average	2 oz	4 oz	Average
1	105	106	106	105.7	104	101	102.5
2	105	107	103	105.0	108	107	107.5
3	102	109	105	105.3	—	107	107.0
4	105	104	104	104.3	105	107	106.0
5	106	104	107	105.7	107	102	104.5
6	110	108	107	108.3	108	107	107.5
7	103	105	105	104.3	102	104	103
8	108	112	114	111.3	113	—	113
Avg.	105.5	106.9	106.4	106.24	106.7	105	106.38
s.d.	2.56	2.75	3.38	2.40	—	—	3.31

The following questions must be answered to pursue the process validation of this product:

1. Are the assays within limits as stated in the in-house specifications?
2. Do the average results differ for the top, middle, and bottom of the bulk? This can be considered as a measure of drug homogeneity. If the results are (statistically or practically) different in different parts of the bulk container, mixing heterogeneity is indicated.
3. Are the average drug concentration and homogeneity in the bulk mix different from the average concentration and homogeneity of the finished product?
4. Are batches in control based on the charting of averages using control charts?

Answers:

Question 1. The in-house specifications call for an average assay between 100% and 120%. All batches pass based on the average results of both the bulk and finished products. Batch 8 has a relatively high assay, but still falls within the specifications.

   Question 2. A two-way analysis of variance (chap. 8) is used to test for equality of means from the top, middle, and bottom of the bulk container. The average results are shown in Table 13.1, and the ANOVA table is shown in Table 13.2. The F test shows lack of significance at the 5% level, and the product can be considered to be homogeneous. The assays of top, middle, and bottom are treated as *replicate assays* for purposes of determining within-batch variability. (Some statisticians may not recommend a two-step procedure where a preliminary statistical test is used to set the conditions for a subsequent test. However, in this case for purposes of validation in the absence of true replicates, there is little choice.) Note that if the average results of top, middle, and bottom showed significant differences (from both a practical and statistical point of view), a mixing problem would be indicated. This would trigger a study to optimize the mixing procedure and/or equipment to produce a relatively homogeneous product. We understand that a heterogeneous system, as exemplified by an ointment, can never be perfectly homogeneous. The aim is to produce a product that has close to the same concentration of material in each part. From Table 13.2, the within-batch variation is obtained by pooling the between position (top, middle, bottom) sum of squares and the error sum of squares. The within-batch error (variance) estimate is $64.67/16 = 4.04$ with 16 d.f. The standard deviation

Table 13.2 ANOVA for Top, Middle, and Bottom of Bulk

Source	d.f.	SS	MS	F
Batches	7	121.8	17.4	—
Top–middle–bottom	2	7.75	3.88	0.95
Error	14	56.92	4.07	—
Total	23	186.5		

Table 13.3   Paired t Test for Comparison of 2- and 4-oz
Containers (Omit Batches 3 and 8)

Average of 2 oz	=	105.67
Average of 4 oz	=	104.67
t	=	$1/(2.76\sqrt{1/6}) = 0.89$

is the square root of the variance, 2.01. This would be the same error that would have been obtained had we considered this a one-way ANOVA with eight batches and disregarded the "top–middle–bottom" factor. Again, statistical analysis of the data is useful. However, statistics should be used to help understand the data only. Hypothesis testing may not be useful because of power considerations, and scientific judgment should always prevail.

Question 3. The comparison of the variability in the bulk and finished product would be a test of change in homogeneity due to handling from the bulk to the finished product. Although this may not be expected in a viscous, semisolid product such as an ointment, a test to confirm the homogeneity of the finished product should be carried out if possible. In powdered mixes such as may occur in the bulk material for tablets, a disruption of homogeneity during the transformation of bulk material into the final tablet is not an unlikely occurrence. For example, movement of the material in the containers during transport, or vibrations resulting in the settling and sifting of particles in the storage containers prior to tableting, may result in preferential settling of the materials comprising the tablet mix.

In order to compare the within-batch variability of the bulk and finished product, a within-batch error estimate for the finished product is needed. We can use a similar approach to that used for the bulk. Compare the average results for the two different containers (when both sizes are manufactured) and if there is no significant difference, consider the results for the two finished containers as duplicates. The analysis comparing the average results for the 2- and 4-oz containers for the six batches where both were manufactured is shown in Table 13.3. The paired t test shows no significant difference ($p > 0.05$). The within-batch variation is obtained by pooling the error from each of the six pairs, considering each pair a duplicate determination.

$$\text{Within} - \text{mean square} = \frac{(104 - 101)^2 + (108 - 107)^2 + \ldots + (102 - 104)^2}{2 \times 6} = 3.67.$$

This estimate of the within-batch variation is very close to that observed for the bulk material (3.67 vs. 4.04). If a doubt concerning variability exists, a formal statistical test may be performed to compare the within-batch variance of the bulk and finished products for the six batches (F test; sect. 5.3) where estimates of the variability of both the bulk and finished product are available. (We can assume that all variance estimates are independent for purposes of this example.) The results show no evidence of a discrepancy in homogeneity between the bulk and finished product. Although this approach may seem complex and circuitous, in retrospective, undesigned experiments, one often must make do with what is available, making reasonable and rational use of the available data.

The average results of the bulk and finished product can be compared using a paired t test. For this test, we first compute the average result of the bulk and finished material for each batch. The average results are shown in Table 13.1. The t test (Table 13.4) shows no significant difference between the average results of the bulk and finished material.

If either or both of the tests comparing bulk and finished product (average result or variance) show a significant difference, the data should be carefully examined for outliers or

Table 13.4   Paired t Test Comparing the Average of the Bulk
and Finished Product

Average of bulk	=	106.24
Average of finished	=	106.38
t	=	$0.14/1/(2.03\sqrt{1/8}) = 0.20$

Table 13.5 Computations for the Moving Average
Control Chart Shown in Figure 13.2

Moving average ($N = 3$)	Moving range
105.33	0.7
104.87	1
105.1	1.4
106.1	4
106.1	4
107.97	7
Av. 105.91	3.02

obviously erroneous data, or research should be initiated to find the cause. In the present example, where data for only eight batches are available, if the cause is not apparent, further data may be collected as subsequent batches are manufactured to ensure that conclusions based on the eight batches remain valid.

Question 4. A control chart can be constructed to monitor the process based on the data available. This chart is preliminary (only eight batches) and should be updated as new data become available. In fact, after a few more batches are produced, the estimates and comparisons described above should also be repeated to ensure that bulk and finished product assays are behaving normally. The usual Shewhart control chart for averages uses the within-batch variation as the estimate of the variance (see chap. 12). Sometimes in the case of naturally heterogeneous products, such as ointments, tablets, and so on, a source of variation between batches is part of the process that is difficult to eliminate. In these cases, we may wish to use between-batch variation as an estimate of error for construction of the control chart. As long as this approach results in limits that are reasonable in view of official and/or in-house specifications, we may feel secure. However, to be prudent, one would want to find reasons why batches cannot be more closely reproduced. The within-batch variation for the bulk material was estimated as (s.d.) 2.01. A control chart with 3 sigma limits could be set up as $\bar{X} \pm 3 \times 2.01/\sqrt{3} = 106.2 \pm 3.5$ based on the average of the top, middle, and bottom assays. Because of the presence of between-lot variability, a moving average control chart may be appropriate for this data. This chart is constructed from the averages of the three bulk assays for the eight batches (Table 13.1) using a control chart with a moving average of size 3. Table 13.5 shows the calculations for this chart.

For samples of size 3, from Table IV.10, the control chart limits are $105.91 \pm 1.02(3.02) = 105.91 \pm 3.08$. Figures 13.1 and 13.2 show the control charts based on within-batch variation and that based on the moving average. Note that the moving average chart show no out-of-control values and would include batch number 8 within the average control chart limits. The control chart based on within-batch variation finds batch number 8 out of limits.

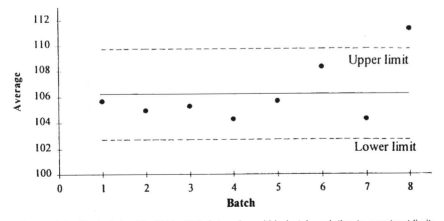

Figure 13.1 Control chart for Table 13.1 data using within-batch variation to construct limits.

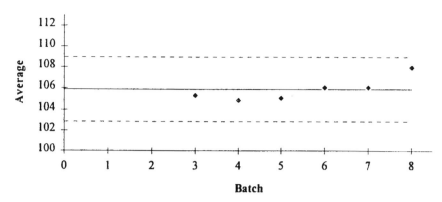

Figure 13.2   Moving average control chart for data of Table 13.1.

Within-batch variation appears to underestimate the inherent variation that includes between-batch variability. Until other sources of variability can be discovered, the moving average chart, which includes between-batch variation, appears to accomplish the objective, that is, to set up a control chart that allows monitoring of the average result of the manufacturing process.

A control chart for the moving range can be constructed using the factor for samples of size 3 in Table IV.10. The upper limit is $2.57 \times 3.02 = 7.76$.

Another control chart of interest is a "range chart" that monitors within-batch variability. If top, middle, and bottom assays are considered to be replicates, we can chart the range of assays within each batch, a monitoring of product homogeneity. The construction of range charts is discussed in chapter 12. Figure 13.3 shows the range chart for the bulk data from Table 13.1 (see also Exercise Problem 1).

A control chart for the finished product is less easily conceived. Different batches may have a different number of assays depending on whether one container or two different size containers are manufactured. There are several alternatives here including the possibility of using (1) separate control charts for the two different sizes, (2) a control chart based on an average result, or (3) a chart with varying limits that depend on the sample size. Note that only a single assay was performed for each finished container. If separate control charts are used for each product, one may wish to consider assays from duplicate containers for each size container so that a range chart to monitor within-batch variability can be constructed. In the present case, limits for the average control chart for the finished product would be wider than that for the bulk average chart since each value is derived from a single (or duplicate) reading rather than the three readings from the bulk. (Note that if within variation is appropriate for construction of the control chart, as may occur with other products, one might use the pooled within variation from both the finished and bulk assays as an estimate of the variance to construct limits.) Exercise Problem 2 asks for the construction of a control chart for finished containers.

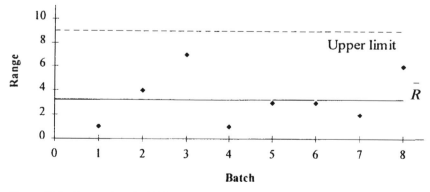

Figure 13.3   Range chart for Table 13.1 data.

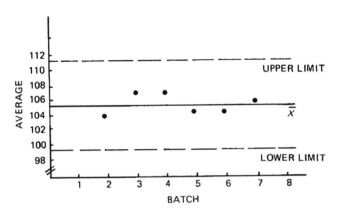

Figure 13.4 Moving average chart for a 4-oz container from Table 13.1 data.

A preliminary control chart using a moving average of size 2 is shown for the 4-oz container in Figure 13.4.

Should any values fall outside the control chart limits, appropriate action should be taken. Refer to the discussion on control charts in chapter 12.

Example 2: An example of a prospective validation study. In this example, a new manufacturing process is just underway for a tablet formulation of either (1) a new drug entity or (2) reformulation of an existing product. Since it would be difficult to generate data from many batches in a reasonable period of time, a recommended procedure is to carefully collect and analyze data from at least three consecutive batches. Of course, this procedure does not negate the necessity of keeping careful in-process and finished-product quality control records to ensure that the quality of the product is maintained.

Prior to the design of the validation procedure, a review of the critical steps in the manufacturing process is necessary. The critical steps will vary from product to product. For the manufacture of tablets, critical steps would include (1) homogeneity and potency after mixing and/or other processes in the preparation of bulk powder prior to tableting, (2) maintenance of homogeneity after storage of the bulk material prior to tableting, (3) the effect of the tableting process on potency as well as other important tablet characteristics such as content uniformity, hardness, friability, disintegration, and dissolution.

In this example, we consider a product in which potency and homogeneity are to be examined as indicators of the validation of the manufacturing process. To this end, both the bulk material and final product are to be tested. We will assume that the critical steps have been identified as (1) the mixing or blending step prior to compression and (2) the manufacture of the finished tablet. Therefore, the product will be sampled both prior to compression in the mixing equipment and after compression, the final manufactured tablet. Three mixing times will be investigated to determine the effect of mixing time on the homogeneity of the mix.

If many variables are considered to be critical, the number of experiments needed to test the effects of these variables may not be feasible from an economic point of view. In these cases, one can restrict the range of many of the variables based on a "knowledge" of their effects from experience. Other options include the use of fractional factorial designs or other experimental screening designs [6].

The question of how many samples to take, as well as where and how to sample is not answered easily. The answer will depend on the nature of the product, the manufacturing procedure, as well as a certain amount of good judgment and common sense. We are interested in taking sufficient samples to answer the questions posed by the validation process.

1. Does the process produce tablets that are uniform?
2. Does the process produce tablets that have the correct potency?
3. Does the variability of the final tablet correspond to the variation in the precompression powdered mix?

Table 13.6 Analysis of Bulk Mix in Blender and Final Tablets

Location	1	2	3	4	5	6
			5 minutes mixing time			
	101	104	101	104	101	109
	93	110	104	100	105	103
	102	106	96	94	99	105
Average	98.7	106.7	100.3	99.3	101.7	105.7
s.d.	4.93	3.06	4.04	5.03	3.06	3.06
			10 minutes mixing time			
	101	105	100	104	99	103
	103	102	99	100	103	104
	103	104	103	101	102	103
Average	102.3	103.7	100.7	101.7	101.3	103.3
s.d.	1.15	1.53	2.08	2.08	2.08	0.58
			20 minutes mixing time			
	102	100	101	99	101	103
	101	102	104	100	101	98
	104	103	100	102	105	102
Average	102.3	101.7	101.7	100.3	102.3	101.0
s.d.	1.53	1.53	2.08	1.53	2.31	2.65

	Final tablets		
	Beginning	Middle	End
	102	99	102
	98	100	103
	103	105	100
	100	101	100
	103	97	104
	103	102	102
	101	98	100
	100	103	97
	99	102	105
	104	100	101
Average	101.3	100.7	101.4
s.d.	2.00	2.41	2.32

Usually, samples are taken directly from the mixing equipment to test for uniformity. Samples may be taken from different parts of the mixer depending on its geometry and potential trouble spots. For example, some mixers, such as the ribbon mixer, are known to have "dead" spots where mixing may not be optimal. Such "dead" spots should be included in the samples to be analyzed. The finished tablets can be sampled at random from the final production batch, or sampled as production proceeds. In the present example, 10 samples (tablets) will be taken at each of the beginning, middle, and end of the tableting process.

Data for the validation of this manufacturing process are shown in Table 13.6. Triplicate assay determinations were made at six different locations in the mixer after 5, 10, and 20 minutes of mixing. In this example, six locations were chosen to represent different parts of the mixture. In other examples, samples may be chosen by a suitable random process. For example, the mixer may be divided into three-dimensional sectors, and samples taken from a suitable number of sectors at random. In the present case, each sample assayed from the bulk mix was approximately the same weight as the finished tablet. During tablet compression, 10 tablets were chosen at three different times in the tablet production run and drug content measured on individual tablets. This procedure was repeated for three successive batches to ensure that the process continued to show good reproducibility. We will discuss the analysis of the results of a single batch.

Table 13.7 ANOVA for Table 13.6

Description	Source	d.f.	MS	F
5-minute mix	Between	5	33.79	2.16
	Within	12	15.67	—
10-minute mix	Between	5	4.10	1.45
	Within	12	2.83	—
20-minute mix	Between	5	1.82	0.46
	Within	12	3.94	—
Tablets	Between	2	1.43	0.28
	Within	27	5.06	—

Analysis of variance can be used to estimate the variability and to test for homogeneity of sample averages from different parts of the blender or from different parts of the production run (Table 13.7).

For the bulk mix, none of the F ratios for between sampling locations mean squares are significant. This suggests that drug is dispersed uniformly to all locations after 5, 10, and 20 minutes of mixing. However, the within-MS is significantly larger in the 5-minute mix compared to the 10- and 20-minute mixes. A test of the equality of variances can be performed using Bartlett's test or a simple F test, whichever is appropriate (see Exercise Problem 3). The data suggest a minimum mixing time of 10 minutes. The homogeneity of the finished tablets is not significantly different from the bulk mixes at 10 and 20 minutes as evidenced by the within-MS error term. The tablet variance is somewhat greater than that in the mix (5.06 compared to 2.83 and 3.94 in the 10- and 20-minute bulk mixes). This may be expected, a result of moving and handling of the mix subsequent to the mixing and prior to the tableting operation.

The average results and homogeneity of the final tablets appear to be adequate. Nevertheless, it would be prudent to continue to monitor the average results and the within variation of both the bulk mix and finished tablet during production batches using appropriate control charts. Again, a moving average chart, where between-batch rather than within-batch variance is the measure of variability, may be necessary in order to keep results for the average chart within limits.

13.2 ASSAY VALIDATION

Validation is an important ingredient in the development and application of analytical methodology for assaying potency of dosage forms or drug in body fluids. Assay validation must demonstrate that the analytical procedure is able to accurately and precisely predict the concentration of unknown samples. This consists of a "documented program which provides a high degree of assurance that the analytical method will consistently result in a recovery and precision within predetermined specifications and limits." To accomplish this, several procedures are usually required. A calibration "curve" is characterized by determining the analytical response (optical density, area, etc.) over a suitable range of known concentrations of drug. Unknown samples are then related to the calibration curve to estimate their concentrations. During the validation procedure, calibration curves may be run in duplicate for several days to determine between- and within-day variation. In most cases, the calibration curve is linear with an intercept close to 0. The proof of the validity of the calibration curve is that known samples, prepared independently of the calibration samples, and in the same form as the unknown samples (tablets, plasma, etc.), show consistently good recovery based on the calibration curve. By "good," we mean that the known samples show both accurate and precise recovery. These known samples are called quality control (QC) samples and are used in both the assay validation and in real studies where truly unknown samples are to be assayed. Typically, the QC samples are prepared in three concentrations that cover the range of concentrations expected in the unknown samples, and are run in duplicate. The QC samples are markers and as long as they show good recovery, the assay is considered to be performing well, as intended.

In general, specific statistical procedures are not recommended by the FDA. This is not necessarily negative as judgment is needed for the many different scenarios that are possible when developing new assays. For example, linearity "should be evaluated by visual inspection." If linearity is accepted, then standard statistical techniques can be applied, such as fitting a regression line by least squares (see sect. 7.5). Transformations to achieve linearity are encouraged. "The correlation coefficient, y-intercept, slope of the regression line and residual sum of squares should be submitted." An analysis of residuals is also recommended.

Some definitions used in assay methodology and validation follow:

Accuracy: Closeness of an analytical procedure result to the true value.

Precision: Closeness of a series of measurements from the same homogeneous sample.

Repeatability: Closeness of results under the same conditions over a short period of time (intra-assay precision).

Interlaboratory (collaborative studies): Studies comparing results from different laboratories. This is not recommended for approval of marketing. This is used more for defining standardization of official assays.

Detection limit: Lowest level that can be detected but not necessarily quantified. The signal-to-noise ratio is used when there is baseline noise. Compare low concentration samples with a blank. "Establish the minimum concentration at which the analyte can be reliably detected." A signal-to-noise ratio of 2/1 or 3/1 is considered acceptable. The detection limit may be expressed as

$$DL = \frac{3.3\,\text{sigma}}{S},$$

where sigma is the s.d. of response and S is the slope of calibration curve.

Quantitation limit (QL): The QL is determined with "known concentrations of analyte, and by establishing the minimum level at which the analyte can be quantified with acceptable accuracy and precision."

A typical calculation of QL is

$$QL = \frac{10\,\text{sigma}}{\text{Slope}}.$$

Good experimental design should be carefully followed in the validation procedure. Careful attention should be paid to the use of proper replicates and statistical analyses. In the following example, the calibration curve consists of five concentrations and is run on three days. Separate solutions are freshly prepared each day for construction of the calibration curve. A large volume of a set of QC samples at three concentrations is prepared from the start to be used throughout the validation and subsequent analyses. A complete validation procedure can be rather complicated in order to cover the many contingencies that may occur to invalidate the assay procedure. In this example, only some of the many possible problems that arise will be presented. The chief purpose of this example is to demonstrate some of the statistical thinking needed when developing and implementing assay validation procedures.

The results of the calibration curves run in duplicate on three days are shown in Table 13.8 and Figure 13.5.

As is typical of analytical data, the variance increases with concentration. For the fitting and analysis of regression lines, a weighted analysis may be used with each value weighted by $1/X^2$, where X is the concentration. For analysis of variance, either a weighted analysis or a log transformation of the data can be used to get rid of the variance heterogeneity (heteroscedasticity). Analyses will be run to characterize the reproducibility and linearity of these data. The calibration lines are at the heart of the analytical procedure as these are used to estimate the unknown samples during biological (e.g., clinical) studies or for QC.

ANOVA: Table 13.9 shows the analysis of variance for the data of Table 13.8 after a log (ln) transformation. The analysis is a three-way ANOVA with factors days (random), replicates (fixed), and concentration (fixed). The two replicates from Table 13.8 are obtained by running all concentrations at the beginning of the day's assays and repeating the procedure at the end of the day. Although the ANOVA for three factors has not been explained in any detail in this

Table 13.8   Calibration Curve Data for Validation (Peak Area)

Day	Concentration	Replicate 1	Replicate 2	Average
1	0.05	0.003	0.004	0.0035
	0.20	0.016	0.018	0.017
	1.00	0.088	0.094	0.092
	10.0	0.920	0.901	0.9105
	20.0	1.859	1.827	1.843
2	0.05	0.006	0.004	0.005
	0.20	0.024	0.020	0.022
	1.00	0.108	0.116	0.112
	10.0	1.009	1.055	1.032
	20.0	2.146	2.098	2.122
3	0.05	0.005	0.008	0.0065
	0.20	0.019	0.023	0.021
	1.00	0.099	0.105	0.102
	10.0	1.000	0.978	0.989
	20.0	1.998	2.038	2.018

book, the interpretation of the ANOVA table follows the same principles presented in chapters 8 and 9, "Analysis of Variance" and "Factorial Designs," respectively.

    The terms of interest in Table 13.9 are replicates and replicate × concentration (BC) interaction. If the assay is performing as expected, neither of these terms should be significant. A significant replicate term indicates that the first replicate is giving consistently higher (or lower) results than the second. This suggests some kind of time trend in the analysis and should be corrected or accounted for in an appropriate manner. A replicate × concentration interaction suggests erroneous data or poor procedure. This interaction may be a result of significant differences between replicates in one direction at some concentrations and opposite differences at other concentrations. For example, if the areas were 1.0 and 1.2 for replicates 1 and 2, respectively, at a concentration of 10.0, and 2.3 and 2.1 at a concentration of 20.0, a significant interaction may be detected. Under ordinary conditions, this interaction is unlikely to occur.

    A least squares fit should be made to the calibration data to check for linearity and outliers. A weighted regression is recommended as noted above (see also sect. 7.7). This analysis is performed if the ANOVA (Table 13.9) shows no problems. A single analysis may be performed

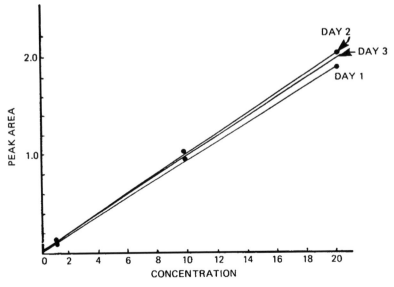

Figure 13.5   Calibration curves from Table 13.8 (weighted least squares fits).

Table 13.9 Analysis of Variance for Calibration Data (Log Transformation)

Source	d.f.	SS	MS	F
Days (A)	2	0.3150	0.1575	—
Replicates (B)	1	0.01436	0.01436	0.36
Concentrations (C)	4	155.78	38.945	1528*
AB	2	0.0803	0.04016	—
AC	8	0.2038	0.0255	—
BC	4	0.0155	0.0387	0.18
ABC	8	0.1742058	0.02178	—
Total	29	156.5834		

$*p < 0.01$.

for all three (days) calibration curves, but experience suggests that calibration curves may often vary from day to day. (This is the reason for the use of QC samples, to check the adequacy of each calibration curve.) In the present case, regression analysis is performed separately for each day's data. Table 13.10 shows the analysis of variance for the weighted least squares fit for the calibration data on day 1 (weight $= 1/X^2$). Each concentration is run in duplicate. The computations for the analysis are lengthy and are not given here. Rather, the interpretation of the ANOVA table (Table 13.10) is more important.

The important feature of the ANOVA is the test of deviations from regression (deviations). This is an F test (deviation-MS/within-MS) with 3 and 5 d.f. The test shows lack of significance (Table 13.10) indicating that the calibration curve can be taken as linear. This is the usual, expected conclusion for analytical procedures. If the F test is significant, the regression plot (Fig. 13.5) should be examined for outliers or other indications that result in nonlinearity (e.g., residual plots, chap. 7). Sometimes, even with a significant F test, examination of the plot will reveal no obvious indication of nonlinearity. This may be due to a very small within-MS error term, for example, and in these cases, the regression may be taken as linear if the other days' regressions show linearity. If curvature is apparent as indicated by inspection of the plot and a significant F test, the data should be fit to a quadratic model, or an appropriate transformation applied to linearize the concentration–response relationship. The test for linearity is discussed further in Appendix II.

A control chart may also be constructed for the slope and intercept of each day's calibration curve, starting with the validation data. This will be useful for detecting trends or outlying data.

A critical step in the assay validation procedure is the analysis of the performance of the QC samples. These samples provide a constant standard from day to day to challenge the validity of the calibration curve. In the simplest case, large volumes of QC samples at three concentrations are prepared to be used both in the validation and in the real studies. The concentrations cover the greater part of the concentration range expected for the unknown samples. The QC samples are run in duplicate (a total of six samples) throughout each day's assays. Usually, the samples will be run at evenly spaced intervals throughout the day with the three concentrations (low, medium, and high) run during the first part of the day and then run again during the latter part of the day. Each set of three should be run in random order. For example, the six QC samples may be interspersed with the unknowns in the following random order:

Medium ... Low ... High ... Low ... High ... Medium

Table 13.10 ANOVA for Regression Analysis for Calibration Data from Day 1

Source	d.f.	SS	MS	F
Slope	1	0.056889	0.056889	1653.0
Error	8	0.000275	0.0000344	—
Deviations from regression	3	0.000004	0.0000013	0.02
Within (duplicates)	5	0.000271	0.0000542	—
Total	9	0.057164		

Slope (weighted regression) $= 0.09153$
Intercept $= -0.00109$

Table 13.11   Data for Quality Control Samples (% Recovery)

Day	Concentration	Replicate 1	Replicate 2	Average
1	0.50	106.5	103.9	105.2
	1.50	97.8	102.4	100.1
	15.0	101.6	97.2	99.4
2	0.50	99.4	107.6	103.5
	1.50	104.0	105.4	104.7
	15.0	96.9	100.7	98.8
3	0.50	97.4	100.2	98.8
	1.50	100.6	99.2	99.9
	15.0	104.2	101.8	103.0

Table 13.11 shows the results for the QC samples, in terms of percent accuracy, during the validation procedure. Percent accuracy is used to help equalize the variances for purposes of the statistical analysis. The first step is to perform an ANOVA for the QC results using all of the data. In this example, the factors in the ANOVA are days (3 days), concentrations (3 concentrations), and replicates (2, beginning of run vs. end of run). The ANOVA table is shown in Table 13.12.

Table 13.12 should not indicate problems if the assay is working as expected. No effect should be significant. A significant replicates effect indicates a trend from the first set of QC samples (beginning of run) to the second set. A significant replicate × concentration interaction is also cause for concern, and the data should be examined for errors, outliers, or other causes. Table 13.12 shows no obvious evidence of assay problems.

To test that the assay is giving close to 100% accuracy, a t test is performed comparing the overall average of all the QC samples versus 100%. This is a two-sided test

$$t = \frac{|\text{Overall average} - 100|}{\sqrt{\text{Days MS}/3}}, \tag{13.1}$$

where 3 = number of days. This is a weak test with only 2 d.f. If no significant effects are obvious in the ANOVA, one may perform the t test on all the data disregarding days and replicates ($N = 18$), and the t test would be

$$t = \frac{|\text{Overall average} - 100|}{\sqrt{S^2/18}}. \tag{13.2}$$

Table 13.12   Analysis of Variance for Quality Control Samples

Source	d.f.	SS	MS	F
Days (A)	2	9.418	4.709	—
Replicates (B)	1	5.556	5.556	0.44
Concentrations (C)	2	13.285	6.642	0.31
AB	2	25.498	12.749	—
AC	4	84.675	21.169	—
BC	2	11.231	5.616	0.73
ABC	4	30.956	7.739	—
Total	17	180.618		

The interpretation of this test should be made with caution because of the assumption of the absence of day, replicate, concentration, and interaction effects. For the data of Table 13.11, the t tests [Eqs. (13.1) and (13.2)] are

$$t = \frac{|101.489 - 100|}{\sqrt{4.709/3}} = 1.188 \tag{13.3}$$

$$t = \frac{|101.489 - 100|}{\sqrt{(180.618/17)/18}} = 1.938. \tag{13.4}$$

We can conclude that the assay is showing close to 100% accuracy. Should the t test show significance, at least one of the three QC concentrations is showing low or high assay accuracy. The data should be examined for errors or outliers, and if necessary, each concentration analyzed separately. The t tests would proceed as above but the data for a single concentration would be used. For the low concentration in Table 13.11, the t test (ignoring the day and replicate effects), would be

$$t = \frac{|102.5 - 100|}{4.12\sqrt{1/6}} = 1.486.$$

To monitor the assay performance, control charts for QC samples may be constructed starting with the results from the validation data. Control charts may be used for each QC concentration separately or, if warranted, all QC concentrations during a day's run can be considered replicates. In the example to follow, we examine the control chart for each QC concentration separately and use the medium concentration as an example. Probably, the best approach is to use a control chart for individuals or a moving average chart (see chap. 12). The validation data cover only three days. Following the validation, data were available for six more days using unknown samples from a clinical study. The data for the medium QC sample from the three validation days and the six clinical study days are shown in Table 13.13.

The average moving range is 2.62 based on samples of size 2. The overall average (of the "average" column in Table 13.13) is 101.17. The 3 sigma limits are $101.17 \pm 3(2.62/1.128) = 101.17 \pm 6.97$. The control chart is shown in Figure 13.6. All the results fall within the control chart limits. Another control chart can be constructed for the range for the duplicate assays performed each day. The average range is 2.38. The upper limit for the range chart is 7.78 (see Exercise Problem 4). As for all control charts, the average and limits should be updated as more data become available.

The control chart for the individual daily averages of the QC samples and the control charts for the slope and intercept, if desired, are used to monitor the process for the analysis of the unknown samples submitted during the clinical studies or for QC. If QC samples fall out of

Table 13.13   Data for Medium QC Sample (Concentration $= 1.50$) for Control Chart

Day	Replicate 1	Replicate 2	Average	Moving range
1	97.8	102.4	100.1	—
2	104.0	105.4	104.7	4.6
3	100.6	99.2	99.9	4.8
4	99.3	97.8	98.55	1.35
5	103.8	101.4	102.6	4.05
6	103.4	103.0	103.2	0.60
7	99.6	102.4	101.0	2.2
8	99.4	103.8	101.6	0.6
9	100.1	97.6	98.85	2.75

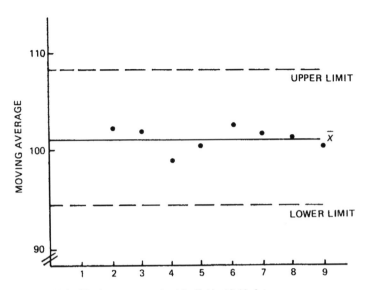

Figure 13.6   Moving average chart for Table 13.13 data.

limits and no obvious errors can be found, the analyses of the samples during that run may be suspect.

A detailed description of analytical validation for blood assays in bioavailability studies has been published by Lang and Bolton [7,8].

For further discussion of assay validation, see Ref. [9].

13.3   CONCLUDING REMARKS

In this chapter, some examples of statistical analysis and design of validation studies have been presented. As we have noted, statistical input is a necessary part of the design and analysis of validation procedures. The statistical procedures that may be used to analyze such data are not limited to the examples given here, but are dependent on the design of the procedures and the characteristics of the data resulting from these experiments. The design of the experiments needed to validate processes will be dependent on the complexity of the process and the identification of critical steps in the process. This is a most important part of validation and the research scientist should be very familiar with the nature of the process, for example, a manufacturing process or assay procedure [1,2,4]. The steps in the validation and statistical analysis are best implemented with the cooperation of a scientist familiar with the physical and chemical processes and a statistician. This is one of the many areas where such a joint venture can greatly facilitate project completion.

KEY TERMS

Assay validation	Process validation
Average control chart	Prospective validation
Calibration curve	Quality control samples
Control chart	Range control chart
Critical steps	Refractive validation
Moving average control chart	Weighted analysis
Moving range control chart	

EXERCISES

1. Construct the range chart using within-batch variation for the bulk material in Table 13.1. Assume that the 3 readings within each batch are true replicates.
2. Construct a moving average chart ($n = 3$) for the 2-oz finished container in Example 2, Table 13.1.

3. Compare the variances during the mixing stage in Example 2 using Bartlett's test. (The variances are estimated from the within-MS terms in the ANOVAs in Table 13.7.)
4. Construct a range chart for the data of Table 13.12. Use the range of the daily duplicates to construct this chart.
5. Construct a control chart for individuals based on the data for three days for the low QC concentration from Table 13.11.

REFERENCES

1. Guideline of General Principles of Process Validation. Rockville, MD: FDA, 1991.
2. Berry IR. Process validation: practical applications to pharmaceutical products. Drug Dev Ind Pharm 1988; 14:377.
3. Berry IR, Nash R. Pharmaceutical Process Validation. New York: Marcel Dekker, 1993.
4. Nash R. In: Lieberman HA, Lachman L, Schwartz J, eds. Pharmaceutical Dosage Forms: Tablets, Vol 3, 2nd ed. New York: Marcel Dekker, 1990.
5. United States v. Barr Labs., Inc., Consolidated Docket No. 92–1744 (AMW) (Court Ruling 2/4/93).
6. Plackett RL, Burman J P. The design of optimum multifactorial experiments. Biometrika 1946; 33:305–325.
7. Lang JR, Bolton S. A comprehensive method validation strategy for bioanalytical applications in the pharmaceutical industry, Part I. J Pharm Biomed Anal 1991; 9:357.
8. Lang JR, Bolton S. A comprehensive method validation strategy for bioanalytical applications in the pharmaceutical industry–2. Statistical analyses. J Pharm Biomed Anal 1991; 9:435.
9. Schofield T. Assay validation. In: Chow S-C, ed. Encyclopedia of Pharmaceutical Statistics. New York: Marcel Dekker, 2000:21–30.

14 | Computer-Intensive Methods

The widespread availability of powerful computers has revolutionized society. The field of statistics is no exception. Twenty years ago, the typical statistician had a collection of tables and charts, without which he or she would have been lost. Today one can perform complex statistical analyses without ever referring to a printed table, relying instead on the statistical functions available in many standard software packages. The ubiquitous availability of personal computing power and sophisticated programming packages permit us to approach statistics today in a less mathematical, more intuitive manner than is possible with the traditional formula-based approach.

The study of statistics by those lacking a strong mathematical background can be a daunting task. The traditional approach usually begins with the introduction of basic probability theory followed by a presentation of standard statistical distributions. To this point, nothing beyond algebra is required. Unfortunately, the progression to real-life problems and the development of inferential methods often involve the derivation of formulas. In many cases, this is accomplished through application of the calculus. The resulting formulas are generally neither intuitive nor simple to comprehend. Too often, the study of statistics is relegated to a process of memorization of these formulas that are then used in cookbook fashion. While the formula-based method of problem solving has an important place in statistics, it is often intimidating to the nonstatistician. For the statistician, this standard approach can become so automatic that the art of data analysis is lost and important characteristics of the data may go unrecognized. Using computer-intensive methods, we approach the solution of statistical problems through a logical application of basic principles applied to a computer-based experiment.

Computer simulations can let us explore the behavior of probability-based processes without becoming overly concerned about the underlying mathematics. When a real-life process can be formulated to follow, or to approximately follow, a known statistical distribution, its characteristics can be explored using Monte Carlo simulation. The Bootstrap Method [1] is a form of computer simulation that is applied to a specific set of data (a sample) without assuming any specific underlying statistical distribution. Bootstrap methods complement standard nonparametric statistical analyses. These are used when we do not know, or do not want to assume, what underlying statistical distribution is operative.

14.1 MONTE CARLO SIMULATION

Monte Carlo simulation enables exploration of complex, probability-based processes, many of which would be difficult to understand by even the most astute statistician using standard formula-based methods. In simulation, the computer performs a large number of experiments, such as the random drawing of balls from an urn, the tossing of a fair or biased coin, or the drawing of random samples from a Normal distribution. Solving a problem using computer simulation involves reducing it to a simple probability-based model, designing a sampling experiment based on the model, and then conducting the experiment, via the computer, a large number of times. The cumulative frequency distribution of the experimental outcomes is viewed as the cumulative probability distribution for the outcomes.

A simple example of how Monte Carlo simulation can be used instead of, or to complement, formula-based methods can be demonstrated using the antibiotic example of chapter 3. In this example, the cure rate for an antibiotic treatment is stated to be 0.75. The question posed is, what is the probability that three of four treated patients will have a cure? The analysis tool add-in of Microsoft Excel provides a convenient way to simulate an answer to this question. To activate this Excel option, if it is not already available in your installation, choose the

Tools option from the Main Menu and then select Add-ins. From the choices in the drop-down Add-ins menu, select (click on) both the Analysis ToolPak and the Analysis ToolPak-VBA. Both choices should show a check mark in their respective boxes.

To answer our antibiotic question, open a new Excel worksheet.

	A	B	C	D	E	F	G
1							
2							
3							
4							
5							
6							
7							
8							

Execute the following commands to simulate 30,000 flips of a biased coin, expected to land on heads 75% of the time and on tails 25% of the time:

From the Main Menu bar, choose Tools.
From the options listed under Tools, choose Data Analysis.
From the Data Analysis options, choose the Random Number Generator.

In the drop-down Dialog Box, enter the following:

For Number of Variables, enter 6.
For Number of Random Numbers, enter 30000.
For Distribution, select Binomial from the choices in the pop-up menu.
Enter 0.75 for the p value and 4 for the Number of Trials.
Enter 12345 for the Random Seed. (Any random number can be used for the seed.)
Click on Output Range and enter A1 in the area to the right of this option.
Click OK to start the simulation.

The commands instructed Excel to generate entries in the cells of the first six columns of the first 30,000 rows of the worksheet. The entry in each of these 180,000 cells represents the simulated number of successes (heads) observed in four independent Bernoulli trials (four flips of a biased coin). The possible outcome of each trial (flip) is either a 0 (tail) or a 1 (head), with the probability of getting a 1 (success) being 0.75 and the probability of getting a 0 (failure) being 0.25. (The coin is biased toward heads.) We might also have flipped a balanced tetrahedron with three sides labeled success and one labeled failure. The possible cell values are 0, 1, 2, 3, or 4 (number of heads in four flips of the coin). The following shows partial results of one simulation and the set of commands used to obtain these results.

	A	B	C	D	E	F	G
1	3	3	3	4	3	4	
2	4	3	3	3	3	3	
3	3	4	3	3	3	3	
4	2	3	4	1	3	2	
5	4	1	4	3	4	4	
6	2	3	3	3	2	4	
7	3	1	3	3	4	4	
8	3	1	1	4	4	3	

Commands in Simulation

Main Menu	Tools → Data Analysis → Random Number Generator

Dialog Box

Number of Variables	Enter 6	Generate 6 variables
Number of Random Numbers	Enter 30,000	Generate 6 variables 30,000 times
Distribution	Select Binomial	Simulate flips of a coin
p Value	Enter 0.75	Coin comes up heads 3 out of 4 flips
Number of Trials	Enter 4	Each variable is # heads in 4 flips
Random Seed	Enter 12345	Can enter desired value here
Output Range	Enter A1	Simulated values in cells A1 − F30000
OK	Click on this to perform the random numbers generation	

We need to determine the proportion of the 180,000 cells that have a simulated value of exactly 3 (three heads from four flips of the coin). The final set of commands to obtain a solution to our question is

Final Commands in Simulation

Into:	*Enter:*	*Result:*
Cell H1	$= IF(A1 = 3,1,0)$	Places a 1 if A1 is a 3, 0 otherwise
Cells I1 through M1	Copy the formula from H1	Determines if 3 heads are in cells B1:F1
Cells H2 through M30000	Copy formulas from row 1 to rows 2 through 30,000	Determines where 3 heads occur in remaining cells
Cell G1	$= AVERAGE(H1: M30000)$	Proportion (probability) of 3 heads in 4 flips

	G	H	I	J	K	L	M
1	0.4235	1	1	1	0	1	0
2		0	1	1	1	1	1
3		1	0	1	1	1	1
4		0	1	0	0	1	0
5		0	0	0	1	0	0

The probability, 0.4235, observed in the simulation, compares favorably to the exact value, 0.4219, calculated using the formula for the binomial expansion. We can increase the accuracy of the simulation estimate by including more columns in the simulation or repeating it a number of times and using the average result of all the simulations. Performing the simulation, with the same seed, 12345, but using 10 columns (variables) instead of 6 gave a probability of 0.4222.

The Central Limit Theorem, used extensively in statistics, indicates that the shape of the distribution of sample means tends toward normality as the sample size increases. This occurs regardless of the underlying statistical distribution from which the sample is drawn. This important concept is not particularly intuitive. Computer simulation is a simple way to demonstrate the impact that the Central Limit Theorem has on the sampling process.

We use Excel to simulate samples drawn from the Uniform distribution, whose shape is markedly different from that of the Normal distribution. In the Uniform distribution, every value has an equal probability of occurrence. A histogram of independent, single samples (sample size of 1) is expected to be represented by a series of bars of equal height (frequency). As a result of the Central Limit Theorem, a histogram of the sample means, where the sample consists of a sufficient number of values drawn from the Uniform distribution, should have a pattern approximating the familiar bell-shaped curve of the Normal distribution. We can show this by performing a Monte Carlo simulation. We simulate the sampling of six values randomly and independently drawn from the Uniform distribution with range 0 to 1. We then determine the mean of the six values in the sample. The sampling is then repeated a large number of times. Histograms are constructed for both the first value from each set of six independent values and for the mean of the six independent values in each sample. The histogram of the single values shows how distinctly different the shape of the Uniform distribution is from that of the Normal distribution. The histogram of the sample mean demonstrates the power of the Central Limit Theorem, even when dealing with a relatively small number of values, only six, sampled from a distribution whose shape is extremely non-normal.

Open an Excel Worksheet and enter the labels shown in row 1

Commands in the Simulation

Main Menu	Tools → Data Analysis → Random Number Generator	
Dialog Box		
Number of Variables	6	Generate 6 variables in each trial
Number of Random Numbers	1000	Generate 1000 trials
Distribution	Uniform	Simulate the Uniform distribution
Between	0 and 1	Distribution range
Random Seed	12345	Can enter a different seed value if desired
Output Range	A2	Simulated values placed in cells A2–F1001
OK		Click to perform simulation
Cell G2	= Average(A2:F2)	Calculate mean of simulated values, trial 1
Cells G3:G1001	Copy G2 formula	Calculate mean for remaining trials
Cells H2:H12	0,0.1,0.2,. . ., 0.9,1.0	Bins for histogram bars
Cells I2:I14	0.20,0.25,. . ., 0.75,0.80	Bins for histogram
Main Menu	Tools → Data Analysis → Histogram	
Dialog Box		
Input Range	A2:A1001	Use variable 1 values
Bin Range	H2:H12	Bin range
New Worksheet Ply	Check this option	
Chart Output	Check this option	
OK	Click to create histogram	
In New Worksheet Click on Histogram Chart		
Main Menu	Chart → Location	
As new sheet	Click this option and enter "Graph 1"	
Double Click on one of the histogram bars		
Options Tab	Click to open	
Gap Width	10	
Sheet 1	Click on this to return to simulation results	

Main Menu   Tools → Data Analysis → Histogram
Dialog Box
  Input Range   G2:G1001   Use mean of the 6 values in
                                                           each trial
  Bin Range   I2:I14   Bin range
  New Worksheet Ply   Check this option
  Chart Output   Check this option
  OK   Click to create histogram

In New Worksheet Ply Click on Histogram Chart
Main Menu   Chart → Location
As new sheet   Click this option and enter
                                             "Graph Mean"

Double Click on a Bar
Options Tab   Click to open
Gap Width   10

	A	B	C	D	E	F	G	H	I
1	Var 1	Var 2	Var 3	Var 4	Var 5	Var 6	Mean	Bins1	Bins Mean
2	0.23	0.58	0.79	0.68	0.18	0.71	0.53	0.0	0.20
3	0.84	0.17	0.89	0.71	0.46	0.45	0.59	0.1	0.25
4	0.62	0.66	0.29	0.43	0.67	0.31	0.50	0.2	0.30
5	0.80	0.12	0.31	0.25	0.34	0.47	0.38	0.3	0.35
6	0.08	0.32	0.65	0.11	0.04	0.84	0.34	0.4	0.40
7	0.65	0.06	0.40	0.97	0.14	0.69	0.48	0.5	0.45
8	0.20	0.64	0.48	0.87	0.46	0.42	0.51	0.6	0.50
9	0.68	0.78	0.38	0.89	0.05	0.23	0.50	0.7	0.55
10	0.09	0.55	0.75	0.86	0.57	0.23	0.51	0.8	0.60
11	0.56	0.22	0.11	0.81	0.11	0.05	0.31	0.9	0.65
12	0.03	0.14	0.81	0.72	0.02	0.28	0.33	1.0	0.70
13	0.85	0.24	0.76	0.54	0.46	0.67	0.59		0.75
14	0.26	0.80	0.86	0.45	0.57	0.44	0.56		0.80

One of the most useful applications of computer simulation is in dealing with a complex probability problem. This can be demonstrated by an example based on FDA's guidance for industry entitled "Bioanalytical Method Validation," May 2001, copies of which are available at http://www.fda.gov/Drugs/GuidanceComplianceRegulatoryInformation/Guidances/ucm064964.htm. The prescribed procedure for monitoring the accuracy and precision of a validated bioanalytical method, in routine use, involves the measurement of quality control (QC) samples, processed in duplicate, at each of three different concentrations. The QC samples are prepared in the same matrix (serum, plasma, blood, urine, etc.) as the samples with unknown concentrations to be analyzed. The three concentration levels of the QC samples cover the working range of the bioanalytical method, one in the lower region, a second at midrange, and the third in the upper region of the standard curve. QC samples are to be analyzed with each batch run of unknown samples. The run is acceptable if at least four of the six QC sample values are within 20% of their nominal concentrations. Two of the six samples may be outside the ± 20% acceptance region, but not two at the same concentration level.

Assume that we have QC levels at 10, 250, and 750 ng/mL. Our assay method has a 15% CV (% relative standard deviation) over its entire working range. What proportion of batch runs do we expect to reject when the assay is running as validated? Also assume that we have accurately prepared our QC samples and that any deviations in their assayed values are random errors that follow a Normal distribution (i.e., the mean deviation = 0%, standard deviation = CV% of the assay).

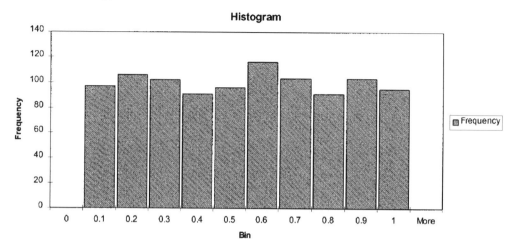

Histogram of simulated values (sample of size 1) from the Uniform distribution.

The traditional formula-based calculations rely on the known properties of the Normal and Binomial distributions. The probability that a single QC value will be within the acceptance region is equal to the proportion of the Standard Normal distribution, which lies between $-Z$ and $+Z$, where $Z = 20\%/CV\%$. With a CV% equal to 15%, the probability that any *single* QC value will be acceptable is the proportion of the Standard Normal distribution that lies between Z values of -1.33 and $+1.33$, or $p = 0.8176$. [$Z = (X - \mu)/\sigma = (20 - 0)/15 = 1.33$; see chap. 3]. The probability that a single QC value will fail to be accepted is $1 - P$ or $q = 0.1824$. The batch run is acceptable if all six QC values pass the criteria, five of six pass, or four of six pass. According to the binomial expansion, this probability is $p^6q^0 + 6p^5q^1 + 15p^4q^2$ (see chap. 3). However, three of the 15 ways that four QC values pass involve two failures at the same concentration level. This is not permitted by the FDA acceptance criteria. Therefore, this reduces the 15 possible ways of 4 QC values passing to 12 ways. The probability of run acceptance, based on the QC results, is $p^6 + 6p^5q^1 + 12p^4q^2$, or 0.88. We expect that 12% of our runs ($1 - 0.88 = 0.12$) will fail due to random error alone.

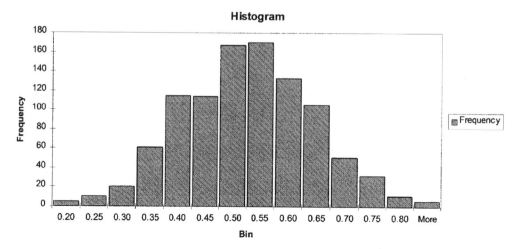

Histogram of the sample mean (n = 6) simulated from the Uniform distribution.

The simulation to evaluate this same question is easily accomplished using Excel. Open an Excel Worksheet and place the labels in the cells as shown in row 1.

	A	B	C	D	E	F	G
1	QC1	QC1	QC2	QC2	QC3	QC3	Prob. Pass
2							
3							
4							
5							

Commands in Simulation

Main Menu	Tools → Data Analysis → Random Number Generator	
Dialog Box		
Number of Variables	6	Generate 6 QC values for each run (row)
Number of Random Numbers	5000	Generate the values for 5000 runs
Distribution	Normal	Sample is from the Normal Dist.
Mean	0	True QC deviation is 0% (100% accurate)
Standard Deviation	15	CV for QC deviation is 15%
Random Seed	Enter 12345	Can enter a different seed if desired
Output Range	Enter A2	Variable values in cells A2 − F5001
Click OK		
Cell H2	= IF(ABS(A2) < 20,1,0)	If QC1 deviation is < 20%, it passes (1)
Cells I2, H3:I5001	Copy H2 formula	Evaluates remaining QC1 values
Cell J2	= IF((H2 + I2) >0,1,0)	QC1 passes (1) if either replicate passes
Cells J3:J5001	Copy J2 formula	Evaluates runs 2–5000 for QC1 passing
Cell K2	= IF(ABS(C2)<20,1,0)	If QC2 deviation is <20%, it passes (1)
Cells L2, K3:L5001	Copy K2 formula	Evaluates remaining QC2 values
Cell M2	= IF((K2 + L2)>0,1,0)	QC2 passes (1) if either replicate passes
Cells M3:M5001	Copy M2 formula	Evaluates runs 2–5000 for QC2 passing
Cell N2	= IF(ABS(E2)<20,1,0)	If QC3 deviation is <20%, it passes (1)
Cells O2, N3:O5001	Copy N2 formula	Evaluates remaining QC3 values
Cell P2	= IF((N2 + O2)>0,1,0)	QC3 passes (1) if either replicate passes
Cells P3:P5001	Copy P2 formula	Evaluatea runs 2–5000 for QC3 passing
Cell Q2	= J2*M2*P2	Flag is 1 if each QC level passes
Cells Q3:Q5001	Copy Q2 formula	Evaluates runs 2–5000 for passing each level
Cell R2	= IF((H2 + I2 + K2 + L2 + N2 + O2) >3,1,0	Flag is 1 if ≥4 QC passing

Cells R3:R5001	Copy R2 formula	Evaluates runs 2–5000 for ≥ 4 passing
Cell S2	= Q2*R2	Flag value is 1 if all QC criteria are met
Cells S3:S5001	Copy S2 formula	Evaluates runs 2—5000 for meeting criteria
Cell G1	= Average(S2:S5001)	Probability of run passing (here, 0.8754)

	A	B	C	D	E	F	G
1	QC1	QC1	QC2	QC2	QC3	QC3	Prob. Pass
2	−11.01	3.22	11.95	6.82	−13.85	8.49	0.8754
3	14.83	−14.60	18.18	8.49	−1.64	−1.82	
4	4.74	6.32	−8.26	−2.66	6.61	−7.36	
5	12.40	−17.59	−7.36	−9.93	−6.22	−1.26	
6	−21.03	−7.02	5.91	−18.44	−26.17	15.19	
7	5.74	−22.78	−3.95	28.12	−16.39	7.45	
8	−12.51	5.25	−0.90	16.61	−1.51	−2.88	
9	7.18	11.53	−4.59	18.02	−25.29	−11.08	
10	−20.20	1.84	9.95	16.37	2.66	−11.17	

In this simulation of 5000 runs, 87.5% passed (probability = 0.8754) and 12.5% failed. These results are in close agreement with the theoretical values of 88% passing and 12% failing.

In the QC example, it would have been easier to apply the normal and binomial formulas rather than conducting the Excel simulation to answer our question. Had we wanted to investigate a more complex and perhaps more realistic situation, a simulation approach might be far simpler, and considerably more intuitive, than the formula-based approach. As an example, consider the situation where the standard deviation is 18% at the lowest concentration QC, 15% at the next higher concentration, and only 12% at the highest concentration. In addition, if the highest QC value exceeds the highest standard curve concentration, it cannot be reported so it is considered a failing value. It would be difficult to deal with this using the formula-based approach, but only marginally more difficult than our previous example if solved by simulation. The more complicated (realistic) our scenario, the more likely it is that computer simulation will prove to be the easier methodology to implement.

	H	I	J	K	L	M	N	O	P	Q	R	S
1	P1_1	P1_2	Pass1	P2_1	P2_2	Pass2	P3_1	P3_2	Pass3	No 2	Pass4	Run
2	1	1	1	1	1	1	1	1	1	1	1	1
3	1	1	1	1	1	1	1	1	1	1	1	1
4	1	1	1	1	1	1	1	1	1	1	1	1
5	1	1	1	1	1	1	1	1	1	1	1	1
6	0	1	1	1	1	1	0	1	1	1	1	1
7	1	0	1	1	0	1	1	1	1	1	1	1
8	1	1	1	1	1	1	1	1	1	1	1	1
9	1	1	1	1	1	1	0	1	1	1	1	1
10	0	1	1	1	1	1	1	1	1	1	1	1

Monte Carlo simulation also offers an intuitive approach to hypothesis testing. In Table 5.9 of Chapter 5, the percent dissolutions after 15 minutes for two different tablet formulations, *A* and *B*, are listed. The distributions of the mean values for the two samples (10 values for each formulation) are assumed to follow Normal distributions. Is the average dissolution of formulation A at 15 minutes different from that of formulation B? The formula-based approach relies on the application of the t test for the difference between two independent means as described in section 5.2.2. The calculated t statistic, 1.99, indicates that the probability of seeing a difference as large as that observed for these two formulations, if the two formulations are actually equivalent, is 0.062. The simulation approach requires applying only our knowledge that the variance of the sample mean is equal to the variance of the individual values divided by n, the size of the sample. The square root of this variance is the standard error of the mean. According to the Central Limit Theorem, the sample mean will tend to be normally distributed about its true mean value with a variability equal to the standard error.

Our question is formulated for a simulation solution by the following null hypothesis and its alternative:

H_0: The difference actually observed between the A and B means, 5.7, occurs by chance at least 5% of the time from two independent samples, each of size 10, taken from the same Normal distribution with mean and standard deviation equivalent to those in the combined (A + B) sample.

H_a: The difference observed between the sample means, 5.7, occurs less than 5% of the time by chance, indicating that it is unlikely that the two formulations represent the same population (i.e., their means are not equal).

The following is a simulation to evaluate our hypotheses:

	A	B	C	D	E	F	G	H
1	Form	Percent		Sim A	Sim B	Abs(Diff)	GE 5.7	Prob. (Diff GE 5.7)
2	A	68		72.7	74.7	2.1	0	0.065
3	A	84		76.0	75.2	0.7	0	
4	A	81		72.2	75.5	3.2	0	
5	A	85		76.4	72.1	4.3	0	
6	A	75		76.9	75.5	1.4	0	
7	A	69		74.0	74.0	0.0	0	
8	A	80		74.9	75.2	0.2	0	
9	A	76		73.1	73.9	0.8	0	
10	A	79		75.2	73.2	2.0	0	
11	A	74		76.1	71.7	4.4	0	
12	B	74		73.2	72.8	0.4	0	
13	B	71		73.3	74.1	0.7	0	
14	B	79		71.2	73.2	2.0	0	
15	B	63		75.1	71.6	3.5	0	
16	B	80		70.4	76.5	6.0	1	
17	B	61		75.1	70.9	4.1	0	
18	B	69		73.7	78.3	4.7	0	
19	B	72		71.9	75.3	3.5	0	
20	B	80		72.4	75.0	2.6	0	
21	B	65		74.1	76.7	2.5	0	
22				74.0	73.8	0.2	0	
23	Mean	74.25		75.3	75.9	0.6	0	
24	Variance	47.46053		73.6	76.9	3.3	0	
25	Stderr 10	2.178544		70.6	72.6	2.1	0	

Commands in Simulation

Cells B2–B21	Enter the 15-minute dissolution values from Table 5.9	
Cell B23	= AVERAGE(B2:B21)	Mean of combined A and B values
Cell B24	= VAR(B2:B21)	Variance of combined values
Cell B25	= SQRT(B24/10)	Standard error of mean for $n = 10$ values
Main Menu	Tools → Data Analysis → Random Number Generator	
Dialog Box		
Number of Variables 2		Generate simulated means for two samples
Number of Random Numbers	30,000	Perform 30,000 simulations
Distribution	Normal	The means are Normally distributed.
Mean	74.25	Actual mean from combined $A + B$ sample
Standard Deviation	2.178544	Standard error of a mean for a sample ($n = 10$)
Random Seed	Enter 12345	Can enter a different seed if desired
Output Range	Enter D2	Simulated means in cells D2–E30001
Click OK		
Cell F2	= ABS(D2–E2)	Absolute value of difference between the two simulated means
Cells F3-F30001	Copy formula from F2	
Cell G2	= IF(F2 < 5.7, 0, 1)	If simulated means differ as much as what we saw for the actual sample then value is 1, otherwise it is 0
Cells G3-G30001	Copy formula from G2	
Cell H2	= AVERAGE(G2: G30001)	Proportion of times that results are as extreme as what we saw with actual sample (probability)

Our estimated probability for the difference between the formulation A and B means is 0.065, which is very similar to the result obtained with the t test, $p = 0.062$. We can further refine our estimate by repeating the simulation multiple times (using different seed values each time) and using the average probability. The results from a second simulation using a seed value 5555 gave a probability estimate of 0.062. The estimated probability obtained from averaging those from the two simulation estimates $p = 0.0635$.

The next example again uses the data in Table 5.9. Having observed a difference of 5.7 between the mean 15-minute dissolution values of formulations A and B, what is the 95% confidence interval for the true mean difference between the formulations? Using Monte Carlo simulation, the answer can be obtained in a very intuitive way. Assume that the means from the two samples are normally distributed, a reasonable assumption given the Central Limit theorem. The variance of the difference between two sample means is the sum of the two samples' variances divided by the number of observations (n) in the samples. It is assumed that there is a common variance (VAR) for the two formulations. The variance for the difference between the sample means is $(VAR/n_a + VAR/n_b)$, where n_a and n_b are the number of values in the *A* and *B* samples, respectfully. As both samples consist of 10 values, the variance for the difference between means is equal to (2 × VAR/10). The standard error for the difference is equal to the square root of this value.

Applying the Central Limit Theorem, we can assume that the difference between the two means will be approximately normally distributed with $\mu = 5.7$ (our observed mean difference) and standard deviation equal to our estimated standard error. Simulating 30,000 mean differences, we can easily estimate the lower and upper 95% confidence limits. The 95% confidence limits encompass values between the 2.5th and 97.5th percentiles of the distribution describing the mean difference between the two samples (see chap. 5). These limits, for our Monte Carlo simulation of 3000 mean differences, are simply the 750th sorted value (2.5th percentile) and the 29,250th sorted value (97.5th percentile). The confidence interval obtained from the simulation, −0.34% to 11.79%, is comparable to that calculated using the *t*-distribution method, −0.32% to 11.72% (see chap. 5 for a description of how to apply the *t*-distribution method).

	A	B	C	D	E	F	G
1	**Form**	**Percent**	**Sim Diff**	**Sorted**	**Position**	**95% CI Lo**	**95% CI Hi**
2	*A*	68	2.14	−23.68		−0.34	11.79
3	*A*	84	−1.95	−23.68	**2.5%**		
4	*A*	81	4.83	−6.65	751		
5	*A*	85	3.79	−6.14			
6	*A*	75	7.35	−5.83	**97.5%**		
7	*A*	69	3.48	−5.83	29251		
8	*A*	80	5.44	−5.60			
9	*A*	76	7.76	−5.15			
10	*A*	79	2.14	−5.15			
11	*A*	74	4.97	−5.04			
12	*B*	74	6.55	−4.95			
13	*B*	71	5.83	−4.78			
14	*B*	79	7.84	−4.64			
15	*B*	63	2.10	−4.64			
16	*B*	80	4.55	−4.51			
17	*B*	61	5.04	−4.41			
18	*B*	69	3.15	−4.31			
19	*B*	72	7.07	−4.26			
20	*B*	80	0.07	−4.18			
21	*B*	65	4.57	−4.03			
22			5.96	−3.96			
23	**Mean *A***	77.1	11.29	−3.93			
24	**Mean *B***	71.4	1.74	−3.90			
25	**Difference**	5.7	10.80	−3.79			
26	**Variance**	47.46053	5.64	−3.59			
27	**Stderr Diff**	3.080926	5.70	−3.59			

Commands in Simulation

Cells B2–B21	Enter the formulation A and B dissolution values from Table 5.9	
Cell B23	= AVERAGE(B2:B11)	Mean of formulation A values
Cell B24	= AVERAGE(B12:B21)	Mean of formulation B values
Cell B25	= B23–B24	Difference between A and B means
Cell B26	= VAR(B2:B21)	Variance of combined A and B values
Cell B27	= SQRT(2*B26/10)	Standard error for difference between means
Main Menu	Tools $\rightarrow$ Data Analysis $\rightarrow$ Random Number Generator	

Dialog Box

Number of Variables 1	Generate a simulated difference between means	
Number of Random 30,000 Numbers	Generate 30,000 mean differences	
Distribution	Normal	Value is from the Normal dist.
Mean	5.7	Observed difference between A and B means
Standard Deviation	3.080926	Standard error of the difference
Random Seed	Enter 12345	Can enter a different seed if desired
Output Range	Enter C2	Simulated differences in cells C2–C30001

Click OK

Select column C by clicking at the top of the column and then from Main Menu choose Edit $\rightarrow$ Copy

Click at the top of column D and from Main Menu choose Edit $\rightarrow$ Paste

Cell D1	Change label to Sorted	
Click on column D to select it		
Main Menu	Data $\rightarrow$ Sort	
	Choose to sort only the selection in ascending order.	
Cell E4	= 1 + 0.25*30,000	Column D cell with 2.5th percentile value
Cell E7	= 1 + 0.975*30,000	Column D cell with 97.5th percentile value
Cell F2	= D751	Lower 95% confidence limit value
Cell G2	= D29251	Upper 95% confidence limit value

The next example comes from section 5.2.6. In two groups of patients, the incidences of headaches are evaluated to obtain a 95% confidence interval on the true difference in headache rates between the groups. In Group 1, there were 46 patients with headaches among the 196 patients, for a rate (proportion) of 0.2347. In the second group, 35 of the 212 patients experienced headaches, for a rate of 0.1651. The following Excel worksheet shows how to obtain the 95% confidence interval on the difference between the incidence proportions in the two groups by simulation.

We start by generating random values of the number of headaches in the two groups based on the binomial distribution. For Group 1, $N = 196$ and $p = 0.2347$. For Group 2, $N = 212$ and $p = 0.1651$. For each generated number of headaches for the two groups (simulated trial), we calculate the proportion of headaches "observed" in the groups and then find the difference between these proportions (Groups 1–Group 2). Thus, we generate 30,000 trials (see below). From the distribution of the Group 1-to-2 differences in theses trials, we find the 2.5th and 97.5th percentiles as in the previous example. This is the 95% confidence interval.

	A	B	C	D	E	F	G	H	I	J
	A	**B**	**C**	**D**	**E**	**F**	**G**	**H**	**I**	**J**
1	S1	S2	Sim S1	Sim S2	**p1**	**p2**	**Diff**	**Sorted**	**Position**	**95% CI**
2	46	35	47	41	0.240	0.193	0.046	−0.079	2.5%	Low
3			51	40	0.260	0.189	0.072	−0.073	751	−0.009
4	N1	N2	33	30	0.168	0.142	0.027	−0.072		
5	196	212	39	27	0.199	0.127	0.072	−0.072	97.5%	Hi
6			50	38	0.255	0.179	0.076	−0.070	29251	0.147
7	P1	P2	40	30	0.204	0.142	0.063	−0.070		
8	0.2347	0.1651	48	47	0.245	0.222	0.023	−0.069		
9			55	33	0.281	0.156	0.125	−0.064		
10			51	40	0.260	0.189	0.072	−0.063		

Commands in Simulation

Cells A2 and B2	Enter the number of patients in Group 1 and 2, respectively, who experienced headaches
Cells A5 and B5	Enter the total number of patients in Groups 1 and 2, respectively
Cell A8	= A2/A5 Group 1 observed proportion of headaches
Cell B8	= B2/B5 Group 2 observed proportion of headaches
Main Menu	Tools → Data Analysis → Random Number Generator

Dialog Box

Number of Variables	1	Simulate Group 1 headache number
Number of Random	30,000	Generate 30,000 simulated trials
Numbers Distribution	Binomial	Simulated number is from binomial distribution
p Value	0.234694	Group 1 observed proportion of headaches
Total Number of Trials	196	Number of patients in Group 1
Random Seed	Enter 12345	Enter a different seed if desired
Output Range	Enter C2	Simulated headache numbers in C2–C30001

Click OK

Main Menu   Tools → Data Analysis → Random Number Generator

Dialog Box

Number of Variables	1	Simulate Group 2 headache number
Number of Random	30,000	Generate 30,000 simulated trials
Numbers Distribution	Binomial	Simulated number is from binomial distribution
p Value	0.165094	Group 2 observed proportion of headaches
Total Number of Trials	212	Number of patients in Group 2

Random Seed	Enter 12345	Enter a different seed if desired
Output Range	Enter D2	Simulated headache numbers in D2–D30001
Click OK		
Cell E2	= C2/196	Proportion of Group 1 simulated headaches in trial 1
Cell F2	= D2/212	Proportion of Group 2 simulated headaches in trial 1
Cell G2	= E2–F2	Difference between group headache rates for 1st trial
Cells E3:G30001	Copy E2:G2	Calculates proportions and differences for other trials
Cells H2:H30001	Copy column G values, using the Paste Special, Values method	
	Click at top of column H to choose it (column is highlighted)	
Main Menu	Tools → Data → Sort	
	Click option to continue without expanding current selection	
	Click OK to sort the column with a header, in ascending order	
Cell I3	= 1 + 0.025*30,000	Row with the 2.5th percentile difference
Cell I6	= 1 + 0.975*30,000	Row with the 97.5th percentile difference
Cell J3	= H751	95% CI low limit = 2.5th percentile value
Cell J6	= H29251	95% CI hi limit = 97.5th percentile value

The 95% confidence interval limits based on the simulations, -0.009 to 0.147, are in close agreement with the limits of -0.008 to 0.148 calculated using normal approximation methods and with the limits -0.012 to 0.152 obtained using the more conservative continuity-corrected, normal approximation. Running the simulation a number of times, using different seed values each time, and then averaging the results should provide values closer to the exact limits.

One area where Monte Carlo methods are extremely useful is in determining the sizes of samples needed to obtain a desired power for a given statistical evaluation. A number of formulas are presented in chapter 6 that can be used for these calculations. In many situations, while the formulas are easily applied, their derivations are not so easily understood. Simulation provides an extremely intuitive approach in this area. As discussed in chapter 6, to determine the sample size needed for a given study we need to state the alpha level (e.g., 0.05), the beta level (e.g., 0.2 = power of 0.8), and a difference between treatments of a specified magnitude (usually a difference of practical significance). To determine the probability of obtaining a given outcome from a particular statistical test (e.g., the probability of getting a P value ≤ 0.05 in an independent group t test) we simply simulate a large number of random samples, calculate the statistic for each simulation, and then determine the proportion of times the statistic had the desired outcome. The more complicated the problem, the more intuitive and useful is the simulation method.

For sample size determination, we usually calculate the proportion of times we get a significant difference under the null hypothesis, which causes us to reject it in favor of the alternative hypothesis where the meaningful difference is specified. The following example uses both this approach and a modification needed when we want to test for noninferiority (or equivalence) rather than testing for a difference.

In this example, we want to conduct a clinical trial on a new drug developed to treat a certain disease. Preliminary animal studies indicate that the new drug will be at least as effective as the current treatment for the disease and is likely to have fewer serious side effects. The FDA has indicated that it wants to see a placebo-controlled, noninferiority trial. This trial will compare the new Drug, A, the current treatment B, and placebo, in the treatment of subjects with the disease. The primary efficacy measure will be the proportion of subjects who show improvement. We intend to show that the new drug is at least as effective as (noninferior to) the current Drug B. To demonstrate noninferiority we must construct a 95% confidence interval for the difference between the Drug A and Drug B proportions and then show that this difference is no worse than 20% (i.e., Drug A is no more inferior to Drug B than 20%). In addition to showing noninferiority, we must simultaneously demonstrate that the clinical trial had adequate sensitivity to detect true differences in efficacy had they existed. This is established by showing that both Drugs A and B have superior efficacy to that of the placebo.

From prior experience, we know that 25% of patients left untreated will improve spontaneously (placebo success proportion is expected to be 0.25) and improvement is seen in 45% of those treated with Drug B (success rate for B is expected to be 0.45). We believe that the new drug will be successful in treating at least 50% of the patients (cure rate for Drug A is conservatively set at 0.50).

The statistical evaluation comparing Drug A to Drug B involves the construction of the 95%, continuity-corrected, confidence interval on the difference between their success proportions. If the lower limit of this confidence interval is greater than -0.20, then noninferiority of A to B will be established. Note that while our interest is only with the lower confidence interval limit (i.e., it is one-sided), the FDA usually requires the use of the more conservative, two-sided, confidence interval (critical Z value of 1.96 is used instead of 1.645). Had our intention been to show therapeutic equivalence of Drug A to Drug B, rather than noninferiority, then we would need to show that the entire confidence interval falls within the equivalence interval -0.20 to $+0.20$. For the trial to be successful, we must also show that the two-sided, continuity-corrected, Z tests on the differences between the success proportions for Drug A compared to placebo and for Drug B compared to placebo show statistical superiority for the active treatments (i.e., differences > 0 and $p < 0.05$). The following equations will be used (see chap. 5):

$$95\% \, CI = (p_a - p_b) \pm \left[1.96 * \left(\frac{p_a * q_a}{n_a} + \frac{p_b * q_b}{n_b} \right)^{1/2} + 0.5 * \left(\frac{1}{n_a} + \frac{1}{n_b} \right) \right]$$

$$Z \, test \, 1 = \frac{\left[(p_a - p_p) - 0.5 * \left((1/n_a) + (1/n_p) \right) \right]}{\left[(p_0 * q_0) \left((1/n_a) + (1/n_p) \right) \right]^{1/2}}$$

$$Z \, test \, 2 = \frac{\left[(p_b - p_p) - 0.5 * \left((1/n_b) + (1/n_p) \right) \right]}{\left[(p_0 * q_0) \left((1/n_b) + (1/n_p) \right) \right]^{1/2}}$$

where p_a is the observed success rate for Drug A, $q_a = 1 - p_a$ failure rate for Drug A, p_b the observed success rate for Drug B, $q_b = 1 - p_b$ failure rate for Drug B, p_p the observed success rate for placebo, and $q_p = 1 - p_p$ failure rate for placebo; n_y the number of patients receiving treatment Y; $Y = A$, B, or placebo, $p_0 = (n_y * p_y + n_p * p_p)/(n_y + n_p)$, pooled success rate; $Y = A$ for Z Test 1, B for Test 2; $q_0 = 1 - p_0$ pooled failure rate for Z test.

We will determine our sample size by trial and error. First we specify a given sample size. We assume that the two active products' success rates actually differ by 5% (0.50 vs. 0.45) and that the placebo success rate is that known to occur in untreated patients (0.25). We then randomly generate (simulate) success/failure results for treating patients with Drug A, Drug B, and placebo. From these results, we calculate the proportions of patients with success in each treatment group and calculate the above statistics. Our probability of trial success (power) is the proportion of times that our simulated samples meet the criteria for noninferiority and superiority.

Our initial evaluation uses a 2:2:1 randomization (A:B:placebo) in about 350 patients (a number consistent with our initial budget allocation). We propose to use 340 patients, 136 in each active treatment group and half that number, 68, in the placebo group. We want to estimate the probability that our trial will show both noninferiority of Drug A compared to Drug B, and superiority of both A and B over placebo. We determine this easily using Monte Carlo simulation.

As shown in the following Excel worksheet, we simulate the results for 30,000 trials each involving the treatment of 136 patients for Drugs A and B, and 68 patients for placebo. The number of successfully treated patients for Drug A is placed in column A, for Drug B in column B, and for placebo in column C. Columns D, E, and F contain the

	A	B	C	D	E	F	G	H	I	J	K
	Drug A	Drug B	Placebo	n_a	n_b	n_p	p_a	p_b	p_p	$p01$	$p02$
1											
2	66	61	10	136	136	68	0.485	0.449	0.147	0.373	0.348
3	65	57	13	136	136	68	0.478	0.419	0.191	0.382	0.343
4	63	59	17	136	136	68	0.463	0.434	0.250	0.392	0.373
5	66	62	18	136	136	68	0.485	0.456	0.265	0.412	0.392
6	66	57	14	136	136	68	0.485	0.419	0.206	0.392	0.348
7	63	60	16	136	136	68	0.463	0.441	0.235	0.387	0.373
8	67	63	14	136	136	68	0.493	0.463	0.206	0.397	0.377
9	67	66	18	136	136	68	0.493	0.485	0.265	0.417	0.412
10	73	61	20	136	136	68	0.537	0.449	0.294	0.456	0.397
11	73	70	19	136	136	68	0.537	0.515	0.279	0.451	0.436
12	82	61	20	136	136	68	0.603	0.449	0.294	0.500	0.397

total number of treated patients (136, 136, and 68) for Drug A, Drug B, and placebo, respectively. The calculated success proportions for Drug A, Drug B, and placebo are placed in columns G, H, and I, respectively. The pooled success proportions for the Drug A and placebo comparisons and for the Drug B and placebo comparisons, under the null hypothesis of no difference between treatment success proportions, are placed in columns J and K, respectively. A portion of the worksheet with these results is shown above along with the following commands used to obtain them.

Commands in Simulation

Main Menu	Tools → Data Analysis → Random Number Generator
Dialog Box	

Number of Variables	1	Simulate number of successes for Drug A
Number of Random Numbers Distribution	30,000 Binomial	Generate 30,000 simulated trials Numbers come from binomial distribution
p Value Total Number of Trials	0.50 136	Expected Drug A success proportion Number of patients in treatment group
Random Seed Output Range	Enter 1234 Enter A2	Enter a different seed if desired Drug A number of successes in A2–A30001
Click OK		

Main Menu	Tools → Data Analysis → Random Number Generator
Dialog Box	

Number of Variables	1	Simulate number of successes for Drug B
Number of Random Numbers Distribution	30,000 Binomial	Generate 30,000 simulated trials Numbers come from binomial distribution

p Value	0.45	Expected Drug B success proportion
Total Number of Trials	136	Number of patients in treatment group
Random Seed	Enter 2341	Enter a different seed if desired
Output Range	Enter B2	Drug B number of successes in B2−B30001

Click OK

Main Menu   Tools→ Data Analysis → Random Number Generator

Dialog Box

Number of Variables	1	Simulate number of successes for placebo
Number of Random	30,000	Generate 30,000 simulated trials
Numbers Distribution	Binomial	Numbers come from binomial distribution
p Value	0.25	Expected placebo success proportion
Total Number of Trials	68	Number of patients in treatment group
Random Seed	Enter 3412	Enter a different seed if desired
Output Range	Enter C2	Placebo number of successes in C2–C30001

Click OK		
Cells D2, E2, F2		Enter number of patients in treatment groups A, B and placebo
Cell G2	= A2/D2	Proportion of successes for Drug A
Cell H2	= B2/E2	Proportion of successes for Drug B
Cell I2	= C2/F2	Proportion of successes for placebo
Cell J2	= (A2 + C2)/(D2 + F2)	Pooled proportion for A and placebo
Cell K2	= (B2 + C2)/(E2 + F2)	Pooled proportion for B and placebo
Cells D3: K30001		Copy formulas from cells D2 through K2

Next we calculate the continuity correction, $0.5 \times (1/n_a + 1/n_b)$, for the noninferiority calculation and place it in column L. We do the same for the superiority comparisons of Drug A to placebo and Drug B to placebo, and place these values in columns M and N. The 90% confidence interval lower limit for each trial (row) is calculated and placed in column O. The Z test value for the comparison of Drug A to placebo is calculated and placed in column P and that for Drug B to placebo is placed in column Q. Flags in columns R, S, and T are set to 1 if we pass the noninferiority test, the A-to-placebo superiority test, and the B-to-placebo superiority test, respectively. If all three tests are passed, then a 1 is placed in column U indicating that the trial was successful. A failed test is designated by a flag value of 0 placed in its respective column.

	L	M	N	O	P	Q	R	S	T	U
1	CCAB	CC1	CC2	95%CI LO	Z test1	Z test2	Flag1	Flag2	Flag3	Flag All
2	0.007	0.011	0.011	−0.089	4.557	4.105	1	1	1	1
3	0.007	0.011	0.011	−0.067	3.820	3.076	1	1	1	1
4	0.007	0.011	0.011	−0.096	2.789	2.406	1	1	1	1
5	0.007	0.011	0.011	−0.097	2.867	2.484	1	1	1	1
6	0.007	0.011	0.011	−0.059	3.701	2.858	1	1	1	1
7	0.007	0.011	0.011	−0.104	2.998	2.714	1	1	1	1
8	0.007	0.011	0.011	−0.097	3.794	3.421	1	1	1	1

9	0.007	0.011	0.011	−0.119	2.962	2.867	1	1	1	1
10	0.007	0.011	0.011	−0.037	3.131	1.973	1	1	1	1
11	0.007	0.011	0.011	−0.104	3.333	3.045	1	1	1	1
12	0.007	0.011	0.011	0.030	4.010	1.973	1	1	1	1
13	0.007	0.011	0.011	−0.082	3.888	3.328	1	1	1	1
14	0.007	0.011	0.011	−0.029	3.397	2.161	1	1	1	1
15	0.007	0.011	0.011	−0.163	3.491	3.960	1	1	1	1
16	0.007	0.011	0.011	−0.082	3.169	2.598	1	1	1	1
17	0.007	0.011	0.011	−0.015	2.139	0.664	1	1	0	0
18	0.007	0.011	0.011	−0.148	3.678	3.960	1	1	1	1

Commands in Simulation (Continued)

Cell L2	$= 0.5*(1/D2 + 1/E2)$	Continuity correction (A vs. B)
Cell M2	$= 0.5*(1/D2 + 1/F2)$	Continuity correction (A vs. placebo)
Cell N2	$= 0.5*(1/E2 + 1/F2)$	Continuity correction (B vs. placebo)
Cell O2	$= (G2{-}H2) - ((1.96*SQRT(G2*(1{-}G2)/D2 + H2*(1{-}H2)/E2) + L2))$	
Cell P2	$= (G2{-}I2 - M2)/SQRT(J2*(1{-}J2)*(1/D2 + 1/F2))$	
Cell Q2	$= (H2{-}I2 - N2)/SQRT(K2*(1{-}K2)*(1/E2 + 1/F2))$	
Cell R2	$= IF(O2{>}{-}0.2,1,0)$	Flag $= 1$ if 95% CI is above −0.20
Cell S2	$= IF(P2{>}1.96,1,0)$	Flag $= 1$ if A versus placebo Z test is significant
Cell T2	$= IF(Q2{>} 1.96,1,0)$	Flag $= 1$ if B versus placebo Z test is significant
Cell U2	$= R2*S2*T2$	Flag $= 1$ if all three tests pass
Cells L3:U30001	Copy formulas from cells L2 through U2	

Our probability (power) of showing noninferiority, superiority, or passing all three required tests is simply the average of the 0/1 entries in the corresponding flag column, the proportion of simulated trials in which we observed a successful (1) outcome.

	V	W	X	Y
1	p (noninf)	p (super$A$)	p (super$B$)	p (trial)
2	0.9830	0.9206	0.7668	0.7353

Final Commands in Simulation

Cell V2 $=$ Average(R2:R30001)	Proportion where A was noninferior to B
Cell W2 $=$ Average(S2:S30001)	Proportion where A was superior to placebo
Cell X2 $=$ Average(T2:T30001)	Proportion where B was superior to placebo
Cell Y2 $=$ Average(U2:U30001)	Proportion where A was noninferior to B and both A and B were each superior to placebo (overall probability of success)

The probability of showing noninferiority, 0.983, and the probability of showing the superiority of Drug *A* over placebo, 0.921, are both high with assumed proportions of 0.5 and 0.25 for Drug *A* and placebo, respectively. The probabilities of showing superiority of Drug *B* (assumed proportion 0.45) over placebo, 0.767, and for the overall success of the trial, 0.735, are unacceptably low.

We would like to know if there is a way to increase the probability of overall success without increasing our costs (i.e., patient numbers). We decide to explore the question by looking to a different randomization scheme. Perhaps a 1:1:1 randomization would increase the probability of trial success. We will evaluate using an equivalent number of patients in each treatment group to see if this improves our expected outcome. By setting our sample sizes to 110 patients in each treatment (330 total) and performing the simulation and calculations again, we find that the probability of showing noninferiority decreases slightly to 0.947, the probability of showing superiority of Drug *A* over placebo increases slightly to 0.963, and the probability of showing Drug *B* to be superior to placebo significantly increases to 0.849. The overall effect is that the probability of a successful trial is now increased to 0.789. By adding a few more patients to each treatment group and using the 1:1:1 randomization scheme, we can bring the overall probability of trial success to 0.80, a typical level of power used in designing a clinical trial. This is accomplished by using essentially the same number of subjects that would provide only a 0.74 probability of trial success with the 2:2:1 randomization scheme. Using computer simulations, these types of what-if evaluations are easy to conduct and to understand.

Another important application of Monte Carlo simulation is estimating the properties of a certain statistic when there is no known formula for doing so. For example, we might want to determine the probability distribution for the difference between two sample medians when the samples are drawn from similar, or dissimilar, statistical distributions. When there are no standard formulas to evaluate the distributional properties of a complex or unusual statistic, computer simulation is often the only tool available.

14.2 BOOTSTRAPPING

Bootstrapping (sometimes called resampling) encompasses a group of computer simulation methods in which samples are repeatedly drawn not from some hypothesized statistical distribution, but from the set of values that come from an actual sample obtained from some real population. These methods typically assume only that the sample was randomly selected from the population, thereby ensuring that it is likely to be representative of the population from which it was drawn. The theory behind Bootstrap methods proposes that the probabilistic information contained in the sample is reflective of corresponding information contained in the actual population. This same assumption is also required for most standard inferential methods. The primary difference between bootstrapping and standard inferential methods is how we use this information contained in the sample.

Standard inferential methods rely on our knowledge of the distribution of a statistic or parameter (e.g., mean, standard deviation, etc.) that we calculate from a sample collected from a population with some assumed statistical distribution. As an example, it is known that the average value calculated from a sample whose underlying population is assumed to be normally distributed with mean μ and variance σ^2 will follow a Normal distribution with mean μ and variance σ^2/n, regardless of the size of the sample, n. When neither μ nor σ is known, we estimate these parameters from the sample average and its standard deviation. A 95% confidence interval on μ is calculated using the standard equation: average $\pm t_{\alpha/2,n-1} \times$ SE, where SE is the sample standard deviation divided by the square root of n. The value $t_{\alpha/2,n-1}$ is obtained from student's t distribution. In the standard method, using statistics calculated from the sample (e.g., mean and standard deviation) we infer back to the values of the unknown parameters (e.g., μ and σ) of the underlying population.

In bootstrapping, we make no assumptions about the statistical distribution of the population from which the sample was collected or about the distributional properties of the sample itself. Instead, we treat the sample as if it was the population and repetitively take samples (resamples) from it using computer simulation. The distribution of statistics calculated from these computer-generated samples theoretically mimics the distribution in the population. Using the frequency distribution of the statistic in the computer-generated samples, we make inferences about the corresponding distribution in the underlying population. One of the simplest bootstrapping methods will be used to provide a brief introduction to these powerful, computer-intensive simulation methods. The method is known as the percentile method and is one of the most intuitive ones available.

Table 5.1 shows the assay results for 10 randomly selected tablets. The average value for these results is 103.0 mg and the standard deviation is 2.218. If we assume that the sample comes from a population that is normally distributed, or that based on the Central Limit Theorem the sample average is normally distributed, then we can calculate a 95% confidence interval on μ. This interval is determined to be 101.4 to 104.6, as shown in chapter 5. If we do not want to make distributional assumptions about the underlying population or about the sample, then a Bootstrap method can be used to obtain a confidence interval on μ (the population average value) as shown below.

	A	B	C	D	E	F	G	H	I	J	K	L	M
1	Mean	Stdev	Number:	1	2	3	4	5	6	7	8	9	10
2	103.0	2.218	Sample:	101.8	102.6	99.8	104.9	103.8	104.5	100.7	106.3	100.6	105.0

In row 1, columns D to M, we enter the numbers 1 to 10 to identify each observed assay value in the sample. The observed sample values are entered into row 2, immediately below their corresponding identification numbers. The mean (cell A2) and standard deviation (cell B2) of the values are calculated using the Excel formulas = AVERAGE(D2:M2) and = STDEV(D2:M2). We now go to column Y, reserving columns N to W for use later. We next generate 10 random numbers from the Uniform distribution for each of our 3001 simulated trials (rows) and place these numbers in columns Y to AH. The numbers will be rounded to integer values, and placed into columns N to W, to be used in obtaining our Bootstrap sample for each simulated trial.

	Y	Z	AA	AB	AC	AD	AE	AF	AG	AH
1										
2										
3	3.08	6.26	8.09	7.08	2.60	7.43	8.55	2.49	8.99	7.43
4	5.11	5.07	6.62	6.97	3.62	4.87	7.03	3.81	8.16	2.08
5	3.81	3.29	4.05	5.20	1.72	3.88	6.88	1.99	1.36	8.60

Commands in Simulation

Main Menu	Tools → Data Analysis → Random Number Generator

Dialog Box

Number of Variables	10	Simulate 10 values for each trial
Number of Random Numbers	3001	Perform the simulation for 3001 trials
Distribution	Uniform	Values come from the Uniform distribution
Parameters Between	1,10	Generate equally probable values between 1 and 10
Random Seed	12345	Enter a different seed if desired
Output Range	Y3:AH3003	Place values in cells Y3 through AH3003

Click OK

To convert the simulated Uniform distribution values into integer, index numbers, enter the following equation into cell N3: = ROUND(Y3,0). Next, copy this formula to all cells within the range N3: W3003.

	N	O	P	Q	R	S	T	U	V	W
1	R1	R2	R3	R4	R5	R6	R7	R8	R9	R10
2										
3	3	6	8	7	3	7	9	2	9	7
4	5	5	7	7	4	5	7	4	8	2
5	4	3	4	5	2	4	7	2	1	9

Using one of Excel's table lookup functions (HLOOKUP), we select values (resample) from the original sample whose assigned numbers in row 1 of columns D to M match the corresponding index values found in cells N3–W3003. In this way, each of the 3001 rows, representing a trial, contains a computer-generated sample of size 10 drawn from the original sample. As each original value can appear more than once in the Bootstrap sample, the method involves sampling with replacement. For each of the 3001 Bootstrap samples (rows 3–3003), we calculate the mean and standard deviation for its 10 values in columns D to M. These are the Bootstrap sample means and standard deviations whose frequency distributions will be used to make inferences to the characteristics of the underlying population from which our original sample was obtained.

	A	B	C	D	E	F	G	H	I	J	K	L	M
1	Mean	Stdev	Number	1	2	3	4	5	6	7	8	9	10
2	103.0	2.2181	Sample	101.8	102.6	99.8	104.9	103.8	104.5	100.7	106.3	100.6	105.0
3	101.6	2.17		99.8	104.5	106.3	100.7	99.8	100.7	100.6	102.6	100.6	100.7
4	103.2	1.99		103.8	103.8	100.7	100.7	104.9	103.8	100.7	104.9	106.3	102.6
5	102.7	1.93		104.9	99.8	104.9	103.8	102.6	104.9	100.7	102.6	101.8	100.6

Commands in Simulation

Cell D3	= HLOOKUP (N3,$D$1:$M$2,2)	From the sample values in row 2, section D1 to M2, select the value whose number in row 1 matches the random index number in N3
Cells E3:M3	Copy formula from cell D3	Generate first bootstrap sample
Cells D4:M3003	Copy formulas from cells D3:M3	Generate the remaining 3000 samples
Cells A3:A3003	Copy formula from cell A2	Calculate the Bootstrap samples' means
Cells B3:B3003	Copy formula from cell B2	Calculate the samples' standard deviations

Now we will estimate a 95% confidence interval on μ and test the hypothesis that μ is less than 102. These results will be compared to those obtained by standard formulas that assume that the sample is normally distributed. The procedure that we use in the percentile method is similar to that shown in previous examples, but here we apply them to the Bootstrap samples rather than samples simulated from some underlying assumed statistical distribution.

We start by opening a new worksheet and transferring the mean (sample average) values from our existing column A to column A in the new worksheet. Wanting to transfer only the values and not the formulas, we use the Edit → Copy → Edit → Paste Special → Values sequence in Excel. We delete the first entry in the new column A, as this is the mean for the original sample, not for a Bootstrap sample. All remaining values in the column shift upwards. The Bootstrap estimate of μ is the average of the Bootstrap sample means. The 95% confidence interval lower limit is the 2.5th percentile sorted mean and the upper limit is the 97.5th percentile sorted mean. The probability that $\mu < 102$ is simply the frequency that we have a Bootstrap mean value that is less than 102. The analyses and the commands to conduct the analyses are shown in the following.

	A	B	C	D
1	**Mean**	**<102**		
2	100.78	1	**Bootstrap**	
3	100.78	1	Mean	103.0
4	100.85	1	2.5% observation	76
5	101.00	1	97.5% observation	2927
6	101.00	1	95% CI Lower	101.7
7	101.03	1	95% CI Upper	104.2
8	101.10	1	Prob(mu<102)	0.076
9	101.10	1		
10	101.14	1		
11	101.14	1	**Normality Assumed**	
12	101.19	1	95% CI Lower	101.4
13	101.29	1	95% CI Upper	104.6
14	101.29	1	Z < 102	−1.426
15	101.29	1	Prob(μ <102)	0.077

Commands in Simulation

New Column A	Data → Sort	Sort the Bootstrap means in ascending order
Cell B2	= IF(A2 < 102,1,0)	1 if bootstrap mean is < 102, 0 otherwise
Cells B3:B3002	Copy formula from B2	
Cell D3	= AVERAGE(A2:A3002)	Bootstrap estimate of μ
Cell D4	= 1 + 0.025*3001	Row number for 2.5th percentile mean value
Cell D5	= 1 + 0.975*3001	Row number for 97.5th percentile value
Cell D6	= A76	2.5th percentile value is Bootstrap CI lower limit
Cell D7	= A2927	97.5th percentile value is Bootstrap CI upper limit
Cell D8	= AVERAGE(B2:B3002)	Proportion of the means that are < 102
Cells D12 and D13	95% confidence limits	Obtain from text assuming Normal distribution
Cell D14	= (102 − 103)/(2.2181/SQRT(10))	Z value for standard test of $\mu <$ 102
Cell D15	= Normsdist(D14)	Probability from Standard Normal distribution

A similar evaluation for standard deviation can be conducted from the Bootstrap sample values. Copy the standard deviation values from column B of our original worksheet into column A of a new worksheet. Delete the standard deviation value for the original sample, leaving only those for the Bootstrap samples. As with the analysis of the means, we sort the column of values. The average of the Bootstrap values is our estimate of σ. The 2.5th and 97.5th percentile values are our lower and upper 95% confidence interval limits.

For the standard method, we rely on the Chi-square distribution (see chap. 5) as the assumed statistical distribution of the variance. The square root of the variance, the standard deviation of the original sample values, is the estimate of σ. By using the 2.5th percentile and 97.5th percentile critical Chi-square values, we can construct a 95% confidence interval on σ using standard formulas.

The average standard deviation value of our Bootstrap samples is 2.122. The 2.5th percentile and the 97.5th percentile standard deviations are the 76th and 2927th sorted values. The Bootstrap 95% confidence interval limits are 1.4 and 2.7.

The 95% confidence interval limits based on the Chi-square distribution are derived from the distribution's critical values of 2.70 (0.025 probability level, 9 d.f.) and 19.02 (0.975 probability level, 9 d.f.). The lower limit, 1.5, is calculated as $SQRT((2.218^2 \times 9)/19.02)$ and the upper limit, 4.0, is calculated as $SQRT((2.218^2 \times 9)/2.70)$.

It is notable that while the lower limits from both the Bootstrap method and the formula-based method are quite similar, those for the upper limit are not. This may be due to the small size of the original sample, resulting in a biased bootstrap estimate of variability. If this was the case, then taking a second sample and combining it with the original sample, then repeating the Bootstrap process, might improve the estimate. It is also possible that the actual underlying statistical distribution for the sample variance is not that of the assumed Chi-square distribution. In this case, the Bootstrap confidence interval may be closer to reality than that obtained by the standard formulas. Only additional actual sampling would help us evaluate the cause of the discrepancy between the two estimates.

Both Monte Carlo simulation and bootstrapping methods are powerful tools for solving problems. Monte Carlo simulation, carrying out repeated computer-simulated experiments based on simple statistical principles, is a process that has an intuitive appeal to many scientists.

	A	C	D
1	Stdev		
2	0.703	Bootstrap	
3	0.811		
4	0.891	stdev	2.122
5	0.982	2.5% obs	76
6	1.001	97.5% obs	2927
7	1.001	95% CI Low	1.4
8	1.020	95% CI Hi	2.7
9	1.020		
10	1.105		
11	1.116	Chi-Sq on sample	
12	1.116		
13	1.128	95% CI Low	1.5
14	1.132	95% CI Hi	4.0
15	1.132		
16	1.160		
17	1.171		

While less intuitive in its theoretical underpinnings, Bootstrapping provides a simple non-parametric method for solving problems when we are unable to make assumptions about the underlying statistical properties that govern the process of interest.

    While the examples presented have relied upon the computing power of Microsoft Excel, there are other packages that may provide more accessible simulation and bootstrapping capabilities. The author is familiar with two such packages provided by the company Resampling Stats, Inc. (*www.resample.com*). One is marketed as an Excel Add-in [2] that enhances the built-in simulation capabilities in Excel and provides a considerably easier way to perform bootstrapping in Excel. The second, Resampling Stats [3,4] is a self-contained simulation and bootstrapping package with extremely intuitive commands and easy to use programming wizard interface. The reader who wishes to further pursue simulation methods would be well advised to consider one of these computer packages.

REFERENCES

1. Mooney CZ, Duval RD. Bootstrapping: A Nonparametric Approach to Statistical Inference. Newbury Park, CA: Sage University Paper Series on Quantitative Applications in the Social Sciences, Sage Publications, Inc., 1993.
2. Blank S, Seiter C, Bruce P. Resampling Stats in Excel. Arlington, VA: Resampling Stats, Inc., 1999.
3. Simon JL. Resampling: The New Statistics, 2nd ed. Arlington, VA: Resampling Stats, Inc., 1998.
4. Simon JL. Resampling Stats: User's Guide. Arlington, VA: Resampling Stats, Inc., 1999.

15 | Nonparametric Methods

Nonparametric statistics, also known as distribution-free statistics, may be applicable when the nature of the distributions are unknown, and we are not willing to accept the assumptions necessary for the application of the usual statistical procedures. For most of the statistical tests described in this book, we have assumed that data are normally distributed. This assumption, although never exactly realized, is bolstered by the central limit theorem (sect. 3.4.2) when we are testing hypotheses concerning the means of distributions. However, occasions arise in which data are clearly too far from normal to accept the assumption of normality. The data may deviate so much from that expected for a normal distribution that to assume normality, even when dealing with means, would be incorrect. In these situations, a data transformation may be used, chapter 10, or nonparametric methods may be applied for statistical tests. As we shall see, many of the nonparametric tests are easy to compute, and can be used for a quick preliminary approximation of the level of significance when parametric tests may be more appropriate. Although some people believe that any kind of data, no matter what the distribution, can be correctly analyzed using nonparametric methods, a kind of panacea, this is not true. Many if not most nonparametric methods require that the distributions be continuous and symmetrical, and that data be independent, for example. These are among the assumptions underlying parametric analyses, as exemplified by the normal t, and F tests.

15.1 DATA CHARACTERISTICS AND AN INTRODUCTION TO NONPARAMETRIC PROCEDURES

Before proceeding, a review of the different kinds of data that are usually encountered in scientific experiments will be useful for the understanding of the applications of nonparametric methods.

1. Perhaps the most elementary kinds of data are *categorical* or *attribute* measurements. These are also known as *nominal* observations (i.e., the observation is given a *name*). Thus, a person is observed to be a "male" or a "female" or "black," "white," or "yellow." Some other examples are given in Table 15.1. The assignment of a number to such nominal data may be useful to differentiate the categories, perhaps for computer usage. However, actual values, a number assigned to these categories where the numbers have meaning in terms of rank, would not make sense. For example, we could assign the number 1 to a male and 2 to a female, but this does not imply that a female is *larger* (or, for that matter, smaller) than a male. Data that comprise two classes and consist of such attribute measurements may be analyzed using the binomial distribution. As discussed in chapter 5, Chi-square tests may be used to test the significance of differences of the proportion of attributes in comparative groups if the sample size and incidences are sufficiently large. These kinds of data are usually presented in the form of contingency tables, such as the 2 × 2 table for proportions discussed in chapter 5.

2. The next, perhaps more "sophisticated" level of measurement involves data that can be *ranked* in order of magnitude. That is, we can say that one measurement is equal to, less than, or greater than another. These kinds of ordered data are known as *ordinal* measurements. Continuous variables are ordinal measurements according to this definition, but here, we usually think of ordinal data as arising from some arbitrary scale, as constructed for rating scales. For example, patients receiving antidepressant medication, may be rated according to attributes such as "sociability." A high score will be assigned to a patient performing well on this criterion. If the patient shows characteristics of "withdrawal," a low score will result. Intermediary scores reflect various degrees of response. These are ordinal measurements. A

Table 15.1   Examples of Nominal Data

Products categorized as acceptable and unacceptable in quality control
Side effects in a clinical study
Males and females in a clinical study
Various descriptions of "feel" of an ointment preparation, or taste of a product
  (tart, biting, sharp, etc.)
Concomitant diseases or medicaments in a clinical study

patient with a score of zero after one week of medication, and a score of 3 after two weeks of medication can be said to have improved during the period between one and two weeks of treatment. A score of 3 is *better* than a score of zero. Some examples of this kind of data are shown in Table 15.2. Many nonparametric tests are based on *ranking* data. Certainly, data derived from a continuous distribution, such as the normal distribution, can be ranked in order of magnitude. (Ordinal data, by definition, can be ranked.) The nonparametric tests that will be discussed here, which use ranks for the analysis, require that the data have a continuous distribution. One might question the validity of nonparametric tests using data derived from an arbitrary ordinal rating scale such as that described above. If we understand (or assume) that the rating scale has an underlying continuity, the discreteness and arbitrary nature of the scale can be considered acceptable for nonparametric tests. The condition of the "depressed" patient is a continuum. The condition can vary from one extreme to another with infinitely small gradations, in theory. It is not possible practically to measure the subjective condition with its infinite subtleties, and therefore we substitute an ordered scale that approximates the condition of the patient. Controversy exists regarding the analysis of this kind of data. Some people believe that data derived from rating scales, as described above, should not be analyzed by parametric methods such as the *t* test. One reason for this position is that the intervals in these rating scales are not equal in terms of the degree of response; that is, the scores do not represent an equi-interval scale. In fact, the scale points do not precisely correspond to the description of the condition. The points are usually arbitrarily defined. Thus, there is not an exact correspondence of the numbers on the rating scale to the patients' conditions, as defined by an arbitrary description based on an assumed underlying continuous distribution (Fig. 15.1). For example, if a score of 3 represents "marked improvement" in sociability, 2 represents "moderate improvement," and 1 represents "no improvement," one usually cannot say that the difference between scores of 3 and 2 is equal in magnitude to the difference of 2 and 1. Yet the data analysis of such scores usually treats a difference between 3 and 2 as equivalent to a difference between 2 and 1. Perhaps, if the psychological aspects of depression were known to a sufficient extent, and the observer could discern subtle differences, the scoring system could be shown to be better represented by 3, 2.5, and 0.8 for the conditions corresponding to "marked improvement," "moderate improvement," and "no improvement," respectively.

   Although we can and do analyze data from a rating scale using nonparametric methods (as presented below), the typical parametric methods (ANOVA, *t* tests) are also commonly applied to such data. The use of parametric methods to analyze rating scale data is considered to be acceptable by many statisticians, including members of the FDA. Snedecor and Cochran discuss the analysis of this kind of data using a modified *t* test [1].

3. When comparing ages using a "ranking" scale, one person may be said to be older than another without regard to the magnitude of the difference in age. One can also specify the numerical differences with such data (e.g., one person is *two* years older than another). This

Table 15.2   Examples of Ordinal Data

Rating scales for sensory attributes (degree of liking)
Degree of effectiveness of therapeutic agent (pain relief, joint swelling, etc.)
Dichotomization of a continuous variable (underweight and overweight)
Number of anginal attacks in one week
Number of ulcers in skin-diseased patient

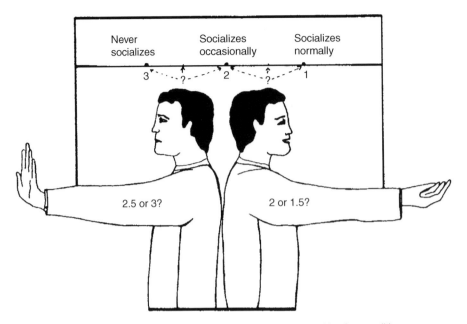

Figure 15.1 Problems with correspondence of a number and a subjective condition.

is an example of numerical data, often encountered in scientific experiments, where the distances between the values representing experimental outcomes have physical meaning. These data have a precise, better-defined meaning than data that are only ranked. Such data are often categorized as *interval* or *ratio* scaled data, depending on whether or not a true "zero point" exists. Age, weight, and concentration are examples of ratio scales. A person who weights 200 pounds is twice as heavy as one who weighs 100 pounds. Temperature does not have a true zero point (according to the concept above) and is an example of an interval scale. A temperature designated as "zero" is an arbitrary position on the scale and does not represent the lack of temperature. We cannot say that 40°C is twice as hot as 20°C. Ratio and interval-type data are the kinds of numbers that usually are subjected to the typical parametric tests. If these data are not normally distributed, they may be appropriately analyzed using nonparametric methods. One should understand that, in general, nonparametric tests can be applied to most of the data that we usually encounter, including that from continuous data distributions. Hence, data that are normally distributed may also be analyzed using these methods. A disadvantage of using nonparametric methods rather than the usual analyses for normally distributed data is that nonparametric methods are less sensitive (i.e., they are less powerful). Nevertheless, some nonparametric methods are surprisingly sensitive and are able to differentiate treatments that are normally distributed with efficiency almost equal to that of the parametric tests.

Nonparametric tests are most effectively used for data that consist of only classified (nominal) variables or ranked variables that are considered to have an underlying continuous distribution. Data derived from continuous distributions are particularly amenable to nonparametric methods when the distributions deviate greatly from normality. The reader should be aware that many nonparametric tests assume a symmetric distribution and equality of variance in the comparative groups. A marked disadvantage of the simpler nonparametric techniques is the lack of flexibility of the design and analysis. Elementary designs may be readily analyzed using nonparametric methods, but more complex designs in which interactions and other ANOVA components are present cannot be simply analyzed with these techniques, particularly when sample sizes are small.

Most of the nonparametric methods for data that are not categorical use *ranking* procedures. The observations in the various treatment groups are ranked according to specific procedures, and the ranks that replace the raw data are then analyzed. These analyses use simpler statistical computations than the corresponding parametric analyses. The transformation to ranks results in simple whole or fractional numbers of relatively small magnitude.

15.2   SIGN TEST

The sign test is probably the simplest of the nonparametric tests. The sign test is a test of the equality of the medians of two comparative groups. This test is used for *paired* data with an underlying continuous distribution, and can be applied to ranked or higher level data such as continuous *interval* and *ratio*-type data. The pairs are matched, and *differences* of the measurements for each pair tabulated. The differences are then categorized only with regard to the *sign* of the difference. That is, we count the number of times one treatment has a higher value than the other. *Ties* are not counted for this test. Ties give no information regarding which treatment has the higher median value. Theoretically, with continuous variables, there should be no ties.* However, with limited measuring instruments or the use of a crude rating scale, ties do occur.

As noted above, the sign test is a test of equal medians. If the test shows "significance," we can say that two comparative populations have different medians at the α level of significance. Under the null hypothesis that the medians of the two comparative distributions are the same, the probability of observing a value for Treatment A being larger or smaller than an observation for Treatment B is *one-half*; that is, the probability that an observation for Treatment A will be greater than a paired observation for Treatment B is one-half. Having recorded the differences, we compute the proportion of observations where the difference of treatment pairs is positive (or negative), disregarding ties (i.e., zero differences).

If positive and negative signs are observed to occur with approximately equal frequency, we can conclude that the treatments have a similar median. If either positive $(+)$ or negative $(-)$ signs predominate, there is evidence that one treatment has a higher median than the other. The statistical test is based on the binomial distribution. When applying two treatments to the same person, there are two possible outcomes: either Treatment A is favored or Treatment B is favored. Under the null hypothesis, the probability of A being favored is one-half; $H_0 : p = 0.5$. We compare the observed proportion to one-half (0.5). With N small and $p = 0.5$, the probabilities of various experimental outcomes can be calculated using computer software, or from the expansion of the binomial [Eq. (3.9)], or from tables of the binomial distribution (Table IV.3). For sample sizes of 6 to 20, inclusive, the number of positive or negative signs needed for significance at the 5% level for the sign test is given in Table IV.12. For sample sizes greater than 20, the normal approximation to the binomial, with a continuity correction, will suffice (see sect. 5.2.4). The normal approximation test is

$$Z = \frac{|p - 0.5| - 1/(2N)}{0.5/\sqrt{N}},$$ (15.1)

where p is the observed proportion and N is the sample size. If Z is greater than 1.96, the treatments differ at the 5% level (two-sided test). The calculation can be simplified as follows:

$$Z = \frac{|\text{number of } + \text{'s} - \text{number of} - \text{'s}| - 1}{\sqrt{\text{number of } + \text{'s} + \text{number of} - \text{'s}}}.$$ (15.2)

Remember that ties are discarded and that N, the sample size, does *not* include ties.

Example 1. Because of its simplicity, the sign test may be used for a fast look at data from comparative experiments before applying a more sensitive parametric test such as the t test (if appropriate). This was the case for the data in Table 15.3, which were obtained to compare the "time to peak" plasma level for two oral formulations of the same drug. These data would usually be analyzed using a more sensitive nonparametric test (see sect. 15.3) or a t test for paired data (or ANOVA for a crossover design). Values were obtained by administering both drugs

---

* With continuous measurements, the probability of two values being identical is zero.

Table 15.3 Paired Data Obtained from the Bioavailability
Experiment: Time to Peak Plasma Concentration

Subject	Time to peak (hr) A	Time to peak (hr) B	Difference B − A
1	2.5	3.5	+ 1
2	3.0	4.0	+ 1
3	1.25	2.5	+ 1.25
4	1.75	2.0	+ 0.25
5	3.5	3.5	0
6	2.5	4.0	+ 1.5
7	1.75	1.5	−0.25
8	2.25	2.5	+ 0.25
9	3.5	3.0	−0.5
10	2.5	3.0	+ 0.5
11	2.0	3.5	+ 1.5
12	3.5	4.0	+ 0.5

to each of 12 persons on two different occasions. Although these data would ordinarily result from a crossover design, and ANOVA techniques might be more appropriate, for the present purposes, we will consider an example where treatments have been assigned in random order. We will, therefore, not analyze "order" effects, and we will assume that no carryover effects are present.

From Table 15.3, tabulation of the differences $(B - A)$ results in nine positive signs and two negative signs. One subject showed no difference between Treatments A and B. Referring to Table IV.12, 10 of 11 positive (or negative) signs are needed to obtain significance at the 5% level. Thus, according to the sign test, the difference just misses significance, although product B appears to take a longer time to peak than does product A.

If the differences can be assumed to have a normal distribution, the paired t test would be a more sensitive test than the sign test. For any given, specific example, one could not predict that the t test would result in a "more significant" difference; but on the average, the t test will be more discriminating. In this example, the t test results in a highly significant difference between the two formulations ($t = 3.02$; see Exercise Problem 1).

15.3 WILCOXON SIGNED RANK TEST

For the comparison of two treatments in a paired design, a more sensitive nonparametric test than the sign test is the Wilcoxon signed rank test. In the Wilcoxon test, the magnitude of the difference between the paired results is taken into consideration in addition to the sign. This feature results in a more powerful test, the sign test still retains its advantage for a very quick assessment of the experimental results.

The Wilcoxon test is based on the assumption that the distributions of the comparative treatments are symmetrical. Therefore, we are testing the equality of the means or the medians; the mean and median are equal in a symmetrical distribution.

The initial calculations are the same as in the sign test. We first take differences between the treatment pairs as in Table 15.3. Again, when the values for a treatment pair are equal (a difference of zero), a tie, these data are discarded for purposes of the test. As in the sign test, a zero difference does not contribute information regarding the differentiation of treatments in the Wilcoxon signed rank test. The differences of the untied pairs are then *ranked* in order of magnitude, *disregarding sign*. For the data in Table 15.3, the comparison of the time to peak plasma concentration for two formulations, A and B, the ranking of the absolute values of the differences is shown in Table 15.4. Differences of equal magnitude (disregarding sign) are given the *average rank*. The three subjects, 4, 7, and 8, all showed a difference (*absolute value*) equal to 0.25. Each of the differences are given a rank of 2, since these are the three smallest differences observed; 2 is the average of ranks 1, 2, and 3.

Table 15.4 Data from Table 15.3: Ranking Differences Without Regard to Sign for the Wilcoxon Signed Rank Test

Subject	Value	Rank	Assigned rank	Assigned rank with sign
7	−0.25	1	2	−2
4	0.25	2	2	2
8	0.25	3	2	2
9	−0.5	4	5	−5
10	0.5	5	5	5
12	0.5	6	5	5
1	1.0	7	7.5	7.5
2	1.0	8	7.5	7.5
3	1.25	9	9	9
6	1.5	10	10.5	10.5
11	1.5	11	10.5	10.5

Ranks with positive signs
2
2
5
5
7.5
7.5
9
10.5
10.5
Sum = 59

Ranks with negative signs
2
5
Sum = 7

After ranking (disregarding sign) is completed, the signs corresponding to the signs of the original differences are reassigned to the ranks. For example, for subject 7 (originally given a rank of 2), the rank is changed to −2, because the difference for this subject was negative. The ranks with *like signs are summed* as shown following Table 15.4. The sum of the positive ranks is 59, and the sum of the negative ranks is 7. These are known as the *rank sums*. Table IV.13 gives the values of the *smaller* of the two rank sums needed for significance at the 5% level for various sample sizes, N (N is the sample size, the number of pairs, less the number of ties). The smaller rank sum must be equal to or less than that designated in Table IV.13 for the two means to be significantly different at the 5% level. In our example, Table IV.13 shows that the means are significantly different. The table shows that a rank sum of 10 or less for the smaller rank sum is significant at the 5% level for $N = 11$. In our example, the smaller rank sum is 7. Therefore the difference is significant at the 0.05 level ($p \sim 0.02$) [2]. This test gives very similar conclusions to that obtained by the t test. The Wilcoxon signed rank test is 95% as efficient as a t test for the comparison of normal populations. This means that a sample size of 100 that is analyzed using the Wilcoxon test would have equal sensitivity to a sample size of 95 using the t test. Considering the less restrictive assumptions of the Wilcoxon test compared to the t test, there is much to recommend it.

For sample sizes larger than those shown in Table IV.13, a normal approximation is available to compare two population means using the Wilcoxon signed rank test

$$Z = \frac{|R - N(N+1)/4|}{\sqrt{[N(N+1/2)(N+1)]/12}},$$ (15.3)

where R is the sum of ranks (either the larger or smaller rank sum can be used) and N is the sample size (disregarding ties). This formula works well also for smaller sample sizes. In our example, $N = 11$ and $R = 59$.

$$Z = \frac{|59 - 11(12)/4|}{\sqrt{[11(11.5)(12)]/12}} = 2.31.$$

From Table IV.2, $p = 0.02$, which is very close to the exact probability, if the data are normally distributed.

15.3.1 Nonparametric Confidence Intervals for Crossover Studies and Bioequivalence

If the assumptions of ANOVA (and t test) are violated, particularly the assumption of normality, a confidence interval can be formed based on a nonparametric approach. The method is based on ordering or ranking the outcomes and is relevant to bioequivalence studies, being introduced in this context. For the analysis of bioequivalence, a controversy concerning the nature of the data distribution recently polarized regulatory agencies. For many years, bioequivalence parameters were analyzed as the raw, untransformed values. For a two-period crossover design, this would be analogous to analyzing the differences of the treatments for each individual in the absence of period and carryover effects. Recently, agreement appears to have been reached, in the spirit of international harmonization, that a log transformation of AUC and C_{max} values is appropriate prior to the statistical analysis. This is analogous (but not the same) to an analysis of the ratio of the estimated parameters. However, one can use a nonparametric test in which the error structure and distribution assumptions are less rigid. A nonparametric confidence interval for ratios (or differences of logs) is given in Hollander and Wolfe [3] and is expounded in a paper by Steinijens and Diletti [4]. In this method, as opposed to parametric techniques, period and sequence (carryover) effects are assumed to be absent, and no adjustment is made for these effects.

The example in Steinijens and Diletti uses logs that would be appropriate in light of current practice. The method is described for N subjects in a two-period crossover design (or paired designs). First, compute the difference for each subject (e.g., test–reference). For the case of a log transformation for AUC, compute the difference of the product responses,

$$\log AUC_t - \log AUC_r = \log \left(\frac{AUC_t}{AUC_r} \right) = R$$

for each subject. (One may also calculate the ratios AUC_t / AUC_r because of the one–one relationship of ranks to the differences of logs. Compute R', the average (geometric mean for ratios) of all possible pairs of the N individual ratios (R), where N is the number of subjects. There are $N(N + 1)/2$ such pairs, including the ratio, R, for the same subject. (This will be clarified in the example below.) The values of R' are then ranked in order from low to high. The lower and upper nonparametric 90% and 95% confidence limits are given in Table 15.5. "C" is defined as the value of the R' that has the rank given in the table. For example, if "C" for the lower limit in Table 15.5 is 11, this means that the 11th ranked R' is given as the lower limit of the confidence interval. The details of the theory and the computations of C are given in Refs. [3,4,5]. In practice, it is not necessary to compute the logs because we are really interested in the ratios of test to reference. If we compute the ratios and use the geometric mean of the $N(N + 1)/2$ pairs for the ranks, we will obtain the confidence interval for the ratio of test/reference directly. Again, this is a result of the monotonic relationship between the ratio and difference of the logs. The following example clarifies the procedure. In this example, both the parametric and nonparametric confidence intervals are calculated for purposes of comparison.

Example 2. Data for 12 subjects comparing two products for C_{max} is shown in Table 15.6. The ratio of the C_{max} for the products (B/A) is also calculated for each subject. In Table 15.7, the geometric mean of each pair of ratios (B/A) is shown in rank order. There are $N(N + 1)/2$ such combinations (pairs) including each ratio with itself. The geometric mean is simply the square root of the product of 2 ratios. Thus, the ratio for subject 1 combined with itself is $102/135 =$

Table 15.5 Nonparametric Confidence Intervals Based on Wilcoxon's Signed Rank Test

Subjects (N)	Rank for lower limit		Rank for upper limit	
	95%	90%	95%	90%
6	1	3	21	19
7	3	4	26	25
8	4	6	33	31
9	6	9	40	37
10	9	11	47	45
11	11	14	56	53
12	14	18	65	61
13	18	22	74	70
14	22	26	84	80
15	26	31	95	90
16	30	36	107	101
17	35	42	119	112
18	41	48	131	124
19	47	54	144	137
20	53	61	158	150
21	59	68	173	164
22	66	76	188	178
23	74	84	203	193
24	82	93	219	208

0.756, and the geometric mean is the square root of $0.756 \times 0.756 = 0.756$. For subject 1 combined with subject 2, the geometric mean is the square root of $0.756 \times 0.821 = 0.788$, and so on.

For 12 subjects, the lower and upper cut-off points for a 95% confidence interval are the values ranked 14 and 65 (Table 15.5). For the data in this example, these values correspond to the ratios 0.800 and 1.247, respectively. The 90% confidence interval refers to the 18th and 61st rankings in Table 15.7, corresponding to an interval of 0.804 to 1.065. The 90% confidence interval would just pass the lower limits of the FDA requirements of 0.8 for the ratio.

Using a parametric analysis of variance (two-way ANOVA, assuming no period or sequence effects), with a log transformation (see Exercise Problem 15), the 90% interval is 0.79 to 1.26. The wider interval observed using the parametric approach is due to the "outlying" ratio for subject 3.

Table 15.6 Results for C_{max} from Bioequivalence Study

Subject	Product		Ratio
	A	B	B/A
1	135	102	0.7555556
2	179	147	0.8212290
3	101	385	3.8118813
4	109	106	0.9724771
5	138	189	1.3695653
6	135	105	0.7777778
7	158	130	0.8227848
8	156	125	0.8012821
9	174	144	0.8275862
10	147	133	0.9047619
11	145	114	0.7862069
12	147	167	1.1360544

Table 15.7 Ranks for Determining Confidence Interval for Bioequivalence Study

Rank	Subjects A, B	R' geometric mean	Rank	Subjects A, B	R' geometric mean
1	1, 1	0.75556	40	2, 4	0.89366
2	1, 6	0.76659	41	4, 7	0.89451
3	1, 11	0.77073	42	4, 9	0.89711
4	6, 6	0.77778	43	10, 10	0.90476
5	1, 8	0.77808	44	1, 12	0.92647
6	6, 11	0.78198	45	4, 10	0.93801
7	11, 11	0.78621	46	6, 12	0.94000
8	1, 2	0.78771	47	11, 12	0.94508
9	1, 7	0.78845	48	8, 12	0.95410
10	6, 8	0.78944	49	2, 12	0.96590
11	1, 9	0.79075	50	7, 12	0.96681
12	8, 11	0.79371	51	9, 12	0.96963
13	2, 6	0.79921	52	4, 4	0.97248
14	6, 7	0.79996	53	10, 12	1.01383
15	8, 8	0.80128	54	1, 5	1.01724
16	6, 9	0.80230	55	5, 6	1.03209
17	2, 11	0.80353	56	5, 11	1.03767
18	7, 11	0.80429	57	5, 8	1.04757
19	9, 11	0.80663	58	4, 12	1.05109
20	2, 8	0.81119	59	2, 5	1.06053
21	7, 8	0.81196	60	5, 7	1.06154
22	8, 9	0.81433	61	5, 9	1.06463
23	2, 2	0.82123	62	5, 10	1.11316
24	2, 7	0.82201	63	12, 12	1.13605
25	7, 7	0.82278	64	4, 5	1.15407
26	2, 9	0.82440	65	5, 12	1.24736
27	7, 9	0.82518	66	5, 5	1.36957
28	1, 10	0.82680	67	1, 3	1.69708
29	9, 9	0.82759	68	3, 6	1.72186
30	6, 10	0.83887	69	3, 11	1.73116
31	10, 11	0.84340	70	3, 8	1.74768
32	8, 10	0.85145	71	2, 3	1.76930
33	1, 4	0.85718	72	3, 7	1.77098
34	2, 10	0.86198	73	3, 9	1.77614
35	7, 10	0.86280	74	3, 10	1.85711
36	9, 10	0.86531	75	3, 4	1.92535
37	4, 6	0.86970	76	3, 12	2.08099
38	4, 11	0.87440	77	3, 5	2.28487
39	4, 8	0.88274	78	3, 3	3.81188

15.4 WILCOXON RANK SUM TEST (TEST FOR DIFFERENCES BETWEEN TWO INDEPENDENT GROUPS)

The sign test and Wilcoxon signed rank test are nonparametric tests for the comparison of paired samples. These data result from designs where each treatment is assigned to the same person or object (or at least subjects that are very much alike). If two treatments are to be compared where the observations have been obtained from two independent groups, the nonparametric *Wilcoxon rank sum* test (also known as the Mann–Whitney U test) is an alternative to the two independent sample *t* test. The Wilcoxon rank sum test is applicable if the data are at least ordinal (i.e., the observations can be ordered). This nonparametric procedure tests the equality of the *distributions* of the two treatments. Thus, this procedure tests for both the location and spread of the distributions.

The calculations for the Wilcoxon rank sum test are similar to those for the signed rank test discussed above. First, the observations from both groups are pooled and ranked, regardless of group designation. Identical observations are given a rank equal to the average of the ranks. In this procedure, the signs of the observations are taken into account for ranking. For example, a

Table 15.8 Results of a Dissolution Test Using the Original Dissolution Apparatus and a Modification: Amount Dissolved in 30 Minutes

Original apparatus		Modified apparatus	
Amount dissolved	Rank	Amount dissolved	Rank
53	3	58	11
61	14	55	5.5
57	9	67	21
50	1	62	15.5
63	17	55	5.5
62	15.5	64	18.5
54	4	66	20
52	2	59	12.5
59	12.5	68	22
57	9	57	9
64	18.5	69	23
		56	7
Sum of ranks	105.5		[170.5]

value of -1 has a lower rank than 0.5, which has a lower rank than 1. After ranking the pooled data, the observations are returned to their respective treatment groups. The observations are then replaced by their corresponding ranks. The *sum of the ranks of the smaller sample* is the basis for the statistical test. If the sample sizes are equal in the two treatment groups, the sum of the ranks in either group can be used as the statistic for the Wilcoxon rank sum test.

Table 15.8 shows tablet dissolution results observed in the original dissolution apparatus and a modification of the apparatus. The objective of this experiment was to compare the performance of the two pieces of apparatus. Twelve individual tablets were used for each "treatment" (apparatus). The amount of drug dissolved in 30 minutes was determined for each tablet. One tablet assay, determined in the original apparatus, is not included in the results (Table 15.5) because of an overt error during the assay procedure for this tablet.

Note how the ranks are obtained. The original apparatus has the four smallest values, 50, 52, 53, and 54, which are ranked 1, 2, 3, and 4, respectively. The next two highest values are from the modified apparatus, both equal to 55. These values are both given the *average* rank of 5 and 6, equal to 5.5. The next value, 56, from the modified apparatus is given a rank of 7. The next highest value is 57, which occurs twice in the original and once in the modified apparatus. These are each given a rank of 9, the average of the three ranks which these values occupy, 8, 9, and 10, and so on.

For moderate-sized samples, the statistical test for equality of the distribution means may be approximated using the normal distribution. This approximation works well if the smaller sample is equal to or greater than 10. For samples less than size 10, refer to Table IV.16 for exact significance levels [2]. The normal approximation is

$$Z = \frac{|T - N_1(N_1 + N_2 + 1)/2|}{\sqrt{N_1 N_2(N_1 + N_2 + 1)/12}}, \tag{15.4}$$

where N_1, is the smaller sample size, N_2 is the larger sample size, and T is the sum of ranks for the smaller sample size. If Z is greater than or equal to 1.96, the two treatments can be said to be significantly different at the 5% level (two-sided test). In our example

$$Z = \frac{|105.5 - 11(11 + 12 + 1)/2|}{\sqrt{(11)(12)(11 + 12 + 1)/12}} = \frac{26.5}{16.25} = 1.63.$$

A value of Z equal to 1.63 is not large enough to show significance in a two-sided test at the 5% level ($p = 0.11$; Table IV.2). Therefore, these data do not provide sufficient evidence to show that the two different pieces of apparatus give different dissolution results.

One should appreciate, as noted previously, that in ranking tests, *ties* result only because of measurement limitations, because the distributions are assumed to be continuous. Too many ties result in erroneous probabilities with regard to the test of significance. The error is on the "conservative" side. For data with many ties (*more than 10% of the data result in ties*, as is the case in our example) statistical tests will tend to give results that overestimate α (i.e., the α error is larger than it should be). Hence, we tend to miss significant differences more often than we should when too many ties appear in the data. A correction for ties is available, but in most applications the difference between the corrected and uncorrected Z value is negligible.

It would also be of interest to compare the two pieces of equipment using the two independent sample t test in order to see how the conclusions might differ. Of course, in general, one cannot determine what would be expected to occur from a single example. The t test is more efficient than the nonparametric rank sum test if the assumptions for the t test are valid (see sect. 5.2.4). Similar to the signed rank test, the Wilcoxon rank sum test is very efficient, approximately 95% compared to the corresponding t test. A two independent groups t test for the data of Table 15.8 results in a t value of 1.84 with 21 d.f. ($p < 0.10$)

$$t = \frac{61.3 - 57.45}{5.05\sqrt{1/12 + 1/11}} = 1.84.$$

The probability level is somewhat less for the t test compared to the Wilcoxon rank sum test in this example. However, the conclusions are similar for the two statistical procedures.

The tests described above may replace the *paired t test* (use the *sign test* or *signed rank test*) or the *two independent groups t test* (use the *rank sum test*) when the assumptions required for the validity of the t tests are questionable. For the comparison of more than two groups, nonparametric tests, analogous to the analysis of variance parametric methods, are available. However, simple nonparametric tests are not available for the analysis of more advanced designs or for tests of interaction. The tests to be described below can be used to test for treatment effects for a simple one-way or two-way analysis of variance. These tests are widely used, and are recommended when ANOVA assumptions regarding normality are suspect and/or cannot be easily tested. The nonparametric tests are useful in experiments where the data consist of values derived from a rating scale with an underlying continuous distribution.

15.4.1 Nonparametric Analysis of Two-Way Crossover (Bioequivalence Designs)

Some people have advocated the use of nonparametric analyses for crossover designs or for pharmacokinetic parameters from bioequivalence studies. As presented earlier (sect. 15.3.1), the reason for this is the less restrictive assumptions of the nonparametric analysis. In particular, this would, apparently, resolve the problem of violations of certain assumptions inherent in ANOVA, for example, linearity, normality, and variance assumptions, although the theory is quite complex. One problem with nonparametric techniques for the analysis of these data is that we cannot account simultaneously for effects due to periods or carryover in the nonparametric model. Cornell [6] has presented a lucid discussion of methods to analyze these data taken from Koch [7]. One can demonstrate that specific sums and differences of observations in the two-way crossover design are equivalent to effects of interest, that is, treatment, period, and carryover effects (see discussion of parametric analysis in sect. 11.4.2). Applying a model that includes treatment, period, carryover, and random effects, the principles of the analysis are as follows:

1. Total the data for each subject over both periods and compare the totals for group (Sequence) I (subjects taking test followed by reference) to the totals for Group II (subjects taking reference followed by test). This comparison is a test for unequal carryover effects (Note that this is

the same procedure as in the parametric analysis, where the sequence effect is confounded with a carryover).

2. Take differences of Period 1 and Period 2 for each subject. Compare the differences for Group 1 to Group 2. This is a test of treatment differences.
3. Take differences of Treatment 1 and Treatment 2 for each subject. Compare the differences for Group 1 to Group 2. This is a test of period differences.

To see how this works, consider the estimate of treatment effects (item 2 above). In Group 1, Treatment 2 follows Treatment 1; in Group 2, Treatment 1 follows Treatment 2. The expected value for each subject is

In Group 1, Period 1 : $\mu + P_1 + T_1$

In Group 1, Period 2 : $\mu + P_2 + T_2 + C_1$

In Group 2, Period 1 : $\mu + P_1 + T_2$

In Group 2, Period 2 : $\mu + P_2 + T_1 + C_2$

where P_i, T_i, and C_i refer to the effects due to period i, treatment i, and carryover due to treatment i, respectively.

The expected values of the differences between Period 1 and 2 for the two groups are

Group 1 : $P_1 + T_1 - P_2 - T_2 - C_1$

Group 2 : $P_1 + T_2 - P_2 - T_1 - C_2$

If carryover has been shown to be nonsignificant, $C_1 = C_2$ (see next paragraph), the difference between the expected values for Groups 1 and 2 is equal to $2(T_1 - T_2)$, or twice the treatment effect. The same approach can be used to demonstrate the results of the calculations for sequence and period effects, items 1 and 3 above (see Exercise Problem 16).

The nonparametric statistical tests may be applied sequentially. If the sequence effect is significant, we may have to use only the Period I data as is the case for the parametric analysis, as has been noted in chapter 11. For bioequivalence studies, real carryover effects are very rare because of the nature of the design (short period of dosing and washout period). Therefore, significant carryover effects may be dismissed if there is no rational or reasonable explanation for their existence. (The FDA has accepted carryover effects as spurious for single dose studies, in some cases, if the sponsor can demonstrate no obvious cause. However, in the nonparametric test, no adjustment is made for the treatment differences in the presence of period or carryover effects. Therefore, one should be cautious when applying these tests in the presence of a significant "carryover.") We can then apply the usual nonparametric tests. The data in Table 15.9, taken from Wallenstein and Fisher [8] as also presented by Cornell, are used to illustrate the procedure.

To test for significance, the Wilcoxon rank sum test is applied to the ranks of each of the differences (Period 1–Period 2 and Treatment 1–Treatment 2) in Table 15.9. Eight subjects are in Sequence I, Treatment 1 followed by Treatment 2. Nine subjects are in Sequence II. The comparison of treatment totals for Sequences I and II is a test for a sequence or carryover effect. The sequence effect is not significant [see Exercise Problem 17 and Eq. (15.4)]. The treatment effect can be tested by comparing the Period I to Period 2 differences for the two sequences [(Treatment 1–Treatment 2)$_1$ − (Treatment 2–Treatment 1)$_2$]. The sum of ranks for Period 1 to

Table 15.9 AUC (log) Data for Crossover Study [11] to Illustrate Nonparametric Analysis

Sequence (Group I)	Pd. 1	Pd. 2	Total	Rank	Pd. 1–Pd. 2	Rank	Tr1–Tr2	Rank
Tr1–Tr2	2.60	2.16	4.76	16	0.44	1.5	0.44	1.5
	2.81	2.53	5.34	5	0.28	4	0.28	4
	3.02	2.69	5.71	1	0.33	3	0.33	3
	2.59	2.50	5.09	9	0.09	9	0.09	12
	2.70	2.45	5.15	8	0.25	5	0.25	5
	2.01	2.49	4.50	17	−0.48	17	−0.48	17
	2.71	2.27	4.98	10	0.44	1.5	0.44	1.5
	2.67	2.55	5.22	7	0.12	8	0.12	10
	Total			73		49		54
Sequence (Group II)								
Tr2–Tr1	2.57	2.38	4.95	11	0.19	6	−0.19	16
	2.36	2.50	4.86	13.5	−0.14	13	0.14	13
	2.73	2.75	5.48	2	−0.02	11	0.02	9
	2.38	2.55	4.93	12	−0.17	14	0.17	8
	2.64	2.75	5.39	3	−0.11	12	0.11	11
	2.52	2.71	5.23	6	−0.19	15	0.19	7
	2.46	2.32	4.78	15	0.14	7	−0.14	15
	2.57	2.79	5.36	4	−0.22	16	0.22	6
	2.46	2.40	4.89	13.5	0.06	10	−0.06	14
	Total			80		104		99

Period 2 for Sequence I is 49. Applying Eq. (15.4),

$$Z = \frac{|49 - 8(8+9+1)/2|}{\sqrt{8 \times 9(8+9+1)/12}} = 2.21(p < 0.05).$$

The test for period effects is based on the comparison of the ranks in the two sequences in the last column of Table 15.9 (see Exercise Problem 17). Of course, the current test for bioequivalence is not based on statistical significance. Nevertheless, the nonparametric approach to this problem is instructive. See section 15.3.1 for an illustration of a nonparametric confidence interval to bioequivalence data.

15.5 KRUSKAL–WALLIS TEST (ONE-WAY ANOVA)

The Kruskal–Wallis test is an extension of the rank sum test to more than two treatments, and is basically a test of the *location* of the distributions, assuming variance symmetry, that is, equality of variances in the different groups. Significant differences can be interpreted as meaning that the averages of at least two of the comparative treatments are different. The computations and analysis will be illustrated using an experiment in which data were obtained from a preclinical experiment in which rats, injected with two doses of an experimental compound and a control (a known sedative), were observed for sedation. The time for the animals to fall asleep after injection was recorded. If an animal did not fall asleep within 10 minutes of the drug injection, the time to sleep was arbitrarily assigned a value of 15 minutes. The experimental results are shown in Table 15.10. One data point was lost from the control group because of an illegible recording, obliterated in the laboratory notebook.

The analysis for treatment differences is not dependent on equal numbers of observations per group, although, as in most experiments, equal sample sizes are most desirable (optional). The analysis consists of first combining all of the data, as in the Wilcoxon rank sum test. To obtain the ranks, one lists all observations in order of magnitude, identifying each value by its group designation. The observations are then reclassified into their original groups, similar to the Wilcoxon rank sum test procedure. The ranks corresponding to each observation are retained and summed for each group as shown in Table 15.10. Note that ties are given the average rank

Table 15.10 "Time to Sleep" for a Control and Two Doses of an
Experimental Compound (minutes)

Control	Rank	Low dose	Rank	High dose	Rank
8	22	10	26	3	10
1	3.5	5	13	4	12
9	24.5	8	22	8	22
		6	15	1	3.5
9	24.5	7	18.5	1	3.5
6	15	7	18.5	3	10
3	10	15	28	1	3.5
15	28	1	3.5	6	15
1	3.5	15	28	2	7.5
7	18.5	7	18.5	2	7.5
Sum of ranks	149.5		191.0		94.5

as in the previously described rank sum test. In addition to the usual analysis, we will present a procedure that corrects the analysis for tied observations [3].

The test statistic for the Kruskal–Wallis test, as described below, is approximately distributed as Chi-square with $k - 1$ d.f., where k is the number of treatments (groups) in the experiment. For small sample sizes, tables to determine the treatment rank sums needed for significance are available [3]. The Chi-square approximation is good if the number of observations in each group is greater than five. The computation of the Chi-square statistic is as follows:

$$\chi^2_{k-1} = \frac{12}{N(N+1)} \left(\sum \frac{R_i^2}{n_i} \right) - 3(N+1), \tag{15.5}$$

where N is the total number of observations in all groups combined, R_i the sums of ranks in ith group, n_i the number of observations in ith group, and k the number of groups.

In our example, $N = 29$, $R_1 = 149.5$, $R_2 = 191$, $R_3 = 94.5$, $n_1 = 9$, $n_2 = 10$, $n_3 = 10$, and $k = 3$. Applying Eq. (15.5), we have

$$\chi^2_2 = \frac{12}{(29)(30)} \left(\frac{149.5^2}{9} + \frac{191^2}{10} + \frac{94.5^2}{10} \right) - 3(29+1) = 6.89.$$

The value of Chi-square with 2 d.f. must be equal to or greater than 5.99 to be significant at the 5% level (Table IV.5). Therefore, the average "time to sleep" differs for at least two of the three treatment groups (control, high dose, and low dose) at the 5% level of significance.

As in the parametric tests, if statistically significant differences among treatments are found, one usually would want to know which treatments are different. For individual (pairwise) comparisons, Table IV.17 tabulates the differences between rank sums needed for significance at the 5% level, given the number of treatments in the design and the sample size [2]. To perform the pairwise treatment comparisons, the number of observations per treatment must be the same. For example, in the case of three treatments, each with a sample size of 10, a difference between the rank sums of two of the treatments (groups) must exceed 92 in order for the two treatments to be considered different at the 5% level. In our example, had the control group had 10 observations instead of 9, we could apply the pairwise test. However, if an additional observation had been included in the control group, the greatest difference between the rank sums of the control group and one of the doses of the experimental drug in this experiment could not exceed 92.[†] The observed difference between the high and low doses is $(191 - 94.5) = 96.5$, which exceeds 92. Thus, the pairwise comparison criterion shows a significant difference

---

[†] The largest difference between the control and one of the experimental drug doses would occur if the tenth value in the control group were the highest observation. The rank sum of the control group would be increased by 30, resulting in a rank sum of 179.5 (Table 15.10). The difference between the rank sums of the control and high-dose groups would be $179.5 - 94.5 = 85$, which is not significant at the 5% level.

between the high and low doses of the experimental drug ($p < 0.05$), agreeing with the significant Chi-square test. For more details concerning multiple comparisons in the Kruskal–Wallis test, see Refs. [2,3].

As in the ranking procedures previously described, tied values are given the average rank. A correction for ties can be used that increases the value of Chi-square. Therefore, if the null hypothesis is rejected (significant treatment differences), the correction only increases the degree of significance. If Chi-square just misses significance, the correction may result in statistically significant differences. The correction is as follows:

$$\text{Correction} = \frac{\chi^2}{1 - \sum (t_i^3 - t_i)/(N^3 - N)},$$

where t_i is the number of tied observations in group i and N is the total number of observations. The calculations are illustrated below. There are eight groups of ties in the data shown in Table 15.10. For example, there are six values equal to 1. For this group of ties, $t^3 - t$ is equal to $6^3 - 6 = 210$. Another group of ties are the two values equal to 2. There are two values of 2 in the data, and for this group, $t = 2$ and $t^3 - t = 6$. The other ties occurred for values of 3, 6, 7, 8, 9, and 15. The reader can verify that the sum of T (where $T = \Sigma(t_i^3 - t_i)$) is 378. The correction for Chi-square is

$$\frac{6.89}{1 - 378/(29^3 - 29)} = \frac{6.89}{0.984} = 7.00.$$

(Note that $N = 29$ in this example.) The correction for ties is usually very small. Of course, in this example, the correction does not change the conclusion of significant differences among treatment means.

15.6 FRIEDMAN TEST (TWO-WAY ANALYSIS OF VARIANCE)

The Friedman test is a nonparametric test applied to data that is, at least, ranked and that is in the form of a two-way ANOVA design (randomized blocks). This test, which may be applied to ranked or interval/ratio-type data, is used when more than two treatment groups are included in the experiment. For two groups in a paired (two-way) design, the rank sum test may be used. In the Friedman test, the treatments are ranked *within each block* (e.g., animal or person), disregarding differences between blocks. The procedure will be illustrated using the data from Table 15.11. These data describe the results of a *validation* experiment to test the performance of four tablet presses, with regard to tablet hardness. The average hardness of 10 tablets was computed for five different tablet products manufactured on four presses. The tablets are a random selection of five typical tablet products. The presses were identically set for the same pressure for each tablet formulation.

The parenthetical values in Table 15.11 are the ranks of the average hardness for each formulation over the four presses. For formulation 1, the lowest value, 6.9, is assigned a rank of 1, and the highest value, 7.5, is assigned a rank of 4. Although no ties occurred in this example, if ties were observed, the average rank would be assigned to the tied observations as discussed in the preceding sections. If one of the presses consistently had the highest (or lowest) rank, one would conclude that the press (treatment) produced harder (or less hard) tablets than the other presses. In our example, tablet press C had the highest hardness value for all formulations with the exception of formulation 1, where it had the next-to-largest value. The test of significance is an objective assessment of whether or not the data of Table 15.11 provide sufficient evidence to say that tablet press C is, indeed, producing harder tablets than the other presses.

Table 15.11 Average Hardness of 10 Tablets for Five Different Tablet Formulations Prepared on Four Presses[a]

Tablet formulation	Tablet press			
	A	B	C	D
1	7.5 (4)	6.9 (1)	7.3 (3)	7.0 (2)
2	8.2 (3)	8.0 (2)	8.5 (4)	7.9 (1)
3	7.3 (1)	7.9 (3)	8.0 (4)	7.6 (2)
4	6.6 (3)	6.5 (2)	7.1 (4)	6.4 (1)
5	7.5 (3)	6.8 (2)	7.6 (4)	6.7 (1)
R_i	14	10	19	7

[a]Parenthetical values are the within-tablet press ranks.

If the sample sizes are sufficiently large, a Chi-square distribution can be used to approximate the test of significance. The Chi-square test is

$$\chi^2_{c-1} = \frac{12}{rc(c+1)} \left(\sum R_i^2 \right) - 3r(c+1),$$

(15.6)

where χ^2_{c-1} is the χ^2 statistic with $c - 1$ d.f., r the number of rows (blocks), c the number of columns (treatments), and R_i the sums of ranks in the ith group (column).

In our example, the Chi-square statistic has 3 d.f.

$$\chi^2_3 = \frac{12}{(5)(4)(4+1)}(14^2 + 10^2 + 19^2 + 7^2) - 3(5)(5) = 9.72.$$

A Chi-square value of 7.81 or larger is needed for significance at the 5% level (Table IV.5). We can conclude that at least two of the tablet presses differ with regard to tablet hardness. Examination of Table 15.11 shows that tablet press C produces harder tablets than those produced by the other presses. Table IV.18 shows that a difference of 11 is needed for significance ($p < 0.05$) for individual comparisons between pairs of means for 4 treatments ($k = 4$) and 5 rows ($n = 5$). Therefore, press C produces significantly harder tablets than press D with a sum of ranks of 19 and 7, respectively.

For small samples, exact probabilities for the Friedman test are given in *Nonparametric Statistical Methods* [3]. This test also describes a test that corrects Chi-square for tied observations.

15.6.1 Modified Friedman Test

Conover [9,10] recommends a statistic that has an approximate F distribution with $(c - 1)$, $(c - 1)(r - 1)$ d.f. (where r is the number of rows and c is the number of columns in the RXC matrix of data). This method of analysis has been shown to be superior to the Chi-square distribution for the Friedman nonparametric analysis (sect. 15.6) of a two-way ANOVA model. The statistic T_2 is calculated as follows:

Compute $A_2 = \sum (x_{ij})^2$, where the x_{ij} are the individual ranks.
A_2 is equal to $cr(c+1)(2c+1)/6$ if there are no ties (ties are given the value of the average rank).
c = number of columns and r = number of rows
Compute $B_2 = (1/r) \sum (C_i)^2$
where C_i is the sum of observations in column i. Then

$$T_2 = \frac{[(r-1)\{B_2 - rc(c+1)^2/4\}]}{A_2 - B_2}$$

Refer T_2 to an F table with $c - 1$ and $(r - 1)(c - 1)$ d.f.

Example 3. The computations for this analysis are shown below for the data from Table 15.11

$$A_2 = \frac{cr(c+1)(2c+1)}{6} = \frac{4 \times 5 \times (4+1)(2 \times 4+1)}{6} = 150$$

$$B_2 = \left(\frac{1}{r}\right) \sum (C_i)^2 = \left(\frac{1}{5}\right)(14^2 + 10^2 + 19^2 + 7^2) = \frac{706}{5} = 141.2$$

$$T_2 = \frac{[(5-1)(141.2 - 4 \times 5(4+1)^2/4)]}{(150 - 141.2)} = 7.364$$

Compare 7.364 to the tabled value of F with 3 and 12 d.f. at the 5% level (App. IV, Table IV.6A)

$$F_{3, 12, 0.05} = 3.49.$$

Since the observed F (7.364) is larger than the tabled F (3.49) at the 5% level, the differences among tablet presses are significant ($p = 0.005$). The usual Friedman test that uses a Chi-square statistic shows a level of 0.02 (sect. 15.6). See Exercise Problem 18 at the end of this chapter for the application of ANOVA to this data.

15.6.1.1 Multiple Comparisons for the Modified Friedman Test
If the null hypothesis of equal treatment means is rejected, the following formula can be used to calculate a least significant difference between pairs of treatments:

$$[C_j - C_i] > t\sqrt{[2r(A_2 - B_2)/[(r-1)(c-1)]}$$

where t is the tabled t value with $(r-1)(c-1)$ d.f. at the specified alpha level.
    Applying this formula to the data in Table 15.11 for tablet press differences at the 5% level.

$$[C_j - C_i] > 2.18\sqrt{[2 \times 5(150 - 141.2)]/[(5-1)(4-1)]} = 5.90.$$

Any difference in rank sums ≥ 5.9 is significant at the 0.05 level. Inspection of the results shown in Table 15.11 shows that Tablet Press C gives higher results ($p < 0.05$) than B and D, and A is higher than D. In this example, we see more significant differences with the modified test compared to the Friedman test described in section 15.6.

15.6.2 Quade Test for Randomized Block Design
Conover [9] presents another test (Quade Test) that is still valid in the presence of many ties. In addition to the usual computations as shown in the Friedman Test, a further computation is needed. The range of values (largest minus smallest value) is calculated for each block (row). The blocks are ranked in order from the smallest to the largest with regard to the range of values within a block. Call these ranks $Q_1 \ldots Q_r$, where r is the number of rows (blocks). Let $R(X_{ij})$ be the rank of each observation, where ranks are within each row or block. Compute for each observation

$$S_{ij} = Q_i[R(X_{ij}) - (k+1)/2],$$

where k is the number of treatments (columns).

Let $S_i = \sum S_{ij}$ for each treatment.
Calculate $A = \sum S_{ij}^2$ (for all observations).
Calculate $B = \sum S_i^2/r$.

For Table 15.11, the calculations for the "Quade" test are as follows:

Range of row 1 = 0.6 (7.5 − 6.9)
Range of row 2 = 0.6 (8.5 − 7.9)
Range of row 3 = 0.7 (8.0 − 7.3)
Range of row 4 = 0.7 (7.1 − 6.4)
Range of row 5 = 0.9 (7.6 − 6.7)

Q_i is the rank of row i.

$Q_1 = 1.5$

$Q_2 = 1.5$

$Q_3 = 3.5$

$Q_4 = 3.5$

$Q_5 = 5$

(Note: as usual, compute the average rank for ties.) As an example, the calculation of S_{11} follows:

S_{11} = value for formulation 1 on press 1 = $1.5[4 − (4 + 1)/2] = 2.25$.

The values of S_{ij} derived from the data in Table 15.11 are shown in Table 15.12.

$$A = \sum (S_{ij})^2 = 270$$

$$B = \sum \frac{S_i^2}{r} = \left[2^2 + (-5.5)^2 + 21^2 + (-17.5^2)\right]5 = 156.3$$

The test statistic is

$$T = \frac{(r − 1)B}{A − B}$$

$$T = \frac{4(156.3)}{270 − 156.3} = 5.499.$$

Refer T to an F distribution with $(c − 1)$ and $(r − 1)(c − 1)$ d.f. at the appropriate alpha level (App. IV, Table IV.6A).

Table 15.12 Table to Aid Computations for Quade Test (S_{ij}) Press

	A	B	C	D	Range	Rank
Formulation						
1	2.25	−2.25	0.75	−0.75	0.6	1.5
2	0.75	−0.75	2.25	−2.25	0.6	1.5
3	−5.25	1.75	5.25	−1.75	0.7	3.5
4	1.75	−1.75	5.25	−5.25	0.7	3.5
5	2.5	−2.5	7.5	−7.5	0.9	5
Sum	2.00	−5.50	21.00	−17.50		

Tabled $F_{3,12} = 3.49$ at the 5% level.

Therefore, at least two of the presses are significantly different ($p = 0.013$). Multiple comparisons can be made if the F test shows significance. The difference between the sums of any two treatments is significant if the absolute value of the difference exceeds

$$t \times \left[\frac{2r(A - B)}{(r - 1)(c - 1)} \right]^{1/2},$$

where t is the appropriate tabled t value at the alpha level with $(r - 1)(c - 1)$ d.f. In this example, $t_{0.05,12} = 2.18$, and the least significant difference is

$$2.18 \left[\frac{10(270 - 156.3)}{(4)(3)} \right]^{1/2} = 21.22.$$

We conclude that press C gives higher results than presses B and D at the 5% level of significance.

This analysis is identical to an analysis of variance on the ranks in Table 15.12. The least significant difference is computed as in Fisher's LSD, based on the analysis of the "adjusted" ranks, as computed from Eq. (8.7) (see Exercise Problem 19).

Three different tests applied to these data give somewhat different overall conclusions. This is caused by the fact that some of the comparisons are close to significant (C vs. A and C vs. B). As always the test to be applied and the level of significance should be clearly defined at the initiation of the experiment.

15.7 NONPARAMETRIC ANALYSIS OF COVARIANCE

Quade has proposed a simple and neat nonparametric analysis of covariance (ANCOVA) [11]. The procedure is described in detail using the data of Table 15.13.

Rank each of X and Y (raw material and product assays, respectively) disregarding treatment. Let the lowest value have rank 1 up to the highest value, rank N, where there are a total of N observations. Correct the rankings so the mean of the ranks $= 0$, by subtracting the average rank, $(N + 1)/2$, from each rank Y. In this example, $N = 8$. The lowest value of Y (product assay) is 95.4 and is given a rank of 1. Subtract $(N + 1)/2 = 9/2 = 4.5$ from 1, resulting in $R_y = 1 - 4.5 = -3.5$. Similarly, the largest assay is 98.5, with an adjusted rank of $8 - 4.5 = +3.5$. The ranks of the raw material assays are calculated similarly. Use average ranks in case of ties.

Table 15.13 Data for Quade Nonparametric Covariance Analysis (ANCOVA)

Final assay Y	Raw material X	Ranks R_y	−4.5 R_x	Predicted	Residual
Method I					
98.00	98.40	2.50	−3.00	1.4451220	1.0548780
97.80	98.60	1.50	−1.00	0.4817073	1.0182927
98.50	98.60	3.50	−1.00	0.4817073	3.0182927
97.40	99.20	−0.50	2.50	−1.2042683	0.7042683
Sum					5.795732
Method II					
97.60	98.70	0.50	0.50	−0.2408537	0.7408537
95.40	99.00	−3.50	1.50	−0.7225610	−2.7774391
96.10	99.30	−2.00	3.50	−1.6859756	−0.3140245
96.10	98.40	−2.00	−3.00	1.4451220	−3.4451220
Sum					−5.795732

Next, perform a regression of the adjusted ranks of $Y(R_y)$ on the adjusted ranks of $Y(R_x)$ for all data, to obtain the residuals. Remember, the residuals are the difference between the observed values of R_y and the predicted values of R_y based on the calculated regression parameters (Table 15.13).

An analysis of variance is performed on the residuals (Group I vs. Group II). Note that there is no correction for the mean because the mean of the residuals is 0.

Quade used the following formula that has an F distribution with $k-1$ and $N-k$ d.f., where k is the number of groups and N is the total number of observations. In our example, we have two groups and eight observations, resulting in an $F_{1,6}$ distribution.

$$F_{k-1,N-k} = \frac{(N-k)\sum (Z_{ij})^2/n_i}{(k-1)[\sum_i \sum_j Z_{ij}^2 - \sum(Z_{ij})^2/n_i]}$$

$N = 8$

$k = 2$

$n_1 = n_2 = 4$

$$\sum \frac{(Z_{ij})^2}{n_i} = \frac{(5.795732^2 + \{-5.795732\}^2)}{4} = 16.795255$$

$$\sum_i \sum_j Z_{ij}^2 = 31.98628$$

$$F_{k-1,N-k} = (8-2)(16.795255/[(2-1)(31.98628 - 16.795255)]$$

$$F_{1,6} = 6.634$$

$p = 0.042$ (this result may be compared to $p = 0.037$ using a parametric analysis, sect. 8.6).

An assumption for this test is that the variables be on an ordinal scale, not necessarily continuous (dichotomous variables may be used). We do not have to assume normality or linearity of y on x. However, the distribution of X should be the same in each group, a requirement not needed for the parametric analysis.

15.8 RUNS TEST FOR RANDOMNESS

When performing an experiment (or observing a process) where values are observed sequentially, it may be of interest to determine whether the observations are randomly varying about the central value (i.e., the median). If the process is not random, we might expect to see trends in the data, perhaps a consecutive series of high or low values, which are unlikely to occur by chance. The *runs* test is a simple method of investigating the "random" nature of such a process. Tests for runs were introduced in section 12.2.5, the discussion of control charts. A run is a series of *uninterrupted, like* observations. For example, suppose that the median weight of 20 tablets, sequentially taken during a batch run, is 200 mg. Twenty consecutive tablets were weighted with the following results:

The first six tablets weighed more than 200 mg.
The next five tablets weighed less than 200 mg.
The next four tablets weighed more than 200 mg.
The next (remaining) five tablets weighed less than 200 mg.

If we designate tablet weights less than 200 mg by a minus (−), and tablet weights more than 200 mg by a plus (+), the 20 weights can be described by the following sequence:

200 mg → $\underline{++++++}$ $\underline{-----}$ $\underline{++++}$ $\underline{-----}$

The first six values, +'s, represent a *run*. Each time that a series of like signs change, a new run begins. There are *four* runs in these data: *six* pluses, *five* minuses, *four* pluses, and *five* minuses. If the tablet weights follow a random process, one might suspect that the sequence of values described above is unlikely. It appears that the pluses and minuses come in "bunches." One might guess that the sequence of pluses and minuses could have been due to too-frequent weight adjustments on the tablet press. For example, the first tablets sampled were over the median weight of 200 mg. The tablet press may then have been adjusted down, more than necessary, resulting in too-low tablet weights (the next five tablets were underweight), and so on.

To test for randomness for sample sizes as large as 40, we can refer to Table IV.14. The table gives the lower and upper limits for the number of runs that would be expected to occur in a random process in a sample of size N. An observed number of runs equal to or less than the critical lower number or greater than the critical upper number shown in Table IV.14 is an indication that the process is not random at the 5% level. The runs test is usually a two-sided test; either too few or too many runs lead to significance (nonrandomness). In some cases, for example, control charts, only relatively few runs may be considered to suggest problems with a process. In these situations, critical values for a one-sided test as shown in Table IV.14 are appropriate. According to Table IV.14, for a sample size of 20, between 7 and 16 runs would be expected to occur if the null hypothesis of randomness is true. We observed four runs in the sample of 20 tablets ($N = 20$) in our example. Therefore, we conclude that the process is not random ($p < 0.05$). The clusters of high and low values are probably due to some malfunctioning of the tableting process.

Consider the following as a further example of an application of the *runs* test. A *standard* is analyzed every 20th sample in an automated analytical procedure. A record of the readings for the standard in chronological order derived from one day's assay results are shown in Table 15.14. The median value for the data in the table is 0.7985 (the 20th and 21st ordered values are 0.798 and 0.799). As in the previous example, we label values greater than the median as + and values less than the median as −. The sequence of pluses and minuses is as follows (Samples 1 and 2 are below the median; 3 and 4 are above the median, etc.):

$\underline{--}\underline{++}\underline{-}\underline{+}\underline{-}\underline{+}\underline{-}\underline{+}\underline{---}\underline{+++++++}\underline{---}$

$\underline{+++++}\underline{--}\underline{+++}\underline{--------}$

The runs are underlined in the previous sequence. There are 15 runs. For sample sizes of 40 or more, a normal approximation to the distribution of runs is available, under the null hypothesis that the observed values occur in a random manner.

$$Z = \frac{|r - (N/2 + 1)|}{\sqrt{N(N-2)/4(N-1)}},\tag{15.7}$$

where r is the number of runs and N is the sample size.

Values of Z equal to or greater than 1.96 are unlikely ($p \leq 0.05$) if the observations are random. In our example $N = 40$ and $r = 15$. Therefore,

$$Z = \frac{|15 - (40/2 + 1)|}{\sqrt{40(40-2)/4(40-1)}} = \frac{6}{3.12} = 1.92.$$

The value of Z is not quite large enough for the data to be considered nonrandom at the 5% level. Table IV.14 shows that for a sample of size 40, an observation of between 15 and 26 runs leads to acceptance of the null hypothesis of randomness, agreeing with the conclusion of

Table 15.14 Readings of a Standard Solution in Chronological Order (Optical Density)

Sample	Reading
1	0.795
2	0.796
3	0.804
4	0.801
5	0.792
6	0.816
7	0.791
8	0.819
9	0.796
10	0.815
11	0.782
12	0.795
13	0.798
14	0.800
15	0.800
16	0.802
17	0.799
18	0.805
19	0.820
20	0.802
21	0.796
22	0.797
23	0.795
24	0.802
25	0.800
26	0.801
27	0.802
28	0.820
29	0.788
30	0.780
31	0.813
32	0.804
33	0.801
34	0.793
35	0.790
36	0.791
37	0.784
38	0.791
39	0.788
40	0.794

the normal approximation [Eq. (15.7)]. Had 14 runs been observed, we would have concluded that the data were not random ($p < 0.05$).

15.9 CONTINGENCY TABLES

Chi-square tests for contingency tables (e.g., 2×2 tables) are often categorized as nonparametric tests. The analysis of 2×2 tables using a Chi-square test was described in section 5.2.5. The Chi-square test can be applied to nominal or categorical data that cannot be analyzed using the ranking techniques discussed above. These data cannot be ordered (the data are not ordinal or on an interval/ratio scale). Nominal data are usually available in the form of *counts*, such as 25 males and 12 females entered into a clinical study; or the *number* of tablets categorized as acceptable, chipped, cracked, and so on. For large samples, Chi-square methods can be used to compare "statistically" the relative frequency of such events that occur in two or more groups. Here we will briefly expand the case of the fourfold table, discussed in Chapter 5, to the analysis

Table 15.15 Examples of $R \times C$ Tables

2 × 2 table (Fourfold table)		
Treatment	Cured	Not cured
A		
B		

2 × 3 table			
Treatment	Unsuccessful	Moderately successful	Successful
A	25	10	40
B	27	23	25

$R \times C$ table

Treatment	Outcome
	1 2 3 ⋯ C
1	
2	
3	
.	
.	
R	

of $R \times C$ tables, R rows and C columns. We will then examine the case of 2 × 2 tables with small expected frequencies, followed by different tests of hypotheses for fourfold tables.

15.9.1 $R \times C$ Tables

In the binominal case, data are dichotomized, resulting in the 2 × 2 table, for example, comparison of success rates of two treatments as shown in Table 15.15. When experiments consist of more than two comparative groups and/or more than two possible outcomes, we are, in general, confronted with an $R \times C$ table (Table 15.15).

In the experiments involving contingency tables, we are usually interested in testing group differences with regard to proportions or the distribution of counts in the various outcome categories. Consider the data in the 2 × 3 table in Table 15.15. Two treatments have been compared where the outcomes are categorized as "unsuccessful," "moderately successful," and "successful." Inspection of the data indicates that Treatment A has a greater incidence of "successful" events and less "moderately successful" events than Treatment B.

Equivalently the hypothesis in contingency tables is often stated in terms of the relationship between rows and columns. "Acceptance" of the null hypothesis suggests that the rows and columns are independent. For example, in the 2 × 3 contingency table in Table 15.15, lack of rejection of the null hypothesis would be interpreted, in this context, as meaning that the experimental outcomes are independent of the treatment (i.e., the treatments do not differ with respect to the experimental outcome).

The relationship of the rows and columns in an $R \times C$ contingency table may be tested by means of the Chi-square distribution with $(R - 1)(C - 1)$ d.f. Note that for a 2 × 2 table, we have 1 d.f., agreeing with the analysis of 2 × 2 tables described in chapter 5. The Chi-square statistic is calculated as

$$\chi^2_{(R-1)(C-1)} = \sum \frac{(O - E)^2}{E}, \tag{15.8}$$

where O is the observed count and E is the expected count. The summation in Eq. (15.8) is for all $R \times C$ cells in the contingency table.

The Chi-square test is an approximate test and should be used only when the expected values are sufficiently large. The usually recommended minimum expected value of five for each cell, as described in section 5.2.5, is conservative [1]. If most of the cells have an expected value of five or more, the test should be reliable. If there is doubt about using the Chi-square test, the exact test (multinomial) may be computed [12]. The calculations for the exact test solution are usually very tedious.

Table 15.16 Patients Categorized by Severity of Disease Entered into Two Treatment Groups in a Clinical Study

	Very severe	Moderately severe	Mildly severe	Total
Treatment A	13	24	18	55
Treatment B	19	20	12	51
Total	32	44	30	106

Table 15.16 shows data from a clinical study in which patients entering the study were categorized according to the severity of disease. Severity was divided into three classes: very severe, moderately severe, and mildly severe. The categorization was made to ensure that the severity of disease was similar for patients in the two treatment groups. Thus, the question addressed by these data is "Is the severity of disease similar for patients entered into the two treatment groups?" or "Is there a relationship between 'treatment' and 'severity of disease'?" In a sense, this test is a confirmation of the randomization procedure used to assign patients to the two treatment groups. We would expect that, "on the average," the severity would be similar in Groups A and B.

The Chi-square calculation is similar to that for the fourfold (2 × 2) table (chapter 5). The *expected* values for each cell are obtained by multiplying the row and column totals corresponding to the cell, and dividing this result by the grand total (row total × column total/grand total). In the example in Table 15.16, this calculation needs to be done for only two cells (note the 2 d.f.), because the remaining four expected values can be obtained by subtraction from the fixed row and column totals. The sum of the expected values must equal the row and column totals of the raw data. In the table the expected value for the cell with 13 patients (Treatment A, very severe) is $(32)(55)/(106) = 16.60$. For the cell defined by Treatment A, moderately severe, the expected value is $(44)(55)/(106) = 22.83$. The expected values are shown in Table 15.17.

The Chi-square statistic is calculated according to Eq. (15.8).

$$\chi_2^2 = \frac{(13 - 16.60)^2}{16.60} + \frac{(24 - 22.83)^2}{22.83} + \frac{(18 - 15.57)^2}{15.57} + \frac{(19 - 15.4)^2}{15.40} + \frac{(20 - 21.17)^2}{21.17} + \frac{(12 - 14.43)^2}{14.43} = 2.54.$$

For significance at the 5% level, a value of 5.99 is needed for Chi-square with 2 d.f. (Table IV.5). Since the observed Chi-square is 2.54, we conclude that there is not sufficient evidence to show that severity and treatment are related; that is, the two treatment groups cannot be shown to differ with regard to the distribution of severity of disease.

Another example of an $R \times C$ table is shown in Table 15.18. This differs from the previous example in that we have three treatments each with a dichotomous outcome, rather than two treatments with three categories of outcome. The analysis tests for differences among the three treatments. This data is derived from a clinical study in which three treatments were randomly assigned to 60 patients. Only 54 patients successfully completed the study. Patients were classified as success or failure, depending on their response to treatment.

The analysis proceeds exactly as in the preceding example. The value of Chi-square with 2 d.f. is 7.76. Since the table Chi-square with 2 d.f. is 5.99, the treatments are significantly different. To test for differences suggested by the data (a posteriori tests), perform a Chi-square test for two

Table 15.17 Expected Values for the Data of Table 15.10

	Very severe	Moderately severe	Mildly severe	Total
Treatment A	16.60	22.83	15.57[a]	55
Treatment B	15.40[a]	21.17[a]	14.43[a]	51
Total	32	44	30	106

[a]Obtained by subtraction from total; see the text (e.g., 55 − 16.60 − 22.83 = 15.57).

Table 15.18 Number of Successes and Failures Following Three Treatments

Treatment	Successes	Failures	Total
A	9	6	15
B	8	11	19
C	17	3	20
Total	34	20	54

treatments (a 1 d.f. test), but use the Chi-square cut-off point for 2 d.f., 5.99, for significance. For example, the Chi-square value for the comparison of Treatments B and C is 7.79, and Treatments B and C are significantly different (see Exercise Problem 13).

For a further discussion of multiple comparisons and other topics in the analysis of categorical data, the book *Statistical Methods for Rates and Proportions* by Fleiss [13] is highly recommended.

15.9.2 Fisher's Exact Test

In the Chi-square analysis of 2×2 contingency tables, if the expected values are too small, the Chi-square test may not be appropriate. Dichotomous data with small expected values are commonly encountered in pharmaceutical research, particularly in preclinical toxicology studies. For example, in preclinical animal carcinogenic studies, when comparing control and treatment groups with respect to some characteristic that occurs infrequently, the comparison of the frequencies may not be amenable to a Chi-square analysis. Fisher's exact test for 2×2 tables can be used to compute the exact probabilities. This test can be used, for example, to compare proportions for two independent groups (treatments), a binomial test, where expected values are very small.

Fisher's exact test makes use of the fact that the probability of a given configuration in a fourfold table with *fixed margins*[‡] can be computed using the *hypergeometric* distribution. The probability calculation will be described with reference to the notation in Table 15.19 to help clarify the procedure.

The probability of the values found in Table 15.19, given the four *fixed* margins, $(A + C)$, $(B + D)$, $(A + B)$, and $(C + D)$, is

$$\frac{(A + B)!(C + D)!(A + C)!(B + D)!}{N!A!B!C!D!} \tag{15.9}$$

The numerator of Eq. (15.9) is obtained by multiplying the factorials of the marginal totals. The denominator is the product of the factorials of the individual cells of the fourfold table, multipled by $N!$, the factorial of the total number of observations.

Table 15.20 shows data typically analyzed using the Fisher's exact test. One group of animals was administered a placebo preparation consisting of all components of the drug formulation with the exception of the active ingredient (placebo group). Another group of animals (drug group) was administered the drug formulation. After a fixed period of time, the incidence of a particular type of carcinoma was noted. The probability of the fourfold table shown in Table 15.20 with fixed margins (12, 14, 5, and 21) is calculated using Eq. (15.9).

$$\frac{5!21!12!14!}{26!1!4!11!10!} = 0.183.$$

[‡] Theoretically, Fisher's exact test is appropriate when marginal totals are fixed. In the example in Table 15.20, this means that before the initiation of the experiment, we decided to use 12 animals on placebo and 14 animals on drug; a total of five carcinomas will be observed in both groups. The latter result is clearly not under our control (although in some experiments, the marginal totals can be controlled). There exists some controversy whether data, in which two independent groups are to be compared (as in Table 15.19), where the margins are not fixed, are appropriate for Fisher's exact test. However, the test is commonly used to analyze such data.

Table 15.19 Fourfold Table as an Aid to the Calculation of Fisher's Exact Test

		Column		
		I	II	Total
Row	I	A	C	$A + C$
	II	B	D	$B + D$
Total		$A + B$	$C + D$	$A + B + C + D = N$

Thus, the probability of the results shown in Table 15.14 are *not* very unlikely. However, this is not the entire statistical test. In Fisher's test, we compute the probability of the observed configuration *plus* the probabilities of *all less likely* configurations (a cumulative probability). If the sum of the observed configuration plus all less likely configurations is *less than* α (0.05, for example), we conclude that the rows and columns (treatment and carcinoma) are *not* independent; that is, the treatments differ with respect to the incidence of carcinomas. If the sum of these probabilities exceeds α (0.05, for example), we accept the null hypothesis of independence, concluding that the evidence is not sufficient to conclude that the treatments differ. In the example (Table 15.20), the sum of probabilities obviously exceeds 0.183. (The probability of the observed table is 0.183). Therefore, there is insufficient data to show conclusively that the incidence of carcinoma is greater in the drug group compared to the placebo group.

To clarify this procedure further, we will work out an example in more detail based on the data shown in Table 15.21. These data are similar to that in Table 15.20, except that no carcinomas were observed in the placebo group and five were observed in the drug group. Thus, the marginal totals are the same in Tables 15.20 and 15.21. The probability of Table 15.21 is calculated as before, using Eq. (15.9).

$$\frac{5!21!12!14!}{26!0!5!12!9!} = 0.03043.$$

In order to assess the possible "statistical" significance of this table, we must compute the probability of all less likely configurations as discussed above. What constitutes less likely tables is not always obvious without some "trial and error" calculations. With experience, good, educated guesses can be made as to what constitutes a less likely table.

If a configuration is mistakenly chosen with a higher probability than the observed table, the calculation is discarded. Possible "less likely" tables are shown in Table 15.22 with the probability of each table. The only table with a lower probability than the observed table (Table 15.21) is the one with all five carcinomas appearing in the placebo group.

$$\frac{5!21!12!14!}{26!5!0!7!14!} = 0.01204.$$

Table 15.20 Incidence of Carcinoma in Drug- and Placebo-Treated Animals: Example 1

	Number of animals with:		
	Carcinoma	No carcinoma	Total
Placebo	1	11	12
Drug	4	10	14
Total	5	21	26

Table 15.21 Incidence of Carcinoma in Drug- and Placebo-Treated Animals: Example 2

	Carcinoma present	Carcinoma absent	Total
Placebo	0	12	12
Drug	5	9	14
Total	5	21	26

The sum of the probabilities of the observed table and all less likely (or equally likely) tables is $0.03043 + 0.01204 = 0.0425$. Therefore, Table 15.15 is "significant" at the 5% level ($p < 0.05$); the drug appears to result in an increased incidence of carcinomas.

Note that Fisher's exact test requires that the probabilities of tables with fixed margins be computed for all possible configurations. If we calculate all possible configurations, the sum of the probabilities of the different tables would be equal to 1. Among all of these probabilities will be the probability of the observed table, in addition to possible probabilities equal to or smaller than that of the observed table. If the sum of these probabilities is less than or equal to 0.05, for example, the treatments are said to be "significantly" different at the 5% level, in the context of the present example.

The computations are often very tedious. For cases where the computations are unduly long and tedious, the use of computer programs or tables to determine significance points in fourfold tables are recommended [14].

15.9.3 Fourfold Tables with Related Samples

The examples of 2×2 contingency tables previously discussed in this chapter and chapter 5 have involved the comparison of proportions or frequencies in two or more *independent* groups. A similar problem that occurs less frequently in pharmaceutical research is the comparison of two groups where the observations are *related*, also known as matched pairs. For example, Table 15.23 shows the results of two versions of an allergy test, A and B, applied to 50 persons. The test reagents were applied at the same time at different sites for each subject, and either a positive or negative reaction was observed. In this design, the total sample size is specified in advance, but the marginal totals are not fixed. We cannot anticipate the total positive and negative for test B in Table 15.23, for example. In the previous example, the size of the two treatment groups can be fixed in advance. Note that in this example each person is subjected to both treatments (allergy tests). In the previous examples of fourfold tables, each person is subjected to a single treatment and a dichotomous response is observed (e.g., *cured* or *not cured*).

The objective of this experiment is contained in the question: "Does the proportion of positive reactions for test A differ from that for test B?" (i.e., $H_0: p_a = p_b$, where p_a and p_b are the proportion of positive reactions in tests A and B, respectively). Note that test A has 32 positive reactions $(23 + 9)$, and test B has 29 positive reactions $(23 + 6)$. It can be shown that the statistical test for the equality of positive reactions for the two tests is equivalent to the test for the equality of the counts in the diagonal cells designated by an *a* in Table 15.23 (9 and 6) [13]. The counts (or proportions) in these two cells represent the *untied* responses (*positive A* and *negative B*, and *negative A* and *positive B*, 9 and 6, respectively). The counts in the other two cells do not differentiate the two allergy tests. For example, the upper left-hand cell shows the 23 patients who were positive on *both* tests.

Table 15.22 Some "Unlikely" Tables with Margins Identical to Table 15.21

	Carcinoma present	Carcinoma absent	Total	Carcinoma present	Carcinoma absent	Total
Placebo	5	7	12	4	8	12
Drug	0	14	14	1	13	14
Total	5	21	26	5	21	26
	Probability $= 0.0120$			Probability $= 0.1054$		

Table 15.23 Frequency of Positive and Negative Reactions to Two Allergy Tests Applied to Two Sites in 50 Persons

		Test B		
		Positive	Negative	Total
Test A	Positive	23	9[a]	32
	Negative	6[a]	12	18
	Total	29	21	50

[a] Patients who were positive on one test and negative on the other test.

Under the null hypothesis that the probability of a positive reaction is equal for both tests, the diagonal counts, 9 and 6, should be equal. The test of significance is a binomial test, as in the sign test (sect. 15.2). In the latter procedures, the observed proportion is compared to 0.5, the expected proportion if both treatment groups have an equal probability of being positive. The statistical test in this example makes use of the normal approximation to the binomial distribution [Eq. (15.1)].

$$Z = \frac{|\text{observed proportion} - 0.5| - 1/(2N)}{\sqrt{(0.5)(0.5)/N}}. \qquad (15.1)$$

If Z is greater than 1.96, the difference is significant at the 5% level and we conclude that the probability of a positive response is different for the comparative treatments. (As in other examples where the normal approximation to the binomial is used, the sample size should be sufficiently large, approximately 10 for this test.) The observed proportion in the example in Table 15.23 is $9/15 = 0.60$. $N = 15$, the number of untied pairs. Therefore,

$$Z = \frac{|0.60 - 0.5| - 1/30}{\sqrt{(0.5)(0.5)/15}} = 0.52.$$

Since Z is not equal to or greater than 1.96, the difference is not significant at the 5% level. The difference is not sufficiently large to conclude that the two tests differ with regard to the frequency (proportion) of positive responses. This test is also known as McNemar's test.

The data shown in Table 15.23 can also answer a different question that requires a different analysis. In the previous example, we inquired if the proportion of positive reactions was different in the two tests. Another question that is often relevant to such data is: "Are the allergy tests independent, that is, is the probability of a positive response for test B independent of the outcome for test A?" This question implies that if A and B are independent, there should be an equal proportion of positive results to test A in both patients with a positive test to B and in patients with a negative test to B. Table 15.24 shows the *expected results* if, in fact, tests A and B are independent. Note that the expected proportion of positive A's in patients who had a positive test for B is 0.64, 18.56/29. This is the same expected proportion of positive A's as that for patients who had a negative test for B, 13.44/21.

Table 15.24 Expected Values from Table 15.17 if Allergy Tests A and B Are Independent

		Test B		
		Positive	Negative	Total
Test A	Positive	18.56	13.44	32
	Negative	10.44	7.56	18
	Total	29	21	50

Table 15.25 Fourfold Table for Treatment and Placebo

	Improvement		
Treatment	None	Some or marked	Total
Active	13	28	41
Placebo	29	14	43
Total	42	42	84

The test for independence is the same Chi-square test as that used for the comparison of proportions in two independent samples, although the question to be answered is different (see sect. 5.2.5). We apply Eq. (15.8)

$$\chi_1^2 = \sum \frac{(O - E)^2}{E},$$ (15.8)

where O is the observed count and E is the expected count. The expected values for the Chi-square test are shown in Table 15.24 (see sect. 5.2.5 for calculation of expected values). Applying Eq. (15.8) to the data of Tables 15.23 and 15.24 (including the continuity correction discussed in sect. 5.2.5), we have

$$\chi_1^2 = \frac{(4)^2}{13.44} + \frac{(4)^2}{18.56} + \frac{(4)^2}{7.56} + \frac{(4)^2}{10.44} = 5.70.$$

To obtain significance at the 5% level, a Chi-square value of 3.84 is needed (Table IV.5). Clearly, the test is significant and we conclude that the results of this test warrant rejection of the null hypothesis (i.e., the results of tests A and B are dependent). This significant result suggests that tests A and B are related; a positive test for A is associated with a positive test for B; and a negative test for A is associated with a negative test for B.

15.9.4 Analysis of Combined Sets of 2 × 2 Tables

Two situations may arise in which the analysis of combined fourfold tables is needed. Consider a clinical study in which two treatments are to be compared with regard to a dichotomous variable where the data are collected from more than one center. Rather than pooling all the data to form one combined table, the analysis is performed with the data stratified by center. In a second example, a study may be performed at a single center, but there may be a variable within the center that needs further clarification with respect to interpretation of the results. The data are then stratified by this variable. Koch and Edwards [15] give an example of a clinical study at a single center comparing a test drug and placebo in a study of arthritis. The outcome of the treatment is dichotomized into either no improvement or (some or marked) improvement. The overall results are shown in Table 15.25. Table 15.26 stratifies Table 15.25 into two groups, results for males and females. The following discussion summarizes part of their presentation (for more detail, see Ref. [15]).

Note that males appear to be less responsive than females to both active drug and placebo. If the distribution of males and females to treatment groups is unbalanced, the experimental results can be biased.

The Chi-square test for significance for the data of Table 15.25 is 10.7 with a correction factor, and 12.3 without the correction factor. The Mantel–Haenszel method [16] tests for significance, taking into account the sex-adjusted response (Table 15.26).

If treatments are equally effective, the expected value of n_{111} and n_{211} in Table 15.25 are

$$E(n_{111}) = m_{111} = \frac{(n_{11+})(n_{1+1})}{n_1}$$

$$E(n_{211}) = m_{211} = \frac{(n_{21+})(n_{2+1})}{n_2}.$$

Table 15.26   Table 15.25 with Two Subgroups

| Sex | Treatment | Improvement | | Total |
		None	Some or marked	
Female	Test drug	$n_{111} = 6$	$n_{112} = 21$	$n_{11+} = 27$
Female	Placebo	$n_{121} = 19$	$n_{122} = 13$	$n_{12+} = 32$
Female total		$n_{1+1} = 25$	$n_{1+2} = 34$	$n_1 = 59$
Male	Test drug	$n_{211} = 7$	$n_{212} = 7$	$n_{21+} = 14$
Male	Placebo	$n_{221} = 10$	$n_{222} = 1$	$n_{22+} = 11$
Male total		$n_{2+1} = 17$	$n_{2+2} = 8$	$n_2 = 25$

The variances of n_{111} and n_{211} are

$$\text{Var}(n_{111}) = \frac{n_{11+}n_{12+}n_{1+1}n_{1+2}}{[(n_1^2)(n_1 - 1)]}$$
$$= \frac{(27)(32)(25)(34)}{(59)^2(58)} = 3.63748$$

$$\text{Var}(n_{211}) = \frac{n_{21+}n_{22+}n_{2+1}n_{2+2}}{[(n_2^2)(n_2 - 1)]}$$
$$= \frac{(14)(11)(17)(8)}{(25)^2(24)} = 1.39627.$$

The Mantel–Haenszel statistic is calculated as

$$\frac{\left[\sum_{h=1}^{2}(n_{h1+}n_{h2+}/n_h)(p_{h11} - p_{h21})\right]^2}{\sum_{h=1}^{2}v_{h11}}, \tag{15.9}$$

where $h = 1, 2$ and $p_{hi1} = (n_{hi1}/n_{hi+})$ the proportion of patients in each sex and treatment group who show no improvement.

$$p_{111} = 6/27$$

$$p_{211} = 7/14$$

$$p_{121} = 19/32$$

$$p_{221} = 10/11$$

For the data of Table 15.26, the calculation of the Mantel–Haenszel statistic (eq. 15.9) is

$$Q_{MH} = \frac{[\{(27)(32)/59\}(6/27 - 19/32) + \{(14)(11)/25\}(7/14 - 10/11)]^2}{3.63748 + 1.39627} = 12.59.$$

Q_{MH} is distributed approximately as Chi-square with 1 d.f. Therefore, the conclusion is that after adjustment for sex differences, the treatments are significantly different ($p < 0.05$).

This analysis summarizes an elementary but common occurrence in the analysis of clinical studies. For more detail of the application of the Mantel–Haenszel statistic, see Refs. [13,15].

Table 15.27 Treatments with Binomial
Outcome in Randomized Blocks[a]

Subject	Treatment I	II	III	B_j
1	1	1	0	2
2	0	0	0	0
3	1	0	1	2
4	1	0	1	2
5	0	1	1	2
6	1	0	1	2
7	1	0	0	1
8	1	0	0	1
9	1	0	1	2
10	1	0	0	1
T_i	8	2	5	15

[a] 1 = success, 0 = failure.

15.9.5 Randomized Blocks with Binomial Outcome

For data with a binomial outcome that is in the form of a randomized block, the following test
to compare treatments [17] may be used:

Compute:

$$Q = \frac{\left[c(c-1)\sum(T_i^2) - (c-1)N^2 \right]}{\left[cN - \sum(B_j^2) \right]},$$

where c is the number of treatments (columns), T_i the total for treatment i, B_j the total for block
j, and N is the grand total.

For large samples, Q has an approximate Chi-square distribution with $c-1$ d.f.

Example 3. Ten subjects were treated with a topical product for a fungus infection. Subjects
were evaluated as cured (1) or not cured (0) (Table 15.27).

$$Q = \frac{\left[c(c-1)\sum(T_i^2) - (c-1)N^2 \right]}{\left[cN - \sum(B_j^2) \right]}$$

$$Q = \frac{\left[3(2)(64 + 4 + 25) - 2(15)^2 \right]}{(3 \times 15 - 27)} = \frac{108}{18} = 6.$$

The tabled value of Chi-square with 2 d.f. at the 5% level is 5.99. Therefore, we can
conclude that the differences are significant at the 0.05 level. (Treatment 1 is different from
Treatment 2.)

15.10 NONPARAMETRIC TOLERANCE INTERVAL

If the data set appears to be non-normal, the usual tolerance interval calculation assuming a
normal distribution may be inappropriate (see sect. 5.1). In this case, a nonparametric tolerance
interval can be constructed. The nonparametric interval can be considered conservative, and
would be wider, on average, than that usually calculated assuming normality, if the data truly
are normal. The computation quantifies the intuition [18] that most future observations would
lie within the minimum and maximum of an observed sample from the distribution. The
calculation is as follows [18]. Given a sample of size n from a distribution, we can state the
probability of the proportion of samples that are within the minimum and maximum values

observed. Let $p =$ the proportion of samples within the maximum and minimum. Let Q be the probability space covered by min, max; n is the sample size.

$$P(Q > p) = 1 - n(p)^{n-1} + (n-1)p^n$$

$P(Q > p)$ is the probability that the minimum and maximum values will cover a proportion p of all values in the distribution.

An example should clarify the calculation. Suppose that we assay 50 individual tablets randomly chosen from a batch, with a minimum of 96% and a maximum of 103%. Furthermore, we have reason to believe that the data are not normally distributed. We wish to compute a tolerance interval that will give a probability that p proportion of the tablets assay between 96% and 103%. In this example, $n = 50$. Suppose, we wish to set the proportion of tablets within 96 to 103 to be 95%. We then calculate the probability.

$$P(Q > p) = 1 - 50(0.95)^{50-1} + (50-1)0.95^{50} = 0.72.$$

Therefore, we say that the probability is 0.72 (72%) that at least 95% of the tablets in the batch are within 96% to 103%.

We might want to compare this result with the tolerance interval assuming a normal distribution (sect. 5.1). The average is 100.2% and the standard deviation of the 50 tablets is 2.5%. Referring to Table IV.19 in the appendix, a 75% probability tolerance interval containing 95% of the tablets is

$$100.2 \pm 2.138 \times 0.025 = 100.2 \pm 5.3 = 94.9 - 105.5.$$

That is, the probability is 75% that at least 95% of the tablets are between 94.9% and 105.5%.

KEY TERMS

ANCOVA	Multinominal distribution
Attribute	Nominal data
Categorical data	Normal approximation
Confidence interval	Ordered data
Contingency table ($R \times C$ table)	Ordinal data
Continuous data	Quade test
Distributions	Rating scale
Efficiency	Run
Fisher's exact test	Runs test
Friedman's test	Sensitive
Hypergeometric distribution	Sign test
Independence	Ties
Interval or ratio scale	Tolerance interval
Kruskal–Wallis test	Wilcoxon rank sum test
Mantell–Haenszel test	Wilcoxon signed rank test
McNemar's test	

EXERCISES

1. Perform a t test to compare treatments for the data from Table 15.3. Compare the results of this test to the nonparametric test presented in the text.

2. The following data were observed comparing two assays using 12 batches of material:
   (a) Use the sign test to determine if the two tests are different.
   (b) Compare the two tests (A and B) using the t test.

Batch	Test A	Test B
1	8.1	9.0
2	9.4	9.9
3	7.2	8.0
4	6.3	6.0
5	6.6	7.9
6	9.3	9.0
7	7.6	7.9
8	8.1	8.3
9	8.6	8.2
10	8.3	8.9
11	7.0	8.3
12	7.7	8.8

3. Use the Wilcoxon signed rank test to compare the two assay methods to determine if the methods are significantly different for the data in Exercise Problem 2. Use Table IV.13 and the normal approximation.

4. Blood glucose uptake for corresponding halves of rat diaphragms for compounds A and B are as follows (adapted from Ref. [2]):

					Rat				
	1	2	3	4	5	6	7	8	9
A	9	9.5	5.7	3.9	6.7	5	8.6	3	8
B	8	9.7	5.1	3.6	7.1	5	8.4	4.2	7.1

Use a nonparametric procedure to compare the two compounds.

5. Twenty patients were randomly allocated to two treatment groups, 10 patients per group. The following data are the change in serum chloride after treatment.

Treatment A	Treatment B
4.3	6.1
6.2	0.9
4.4	0.7
8.2	0.8
0.5	1.3
2.6	3.1
4.2	1.9
4.1	3.9
5.6	2.1
3.4	0.1

Test for treatment differences using a nonparametric test and a t test.

6. Dissolution is compared for three experimental batches with the following results (each point is the time in minutes to 50% dissolution for a single tablet):
   Batch 1: 15, 18, 19, 21,23, 26
   Batch 2: 17, 18, 24, 20
   Batch 3: 13, 10, 16, 11,9
   Is there a significant difference among batches?

7. A bioavailability study was conducted in which three products were compared: a standard product and two new formulations, A and B. The peak blood concentrations were as follows:

Subject	Standard	A	B
1	14	12	17
2	12	18	9
3	11	17	8
4	17	15	14
5	20	16	16
6	16	12	13
7	14	11	10
8	16	16	10
9	18	17	19
10	15	10	8
11	22	15	15
12	14	13	14

Use Friedman's test to determine if there is a difference among the three treatments.

8. In a test for pain relief, two drugs are compared where the outcome is 0, 1, or 2, where 0 = no relief, 1 = partial relief, 2 = complete relief. With drug A, 50 had a score of 0, 50 scored 1, and 75 scored 2. With drug B, 20 had a score of 0, 60 scored 1, and 60 scored 2. Use a Chi-square test to compare drugs A and B. How would you interpret a significant effect?

9. The following fourfold table was constructed from data for inspection of 1000 tablets in quality control.
   (a) Are "specks" and "capping" independent?
   (b) Are the proportion of tablets specked and capped different in this batch of tablets?

		Capped	
		Yes	No
Specked	Yes	13	45
	No	18	924

$§$10. In a preclinical study, the following incidence of tumors was observed in control and treated animals:
   Controls: 0 of 12 animals
   Treated: 5 of 14 animals
   Use Fisher's exact test to determine if the incidence is significantly different in the two groups. Compare the results to a Chi-square test with continuity correction.

11. The following assay results were observed from sequential readings from a control chart. Using the runs test, determine if these values conform to a "random" sequence. Use a two-sided test. What would be your conclusion if the test were one-sided? 300.1, 300.5, 300.7, 308.2, 304.4, 303.9, 302.1, 303.1, 300.9, 303.4, 305.6, 306.2, 304.1, 306.1, 306.8, 301.3, 304.3, 301.9, 304.2, 302.6

12. Confirm that the corrected χ^2 is 7.0 by computing the correction for ties for the analysis of the data in Table 15.10.

13. For the 3×2 table at the end of section 15.9.1 (Table 15.18), compute the χ^2 value for the entire table and for the comparison of Treatments B and C.

$§§$ Optional, more advanced problems.

14. Analyze the following data, using the combined data from two centers. Use the Mantel–Haenszel test.

	Success	Failure	Total
Center I			
Drug *A*	12	6	18
Drug *B*	9	9	18
Total	21	15	36
Center II			
Drug *A*	14	3	17
Drug *B*	9	11	20
Total	23	14	37

Are the two treatments significantly different?

15. Compute the parametric two-way ANOVA and confidence intervals for the data of Table 15.6, using a log transformation.

16. In section 15.4.1, show that the comparison of Treatment 1 with Treatment 2 for Sequence 1 and Sequence 2 is equal to twice the period effect (no carryover).

17. Compute the tests for sequence and period effects for Table 15.9; use Eq. (15.4).

18. Perform a two-way analysis of variance on the data of Table 15.11. Assuming no interaction, what is the probability associated with the *F* test (tablet press MS/error MS).

19. Perform a two-way ANOVA on ranks in Table 15.11. Show that Fisher's LSD is the same as the nonparametric multiple comparison.

REFERENCES

1. Snedecor GW, Cochran WG. Statistical Methods, 7th ed. Ames, IA: State University Press, 1980.
2. Wilcoxon F, Wilcox RA. Some Rapid Approximate Statistical Procedures. Pearl River, NY: Lederle Laboratories, 1964.
3. Hollander M, Wolfe DA. Nonparametric Statistical Methods. New York: Wiley, 1973.
4. Steinijens VW, Diletti E. Statistical analysis of bioavailability studies: parametric and nonparametric confidence intervals. Ear J Clin Pharmacol 1983; 24:127–136.
5. Jones B, Kenward MC. Design and Analysis of Cross-Over Trials. London: Chapman and Hall, 1989.
6. Cornell RG. Evaluation of bioavailability data using nonparametric statistics. In: Albert KS, ed. Drug Absorption and Disposition, Statistical Considerations. Washington, DC: American Pharmaceutical Association, Academy of Pharmaceutical Sciences, 1980:51.
7. Koch G. The use of nonparametric methods in the statistical analysis of the two period change-over design. Biometrics 1972; 28:577–584.
8. Wallenstein S, Fisher AC. The analysis of the two-period repeated measurements crossover design with application to clinical trails. Biometrics 1977; 33:261.
9. Conover W. Practical Nonparametric Statistics, 2nd ed. New York: John Wiley and Sons, 1980.
10. Sprent P. Applied Nonparametric Statistical Methods. New York: Chapman and Hall, 1989.
11. Quade D. J Am Stat Assoc 1967:1187.
12. Sokal RR, Rohlf FJ. Biometry. San Francisco: W.H. Freeman, 1969.
13. Fleiss JL. Statistical Methods for Rates and Proportions. New York: Wiley, 1980.
14. Dixon WJ, Massey FJ Jr. Introduction to Statistical Analysis, 3rd ed. New York: McGraw-Hill, 1969.
15. Koch GG, Edwards S. Summarization analysis, and monitoring of adverse experiences. In: Peace KE, ed. Statistical Issues in Drug Research and Development. New York: Marcel Dekker, 1990: 19–161.
16. Mantel N, Haenszel W. Statistical aspects of the analysis of data from retrospective studies of disease. J Natl Canc Inst 1959; 22:719.
17. Sprent P. Applied Nonparametric Statistical Methods. New York: Chapman and Hall, 1989:128.
18. Rice JA. Mathematical Statistics and Data Analysis. Pacific Grove, CA: Wadsworth and Brooks/Cole, Statistics/Probability Series, 1988.

16 | Optimization Techniques and Screening Designs*

The* optimization of pharmaceutical formulations with regard to one or more attributes has always been a subject of importance and attention for those engaged in formulation research. Product formulation is often considered an art, the formulator's experience and creativity providing the "raw material" for the creation of a new product. Given the same active ingredient and a description of the final marketed product, two different scientists will very likely concoct different formulations. Certainly, human input is an essential ingredient of the creative process. In addition to the *art* of formulation, techniques are available that can aid the scientist's choice of formulation components that will optimize one or more product attributes. These techniques have been traditionally applied in the chemical and food industries, for example, and in recent years have been applied successfully to pharmaceutical formulations. In this chapter, we describe the application of factorial designs (and modified factorials) and simplex lattice designs to formulation optimization. When the effects of factors on a pharmaceutical process or response are unknown, the use of screening designs to estimate factor effects may be indicated.

16.1 INTRODUCTION

The pharmaceutical scientist has the responsibility to choose and combine ingredients that will result in a formulation whose attributes conform with certain prerequisite requirements. Often, the choice of the nature and quantities of additives (excipients) to be used in a new formulation is based on experience, for example, similar products previously prepared by the scientist or his or her colleagues. To break habits based on experience and tradition is difficult. Although there is much to be said for the practical experience of many years, we often become caught in the web of the past. The application of formulation optimization techniques is relatively new to the practice of pharmacy. When used intelligently, with common sense, these "statistical" methods will broaden the perspective of the formulation process.

Although several optimization procedures are available to the pharmaceutical scientist, a few frequently used methods will be presented in this chapter. The objective is to produce a mathematical model that describes the responses. In general, the procedure consists of preparing a series of formulations, varying the concentrations of the formulation ingredients in some systematic manner. These formulations are then evaluated according to one or more attributes, such as hardness, dissolution, appearance, stability, taste, and so on. Based on the results of these tests, a particular formulation (or series of formulations) may be predicted to be optimal. The "proof of the pudding," however, is actually to prepare and evaluate the predicted *optimal* formulation.

If the formulation is optimized according to a single attribute, the optimization procedure is relatively uncomplicated. To optimize on the basis of two or more attributes, dissolution and hardness, for example, may not be possible. The formulation that is optimal for one attribute very well may be different from the formulation needed to optimize other attributes. In these cases, a compromise must be made, depending on the relative importance of each attribute. The final formulation, therefore, is suitably modified to attain an acceptable performance of all relevant attributes, if possible. We will discuss the optimization procedure based on a single attribute. More complex situations may require more complex designs, and the advice of an experienced statistician is recommended in these cases. Therefore, the use of the term, "optimization" may be a misnomer. An optimal response may not be a single response, but a region of responses that

---

* This is an advanced topic.

satisfy the requirements of the formulation. Once such a region is defined, the desired response may be defined using a range of factors.

In general, an advanced understanding of statistics is not necessary. One should be familiar with the following concepts as described elsewhere in this book.

16.1.1 Planning Experiments

Common sense should prevail. Design and choice of variables are discussed later in this chapter. In most cases, we have a reasonable idea of which variables are important, and their effective ranges. But, we may be surprised. If everything were known, we would not have to experiment. Also, we should be careful not to neglect potentially important variables. Screening designs may be useful if little is known of the system.

16.1.2 Variables

Variables may be considered as Independent and Dependent (X, Y). Dependent variables (Y) are outcome variables (e.g., dissolution). Independent variables (X) are set in advance (e.g., lubricant level). Variables can be continuous or discrete. The number of experiments should be kept at a reasonable level. The more variables used, the more knowledge is gained, but expense and time should be taken into consideration.

16.1.3 Variability or Experimental Error

It is important to have an idea about variability of response (Y) and/or "predicted response." Replication is typically needed to estimate variability, but this adds time and cost to the study. Estimates of variance can be obtained from replication, from ANOVA or from experience.

16.1.4 Regression

For our purposes, regression is used to predict Responses, and/or to describe relationships. Either simple linear or multiple regression may be used to obtain optimized systems. We derive a response equation from the data (as described in this chapter), and predict a response within the bounds of the fixed independent variables, X. Prediction outside of the bounds of the independent variables is unreliable. Consider the following example.

Suppose that the theoretical response relationship (Y as a function of X_1 and X_2, where we have two independent variables) is $Y = 5 + 6X_1 + 7X_1^2 + 3X_2$. We obtain six values of Y as follows:

X_1	X_2	Y
1	1	21
2	1	48
1	2	24
2	2	57
3	1	89
1	3	45

Using multiple regression, we obtain the following equation relating Y to the independent variables.

$$Y = -7 + 7.2X_1 + 7X_1^2 + 11.4X_2$$

This works well within the experimental space. But predictions outside are questionable. For example, if $X_1 = 4$ and $X_2 = 4$

Predicted $= 179.4$
Actual $= 153$

16.2 OPTIMIZATION USING FACTORIAL DESIGNS

The basic principles of factorial designs have been presented in chapter 9. In factorial designs, levels of factors are independently varied, each factor at two or more levels. The effects that can be attributed to the factors and their interactions are assessed with maximum efficiency in factorial designs. Also, factorial designs allow for the estimation of the effects of each factor and interaction, unconfounded by the other experimental factors. Thus, if the effect of increasing stearic acid by 1 mg is to decrease the dissolution by 10%, in the absence of interactions, this effect is independent of the levels of the other factors. This is an important concept. If the levels of factors are allowed to vary haphazardly, as in an undesigned experiment, the observed effect due to any factor is dependent on the levels of the other varying factors. Generalities, or predictions, based on results of an undesigned experiment will be less reliable than those that would be obtained in a designed experiment, in particular, a factorial design. Screening designs use less runs, and estimate the main effects of factors. The latter part of this chapter will introduce screening designs. These designs are useful when a relatively large number of factors may affect the response or process. From a regulatory viewpoint, the data derived from factorial designs can be useful to predict responses when confronted with formulation or manufacturing modifications.

The optimization procedure is facilitated by construction of an equation that describes the experimental results as a function of the factor levels. A *polynomial* equation can be constructed, in the case of a factorial design, where the coefficients in the equation are related to the effects and interactions of the factors. For the present, we will restrict our discussion to factorial designs with factors at only two levels, called 2^n factorials, where n is the number of factors (see chap. 9). These designs are simplest and often are adequate to achieve the experimental objectives. These designs estimate only linear effects. That is, if there is a curved response as a function of factor levels or combination, such effects will be missed. Sometimes, use of these smaller designs is imperative, for the sake of economy. Increasing the number of factor levels dramatically increases the number of formulations that are needed to complete the design. With a large number of factors, even designs where factors are restricted to two levels may result in a very large number of formulations to be prepared and tested. In such cases, *fractional* factorial designs may be used. Some information is lost when using fractional factorial designs, but one-half, one-fourth, or less of the formulations are needed compared to those needed to run a full factorial design. A brief description of fractional factorial designs is presented in section 9.5. The theory and construction of these designs are presented in detail in *The Design and Analysis of Industrial Experiments,* edited by Davies [1]. Also see Ref. [2] for an example of optimization applied to an HPLC analytical method.

As noted above, the optimization procedure is facilitated by the fitting of an empirical polynomial equation to the experimental results. The equation constructed from a 2^n factorial experiment is of the following form:

$$Y = B_0 + B_1 X_1 + B_2 X_2 + B_3 X_3 + \ldots + B_{12} X_1 X_2$$
$$+ B_{13} X_1 X_3 + B_{23} X_2 X_3 + \ldots + B_{123} X_1 X_2 X_3 + \ldots \tag{16.1}$$

where Y is the measured response, X_i is the level (e.g., concentration) of the ith factor, B_i, B_{ij}, B_{ijk}, . . .represent coefficients computed from the responses of the formulations in the design, as will be described below. (B_0 represents the intercept.)

For example, in an experiment with three factors, each at two levels, we have eight formulations, a total of eight responses. The eight coefficients in Eq. (16.1) will be determined from the eight responses in such a way that each of the responses will be exactly predicted by the polynomial equation. For the present, to illustrate this concept we will look at the problem in reverse. Suppose that we already have an equation to predict the experimental results derived from a factorial design as follows:

$$Y = 5 + 2(X_1) + 3(X_2) + X_3 - 0.6(X_1 X_2) - 0.4(X_1 X_3)$$
$$+ 0.7(X_2 X_3) + 0.12(X_1 X_2 X_3) \tag{16.2}$$

From Eq. (16.2), we can reconstruct the original data from the 2^3 experiment. Suppose that the levels (in mg) of the three factors in the design were as follows:

	Low level	High level
X_1 = stearate	0	2
X_2 = colloidal silica	0	1
X_3 = drug	0	5

Based on Eq. (16.2), the formulation with all factors at the low level will have a response of five. All factors are equal to 0, and all terms containing X_1, X_2, or X_3 are equal to 0. If X_1 is at the high level (2 mg), and X_1, and X_3 are at the low level (0), the predicted response is $Y = 5 + 2(X_1) = 5 + 2(2) = 9$. All other terms are equal to 0. If X_1 and X_2 are at the high level, and X_3 is at the low level, the response is

$$5 + 2(X_1) + 3(X_2) - 0.6(X_1 X_2) = 5 + 2(2) + 3(1) - 0.6(2)(1) = 10.8.$$

The results for all eight combinations (formulations) as predicted from Eq. (16.2) are shown in (Table 16.1).

Table 16.1 shows the results of the factorial experiment that were used to construct Eq. (16.2). The practical, more realistic problem is to construct the polynomial equation, given the experimental results. To solve this problem, we find the solution to eight equations with eight unknowns [the unknowns are the eight coefficients in Eq. (16.2)]. For example, in formulation 1 (Table 16.1),

$$X_1 = X_2 = X_3 = 0.$$

Substituting $X_1 = X_2 = X_3 = 0$ into the general equation [Eq. (16.1)] results in

$$Y = B_0 \text{(all other terms are 0).}$$

Since the response (Y) for formulation 1 (where $X_1 = X_2 = X_3 = 0$) is equal to 5,

$$Y = B_0 = 5.$$

This is the simple solution for the first of the simultaneous equations.
In the second formulation, $X_1 = 2$, X_2 and X_3 are equal to 0 and Eq. (16.1) reduces to

$$Y = B_0 + B_1 X_1 \text{(all other terms are 0).} \tag{16.3}$$

Table 16.1 Results of the 2^3 Factorial Experiment That Led to the Construction of the Polynomial Equation (16.2)

Formulation	Factor level			Predicted response, Y
	X_1	X_2	X_3	
1	0	0	0	5
2	2	0	0	9
3	0	1	0	8
4	2	1	0	10.8
5	0	0	5	10
6	2	0	5	10
7	0	1	5	16.5
8	2	1	5	16.5

The response, Y, for formulation 2 is 9 (Table 16.1). We can solve for B_1, using Eq. (16.3) ($B_0 = 5$ and $X_1 = 2$)

$$9 = 5 + B_1(2) \quad B_1 = 2.$$

This procedure is continued, until we solve for all coefficients, B_i, B_{ij}, B_{ijk}, and so on.

In the example above, the solution for the coefficients for the polynomial equation is very simple, because the low level of all factors is zero. In general, the solution would be more difficult if the low level of all factors is not equal to zero. However, the general solution for the polynomial coefficients is not difficult for 2^n factorial designs, because of the independence (orthogonality) inherent in factorial designs. The first step in the solution is to code the levels of the factors so that the high level of each factor is $+1$, and the low level of each factor is -1. This procedure requires a transformation of each of the three variables, X_1, X_2, and X_3 to X_1', X_2', and X_3', respectively, as follows:

For X_1, let $X_1' = X_1 - 1$. Note that when $X_1 = 2$ (the high level), $X_1' = +1$, and when $X_1 = 0$ (the low level), $X_1' = -1$.

For X_2, let $X_2' = 2X_2 - 1$.

For X_3, let $X_3' = \dfrac{2X_3 - 5}{5}$.

In general, the formula for the transformation is

$$\frac{X - \text{the average of the two levels}}{\text{one-half the difference of the levels}}. \tag{16.4}$$

After the transformation, the levels of the factors are as shown in Table 16.2 (see also chap. 9).

Table 16.2 also contains "transformed" values for the interactions, represented by $+1$ or -1. These values are obtained by multiplying the values in the appropriate columns of X_1, X_2, and X_3. For example, in formulation 1, $X_1 X_2$ is represented by $+1$, the product of -1 for X_1 and -1 for X_2 [$X_1 X_2 = (-1)(-1) = +1$]. $X_1 X_2 X_3$ is represented by the product of $(-1)(-1)(-1) = -1$, derived from the values in the columns headed by X_1, X_2, and X_3. (See also chap. 9 to clarify this procedure.) The "total" column contains only the value $+1$, and is used to calculate the intercept, B_0.

The coefficients for the polynomial equation (16.1) are calculated as $\Sigma XY/8\,(\Sigma XY/2^n,$ in general), where X is the value ($+1$ or -1) in the *column* appropriate for the coefficient being calculated, and Y is the response. An example should make the calculation clear. For the coefficient corresponding to X_1 (B_1), the calculation is performed as follows. We multiply each

Table 16.2 Transformed Levels of Factors Showing Signs to Be Used to Determine Effects and Polynomial Coefficients

Formulation	X_1	X_2	X_3	$X_1 X_2$	$X_1 X_3$	$X_2 X_3$	$X_1 X_2 X_3$	Total	Y
1[a]	-1	-1	-1	$+1$	$+1$	$+1$	-1	$+1$	5
2	$+1$	-1	-1	-1	-1	$+1$	$+1$	$+1$	9
3	-1	$+1$	-1	-1	$+1$	-1	$+1$	$+1$	8
4	$+1$	$+1$	-1	$+1$	-1	-1	-1	$+1$	10.8
5	-1	-1	$+1$	$+1$	-1	-1	$+1$	$+1$	10
6	$+1$	-1	$+1$	-1	$+1$	-1	-1	$+1$	10
7	-1	$+1$	$+1$	-1	-1	$+1$	-1	$+1$	16.5
8	$+1$	$+1$	$+1$	$+1$	$+1$	$+1$	$+1$	$+1$	16.5

[a]Note that X_1, X_2, and X_3 are at their low levels (0). Transformed values are -1, -1, and -1.

value in the column headed X_1 (+1 or −1) by the corresponding response, Y. The sum of these products (ΣXY) divided by 8 (2^n) is the coefficient, B_1.

$$[(-1)(5) + (+1)(9) + (-1)(8) + (+1)(10.8) + (-1)(10) + (+1)(10)$$
$$+(-1)(16.5) + (+1)(16.5)] = \frac{6.8}{8} = 0.85.$$

The coefficient, B_2, is calculated using the values (+1 or −1) in the second column, the X_2 column.

$$[(-1)(5) + (-1)(9) + (+1)(8) + (+1)(10.8) + (-1)(10)$$
$$+(-1)(10) + (+1)(16.5) + (+1)(16.5) = \frac{17.8}{8} = 2.225.$$

The coefficient for $X_1 X_2 X_3$ is B_{123}, and is calculated using the values in the column headed by $X_1 X_2 X_3$ as follows:

$$[(-1)(5) + (+1)(9) + (+1)(8) + (-1)(10.8) + (+1)(10)$$
$$+(-1)(10) + (-1)(16.5) + (+1)(16.5)] = \frac{1.2}{8} = 0.15.$$

All of the coefficients are calculated in this manner. B_0 is the sum of all of the observations, Y, divided by 8 (10.725).[†] (Note that all of the values in the "total" column are +1; this column is used to obtain B_0 in the same manner as the other coefficients.) The final polynomial equation for predicting the response, Y, is

$$Y = 10.725 + 0.85(X_1) + 2.225(X_2) + 2.525(X_3)$$
$$-0.15(X_1 X_2) - 0.85(X_1 X_3) + 1.025(X_2 X_3) + 0.15(X_1 X_2 X_3) \tag{16.5}$$

This equation looks entirely different from Eq. (16.2), which also predicts the responses in this experiment. However, the two equations predict the same response. Equation (16.5) uses the transformed levels of X_1, X_2, and X_3 (+1 or −1), and Eq. (16.2) uses the actual, observed, untransformed values. For example, if X_1 and X_2 are at their high levels, and X_3 is at the low level, we can solve for the response, Y, using Eq. (16.5) and the transformed values, +1, +1, and −1 for X_1, X_2, and X_3, respectively.

$$Y = 10.725 + 0.85(+1) + 2.225(+1) + 2.525(-1) - 0.15(+1)(+1)$$
$$-0.85(+1)(-1) + 1.025(+1)(-1) + 0.15(+1)(+1)(-1) = 10.8.$$

The response with X_1 and X_2 at the high level is 10.8, exactly equal to the value obtained from Eq. (16.2), where X_1, X_2, and X_3 are the actual levels, 2, 1, and 0 mg, respectively.

To reiterate, the reason for the transformation (also called coding) is to allow for calculation of the coefficients in the polynomial equation.[‡] The transformation of the high and low factor levels to +1 and −1 also results in easy calculation of the variance of the coefficients. Using the transformed levels, the variance of a coefficient is $\sigma^2 / \Sigma(X - \overline{X})^2 = \sigma^2 / 8$. With an estimate of the variance, S^2, each coefficient can be tested for significance, using a t test. These tests are exactly equivalent to the testing of the effects of the ANOVA of a factorial design as explained in chapter 9. If, for example, the $X_1 X_2$ interaction were found to be nonsignificant in an ANOVA, the coefficient of $X_1 X_2$, −0.15 in this example, will also be nonsignificant. Usually, when constructing the polynomial equation, only those terms that are statistically "significant" are retained. In the

---

[†] $B_0 = \overline{Y}$.
[‡] The coded values also result in orthogonality (independence) of effects.

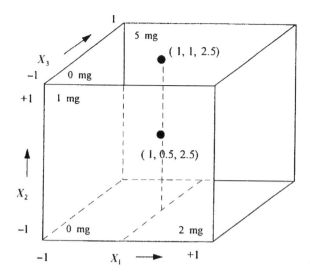

Figure 16.1 Factor space for experiment with factor levels shown in Table 16.1.

experiment above, an estimate of the standard deviation was available from previous similar experiments; s.d. $= 0.32$ with 16 d.f. Therefore, the coefficients B_{12} and B_{123} (0.15) are not significant.

$$t = \frac{|0.15|}{0.32/\sqrt{8}} = 1.3(P > 0.05).$$

Omitting the "nonsignificant" B_{12} and B_{123} terms, the final equation is

$$Y = 10.725 + 0.85(X_1) + 2.225(X_2) + 2.525(X_3) \\ -0.85(X_1 X_3) + 1.025(X_2 X_3). \tag{16.6}$$

An advantage of the transformation described above is that the omission of the two coefficients, B_{12} and B_{123} does not affect the values of the remaining coefficients, that is, recalculation of the polynomial equation results in the same coefficients. This result would not occur if Eq. (16.2) were used to describe the data. Equation (16.2) used the untransformed factor levels and would necessitate extensive computations if some terms were omitted, probably requiring use of a computer as a computing aid. Using the transformed values ensures that the factors are orthogonal. This means that the estimates of the coefficients are independent.

Having derived an equation (16.6) that describes the experimental system based on the results of the experimental formulations, we consider this equation to approximately predict the response within the experimental space. Figure 16.1 shows the space described by this design. The *prediction* of the response, Y, at $X_1 = 1$ mg, $X_2 = 1$ mg, and $X_3 = 2.5$ mg is 12.95 [Eq. (16.6)] (see Exercise Problem 1). How do we know that Eq. (16.6) will be a good predictor for responses other than those included in the factorial design? Without actually testing some "extra-design" formulations, we have no way of knowing that the derived empirical equation will be adequate to predict the results of yet-to-be-tested formulations. If the response is "well behaved," the in-between points should be able to be accurately predicted from the response equation.

Usually, it is a good idea to test at least one formulation, not included in the design, as a *checkpoint*. The observed results of the checkpoint formulation can then be compared to the predicted value to test the equation. In our example, a formulation was prepared with $X_1 = 1$ mg, $X_2 = 0.5$ mg, and $X_3 = 2.5$ mg. The transformed values are equal to zero for the three variables [see the transformation equation (16.4)]. Using Eq. (16.6), the predicted response is 10.725 (only the intercept term is not equal to 0). The factor values for the checkpoint are the average of the low and high levels of the factors (X variables), and lie in the center of the cube in Figure 16.1. This is called a "Center Point." The actual observation made on this formulation was 10.5, very close to the predicted value. Extrapolation of predicted results outside the factor

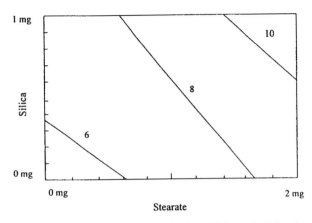

Figure 16.2 Response surface with drug (X_3) constant (low level) [Eq. (16.6)].

space, as shown in Figure 16.1, is not recommended. A two-level design can make predictions only in a linear fashion, usually a gross approximation. If curvature is present, the response may be misrepresented both inside and outside the confines of the design.

Once the polynomial-response equation has been established, an optimum formulation (or a region of optimum formulations) can be found by various techniques. Sometimes, inspection of the experimental results may be sufficient to choose the desired product. In the example above, if large values of the response are desirable, formulations 7 and 8 may be chosen as "best" (Table 16.1). With the use of computers (programmable calculators will often do), a "grid" method may be used to identify optimum regions, and response surfaces may be depicted (Fig. 16.2). The response surface is a geometrical representation of the response and the factor levels, similar to a contour map. For more than two factors, response surfaces cannot be easily represented in two-dimensional space. However, one can take slices of the surface, with all but two factors at fixed levels, as shown in Figure 16.2. A computer can calculate the response, based on Eq. (16.1), at many combinations of the factor levels. The formulation(s) whose response has optimal characteristics based on the experimenter's specifications can then be chosen. To illustrate the grid method, a very rough grid with predicted responses based on Eq. (16.6) is shown in Table 16.3.

The experimental system analyzed above is a very simple example, but is a typical approach to the optimization process. More sophisticated designs may be used, such as the composite designs to be described below (sect. 16.3), or fractional factorial designs. The principles are the same. All of these designs have orthogonal properties to allow for clear and simple estimation of the polynomial coefficients. For these designs, the magnitude of the coefficients is directly related to the magnitude of the response.

The polynomial coefficients may be calculated by techniques such as described here, or by using a multiple regression computer program (see App. III). For two-level experiments

Table 16.3 Grid Solutions for Responses (Y) Based on Eq. (16.6)

X_1[a]	X_2	X_3	Y	X_1	X_2	X_3	Y	X_1	X_2	X_3	Y
−1	−1	−1	5.3	0	−1	−1	7	+1	−1	−1	8.7
−1	−1	0	7.65	0	−1	0	8.5	+1	−1	0	9.35
−1	−1	+1	10	0	−1	+1	10	+1	−1	+1	10
−1	0	−1	6.5	0	0	−1	8.2	+1	0	−1	9.9
−1	0	0	9.875	0	0	0	10.725	+1	0	0	11.575
−1	0	+1	13.25	0	0	+1	13.25	+1	0	+1	13.25
−1	+1	−1	7.7	0	+1	−1	9.4	+1	+1	−1	11.1
−1	+1	0	12.1	0	+1	0	12.95	+1	+1	0	13.8
−1	+1	+1	16.5	0	+1	+1	16.5	+1	+1	+1	16.5

[a] Transformed values.

(2^n factorials), the factor levels should be transformed so that the low level is equal to -1 and the high level equal to $+1$, according to Eq. (16.4). (Experiments with factors at more than two levels should be analyzed with the help of a statistician.) The transformation considerably reduces the complexity of the computations, and aids in the interpretation of the results. Each coefficient may be tested for significance discarding those coefficients that are not significant, although there are no firm rules regarding this procedure. In addition to the statistical criteria, scientific judgment may be used in making decisions about the "significance" of the coefficients. In order to statistically test the coefficients for significance, an estimate of the experimental error is required. This error estimate may be obtained from previous experience, but is best estimated by replicating runs. Replication, however, may result in a large number of experiments, which could be very costly. Replication, accomplished by performing duplicate assays on the same sample, for example, is usually *not* sufficient. The best procedure for replication consists of preparing each formulation or experiment in duplicate (or more), and randomizing the order of the experiments, if all formulations cannot be prepared and tested simultaneously. Methods are available to obtain an estimate of error from an unreplicated factorial experiment (e.g., halfnormal plots [3,4], or from higher order interactions as discussed in chap. 9, but these procedures will not be discussed here).

16.2.1 Replication (Sample Size)

We may only want to find optimum conditions, or we may want to know that effects are real, and not just due to random error. In the latter case, we may want to perform statistical tests (or confidence intervals). To determine the sample size for hypothesis tests, we may use the approximate formula, $N = 4(S^2/\text{delta}^2)(10)$, where N is the sample size for the comparative groups ($N = 4$ for the 2^3 design), where alpha $= 0.05$ and beta $= 0.8$. Usually a sample size between 10 and 20 should be sufficient.

Note that for two-level designs, the variance of an effect is $4S^2/N$, where N is the number of runs.

EXAMPLE:

A difference in response of 2.5 units is meaningful in a 2^3 experiment. The s.d. is expected to be 1.5. What size sample should we use?

$$N = 4(2.25/6.25)(10) = \text{approximately } 16.$$

16.2.2 Extra (Center) Points

Often, it is useful to include an extra run as a "prediction" point, or to estimate curvature. A center point should be equal to the average of the "run" points if there is no curvature. If curvature is present, more runs will be needed to model the data.

The ANOVA for the following data set is shown below to illustrate the analysis of replicated data.

Experiment	A,B	Level P	D	Response
1 (1)	A	1	0.1	5,6
2 P	B	1	0.1	7,11
3 D	A	2	0.1	4,6
4 PD	B	2	0.1	8,11
5 A	A	1	0.2	12,12
6 PA	B	1	0.2	16,21
7 DA	A	2	0.2	11,12
8 PDA	B	2	0.2	24,29
9 Checkpoint	B	1.5	0.15	22

Analysis of variance table

Source term	d.f.	Sum of squares	Mean square	F ratio	Prob. level
P	1	162	162	40.50	0.000380*
D	1	5.555555	5.555555	1.39	0.277097
PD	1	10.88889	10.88889	2.72	0.142947
A	1	304.2222	304.2222	76.06	0.000052*
AP	1	26.88889	26.88889	6.72	0.035802*
AD	1	5.555555	5.555555	1.39	0.277097
APD	1	5.555555	5.555555	1.39	0.277097
S	7	28	4		
Total	14	456.9333			

*$p < 0.05$.

In the absence of replication, there is no proper error term to test significance of the effects. Sometimes we can use an estimate of error from previous experiments or pool the higher order interaction terms. If the runs are replicated, we would have a new term in the ANOVA, residual or error. Then, we can perform F (or t) tests to test for significance.

We could also construct an equation to predict the response (assuming a linear response with factors at two levels). This will be discussed later.

Fractional factorial designs use a fraction of the full factorials (e.g., $^1/_2$, $^1/_4$). The gain is that we use less runs in the experiment. The loss is that we confound some effects. We try to confound effects that we feel are not significant (or very small) with effects that we wish to measure. In this example, the smallest fractional design is a $^1/_2$ replicate, using four of the eight runs. In four runs, we can only measure three effects. The logical choice of effects to measure are A, P, and D. We assume that all interactions are negligible. If our assumption is wrong, the measure of the main effects will be biased.

16.2.3 Optimization of a Combination Drug Product

The following example of a 2^2 factorial experiment is another illustration of the technique of "optimization" using factorial designs. In this experiment, a *combination* drug product was tested to obtain the dose of each drug that would result in an optimal response. The product contained two drugs, $A(X_1)$ and $B(X_2)$. The experiment consists of formulating combinations containing each drug at two dose levels. The doses for A were 5 and 10 mg; B was chosen at doses of 50 and 100 mg. These levels were carefully selected to cover a range of doses that would include an appropriate dose to be chosen as the prime candidate for the final marketed product. The full factorial consists of the four experiments shown in Table 16.4.

The product is a local anesthetic, and the response (Y) is the average time to anesthesia for 12 patients per group. The high and low levels of drug A and drug B are transformed to $+1$ and -1 [Eq. (16.4)]. For drug A, the transformation is

$$\frac{\text{Potency} - 7.5}{2.5}\text{(high level is 10; low level is 5)}.$$

Table 16.4 Factorial Design for the Drug Combination Study

Formulation	Potency (mg) A (X_1)	B (X_2)	Potency (transformed) A (X_1)	B (X_2)	AB (X_1X_2)	Response, Y (min)
1	5	50	−1	−1	+1	9.7
2	10	50	+1	−1	−1	7.2
3	5	100	−1	+1	−1	8.4
4	10	100	+1	+1	+1	4.1

For drug B, the transformation is

$$\frac{\text{Potency} - 75}{25} \text{(high level is 100; low level is 50).}$$

The response equation has the form

$$Y = B_0 + B_1(X_1) + B_2(X_2) + B_{12}(X_1)(X_2). \tag{16.7}$$

The coefficients are computed as described earlier in this section. For example, referring to Table 16.4, B_1 is

Column A (X_1)	Y	X_1Y
−1	9.7	−9.7
+1	7.2	+7.2
−1	8.4	−8.4
+1	4.1	+4.1
		−6.8/4 = −1.7

(B_1 is the sum of $X_1Y/4 = -1.7$.) The polynomial equation is calculated as

$$Y = 7.35 - 1.7(X_1) - 1.1(X_2) - 0.45(X_1 X_2). \tag{16.8}$$

The response, Y, is the time to anesthesia. Formulation 4, which has the high levels of both drugs, has the shortest time to anesthesia, and formulation 1 or 4 would be chosen as optimal if either a long time or a short time to anesthesia is desired. However, an intermediate time might be more desirable. For example, suppose that a time of 5 minutes is the most desirable time based on considerations such as the administration of the product and the type of conditions that are meant to be treated with the aid of the product. Table 16.5 is a rough grid of the predicted responses based on Eq. (16.8). Based on a time to anesthesia of approximately 5 minutes, a formulation containing 0.5 of A and 1 of B would be a candidate. Decoding the values results in a formulation containing 8.75 mg of A and 100 mg of B.

16.3 COMPOSITE DESIGNS TO ESTIMATE CURVATURE

In general, when looking for optimality, the response equation will be more reliable if it contains terms that reflect curvature. Physical systems are less satisfactorily described by empirical equations containing only linear terms. Figure 16.3 shows an example of a single factor, X, at two levels. Clearly, to interpolate the response, Y, at values of X between the low and high levels requires an assumption of linearity. These predictions would be very much in error if the response is curved, as shown in Figure 16.4.

In order to estimate curvature, more than two levels of the factor must be included in the experiment. The presence of curvature would be reflected in the presence of terms with a power

Table 16.5 Predicted Values of Response to Anesthetic Combinations of Drugs A and B Based on Eq. (16.8)

		Dose of drug A[a]				
		−1	−0.5	0	+0.5	+1
Dose of drug B[a]	−1	9.7	9.075	8.45	7.825	6.2
	0	9.05	8.2	7.35	6.5	6.65
	+1	8.4	7.325	6.25	5.17	4.1

[a]Coded values of drug potency.

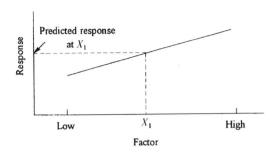

Figure 16.3 Figure showing linear response as a function of a single variable (factor).

greater than 1 (e.g., X_1^2) in the response equation. Such equations are known as polynomials of order 2, and have the following form for a two-factor design:

$$
\begin{aligned}
Y = B_0 + B_1 X_1 + B_{11} X_1^2 + B_2 X_2 \\
+ B_{22} X_2^2 + B_{12} X_1 X_2 + \ldots
\end{aligned}
\tag{16.9}
$$

Composite designs are effective designs to estimate second-order terms. These designs have a number of desirable features. In addition to allowing an estimate of curvature, composite designs give orthogonal estimates of the polynomial coefficients, and allow for the possibility of proceeding with the experiment in a stepwise fashion rather than performing the entire experiment at once. The theory underlying composite designs is beyond the scope of this book. An excellent description of this design and optimization procedure can be found in chapter 11 of Ref. [1].

Although the following discussion is somewhat more advanced than the bulk of material presented in this book, for those who are interested in this subject, an example of a two-factor composite design will be presented to illustrate the technique. A two-factor composite design is identical to a 3^2 factorial design, that is, two factors each at three levels, a total of nine combinations (Table 16.6).

In general, composite designs are not full factorials of the class 3^n, where n is the number of factors. These full factorial designs require a larger number of experiments. For example, a 3^n design with three factors requires 27 runs (27 formulations, for example), 3^3. With more than two factors, composite designs consist of the 2^n design, plus *extra-design* points. The extra points include a *center point* and 2^n extra points, appropriately chosen to maintain orthogonality of the design [1]. The two-factor composite design is shown in Figure 16.5.

The coded values -1, 0, and $+1$ in Table 16.6 for the factor levels represent three *equally spaced* levels of each factor. The coded values in the column headed $X_1 X_2$ are obtained by multiplying the corresponding values in the first two columns (X_1, X_2) as previously described. The values in the columns $X_1^2 - 2/3$ and $X_2^2 - 2/3$ are derived so that the product of corresponding values in any two columns of Table 16.6 sum to zero, resulting in orthogonality (independence) of effects. The special orthogonality obtained by transforming X_i^2 to $X_i^2 - 2/3$ allows for easy

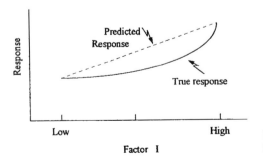

Figure 16.4 Figure showing curved response as a function of a single variable (factor).

Table 16.6 Orthogonal Composite Design with Two Factors (3^2 Design)

Formulation	Coded level					Response, Y	Predicted response
	X_1	X_2	$X_1 X_2$	$X_1^2 - \frac{2}{3}$	$X_2^2 - \frac{2}{3}$		
1	-1	-1	$+1$	$+1/3$	$+1/3$	9.7	9.3
2	-1	0	0	$+1/3$	$-2/3$	9.0	9.4
3	-1	$+1$	-1	$+1/3$	$+1/3$	8.4	8.4
4	0	-1	0	$-2/3$	$+1/3$	5.3	5.6
5	0	0	0	$-2/3$	$-2/3$	4.8	5.0
6	0	$+1$	0	$-2/3$	$+1/3$	3.8	3.3
7	$+1$	-1	-1	$+1/3$	$+1/3$	8.2	8.3
8	$+1$	0	0	$+1/3$	$-2/3$	7.5	6.9
9	$+1$	$+1$	$+1$	$+1/3$	$+1/3$	4.1	4.6

calculation of the coefficients and their variances. With this transformation, Eq. (16.9) is modified to

$$Y = B_0 + B_1 X_1 + B_{11}\left(X_1^2 - \frac{2}{3}\right) + B_2 X_2 + B_{22}\left(X_2^2 - \frac{2}{3}\right)$$
$$+ B_{12} X_1 X_2 + \ldots$$
(16.10)

The data in Table 16.6 consist of the four formulations from Table 16.4 plus five new runs to complete the composite design. The doses of each drug (X_1 and X_2) were chosen such that the three doses are at equally spaced intervals. Thus the third dose, in addition to the two doses chosen for the 2^2 factorial, is 7.5 mg for $X_1(A)$ and 75 mg for $X_2(B)$. The experiment consists of evaluating the nine combinations of doses, 5, 7.5, and 10 mg for $X_1(A)$ and 50, 75, and 100 mg for $X_2(B)$. Note that the *center point* for the composite design is the combination 7.5 and 75 mg of X_1 and X_2, respectively.

The results of the nine runs are shown in Table 16.6. The results are shown schematically in Figure 16.6(A). The plane at the bottom of the figure shows the combinations of X_1 and X_2. The vertical "sticks" are the responses at each combination of X_1 and X_2. We will compute an equation of the form of Eq. (16.10) that represents a smooth curved surface based on the experimental data. In general, the equation can be obtained through the use of a multiple regression computer program.

The coefficients can also be calculated by "hand" (calculator) using the coded values in Table 16.6. The sum of the products of the coded values times the responses divided by the sum of the squared coded values in the column of interest gives the coefficient. For example, the

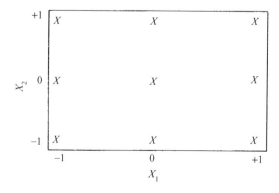

Figure 16.5 Two-factor composite design (3^2 factorial).

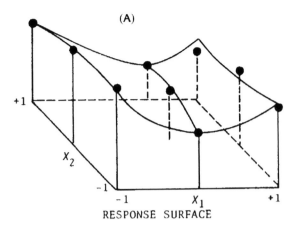

(A)

RESPONSE SURFACE

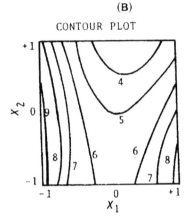

(B)

CONTOUR PLOT

Figure 16.6 Results of composite design experiment from Table 16.6 and response surface computed from Eq. (16.11).

coefficient B_{11} in Eq. (16.10) is calculated as follows:

$X_1'^2 = X_1^2 - 2/3$	Y	$(X_1'^2)(Y)$
+1/3	9.7	3.23
+1/3	9.0	3.00
+1/3	8.4	2.80
−2/3	5.3	−3.53
−2/3	4.8	−3.20
−2/3	3.8	−2.53
+1/3	8.2	2.73
+1/3	7.5	2.50
+1/3	4.1	1.37
$\sum X_1'^2 = 2$		Sum = 6.37

The sum of squared values in the $(X_1^2 - 2/3)$ column is 2. Therefore, the coefficient, B_{11}, is $6.37/2 = 3.18$. The intercept, B_0, is the average of the nine responses, $\overline{Y}$, equal to 6.756. The response equation is

$$Y = 6.756 - 1.22(X_1) + 3.18\left(X_1^2 - \frac{2}{3}\right) - 1.15(X_2)$$

$$-0.52\left(X_2^2 - \frac{2}{3}\right) - 0.7(X_1 X_2)$$

(16.11)

Note that Eq. (16.11) is not an exact fit to the experimental data, as was the case with the polynomial fit described for factorial designs in section 16.2. Had we included three more terms representing various interactions, the equation would exactly fit the data. Equation (16.11) is computed with the assumption that interactions are negligible. Because of the larger number of experiments and the estimation of only six coefficients, we have 2 d.f. for error. Although such an error estimate is not very reliable, it does give us some information, albeit small. The response surface described by Eq. (16.11) is shown in Figure 16.6(A). If this equation does not adequately represent the experimental observations, more terms may be needed in the polynomial equation [Eq. (16.9)] to improve the fit.

The contour plot (similar to contour maps) shown in Figure 16.6(B) allows the selection of combinations of X_1 and X_2 to satisfy given levels of the response. If a maximum response is desired, the X_1, X_2 combinations are limited to a small area of the $X_1 - X_2$ space. If a response of approximately 5 minutes is desired, various combinations of X_1 and X_2 will satisfy the requirements. The ultimate choice will probably depend on other factors, as well, such as cost, toxicity, and so on.

Use of factorial designs in tablet formulation optimization has been presented by Schwartz et al. [5], Fonner et al. [6], and Lindberg et al. [7]. These papers discuss designs somewhat more complex than that presented here. However, for those interested in pursuing this topic further, these papers and the books *The Design and Analysis of Industrial Experiments* [1] and *Statistics for Experimenters* [4] are recommended.

16.4 THE SIMPLEX LATTICE [12]

Response surfaces and optimal regions for *formulation* characteristics are frequently obtained from the application of simplex lattice designs. This class of designs is particularly appropriate in formulation optimization procedures where the *total* quantity of the different ingredients under consideration must be *constant*. Therefore, these are also called "Mixture Designs." For example, suppose that in a liquid formulation, the active ingredient and solvent compose 90% of the product. The remaining 10% of the formulation consists of preservatives, coloring agents, and a surfactant. We wish to prepare a formulation with a certain optimal attribute(s) that is dependent on the relative concentrations of preservative, color, and surfactant. In order to determine optimal regions, we vary the concentrations of these three ingredients in a systematic manner, with the restriction that the total concentration of these ingredients is 10%. This approach differs from the previous procedures (sects. 16.2 and 16.3) in that a constraint is imposed on the total amount of the varying ingredients. In this example, the total amount of the varying components is maintained at 10%. Given the concentration of two of the ingredients, the third ingredient is fixed where in this example $C = 10\% - A - B$.

Implementation of the simplex design consists of preparing various formulations containing different combinations of the variable ingredients. The combinations are prepared in a manner such that the experimental data can be used to predict the responses over the simplex space[§] in a simple and efficient manner. The combinations (formulations) in a simplex design are chosen to cover the space of interest in a symmetrical manner. The experimental results are used to compute a polynomial (simplex) equation that can be used to estimate the response surface. As is true with all optimization and so-called response surface procedures, extrapolation to combinations outside the range included in the experimental design is not recommended. The equation resulting from the experiment, the simplex equation, is an empirical equation that approximately describes the response pattern in the simplex space. There is no reason to believe that the equation has any physical meaning, other than the fact that the complex response patterns resulting from the varying formulations can often be approximated by simple polynomial equations.

Figure 16.7 representing a two-component system (A and B) is useful to help clarify some concepts of simplex designs. One can consider components A and B to be two solvents, which

---

[§] The simplex space is the region enclosed by the various combinations of ingredients chosen for the experiment. See Figure 16.8, for example.

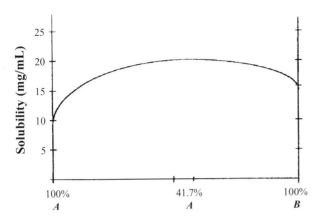

Figure 16.7   Two-component solvent system used to illustrate the simplex approach to optimization.

together comprise the entire solvent system of a drug product. We wish to mix A and B in the correct proportion to optimize the solubility of the drug.

Figure 16.7 is familiar as a solubility phase diagram. This system can also be visualized as an elementary simplex system. The constraint is that the concentrations of A and B must add to 100%. This experiment consists of observing responses (solubility) at three points, *100% A*, *100% B*, and a *50–50 mixture of A and B*, an elementary simplex design. According to Figure 16.7, the solubilities of the drug at the three simplex points, 100% A, 100% B, and 50% A to 50% B, are 10, 15, and 20 mg/mL, respectively. In the simplex approach, we construct an equation of the form

$$Y = B_1(A) + B_2(B) + B_{12}(A)(B), \tag{16.12}$$

where Y is the response (solubility in this example), and (A) and (B) are the concentrations (proportions) of A and B, respectively. The coefficients, B_1, B_2, and B_{12}, are calculated from the experimental observations. The response, Y, can then be predicted for all combinations of A and B, where $(A) + (B) = 1.0$ (100%). (The proportion of each component is usually indicated as a decimal rather than as a percentage.) The form of the simplex design allows for easy calculation of the coefficients. In this example, the coefficients are simply calculated as follows:

$B_1 = $ response at (A) equal to $1.0(100\%) = 10$

$B_2 = $ response at (B) equal to $1.0(100\%) = 15$

$B_{12} = 4$(response at $0.5 – 0.5$ mixture of $A – B$)
$\quad\quad - 2$ (sum of responses at $A = 1.0$ and $B = 1.0$)

$B_{12} = 4(20) - 2(10 + 15) = 30$

The response equation is

$$Y = 10(A) + 15(B) + 30(A)(B). \tag{16.13}$$

The solution above for the three coefficients is a result of the solution of three simultaneous equations:

With $A = 1.0$ and $B = 0$, from Eq. (16.12), $B_1{}^{**} = 10$

With $A = 0$ and $B = 1.0$, from Eq. (16.12), $B_2 = 15$

**The response, Y, with A equal to 1.0 (100%) is 10.

With $A = 0.5$ and $B = 0.5$, from Eq. (16.12),

$$20 = 0.5B_1 + 0.5B_2 + 0.25B_{12} \text{ or } B_{12} = 4(20) - 2(B_1 + B_2) = 30.$$

We will see that in more complex simplex designs, the polynomial coefficients are, similarly, easily calculated as linear combinations of experimental results.

Equation (16.13) exactly predicts the observed points: a fit of a polynomial with three terms to three experimental points. We can always construct an equation with N coefficients that will exactly pass through N points. For example, for the 50–50 mixture,

$$Y = 10(0.5) + 15(0.5) + 30(0.5)(0.5) = 20.$$

The *response equation* predicts responses at extra-design points, those formulations not included in the experiment but that lie within the simplex space, 100% A to 100% B in this example. For example, what solubility would be predicted in a solvent system containing 75% A and 25% B? (Note that $A + B$ must equal 100%.) Applying Eq. (16.13), we have

$$Y = 10(0.75) + 15(0.25) + 30(0.75)(0.25) = 16.875.$$

See also Figure 16.7. The entire response may be sketched in by predicting solubilities along the curve, as shown in the figure.

The primary experimental objective in experiments such as that described above may be the determination of the solvent combination that results in maximum drug solubility. The optimum solubility can be computed by calculating the predicted solubility at many solvent combinations so as to clearly define the response over the solvent mixture continuum. This may seem an indirect and tedious approach, but with the ready availability of computers, this is often the most expeditious route. The maximum solubility is predicted to occur at 41.67% A. In this simple example, the maximum can easily be calculated by setting the first derivative of Eq. (16.13) equal to 0 (see Exercise Problem 6).

In general, the simplex design is usually applied to formulation problems in which a mixture of three or more components is to be investigated. The design is conveniently represented by regular-sided figures, which can be visualized for three- or four-component systems. For more than four components, a single figure cannot be conveniently constructed, but can be theoretically conceived as an N-sided figure in $(N - 1)$-dimensional space. For example, Figure 16.8 shows the three-component system that is represented as an equilateral triangle in two-dimensional space. A regular simplex design for a three-component mixture system consists of six or seven formulations.

Three formulations, one each at each vertex, A, B, and C. These formulations represent formulations with the pure components, A, B, and C, respectively.
Three formulations are prepared with 50–50 mixtures of each pair of components, AB, AC, and BC.
A *seventh* formulation may be prepared with one-third of each component. This lies in the *center* of the design.

An example of a simplex design for four components consisting of 15 formulations is shown in Figure 16.8. The 15 formulations consist of

Four formulations each with 100% of each of the four pure components *Six* formulations of 50–50 mixtures of component pairs (AB, AC, AD, BC, BD, and CD).
Four formulations consisting of one-third mixtures of combinations of three components (ABC, ABD, ACD, BCD).
A mixture containing 25% of each of the four components ($ABCD$).

The simplex design is arranged so that the experimental space is well covered in a symmetrical fashion. In addition, the symmetrical spacing of the points allows for an easy computation

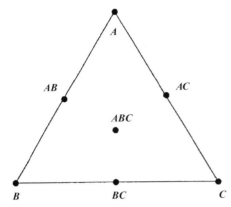

Four-component simplex

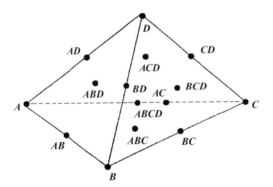

Figure 16.8 Three-component simplex lattice design and four-component simplex lattice design.

of the response equation coefficients. The general equation for the response based on a simplex design contains terms for pure components and all mixtures of components as follows:

$$Y = B_a(A) + B_b(B) + B_c(C) + \ldots + B_{ab}(A)(B) + B_{ac}(A)(C)$$
$$+ B_{bc}(B)(C) + \ldots + B_{abc}(A)(B)(C) + \cdots \tag{16.14}$$

where (A), (B), and (C) are the proportions of components A, B, and C, and $(A) + (B) + (C) + \ldots$ is equal to 1.0.

The subscripted B's (e.g., B_a) are coefficients that can be easily calculated from the responses, Y, or using a multiple regression computer program.

After the coefficients have been calculated, the response equation [Eq. (16.14)] may be used to predict the response of combinations of the N components in the system. With the aid of a computer, responses may be calculated over the simplex space, and contour diagrams printed (see also Fig. 16.6). The contour plot is a graphic description of the response surface resulting from data derived from experimental designs such as the simplex. For the two-component system (Fig. 16.7), the response surface is simply the solubility curve. With three components, a three-dimensional figure would be necessary to show the response surface. A contour plot is a means of illustrating the response on a two-dimensional surface, as is familiar to those who have been exposed to contour maps. A computer may be programmed to produce two-dimensional figures (commercial programs are also available) that are slices through the three-dimensional figure for a three-component system. The slices are taken at a constant concentration of one of the components. In computer outputs, the regions of equal response are indicated by a common symbol, such as a letter or a figure. An example of a contour plot was shown in Figure 16.6. The contour plot will be discussed further in the example that follows. Examination of the contour plot(s) allows the experimenter to choose formulations that have predicted responses of some specified magnitude.

When constructing an empirical response equation based on a limited number of experimental observations, one should understand that predicted values based on the equation may be in error for several reasons. For example, the empirical equation (or model, as it is often called) rarely exactly defines the experimental system. The equation is an approximation to the system. To understand this important concept, note that the same problem would exist if we had only two points in the experimental space. The empirical equation derived from the two points could only relate the observations by a straight line. In-between points could only be predicted on the basis of the straight-line relationship (Figs. 16.3 and 16.4).

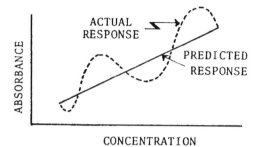

If the true relationship of the X, Y variables were curved, the linear interpolation would be in error. In the simplex design, we used a limited number of points to define a relatively large region of response. Even if the model represented by the empirical equation is a reasonable representation of the true surface, other sources of variation can contribute to error in the prediction equation and predicted responses (e.g., error in measuring the response). Thus, in these systems, we have at least two obvious sources of variability: that due to the empirical model and that due to observational errors.

How can we protect ourselves from inadvertently proceeding with predictions when the derived equation is indeed inaccurate? As insurance against such a possibility, it is a good idea to run one or more extra-design points. These points are not used to estimate the coefficients in the simplex equation [Eq. (16.14)] but will be used as checkpoints. Once the simplex equation is derived, the result at the extra-design checkpoint(s) is predicted based on the equation, and its agreement with the observed value assessed. If the agreement is close, we have increased faith in the predictive power of the response equation (see sect. 16.2). If we have an estimate of error from replication or other means, we may wish to perform a statistical test to test the adequacy of the model (a statistician may be consulted for this calculation).

The calculation of the simplex equation coefficients is easily accomplished using the following formulas. These formulas are an extension of those discussed previously for the two-component system as applied to a three-component system. The general formulas for calculation of coefficients for an N-component system may be found in Ref. [7].

$B_1 = Y_1$, the response at 100% A

$B_2 = Y_2$, the response at 100% B

$B_3 = Y_3$ the response at 100% C

$B_{12} = 4(Y_{12}) - 2(Y_1 + Y_2)$, where Y_{12} is the response at 50 – 50 AB
$B_{13} = 4(Y_{13}) - 2(Y_1 + Y_3)$, where Y_{13}, is the response at 50 – 50 AC
$B_{23} = 4(Y_{23}) - 2(Y_2 + Y_3)$, where Y_{23} is the response at 50 – 50 BC $\hspace{2cm}$ (16.15)
$B_{123} = 27(Y_{123}) - 12(Y_{12} + Y_{13} + Y_{23}) + 3(Y_1 + Y_2 + Y_3)$,
$\hspace{1cm}$ where Y_{123} is the response at $1/3A, 1/3B,$ and $1/3C$

The discussion above has been based on an experimental situation where the components being varied in the simplex design comprise the entire mixture (100%). In pharmaceutical formulations, a more common situation is one in which part of the formulation must remain fixed (e.g., drug concentration in a tablet). The remaining components, which may be varied, therefore do not make up 100% of the mixture. In addition, the lower limit for the varying components is often not equal to 0. For example, some components must be present in some minimal quantity in order that a marketable product can be manufactured. This is known as a design with constraints. For tablets, some minimal amount of a lubricating agent may be necessary in order to obtain an acceptable product. These modifications in the simplex design present no problem, however, because we can restrict the treatment of the simplex to those components that are varied, and with suitable transformations, treat the data in exactly the same way as described above. For example, if the components to be varied make up 60% of the total formulation ingredients, we can appropriately transform the actual percentages of these components so that the transformed percentages total 100%. In a three-component mixture containing 20% of each of three components, each component can be transformed to 33.3% (1/3) for purposes of the simplex analysis. Transformations can also be made where the components have a lower limit greater than 0% and an upper limit less than 100%, as will be explained in the following worked example.

The example presented below is an experiment in which a simplex design was used to obtain a formulation with optimal properties. This example should clarify the concepts and procedures described above. This experiment was prompted by problems with tablet hardness for a large-volume marketed product. Although the reason for the problem was not obvious, the pharmaceutical product development scientists felt that the cause could be traced to three components of the tablets, which we will denote as ingredients A, B, and C. Together, these components consisted of 25% of the original formulation, or 75 mg of the total tablet weight of 300 mg. A careful evaluation of the product ingredients indicated that the three components had to be present in an amount equal to at least 10 mg each in order for the tablet to be satisfactorily compressed. Thus, the recommended simplex design to obtain a satisfactory tablet hardness consisted of varying the three components with the constraint that the sum of the components must be 75 mg, and that each component be present in an amount equal to at least 10 mg.

In order to apply the simplex equation to be derived from this experiment in a convenient manner, the actual concentrations used should be *transformed* such that the minimum concentration (10 mg) corresponds to 0% and the highest concentration corresponds to 100%.[††] In our example, the transformation is the same for all three components because each component is subject to the same restrictions. The minimum quantity is 10 mg and the maximum is 55 mg. (The other two components, each at 10 mg, make up the 20-mg difference, a total of 75 mg.) The transformation is as follows:

$$\text{Transformed proportion} = \frac{\text{Amount used} - \text{minimum}}{\text{maximum} - \text{minimum}}$$

$$= \frac{\text{Amount used} - 10}{55 - 10}.$$

(16.16)

Thus, a formulation prepared with a 50–50 mixture of components A and B would actually contain 32.5 mg of A, 32.5 mg of B, and 10 mg of C. Note that from Eq. (16.16), if a component is at a concentration of 32.5 mg, the transformed proportion is $(32.5 - 10)/(55 - 10) = 0.5$. A formulation with "100%" A would actually contain 55 mg of A, 10 mg of B, and 10 mg of C.

The three-component simplex design was run with one checkpoint, as shown in Table 16.7. The hardness values represent the average hardness of 20 tablets taken at random from the experimental batches. The simplex coefficients are computed as described previously [Eq. (16.15)], resulting in the following equation:

$$Y = 6.1(A) + 7.5(B) + 5.3(C)$$
$$-0.8(A)(B) + 2.8(A)(C) + 2.0(B)(C) + 15(A)(B)(C).$$

(16.17)

[††]If there are no constraints on the upper and lower limits, the highest concentration would ordinarily be 100% and the lowest 0%.

Table 16.7 Results of a Three-Component Simplex System for Tablet Hardness

Formulation components			Transformed proportion			
A	B	C	A	B	C	Average hardness, Y
55	10	10	1.0	0	0	6.1
10	55	10	0	1.0	0	7.5
10	10	55	0	0	1.0	5.3
32.5	32.5	10	0.5	0.5	0	6.6
32.5	10	32.5	0.5	0	0.5	6.4
10	32.5	32.5	0	0.5	0.5	6.9
25	25	25	0.33	0.33	0.33	7.3
32.5[a]	21.25	21.25	0.5	0.25	0.25	7.2

[a]Extra-design checkpoint.

For example, the coefficient B_{123} is calculated as follows:

$$27(7.3) - 12(6.6 + 6.4 + 6.9) + 3(6.1 + 7.5 + 5.3) = 15.$$

(A), (B), and (C) in Eq. (16.17) are the transformed proportions. The extra-design checkpoint (the final formulation in Table 16.7) has a response of 7.2. The predicted value based on Eq. (16.17) is 7.09, very close to the observed value, 7.2. This is some confirmation of the adequacy of Eq. (16.17) as a predictor of tablet hardness. Figure 16.9 shows a contour plot of the results of the experiment based on Eq. (16.17). Tablets with high hardness are found in the region with relatively larger amounts of component B. If a tablet hardness of 7 or more is satisfactory, the pharmaceutical scientist has a choice of formulations. The final composition may then be dependent on other factors, such as cost or other tablet properties.

The following example shows data (Table 16.8) and analysis from a replicated simplex design that gives an estimate of experimental error. The design is a basic three-component (A, B, and C) simplex design with a center point consisting of 1/3 of each of the three components. This example is set up for a computer analysis. Note that the interaction term coefficients are the product of the main effect coefficients. For example for Run #7, the ABC interaction is

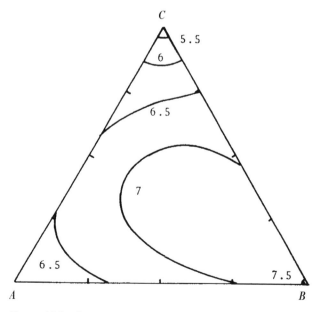

Figure 16.9 Contour plot of three-component simplex system (Table 16.7).

Table 16.8 Example of a Replicated Simplex Design

Run	A	B	C	AB	AC	BC	ABC	Response
1	1	0	0	0	0	0	0	38
2	0	1	0	0	0	0	0	27
3	0	0	1	0	0	0	0	46
4	0.5	0.5	0	0.25	0	0	0	33
5	0.5	0	0.5	0	0.25	0	0	51
6	0	0.5	0.5	0	0	0.25	0	32
7	0.333	0.333	0.333	0.111	0.111	0.111	0.037	48
8	1	0	0	0	0	0	0	42
9	0	1	0	0	0	0	0	28
10	0	0	1	0	0	0	0	41
11	0.5	0.5	0	0.25	0	0	0	35
12	0.5	0	0.5	0	0.25	0	0	47
13	0	0.5	0.5	0	0	0.25	0	32
14	0.333	0.333	0.333	0.111	0.111	0.111	0.037	50

Independent variable	Regression coefficient	Standard error	Lower 95% CL	Upper 95% CL
A	40	1.535299	36.36959	43.63041
B	27.5	1.535299	23.86959	31.13041
C	43.5	1.535299	39.86959	47.13041
AB	1	7.521398	−16.78528	18.78528
AC	29	7.521398	11.21472	46.78528
BC	−14	7.521398	−31.78528	3.78528
ABC	277.1	52.90734	151.9937	402.2056

			Analysis of variance section		
Source	d.f.	Sum of squares	Mean square	Prob. F ratio	Level
Intercept	0	0	0		
Model	7	22461	3208.714	680.6364	0.000000
Error	7	33	4.714286		

$0.333 \times 0.333 \times 0.333 = 0.037$. The computer analysis gives the regression coefficients for the response equation, and an ANOVA to estimate the experimental error. The variance estimate is 4.71.

A checkpoint was run at $A = 0.25$, $B = 0.25$, and $C = 0.5$ with a response of 46. The model predicted 49.2.

In my experience, this approach gives excellent results.

16.5 SEQUENTIAL OPTIMIZATION**

Sequential optimization was developed as a means to optimize a process in a stepwise fashion. Evolutionary operation (EVOP) uses factorial type designs and usually requires a large number of experiments [8]. A relatively simple approach to sequential optimization is a stepwise application of the simplex procedure [9,10]. The procedure consists of first generating data from $n + 1$ experiments where n is the number of independent variables or factors. Based on the $n + 1$ responses and predetermined rules, one result is eliminated from the set and a new experiment is performed. A decision is made as a result of the most recent experiment, generating another new experiment, and so on, eventually terminating the design at an "optimal" response. Thus, each new experiment leads the researcher on a path toward an optimum. The procedure and rules are illustrated in the following example. For further details and illustrations, the reader is encouraged to study Refs. [9–11].

** A more advanced topic.

Table 16.9 Initial Four Experiments for Simplex Experiment

Experiment	Disintegrant	Lubricant	Fill weight	Response
1	+(50)[a]	−(0.2)	−(100)	37
2	−(0)	+(2.2)	−(100)	58
3	−(0)	−(0.2)	+(400)	46
4	+(50)	+(2.2)	+(400)	40

[a]Parenthetical value is the amount of ingredient in the formulation.

16.5.1 An Example of Sequential Simplex Optimization

This example is based on the presentation by Shek et al. [11] using the simplex procedure to optimize properties of a capsule formulation. They were interested in optimizing dissolution and compaction rates as a function of the factors (or variables) drug, disintegrant, lubricant, and fill weight. In this synthetic example, we will look at a single response, dissolution at 30 minutes, as a function of three variables: disintegrant, lubricant, and fill weight.

We start with four experiments (we have three variables). There are no firm rules regarding the design of these experiments, but principles of good experimental design should prevail. For example, a 1/2 replicate of a 2^3 factorial design can be used for the initial four experiments. This requires setting low (−) and high (+) levels for each factor; see Table 16.9.

Let W = vector of worst response
Let S = vector of second worst response
Let B = vector of best response
Let R_w = worst response
Let R_s = second worst response
Let R_b = best response
Let P = average vector after elimination of worst response among formulations under consideration.

Note that since Formula 2 shows the worst response (the longest dissolution time) $\overline{P}$ is the average of experiments 1, 3, and 4 and is equal to (33.3, 0.87, 300). For example, the first vector element refers to the average disintegrant = (+ 50 − 0 + 50)/3 = 33.3.

Procedure:

Step 1. Eliminate W, the vector of the worst response from the data set and compute R [Eq. (16.18) below], the formulation for the new experiment.

$$R = \overline{P} + (\overline{P} - W)$$
$$(33.3,\ 0.87,\ 300) + (33.3,\ -1.33,\ 200) = (66.6,\ -0.46,\ 500). \tag{16.18}$$

In this example, we need 66.6 of disintegrant, −0.46 of lubricant and a fill weight of 500. We will interpret this result after the rules are specified and we proceed with the optimization.

If the response from experiment R, R_r, is better than the second-worst response, R_s, but worse than the best response, retain R_r and proceed to Step 1, evaluating a new formulation with the new set of four formulations.

If the response to R_r is better than the best response, proceed to Step 2.
If the response to R_r is worse than the second-worst response, go to Step 3.
If the response to R_r is worse than the worst response, go to Step 4.

Step 2. Compute E [Eq. (16.19) below] and evaluate R_e.

$$E = \overline{P} + 2(\overline{P} - W) \tag{16.19}$$

If R_r is better than the response to E, R_e, retain R. If R_e is better than R_r, retain E.

Step 3. Compute C_r [Eq. (16.20) below] and evaluate the response to C_r, R_{cr}.

$$C_r = \overline{P} + 0.5(\overline{P} - W) \tag{16.20}$$

Retain C_r. However, if R_{cr} is worse than R_s (the next-to-worst response), then set $R_w = R_s$ and $W = S$. (This means that the worst response is set equal to the next-to-worst response.) Set R_{cr} as the next-to-worst response, that is, $S = C_r$ and $R_s = R_{cr}$.

Step 4. Compute C_w [Eq. (16.21) below] and evaluate R_{cw}. Retain C_w. However, if R_{cw} is worse than R_s (the next-to-worst response), then set $R_{cw} = R_s$ and $W = S$ (this means that the worst response is set equal to the next-to-worst response). Set R_{cw} as the next-to-worst response, that is, $S = C_w$ and $R_s = R_{cw}$.

Summary of calculation of new formulations

1.
$$R = \overline{P} + (\overline{P} - W) \tag{16.18}$$
R_r = The response to formula R

2.
$$E = \overline{P} + 2(\overline{P} - W) \tag{16.19}$$
R_e = The response to formula E

3.
$$C_r = \overline{P} + 0.5(\overline{P} - W) \tag{16.20}$$
R_{cr} = The response to formula C_r

4.
$$C_w = \overline{P} - 0.5(\overline{P} - W) \tag{16.21}$$
R_{cw} = The response to formula C_r

Although this procedure may appear confusing, if one follows the example, the process will be clarified.

We have already calculated the vector for the first new formulation using Step 1 above: $(66.6, -0.46, 500)$. The response to this formulation will replace the worst formulation, W, which is formulation 2. Unfortunately, we cannot prepare this formulation because of the negative quantity of lubricant. We will make a rule that in such impossible situations we consider the response to this new formulation to be worse than the remaining formulations under consideration (formulations 1, 3, and 4).

This sends us to Step 4 according to our rules. The formulations under consideration are 1, 3, 4, and 5 in Table 16.10. According to Eq. (16.21)

$$C_w = (33.3, 0.87, 300) - 0.5(-33.3, 1.33, -200)$$
$$= (50, 0.20, 400).$$

The response, R_{cw}, to C_w is 44. According to Step 4 above, we retain this result. This is shown as experiment 6 in Table 16.9. We now operate on experiments 1, 3, 4, and 6; experiment 3 is the new worst result.

We go to Step 1 and compute our new formulation R from Eq. (16.18)

$$R = (50, 0.87, 300) + (50, 0.67, -100) = (100, 1.54, 200).$$

Table 16.10 Sequential Experiments in Optimization Process

Experiment	Disintegrant	Lubricant	Fill weight	Response
1	50	0.2	100	37
2	0	2.2	100	58(W_1)[a]
3	0	0.2	400	46(W_3)
4	50	2.2	400	40
5	66.6	−0.46	500	(W_2)
6	50	0.20	400	44(W_4)
7	100	1.54	200	42(W_6)
8	83.3	2.42	67	43(W_5)
9	58.4	0.75	316	36
10	8.5	0.07	416	41(W_7)
11	39	0.56	344	44(W_8)
12	56.2	0.8	308	35

[a] W_1 means that this result was eliminated after the first evaluation.

The response R, is 42 (represented by experiment 7 in Table 16.9). This is better than the second worst response (44 for experiment 6) and we retain R_r as directed in Step 1 above. We recompute R for the set of experiments 1, 4, 6, and 7

$$R = (66.7, 1.31, 233) + (16.7, 1.11, -167) = (83.3, 2.42, 67).$$

The response, R_r, is 43. This is worse than the second-to-worst response, 42. Therefore we go to Step 3

$$C_r = \overline{P} + 0.5(\overline{P} - W)$$

$$C_r = (66.7, 1.31, 233) + 0.5(-16.7, -1.11, 167)$$
$$= (58.4, 0.75, 316).$$

The new response (experiment 9) is 36.
According to our rules, we go to Step 2

$$E = \overline{P} + 2(\overline{P} - W)$$

$$E = (69.5, 1.05, 272) + 2(-30.5, -0.49, 72)$$
$$= (8.5, 0.07, 416).$$

The response to E is 41. According to Step 2, we retain R in lieu of E because R gave the better response. We compute a new R from Step 1

$$R = (69.5, 1.05, 272) + (-30.5, -0.49, 72)$$
$$= (39, 0.56, 344).$$

The response is 44. Our new set of four experiments is numbers 1, 4, 9, and 11, with number 11 the worst.
We go to Step 4 and compute C_w because the value of R is worse than R_w

$$C_w = (69.5, 1.05, 272) - 0.5(30.5, 0.49, -72)$$
$$= (54.2, 0.8, 308).$$

The response was 35 (see experiment 12).
The experiments may continue as described above until repeated experiments do not show improvement. We are searching for an optimal response in the presence of variability. In the present case, a formula containing approximately 55 of disintegrant and 0.75 of lubricant with a fill weight of 300 mg appeared to show minimal dissolution time; the study was stopped after experiment 12.

As with other optimization procedures presented in this chapter, studying details in the literature references is essential to understand the procedure and calculations [8–11].

16.6   SCREENING DESIGNS

Usually, we know the factors that we wish to investigate, from our experience. However, in new, unknown, situations, it is possible that we may consider a number of factors to investigate, to see if any of these may affect the response or outcome. If there are only a few such variables (or factors), we may wish to use a factorial or fractional factorial design. If there are many potential factors of interest, screening designs are available that use less runs, but do give us insight into effects of interest. The most popular of such designs are the Plackett–Burman designs.

Screening designs may be useful if little is known of the system. In most cases, one should have a reasonable idea of which variables are important, and their effective ranges. But, we may be surprised. If everything were known, experimentation would not be necessary. Also, one should be careful not to neglect potentially important variables.

Table 16.11 Twelve Run Plackett–Burman Design

Run	X_1	X_2	X_3	X_4	X_5	X_6	X_7	X_8	X_9	X_{10}	X_{11}
1	+	+	−	+	+	+	−	−	−	+	−
2	+	−	+	+	+	−	−	−	+	−	+
3	−	+	+	+	−	−	−	+	−	+	+
4	+	+	+	−	−	−	+	−	+	+	−
5	+	+	−	−	−	+	−	+	+	−	+
6	+	−	−	−	+	−	+	+	+	−	+
7	−	−	−	+	−	+	+	−	+	+	+
8	−	−	+	−	+	+	−	+	+	+	−
9	−	+	−	+	+	−	+	+	+	−	−
10	+	−	+	+	−	+	+	+	−	−	−
11	−	+	+	−	+	+	+	−	−	−	+
12	−	−	−	−	−	−	−	−	−	−	−

Screening designs, in general, are fractional factorials of 2^n designs that estimate main effects, but not interactions. If results of such experiments point to specific factors, one can follow up with more complete designs to evaluate specific interactions.

A 12-run design is shown in Table 16.11. Note that the − and + signs refer to the low and high levels of the factor, respectively. Thus, for example, factor 1 in run 1 is at the high level. (See chap. 9 for further explanation of terminology.) For other designs, for example, higher order or more complex designs, a statistician should be consulted. In general, variability cannot be estimated without replication (run the design in duplicate, for example) or partial replication. This would increase the size and cost of the experiment. As in other design considerations, the cost and time considerations must be weighed against the information gained from expanded experiments. If less factors than runs are used, an estimate of variability can be provided. This is shown in the following example.

An example of a 12-run Plackett–Burman design is shown in Table 16.12. This design estimates the main effects of six variables. This leaves 5 d.f. for estimating the error. The estimates based on columns 7 to 11, inclusive, are only used to compute the variability, and are not related to the six factors in the experiment. An example of an experiment using this design could be as follows. The effect of six variables on the dissolution of a tablet is to be investigated. The six factors are (X_1) hardness, (X_2) level of disintegrant, (X_3) time of mixing granulation, (X_4) level of lubricant, (X_5) type of coating, and (X_6) tablet press pressure. The response is the percent dissolution in 30 minutes. Each factor is set at a low (−1) and a high (+1) level. (Note that "type of coating" is arbitrarily set at −1 and +1.)

The analysis is most easily accomplished using a multiple regression computer program. When designating values for the model in the computer program, it is convenient to input −1 for the low level and +1 for the high level. Table 16.13 shows an example of relevant computer output.

Table 16.12 Example of Twelve Run Plackett–Burman Design

Dissolution	X_1	X_2	X_3	X_4	X_5	X_6	Error	Error	Error	Error	Error
75	1	1	−1	1	1	1	−1	−1	−1	1	−1
104	1	−1	1	1	1	−1	−1	−1	1	−1	1
57	−1	1	1	1	−1	−1	−1	1	−1	1	1
54	1	1	1	−1	−1	−1	1	−1	1	1	−1
46	1	1	−1	−1	−1	1	−1	1	1	−1	1
58	1	−1	−1	−1	1	−1	1	1	−1	1	1
3	−1	−1	−1	1	−1	1	1	−1	1	1	1
98	−1	−1	1	−1	1	1	−1	1	1	1	−1
80	−1	1	−1	1	1	−1	1	1	1	−1	−1
12	1	−1	1	1	−1	1	1	1	−1	−1	−1
100	−1	1	1	−1	1	1	1	−1	−1	−1	1
13	−1	−1	−1	−1	−1	−1	−1	−1	−1	−1	−1

Table 16.13 Multiple Regression Computer Output of Data in Table 16.12

Independent variable	Regression coefficient	T value $(H_0: B = 0)$	Prob. level	Decision (5%)
Intercept	58.33	13.3748	0.000042	Reject H_0
X_1	−0.167	−0.0382	0.970996	Accept H_0
X_2	10.33	2.3692	0.064013	Accept H_0
X_3	12.5	2.8660	0.035158	Reject H_0
X_4	−3.167	−0.7261	0.500353	Accept H_0
X_5	27.5	6.3052	0.001477	Reject H_0
X_6	−2.667	−0.6114	0.567651	Accept H_0

Analysis of variance Source	d.f.	Sum of squares	Mean square	F ratio	Prob. level
Intercept	1	40833.33	40833.33		
Model	6	12437.33	2072.889	9.0810	0.014
Error	5	1141.33	228.267		
Total	11	13578.67	1234.424		

Note that only main effects are estimated. The error term comprises the five columns that were not assigned to factors (columns 7–11). If only five factors were investigated, columns 6 to 11 would be used to estimate error with 6 d.f. The estimate of error allows us to test the main effects for significance. This is a conservative test because the error will be, if anything, estimated on the high side. That is, if any interactions are present, the error estimate will be too high. This means that we may miss some significant effects if interaction is present. In this example, X_2 just misses significance, and X_3 and X_5 are significant. Again, the six factors are (X_1) hardness, (X_2) level of disintegrant, (X_3) time of mixing granulation, (X_4) level of lubricant, (X_5) type of coating, and (X_6) tablet press pressure. Therefore, we might wish to consider the level of disintegrant, time of mixing, and type of coating if we wish to modify the dissolution. The type of coating seems to have the greatest effect.

KEY TERMS

Checkpoint
Coding
Composite designs
Contour plot
Extra-design points
Factorial designs
Fractional factorial designs
Grid
Independence
Model
Model error
Multiple regression

Optimization
Orthogonality
Plackett–Burman
Polynomial equation
Replication
Response equation
Response surface
Screening designs
Sequential optimization
Simplex design
Simplex space
Transformation

EXERCISES

1. Calculate the predicted response from Eq. (16.6) for
   (a) $X_1 = 1$ mg, $X_2 = 1$ mg, $X_3 = 2.5$ mg
   (b) $X_1 = 2$ mg, $X_2 = 1$ mg, $X_3 = 4$ mg
   Note that Eq. (16.6) uses coded values; see Eq. (16.4).] For example, the coded value for $X_1 = 1$ mg is $0 = (1 − 1)/1$.
2. Show that the transformed values of $X_1 = 1$, $X_2 = 0.5$, and $X_3 = 2.5$ are all equal to zero for the three variables in Exercise Problem 1.
3. Calculate the coefficients for the polynomial equation, (16.8). The coefficients are calculated from the data in Table 16.4.

4. Show that decoded values of A and B equal to 0.5 and 1, respectively, are equal to 8.75 mg of A and 100 mg of B, for the data of Table 16.4 and Eq. (16.8). Calculate the expected response of this combination of A and B using Eq. (16.8).

5. A formulation was to be prepared to optimize dissolution time. (The formulation with the dissolution time of approximately 15 minutes is "optimal.") Stearic acid and mixing time were varied according to a 2^2 factorial design with the following results:

		Stearic acid	
		0.25%	1%
Mixing time (min)	15	10	23
	30	21	25

(a) Construct a polynomial response equation [see Eq. (16.8)].
(b) What concentration of stearic acid and mixing time would you choose for the final product?

‡‡6. Calculate the maximum solubility based on Eq. (16.13), using procedures of calculus. [Hint: Set the first derivative equal to 0 after substituting $(1.00 - A)$ for B.]

7. A total of 100 mg of three components, stearic acid (A), starch (B), and DCP (C), are to be added to a tablet formulation. Dissolution time was measured in a simplex design with the following results:

100% A:	292.0 min
100% B:	5.6 min
100% C:	50.4 min
50% A, 50% B:	25.6 min
50% B, 50% C:	15.6 min
50% A, 50% C:	124.5 min
1/3 A, 1/3 B, and 1/3 C:	37.0 min

(a) Compute the simplex equation coefficients.
(b) Give a combination with very fast dissolution.
(c) Give a combination that has a dissolution time of 90 minutes.

REFERENCES

1. Davies OL. The Design and Analysis of Industrial Experiments. New York: Hafner, 1963.
2. Ahmed S, Bolton S. Factorial design in the study of the effects of selected liquid chromatographic conditions on resolution and capacity factors. J Liq Chromatogr 1990; 13:525.
3. Daniel C. Use of half normal plots in interpreting factorial two-level experiments. Technometrics 1959; 1:311.
4. Box GE, Hunter WG, Hunter JS. Statistics for Experimenters. New York: Wiley, 1978.
5. Schwartz JB, Flamholtz JR, Press RH. Computer optimization of pharmaceutical formulations. I. General procedure. J Pharm Sci 1973; 62:1165.
6. Fonner DE Jr, Buck JR, Banker GS. Mathematical optimization techniques in drug product design and process analysis. J Pharm Sci 1970; 59:1587.
7. Lindberg N-O, Jonsson C, Holmquist B. Optimization of disintegration time and crushing strength of a tablet formulation. Drug Dev Ind Pharm 1985; 11(4):931–943.
8. Box GEP, Draper NR. Evolutionary Operations. New York: Wiley, 1969.
9. Spendley W, Hext GR, Himsworth FR. Sequential application of simplex designs in optimization and evolutionary operation. Technometrics 1962; 4:441.
10. Nelder JA, Mead R. A simplex method for function minimization. Comput J 1965; 7:308.
11. Shek E, Ghani M, Jones RE. A new attempt to solve the scale-up problem for granulation using response surface methodology. J Pharm Sci 1980; 69:1135.
12. Gorman JW, Hinman JE. Simplex lattice designs for multicomponent systems. Technometrics 1962; 4:463.

‡‡Optional, more advanced problem.

Glossary

a	calculated intercept in regression
ANCOVA	analysis of covariance
ANOVA	analysis of variance
b	calculated slope in regression
BMS	between mean square
BSS	between sum of squares
C. T.	correction term
CI	confidence interval
CV	coefficient of variation; relative error; relative standard deviation
CXR	column × row interaction
df	degrees of freedom
E	expected number in chi-square table
F	F value for F distribution
Ha	alternative hypothesis
Ho	null hypothesis
In	natural log
LSD	least significant difference
O	observed number in chi-square table
p	estimated proportion (binomial)
p (A)	probability that event will occur
p (A|B)	conditional probability of A given B
Po	true or hypothesized proportion
q	probability of failure in binomial
R	range
r	calculated correlation coefficient
r (Dixon)	computation for outlier analysis
r^2	square of correlation coefficient
RSD	relative standard deviation
S	sample standard deviation
S^2	sample variance
$S^2y.x$	estimated variance from line fitting
t	t value for t distribution
Tn	test for outlier
σ	true standard deviation of distribution
w	weight in weighted least squares
WSS	within sum of squares
X_i	ith observation
Z	normal standard deviate
X^2	chi square
Δ	delta, true change or difference
N	sample size
Σ	sum of observations

α	alpha level or error for null hypothesis; error of first kind
β	beta error (1-power)
δ	observed change or difference
μ	true mean of distribution

Appendix I
Some Properties of the Variance

I.1 POOLING VARIANCES

In many statistical procedures, an estimate of the variance is obtained by "averaging" or *pooling* the variances from more than one group of observations. The pooling of variances is appropriate in cases where samples from separate groups or different experiments provide estimates of the *same* variance. Note that we do not pool or average standard deviations. As we have previously noted, the sample variance, $\sum (X - \bar{X})^2/(N-1)$ [Eq. (1.5)], is an unbiased estimate of the true population variance. The standard deviation, estimated from a *sample*, is a *biased estimate* of the true *population* standard deviation. On the average, the sample standard deviation underestimates the population standard deviation. Estimation and properties of the variance are important considerations in both theoretical and applied statistics.

A common example of a procedure where variance estimates from different groups are pooled is the two-sample independent-groups t test for comparison of means discussed in chapter 5. In this test, the average results of two treatments* (e. g., active drug versus placebo; dissolution behavior of two tablet formulations) are compared. An estimate of the variance of the observations is needed in order to compare the two treatment groups statistically. An important assumption underlying this test is that the variances for each group are equal. The variance is first calculated for each treatment group separately. The variance is more precisely estimated from samples with a larger number of observations, and the pooled variance from both treatment groups is the best estimate of the common variance. For example, suppose that the following variances were observed in a comparative experiment:

Placebo group: $N = 25$ and the variance $(S^2) = 10$
Drug group: $N = 20$ and the variance $(S^2) = 15$

Although we assume that the true variance (the population variance) is the same for each group, different variances are observed in the two groups. If the two groups truly have equal variance, the difference in the observed variance is a consequence of random variation, due in part to the particular samples which were chosen, and measurement errors. The pooling procedure, in general, uses a *weighted average*, where the weights are equal to the degrees of freedom [see Eq. (1.2)].

$$S^2 \text{ pooled} = S_p^2 = \frac{(24)(10) + (19)(15)}{24 + 19} = 12.21.$$

The standard deviation is 3.49 ($\sqrt{12.21}$). The numbers 24 and 19 are the degrees of freedom for the two groups. If variances are to be pooled from more than two groups, the procedure is the same. Use a weighted average of the group variances, weighting the variance in each group by its number of degrees of freedom.

I.2 COMPONENTS OF VARIANCE

Variability of observations usually arise from more than one source. Hence, the variability of observations can often be expressed as the sum of independent sources of error that comprise the

---

* The word "treatment" in statistics does not necessarily mean treatment in the medical sense. Treatments are conditions or combinations of conditions whose effects on an experimental outcome are to be assessed.

total variation. This notion is presented in more detail under the topic of *components of variance* in section 12.4.1. The variance of the average of assay results for three tablets obtained by selecting a single tablet from each of three batches and assaying each tablet is as follows: [*variance* due to mean potency differences among batches (i.e., the batch averages are not identical) + *variance* due to tablet differences within batches[i] + *variance* due to drug assay]/3. Note that this is the variance of a mean of three results (a total of three tablets have been assayed from the three batches). This accounts for the number 3 in the denominator ($S^2 = S^2/N$).

Similarly, the variability of individual cholesterol changes, derived from a group of patients, such as shown in Table 1.1, is the sum of the components that contribute to the overall variability: (a) biological variation as reflected in inherent differences between patients, (b) the day-to-day variability within patients (a single person's cholesterol varies from day to day), and (c) the analytical error, among other sources of error.

I.3 VARIANCE OF LINEAR COMBINATIONS OF INDEPENDENT VARIABLES

The variance of linear combinations of variables, where the variables are independent, can be shown to be

$$\text{Variance}(m X_1 \pm n X_2) = m^2 \, \text{variance}(X_1) + n^2 \, \text{variance}(X_2), \qquad (\text{I.1})$$

where m and n are constants. This important result can be used to derive the variance of the mean of n independent observations, for example. Consider m observations of the variable X. We can represent the observations as $X_1, X_2, X_3, \ldots, X_m$. The mean is

$$\frac{\sum X_i}{m} = \frac{X_1 + X_2 + X_3 + \cdots + X_m}{m}.$$

The variance of each X is σ^2. Therefore, the variance of the mean is

$$\frac{\sigma_1^2 + \sigma_2^2 + \sigma_3^2 + \cdots \sigma_m^2}{m^2} = \frac{m(\sigma^2)}{m^2} = \frac{\sigma^2}{m}$$

Equation (I.1) also demonstrates that the variance of the difference of two independent observations is the *sum* of their variances. An example noted by Mandel [1] that illustrates this concept is the timing of a reaction. A stopwatch is started at the initiation of the reaction and stopped at some end point. The time depends on both the initial and final readings. If errors in the times are independent, the variance of $t_2 - t_1$, the difference between final and initial readings, is the sum of the variances; that is, the error of the difference of the two readings is larger than the error of either reading alone. Consider another example where a procedure calls for 10 mL of solution to be removed from a beaker containing 30 mL. Only 10-mL pipettes are available. The original 30 mL of solution is prepared by pipetting three 10-mL portions into a beaker. A total of 10 mL is then removed. The variance of the volume remaining in the solution is calculated as follows:

$$\text{Variance}(P_1 + P_2 + P_3 - P_4) = \sigma^2 P_1 + \sigma^2 P_2 + \sigma^2 P_3 + \sigma^2 P_4,$$

where P_i ($i = 1, 2, 3, 4$) represents the four pipetting steps. If the variance of a pipetting step is 0.01, the total variance of the remaining solution (with an expected volume of 20 mL) is (4)(0.01) = 0.04.

REFERENCE
1. Mandel J. The Statistical Analysis of Experimental Data, New York: Interscience, 1964.

---

[i] Variation resulting from differences in tablet potency in a randomly chosen sample of tablets which is due to the inherent variability of tablets (a result of the heterogeneity of the tableting process) is also known as "sampling error."

Appendix II

Comparison of Slopes and Testing of Linearity: Determination of Relative Potency

A common problem in bioassay, or when comparing the potency of compounds such as in drug screening programs, is the assessment of the relative potency of the comparative drugs. The problems in this analysis consist of (a) obtaining a function of dose and response that is linear, (b) testing the lines for each compound for parallelism (i.e., equality of slopes), and (c) determining the relative potency. We will discuss some elementary concepts for a comparison of two anti-inflammatory compounds, a standard drug (St) and an experimental compound (Ex). The experiment consists of measuring the reduction in volume after treatment of initially inflamed paws of two animals at each of three doses for each compound. The results are shown in Table II.1 and plotted in Figure II.1. The figure shows that the plot of *log* dose versus response is approximately linear. A *transformation* of dose and/or response is often necessary to achieve linearity in dose-response relationships. The response is usually considered to be a linear function of *log* dose (see chap. 10). Transformations to obtain linearity are desirable because straight-line relationships are more easily analyzed and interpreted than are more complex functions.

How does one determine if the data are represented by a linear function such as a straight line? A known theoretical relationship between X and Y may be sufficient to answer the question. From a statistical point of view, replicate measurements at fixed values of X are needed to test for linearity. Replicate measurements of Y at a fixed X represent S^2_y only, a variance estimate which is independent of the functional form of X and Y. If X and Y are truly related by a straight-line function, deviations of the observed values of Y from the fitted line should be due only to the variability of Y. If the relationship between X and Y is not a straight line, the variance as measured by the deviations of Y from the fitted line will be increased due to "nonlinearity" (see Fig. 7.4b). To test for linearity, we compare the variance due to deviations of Y from the fitted line (deviations from regression) to the variation due only to Y (the pooled error from the Y replicates, the within mean square). The "deviations" mean square is the mean square due to deviations of the averages of Y (at each X) from the fitted line. The statistical test is an F test obtained from an analysis of variance. The concept of this test is illustrated in Figure II.2.

To perform the test, a one-way ANOVA is first performed on the data (Table II.2), duplicate determinations for three doses in the present example. The ANOVA is computed for each of both the standard and experimental drugs. For example, the calculations for the ANOVA for the standard drug are as follows:

$$\text{Total SS} = \sum Y^2 - \frac{\left(\sum Y^2\right)}{N} = 1.674 - 1.4406 = 0.2334$$

$$\text{Between-doses SS} = \frac{0.49^2 + 1.00^2 + 1.45^2}{2} - 1.4406 = 0.2307.$$

The within SS is the difference between the total SS and the between SS (see sec. 8.1).

The between-doses SS is the sum of two components: (a) the SS due to the slope (regression SS) and (b) the SS due to *deviations of the mean values (at each X) of Y from the fitted line. The deviation SS has been discussed* above and is shown in Figure II.2. The easiest way to compute the deviation SS is to divide the between-doses SS into its components as follows. The "regression" SS has 1 degree of freedom and is defined as

$$\text{Regression SS} = b^2 \sum (X - \overline{X})^2. \tag{II.1}$$

Table II.1 Results of the Experiment Comparing Potencies of Two Compounds[a]

Compound	Dose (mg)		
	5	15	45
Standard (St)	0.22	0.51	0.70
	0.27	0.49	0.75
Experimental (Ex)	0.29	0.55	0.76
	0.26	0.54	0.83

[a]Data are relative reduction in paw volume from baseline value.

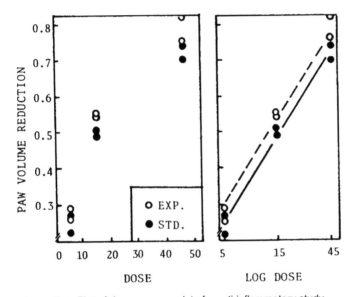

Figure II.1 Plot of dose response data for anti-inflammatory study.

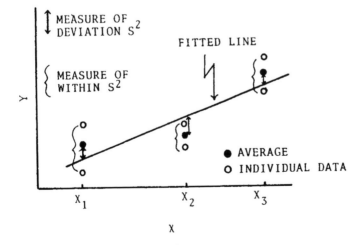

Figure II.2 ANOVA test for linearity.

Table II.2 One-way ANOVA for Data from Standard and Experimental Drugs

	Standard drug			Experimental drug		
Source	d.f.	SS	MS	d.f.	SS	MS
Between doses	2	0.2307	0.1154	2	0.27053	0.1353
Within (doses)	3	0.0027	0.0009	3	0.00295	0.00098
Total	5	0.2334		5	0.27348	

This SS, a result of the slope of the line, will be zero for a line of zero slope ($b = 0$), and will be large for a line with a steep positive or negative slope. For the standard drug, the regression sum of squares is calculated as follows (remember, we are using log dose $= X$):

$$b = 0.503$$
$$b^2 = \sum (X - \overline{X})^2 = 0.503^2(0.9106) = 0.2304.$$

The deviation SS (sometimes called "lack of fit" SS) is equal to the between-doses SS minus the regression SS. Therefore, the deviation SS $= 0.2307 - 0.2304 = 0.0003$.

The results of this calculation for both standard and experimental drugs are shown in Table II.3.

The test for linearity is an F test (deviation MS)/(within MS). For the standard drug, for example, the F ratio is $0.0003/0.0009 = 0.33$, with 1 and 3 d.f., which is not significant (within MS $= 0.0009$, Table II.2). There is no evidence for lack of linearity for both lines.

Usually, in these assays, the deviation mean squares are pooled from both products and compared to the pooled error (within MS), testing linearity of both lines simultaneously. The pooled deviation MS is $(0.000433)/2$ with 2 degrees of freedom. The pooled within **MS** is 0.000942 with 6 degrees of freedom. The F test for linearity is $0.000217/\ 0.00094 = 0.23$ (2 and 6 d.f.), which is clearly not significant. The pooling assumes that the error for both drugs is the same, and that both drugs show a linear response versus log dose.

Another assumption in the analysis of the parallel-line assay is that the two lines are parallel. A test of parallelism is equivalent to a test of equality of slopes. The common slope, calculated from all the data combined, is

$$b = \frac{\sum XY - (\sum X \sum Y)/N}{\sum (X - \overline{X})^2} = 0.5240.$$

The regression SS due to the common slope is

$$b^2 \sum (X - \overline{X})^2 = (0.5240)^2(1.8212) = 0.500.$$

The regression SS of the common slope is subtracted from the pooled *regression* SS for the two drugs to obtain the SS attributed to lack of parallelism of the lines. The pooled regression SS is

Table II.3 Regression and Deviations Sum of Squares for Standard and Experimental Drugs[a]

	Standard drug		Experimental drug	
Source	d.f.	SS	d.f.	SS
Regression	1	0.2304	1	0.2704
Deviations	1	0.0003	1	0.000133
Between doses	2	0.2307	2	0.270533

[a]Degrees of freedom for "regression" in the simple linear regression case is always equal to 1. Degrees of freedom for "deviations" is equal to (number of doses − 2).

0.2304 + 0.2704 = 0.5008. The SS for "parallelism" is 0.0008 (0.5008 − 0.5000). The F test has 1 and 6 d.f., using the pooled error term:

$$F_{1,6} = \frac{0.0008}{0.00094} = 0.851.$$

Since the F value shows lack of significance at the 5% level, we conclude that the lines appear to be parallel within "experimental error."

The test for parallelism for *two* lines can also be done by using a t test with the same results as the F test. (For the case of two lines, the t is the square root of the F value.) For the t test, we compare the two slopes, using the standard deviation of the difference of the two slopes in the denominator of the t ratio. The slopes are 0.5030 and 0.5449 for the standard and experimental drugs, respectively. The variances in both groups are assumed to be equal.

$$t = \frac{|b_1 - b_2|}{\sqrt{S^2 \left[1/\sum_1 (X - \overline{X})^2 + 1\sum_2 (X - \overline{X})^2 \right]}}$$

$$t = \frac{|0.5030 - 0.5449|}{\sqrt{0.00094 \left[1/\sum_1 (X - \overline{X})^2 + 1\sum_2 (X - \overline{X})^2 \right]}},$$

(II.2)

where $\sum_i (X - \overline{X})^2$ represents the sum of squares of the $X's$ for the respective groups. [Note that the variance of a slope equals $S^2/\sum (X - \overline{X})^2$.]

Having satisfied ourselves that the assumptions of the assay have been met (i.e., particularly, linearity and parallelism), we can now estimate the relative potency. The relative potency is the ratio of the comparative drugs that will give the same response. If the lines are parallel, we can choose any response (Y) to estimate the relative potency; the answer will be the same (Fig. II.3).

One can show that the log of the relative potency (log R) is equal to

$$\log R = \log \left[\frac{\text{experimental}}{\text{standard}} \right] = \frac{a_e - a_d}{b},$$

where a_e and a_d are the intercepts for the experimental drug and the standard drug, respectively; b is the common slope (0.524, in our example); and (experimental/standard) is the inverse ratio of doses that gives equal response. For the data of Table II.1,

$$a_d = -0.1262 \quad a_e = -0.0779$$

$$\log R = \frac{-0.0779 - (-0.1262)}{0.5240} = 0.092.$$

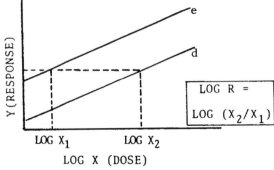

LOG R =

LOG (X_2/X_1)

Figure II.3 Relative potency estimation using parallel dose–response lines; doses equivalent to log X_1 and log X_2 give the same response for products e and d, respectively.

The relative potency is 1.24; that is, the experimental drug is 1.24 times as potent as the standard. This means that 124 mg of the standard is needed to give the same response as 100 mg of the experimental drug, for example.

Confidence limits can be put on the relative potency based on Fieller's theorem (similar to confidence limits for X at a given Y; see chap. 7). The procedure is complicated, and the interested reader is referred to the book by Finney, *Statistical Methods in Biological Assay* [1], for details of the computations.

REFERENCE

1. Finney DJ. Statistical Methods in Biological Assay. New York: Hafner, 1964.

Appendix III
Multiple Regression

Multiple regression is a topic of utmost importance in statistics, analysis of variance being a special case of the more general regression techniques. Multiple regression is an extension of linear regression, in which we wish to relate a response, Y (dependent variable), to more than one independent variable, X_i.

Linear regression: $Y = A + BY$

Multiple regression: $Y = B_0 + B_1 X_1 + B_2 X_2 + \ldots$

The independent variables, X_1, X_2, and so on, generally represent factors that we believe influence the response. Usually, the purpose of multiple regression analysis is to quantitate the relationship between Y and the X_i's by means of an equation, the multiple regression equation. For example, tablet dissolution may be measured as a function of several variables, such as level of disintegrant, lubricant, and drug. In this case, a multiple regression equation would be useful to predict dissolution, at given levels of the independent variables.

$$Y = B_0 + B_1 X_1 + B_2 X_2 + B_3 X_3, \tag{III.1}$$

where Y is the some measure of dissolution, X_i is ith independent variable, and B_i the regression coefficient for the ith independent variable.

Here, X_1, X_2, and X_3 refer to the level of disintegrant, lubricant, and drug. B_1, B_2, and B_3 are the coefficients relating the X_i to the response. These coefficients correspond to the slope (B) in linear regression. B_0 is the intercept. This equation cannot be simply depicted, graphically, as in the linear regression case. With two independent variables (X_1 and X_2), the response surface is a plane (Fig. III.1). With more than two independent variables, it is not possible to graph the response in two dimensions.

Data suitable for multiple regression analysis can be obtained in different ways. Optimal efficiency and interpretation are obtained by using data from "designed" experiments. In designed experiments, the independent variables are carefully chosen and deliberately controlled at preassigned levels. For example, in the dissolution experiment noted above, we may be able to fix the levels of disintegrant, lubricant, and drug according to a factorial design (as described in chap. 9). Table III.1 illustrates a 2^3 factorial design. These data correspond to the eight combinations in the 2^3 design that can be used to construct a multiple regression equation. The procedure for fitting data from a factorial design to a regression equation is given in section 16.2.

The form of the equation and the number of independent variables necessary to define the response adequately depend on a knowledge of the system being investigated. In the example above, there are three independent variables (factors), but interactions of factors may also be needed to define the response. In multiple regression equations, interactions may be represented by "cross-product" terms, such as ($X_1 X_2$) or ($X_1 X_2 X_3$). We usually include only those terms in the equation that probably have a meaningful effect on the response. Suppose, in our example, that the three factors and the lubricant X drug interaction are related to the response, dissolution. We would include terms for X_1, X_2, X_3, and $X_2 X_3$ in the model.

$$Y = B_0 + B_1 X_1 + B_2 X_2 + B_3 X_3 + B_{23} X_2 X_3.$$

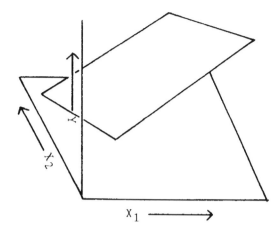

Figure III.1 Representation of the multiple regression equation response, $Y = B_0 + B_1 \times_1 + B_2 X_2$, as a plane.

Data for multiple regression fits are often obtained from undesigned experiments where the levels of the independent variables are not controlled. This less desirable alternative is often a consequence of convenience or cost considerations. Sometimes, the circumstances are such that we have no choice; we get the data in any way that we can. For example, suppose that tableting pressure, temperature, and humidity all affect some particular quality of a finished tablet. Tablets may be conveniently selected for inspection during the manufacturing process, at which time measurements of the pressure, temperature, and humidity are made. After collecting a sufficient quantity of data, these variables may be related to tablet quality using multiple regression techniques.

$$Y = B_0 + B_1(\text{tablet press pressure}) + B_2(\text{temperature}) + B_3(\text{humidity}).$$

In this example, we have no control of the variables; their values are a matter of "happenstance." We take the values as they come. A significant disadvantage of making conclusions based on data of this sort is that a correlation exists among the independent variables, which can be eliminated (or controlled) in a designed experiment. The result of this correlation is that the effects of the variables cannot be clearly separated. What we attribute to one variable, temperature for example, has a component due to humidity and pressure as well. With data derived from a designed experiment, such as the factorial design noted above, the regression equation can be constructed so that the effects of different factors and interactions are represented by the coefficients (B_i) and are independent of other factors.

The computations to determine the coefficients in multiple regression analysis are very tedious, and without the use of computers, analysis of undesigned experiments of reasonable size are virtually impossible. Manipulations of large matrices are often performed in the solution of these problems. Regression equations for orthogonal (designed) factors are much easier to compute. However, with easy access to computers, hand analysis should be done only as a learning tool to gain insight into the analytical process. We will not discuss computational methods in the general multiple regression model. However, because of the importance of multiple regression in optimization procedures discussed in chapter 16, some further introductory concepts will be presented here.

Table III.1 Factorial Design to Be Used as the Source for a Multiple Regression Equation

| | | Disintegrant low level | | Disintegrant high level | |
| | | Drug | | Drug | |
		Low level	High level	Low level	High level
Lubricant	Low level				
	High level				

The technique of fitting a linear model to data consisting of N observations of a response, Y, and one or more independent variables, X_i, is applicable when the number of observations is equal to or greater than the number of parameters to be estimated (the coefficients are the parameters in multiple regression). In simple linear regression, we estimate two parameters in the usual case, the intercept and the slope. Given two X, Y points, the line (slope and intercept) is unambiguously fixed. With more than two points, the best straight line is considered to be the line that minimizes the sum of the squared deviations of the observed values from the fitted least squares line. Multiple regression is just an extension of this procedure. If there are N parameters (coefficients) in the regression model, N observations will result in an exact fit to the model. For example, an equation with six coefficients will be exactly fit to six appropriate experimental values (with certain mathematical restrictions). With more than N observations, the coefficients, B_i, are calculated to minimize the squared deviations of the observations from the least squares regression fit (the same concept as in simple linear regression).

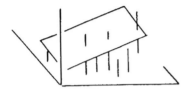

The relationship of the independent variables and the dependent variable in the multiple regression model must be *linear in the coefficients, B_i*, in order to obtain the regression equation by the usual procedures [1]. The general form of the regression equation is given by Eq. (III.1).

$$Y = B_0 + B_1 X_1 + B_2 X_2 + B_3 X_3. \tag{III.1}$$

The X_i's can be "nonlinear" functions such as X^2, log X, or 10^x. However, the coefficients, B_i, cannot be in this nonlinear form. Thus

$$Y = B_1 X_1 + B_2 X_2 + B_3 X_1^2 + B_4 X_1 X_2 \quad \text{is linear in } B_i$$
$$Y = B_0 + B_1 X_1 + X_1^{B_2} \quad \text{is not linear in } B_i.$$

The basic problems in multiple regression analysis are concerned with estimation of the error and the coefficients (parameters) of the regression model. Statistical tests can then be performed for the significance of the coefficient estimates.

When many independent variables are candidates to be entered into a regression equation, one may wish to use only those variables that contribute "significantly" to the relationship with the dependent variable. In designed experiments (e. g., factorial designs) the significance of each factor can be determined using analysis of variance, or, equivalently, by testing the regression coefficients for significance. In an undesigned experiment, where the data come from "uncontrolled" combinations of the variables, the independent variables will inevitably be more or less correlated. Thus, if dissolution is to be related to tablet weight, drug content, and tablet hardness, based on production records, we are obliged to fit an equation with the available data, and some correlation will exist between drug content and weight, for example. This lack of independence presents special problems when deciding which variables are relevant, contributing significantly to the regression relationship. If two of the X variables, X_i and X_j, are highly correlated, inclusion of both in the regression equation will be redundant. Therefore, there may be some X variables that appear to contribute to the regression but which are correlated to other X variables. We must then make a choice regarding their inclusion in the final regression equation. Draper and Smith note: "There is no unique statistical procedure for doing this," and some degree of arbitrariness must be used in making choices [1]. Two methods used to help make such decisions are made possible through the use of computers. One method involves regression fits using all possible combinations of the independent variables (2^k regressions, where k is the number of independent variables). For two independent variables, X_1 and X_2, the four possible regressions are

1. $Y = B_0$
2. $Y = B_0 + B_1X_1$
3. $Y = B_0 + B_2X_2$
4. $Y = B_0 + B_1X_1 + B_2X_2$

The best equation may then be selected based on the fit and the number of variables needed for the fit. The multiple correlation coefficient, R^2, is a measure of the fit. R^2 is the sum of squares due to regression divided by the sum of squares without regression. For example, if R^2 is 0.85 when three variables are used to fit the regression equation, and R^2 is equal to 0.87 when six variables are used, we probably would be satisfied using the equation with three variables, other things being equal. The inclusion of more variables in the regression equation cannot result in a decrease of R^2.

Another method of selecting variables to be included in the regression equation is the popular stepwise procedure, which is considered a better method than the "all possible regressions" approach. Independent variables (X_i) are entered into the equation, one at a time, starting with the independent variable that is most highly correlated to the dependent variable, Y. As each new variable is considered, its inclusion is based on a preassigned statistical test related to its correlation with the dependent variable, as well as its correlation to those independent variables already included in the regression equation.

Probably the biggest pitfall in multiple regression techniques lies in the interpretation of the coefficients. Draper and Smith discuss this problem, and the answer is by no means simple [1].

Interpretation of the meaning of the coefficients in multiple regression equations is much more clear in a designed (orthogonal) experiment. As we have noted previously, in a factorial experiment, the levels of the factors can be controlled, so that the effects of the factors can be independently evaluated. Techniques to describe and optimize pharmaceutical systems by fitting experimental data to regression models using designed experiments are discussed in chapter 16.

An application of regression analysis to physical properties of finished tablets, with compression pressure and various tablet components as independent variables can be found in Ref. [2]. In this paper, the authors considered five independent variables for inclusion in the regression equation. They suggested the following equation as a predictor of dissolution:

$$Y = 69.91 - 37.3X_5 - 17.48X_2 + 4.24X_3, \tag{III.2}$$

where Y is the dissolution, X_5 the magnesium stearate level, X_2 the compression pressure, and X_3 the starch disintegrant.

Magnesium stearate and compression pressure decrease dissolution (negative coefficient). Starch increases dissolution. The authors discuss possible mechanisms for these effects.

Multiple regression equations that relate variables such as those described above are empirical relationships. We do not encounter real systems that can be described so simply, theoretically. The multiple regression equation is a "model" of a real system that must be recognized as being only an approximation of reality. How good an approximation the equation is can be evaluated only by seeing how the equation performs as a predictor of the response in new situations, where the levels of the independent variables are changed. Also, particularly in undesigned systems, placing physical interpretation on the signs and magnitude of the coefficients can be hazardous. As noted previously, the coefficients can give insights into the mechanisms of a process, but great caution is needed before making definitive judgments on this basis. Problems similar to those discussed for prediction in linear regression apply here as well. Error (variability) in the estimation of the coefficients, extrapolating to areas outside the levels of the variables in the experiment, and the choice of an incorrect model all adversely affect the reliability of the predicted value.

In addition to its use as a predictive equation, the regression equation may also be used to help obtain combinations of ingredients that will give a desired (e. g., optimum) response.

This process is discussed in chapters 9 and 16. For those readers who are interested in a more advanced, in-depth discussion of regression, the excellent book by Draper and Smith, *Applied Regression Analysis*, is recommended [1].

REFERENCES
1. Draper NR, Smith H. Applied Regression Analysis, 2 nd ed. New York: Wiley, 1981.
2. Bohidar NR, Restaino FA., Schwartz JB. Selecting Key Pharmaceutical Formulation Factors by Regression Analysis. Drug Dev Ind Pharm 1979; 5: 175.

Appendix IV
Tables

Table IV.1 Random Numbers

44	17	50	92	09	79	27	71	05	07	76	21	95	93	04
83	50	39	13	89	83	45	72	40	94	78	62	93	55	62
28	79	77	81	43	04	54	23	14	80	49	98	32	70	27
55	29	62	11	00	62	65	76	31	83	08	22	02	35	53
88	93	30	81	50	24	43	07	88	45	96	24	60	78	89
46	00	76	13	83	31	98	15	30	74	17	76	73	31	40
99	05	78	83	75	79	52	47	39	12	70	33	42	30	45
24	88	59	45	16	73	64	63	03	16	04	43	81	66	97
14	90	27	33	43	46	37	68	94	35	12	72	70	43	54
50	27	98	87	19	20	15	73	00	94	52	85	80	22	26
55	47	03	77	04	44	22	78	84	26	04	33	46	09	52
59	29	97	68	60	71	91	38	67	54	13	58	18	24	76
48	55	90	65	72	96	57	69	36	10	96	46	92	42	45
66	37	32	20	30	77	84	57	03	29	10	45	65	04	26
68	49	69	10	82	53	75	91	93	30	34	25	20	57	27
83	62	64	11	12	67	19	00	71	74	60	47	21	92	86
06	90	91	47	68	25	49	33	74	02	16	29	35	65	16
33	23	97	78	26	78	26	45	40	19	61	29	53	73	09
47	15	40	15	02	82	06	93	20	01	67	38	02	37	90
79	65	14	62	16	34	96	02	75	82	46	75	43	89	36

Table IV.2 Cumulative Normal Distribution: Cumulative Area Under the Normal Distribution (Less Than or Equal to Z)

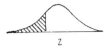

Z	Area	Z	Area	Z	Area	Z	Area
−3.25	0.0006	−1.50	0.0668	0.25	0.5987	2.00	0.9772
−3.20	0.0007	−1.45	0.0735	0.30	0.6179	2.05	0.9798
−3.15	0.0008	−1.40	0.0808	0.35	0.6368	2.10	0.9821
−3.10	0.0010	−1.35	0.0885	0.40	0.6554	2.15	0.9842
−3.05	0.0011	−1.30	0.0968	0.45	0.6736	2.20	0.9861
−3.00	0.0013	−1.25	0.1056	0.50	0.6915	2.25	0.9878
−2.95	0.0016	−1.20	0.1151	0.55	0.7088	2.30	0.9893
−2.90	0.0019	−1.15	0.1251	0.60	0.7257	2.35	0.9906
−2.85	0.0022	−1.10	0.1357	0.65	0.7422	2.40	0.9918
−2.80	0.0026	−1.05	0.1469	0.70	0.7580	2.45	0.9929
−2.75	0.0030	−1.00	0.1587	0.75	0.7734	2.50	0.9938
−2.70	0.0035	−0.95	0.1711	0.80	0.7881	2.55	0.9946
−2.65	0.0040	−0.90	0.1841	0.85	0.8023	2.60	0.9953
−2.60	0.0047	−0.85	0.1977	0.90	0.8159	2.65	0.9960
−2.55	0.0054	−0.80	0.2119	0.95	0.8289	2.70	0.9965
−2.50	0.0062	−0.75	0.2266	1.00	0.8413	2.75	0.9970
−2.45	0.0071	−0.70	0.2420	1.05	0.8531	2.80	0.9974
−2.40	0.0082	−0.65	0.2578	1.10	0.8643	2.85	0.9978
−2.35	0.0094	−0.60	0.2743	1.15	0.8749	2.90	0.9981
−2.30	0.0107	−0.55	0.2912	1.20	0.8849	2.95	0.9984
−2.25	0.0122	−0.50	0.3085	1.25	0.8944	3.00	0.9987
−2.20	0.0139	−0.45	0.3264	1.30	0.9032	3.25	0.9994
−2.15	0.0158	−0.40	0.3446	1.35	0.9115		
−2.10	0.0179	−0.35	0.3632	1.40	0.9192	Z	Area
−2.05	0.0202	−0.30	0.3821	1.45	0.9265	1.282	0.90
						1.645	0.95
−2.00	0.0228	−0.25	0.4013	1.50	0.9332	1.960	0.975
−1.95	0.0256	−0.20	0.4207	1.55	0.9394	2.326	0.99
−1.90	0.0287	−0.15	0.4404	1.60	0.9452	2.576	0.995
−1.85	0.0322	−0.10	0.4602	1.65	0.9505	3.090	0.999
−1.80	0.0359	−0.05	0.4801	1.70	0.9554		
−1.75	0.0401	0	0.5000	1.75	0.9599		
−1.70	0.0446	0.05	0.5199	1.80	0.9641		
−1.65	0.0495	0.10	0.5398	1.85	0.9678		
−1.60	0.0548	0.15	0.5596	1.90	0.9713		
−1.55	0.0606	0.20	0.5793	1.95	0.9744		

Table IV.3 Individual Terms of the Binomial Distribution for $N = 2$ to 10 and $P = 0.2$, 0.5, and 0.7[a]

				$P = 0.2$					
					N				
X	2	3	4	5	6	7	8	9	10
0	0.64	0.512	0.410	0.328	0.262	0.210	0.168	0.134	0.107
1	0.32	0.384	0.410	0.410	0.393	0.367	0.336	0.302	0.268
2	0.04	0.096	0.154	0.205	0.246	0.275	0.294	0.302	0.302
3		0.008	0.026	0.051	0.082	0.115	0.147	0.176	0.201
4			0.002	0.006	0.015	0.029	0.046	0.066	0.088
5				*	0.002	0.004	0.009	0.017	0.026
6					*	*	0.001	0.003	0.006
7						*	*	*	0.001
8							*	*	*
9								*	*
10									*

				$P = 0.5$					
					N				
X	2	3	4	5	6	7	8	9	10
0	0.250	0.125	0.0625	0.031	0.016	0.008	0.004	0.002	0.001
1	0.500	0.375	0.250	0.156	0.094	0.055	0.031	0.018	0.010
2	0.250	0.375	0.375	0.313	0.234	0.164	0.109	0.070	0.044
3		0.125	0.250	0.313	0.313	0.273	0.219	0.164	0.117
4			0.0625	0.156	0.234	0.273	0.273	0.246	0.205
5				0.031	0.094	0.164	0.219	0.246	0.246
6					0.016	0.055	0.109	0.164	0.205
7						0.008	0.031	0.070	0.117
8							0.004	0.018	0.044
9								0.002	0.010
10									0.001

				$P = 0.7$					
					N				
X	2	3	4	5	6	7	8	9	10
0	0.090	0.027	0.008	0.002	0.001	*	*	*	*
1	0.420	0.189	0.076	0.028	0.010	0.004	0.001	*	*
2	0.490	0.441	0.265	0.132	0.060	0.025	0.010	0.004	0.001
3		0.343	0.412	0.309	0.185	0.097	0.047	0.021	0.009
4			0.240	0.360	0.324	0.227	0.136	0.074	0.037
5				0.168	0.303	0.318	0.254	0.172	0.103
6					0.118	0.247	0.296	0.267	0.200
7						0.082	0.198	0.267	0.267
8							0.058	0.156	0.233
9								0.040	0.121
10									0.028

*$P < 0.0005$.

[a] These tables may be used for $P = 0.8$ and $P = 0.3$ as follows. Use the table with $P = 0.2$ to obtain terms for $P = 0.8$; and use the table with $P = 0.7$ to obtain terms for $P = 0.3$. For example, for the probability of 5 ($x' = 5$) successes in 8 trials ($N = 8$) for $P = 0.8$, look in the table for $P = 0.2$, $N = 8$, and $X = N - X' = 8 - 5 = 3$. This is equal to 0.147.

Table IV.4 *t* Distributions

Two-sided: One-sided: d.f.:	40% 20% $t_{0.80}$	20% 10% $t_{0.90}$	10% 5% $t_{0.95}$	5% 2.50% $t_{0.975}$	1% 0.50% $t_{0.995}$
1	1.376382	3.077684	6.313752	12.7062	63.65674
2	1.06066	1.885618	2.919986	4.302653	9.924843
3	0.978472	1.637744	2.353363	3.182446	5.840909
4	0.940965	1.533206	2.131847	2.776445	4.604095
5	0.919544	1.475884	2.015048	2.570582	4.032143
6	0.905703	1.439756	1.94318	2.446912	3.707428
7	0.89603	1.414924	1.894579	2.364624	3.499483
8	0.88889	1.396815	1.859548	2.306004	3.355387
9	0.883404	1.383029	1.833113	2.262157	3.249836
10	0.879058	1.372184	1.812461	2.228139	3.169273
11	0.87553	1.36343	1.795885	2.200985	3.105807
12	0.872609	1.356217	1.782288	2.178813	3.05454
13	0.870152	1.350171	1.770933	2.160369	3.012276
14	0.868055	1.34503	1.76131	2.144787	2.976843
15	0.866245	1.340606	1.75305	2.13145	2.946713
16	0.864667	1.336757	1.745884	2.119905	2.920782
17	0.863279	1.333379	1.739607	2.109816	2.898231
18	0.862049	1.330391	1.734064	2.100922	2.87844
19	0.860951	1.327728	1.729133	2.093024	2.860935
20	0.859964	1.325341	1.724718	2.085963	2.84534
21	0.859074	1.323188	1.720743	2.079614	2.83136
22	0.858266	1.321237	1.717144	2.073873	2.818756
23	0.85753	1.31946	1.713872	2.068658	2.807336
24	0.856855	1.317836	1.710882	2.063899	2.796939
25	0.856236	1.316345	1.708141	2.059539	2.787436
26	0.855665	1.314972	1.705618	2.055529	2.778715
27	0.855137	1.313703	1.703288	2.05183	2.770683
28	0.854647	1.312527	1.701131	2.048407	2.763262
29	0.854192	1.311434	1.699127	2.04523	2.756386
30	0.853767	1.310415	1.697261	2.042272	2.749996
31	0.85337	1.309464	1.695519	2.039513	2.744042
32	0.852998	1.308573	1.693889	2.036933	2.738481
33	0.852649	1.307737	1.69236	2.034515	2.733277
34	0.852321	1.306952	1.690924	2.032244	2.728394
35	0.852012	1.306212	1.689572	2.030108	2.723806
36	0.85172	1.305514	1.688298	2.028094	2.719485
37	0.851444	1.304854	1.687094	2.026192	2.715409
38	0.851183	1.30423	1.685954	2.024394	2.711558
39	0.850935	1.303639	1.684875	2.022691	2.707913
40	0.8507	1.303077	1.683851	2.021075	2.704459
41	0.850476	1.302543	1.682878	2.019541	2.701181
42	0.850263	1.302035	1.681952	2.018082	2.698066
43	0.85006	1.301552	1.681071	2.016692	2.695102
44	0.849867	1.30109	1.68023	2.015368	2.692278
45	0.849682	1.300649	1.679427	2.014103	2.689585
46	0.849505	1.300228	1.67866	2.012896	2.687013
47	0.849336	1.299825	1.677927	2.01174	2.684556
48	0.849174	1.299439	1.677224	2.010635	2.682204
49	0.849018	1.299069	1.676551	2.009575	2.679952
50	0.848869	1.298714	1.675905	2.008559	2.677793
75	0.84644	1.292941	1.665425	1.992102	2.642983
100	0.84523	1.290075	1.660234	1.983971	2.625891
500	0.842341	1.283247	1.647907	1.96472	2.585698
infinity	0.841621	1.281552	1.644855	1.959966	2.575834

Table IV.5 Chi-Square Distributions

Degrees of Freedom	Probability								
	0.01	0.025	0.05	0.1	0.2	0.8	0.9	0.95	0.99
1	6.634897	5.023886	3.841459	2.705544	1.642375	0.064185	0.015791	0.003932	0.000157
2	9.21034	7.377759	5.991465	4.60517	3.218876	0.446287	0.210721	0.102587	0.020101
3	11.34487	9.348404	7.814728	6.251388	4.641628	1.005174	0.584374	0.351846	0.114832
4	13.2767	11.14329	9.487729	7.77944	5.988617	1.648777	1.063623	0.710723	0.297109
5	15.08627	12.8325	11.0705	9.236357	7.289276	2.342534	1.610308	1.145476	0.554298
6	16.81189	14.44938	12.59159	10.64464	8.55806	3.070088	2.204131	1.635383	0.87209
7	18.47531	16.01276	14.06714	12.01704	9.80325	3.822322	2.833107	2.16735	1.239042
8	20.09024	17.53455	15.50731	13.36157	11.03009	4.593574	3.489539	2.732637	1.646497
9	21.66599	19.02277	16.91898	14.68366	12.24215	5.380053	4.168159	3.325113	2.087901
10	23.20925	20.48318	18.30704	15.98718	13.44196	6.179079	4.865182	3.940299	2.558212
11	24.72497	21.92005	19.67514	17.27501	14.63142	6.988674	5.577785	4.574813	3.053484
12	26.21697	23.33666	21.02607	18.54935	15.81199	7.807328	6.303796	5.226029	3.570569
13	27.68825	24.7356	22.36203	19.81193	16.9848	8.633861	7.041505	5.891864	4.106915
14	29.14124	26.11895	23.68479	21.06414	18.15077	9.467328	7.789534	6.570631	4.660425
15	30.57791	27.48839	24.99579	22.30713	19.31066	10.30696	8.546756	7.260944	5.229349
16	31.99993	28.84535	26.29623	23.54183	20.46508	11.15212	9.312236	7.961646	5.812213
17	33.40866	30.19101	27.58711	24.76904	21.61456	12.00227	10.08519	8.67176	6.40776
18	34.80531	31.52638	28.8693	25.98942	22.75955	12.85695	10.86494	9.390455	7.014911
19	36.19087	32.85233	30.14353	27.20357	23.90042	13.71579	11.65091	10.11701	7.63273
20	37.56623	34.16961	31.41043	28.41198	25.03751	14.57844	12.44261	10.85081	8.260398
21	38.93217	35.47888	32.67057	29.61509	26.1711	15.44461	13.2396	11.59131	8.897198
22	40.28936	36.78071	33.92444	30.81328	27.30145	16.31404	14.04149	12.33801	9.542492
23	41.6384	38.07563	35.17246	32.0069	28.42879	17.18651	14.84796	13.09051	10.19572
24	42.97982	39.36408	36.41503	33.19624	29.55332	18.0618	15.65868	13.84843	10.85636
25	44.3141	40.64647	37.65248	34.38159	30.6752	18.93975	16.47341	14.61141	11.52398
26	45.64168	41.92317	38.88514	35.56317	31.79461	19.82019	17.29189	15.37916	12.19815
27	46.96294	43.19451	40.11327	36.74122	32.91169	20.70298	18.1139	16.1514	12.8785
28	48.27824	44.46079	41.33714	37.91592	34.02657	21.58797	18.93924	16.92788	13.56471
29	49.58788	45.72229	42.55697	39.08747	35.13936	22.47505	19.76774	17.70837	14.25645
30	50.89218	46.97924	43.77297	40.25602	36.25019	23.36412	20.59923	18.49266	14.95346
31	52.19139	48.23190	44.98534	41.42174	37.35914	24.25506	21.43356	19.28057	15.65546
32	53.48577	49.48044	46.19426	42.58475	38.46631	25.14779	22.27059	20.07191	16.36222
33	54.77554	50.72508	47.39988	43.74518	39.57179	26.04222	23.1102	20.86653	17.07351
34	56.06091	51.966	48.60237	44.90316	40.67565	26.93827	23.95225	21.66428	17.78915
35	57.34207	53.20335	49.80185	46.05879	41.77796	27.83587	24.79666	22.46502	18.50893
36	58.61921	54.43729	50.99846	47.21217	42.8788	28.73496	25.6433	23.26861	19.23268
37	59.8925	55.66797	52.19232	48.36341	43.97822	29.63547	26.49209	24.07494	19.96023
38	61.16209	56.89552	53.38354	49.51258	45.07628	30.53734	27.34295	24.8839	20.69144
39	62.42812	58.12006	54.57223	50.65977	46.17303	31.44052	28.19579	25.69539	21.42616
40	63.69074	59.34171	55.75848	51.80506	47.26854	32.34495	29.05052	26.5093	22.16426
41	64.95007	60.56057	56.94239	52.94851	48.36283	33.2506	29.90709	27.32555	22.90561
42	66.20624	61.77676	58.12404	54.0902	49.45597	34.15741	30.76542	28.14405	23.65009
43	67.45935	62.99036	59.30351	55.23019	50.54799	35.06534	31.62545	28.96472	24.3976
44	68.70951	64.20146	60.48089	56.36854	51.63892	35.97435	32.48713	29.78748	25.14803
45	187.5299	180.2291	174.101	167.2074	159.1036	130.5082	123.6489	118.1714	108.3451
46	71.2014	66.61653	62.82962	58.64054	53.8177	37.79548	34.21517	31.439	26.65724
47	72.44331	67.82065	64.00111	59.77429	54.90561	38.70752	35.08143	32.26762	27.41585
48	73.68264	69.02259	65.17077	60.90661	55.99258	39.62051	35.94913	33.09808	28.17701
49	74.91947	70.22241	66.33865	62.03754	57.07863	40.53442	36.81822	33.93031	28.94065
50	76.15389	71.4202	67.50481	63.16712	58.1638	41.44921	37.68865	34.76425	29.70668
51	77.38596	72.61599	68.66929	64.2954	59.24811	77.38596	38.56038	35.59986	30.47505
52	78.61576	73.80986	69.83216	65.42241	60.33158	78.61576	39.43339	36.43709	31.24567
53	79.84334	75.00186	70.99345	66.5482	61.41425	79.84334	40.30762	37.27589	32.01849
54	81.06877	76.19205	72.15322	67.67279	62.49613	81.06877	41.18304	38.11622	32.79345
55	82.29212	77.38047	73.31149	68.79621	63.57724	82.29212	42.05962	38.95803	33.57048
56	83.51343	78.56716	74.46832	69.91851	64.65762	83.51343	42.93734	39.80128	34.34952
57	84.73277	79.75219	75.62375	71.03971	65.73727	84.73277	43.81615	40.64593	35.13053
58	85.95018	80.93559	76.7778	72.15984	66.81621	85.95018	44.69603	41.49195	35.91346
59	87.16571	82.11741	77.93052	73.27893	67.89448	87.16571	45.57695	42.33931	36.69825
60	88.37942	83.29768	79.08194	74.39701	68.97207	88.37942	46.45889	43.18796	37.48485
80	112.3288	106.6286	101.8795	96.5782	90.40535	112.3288	64.27785	60.39148	53.54008
100	135.8067	129.5612	124.3421	118.498	111.6667	135.8067	82.35814	77.92947	70.0649
500	576.4928	563.8515	553.1268	540.9303	526.4014	576.4928	459.9261	449.1468	429.3875

Table IV.6A1 Upper 5% Values of the F Distribution

Degrees of freedom in denominator	Degrees of freedom in numerator																											
	1	2	3	4	5	6	7	8	9	10	11	12	13	14	15	16	17	18	19	20	25	30	40	50	60	80	100	inf
1	161.448	199.500	215.707	224.583	230.162	233.986	236.768	238.883	240.543	241.882	242.983	243.906	244.690	245.364	245.950	246.464	246.918	247.323	247.686	248.013	249.260	250.095	251.143	251.774	252.196	252.724	253.041	254.314
2	18.513	19.000	19.164	19.247	19.296	19.330	19.353	19.371	19.385	19.396	19.405	19.413	19.419	19.424	19.429	19.433	19.437	19.440	19.443	19.446	19.456	19.462	19.471	19.476	19.479	19.483	19.486	19.496
3	10.128	9.552	9.277	9.117	9.013	8.941	8.887	8.845	8.812	8.786	8.763	8.745	8.729	8.715	8.703	8.692	8.683	8.675	8.667	8.660	8.634	8.617	8.594	8.581	8.572	8.561	8.554	8.526
4	7.709	6.944	6.591	6.388	6.256	6.163	6.094	6.041	5.999	5.964	5.936	5.912	5.891	5.873	5.858	5.844	5.832	5.821	5.811	5.803	5.769	5.746	5.717	5.699	5.688	5.673	5.664	5.628
5	6.608	5.786	5.409	5.192	5.050	4.950	4.876	4.818	4.772	4.735	4.704	4.678	4.655	4.636	4.619	4.604	4.590	4.579	4.568	4.558	4.521	4.496	4.464	4.444	4.431	4.415	4.405	4.365
6	5.987	5.143	4.757	4.534	4.387	4.284	4.207	4.147	4.099	4.060	4.027	4.000	3.976	3.956	3.938	3.922	3.908	3.896	3.884	3.874	3.835	3.808	3.774	3.754	3.740	3.722	3.712	3.669
7	5.591	4.737	4.347	4.120	3.972	3.866	3.787	3.726	3.677	3.637	3.603	3.575	3.550	3.529	3.511	3.494	3.480	3.467	3.455	3.445	3.404	3.376	3.340	3.319	3.304	3.286	3.275	3.230
8	5.318	4.459	4.066	3.838	3.687	3.581	3.500	3.438	3.388	3.347	3.313	3.284	3.259	3.237	3.218	3.202	3.187	3.173	3.161	3.150	3.108	3.079	3.043	3.020	3.005	2.986	2.975	2.928
9	5.117	4.256	3.863	3.633	3.482	3.374	3.293	3.230	3.179	3.137	3.102	3.073	3.048	3.025	3.006	2.989	2.974	2.960	2.948	2.936	2.893	2.864	2.826	2.803	2.787	2.768	2.756	2.707
10	4.965	4.103	3.708	3.478	3.326	3.217	3.135	3.072	3.020	2.978	2.943	2.913	2.887	2.865	2.845	2.828	2.812	2.798	2.785	2.774	2.730	2.700	2.661	2.637	2.621	2.601	2.588	2.538
11	4.844	3.982	3.587	3.357	3.204	3.095	3.012	2.948	2.896	2.854	2.818	2.788	2.761	2.739	2.719	2.701	2.685	2.671	2.658	2.646	2.601	2.570	2.531	2.507	2.490	2.469	2.457	2.404
12	4.747	3.885	3.490	3.259	3.106	2.996	2.913	2.849	2.796	2.753	2.717	2.687	2.660	2.637	2.617	2.599	2.583	2.568	2.555	2.544	2.498	2.466	2.426	2.401	2.384	2.363	2.350	2.296
13	4.667	3.806	3.411	3.179	3.025	2.915	2.832	2.767	2.714	2.671	2.635	2.604	2.577	2.554	2.533	2.515	2.499	2.484	2.471	2.459	2.412	2.380	2.339	2.314	2.297	2.275	2.261	2.206
14	4.600	3.739	3.344	3.112	2.958	2.848	2.764	2.699	2.646	2.602	2.565	2.534	2.507	2.484	2.463	2.445	2.428	2.413	2.400	2.388	2.341	2.308	2.266	2.241	2.223	2.201	2.187	2.131
15	4.543	3.682	3.287	3.056	2.901	2.790	2.707	2.641	2.588	2.544	2.507	2.475	2.448	2.424	2.403	2.385	2.368	2.353	2.340	2.328	2.280	2.247	2.204	2.178	2.160	2.137	2.123	2.066
16	4.494	3.634	3.239	3.007	2.852	2.741	2.657	2.591	2.538	2.494	2.456	2.425	2.397	2.373	2.352	2.333	2.317	2.302	2.288	2.276	2.227	2.194	2.151	2.124	2.106	2.083	2.068	2.010
17	4.451	3.592	3.197	2.965	2.810	2.699	2.614	2.548	2.494	2.450	2.413	2.381	2.353	2.329	2.308	2.289	2.272	2.257	2.243	2.230	2.181	2.148	2.104	2.077	2.058	2.035	2.020	1.960
18	4.414	3.555	3.160	2.928	2.773	2.661	2.577	2.510	2.456	2.412	2.374	2.342	2.314	2.290	2.269	2.250	2.233	2.217	2.203	2.191	2.141	2.107	2.063	2.035	2.017	1.993	1.978	1.917
19	4.381	3.522	3.127	2.895	2.740	2.628	2.544	2.477	2.423	2.378	2.340	2.308	2.280	2.256	2.234	2.215	2.198	2.182	2.168	2.155	2.106	2.071	2.026	1.999	1.980	1.955	1.940	1.878
20	4.351	3.493	3.098	2.866	2.711	2.599	2.514	2.447	2.393	2.348	2.310	2.278	2.250	2.225	2.203	2.184	2.167	2.151	2.137	2.124	2.074	2.039	1.994	1.966	1.946	1.922	1.907	1.843
21	4.325	3.467	3.072	2.840	2.685	2.573	2.488	2.420	2.366	2.321	2.283	2.250	2.222	2.197	2.176	2.156	2.139	2.123	2.109	2.096	2.045	2.010	1.965	1.936	1.916	1.891	1.876	1.812
22	4.301	3.443	3.049	2.817	2.661	2.549	2.464	2.397	2.342	2.297	2.259	2.226	2.198	2.173	2.151	2.131	2.114	2.098	2.084	2.071	2.020	1.984	1.938	1.909	1.889	1.864	1.849	1.783
23	4.279	3.422	3.028	2.796	2.640	2.528	2.442	2.375	2.320	2.275	2.236	2.204	2.175	2.150	2.128	2.109	2.091	2.075	2.061	2.048	1.996	1.961	1.914	1.885	1.865	1.839	1.823	1.757
24	4.260	3.403	3.009	2.776	2.621	2.508	2.423	2.355	2.300	2.255	2.216	2.183	2.155	2.130	2.108	2.088	2.070	2.054	2.040	2.027	1.975	1.939	1.892	1.863	1.842	1.816	1.800	1.733
25	4.242	3.385	2.991	2.759	2.603	2.490	2.405	2.337	2.282	2.236	2.198	2.165	2.136	2.111	2.089	2.069	2.051	2.035	2.021	2.007	1.955	1.919	1.872	1.842	1.822	1.796	1.779	1.711
26	4.225	3.369	2.975	2.743	2.587	2.474	2.388	2.321	2.265	2.220	2.181	2.148	2.119	2.094	2.072	2.052	2.034	2.018	2.003	1.990	1.938	1.901	1.853	1.823	1.803	1.776	1.760	1.691
27	4.210	3.354	2.960	2.728	2.572	2.459	2.373	2.305	2.250	2.204	2.166	2.132	2.103	2.078	2.056	2.036	2.018	2.002	1.987	1.974	1.921	1.884	1.836	1.806	1.785	1.758	1.742	1.672
28	4.196	3.340	2.947	2.714	2.558	2.445	2.359	2.291	2.236	2.190	2.151	2.118	2.089	2.064	2.041	2.021	2.003	1.987	1.972	1.959	1.906	1.869	1.820	1.790	1.769	1.742	1.725	1.654
29	4.183	3.328	2.934	2.701	2.545	2.432	2.346	2.278	2.223	2.177	2.138	2.104	2.075	2.050	2.027	2.007	1.989	1.973	1.958	1.945	1.891	1.854	1.806	1.775	1.754	1.726	1.710	1.638
30	4.171	3.316	2.922	2.690	2.534	2.421	2.334	2.266	2.211	2.165	2.126	2.092	2.063	2.037	2.015	1.995	1.976	1.960	1.945	1.932	1.878	1.841	1.792	1.761	1.740	1.712	1.695	1.622
31	4.160	3.305	2.911	2.679	2.523	2.409	2.323	2.255	2.199	2.153	2.114	2.080	2.051	2.026	2.003	1.983	1.965	1.948	1.933	1.920	1.866	1.828	1.779	1.748	1.726	1.699	1.681	1.608
32	4.149	3.295	2.901	2.668	2.512	2.399	2.313	2.244	2.189	2.142	2.103	2.070	2.040	2.015	1.992	1.972	1.953	1.937	1.922	1.908	1.854	1.817	1.767	1.736	1.714	1.686	1.669	1.594
33	4.139	3.285	2.892	2.659	2.503	2.389	2.303	2.235	2.179	2.133	2.093	2.060	2.030	2.004	1.982	1.961	1.943	1.926	1.911	1.898	1.844	1.806	1.756	1.724	1.702	1.674	1.657	1.581
34	4.130	3.276	2.883	2.650	2.494	2.380	2.294	2.225	2.170	2.123	2.084	2.050	2.021	1.995	1.972	1.952	1.933	1.917	1.902	1.888	1.833	1.795	1.745	1.713	1.691	1.663	1.645	1.569
35	4.121	3.267	2.874	2.641	2.485	2.372	2.285	2.217	2.161	2.114	2.075	2.041	2.012	1.986	1.963	1.942	1.924	1.907	1.892	1.878	1.824	1.786	1.735	1.703	1.681	1.652	1.635	1.558
36	4.113	3.259	2.866	2.634	2.477	2.364	2.277	2.209	2.153	2.106	2.067	2.033	2.003	1.977	1.954	1.934	1.915	1.899	1.883	1.870	1.815	1.776	1.726	1.694	1.671	1.643	1.625	1.547
37	4.105	3.252	2.859	2.626	2.470	2.356	2.270	2.201	2.145	2.098	2.059	2.025	1.995	1.969	1.946	1.926	1.907	1.890	1.875	1.861	1.806	1.768	1.717	1.685	1.662	1.633	1.615	1.537
38	4.098	3.245	2.852	2.619	2.463	2.349	2.262	2.194	2.138	2.091	2.051	2.017	1.988	1.962	1.939	1.918	1.899	1.883	1.867	1.853	1.798	1.760	1.708	1.676	1.653	1.624	1.606	1.527
39	4.091	3.238	2.845	2.612	2.456	2.342	2.255	2.187	2.131	2.084	2.044	2.010	1.981	1.954	1.931	1.911	1.892	1.875	1.860	1.846	1.791	1.752	1.700	1.668	1.645	1.616	1.597	1.518
40	4.085	3.232	2.839	2.606	2.449	2.336	2.249	2.180	2.124	2.077	2.038	2.003	1.974	1.948	1.924	1.904	1.885	1.868	1.853	1.839	1.783	1.744	1.693	1.660	1.637	1.608	1.589	1.509
50	4.034	3.183	2.790	2.557	2.400	2.286	2.199	2.130	2.073	2.026	1.986	1.952	1.921	1.895	1.871	1.850	1.831	1.814	1.798	1.784	1.727	1.687	1.634	1.599	1.576	1.544	1.525	1.438
60	4.001	3.150	2.758	2.525	2.368	2.254	2.167	2.097	2.040	1.993	1.952	1.917	1.887	1.860	1.836	1.815	1.796	1.778	1.763	1.748	1.690	1.649	1.594	1.559	1.534	1.502	1.481	1.389
80	3.960	3.111	2.719	2.486	2.329	2.214	2.126	2.056	1.999	1.951	1.910	1.875	1.845	1.817	1.793	1.772	1.752	1.734	1.718	1.703	1.644	1.602	1.545	1.508	1.482	1.448	1.426	1.325
100	3.936	3.087	2.696	2.463	2.305	2.191	2.103	2.032	1.975	1.927	1.886	1.850	1.819	1.792	1.768	1.746	1.726	1.708	1.691	1.676	1.616	1.573	1.515	1.477	1.450	1.415	1.392	1.283
200	3.888	3.041	2.650	2.417	2.259	2.144	2.056	1.985	1.927	1.878	1.837	1.801	1.769	1.742	1.717	1.694	1.674	1.656	1.639	1.623	1.561	1.516	1.455	1.415	1.386	1.346	1.321	1.189
infinity	3.841	2.996	2.605	2.372	2.214	2.099	2.010	1.938	1.880	1.831	1.789	1.752	1.720	1.692	1.666	1.644	1.623	1.604	1.587	1.571	1.506	1.459	1.394	1.350	1.318	1.274	1.244	1.000

Table IV.6A2 Upper 10% Values of the F Distribution

Degrees of freedom in numerator

Degrees of freedom in denominator	1	2	3	4	5	6	7	8	9	10	11	12	13	14	15	16	17	18	19	20	25	30	40	50	60	80	100	inf
1	39.863	49.500	53.593	55.833	57.240	58.204	58.906	59.439	59.858	60.195	60.473	60.705	60.903	61.073	61.220	61.350	61.464	61.566	61.658	61.740	62.055	62.265	62.529	62.688	62.794	62.927	63.007	63.328
2	8.526	9.000	9.162	9.243	9.293	9.326	9.349	9.367	9.381	9.392	9.401	9.408	9.415	9.420	9.425	9.429	9.433	9.436	9.439	9.441	9.451	9.458	9.466	9.471	9.475	9.479	9.481	9.491
3	5.538	5.462	5.391	5.343	5.309	5.285	5.266	5.252	5.240	5.230	5.222	5.216	5.210	5.205	5.200	5.196	5.193	5.190	5.187	5.184	5.175	5.168	5.160	5.155	5.151	5.147	5.144	5.134
4	4.545	4.325	4.191	4.107	4.051	4.010	3.979	3.955	3.936	3.920	3.907	3.896	3.886	3.878	3.870	3.864	3.858	3.853	3.849	3.844	3.828	3.817	3.804	3.795	3.790	3.782	3.778	3.761
5	4.060	3.780	3.619	3.520	3.453	3.405	3.368	3.339	3.316	3.297	3.282	3.268	3.257	3.247	3.238	3.230	3.223	3.217	3.212	3.207	3.187	3.174	3.157	3.147	3.140	3.132	3.126	3.105
6	3.776	3.463	3.289	3.181	3.108	3.055	3.014	2.983	2.958	2.937	2.920	2.905	2.892	2.881	2.871	2.863	2.855	2.848	2.842	2.836	2.815	2.800	2.781	2.770	2.762	2.752	2.746	2.722
7	3.589	3.257	3.074	2.961	2.883	2.827	2.785	2.752	2.725	2.703	2.684	2.668	2.654	2.643	2.632	2.623	2.615	2.607	2.601	2.595	2.571	2.555	2.535	2.523	2.514	2.504	2.497	2.471
8	3.458	3.113	2.924	2.806	2.726	2.668	2.624	2.589	2.561	2.538	2.519	2.502	2.488	2.475	2.464	2.455	2.446	2.438	2.431	2.425	2.400	2.383	2.361	2.348	2.339	2.328	2.321	2.293
9	3.360	3.006	2.813	2.693	2.611	2.551	2.505	2.469	2.440	2.416	2.396	2.379	2.364	2.351	2.340	2.329	2.320	2.312	2.305	2.298	2.272	2.255	2.232	2.218	2.208	2.196	2.189	2.159
10	3.285	2.924	2.728	2.605	2.522	2.461	2.414	2.377	2.347	2.323	2.302	2.284	2.269	2.255	2.244	2.233	2.224	2.215	2.208	2.201	2.174	2.155	2.132	2.117	2.107	2.095	2.087	2.055
11	3.225	2.860	2.660	2.536	2.451	2.389	2.342	2.304	2.274	2.248	2.227	2.209	2.193	2.179	2.167	2.156	2.147	2.138	2.130	2.123	2.095	2.076	2.052	2.036	2.026	2.013	2.005	1.972
12	3.177	2.807	2.606	2.480	2.394	2.331	2.283	2.245	2.214	2.188	2.166	2.147	2.131	2.117	2.105	2.094	2.084	2.075	2.067	2.060	2.031	2.011	1.986	1.970	1.960	1.946	1.938	1.904
13	3.136	2.763	2.560	2.434	2.347	2.283	2.234	2.195	2.164	2.138	2.116	2.097	2.080	2.066	2.053	2.042	2.032	2.023	2.014	2.007	1.978	1.958	1.931	1.915	1.904	1.890	1.882	1.846
14	3.102	2.726	2.522	2.395	2.307	2.243	2.193	2.154	2.122	2.095	2.073	2.054	2.037	2.022	2.010	1.998	1.988	1.978	1.970	1.962	1.933	1.912	1.885	1.869	1.857	1.843	1.834	1.797
15	3.073	2.695	2.490	2.361	2.273	2.208	2.158	2.119	2.086	2.059	2.037	2.017	2.000	1.985	1.972	1.961	1.950	1.941	1.932	1.924	1.894	1.873	1.845	1.828	1.817	1.802	1.793	1.755
16	3.048	2.668	2.462	2.333	2.244	2.178	2.128	2.088	2.055	2.028	2.005	1.985	1.968	1.953	1.940	1.928	1.917	1.908	1.899	1.891	1.860	1.839	1.811	1.793	1.782	1.766	1.757	1.718
17	3.026	2.645	2.437	2.308	2.218	2.152	2.102	2.061	2.028	2.001	1.978	1.958	1.940	1.925	1.912	1.900	1.889	1.879	1.870	1.862	1.831	1.809	1.781	1.763	1.751	1.735	1.726	1.686
18	3.007	2.624	2.416	2.286	2.196	2.130	2.079	2.038	2.005	1.977	1.954	1.933	1.916	1.900	1.887	1.875	1.864	1.854	1.845	1.837	1.805	1.783	1.754	1.736	1.723	1.707	1.698	1.657
19	2.990	2.606	2.397	2.266	2.176	2.109	2.058	2.017	1.984	1.956	1.932	1.912	1.894	1.878	1.865	1.852	1.841	1.831	1.822	1.814	1.782	1.759	1.730	1.711	1.699	1.683	1.673	1.631
20	2.975	2.589	2.380	2.249	2.158	2.091	2.040	1.999	1.965	1.937	1.913	1.892	1.875	1.859	1.845	1.833	1.821	1.811	1.802	1.794	1.761	1.738	1.708	1.690	1.677	1.660	1.650	1.607
21	2.961	2.575	2.365	2.233	2.142	2.075	2.023	1.982	1.948	1.920	1.896	1.875	1.857	1.841	1.827	1.815	1.803	1.793	1.784	1.776	1.742	1.719	1.689	1.670	1.657	1.640	1.630	1.586
22	2.949	2.561	2.351	2.219	2.128	2.060	2.008	1.967	1.933	1.904	1.880	1.859	1.841	1.825	1.811	1.798	1.787	1.777	1.768	1.759	1.726	1.702	1.671	1.652	1.639	1.622	1.611	1.567
23	2.937	2.549	2.339	2.207	2.115	2.047	1.995	1.953	1.919	1.890	1.866	1.845	1.827	1.811	1.796	1.784	1.772	1.762	1.753	1.744	1.710	1.686	1.655	1.636	1.622	1.605	1.594	1.549
24	2.927	2.538	2.327	2.195	2.103	2.035	1.983	1.941	1.906	1.877	1.853	1.832	1.814	1.797	1.783	1.770	1.759	1.748	1.739	1.730	1.696	1.672	1.641	1.621	1.607	1.590	1.579	1.533
25	2.918	2.528	2.317	2.184	2.092	2.024	1.971	1.929	1.895	1.866	1.841	1.820	1.802	1.785	1.771	1.758	1.746	1.736	1.726	1.718	1.683	1.659	1.627	1.607	1.593	1.576	1.565	1.518
26	2.909	2.519	2.307	2.174	2.082	2.014	1.961	1.919	1.884	1.855	1.830	1.809	1.790	1.774	1.760	1.747	1.735	1.724	1.715	1.706	1.671	1.647	1.615	1.594	1.581	1.562	1.551	1.504
27	2.901	2.511	2.299	2.165	2.073	2.005	1.952	1.909	1.874	1.845	1.820	1.799	1.780	1.764	1.749	1.736	1.724	1.714	1.704	1.695	1.660	1.636	1.603	1.583	1.569	1.550	1.539	1.491
28	2.894	2.503	2.291	2.157	2.064	1.996	1.943	1.900	1.865	1.836	1.811	1.790	1.771	1.754	1.740	1.726	1.715	1.704	1.694	1.685	1.650	1.625	1.592	1.572	1.558	1.539	1.528	1.478
29	2.887	2.495	2.283	2.149	2.057	1.988	1.935	1.892	1.857	1.827	1.802	1.781	1.762	1.745	1.731	1.717	1.705	1.695	1.685	1.676	1.640	1.616	1.583	1.562	1.547	1.528	1.517	1.467
30	2.881	2.489	2.276	2.142	2.049	1.980	1.927	1.884	1.849	1.819	1.794	1.773	1.754	1.737	1.722	1.709	1.697	1.686	1.676	1.667	1.632	1.606	1.573	1.552	1.538	1.519	1.507	1.456
31	2.875	2.482	2.270	2.136	2.042	1.973	1.920	1.877	1.842	1.812	1.787	1.765	1.746	1.729	1.714	1.701	1.689	1.678	1.668	1.659	1.623	1.598	1.565	1.543	1.529	1.509	1.498	1.446
32	2.869	2.477	2.263	2.129	2.036	1.967	1.913	1.870	1.835	1.805	1.780	1.758	1.739	1.722	1.707	1.694	1.682	1.671	1.661	1.652	1.616	1.590	1.556	1.535	1.520	1.501	1.489	1.437
33	2.864	2.471	2.258	2.123	2.030	1.961	1.907	1.864	1.828	1.799	1.773	1.751	1.732	1.715	1.700	1.687	1.675	1.664	1.654	1.645	1.608	1.583	1.549	1.527	1.512	1.493	1.480	1.428
34	2.859	2.466	2.252	2.118	2.024	1.955	1.901	1.858	1.822	1.793	1.767	1.745	1.726	1.709	1.694	1.680	1.668	1.657	1.647	1.638	1.601	1.576	1.541	1.520	1.505	1.485	1.473	1.419
35	2.855	2.461	2.247	2.113	2.019	1.950	1.896	1.852	1.817	1.787	1.761	1.739	1.720	1.703	1.688	1.674	1.662	1.651	1.641	1.632	1.595	1.569	1.535	1.513	1.497	1.478	1.465	1.411
36	2.850	2.456	2.243	2.108	2.014	1.945	1.891	1.847	1.811	1.781	1.756	1.734	1.715	1.697	1.682	1.669	1.656	1.645	1.635	1.626	1.589	1.563	1.528	1.506	1.491	1.471	1.458	1.404
37	2.846	2.452	2.238	2.103	2.009	1.940	1.886	1.842	1.806	1.776	1.751	1.729	1.709	1.692	1.677	1.663	1.651	1.640	1.630	1.620	1.583	1.557	1.522	1.500	1.484	1.464	1.452	1.397
38	2.842	2.448	2.234	2.099	2.005	1.935	1.881	1.838	1.802	1.772	1.746	1.724	1.704	1.687	1.672	1.658	1.646	1.635	1.624	1.615	1.578	1.551	1.516	1.494	1.478	1.458	1.445	1.390
39	2.839	2.444	2.230	2.095	2.001	1.931	1.877	1.833	1.797	1.767	1.741	1.719	1.700	1.682	1.667	1.653	1.641	1.630	1.619	1.610	1.573	1.546	1.511	1.488	1.473	1.452	1.439	1.383
40	2.835	2.440	2.226	2.091	1.997	1.927	1.873	1.829	1.793	1.763	1.737	1.715	1.695	1.678	1.662	1.649	1.636	1.625	1.615	1.605	1.568	1.541	1.506	1.483	1.467	1.447	1.434	1.377
50	2.809	2.412	2.197	2.061	1.966	1.895	1.840	1.796	1.760	1.729	1.703	1.680	1.660	1.643	1.627	1.613	1.600	1.588	1.578	1.568	1.529	1.502	1.465	1.441	1.424	1.402	1.388	1.327
60	2.791	2.393	2.177	2.041	1.946	1.875	1.819	1.775	1.738	1.707	1.680	1.657	1.637	1.619	1.603	1.589	1.576	1.564	1.553	1.543	1.504	1.476	1.437	1.413	1.395	1.372	1.358	1.291
80	2.769	2.370	2.154	2.016	1.921	1.849	1.793	1.748	1.711	1.680	1.653	1.629	1.609	1.590	1.574	1.559	1.546	1.534	1.523	1.513	1.472	1.443	1.403	1.377	1.358	1.334	1.318	1.245
100	2.756	2.356	2.139	2.002	1.906	1.834	1.778	1.732	1.695	1.663	1.636	1.612	1.592	1.573	1.557	1.542	1.528	1.516	1.505	1.494	1.453	1.423	1.382	1.355	1.336	1.310	1.293	1.214
200	2.731	2.329	2.111	1.973	1.876	1.804	1.747	1.701	1.663	1.631	1.603	1.579	1.558	1.539	1.522	1.507	1.493	1.480	1.468	1.458	1.414	1.383	1.339	1.310	1.289	1.261	1.242	1.144
infinity	2.706	2.303	2.084	1.945	1.847	1.774	1.717	1.670	1.632	1.599	1.571	1.546	1.524	1.505	1.487	1.471	1.457	1.444	1.432	1.421	1.375	1.342	1.295	1.263	1.240	1.207	1.185	1.000

Table IV.6B Upper 1% Values of the F Distribution

Degrees of freedom in denominator	1	2	3	4	5	6	7	8	9	10	11	12	13
											Degrees of Freedom in numerator		
1	4052.181	4999.500	5403.352	5624.583	5763.650	5858.986	6125.865	6125.865	6022.473	6055.847	6083.317	6106.321	6125.865
2	98.503	99.000	99.166	99.249	99.299	99.333	99.422	99.374	99.388	99.399	99.408	99.416	99.422
3	34.116	30.817	29.457	28.710	28.237	27.911	26.983	27.489	27.345	27.229	27.133	27.052	26.983
4	21.198	18.000	16.694	15.977	15.522	15.207	14.307	14.799	14.659	14.546	14.452	14.374	14.307
5	16.258	13.274	12.060	11.392	10.967	10.672	9.825	10.289	10.158	10.051	9.963	9.888	9.825
6	13.745	10.925	9.780	9.148	8.746	8.466	7.657	8.102	7.976	7.874	7.790	7.718	7.657
7	12.246	9.547	8.451	7.847	7.460	7.191	6.410	6.840	6.719	6.620	6.538	6.469	6.410
8	11.259	11.259	7.591	7.006	6.632	6.371	5.609	6.029	5.911	5.814	5.734	5.667	5.609
9	10.561	8.022	6.992	6.422	6.057	5.802	5.055	5.467	5.351	5.257	5.178	5.111	5.055
10	10.044	7.559	6.552	5.994	5.636	5.386	4.650	5.057	4.942	4.849	4.772	4.706	4.650
11	9.646	7.206	6.217	5.668	5.316	5.069	4.342	4.744	4.632	4.539	4.462	4.397	4.342
12	9.330	6.927	5.953	5.412	5.064	4.821	4.100	4.499	4.388	4.296	4.220	4.155	4.100
13	9.074	6.701	5.739	5.205	4.862	4.620	3.905	4.302	4.191	4.100	4.025	3.960	3.905
14	8.862	6.515	5.564	5.035	4.695	4.456	3.745	4.140	4.030	3.939	3.864	3.800	3.745
15	8.683	6.359	5.417	4.893	4.556	4.318	3.612	4.004	3.895	3.805	3.730	3.666	3.612
16	8.531	6.226	5.292	4.773	4.437	4.202	3.498	3.890	3.780	3.691	3.616	3.553	3.498
17	8.400	6.112	5.185	4.669	4.336	4.102	3.401	3.791	3.682	3.593	3.519	3.455	3.401
18	8.285	6.013	5.092	4.579	4.248	4.015	3.316	3.705	3.597	3.508	3.434	3.371	3.316
19	8.185	5.926	5.010	4.500	4.171	3.939	3.242	3.631	3.523	3.434	3.360	3.297	3.242
20	8.096	5.849	4.938	4.431	4.103	3.871	3.177	3.564	3.457	3.368	3.294	3.231	3.177
21	8.017	5.780	4.874	4.369	4.042	3.812	3.119	3.506	3.398	3.310	3.236	3.173	3.119
22	7.945	5.719	4.817	4.313	3.988	3.758	3.067	3.453	3.346	3.258	3.184	3.121	3.067
23	7.881	5.664	4.765	4.264	3.939	3.710	3.020	3.406	3.299	3.211	3.137	3.074	3.020
24	7.823	5.614	4.718	4.218	3.895	3.667	2.977	3.363	3.256	3.168	3.094	3.032	2.977
25	7.770	5.568	4.675	4.177	3.855	3.627	2.939	3.324	3.217	3.129	3.056	2.993	2.939
26	7.721	5.526	4.637	4.140	3.818	3.591	2.904	3.288	3.182	3.094	3.021	2.958	2.904
27	7.677	5.488	4.601	4.106	3.785	3.558	2.871	3.256	3.149	3.062	2.988	2.926	2.871
28	7.636	5.453	4.568	4.074	3.754	3.528	2.842	3.226	3.120	3.032	2.959	2.896	2.842
29	7.598	5.420	4.538	4.045	3.725	3.499	2.814	3.198	3.092	3.005	2.931	2.868	2.814
30	7.562	5.390	4.510	4.018	3.699	3.473	2.789	3.173	3.067	2.979	2.906	2.843	2.789
31	7.530	5.362	4.484	3.993	3.675	3.449	2.765	3.149	3.043	2.955	2.882	2.820	2.765
32	7.499	5.336	4.459	3.969	3.652	3.427	2.744	3.127	3.021	2.934	2.860	2.798	2.744
33	7.471	5.312	4.437	3.948	3.630	3.406	2.723	3.106	3.000	2.913	2.840	2.777	2.723
34	7.444	5.289	4.416	3.927	3.611	3.386	2.704	3.087	2.981	2.894	2.821	2.758	2.704
35	7.419	5.268	4.396	3.908	3.592	3.368	2.686	3.069	2.963	2.876	2.803	2.740	2.686
36	7.396	5.248	4.377	3.890	3.574	3.351	2.669	3.052	2.946	2.859	2.786	2.723	2.669
37	7.373	5.229	4.360	3.873	3.558	3.334	2.653	3.036	2.930	2.843	2.770	2.707	2.653
38	7.353	5.211	4.343	3.858	3.542	3.319	2.638	3.021	2.915	2.828	2.755	2.692	2.638
39	7.333	5.194	4.327	3.843	3.528	3.305	2.624	3.006	2.901	2.814	2.741	2.678	2.624
40	7.314	5.179	4.313	3.828	3.514	3.291	2.611	2.993	2.888	2.801	2.727	2.665	2.611
50	7.171	5.057	4.199	3.720	3.408	3.186	2.508	2.890	2.785	2.698	2.625	2.562	2.508
60	7.077	4.977	4.126	3.649	3.339	3.119	2.442	2.823	2.718	2.632	2.559	2.496	2.442
80	6.963	4.881	4.036	3.563	3.255	3.036	2.361	2.742	2.637	2.551	2.478	2.415	2.361
100	6.895	4.713	3.984	3.513	3.206	2.988	2.313	2.694	2.590	2.503	2.430	2.368	2.313
200	6.763	4.713	3.881	3.414	3.110	2.893	2.220	2.601	2.497	2.411	2.338	2.275	2.220
infinity	6.635	3.912	3.782	3.319	3.017	2.802	2.130	2.511	2.408	2.321	2.248	2.185	2.130

14	15	16	17	18	19	20	25	30	40	50	60	80	100	inf
6142.674	6157.285	6125.865	6181.435	6191.529	6200.576	6208.730	6239.825	6260.649	6286.782	6302.517	6313.030	6326.197	6334.110	6365.861
99.428	99.433	99.437	99.440	99.444	99.447	99.449	99.459	99.466	99.474	99.479	99.482	99.487	99.489	99.499
26.924	26.872	26.827	26.787	26.751	26.719	26.690	26.579	26.505	26.411	26.354	26.316	26.269	26.240	26.125
14.249	14.198	14.154	14.115	14.080	14.048	14.020	13.911	13.838	13.745	13.690	13.652	13.605	13.577	13.463
9.770	9.722	9.680	9.643	9.610	9.580	9.553	9.449	9.379	9.291	9.238	9.202	9.157	9.130	9.020
7.605	7.559	7.519	7.483	7.451	7.422	7.396	7.296	7.229	7.143	7.091	7.057	7.013	6.987	6.880
6.359	6.314	6.275	6.240	6.209	6.181	6.155	6.058	5.992	5.908	5.858	5.824	5.781	5.755	5.650
5.559	5.515	5.477	5.442	5.412	5.384	5.359	5.263	5.198	5.116	5.065	5.032	4.989	4.963	4.859
5.005	4.962	4.924	4.890	4.860	4.833	4.808	4.713	4.649	4.567	4.517	4.483	4.441	4.415	4.311
4.601	4.558	4.520	4.487	4.457	4.430	4.405	4.311	4.247	4.165	4.115	4.082	4.039	4.014	3.909
4.293	4.251	4.213	4.180	4.150	4.123	4.099	4.005	3.941	3.860	3.810	3.776	3.734	3.708	3.602
4.052	4.010	3.972	3.939	3.909	3.883	3.858	3.765	3.701	3.619	3.569	3.535	3.493	3.467	3.361
3.857	3.815	3.778	3.745	3.716	3.689	3.665	3.571	3.507	3.425	3.375	3.341	3.298	3.272	3.165
3.698	3.656	3.619	3.586	3.556	3.529	3.505	3.412	3.348	3.266	3.215	3.181	3.138	3.112	3.004
3.564	3.522	3.485	3.452	3.423	3.396	3.372	3.278	3.214	3.132	3.081	3.047	3.004	2.977	2.868
3.451	3.409	3.372	3.339	3.310	3.283	3.259	3.165	3.101	3.018	2.967	2.933	2.889	2.863	2.753
3.353	3.312	3.275	3.242	3.212	3.186	3.162	3.068	3.003	2.920	2.869	2.835	2.791	2.764	2.653
3.269	3.227	3.190	3.158	3.128	3.101	3.077	2.983	2.919	2.835	2.784	2.749	2.705	2.678	2.566
3.195	3.153	3.116	3.084	3.054	3.027	3.003	2.909	2.844	2.761	2.709	2.674	2.630	2.602	2.489
3.130	3.088	3.051	3.018	2.989	2.962	2.938	2.843	2.778	2.695	2.643	2.608	2.563	2.535	2.421
3.072	3.030	2.993	2.960	2.931	2.904	2.880	2.785	2.720	2.636	2.584	2.548	2.503	2.475	2.360
3.019	2.978	2.941	2.908	2.879	2.852	2.827	2.733	2.667	2.583	2.531	2.495	2.450	2.422	2.305
2.973	2.931	2.894	2.861	2.832	2.805	2.781	2.686	2.620	2.535	2.483	2.447	2.401	2.373	2.256
2.930	2.889	2.852	2.819	2.789	2.762	2.738	2.643	2.577	2.492	2.440	2.403	2.357	2.329	2.211
2.892	2.850	2.813	2.780	2.751	2.724	2.699	2.604	2.538	2.453	2.400	2.364	2.317	2.289	2.169
2.857	2.815	2.778	2.745	2.715	2.688	2.664	2.569	2.503	2.417	2.364	2.327	2.281	2.252	2.131
2.824	2.783	2.746	2.713	2.683	2.656	2.632	2.536	2.470	2.384	2.330	2.294	2.247	2.218	2.097
2.795	2.753	2.716	2.683	2.653	2.626	2.602	2.506	2.440	2.354	2.300	2.263	2.216	2.187	2.064
2.767	2.726	2.689	2.656	2.626	2.599	2.574	2.478	2.412	2.325	2.271	2.234	2.187	2.158	2.034
2.742	2.700	2.663	2.630	2.600	2.573	2.549	2.453	2.386	2.299	2.245	2.208	2.160	2.131	2.006
2.718	2.677	2.640	2.606	2.577	2.550	2.525	2.429	2.362	2.275	2.220	2.183	2.135	2.106	1.980
2.696	2.655	2.618	2.584	2.555	2.527	2.503	2.406	2.340	2.252	2.198	2.160	2.112	2.082	1.956
2.676	2.634	2.597	2.564	2.534	2.507	2.482	2.386	2.319	2.231	2.176	2.139	2.090	2.060	1.933
2.657	2.615	2.578	2.545	2.515	2.488	2.463	2.366	2.299	2.211	2.156	2.118	2.070	2.040	1.911
2.639	2.597	2.560	2.527	2.497	2.470	2.445	2.348	2.281	2.193	2.137	2.099	2.050	2.020	1.891
2.622	2.580	2.543	2.510	2.480	2.453	2.428	2.331	2.263	2.175	2.120	2.082	2.032	2.002	1.872
2.606	2.564	2.527	2.494	2.464	2.437	2.412	2.315	2.247	2.159	2.103	2.065	2.015	1.985	1.854
2.591	2.549	2.512	2.479	2.449	2.421	2.397	2.299	2.232	2.143	2.087	2.049	1.999	1.968	1.837
2.577	2.535	2.498	2.465	2.434	2.407	2.382	2.285	2.217	2.128	2.072	2.034	1.984	1.953	1.820
2.563	2.522	2.484	2.451	2.421	2.394	2.369	2.271	2.203	2.114	2.058	2.019	1.969	1.938	1.805
2.461	2.419	2.382	2.348	2.318	2.290	2.265	2.167	2.098	2.007	1.949	1.909	1.857	1.825	1.683
2.394	2.352	2.315	2.281	2.251	2.223	2.198	2.098	2.028	1.936	1.877	1.836	1.783	1.749	1.601
2.313	2.271	2.233	2.199	2.169	2.141	2.115	2.015	1.944	1.849	1.788	1.746	1.690	1.655	1.494
2.265	2.223	2.185	2.151	2.120	2.092	2.067	1.965	1.893	1.797	1.735	1.692	1.634	1.598	1.427
2.172	2.129	2.091	2.057	2.026	1.997	1.971	1.868	1.794	1.694	1.629	1.583	1.521	1.481	1.279
2.082	2.039	2.000	1.965	1.934	1.905	1.878	1.773	1.697	1.592	1.523	1.473	1.404	1.358	1.000

Table IV.7A Upper 5% Points in the Studentized Range

					Number of treatments, k						
d.f. (error)	2	3	4	5	6	7	8	9	10	15	20
2		8.33	9.80	10.89	11.73	12.43	13.03	13.54	13.99	15.65	16.77
4		5.04	5.76	6.29	6.71	7.06	7.35	7.60	7.83	8.67	9.24
5	3.64	4.60	5.22	5.67	6.03	6.33	6.58	6.80	6.99	7.72	8.21
6	3.46	4.34	4.90	5.31	5.63	5.89	6.12	6.32	6.49	7.14	7.59
8	3.26	4.04	4.53	4.89	5.17	5.40	5.60	5.77	5.92	6.48	6.87
10	3.15	3.88	4.33	4.66	4.91	5.12	5.30	5.46	5.60	6.12	6.47
12	3.08	3.77	4.20	4.51	4.75	4.95	5.12	5.27	5.40	5.88	6.21
14	3.03	3.70	4.11	4.41	4.64	4.83	4.99	5.13	5.25	5.72	6.03
16	3.00	3.65	4.05	4.34	4.56	4.74	4.90	5.03	5.15	5.59	5.90
18	2.97	3.61	4.00	4.28	4.49	4.67	4.83	4.96	5.07	5.50	5.79
20	2.95	3.58	3.96	4.24	4.45	4.62	4.77	4.90	5.01	5.43	5.71
24	2.92	3.53	3.90	4.17	4.37	4.54	4.68	4.81	4.92	5.32	5.59
30	2.89	3.48	3.84	4.11	4.30	4.46	4.60	4.72	4.83	5.21	5.48
40	2.86	3.44	3.79	4.04	4.23	4.39	4.52	4.63	4.74	5.11	5.36
60	2.83	3.40	3.74	3.98	4.16	4.31	4.44	4.55	4.65	5.00	5.24
120	2.80	3.36	3.69	3.92	4.10	4.24	4.36	4.47	4.56	4.90	5.13
∞	2.77	3.32	3.63	3.86	4.03	4.17	4.29	4.39	4.47	4.80	5.01

Table IV.7B Values of t' for Dunnett's Comparison of Several Treatments and a Control ($\alpha = 0.05$)

			Number of treatments			
d.f.	2	3	4	5	6	7
5	3.03	3.39	3.66	3.88	4.06	4.22
6	2.86	3.18	3.41	3.60	3.75	3.85
7	2.75	3.04	3.24	3.41	3.54	3.66
8	2.67	2.94	3.13	3.28	3.40	3.51
9	2.61	2.86	3.04	3.18	3.29	3.39
10	2.57	2.81	2.97	3.11	3.21	3.31
11	2.53	2.76	2.92	3.05	3.15	3.24
12	2.50	2.72	2.88	3.00	3.10	3.18
13	2.48	2.69	2.84	2.96	3.06	3.14
14	2.46	2.67	2.81	2.93	3.02	3.10
15	2.44	2.64	2.79	2.90	2.99	3.07
20	2.38	2.57	2.70	2.81	2.89	2.96
24	2.35	2.53	2.66	2.76	2.84	2.91
30	2.32	2.50	2.62	2.72	2.79	2.86
40	2.29	2.47	2.58	2.67	2.75	2.81
60	2.27	2.43	2.55	2.63	2.70	2.76
120	2.24	2.40	2.51	2.59	2.66	2.71
∞	2.21	2.37	2.47	2.55	2.62	2.67

Table IV.8 Dixon's Criteria for Rejecting Outliers

		Significance level	
k		5%	1%
3	$r_{10} = (X_2 - X_1)/(X_k - X_1)$ if smallest value is suspected;	0.941	0.988
4		0.765	0.889
5	$= (X_k - X_{k-1})/(X_k - X_1)$ if largest value is suspected	0.642	0.780
6		0.560	0.698
7		0.507	0.637
8	$r_{11} = (X_2 - X_1)/(X_{k-1} - X_1)$ if smallest value is suspected;	0.554	0.683
9		0.512	0.635
10	$= (X_k - X_{k-1})/(X_k - X_2)$ if largest value is suspected	0.477	0.597
11	$r_{21} = (X_3 - X_1)/(X_{k-1} - X_1)$ if smallest value is suspected;	0.576	0.679
12		0.546	0.642
13	$= (X_k - X_{k-2})/(X_k - X_2)$ if largest value is suspected	0.521	0.615
14	$r_{22} = (X_3 - X_1)/(X_{k-2} - X_1)$ if smallest value is suspected;	0.546	0.641
15		0.525	0.616
16	$= (X_k - X_{k-2})/(X_k - X_3)$ if largest value is suspected	0.507	0.595
17		0.490	0.577
18		0.475	0.561
19		0.462	0.547
20		0.450	0.535
21		0.440	0.524
22		0.430	0.514
23		0.421	0.505
24		0.413	0.497
25		0.406	0.489

Table IV.9 Critical Values of T for a Two-Sided Test at the 5% Level of Significance (Test for Outliers)

Sample size	T
3	1.155
4	1.481
5	1.715
6	1.887
7	2.020
8	2.126
9	2.215
10	2.290
11	2.355
12	2.412
13	2.462
14	2.507
15	2.549
16	2.585
17	2.620
18	2.651
19	2.681
20	2.709
25	2.822
30	2.908
35	2.979
40	3.036
50	3.128
100	3.383

Table IV.10 Factors for Determining Upper and Lower 3σ Limits for Mean ($\overline{X}$) and Range (R) Charts, and for Estimating σ from $\overline{R}$

		Factors for range chart		
Sample size of subgroup, N	A: Factor for $\overline{X}$ chart	D_L for lower limit	D_U for upper limit	$\sigma = \dfrac{\overline{R}}{d_2}$ d_2
2	1.88	0	3.27	1.128
3	1.02	0	2.57	1.693
4	0.73	0	2.28	2.059
5	0.58	0	2.11	2.326
6	0.48	0	2.00	2.534
7	0.42	0.08	1.92	2.704
8	0.37	0.14	1.86	2.847
9	0.34	0.18	1.82	2.970
10	0.31	0.22	1.78	3.078
15	0.22	0.35	1.65	3.472
20	0.18	0.41	1.59	3.735

Example: If $\overline{X} = 100$ and $\overline{R}$ (the average range) $= 5$, and $N = 6$, the upper and lower limits for the $\overline{X}$ chart are

$\overline{X} \pm A\overline{R} = 100 \pm 0.48(5) = 100 \pm 2.4 = (102.4, 97.6)$.

The upper limit for the range chart is $D_U\overline{R} = 2.0(5) = 10$. The lower limit for the range chart is $D_L\overline{R} = 0(5) = 0$. .

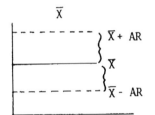

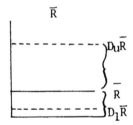

For samples of size 4, $\sigma = \dfrac{\overline{R}}{2.059}$.

If $R = 5, \sigma = \dfrac{5}{2.059} = 2.43$.

Table IV.11 Number of Correct Guesses Needed for Significance in the Triangle Test[a]

Panel size	Correct guesses for significance	
	5% Level	1% Level
6	5	6
7	5	6
8	6	7
9	6	7
10	7	8
11	7	8
12	8	9
13	8	9
14	9	10
15	9	10
16	9	11
17	10	11
18	10	12
19	11	12
20	11	13
21	12	13
22	12	14
23	12	14
24	13	15

[a]Pick-up Table from 3rd ed.

Table IV.12 Number of Positive or Negative Signs Needed for Significance for the Sign Test

Sample size	Number of positive or negative signs for significance[a]	
	5% Level	1% Level
6	6	—
7	7	—
8	8	8
9	8	9
10	9	10
11	10	11
12	10	11
13	11	12
14	12	13
15	12	13
16	13	14
17	13	15
18	14	15
19	15	16
20	15	17

[a]This is a two-sided test. Choose positive or negative signs, whichever is larger.

Table IV.13 Values Leading to Significance for the Wilcoxon Signed Rank Test (Two-Sided Test)

Sample size, N	5% Level[a]	1% Level
6	0	—
7	2	—
8	3	0
9	5	1
10	8	3
11	10	5
12	13	7
13	17	10
14	21	13
15	25	16
16	30	19
17	35	23
18	40	28
19	46	32
20	52	37

[a] If the smaller rank sum is less than or equal to the table value, the comparative groups are different at the indicated level of significance.

Table IV.14 Critical Values for Number of Runs at the 5% Level of Significance

Sample size, N	Two-sided test Lower number[a]	Upper number	One-sided test Lower number
10	2	9	3
12	3	10	3
14	3	12	4
16	4	13	5
18	5	14	6
20	6	15	6
22	7	16	7
24	7	18	8
26	8	19	9
28	9	20	10
30	10	21	11
32	11	22	11
34	11	24	12
36	12	25	13
38	13	26	14
40	14	27	15

[a] If the number of runs is less than or equal to the lower number or greater than or equal to the upper value, the sequence is considered nonrandom at the 5% level of significance. The sample size (N) is the number of values above and below the median. For odd-size samples where one value is the median, use the next smaller sample size for the critical values.

Table IV.15 Probability of Getting at Least One Run of Given Size for N Samples

N	5% Level	1% Level
10	5	—
20	7	8
30	8	9
40	9	10
50	10	11

Table IV.16 Critical Values for Wilcoxon Rank Sum Test[a] ($\alpha = 0.05$)

Size of larger sample	Size of smaller sample (M)						
	M = 3	4	5	6	7	8	9
M	5,16	11,25	18,37	26,52	37,68	49,87	63,108
M + 1	6,18	12,28	19,41	28,56	39,73	51,93	66,114
M + 2	6,21	12,32	20,45	29,61	41,78	54,98	68,121
M + 3	7,23	13,35	21,49	31,65	43,83	56,104	71,127
M + 4	7,26	14,38	22,53	32,70	45,88	58,110	74,133
M + 5	8,28	15,41	24,56	34,74	46,94	61,115	77,139
M + 6	8,31	16,44	25,60	36,78	48,99	63,121	79,146
M + 7	9,33	17,47	26,64	37,83	50,104	65,127	82,152
M + 8	10,35	17,51	27,68	39,87	52,109	68,132	85,158
M + 9	10,38	18,54	29,71	41,91	54,114	70,138	88,164
M + 10	11,40	19,57	30,75	42,96	56,119	72,144	90,171
M + 15	13,53	24,72	36,94	50,118	66,144	84,172	104,202
M + 20	16,65	28,88	42,113	58,140	76,169	96,200	118,223
M + 25	18,78	32,104	48,132	66,162	86,194	108,228	132,264

[a]From Wilcoxon F, and Wilcox RA. Some Rapid Approximate Statistical Procedures. Pearl River, NY: Lederle Laboratories, 1964. If rank sum of smaller sample is equal to or lower than smaller numbers in table or equal to or larger than larger number, groups are significantly different at 0.05 level.

Table IV.17 Critical Difference for Significance ($\alpha = 0.05$) Comparing All Possible Pairs of Treatments for Nonparametric One-Way ANOVA[a]

N (for each treatment)	Number of treatments				
	3	4	5	6	7
3	15	23	30	37	45
4	24	35	46	57	69
5	33	48	63	79	96
6	43	63	83	104	125
7	54	79	105	131	158
8	66	96	128	160	192
9	79	115	152	190	229
10	92	134	178	223	268
11	106	155	205	257	309
12	121	176	233	292	352
13	136	199	263	329	397
14	152	222	294	368	444
15	169	246	326	408	492
16	186	271	359	449	542
17	203	296	393	492	593
18	221	323	428	536	646
19	240	350	464	581	700
20	259	378	501	627	756
21	278	406	538	674	814
22	298	435	577	723	872
23	319	465	617	773	932
24	340	496	657	824	994
25	361	527	699	875	1056

[a]From Wilcoxon F, Wilcox RA. Some Rapid Approximate Statistical Procedures. Pearl River, NY: Lederle Laboratories, 1964.

Table IV.18 Critical Differences for Significance ($\alpha = 0.05$) Comparing All Possible Pairs of Treatments for Nonparametric Two-Way ANOVA[a]

N (for each treatment)	Number of treatments				
	3	4	5	6	7
3	6	8	10	13	15
4	7	10	12	15	18
5	8	11	14	17	20
6	9	12	15	18	22
7	9	13	16	20	24
8	10	14	17	21	25
9	10	14	18	23	27
10	11	15	19	24	28
11	11	16	20	25	30
12	12	16	21	26	31
13	12	17	22	27	32
14	13	18	23	28	34
15	13	18	24	29	35
16	13	19	24	30	36
17	14	19	25	31	37
18	14	20	26	32	38
19	14	20	27	33	39
20	15	21	27	34	40
21	15	21	28	35	41
22	16	22	29	35	42
23	16	22	29	36	43
24	16	23	30	37	44
25	17	23	31	38	45

[a]From Wilcoxon F, Wilcox RA. Some Rapid Approximate Statistical Procedures. Pearl River, NY: Lederle Laboratories, 1964.

Table IV.19 Factors for Two-Sided Tolerance Limits for Normal Distributions[a]

| | | | $\gamma = 0.75$ | | | | | $\gamma = 0.90$ | | |
| | | | p | | | | | p | | |
n	0.75	0.90	0.95	0.99	0.999	0.75	0.90	0.95	0.99	0.999
2	4.498	6.301	7.414	9.531	11.920	11.407	15.978	18.800	24.167	30.227
3	2.501	3.538	4.187	5.431	6.844	4.132	5.847	6.919	8.974	11.309
4	2.035	2.892	3.431	4.471	5.657	2.932	4.166	4.943	6.440	8.149
5	1.825	2.599	3.088	4.033	5.117	2.454	3.494	4.152	5.423	6.879
6	1.704	2.429	2.889	3.779	4.802	2.196	3.131	3.723	4.870	6.188
7	1.624	2.318	2.757	3.611	4.593	2.034	2.902	3.452	4.521	5.750
8	1.568	2.238	2.663	3.491	4.444	1.921	2.743	3.264	4.278	5.446
9	1.525	2.178	2.593	3.400	4.330	1.839	2.626	3.125	4.098	5.220
10	1.492	2.131	2.537	3.328	4.241	1.775	2.535	3.018	3.959	5.046
11	1.465	2.093	2.493	3.271	4.169	1.724	2.463	2.933	3.849	4.906
12	1.443	2.062	2.456	3.223	4.110	1.683	2.404	2.863	3.758	4.792
13	1.425	2.036	2.424	3.183	4.059	1.648	2.355	2.805	3.682	4.697
14	1.409	2.013	2.398	3.148	4.016	1.619	2.314	2.756	3.618	4.615
15	1.395	1.994	2.375	3.118	3.979	1.594	2.278	2.713	3.562	4.545
16	1.383	1.977	2.355	3.092	3.946	1.572	2.246	2.676	3.514	4.484
17	1.372	1.962	2.337	3.069	3.917	1.552	2.219	2.643	3.471	4.430
18	1.363	1.948	2.321	3.048	3.891	1.535	2.194	2.614	3.433	4.382
19	1.355	1.936	2.307	3.030	3.867	1.520	2.172	2.588	3.399	4.339
20	1.347	1.925	2.294	3.013	3.846	1.506	2.152	2.564	3.368	4.300
21	1.340	1.915	2.282	2.998	3.827	1.493	2.135	2.543	3.340	4.264
22	1.334	1.906	2.271	2.984	3.809	1.482	2.118	2.524	3.315	4.232
23	1.328	1.898	2.261	2.971	3.793	1.471	2.103	2.506	3.292	4.203
24	1.322	1.891	2.252	2.959	3.778	1.462	2.089	2.489	3.270	4.176
25	1.317	1.883	2.244	2.948	3.764	1.453	2.077	2.474	3.251	4.151
26	1.313	1.877	2.236	2.938	3.751	1.444	2.065	2.460	3.232	4.127
27	1.309	1.871	2.229	2.929	3.740	1.437	2.054	2.447	3.215	4.106
30	1.297	1.855	2.210	2.904	3.708	1.417	2.025	2.413	3.170	4.049
35	1.283	1.834	2.185	2.871	3.667	1.390	1.988	2.368	3.112	3.974
40	1.271	1.818	2.166	2.846	3.635	1.370	1.959	2.334	3.066	3.917
45	1.262	1.805	2.150	2.826	3.609	1.354	1.935	2.306	3.030	3.871
50	1.255	1.794	2.138	2.809	3.588	1.340	1.916	2.284	3.001	3.833
55	1.249	1.785	2.127	2.795	3.571	1.329	1.901	2.265	2.976	3.801
60	1.243	1.778	2.118	2.784	3.556	1.320	1.887	2.248	2.955	3.774
65	1.239	1.771	2.110	2.773	3.543	1.312	1.875	2.235	2.937	3.751
70	1.235	1.765	2.104	2.764	3.531	1.304	1.865	2.222	2.920	3.730
75	1.231	1.760	2.098	2.757	3.521	1.298	1.856	2.211	2.906	3.712
80	1.228	1.756	2.092	2.749	3.512	1.292	1.848	2.202	2.894	3.696
85	1.225	1.752	2.087	2.743	3.504	1.287	1.841	2.193	2.882	3.682
90	1.223	1.748	2.083	2.737	3.497	1.283	1.834	2.185	2.872	3.669
95	1.220	1.745	2.079	2.732	3.490	1.278	1.828	2.178	2.863	3.657
100	1.218	1.742	2.075	2.727	3.484	1.275	1.822	2.172	2.854	3.646
110	1.214	1.736	2.069	2.719	3.473	1.268	1.813	2.160	2.839	3.626
120	1.211	1.732	2.063	2.712	3.464	1.262	1.804	2.150	2.826	3.610
130	1.208	1.728	2.059	2.705	3.456	1.257	1.797	2.141	2.814	3.595
140	1.206	1.724	2.054	2.700	3.449	1.252	1.791	2.134	2.804	3.582
150	1.204	1.721	2.051	2.695	3.443	1.248	1.785	2.127	2.795	3.571
160	1.202	1.718	2.047	2.691	3.437	1.245	1.780	2.121	2.787	3.561
170	1.200	1.716	2.044	2.687	3.432	1.242	1.775	2.116	2.780	3.552
180	1.198	1.713	2.042	2.683	3.427	1.239	1.771	2.111	2.774	3.543
190	1.197	1.711	2.039	2.680	3.423	1.236	1.767	2.106	2.768	3.536
200	1.195	1.709	2.037	2.677	3.419	1.234	1.764	2.102	2.762	3.529
250	1.190	1.702	2.028	2.665	3.404	1.224	1.750	2.085	2.740	3.501
300	1.186	1.696	2.021	2.656	3.393	1.217	1.740	2.073	2.725	3.481
400	1.181	1.688	2.012	2.644	3.378	1.207	1.726	2.057	2.703	3.453
500	1.177	1.683	2.006	2.636	3.368	1.201	1.717	2.046	2.689	3.434
600	1.175	1.680	2.002	2.631	3.360	1.196	1.710	2.038	2.678	3.421
700	1.173	1.677	1.998	2.626	3.355	1.192	1.705	2.032	2.670	3.411
800	1.171	1.675	1.996	2.623	3.350	1.189	1.701	2.027	2.663	3.402
900	1.170	1.673	1.993	2.620	3.347	1.187	1.697	2.023	2.658	3.396
1000	1.169	1.671	1.992	2.617	3.344	1.185	1.695	2.019	2.654	3.390

(Continued)

Table IV.19 *(Continued)*

n	γ = 0.95 p					γ = 0.99 p				
	0.75	0.90	0.95	0.99	0.999	0.75	0.90	0.95	0.99	0.999
∞	1.150	1.645	1.960	2.576	3.291	1.150	1.645	1.960	2.576	3.291
2	22.858	32.019	37.647	48.430	60.573	114.363	160.193	188.491	242.300	303.054
3	5.922	8.380	9.916	12.861	16.208	13.378	18.930	22.401	29.055	36.616
4	3.779	5.369	6.370	8.299	10.502	6.614	9.398	11.150	14.527	18.383
5	3.002	4.275	5.079	6.634	8.415	4.643	6.612	7.855	10.260	13.015
6	2.604	3.712	4.414	5.775	7.337	3.743	5.337	6.345	8.301	10.548
7	2.361	3.369	4.007	5.248	6.676	3.233	4.613	5.488	7.187	9.142
8	2.197	3.136	3.732	4.891	6.226	2.905	4.147	4.936	6.468	8.234
9	2.078	2.967	3.532	4.631	5.899	2.677	3.822	4.550	5.966	7.600
10	1.987	2.839	3.379	4.433	5.649	2.508	3.582	4.265	5.594	7.129
11	1.916	2.737	3.259	4.277	5.452	2.378	3.397	4.045	5.308	6.766
12	1.858	2.655	3.162	4.150	5.291	2.274	3.250	3.870	5.079	6.477
13	1.810	2.587	3.081	4.044	5.158	2.190	3.130	3.727	4.893	6.240
14	1.770	2.529	3.012	3.955	5.045	2.120	3.029	3.608	4.737	6.043
15	1.735	2.480	2.954	3.878	4.949	2.060	2.945	3.507	4.605	5.876
16	1.705	2.437	2.903	3.812	4.865	2.009	2.872	3.421	4.492	5.732
17	1.679	2.400	2.858	3.754	4.791	1.965	2.808	3.345	4.393	5.607
18	1.655	2.366	2.819	3.702	4.725	1.926	2.753	3.279	4.307	5.497
19	1.635	2.337	2.784	3.656	4.667	1.891	2.703	3.221	4.230	5.399
20	1.616	2.310	2.752	3.615	4.614	1.860	2.659	3.168	4.161	5.312
21	1.599	2.286	2.723	3.577	4.567	1.833	2.620	3.121	4.100	5.234
22	1.584	2.264	2.697	3.543	4.523	1.808	2.584	3.078	4.044	5.163
23	1.570	2.244	2.673	3.512	4.484	1.785	2.551	3.040	3.993	5.098
24	1.557	2.225	2.651	3.483	4.447	1.764	2.522	3.004	3.947	5.039
25	1.545	2.208	2.631	3.457	4.413	1.745	2.494	2.972	3.904	4.985
26	1.534	2.193	2.612	3.432	4.382	1.727	2.469	2.941	3.865	4.935
27	1.523	2.178	2.595	3.409	4.353	1.711	2.446	2.914	3.828	4.888
30	1.497	2.140	2.549	3.350	4.278	1.668	2.385	2.841	3.733	4.768
35	1.462	2.090	2.490	3.272	4.179	1.613	2.306	2.748	3.611	4.611
40	1.435	2.052	2.445	3.213	4.104	1.571	2.247	2.677	3.518	3.493
45	1.414	2.021	2.408	3.165	4.042	1.539	2.200	2.621	3.444	3.399
50	1.396	1.996	2.379	3.126	3.993	1.512	2.162	2.576	3.385	4.323
55	1.382	1.976	2.354	3.094	3.951	1.490	2.130	2.538	3.335	4.260
60	1.369	1.958	2.333	3.066	3.916	1.471	2.103	2.506	3.293	4.206
65	1.359	1.943	2.315	3.042	3.886	1.455	2.080	2.478	3.257	4.160
70	1.349	1.929	2.299	3.021	3.859	1.440	2.060	2.454	3.225	4.120
75	1.341	1.917	2.285	3.002	3.835	1.428	2.042	2.433	3.197	4.084
80	1.334	1.907	2.272	2.986	3.814	1.417	2.026	2.414	3.173	4.053
85	1.327	1.897	2.261	2.971	3.795	1.407	2.012	2.397	3.150	4.024
90	1.321	1.889	2.251	2.958	3.778	1.398	1.999	2.382	3.130	3.999
95	1.315	1.881	2.241	2.945	3.763	1.390	1.987	2.368	3.112	3.976
100	1.311	1.874	2.233	2.934	3.748	1.383	1.977	2.355	3.096	3.954
110	1.302	1.861	2.218	2.915	3.723	1.369	1.958	2.333	3.066	3.917
120	1.294	1.850	2.205	2.898	3.702	1.358	1.942	2.314	3.041	3.885
130	1.288	1.941	2.194	2.883	3.683	1.349	1.928	2.298	3.019	3.857
140	1.282	1.833	2.184	2.870	3.666	1.340	1.916	2.283	3.000	3.833
150	1.277	1.825	2.175	2.859	3.652	1.332	1.905	2.270	2.983	3.811
160	1.272	1.819	2.167	2.848	3.638	1.326	1.896	2.259	2.968	3.792
170	1.268	1.813	2.160	2.839	3.527	1.320	1.887	2.248	2.955	3.774
180	1.264	1.808	2.154	2.831	3.616	1.314	1.879	2.239	2.942	3.759
190	1.261	1.803	2.148	2.823	3.606	1.309	1.872	2.230	2.931	3.744
200	1.258	1.798	2.143	2.816	3.597	1.304	1.865	2.222	2.921	3.731
250	1.245	1.780	2.121	2.788	3.561	1.286	1.839	2.191	2.880	3.678
300	1.236	1.767	2.106	2.767	3.535	1.273	1.820	2.169	2.850	3.641
400	1.223	1.749	2.084	2.739	3.499	1.255	1.794	2.138	2.809	3.589
500	1.215	1.737	2.070	2.721	3.475	1.243	1.777	2.117	2.783	3.555
600	1.209	1.729	2.060	2.707	3.458	1.234	1.764	2.102	2.763	3.530
700	1.204	1.722	2.052	2.697	3.445	1.227	1.755	2.091	2.748	3.511
800	1.201	1.717	2.046	2.688	3.434	1.222	1.747	2.082	2.736	3.495
900	1.198	1.712	2.040	2.682	3.426	1.218	1.741	2.075	2.726	3.483
1000	1.195	1.709	2.036	2.676	3.418	1.214	1.736	2.068	2.718	3.472
∞	1.150	1.645	1.960	2.576	3.291	1.150	1.645	1.960	2.576	3.291

[a]Factors t' such that the probability is γ that at least a proportion P of the distribution will be included between $\overline{X} \pm t's$ where $\overline{X}$ and s are estimates of the mean and the standard deviation computed from a sample size of n.

Table IV.20 Test for Outliers (Upper Band for Critical Values for Studentized Residual)

$(\alpha = .10)$

n	1	2	3	4	5	6	8	10	15	25
					q					
5	1.87									
6	2.00	1.89								
7	2.10	2.02	1.90							
8	2.18	2.12	2.03	1.91						
9	2.24	2.20	2.13	2.05	1.92					
10	2.30	2.26	2.21	2.15	2.06	1.92				
12	2.39	2.37	2.33	2.29	2.24	2.17	1.93			
14	2.47	2.45	2.42	2.39	2.36	2.32	2.19	1.94		
16	2.53	2.51	2.50	2.47	2.45	2.42	2.34	2.20		
18	2.58	2.57	2.56	2.54	2.52	2.50	2.44	2.35		
20	2.63	2.62	2.61	2.59	2.58	2.56	2.52	2.46	2.11	
25	2.72	2.72	2.71	2.70	2.69	2.68	2.66	2.63	2.50	
30	2.80	2.79	2.79	2.78	2.77	2.77	2.75	2.73	2.66	2.13
35	2.86	2.85	2.85	2.85	2.84	2.84	2.82	2.81	2.77	2.55
40	2.91	2.91	2.90	2.90	2.90	2.89	2.88	2.87	2.84	2.72
45	2.95	2.95	2.95	2.95	2.94	2.94	2.93	2.93	2.90	2.82
50	2.99	2.99	2.99	2.99	2.98	2.98	2.97	2.95	2.89	
60	3.06	3.06	3.05	3.05	3.05	3.05	3.05	3.04	3.03	3.00
70	3.11	3.11	3.11	3.11	3.11	3.11	3.10	3.10	3.09	3.07
80	3.16	3.16	3.16	3.15	3.15	3.15	3.15	3.15	3.14	3.12
90	3.20	3.20	3.19	3.19	3.19	3.19	3.19	3.19	3.18	3.17
100	3.23	3.23	3.23	3.23	3.23	3.23	3.23	3.22	3.22	3.21

$(\alpha = .05)$

n	1	2	3	4	5	6	8	10	15	25
					q					
5	1.92									
6	2.07	1.93								
7	2.19	2.08	1.94							
8	2.28	2.20	2.10	1.94						
9	2.35	2.29	2.21	2.10	1.95					
10	2.42	2.37	2.31	2.22	2.11	1.95				
12	2.52	2.49	2.45	2.39	2.33	2.24	1.96			
14	2.61	2.58	2.55	2.51	2.47	2.41	2.25	1.96		
16	2.68	2.66	2.63	2.60	2.57	2.53	2.43	2.26		
18	2.73	2.72	2.70	2.68	2.65	2.62	2.55	2.44		
20	2.78	2.77	2.76	2.74	2.72	2.70	2.64	2.57	2.15	
25	2.89	2.88	2.87	2.86	2.84	2.83	2.80	2.76	2.60	
30	2.96	2.96	2.95	2.94	2.93	2.93	2.90	2.88	2.79	2.17
35	3.03	3.02	3.02	3.01	3.00	3.00	2.98	2.97	2.91	2.64
40	3.08	3.08	3.07	3.07	3.06	3.06	3.05	3.03	3.00	2.84
45	3.13	3.12	3.12	3.12	3.11	3.11	3.10	3.09	3.06	2.96

$(\alpha = .05)$

n	1	2	3	4	5	6	8	10	15	25
					q					
50	3.17	3.16	3.16	3.16	3.15	3.15	3.14	3.14	3.11	3.04
60	3.23	3.23	3.23	3.23	3.22	3.22	3.22	3.21	3.20	3.15
70	3.29	3.29	3.28	3.28	3.28	3.28	3.27	3.27	3.26	3.23
80	3.33	3.33	3.33	3.33	3.33	3.33	3.32	3.32	3.31	3.29
90	3.37	3.37	3.37	3.37	3.37	3.37	3.36	3.36	3.36	3.34
100	3.41	3.41	3.40	3.40	3.40	3.40	3.40	3.40	3.39	3.38

(Continued)

Table IV.20   *(Continued)*

					$(\alpha = .01)$					
					q					
n	1	2	3	4	5	6	8	10	15	25
5	1.98									
6	2.17	1.98								
7	2.32	2.17	1.98							
8	2.44	2.32	2.18	1.98						
9	2.54	2.44	2.33	2.18	1.99					
10	2.62	2.55	2.45	2.33	2.18	1.99				
12	2.76	2.70	2.64	2.56	2.46	2.34	1.99			
14	2.86	2.82	2.78	2.72	2.65	2.57	2.35	1.99		
16	2.95	2.92	2.88	2.84	2.79	2.73	2.58	2.35		
18	3.02	3.00	2.97	2.94	2.90	2.85	2.75	2.59		
20	3.08	3.06	3.04	3.01	2.98	2.95	2.87	2.76	2.20	
25	3.21	3.19	3.18	3.16	3.14	3.12	3.07	3.01	2.78	
30	3.30	3.29	3.28	3.26	3.25	3.24	3.21	3.17	3.04	2.21
35	3.37	3.36	3.35	3.34	3.34	3.33	3.30	3.28	3.19	2.81
40	3.43	3.42	3.42	3.41	3.40	3.40	3.38	3.36	3.30	3.05
45	3.48	3.47	3.47	3.46	3.46	3.45	3.44	3.43	3.38	3.23
50	3.52	3.52	3.51	3.51	3.51	3.50	3.49	3.48	3.45	3.34
60	3.60	3.59	3.59	3.59	3.58	3.58	3.57	3.56	3.54	3.48
70	3.65	3.65	3.65	3.65	3.64	3.64	3.64	3.63	3.61	3.57
80	3.70	3.70	3.70	3.70	3.69	3.69	3.69	3.68	3.67	3.64
90	3.74	3.74	3.74	3.74	3.74	3.74	3.73	3.73	3.72	3.70
100	3.78	3.78	3.78	3.77	3.77	3.77	3.77	3.77	3.76	3.74

n = number of observations
q = number of independent variables (including count for intercept if fitted)
Source: Lund, *Technometrics 17*(4), Nov. 1975.

Appendix V
Outlier Tests and Chemical Assays

V.1 INTRODUCTION

In a recent landmark decision resulting from a trial involving the Federal Government and Barr Laboratories, Judge Wolin made many judgments based on his constant probing and the testimony of expert witnesses [1]. Remarkably, most of what he had to say was clear, correct, and to the point, despite his sparse background in the subject material. Much of the Decision related to testing drug products during their production when failing results (out of specification) were observed. A summary of the Decision is available from the FDA [2]. A previous paper by this author [3] presented some alternatives to retesting when a single out of specification result was observed for which no obvious cause was apparent, a situation that is common in my experience. This paper discusses some issues related to the elimination of an out of specification (OOS) result with no obvious cause, based on an outlier test. The Judge, in his Decision, stated that tests for outliers that can be used to exclude an aberrant observation are not appropriate for chemical tests. His reasoning was that the USP includes tests for outliers, but presents these tests only in the context of biological assays, which tend to be very variable. This, he suggests, is appropriate because of the large variability of these kinds of procedures. Judge Wolin further suggests in his Decision that such outlier analyses should not be used for chemical assays, because if they were appropriate, the USP would have recommended the procedure for chemical assays. Thus, the judgment is that, by default, outlier tests for chemical assays should not be used. All of this raises several questions, including (a) Was it the USP's intention to exclude outlier tests for chemical assays? (b) Was this an oversight or was it intentional? (c) Does the USP not discuss outlier tests for chemical assays because the issue is complex with many possible alternatives?

I do not believe that it was Judge Wolin's intention that his Decision should result in nonscientifically based procedures by pharmaceutical firms. I also believe that he would be disturbed if his Decision and FDA's interpretation of his Decision would lead to increased costs because good judgment was cast aside in lieu of fear of a "483 citation." For example, one firm discarded a batch of product because a single content uniformity value failed, despite the fact that 100 individual repeat assays all yielded results between 85% and 115%. Another firm assayed the blend for a capsule product more than 50 times using single dose unit samples during a validation study (because the recommended 3 dosage units were not feasible), with one value being at 119%. All other values were between 90% and 110%. For fear of a "483," the company was reluctant to release the batch. They would have been equally fearful, had the OOS value been 111%, because they interpreted the Decision to impose limits of 90% to 110% for 3 dosage unit weight assays at the blend stage. The final product passed with all content uniformity values between 90% and 110% and an RSD of 2%. Would this firm have been better off performing an absolute minimum number of assays to validate the batch in order to decrease the probability of a failing assay, or to proceed as they did to ensure a thorough validation with increased risk of failure? Once more, I cannot believe that it was Judge Wolin's intention to impose such irrational hardships on the industry. Thus, part of the incentive for this paper (and one previously published. Ref. [2]) is to propose some rational alternatives in the spirit of the Judge's Decision.

V.2 CAN OUTLIER TESTS BE JUSTIFIED?

In fact, the outlier problem remains perplexing, whether applied to questions of fundamental science or problems of more direct practical application. Stories abound in the history of science about how a single outlier, discarded, was eventually found to have contained important information. Similarly, anecdotes exist about outliers not discarded obscuring the truth. Thus, scientists understand that there is no one answer to the problem, and that there are risks associated with making decisions about how to handle outlying observations. Although the question of how to deal with apparent outliers resulting from chemical assays cannot be resolved easily, the use of outlier tests is ubiquitous in both practical laboratory SOPs as well as the chemical and statistical literature. Pages could be filled with references on this subject, including many from scientists associated with the National Bureau of Standards, for example, the prominent statisticians, Drs. Youden [4] and Mandell [5]. Dr. Youden [4] commented that the experimenter is better equipped to detect outliers than the statistician when a small number of values (e.g., 3) are observed. In fact, with only 3 observations, a value must appear to be extremely divergent before it could be considered an outlier. Thus, he suggests that the experienced experimenter probably would be less conservative than the statistician in finding an observation suspect (the statistical test may be considered conservative in the decision to reject an outlier). Natrella [6] discusses this problem, noting that "There have been many criteria proposed for guiding the rejection of observations." She also states that "no available criteria are superior to the judgment of an experienced investigator . . . ." She gives several statistical procedures for identifying outliers.

It is obvious that there is both theoretical and practical interest in this problem. Again, scientific judgment appears dominant in approaching such problems. Judgment can be defined to be a result of education, knowledge, experience, and common sense. All of these must come into play, and we can be 100% sure that there will never be unanimous agreement on controversial issues. However, because many statistical and chemical treatises discuss the outlier problem, I do not believe that its use can be dismissed out of hand, only because the USP lacks a specific recommendation. Other often used references and documents (OAOC, etc.), including some that are government sponsored, recommend use of outlier tests, when appropriate, for all kinds of data, in particular chemical assays. Virtually every well meaning, knowledgeable scientist would probably entertain the possibility of excluding an outlying value from a set of experimental data. One could give an example of a single assay showing zero drug content, an extreme case, in which it would be absurd not to follow-up with further testing, even if no cause for the "erroneous" result could be found. Similarly, if 3 assays were performed on a relatively homogeneous blend such as a 20 tablet composite, with results of 99, 101, and 0, the null assay would have to be considered suspect. Of course, most situations that might provoke use of an outlier test are less extreme, and probably would need the application of judgment. Certainly, excessive use of outlier tests would suggest some persistant problem that needs to be resolved, unrelated to the assay. Perhaps, there exists a compromise that could satisfy both the conservative (never apply an outlier test) and more liberal (always apply an outlier test and discard the outlying value if present) critic?

V.3   WHY IS THERE NOT A USP TEST FOR OUTLIERS FOR CHEMICAL ASSAYS?

The answer as to why outlier tests are not specifically recommended for chemical assays in the USP is not entirely clear, but I can conjure up a possible scenario. Because of the variability of biological assays, to obtain a more precise estimate of drug content, replicate assays are frequently employed. This is a good scientific approach. The average of replicate assays always gives a better estimate of the true average drug content than a single assay. For very variable assays, a single result may fail because of the large assay variability, not related to the true drug content. For chemical assays, the assay variability is usually relatively small, and a single assay may give a good estimate of the true drug content of the batch. On the other hand, chemical assays with large variability should use replicate assays, with the average result representing the true drug content. Thus, the USP may not want to commit to any specific assay scheduling.

The USP does not comment on the number of assays to be performed, and, in particular, does not suggest multiple assays on a single "homogeneous" portion of material, such as a composite mixture or solution. The number of assays to be performed would appear to be a matter of judgment, each laboratory using its own criteria. This seems reasonable and appropriate. Clearly, in any event, if a single assay or duplicate assays (with no previous estimate of the standard deviation) are performed, outlier tests cannot be applied. At least three assays are needed for an independent application of an outlier test. Thus, the USP cannot apply outlier tests to chemical assays unless at least three assays are performed. In my experience, only one or two assays are routinely performed for the chemical analysis of composite material. Therefore, for the USP to have an outlier test for chemical assays, at least three assays must be performed. As previously noted, the USP makes no such recommendation. In fact, if a firm is considering multiple assays on a composite, for example, and no provision is made for an outlier test, a decision to perform a single assay would probably cause the least problems, and would be the most prudent from an economic point of view. The more assays that are performed on good material, the greater the chance that at least one of the assays will fail. Yet, from a scientific viewpoint, performing multiple assays and using the average result as a measure of the batch parameter is clearly superior to a single assay. This important point is discussed in more detail later and is also exemplified by the multiple assays performed in a validation batch noted earlier.

V.4 SOME COMMENTS ON THE NATURE OF OUTLIERS AND OUTLIER TESTS, AND OTHER INCONSISTENCIES IN THE DECISION THAT OUTLIER TESTS BE USED FOR BIOLOGICAL ASSAYS BUT NOT FOR CHEMICAL ASSAYS

When performing multiple assays on a single source of material, such as a relatively homogeneous mix or a solution, there is a reasonable probability that one of the replicates may be deviant due to chance or due to an outright error. Whether or not a cause for the deviant assay is documented, the USP suggests (for biological assays) the value may be excluded if an outlier test confirms that the observation is deviant at the 4% level (the chance that the value will be incorrectly excluded is less than 1 in 25). The USP makes it clear that outlier tests should be used sparingly, when unavoidable. Certainly, the situation that is "unavoidable" is open to interpretation or judgment. It would appear to me that one situation that might fit the USP's definition is where inclusion of the outlier would cause the batch to fail and no cause can be found for the outlying value following a suitable investigation. Since such general statements need some interpretation, one would want to know the relevant batch history as well as other measures of the batch performance as part of the justification for performing an outlier test and discarding the outlying result. According to my experience, exclusion of biological assay results based on the outlier test is rarely questioned. This situation should be considered carefully in light of the potential 100% exclusion of outlier tests for chemical assays under all circumstances.

What is the nature of an outlier test? Very important in any such test is an assumption about the underlying distribution of the population data, the distribution of analytical results that might arise from the analysis of a sample, in our context. If we consider the assay results to have an approximate normal distribution, then the outlier test recommended in the USP is appropriate. We probably would be not too far wrong using this assumption for the analytical results derived from a single homogeneous sample. The outlier test recommended in the USP compares the ranges of values in order to assess if the extreme value is far enough removed to be considered discordant relative to the rest of the data. The assumption is that the data are normally distributed and if the probability that the extreme value comes from the distribution is less than 1 in 25, then the value may be considered discordant. It is extremely important to understand that this test is not dependent on the absolute variability of data, but rather on the distance of the suspected outlier from the rest of the data relative to the dispersion of the remaining data. Thus, this test will reject an outlier with the same probability no matter what the variance of the data. The following example may clarify this concept. In a microbiological assay, the following three values were obtained for potency based on three replicate assays: 52.3, 99.9, 101.9. The USP outlier test would be just satisfied, that is, we could exclude the outlier, 52.3. Note that for 3 assays, the outlier must be very far (and obviously) removed from the other two values in order to be discarded. In a chemical assay, the following three values were observed, 86.14, 97.64, 97.87. Again, the value of 86.14 is found to be an outlier. The higher precision of the chemical assay as suggested by the two values close together allows a less distant outlier to be detected. Note that if two of three assays are identical (which may occur if rounding results in identical assays), the third result will always be an outlier. The important point to remember is that the probability of incorrectly eliminating the outlier is less than 1 in 25 for both of these examples. Thus, the risk of incorrectly eliminating an outlier is not dependent on the underlying variability of the normal distribution associated with the assay data.

As suggested previously, if a testing recommendation is not scientifically sound, less valid testing situations will be used to satisfy the recommendations in lieu of more valid approaches. The exclusion of outlier testing for chemical assays may promote less good testing procedures, in my opinion.

V.5 WHAT IS THE PURPOSE OF PERFORMING REPLICATE ASSAYS AND WHEN IS AVERAGING APPROPRIATE?

Although the Barr Decision suggests that averaging is not correct in some circumstances, averaging is appropriate in the situation where multiple assays are used to obtain a better estimate of the true parameter (which is the case for biological assays as well as chemical assays). The reason for performing multiple assays is not to detect nonuniformity, but rather to obtain a better estimate of a parameter, the true drug content. The more assays performed the better the estimate based on the average. This would apply to any assay, but would be more important for variable assays. For chemical assays that are usually (but not always) more precise than biological assays, a single assay may be sufficient to get a good estimate of the drug content. An important consideration is that the average result is what is needed in this circumstance. Still, as noted above, a single assay among the replicates that is found to be OOS may suggest further testing, depending on circumstances (e.g., as noted in the Decision, assays of 91, 91, and 89). It would appear perfectly reasonable to me that if replicate chemical assays (3 or more) are to be performed on a sample (a priori as specified in an SOP), that the same considerations be given to outliers in this situation as is given to biological assays. (Due to its far-ranging implications, perhaps the USP can look further into this very important question.)

The more difficult question to answer is how to apply outlier tests when retesting or resampling is considered to be appropriate. This has been addressed briefly in a previous publication [3], but I will pursue this further here.

V.6 IN WHAT SITUATIONS MIGHT OUTLIER TESTS BE APPLICABLE?

V.6.1 Homogeneous Sample (Solution or Composite Powder)

When performing replicate assays on the same portion of material, the average result is typically used as representing the batch parameter. However, although not specifically recommended in official documents, if one of the replicate values is outside of official specifications (whether an outlier or not), a prudent manufacturer may decide to perform further analyses [3]. In particular, in my opinion, if replicate assays (at least 3) are performed based on SOPs, an outlier test is appropriate. Another application of outlier analysis may occur when a single assay fails and no cause is found. In this case, I recommend further sampling as discussed in the previous paper [3]. If the further assays indicate that the original result is an outlier, then it may be discarded in the calculation of the average. The calculation of the number of samples to be reassayed is also discussed in Ref. [2]. For example, consider the following hypothetical scenario. The original assay is 75% from a composite that should have a mean of 100%. Three new samples are assayed from the same composite with results of 98%, 99% and 100%. The lower limit for passing is 93%. Would you accept or reject this test? (The value of 75% tests as an outlier). If the original OOS result does not meet the outlier criterion, then scientific judgment is needed. If the average passes including the outlier, a prudent manufacturer will examine other batch records and batch history to aid in a decision. For example, with no evidence of batch failure, a passing average in this example may be considered to represent the batch. The same considerations may apply if the average does not pass when the original result is included. If the original result was 75% and three reassays were 93%, 96% and 105%, the 75% value would not be a significant outlier. Considerable judgment would be required here. Should the batch be rejected based only on this evidence? Is further testing appropriate? According to the Court, a product should not be tested into compliance, certainly a reasonable and prudent decision. On the other hand, the hypothetical situation presented here begs for further testing, in my opinion. I would hesitate to make any specific recommendations for this case, but further information about the product would be needed to come to any decision. Again, to establish inflexible rules for every situation does not seem to be a good substitute for scientific judgment. That is not to say that reasonable guidelines are not needed and are not important. For a further discussion, see Ref. [3].

V.6.2 Outlier Tests for Destructive Testing

A particularly difficult situation for the application of outlier tests is testing where the sample, once analyzed, is no longer available. This situation is most prevalent in the context of Quality Control testing of content uniformity and dissolution results. Similar situations may arise in stability testing. Another controversial area in which this problem has been extensively discussed is in bioequivalence testing, where the outlying subject is either not available or has changed since the original observation. Another situation that may be included here is when a large sample of homogeneous material presented for analysis continues to fail after multiple testing, and the possibility exists (but undocumented) that the sample does not truly represent the batch, perhaps due to mishandling or an error in preparation. In these cases, further testing may be indicated, and this has been discussed in Ref. [3]. Because we cannot retest the original material, we can never be certain whether the original analysis is correct. In particular, if the result is a failure, we will never know the truth unless an obvious cause is discovered. This would be the case in content uniformity (CU) testing where a single value outside the range of 75% to 125% is observed. This single value would almost certainly be tested as an outlier. If not, the batch would be suspect. Before discussing this situation, we might try to gain some insight into the nature of the CU test. The CU test does not say that OOS values do not exist in the batch. For example, if 0.1% (1/1000) of the tablets in a batch were outside 75% to 125%, assuming a normal distribution, about 94% of the tablets would be between 85% and 115% and about 6% between 75% to 85% and 115% to 125%. The chances of finding one of these OOS tablets in a random sample of 10 is about 1 in a 100, a very small probability. Yet, 1 in every 1000 tablets is OOS. The probability that the CU test would pass based on the first 10 tablets is >0.88. The probability that the CU test would pass based on the second tier testing is >0.94. Therefore, the CU test is not very discriminating in finding OOS tablets. We would have to have at least 1% of the tablets OOS (less than 75% or greater than 125%) before the CU test would have a good chance (about 50–50) of failing. Thus, the CU test can be considered as a screening test, but relatively nondiscriminating in finding tablets OOS if there are less than 1% in the batch. If we observe a tablet outside 75% to 125%, which tests as an outlier with no obvious cause, should the batch be rejected? There is no way of knowing with certainty whether the value is real or due to some malfunction during the assay, or if real was only a chance observation of an event that has very small probability. I propose that in such situations, following a failure investigation, if appropriate, that a sufficient number of tablets be assayed to give high assurance that the proportion of OOS tablets in the batch is small. Remembering that we cannot ever know with certainty that such tablets do not exist in the batch and that the CU test does not discriminate against a small percentage of such tablets, this seems a prudent approach. This problem has also been addressed in the previous publication where in most cases (small RSD and average potency near 100%) with a sufficient number of passing reassays, we can have high confidence that more than 99.9% of the tablets are within 85% to 115% [3]. It would seem to me that such a probability statement is stronger and carries more information than the usual USP test with regard to tablet uniformity. As suggested in the decision [1], resampling should be conducted using the original sample if possible. Thus, in the case of CU testing or a composite sample (which continually fails), the new samples should be taken from the larger sample of product submitted for analysis by Quality Control Personnel. For example, if the CU test is conducted on tablets taken from a bottle of 1000 tablets submitted by QC, the resampling should be from the remaining tablets.

The approach to demonstrating the validity of data presented here is only one way of coping with a difficult problem. However, any method that is backed by scientific reasoning and common sense should certainly be an improvement over arbitrary approaches. In a sense, the application of this kind of reasoning to such methods may be compared to the application of probability and statistical reasoning substantiating or defining findings in criminal court decisions.

REFERENCES

1. United States v. Barr Laboratories, Inc., Consolidated Docket No. 92-1744 (AMW) (Court Ruling, 2/4/93).

2. Judge Wolin's Interpretation of GMP Issues, Document submitted by Mary Ann Danello, Director of Small Business, Scientific and Trade Affairs, FDA, Rockville, MD 20857, May 14, 1993.
3. Bolton S. Should a single unexplained failing assay be reason to reject a batch? Clin Res Regul Aff., 1993; 10(3): 159–176.
4. Youden WJ. Statistical Methods for Chemists. New York: Wiley, 1951.
5. Mandell J. The Statistical Analysis of Experimental Data. New York: Wiley, 1967.
6. Natrella MG. Experimental Statistics, National Bureau of Standards Handbook 91, Superintendent of Documents. Washington, DC: U.S. Government Printing Office, 20402. 1966.

Appendix VI
Should a Single Unexplained Failing Assay be Reason to Reject a Batch?

The problem of what to do with data that appear to be erroneous, but for which no cause is apparent, has puzzled scientists for as long as data have been collected and evaluated. These data can be characterized as outliers, not appearing to be of the same kind as other data collected under the same circumstances. One might suppose that situations exist where such outliers can be considered absurd, for example, nobody with any knowledge of the process could conceive that such a value could exist. For example, if an automatic device for weighing individual tablets would record a zero, we would be "certain" that the result was not due to a weightless tablet, but rather due to some malfunction of the process. However, in the great majority of cases, the cause for an outlying result cannot be ascertained. In the case of scientific experiments for research purposes, the outlier appears among other experimental results, and the scientist can freely hypothesize reasons and explanations for its presence. Thus, the scientist can make a case for exclusion or inclusion of the outlier, and discuss reasons, implications, etc., with impunity. The future will demonstrate the correctness of his evaluation and judgment; "Time will tell." In a regulatory environment, time is of the essence. We cannot wait for time to prove a hypothesis about an outlying observation, correct or not. Usually, a decision must be made quickly. Although there is no absolute right or wrong way to proceed, "judgment" seems to be a key word. Under a given set of circumstances, what is to be done with the "outlier" is not easy to answer. These problems were at the heart of a recent litigation involving the Federal Government (FDA) and a generic company (Barr Labs, Inc.) [1] that involved testing of solid dosage forms or products for reconstitution. Much of the government's case against Barr related to the passing of batches in which a single failing or outlying assay was observed. The government suggested that if a single assay was not within specifications, in the face of all other tests performed on the batch, the product should be rejected. This "outlying" result or test failure could occur as a result of in-process testing or final product testing, either situation resulting in the rejection of the batch. This was the point of much of the trial proceedings, with a willing judge looking for the truth. In fact, there is no truth. What is to be done is a matter of judgment and common sense, grounded in experience, knowledge, and scientific know-how. Nevertheless, it is certainly possible that two knowledgeable and intelligent experts might disagree on what to do in any given situation. Good judgment does not necessarily lead to a single universal truth. Thus, the procedures recommended in this paper represent my judgment and experience.

In my opinion, a single outlying or failing result among many test results accumulated during the manufacture of a batch of product does not necessarily mean that the batch is unacceptable. In fact, I would think quite the contrary, that if all measures of batch quality other than the "outlier" suggests that the batch is acceptable, indeed the batch probably represents an acceptable product. In any event, the decision as how to proceed should consider other measurements observed during production as well as the product history. If a product has a history of problems, then failing results must be taken very seriously, and the onus of quality falls heavily on the product. On the other hand, if the product has a history of good quality, the outlier may not be due to the product, but rather due to a human or equipment malfunction. Thus, the data should be taken in context. Data available for the batch under consideration and past batches consist of, for example, raw material and blend assays during production, dissolution, content uniformity, final product assay, weight variation, hardness, thickness, and friability. Nevertheless, judgment is difficult to document, and who is to say what person has the qualities to make the correct decision. We can only hope to make a decision that is sensible under the circumstances, knowing that all circumstances differ.

As stated in the "Opinion" [2], "The goal is to distinguish between an anomaly and a reason to reject the batch." If a single assay fails, and all other evidence indicates "quality," the manufacturer has the responsibility to demonstrate that the failing result does not represent the product. If the data were observed in a scientific experiment, the researcher could hypothesize reasons for accepting the bulk of the evidence, with possible justifications for the aberrant result, as noted previously. No harm is done. In a manufacturing environment where GMPs dictate procedures, explanations, no matter how rational or scientifically rigorous, are useless, if a judgment is made by an FDA inspector that the result impugns the quality of the product. There is no unanimity concerning the procedure of evaluating an outlier. This small paper discusses approaches in a few commonly encountered situations in the presence of a failing result or outlier. The discussion presupposes that a cause for the aberrant data is not apparent. *Clearly, if a cause can be identified, for example, analyst mistake, instrument malfunction, or sample preparation error, then a reassay on the same or a new sample (as appropriate) according to the original procedure, would be a reasonable procedure to follow.*

VI.1   CASE 1
The original material from which the failing result or outlier was observed is still available and is (relatively) homogeneous. For example, this would occur in the case of an assay of a blend composite or the assay of a composite of 20 tablets for the final product assay. We assume relatively good homogeneity. The same situation would apply for the assay of a solution when some sample is still available after the assay.

VI.2   CASE 1A
A single assay is reported and fails, for example, outside the 90% to 110% release limits. No cause can be determined. How many reassays are necessary to discredit (or verify) the original assay and ensure the integrity of the batch? The Court's "Opinion" [2] suggests that 7 of 8 passing results may possibly suffice. The recommendation is subjective, although not altogether unreasonable. The number of samples to be retested may be quantified in an objective way, but the final decision still requires "judgment." Although the following analysis could apply to any of the situations described above, I will use the example of a final composite assay for tablets (a homogeneous mix of 20 tablets) to illustrate one possible approach. Thus, when failing or aberrant data with no obvious cause are observed, a reasonable sample size for reassay could be calculated as follows:

Estimate the true batch average and RSD from other data compiled during the batch testing, in particular content uniformity (CU). (We assume that CU data have passed. If not, a failure investigation is warranted.) Assay a sufficient number of new samples so that the 99% confidence interval for the average result, calculated from the available assays on the composite sample, is within specifications. In order to make the calculation for the number of samples to be reassayed, we need to estimate both assay and tablet variability. We can either assume that assay variability is considerably larger than tablet variability (use the RSD from CU); or estimate assay and tablet content variability separately from other available data (previous lots, assay data, etc.) in order to make a more realistic estimate. Bolton has discussed how this may be done in a previous publication [3]. For simplicity, estimate the average tablet content and RSD from the CU data. Note that the RSD estimated from the CU data will be an overestimate of the RSD for the composite ($S^2[CU] = S^2[assay] + S^2[tablet\ uniformity]$; $S^2[composite] = S^2[assay] + S^2[tablet\ uniformity/20]$), so that the sample size for the reassay will be overestimated, a maximum estimate. (We assume that assay variance is large compared to tablet variance.) The confidence interval depends on the sample size and d.f., and we can estimate a sample size iteratively. Use Table VI.1 for the estimate of number of samples to be reassayed from the composite (or original sample) as a function of mean potency and RSD. Use a slightly larger sample if in doubt. This table is based on a one-sided confidence interval. Typically, we are concerned about an out-of-specification result that is either too low or too high. Note that the numbers in Table VI.1 are based on the sample having the mean and RSD shown in the table. Therefore, the a priori estimate of the sample mean and RSD should be made with care. If in doubt, choose a sample somewhat larger than given in the table. On the other hand, if estimates

Table VI.1   Estimate of Approximate Number of Samples to Be Reassayed Based on Estimate of Mean and RSD for One-Sided 99% Confidence Limit (See Text)

RSD (%)	Mean potency (%)			
	94	96	98	100[a]
1	4	3	3	3
2	5	4	3	3
3	7	5	4	4
4	9	6	5	4
5	12	7	6	5
6	16	9	7	6

[a]For estimates greater than 100%, use the 98% column for 102%, etc.

of RSD are made from CU data, the estimate is apt to be too large, and this would tend to make the choice of sample size conservative.

An example should make this clear: Specifications for an active ingredient are 90% to 110%. A single assay of 89% is observed on a composite sample of 20 tablets. From CU data, the average result is 97% with RSD $= 4$. From Table VI.1, $N \approx 6$. If RSD is 3 in this example, $N \approx 5$.

The number of reassayed samples is sensible. If the average is close to 100% and the RSD is small, only a few samples need to be reanalyzed. If the RSD is large and the average is close to the limits, a larger sample is necessary. *Note that if the sample size, mean potency, and RSD match the values in Table VI.1, the one-sided 99% CI will be within specifications (90–110).* Finally, one may want to know if the original "outlier" or failing result should be included in the calculation of the average and standard deviation. I would recommend applying the USP test for outliers (Dixon's test) [4] to make a decision as to whether the original outlying observation should be included (see Note 1 on Court Opinion at the end of this paper.) For example, if the original assay is 85%, but we believe that the average potency should be 98% with RSD of 2%, we would assay (at least) three more samples from the same composite (from Table VI.1). If the observed reassay values are 96%, 98%, and 99%, the original assay of 85% is an outlier (Dixon test), and only the 3 reassay values are used in the calculation. The mean is 97.7% and the RSD is 1.55%. The 99% (one-sided) CI is $97.7 - 6.23 = 91.47$, which is within the 90% to 100% limits, and passes. A sample size of 4 or more would give a "comfort" zone.

Note that the Court recommendation of 7 of 8 passing results could be overly conservative in some cases, but less than adequate in other cases. In fact, with moderate variation, 8 samples would be a good number if the average observed potency is close to the specification limits. If the observed potency is close to 100% with moderate variability, less samples are needed.

Also, one might be concerned that if more than one assay fails, the product may still pass (i.e., the average is within limits and the 99% CI is within limits). This would seem to be a most unlikely occurrence, because the inclusion of a failing result would increase the variance considerably if the rest of the values were well within the specification limits. For example, the six assays, 88% 89% 97% 98% 97% and 101%, have an average of 95% and a s.d. of 5.25. The confidence limit would be below 90%.

VI.3   CASE 1B

Replicate assays are performed and the average of the assays is within limits, but one assay fails. No cause can be found. For example, three assays of a homogeneous blend show results of 88%, 95%, and 98%. The average is within 90% to 100%, but one assay is out of limits. One could accept the batch based on the average result (93.7%), but prudence may dictate further testing. We would like to establish a reasonable retesting procedure. Based on the discussion above, it would seem reasonable to assay new samples according to Table VI.1 based on an estimate of the average and RSD. This estimate should be made based on all information available, for example, CU results, not only the results of the assays in question. One could further determine that the passing assays be part of the retesting if there is evidence that one of the values is in error, for example, based on other batch data. Thus, in this case, if a sample of size 4 is called for, only two samples could be tested and combined with the remaining data (2 passing values). Thus, judgment is critical. But, the rationale for retesting should be recorded and made clear. The procedure could be part of SOPs for retesting. For example, in this example, CU data may have shown an average of 98% and an RSD of 3%. In the current example, the value of 88% appears to be an outlier and the retesting plan would be based on the CU data. On the other hand, if the CU data showed an average of 94% and an RSD of 4%, one might believe that the 88% value may be a legitimate value to be included in the average. In this case, the RSD of the three original assays may be factored into the decision of how much retesting is to be done. Consider the following example to help clarify this decision-making process.

Two assays are performed on a composite sample (2 portions of the same composite), with assay results of 90% and 98%. Note that, in the absence of an outright analytical error, differences in results of such replicates can be, at first, attributed to assay variability. The variability (RSD) of such duplicates based on retrospective data (accumulated from past lots, e.g., from control charts) is determined to be 2%. This suggests that the difference between the two assays (8%) is excessive and probably due to an analytical error. Also, the CU data show an average of 97% and an RSD of 3.5%. From Table VI.1, a sample of size 5 is recommended. Include the 98% observation, but not the 90% value, as one of the 5 samples. (Of course, there is nothing wrong with taking a conservative approach and reassaying 6 new samples.) Note again that one is penalized (more samples to be assayed) when a product is either very variable, not close to 100% in potency, or both.

VI.4   CASE 2

The material from which the failing result or outlier was observed is no longer available. This could occur, for example, for single tablet assays where the test is destructive, or for assays where stability is an issue, and a repeat assay on the same material may not be indicative of the original assayed material. This situation may also occur if repeated testing of a sample shows failure, but where the failure is not necessarily indicative of the quality of the product. An example of this latter situation is repeated failures on a single composite, where the failures could be possibly attributed to an error in preparation of the composite. The process of testing further samples is termed "resampling" (as opposed to "retesting" in the Opinion).

VI.5   CASE 2A

Specific examples of the situation described in CASE 2 above may be considered for the cases of dissolution and content uniformity. In these cases, the original material is not present, and multiple units have been assayed. Outliers may be observed more frequently in these cases because of the multiplicity of assays. Clearly, the more assays performed, the greater the probability of an analytical "error" causing an outlier, or the higher the probability of including an occasional aberrant tablet among those items assayed. For example, one could reasonably argue that in a large batch of tablets or capsules, there is a high probability that the batch contains one or more unusually low and/or high potency units. The chances that such aberrant units will be contained in the sample tested (from 6 to 30 units, for example) are very small if only a few of these outliers exist in the batch. Thus, if an outlying value is observed without any obvious cause, we have no way of knowing the true situation. A very conservative view would be to throw out the batch, no matter if all other tests are within specifications (the "FDA" position in the Barr Case). From a practical (cost) and scientific point of view, throwing out the batch based on such an event seems severe. If we decide that further testing should be done to assess the true nature of the batch, in terms of doing the right thing, we want to be "sure" that the observed outlier is not representative of the batch. Of course, we can never be 100% sure. The degree of assurance should be high and would be difficult to quantify. However, it seems fair to say that if there were any sense that the failure could represent a public health hazard, the desired degree of assurance should be greater.

At the present time, there is no unanimity on what is to be done. For example, in a content uniformity test, a single failing result of 70% is observed for a tablet assay. In one instance, at least, I know that a firm assayed 100 additional tablets (all of which were between 85% and 115%), and nevertheless, the batch was rejected. [The reason for the excessive testing was to

Table VI.2A   Minimum Number of Tablets Needed for Various Observed Values of Mean Potency and RSD for Product to Be Acceptable (99% Tolerance Interval)

RSD	Mean potency					
	95%	**96%**	**97%**	**98%**	**99%**	**100%**
1%	6(7)	5(6)	5(6)	5(6)	5(5)	4(5)
2%	13 (25)	11(18)	9(15)	8(12)	8(11)	7(10)
3%	60 (>1000)	35 (250)	22 (90)	18 (50)	15(35)	13 (25)
4%	Fails	700 (Fails)	140 (Fails)	70 (Fails)	45 (800)	30 (190)
5%	Fails	Fails	Fails	Fails	500 (Fails)	140 (Fails)
6%	Fails	Fails	Fails	Fails	Fails	Fails

99% assurance that 99% of tablets within 85% to 115% (99% assurance that 99.9% of tablets within 85% to 115% for potent drugs).

meet GMP requirements, according to one defensive (my opinion) interpretation of a failure investigation.]

The question is how much more testing should be done to give a given degree of assurance. To come upon such a number, we need a measure of the "degree of assurance." One reasonable measure is to have assurance that the great majority of units (tablets) are within 85% to 115%. For example, we may want 99% assurance that 99% of the tablets are within 85% to 115%. From my point of view, such a conclusion would be satisfactory for most products. For very potent products, we may want to have 99% assurance that 99.9% of the tablets are within 85% to 115%. If we assume that the tablet drug content is normally distributed, tolerance intervals can be calculated based on assay results. I would propose that further testing be done in cases of a failing result caused by a single outlier (where no cause can be found), and the mean (% of label) and RSD calculated from the reassays.

In this example (CU), all reassays should be within 85% to 115%, with the exception that not more than 1 (3 in the case of capsules) in every 30 could be within 75% to 125%, as defined for CU limits in the USP [5 If one or more items among the new values assay outside 75% to 125%, a full investigation is warranted and indicated. With an estimate of the mean (%) and RSD from the assayed samples, the tolerance interval can be calculated, that is, we can say with 99% assurance that p percent of the tablets are within some upper and lower limit. Tables VI.2A and VI.2B show some possible scenarios of extended testing in these situations. The number of tablets (capsules) to be reassayed are given for 95% and 99% tolerance probabilities. Note that in all these cases, there is very high assurance that practically 100% of the tablets will be within 75% to 125% of label. This plan certainly seems reasonable. Products with a large RSD (e.g., 5%) must be very close to 100% in order to have any chance of passing. If such products contain potent drugs (a matter of judgment), then a product that shows 5% RSD cannot pass if an outlier is observed (a full failure investigation is indicated.) Thus, the product must exhibit moderate or low variability and be close to 100% in order to give assurance that the product is acceptable. As noted previously, one must understand that the average result and RSD are not

Table VI.2B   Minimum Number of Tablets Needed for Various Observed Values of Mean Potency and RSD for Product to Be Acceptable (95% Tolerance Interval)

RSD	Mean potency					
	95%	**96%**	**97%**	**98%**	**99%**	**100%**
1%	3 (4)	3(3)	3(3)	2(3)	2(3)	2(3)
2%	8 (15)	7(11)	6(9)	6(8)	5(7)	5(6)
3%	35 (>1000)	19(140)	14(50)	11(30)	9(19)	8(15)
4%	Fails	400 (Fails)	80 (Fails)	35 (>1000)	24 (400)	18(100)
5%	Fails	Fails	Fails	Fails	250 (Fails)	80 (Fails)
6%	Fails	Fails	Fails	Fails	Fails	Fails

95% assurance that 99% of tablets within 85% to 115% (95% assurance that 99.9% of tablets within 85% to 115% for potent drugs).

known until the assays are completed. (The RSD and mean potency are determined from the assay results.) These values should be estimated in advance in order to determine the sample size needed for reassay. These values can be estimated from the batch assays. (Use the passing content uniformity data or past batch data for this estimate.) Clearly, the failing value, the suspected faulty result, should not be included in sample size calculations. If unsure about the number of samples to be reassayed, one should estimate conservatively, that is, a larger number of reassays.

For example, a CU test showed 9 passing results (85–115%) and one value less than 75%. The 9 passing values showed a mean of 97% with an RSD of 3%. According to Table VI.2A, 22 more tablets are assayed. If the average of these 22 tablets is close to 97% with RSD approximately equal to 3%, we would have 99% confidence that 99% of the tablets are between 85% and 115%. If the number of tablets to be reassayed based on Table VI.2A is less than 20, reassay at least 20 according to USP CU test specifications [4] the outlier occurred during the first stage of CU testing. If the outlier (<75% or >125%) occurred during the second stage of testing (a total of 30 tablets have been tested), then the numbers in Table VI.2A can be used directly as is.

An important point to be emphasized once more is that the sample sizes in Tables VI.2A and VI.2B will give the indicated tolerance interval if the observed mean and RSD are as indicated in the table. The values of the mean and RSD are not known until the assays are completed. Thus, the numbers in Tables VI.2A and VI.2B are based on a good guess of the expected mean and RSD. A conservative approach would use larger sample sizes than indicated to protect against a bad estimate or chance outcomes. How many more samples to use is strictly a matter of judgment and cost considerations.

A similar table can be constructed for dissolution. This is generally one-sided, in that low values result in failures. For example if the lower limit is 80% dissolution in 30 minutes, the number of retests should result in 95% assurance that 99% of the tablets have a dissolution above 80% in 30 minutes.

One potential cause for product failure is the observation of a large RSD in the CU test. If a product passes based on the individual observations, but fails the RSD test, the individual observations should be evaluated for possible outliers. If a single outler is observed as a possible cause, reassay using the sample size given in Tables VI.2A and VI.2B. If the removal of a single value still results in a failing RSD, a full batch investigation is warranted. For example, suppose that 10 tablets are assayed and 9 have results between 101% and 103%, one value is at 109%, and one value is 86%. Suppose the calculated RSD is greater than 6% (a failure). A reasonable approach would be to reassay, assuming that the 86% value was an outlier. The remaining 9 values have an average result of 103% and RSD of 2.5%. From Table VI.2A, about 15 to 20 tablets would be reassayed. In this example, if the tablets were evenly spread from 85% to 115%, it is possible that elimination of a single tablet would not bring the RSD within specifications. In this case a full investigation would be required. In my experience, this situation would be very unlikely to occur.

VI.6   CASE 2B

Another somewhat different example would be a situation where a single assay fails (or is borderline) and the original sample is no longer available or has been compromised. Again, no cause for the result is obvious, and we cannot differentiate between a true failing result or an analytical error. We need high assurance that the original value does not represent the batch. We could follow the previous example, and estimate the resampling size from Tables VI.2A and VI.2B. However, in these situations, often the material available may be limited. For example, with stability samples, insufficient material may be available for reassay. Another situation that may be considered similar is the case where a composite sample shows consistent failing results and no cause is obvious. The result may have been caused by faulty preparation of the composite. In both of these cases, new samples need to be prepared to verify the integrity of the batch (or stability). In these situations, repeat assay on a new *single* sample (new composite of 20 tablets or new bottle of liquid product) would not be sufficient to assure product quality. One conceivable approach to this problem, if material is lacking, is to take sufficient samples according to Table VI.1, so the results would give a 99% confidence interval for the true potency. The new sample, in this example, would consist of new composites (each individual sample is a 20 tablet composite) or new bottles of liquid on stability (if available). Consider the following example: A composite assay shows 80% potency after 4 assays. Evidence from CU and other batch data suggest that there is an analytical or preparation error. Note that the composite is an average of at least 20 random tablets, and this low observed value is almost surely not due to lack of mixing (heterogeneous mix). The average potency appears to be about 99% with an RSD of 2% based on other available data. Table VI.1 indicates that three new samples should be taken. Three new composites of 20 tablets each are prepared and assayed, and a 99% confidence interval calculated (one-sided). If the confidence limit is contained in the release specifications, the product is considered to be acceptable. If this were a liquid product (which continues to fail upon reassay), we would need to sample three new bottles. (If three stability samples are not available, one might consider sampling from the field.)

VI.7 CONCLUSION

In my opinion, a single failing or outlying test result (with no documentable cause) is not sufficient to fail a batch of product if other test results for the batch indicate no problems. In these cases, a sufficient amount of further testing should be performed so that the product quality can be assured with high probability. This paper proposes one way of approaching the question of "what is the sufficient number of samples to reassay?"

Notes on the Court's Opinion

1. The Opinion [6] suggests that the fact that the outlier test in the USP is directed toward biological assays, and no mention is made of chemical assays, means that the test is not applicable to chemical assays. It is unfortunate that this inference is made. Perhaps the USP, inadvertently, is at fault, for lack of further explanation when describing the test. In addition, the Opinion further states the reason for the omission of chemical assays with regard to testing for outliers is due to the "innate variability of microbiological assays," "... subject to the whims of microorganisms." In fact, the legitimacy of tests for outliers is not dependent on inherent variability in the sense that the variability is taken into account in the test. Thus, an assay with large variability, such as a microbiological assay, would have to show considerable divergence due to the suspected outlier for the value to be rejected. Because of lower variability, testing for an outlier in a chemical assay might reject a less distant observation. Also, there are surely some chemical assays that are more variable than some biological assays. Thus, the use of an outlier test should not be judged based on the variability of the observation, but, rather on other criteria, for example, the nature of the distribution of results or, perhaps, on philosophical grounds.

2. On pages 74 to 75 of the Opinion [7], the following statement appears: "Unless a firm with certainty establishes grounds to reject the tablet falling outside the 75 to 125 range, the batch should not be released." There is no way to be 100% certain (certainty) in this situation (or any situation for that matter). If the tablet is no longer available for assay and no cause for the outlying result can be found, one can never resurrect the original scenario with any confidence. I believe that if we replace the words, "with certainty", with "with a high degree of assurance", that the methods proposed in this paper fulfill the latter definition.

REFERENCES

1. United States v. Barr Laboratories, Inc., Consolidated Docket No. 92-1744 (AMW).
2. United States v. Barr Laboratories, Inc., Consolidated Docket No. 92-1744 (AMW), p. 26.
3. Bolton S. Computation of in-house quality control limits for pharmaceutical dosage forms based on product variability. J Pharm Sci 1983; 72: 405.
4. USP XXII, USP Convention, Inc., Rockville, MD 20852, 1990, p. 1503.
5. USP XXII, USP Convention, Inc., Rockville, MD 20852, 1990, p. 1618CU test.
6. United States v. Barr Laboratories, Inc., Consolidated Docket No. 92-1744 (AMW), pp. 17, 44.
7. United States v. Barr Laboratories, Inc., Consolidated Docket No. 92-1744 (AMW), pp. 74, 75.

Appendix VII
When is it Appropriate to Average and its Relationship to the Barr Decision

VII.1 BACKGROUND: ASSAY AND CONTENT UNIFORMITY TESTS

Analytical procedures to determine the drug content of pharmaceutical dosage forms are of two kinds. One is to estimate the true average drug content of the product (e. g., mg/tablet or mg/mL), and the other is to determine the uniformity of the product, that is, to assess the degree to which different dosage units may differ. For true solutions, the question of uniformity is mute, because solutions are homogeneous by definition (In certain cases, it may be desirable to check uniformity for large volumes of solutions to ensure dissolution and adequate mixing prior to transfer). For solid dosage forms, uniformity is determined by assaying different portions of the powdered blend at the initial stages of the process, and individual finished tablets at the final stage. For assessing uniformity, there are no "official" regulations for conformance for blends. The finished product content uniformity test is defined in the USP. Release limits for blend testing for uniformity is at the discretion of the pharmaceutical firm, and should have a scientific as well as practical basis. The subject of blend testing was an important issue in the Barr Trial and Judge Wolin's Decision [1]. In particular, Judge Wolin condemned the averaging of different samples of powdered blend when the purpose of the test was to determine uniformity. This is obvious to the pharmaceutical scientist. Not that it is wrong to average the results (we are always interested in the average), but we do not want to obscure the variability by mixing heterogeneous samples and then reporting only an average, when the purpose of the test is to assess that variability. Therefore, procedures for assessing and reporting variability are clear, although the regulations for blend testing and interpretation of data are not "official" and need scientific judgment. (A further dilemma here is that some pharmaceutical firms do not perform blend testing on some products, at their discretion.)

VII.2 AVERAGING REPLICATES FROM A HOMOGENEOUS SAMPLE

The problem that I want to present here is: When is averaging appropriate and correct, and how do we deal with the individual values that make up the average in these circumstances? This can be simplified by limiting this question to one particular situation:

AVERAGING IS APPROPRIATE AND CORRECT WHEN MULTIPLE ASSAYS ARE PERFORMED ON THE SAME SAMPLE, OR ON REPLICATE SAMPLES FROM THE SAME HOMOGENOUS MIX, FOR PURPOSES OF DETERMINING THE TRUE AVERAGE CONTENT.

I do not believe that any knowledgeable scientist would argue or contradict this. It is a scientific, statistical fact that the average of multiple assays on the same material will give a better estimate of the true content than single assays (the more assays, the better the estimate). Thus, a pharmaceutical firm would better fulfill its obligation of supplying conforming material to the public by performing multiple assays. Nevertheless, the number of assays performed for purposes of estimating the true drug content is not fixed by law, and many companies perform a single assay, whereas other companies may perform three or more assays. In fact, the manner in which the replicates are performed may differ among companies. For example, a replicate assay may be defined as coming from replicate analyses of the same final solution prepared from a single portion of material, such as replicate HPLC injections from the same solution. The variability among the replicate readings in this case represents instrumental variability rather than product variability. If we are dealing with a solution or a homogenized composite of 20 tablets, there are other sources of variability that are not accounted for in such a replicate scheme. In particular, the variability arising from the sample preparation for analysis is neglected in the former scheme because only one sample has been analyzed. Sample preparation variability would include weighing variability as well as variability during the various steps of preparing the product for the analysis. Therefore, the average of replicates using different sample preparations will give a better estimate of the true drug content than the same number of replicate analyses on the same sample. The latter gives a good estimate of a single sample, whereas the former better estimates the batch. Again, this is a scientific, statistical fact. We can define the variability of such an assay measurement as the sum of independent variances

Variance (assay) = variance(I) + variance(P) + variance(O),

where I = instrumental, P = preparation and O = other sources of variation.

The variance of the average of 3 replicates where the replicates are multiple injections from the same sample is

$$\frac{\text{variance(I)}}{3} + \text{variance(O)} + \text{variance(P)}.$$

The variance of the average of 3 replicates where the replicates are multiple preparations from the same sample is:

$$\frac{\text{variance(I)} + \text{variance(O)} + \text{variance(P)}}{3}.$$

Therefore, given a choice, to obtain a more precise estimate of the average drug content of a batch, assaying multiple preparations from the same homogeneous sample is a more desirable procedure than assaying multiple injections from a single preparation. This would apply for both solutions and homogeneous powders. Thus, there is little doubt as to what constitutes a better testing procedure for estimating drug content

USE MORE INDEPENDENT SAMPLES!

Again, there are no official regulations on how many samples to use. Assaying a single sample may be acceptable in this respect.

VII.3 HOW DO WE DEAL WITH SINGLE OOS RESULTS WHEN THE AVERAGE CONFORMS?

What, then, is the problem? The problem is that there is confusion as to how to handle the individual observations that make up the average in certain situations. There should be no argument as to when it is appropriate to average. As emphasized throughout this discussion, averaging multiple observations is appropriate when the purpose is to estimate the average drug content. If all of the individual observations fall within release limits, there is no ambiguity. The question is, "What do we do if one of the individual observations falls outside of the release limits?"

Although not explicitly stated, official limits are absolute. A product either does or does not pass. The official limits for drug content, as stated in the USP, for example, are based on the average drug content. Clearly, some individual units may lie outside these limits as defined in the content uniformity test. From a legal point of view, it appears that if the measure of the average content falls within limits, the product is acceptable. Thus, an average result of 90.5 based on a single assay or duplicates of 89.5 and 91.5 is within limits. On the other hand, such a result suggests that the true average may be below 90 with substantial probability. A prudent manufacturer would want more assurance that the product is truly within specifications. In-house limits such as 95 to 105 are constructed to give such assurance. These limits are usually computed so that there is high assurance that the product truly meets official specifications if an analytical result falls within these limits. The in-house specifications are not legal limits, but, rather, are computed, conservative limits to ensure that the legal limits will be met. The construction of such limits should include all sources of variability including analytical error. Thus, a single assay of 95.5% should be sufficient to release the product if the in-house limits are computed correctly. In this situation, there is no question about the decision, the product passes or does not pass. Suppose, that a company wants to improve this assessment of lot performance by performing triplicate assays in this same situation. Because the single assay is close to the in-house limit, repeat assays are apt to give values below 95. For example, triplicate assays may give values of 94.5, 95.5, and 96.5, with an average of 95.5. In this case, the average result is definitive and the single value below 95 should not invalidate the average. Otherwise, we would be saying that a single assay of 95.5 is a better indicator of batch quality than triplicate assays that average 95.5. Clearly, this is contradictory to scientific and statistical fact. If we act otherwise, we would be defeating the intent and purpose of scientific QC analytical techniques.

How do we account for the fact that an average may fall within limits, but a single assay may fall outside the limits (without obvious cause)? It is a well-known statistical fact that the more observations we make, the greater the likelihood of seeing extreme observations because of inherent variability in the observations. The variability has a probability distribution, say approximately normal. Every observation has some probability of falling outside the release limits due to extreme errors (variability) that can occur during an analysis. These extreme observations are apt to happen from time to time, by chance. If we are unlucky enough to see such an observation, is this irrevocable? Does this mean the batch is not good? The answer requires scientific judgment. In the absence of a definitive mistake, examination of batch records and product history, as well as the nature of the assay and release limits should lead to either acceptance of the batch or further testing (according to SOPs). Further testing should help to assess the true nature of the data, that is, to differentiate a failure from an anomalous result.

Unfortunately, Judge Wolin, in his decision (Barr Decision), excluded outlier tests from chemical assays (this ruling is controversial and will almost certainly be modified in the near future). But, even if a single failing value is not an outlier, is this cause for rejection, when the average is the objective of the test? Certainly, some scientific judgment is needed here. Otherwise, we will be throwing out much good material at the expense of the manufacturer and taxpayer, and we will be condoning nonscientific, suboptimal testing techniques. If, in fact, there is no give or compromise in this dilemma, companies will do an absolute minimal amount of testing to reduce the probability of out-of-specification (OOS) results.

So the question remains as to how to handle this perplexing problem, "What do we do about a single OOS result among replicates that are meant to be averaged?" I do not believe that there can be a single inflexible rule. Scientific judgment and common sense are needed. I will give a couple of examples.

Example 1. The official limits for a product are 90 to 110. In-house limits are set at 95 to 105. The in-house limits are based on the variability of the product, that is, the manufacturer believes that based on the variability inherent in measuring the drug content of the product (perhaps including assay error, stability, uniformity, etc.) that the average content when the product is released based on a 20 tablet composite should be between 95 and 105. Thus, the manufacturer is prepared to release the product if the average composite assay is 95 to 105. Triplicate analyses yield results of 99, 98, and 94.5, an average of 97.17, which passes. However, one assay is below 95 (note the triple jeopardy incurred by the triplicate determinations). Should this product be released? Note that the release limits of 95 to 105 are based on inherent variability of the product, including its measurement. On this basis, the product should pass, because it is the average in which we are interested. If there is any doubt, I would want to look at other product characteristics and batch history. Certainly, if there were no suggestion of a problem based on other relevant data, release of this batch would be indicated. Another scientific contradiction here concerns in-house limits that apparently are not subject to regulations. Firms that use in-house limits for release, certainly a better and more conservative approach to releasing material than using the absolute official limits, may be penalized for using a more scientific approach to drug testing. Also, I believe that there is a qualitative difference for single OOS results when applying "Official" and "in-house" release limits. "Official" limits are irrevocable, set by "law" without a truly scientific basis. An average of 89.9 for a product with official limits of 90 to 110 cannot be released! In-house limits are set by individual companies based on scientific "know-how" and have built-in allowances for variability. Thus, a single replicate falling slightly below the "Official" limit should probably be treated with greater concern than-the single value outside in-house limits but within official limits as observed in this example.

   Example 2. Consider the situation where the Official release limits are 95 to 105 and the three assays are 96.5, 95.5, and 94.5. The average is 95.5 that passes. All other data are conforming. In this case, although it still may be argued convincingly that the product passes, I would suggest additional testing. I believe that this is appropriate even if no cause can be found for the low result. This question was raised in the Barr trial, in which results of 89, 89, 92 were contemplated for a product with release limits of 90 to 110 (paragraph 49, Barr Decision). Further testing was recommended by the witness, and the judge seemed to be satisfied with this approach. The real problem here, is not the problem of averaging, or retesting, but of "retesting into compliance." Clearly, the latter approach is not satisfactory, and should be addressed in SOPs. The SOP should recommend the number of retests to be performed when there is reasonable doubt about the quality of the batch as suggested in this example.

VII.4   DISCUSSION

Because of the lack of specific regulations concerning averaging of data, scientific judgment and common sense should prevail. Certainly, situations exist where averages are the optimal way of treating and reporting data. In particular, replicate measures based on a homogeneous sample are meant to be averaged. Procedures for averaging data and retesting should be contained in the company's SOPs.

   The question of what to do if a single OOS result is observed is addressed to some extent in the Barr Decision. A single OOS result that cannot be attributed to the process or to an operator error, as opposed to a laboratory error, is not labeled as a failure. According to Inspector Mulligan of the FDA (Barr Decision, paragraph 21), an OOS result overcome by retesting is not a failure. "The inability to identify an error's cause with confidence affects retesting procedures, see paragraph 38–39. . ." (Barr Decision, paragraph 28). Paragraphs 38 and 39 suggest that retesting is part of the failure investigation. "A retest is similarly acceptable when review of the analyst's work is inconclusive." Thus, retesting is not disallowed when the retests are used to isolate the cause or nature of the outlying result. The amount of retesting should be sufficient to differentiate an anomaly and a reason to reject a batch (paragraph 39). Thus, according to the decision, retesting may be done with discretion (based on SOPs) to help identify a cause for OOS results.

   An important consideration is that good testing procedures should not be penalized. As noted in the examples above, a single OOS result contained in an average that passes specifications should not be reason to reject a batch in general without further testing. Otherwise, firms will be forced into performing single assays to reduce the risk of failure. This is based on the fact that the penalty for an OOS result would be the same for both (a) one of several assays OOS or (b) a single assay OOS. Biological assays are often based on the average of triplicates, in which the average result is the basis for release, regardless of the individual values. In principal, chemical assays should be treated in a similar manner, with scientific judgment always in mind.

REFERENCE

1.   Barr Decision, Civil Action No. 92–1744, OPINION, United States District Court for the District of New Jersey, Judge Alfred M. Wolin, February, 1993.

Appendix VIII
Excel Workbooks and SAS Programs

Excel Workbooks

Microsoft *Excel* provides a powerful package to solve many statistical problems. The following Workbooks are provided as examples of how this package can be used to solve problems presented in this book. It is hoped that the reader will be able to apply the principles illustrated in these examples to the real-life statistical problems that he or she encounters. It is anticipated that the reader has some familiarity with Excel and the basic mathematical functions available in Excel. The reader should also be familiar with the basic methods to copy and paste values and formulas from one cell or group of cells to another.

Many of the examples use Excel's built-in statistical modules. These are available in the Statistical Analysis ToolPak add-in. If this feature is activated in your installation of Excel, you will see it by choosing Tools in the main menu of Excel. If you find the Data Analysis option, the add-in is activated. If not, choose Tools and then select Add-Ins. From the choice of Add-ins, select both the Analysis ToolPak and the Analysis ToolPak-VBA options. This will install the package.

In the following examples, sequences of Excel commands will be presented to accomplish the data analyses. The Main Menu bar, in the following illustration, is just below the Microsoft Excel heading. It has the headings of File, Edit, View, Insert, Format, Tools, Data, Window and Help.

The command sequence:

Main Menu   Tools → Data Analysis → Descriptive Statistics

Refers to the steps:

1. Choose Tools from the Main Menu
2. Select Data Analysis under the Tools menu
3. Move the highlight down to Descriptive Statistics
4. Click OK

The first example is based on the Serum Cholesterol Changes for 156 Patients shown in Table 1.1. Workbook 1.1 shows how to perform descriptive analyses of the data and how to obtain a cumulative frequency distribution.

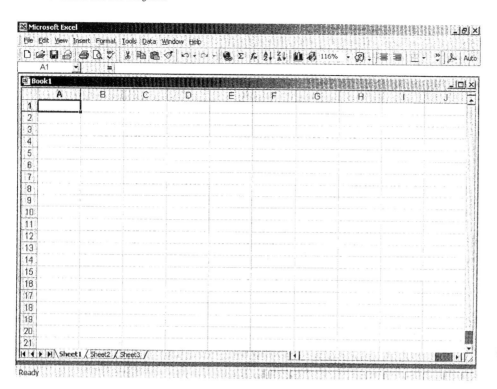

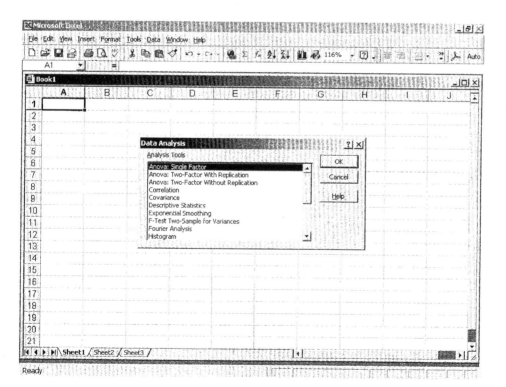

Workbook 1.1 Descriptive Analyses and Cumulative Frequency Distribution (partial workbook shown)

	A	B	C	D	E	F	G
1	*Change*	*Change*		*Point*	*Change*	*Rank*	*Percent*
2	17			125	55	1	100.00%
3	−12	Mean	−10.6218	91	46	2	99.30%
4	25	Standard Error	2.216327	60	40	3	98.70%
5	−37	Median	−9.5	105	39	4	98.00%
6	−29	Mode	17	109	38	5	97.40%
7	−39	Standard Deviation	27.68191	50	35	6	96.10%
8	−22	Sample Variance	766.2883	88	35	6	96.10%
9	0	Kurtosis	−0.16183	11	34	8	94.80%
10	−22	Skewness	−0.28357	126	34	8	94.80%
11	−63	Range	152	18	33	10	94.10%
12	34	Minimum	−97	97	27	11	93.50%
13	−31	Maximum	55	113	26	12	92.90%
14	−64	Sum	−1657	3	25	13	92.20%
15	−12	Count	156	37	24	14	90.90%
16	−49			92	24	14	90.90%
17	5			98	23	16	90.30%

Commands in Analyses

Cells A1 – A157	Enter "Change", then in A2-A157 the 156 change values from Table 1.1.
Main Menu	Tools → Data Analysis → Descriptive Statistics
Dialog Box	
Input Range:	Highlight or enter A1:A157
Grouped By:	Click on Columns option
Labels in First Row:	Click on this option
Output Range	Click on Column B or enter B1
Summary Statistics	Click on this option
OK	Click to calculate
Main Menu	Tools → Data Analysis → Rank and Percentile
Dialog Box	
Input Range:	Highlight or enter A1:A157
Grouped By:	Click on Columns option
Labels in First Row:	Click on this option
Output Range	Click on Column D or enter D1
OK	Click to Calculate

Notes on Analyses Interpretation:
Columns C lists the value of the sample statistic referenced in Column B
The statistic "Mode" (most frequent value) is not a unique value in this data set.

Workbook 1.4 Entry of Tablet Potencies When Frequency Distribution Is Given (partial worksheet shown)

	A	B	C	D	E	F	G	H	I
1	90	92	93	94	95	96	97	98	99
2		92		94		96	97	98	99
3				94			97	98	99
4				94			97	98	99
5				94			97	98	99
6							97	98	99
7							97	98	99
8								98	99
9								98	
10								98	

Column D lists the observation number, in Column A, for the Change value in Column E
Column F lists the rank (highest to lowest) for the Change value shown in Column E
Column G lists the cumulative frequency percentile for the Change value in column E
The next example creates a histogram and a cumulative frequency plot from the tablet potency values presented in Table 1.4.

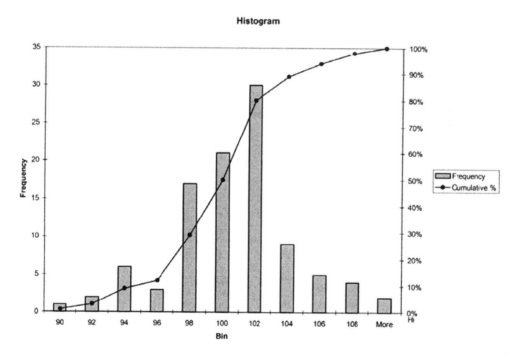

Each Xi value is entered into a separate worksheet column, the number of replicate entries of a value is given by its frequency, Wi, in Table 1.4. Entering a value once and then copying it through the range of desired cells simplifies the process.

The Xi values in each column are then copied to a new worksheet to create a single column of all 100 tablet potencies, as shown in the following partial worksheet.

	A
1	Potency
2	90
3	92
4	92
5	93
6	94
7	94
8	94
9	94
10	94
11	95
12	96
13	96
14	97
15	97
16	97
17	97
18	97
19	97
20	97

Descriptive analyses can now be conducted on the values in Column A of this second worksheet (e.g. creation of histogram and cumulative % plots).

Commands in Analyses

Main Menu Bar	Tools → Data Analysis → Histogram
Dialog Box	
Input Range:	Highlight or enter A1:A101
Labels	Click on this option
Output	Click on New Worksheet Ply
Cumulative Percentage	Click on this option
Chart Output	Click on this option
OK	Click to plot histogram
Click on Histogram	
Main Menu Bar	Chart → Location
Dialog Box	
As New Sheet:	Click on this option and enter "Histogram" in box to right
Click on cumulative	
percentage line:	Format symbol and colors as desired
Click on y-axis	Format scale, font, number as desired
Click on histogram Bars	Format color, patterns, fill effects as desired

Note: If it were necessary to format the x-axis (Bin) values, this is done by changing the format of the Bin values column in the worksheet containing these values.

Next the tablet assay results shown in Table 5.1 are used to demonstrate construction of a 95% confidence interval for the sample mean under the assumption that the data are normally distributed. The mean, μ, and the standard deviation, σ, for the population are unknown and must be estimated from the data. As such, the t-distribution is used to obtain the confidence interval limits.

Workbook 5.1   Confidence Interval When Mean and Sigma Are Unknown

	A	B	C	D	E	F
	Potency	n	Mean	S	alpha	Confidence
1	Potency	n	Mean	S	alpha	Confidence
2	101.8	10	103.0	2.22	0.05	0.95
3	102.6					
4	99.8	df	t-value	CI Lower	CI Upper	
5	104.9	9	2.26	101.41	104.59	
6	103.8					
7	104.5					
8	100.7					
9	106.3					
10	100.6					
11	105.0					
12						
13						

Commands in Analysis (commands for up to 100 entries in column A):

Cells in Column A	Enter tablet potency results from Table 5.1	
Cell B2	= COUNT(A2:A101)	Total number of potency values
Cell B5	= B2–1	Degrees of freedom (df) = n-1
Cell C2	= AVERAGE(A2:A101)	Arithmetic mean of potency values
Cell D2	= STDEV(A2:A101)	Sample standard deviation for values
Cell E2	Enter alpha level	0.05 for 95% CI, 0.10 for 90% CI, etc.
Cell F2	= 1-E2	Confidence Interval coverage
Cell C5	= TINV(E2,B5)	Critical t-value for alpha & df
Cell D5	= C2-C5*D2/SQRT(B2)	95% Confidence Interval lower limit
Cell E5	= C2 + C5*D2/SQRT(B2)	95% Confidence Interval upper limit

The following uses the percent dissolution values of Table 5.9 to demonstrate how to use Excel's built in statistical tools to conduct an independent sample t-test.

Workbook 5.9   Two Independent Sample t-Test

	A	B	C	D	E
1	**FORM A**	**FORM B**	t-Test: Two-Sample Assuming Equal Variances		
2	68	74			
3	84	71		*FORM A*	*FORM B*
4	81	79	Mean	77.1	71.4
5	85	63	Variance	33.43333333	48.71111111
6	75	80	Observations	10	10
7	69	61	Pooled Variance	41.07222222	
8	80	69	Hypothesized Mean Diff	0	
9	76	72	Df	18	
10	79	80	t Stat	1.988775482	
11	74	65	P(T<=t) one-tail	0.031073458	
12			t Critical one-tail	1.734063062	
13			P(T<=t) two-tail	0.062146917	
14			t Critical two-tail	2.100923666	

Commands in Analyses
Columns A & B   Enter Form A and Form B values from
                                 Table 5.9
Main Menu Bar   Tools → Data Analysis → t-Test:
                                 Two-Sample Assuming Equal
                                 Variances
Dialog Box
   Variable 1 Range: Highlight or enter A1:A11
   Variable 2 Range: Highlight or enter B1:B11
   Hypothesized Mean Diff: Enter the null hypothesis difference
                                 between means, 0
   Labels: Click on this option
   Alpha: Enter desired alpha level for t-Test, 0.05
   Output Range   Highlight cell C1 or enter C1.
   OK   Click to perform calculations.
                                 Results appear in Columns C-E.

The next workbook performs the analysis for a paired sample t-test as shown in Table 5.11. The comparison of the Areas under the blood-level curve calculated for six animals dosed in a bioavailability study with both a new drug formulation (A) and the marketed formulation (B) is easily performed using Excel's built-in statistical program.

Workbook 5.11   Paired Sample *t*-Test

	A	B	C	D	E	F	G
1	**Animal**	**FORM A**	**FORM B**	**Ratio**	**Expected**		
2	1	136	166	0.82	1		
3	2	168	184	0.91	1		
4	3	160	193	0.83	1		
5	4	94	105	0.90	1		
6	5	200	198	1.01	1		
7	6	174	197	0.88	1		
8							
9	t-Test: Paired Two Sample for Means				t-Test: Paired Two Sample for Means		
10							
11		*FORM A*	*FORM B*			*Ratio*	*Expected*
12	Mean	155.33333	173.83333		Mean	0.891654	1
13	Variance	1332.2667	1278.1667		Variance	0.004747	0
14	Observations	6	6		Observations	6	6
15	Pearson Correlation	0.9354224			Pearson Correlation	#DIV/0!	
16	Hypothesized Mean Difference	0			Hypothesized Mean Difference	0	
17	Df	5			df	5	
18	t Stat	−3.484781			t Stat	−3.85212	
19	P(T<=t) one-tail	0.0087842			P(T<=t) one-tail	0.005988	
20	t Critical one-tail	2.0150492			t Critical one-tail	2.015049	
21	P(T<=t) two-tail	0.0175684			P(T<=t) two-tail	0.011975	
22	t Critical two-tail	2.5705776			t Critical two-tail	2.570578	

Commands in Analyses

Columns A, B, C & D	Enter values from Table 5.11.
Column E	Enter value of 1 for each entry in Column D (for analysis of ratios)
Main Menu Bar	Tools Data → Analysis → t-Test: Paired Two-Sample for Means
Dialog Box	
Variable 1 Range:	Highlight or enter B1:B7
Variable 2 Range:	Highlight or enter C1:C7
Hypothesized Diff:	Enter the null hypothesis difference between means, 0
Labels:	Click on this option
Alpha:	Enter desired alpha level for t-test, 0.05
Output Range	Click on cell or enter A9.
OK	Click to perform calculations

Note: To obtain an analysis of the Form A/Form B ratios, perform the same sequence of operations using the Ratio values (D1:D7) as Variable 1 and the Expected values (E1:E7) as Variable 2. Choose output Range as E9.

Section 5.2.6 discusses how to construct a 95% confidence interval on the difference between the proportions of headaches observed in two different groups of patients. The calculation uses a normal approximation and incorporates a continuity correction.

The following Excel workbook shows how to carry out the calculations.

Workbook 5.2.6   Continuity-Corrected 95% Confidence Interval

	A	B	C	D	E	F
1		**Group I**	**Group II**	**alpha**	**Z-value**	**correction**
2	**Headaches**	35	46	0.05	1.96	0.00491
3						
4	**N**	212	196	**difference**	**se**	**Z*se**
5	**P**	0.165	0.235	0.070	0.03958	0.077575
6	**Q**	0.835	0.765			
7				**Cl_low**	**Cl_high**	
8				−0.013	0.152	

Commands in Analysis

Data Entry:	Enter Section 5.2.6 values into cells B2, C2, B4, C4, D2	
Cell B5:	$= B2/B4$	$p = \#/n$
Cell C5:	$= C2/C4$	
Cell B6:	$= 1-B5$	$q = 1 - p$
Cell C6:	$= 1-C5$	
Cell D5:	$= C5-B5$	difference between p values (group I-II)
Cell E2:	$= NORMSINV(1-D2/2)$	Critical Z- value for 95% confidence interval
Cell E5:	$= SQRT(B5*B6/B4 + C5*C6/C4)$	$se = (\Sigma(pq/n))^{1/2}$
Cell F2:	$= 0.5*(1/B4 + 1/C4)$	continuity correction $= 0.5\,(1/n_I + 1/n_{II})$
Cell F5:	$= E2*E5$	
Cell D8:	$= D5 - (F5 + F2)$	CI low $=$ diff $-$ [se*Z $+$ correction]
Cell E8:	$= D5 + (F5 + F2)$	CI high $=$ diff $+$ [se* Z $+$ correction]

Excel has utilities for performing linear regression analyses and creating graphs of the results of such analyses. The power of these utilities can be seen in this next example which uses tablet assay results from a stability study (Table 7.5).

In this workbook, linear regression is used to model the stability of tablet potency over time. A 95% confidence interval about the stability line is constructed and the results are graphically illustrated using Excel's Chart Wizard.

Commands in Analyses

Columns A & B	Enter Month and Assay values from Table 7.5	
Main Menu Bar	Tools $\rightarrow$ Data Analysis $\rightarrow$ Regression	
Dialog Box		
Input Y Range:	Highlight or enter B1:B19	
Input X Range:	Highlight or enter A1:A19	
Labels:	Click on this option	
Output Range	Click on cell C1 or enter C1.	Results start in Column C.
OK	Click to perform calculations	
Cell D16	$= AVERAGE(A2:A19)$	Mean value for the X values
Cell D17	$= 18*(VARP(A1:A19))$	equal to $\Sigma(X_i - mean)^2$

Open a second worksheet in this workbook. This sheet will be used to calculate the predicted values for the stability regression line and the 95% confidence interval band around the line. The measured potency values from Worksheet 1 and the predicted values and their confidence bounds from this new worksheet (Worksheet 2) will be used to create a stability trending graph.

Commands in Analyses

Column A	Enter 0 & 1 into Cells A2 & A3, highlight & drag through Cell A62 to obtain Month numbers 0 through 60.
Cells E2 and F2	Copy Slope and Intercept values from Worksheet 1
Cell E5	Enter 16, the residual df from ANOVA in Worksheet 1 equal to N-2
Cell F5	Enter or copy the SSQ__Diff value from Worksheet 1
Cell F8	Enter or copy the Month Mean value from Worksheet 1
Cell E8	= TINV(0.05,E5), t-value for two-sided, 95% confidence interval
Cell E11	= SQRT(1.825), square root of residual MS from ANOVA in Worksheet 1
Cell B2	= $F$2 + A2*$E$2, intercept + month * slope
Cells B3-B62	Copy formula from B2 into these cells to obtain predicted values
Cell C2	= $B2-$E$8*$E$11*SQRT(1/($E$5 + 2) + POWER(($A2-$F$8),2)/$F$5)
Cells C3-C62	Copy formula from C2 to obtain 95% Conf. Interval lower bound
Cell D2	= $B2 + $E$8*$E$11*SQRT(1/($E$5 + 2) + POWER(($A2-$F$8),2)/$F$5)
Cells D3-D62	Copy formula from D2 into these cells to obtain 95% Conf. Interval upper bound

Create graph

Highlight cells A1:B62, click Chart Wizard icon and choose XY scatter plot.
Click Next and choose the series tab.
Click on ADD.
Click in the Name box and enter *95% CI*.
For X-values, choose A1:A62.
For Y-values choose Cl:C62.
Repeat the process to add the graph of the 95% CI upper limits (D1:D62 values).
Next, repeat the process for the Month (X) and Assay (Y) values from Worksheet 1.
Click on Next and enter the title and axes labels for the graph.
Click on finish.

From the Main Toolbar Menu, choose Chart and then under that choose Location
Enter a Name so that the graph is placed as a chart separate from Worksheet 2.
The lines on the graph can now be edited by double clicking on each one.
Edit the predicted line to be solid with no symbols.
Edit the confidence interval curves to be smoothed, no-symbol, dashed.
The y and x axes can be edited (double click on each) to change the range of the Scale.

Workbook 7.5 Linear Regression of Tablet Stability Results (Worksheet 1)

	A	B	C	D	E	F	G	H	I	J
	Month	**Assay**								
1			**SUMMARY OUTPUT**							
2	0	51								
3	0	51	*Regression Statistics*							*P-value*
4	0	53	Multiple R	0.778		Intercept	51.8	0.535524	96.7277	1.42E-23
5	3	51	R Square	0.605		Month	−0.267	0.053822	−4.9546	0.000143
6	3	50	Adjusted R Sq	0.581						
7	3	52	Standard Err	1.351						
8	6	50	Observations	18						
9	6	52								
10	6	48	ANOVA							
11	9	49		*Df*	*SS*	*MS*	*F*		*Significance F*	
12	9	51	Regression	1	44.8	44.8	24.548	0.0001434		
13	9	51	Residual	16	29.2	1.825				
14	12	49	Total	17	74					
15	12	48								
16	12	47	**Month Mean**	8						
17	18	47	**SSQ_diff**	630						
18	18	45								
19	18	49								

Workbook 7.5   Linear Regression of Tablet Stability Results (Worksheet 2) (Listing of first 14 rows of the 62-row worksheet)

	A	B	C	D	E	F
1	**Month**	**Predicted**	**95% CI Low**	**95% CI Hi**	**slope**	**intercept**
2	0	51.8	50.7	52.9	−0.26667	51.8
3	1	51.5	50.5	52.6		
4	2	51.3	50.3	52.2	**Df**	**SSQDx**
5	3	51.0	50.1	51.9	16	630
6	4	50.7	49.9	51.5		
7	5	50.5	49.7	51.2	**t-val**	**Meanx**
8	6	50.2	49.5	50.9	2.12	8
9	7	49.9	49.2	50.6		
10	8	49.7	49.0	50.3	**S_yx**	
11	9	49.4	48.7	50.1	1.351	
12	10	49.1	48.4	49.8		
13	11	48.9	48.1	49.6		
14	12	48.6	47.8	49.4		

Stablility of Tablet Assays

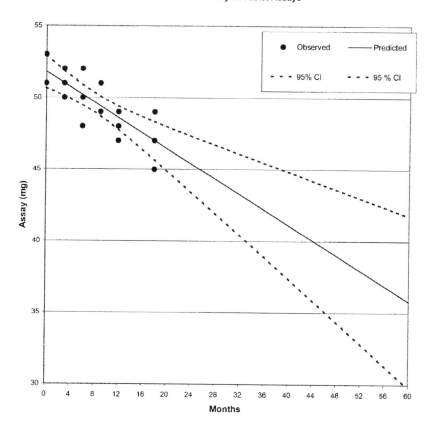

The following Workbook uses the spectrophotometric calibration curve results of Table 7.6. While it employs only the basic mathematical functions of Excel, it provides a powerful method for performing weighted linear regression analysis which can be used in situations where a straight-line model is appropriate. In this example, the weight is the inverse of concentration squared $(1/X^2)$, but the method can be easily adapted to other appropriate weights (some function that is inversely proportional to the variance in y).

Commands in Analyses

Cells A2:B11	Enter the X and y values from Table 7.6	
Cell C2	$= 1/(A2^2)$ Weight, w, is inverse of concentration squared	
Cells C3:C11	Copy the formula from Cell C2	
Cell D2	$= C2^* A2^* B2$	wXy
Cells D3:D11	Copy the formula from Cell D2	
Cell E2	$= C2^* A2$	wX
Cells E3:E11	Copy the formula from Cell E2	
Cell F2	$= C2^* B2$	wy
Cells F3:F11	Copy the formula from Cell F2	
Cell G2	$= C2^* A2^2$	wX^2
Cells G3:G11	Copy the formula from Cell G2	
Cell A13	$= SUM(A2:A11)$	ΣX
Cells B13:G13	Copy formula from Cell A13	$\Sigma y, \Sigma w, \Sigma wXy, \Sigma wX, \Sigma wy$ & ΣwX^2
Cell B15	$= (D13-E13^*F13/C13)/(G13 = (E13^2)/C13)$	slope
Cell B17	$= (F13/C13)-B15^*(E13/C13)$	intercept

Workbook 7.6 Weighted $(1/X^2)$ Linear Regression Analysis

	A	B	C	D	E	F	G
1	Conc (X)	OD (y)	1/X^2	wXy	wX	wy	wX^2
2	5	0.105	0.04	0.021	0.2	0.0042	1
3	5	0.098	0.04	0.0196	0.2	0.00392	1
4	10	0.201	0.01	0.0201	0.1	0.00201	1
5	10	0.194	0.01	0.0194	0.1	0.00194	1
6	25	0.495	0.0016	0.0198	0.04	0.000792	1
7	25	0.508	0.0016	0.02032	0.04	0.0008128	1
8	50	0.983	0.0004	0.01966	0.02	0.0003932	1
9	50	1.009	0.0004	0.02018	0.02	0.0004036	1
10	100	1.964	0.0001	0.01964	0.01	0.0001964	1
11	100	2.013	0.0001	0.02013	0.01	0.0002013	1
12	Sum(X)	Sum(y)	Sum(w)	Sum(wXy)	Sum(wX)	Sum(wy)	Sum(Wx^2)
13	380	7.57	0.1042	0.19983	0.74	0.0148693	10
14							
15	Slope (b) =	0.01986					
16							
17	Intercept (a) =	0.00166					

The next example shows how to use Excel to perform a series of calculations iteratively across different parameter values of a function to determine which values give the best fit to the observed values. In this example, using the method of least-squares, the best estimates for the parameters (slope and intercept) of the function (regression line) occur at the minimum sum of squares for the difference between the predicted values of the regression line and the observed values. The regression line is defined by its slope (K) and its intercept (C0) and there are three observed time (hour)-concentration values (mg/L): (1,63), (2,34), and (3,22). These data are the stability results shown in Table 7.8. It is first necessary to determine a plausible range of values for C0 and K. For C0, this could be done graphically by plotting the data and then extrapolating the curve back to 0 time. A wide range of values should be selected around this estimate for the first iteration. In the first worksheet, a range of 50–400 was chosen. An initial range of estimates for K can be obtained in several ways: by using the estimate of C0 and then solving the equation $C = C0*Exp(-K*t)$ for each time (t)-concentration (C) pair in the data set. Alternatively, the natural logarithm of each concentration can be plotted against time. The slope of the line through the plotted points is an estimate of $- K$. In this example, K was found to be close to 0.5 and a range of 0.1–0.7 was chosen for evaluation. The analysis requires the calculation of the sum of squares (SSQ) of the deviations (DEV = observed-predicted) for each of the three data points based on all combinations of the chosen C0 and K values. The C0 and K values that result in the minimum SSQ represent the least-squares estimates.

Workbook 7.8 Nonlinear Fit of Stability Data by the Method of Least-Squares (first iteration)

	A	B	C	D	E	F	G	H	I	J
1	C0	K	63.0	34.0	22.0	Dev_1	Dev_2	Dev_3	SSQ	
2	400	0.1	361.9	327.5	296.3	−298.9	−293.5	−274.3	250755.3	
3	200	0.1	181.0	163.7	148.2	−118.0	−129.7	−126.2	46667.7	
4	100	0.1	90.5	81.9	74.1	−27.5	−47.9	−52.1	5759.7	
5	50	0.1	45.2	40.9	37.0	17.8	−6.9	−15.0	589.7	
6	400	0.3	296.3	219.5	162.6	−233.3	−185.5	−140.6	108637.2	
7	200	0.3	148.2	109.8	81.3	−85.2	−75.8	−59.3	16510.9	
8	100	0.3	74.1	54.9	40.7	−11.1	−20.9	−18.7	906.9	
9	50	0.3	37.0	27.4	20.3	26.0	6.6	1.7	719.7	
10	400	0.5	242.6	147.2	89.3	−179.6	−113.2	−67.3	49586.7	
11	200	0.5	121.3	73.6	44.6	−58.3	−39.6	−22.6	5477.8	
12	100	0.5	60.7	36.8	22.3	2.3	−2.8	−0.3	13.4	**
13	50	0.5	30.3	18.4	11.2	32.7	15.6	10.8	1428.7	
14	400	0.7	198.6	98.6	49.0	−135.6	−64.6	−27.0	23302.8	
15	200	0.7	99.3	49.3	24.5	−36.3	−15.3	−2.5	1559.8	
16	100	0.7	49.7	24.7	12.2	13.3	9.3	9.8	360.4	
17	50	0.7	24.8	12.3	6.1	38.2	21.7	15.9	2178.7	
18								MIN	13.4	

Additional iterations are performed to refine the estimates to the desired level of precision. In this example, precision to one decimal place for C0 and to three decimal places for K was considered appropriate.

Commands in Analyses (Commands are repeated for each iteration)

Columns A and B		Enter all possible combinations of the selected C0 and K values.
Cell C2	$= A2*EXP(-B2*1)$	Predicted Concentration at 1 hour
Cell D2	$= A2*EXP(-B2*2)$	Predicted Concentration at 2 hour
Cell E2	$= A2*EXP(-B2*3)$	Predicted Concentration at 3 hour
Cell F2	$= 63-C2$	1 hour deviation (observed-predicted)
Cell G2	$= 34-D2$	2 hour deviation
Cell H2	$= 22-E2$	3 hour deviation
Cell I2	$= SUMSQ(F2,G2,H2)$	Sum of squared deviations (SSQ)

Workbook 7.8 Nonlinear Fit of Stability Data by the Method of Least-Squares (section of the worksheet to refine the estimates)

A	B	C	D	E	F	G	H	I	
C0	**K**	**63.0**	**34.0**	**22.0**	**Dev_1**	**Dev_2**	**Dev_3**	**SSQ**	
130	0.4	87.1	58.4	39.2	−24.1	−24.4	−17.2	1473.1	
115	0.4	77.1	51.7	34.6	−14.1	−17.7	−12.6	670.5	
100	0.4	67.0	44.9	30.1	−4.0	−10.9	−8.1	201.7	
85	0.4	57.0	38.2	25.6	6.0	−4.2	−3.6	66.8	
130	0.5	78.8	47.8	29.0	−15.8	−13.8	−7.0	491.4	
115	0.5	69.8	42.3	25.7	−6.8	−8.3	−3.7	128.0	
100	0.5	60.7	36.8	22.3	2.3	−2.8	−0.3	13.4	
85	0.5	51.6	31.3	19.0	11.4	2.7	3.0	147.6	
130	0.6	71.3	39.2	21.5	−8.3	−5.2	0.5	96.5	
115	0.6	63.1	34.6	19.0	−0.1	−0.6	3.0	9.4	**
100	0.6	54.9	30.1	16.5	8.1	3.9	5.5	110.9	
85	0.6	46.6	25.6	14.1	16.4	8.4	7.9	401.1	
130	0.7	64.6	32.1	15.9	−1.6	1.9	6.1	43.2	
115	0.7	57.1	28.4	14.1	5.9	5.6	7.9	129.2	
100	0.7	49.7	24.7	12.2	13.3	9.3	9.8	360.4	
85	0.7	42.2	21.0	10.4	20.8	13.0	11.6	736.6	
							MIN	9.4	

Workbook 7.8 Nonlinear Fit of Stability Data by the Method of Least-Squares (further refining of the estimates) (C0 range examined was 100–130, K range 0.50–0.70; only section with minimum is shown)

A	B	C	D	E	F	G	H	I	
C0	**K**	**63.0**	**34.0**	**22.0**	**Dev_1**	**Dev_2**	**Dev_3**	**SSQ**	
104	0.53	61.2	36.0	21.2	1.8	−2.0	0.8	7.9	
105	0.53	61.8	36.4	21.4	1.2	−2.4	0.6	7.4	
106	0.53	62.4	36.7	21.6	0.6	−2.7	0.4	7.9	
105	0.54	61.2	35.7	20.8	1.8	−1.7	1.2	7.5	
106	0.54	61.8	36.0	21.0	1.2	−2.0	1.0	6.5	
107	0.54	62.4	36.3	21.2	0.6	−2.3	0.8	6.6	
107	0.55	61.7	35.6	20.5	1.3	−1.6	1.5	6.3	
108	0.55	62.3	36.0	20.7	0.7	−2.0	1.3	5.9	
109	0.55	62.9	36.3	20.9	0.1	−2.3	1.1	6.4	
108	0.56	61.7	35.2	20.1	1.3	−1.2	1.9	6.8	
109	0.56	62.3	35.6	20.3	0.7	−1.6	1.7	5.8	**
110	0.56	62.8	35.9	20.5	0.2	−1.9	1.5	5.8	
111	0.56	63.4	36.2	20.7	−0.4	−2.2	1.3	6.8	
110	0.57	62.2	35.2	19.9	0.8	−1.2	2.1	6.5	
111	0.57	62.8	35.5	20.1	0.2	−1.5	1.9	6.0	
112	0.57	63.3	35.8	20.3	−0.3	−1.8	1.7	6.5	

Workbook 7.8 Nonlinear Fit of Stability Data by the Method of Least-Squares (C0 range evaluated was 108.0–110.0 by 0.2; K was 0.550–0.570 by 0.002)

A	B	C	D	E	F	G	H	I	
C0	K	63.0	34.0	22.0	Dev_1	Dev_2	Dev_3	SSQ	
108.0	0.552	62.2	35.8	20.6	0.8	−1.8	1.4	5.838	
108.2	0.552	62.3	35.9	20.7	0.7	−1.9	1.3	5.804	
108.4	0.552	62.4	35.9	20.7	0.6	−1.9	1.3	5.807	
108.6	0.552	62.5	36.0	20.7	0.5	−2.0	1.3	5.850	
108.4	0.554	62.3	35.8	20.6	0.7	−1.8	1.4	5.772	
108.6	0.554	62.4	35.9	20.6	0.6	−1.9	1.4	5.756	
108.8	0.554	62.5	35.9	20.6	0.5	−1.9	1.4	5.779	
108.6	0.556	62.3	35.7	20.5	0.7	−1.7	1.5	5.766	
108.8	0.556	62.4	35.8	20.5	0.6	−1.8	1.5	5.732	
109.0	0.556	62.5	35.9	20.6	0.5	−1.9	1.4	5.735	
109.2	0.556	62.6	35.9	20.6	0.4	−1.9	1.4	5.777	
109.0	0.558	62.4	35.7	20.4	0.6	−1.7	1.6	5.733	
109.2	0.558	62.5	35.8	20.5	0.5	−1.8	1.5	5.718	**
109.4	0.558	62.6	35.8	20.5	0.4	−1.8	1.5	5.741	
109.2	0.560	62.4	35.6	20.4	0.6	−1.6	1.6	5.761	
109.4	0.560	62.5	35.7	20.4	0.5	−1.7	1.6	5.727	
109.6	0.560	62.6	35.8	20.4	0.4	−1.8	1.6	5.731	

A	B	C	D	E	F	G	H	I	
C0	K	63.0	34.0	22.0	Dev_1	Dev_2	Dev_3	SSQ	
109.1	0.557	62.5	35.8	20.5	0.5	−1.8	1.5	5.723	
109.2	0.557	62.6	35.8	20.5	0.4	−1.8	1.5	5.735	
109.3	0.557	62.6	35.9	20.6	0.4	−1.9	1.4	5.755	
109.1	0.558	62.4	35.7	20.5	0.6	−1.7	1.5	5.721	
109.2	0.558	62.5	35.8	20.5	0.5	−1.8	1.5	5.718	**
109.3	0.559	62.5	35.7	20.4	0.5	−1.7	1.6	5.719	
109.1	0.559	62.4	35.7	20.4	0.6	−1.7	1.6	5.744	
109.2	0.559	62.4	35.7	20.4	0.6	−1.7	1.6	5.727	
109.3	0.559	62.5	35.7	20.4	0.5	−1.7	1.6	5.719	

Columns C through I	Copy Row 2 formulas through rows 3–17	
Cell I18	= MIN(I2:I17)	Minimum of SSQ values
Cell J2	= IF(I2 = I$18,"**","")	Flags row if it contains minimum SSQ
Cell J3-J17	Copy formula from Cell J2	Flags row with best C0 and K estimates

Based on these results, it appears that the best estimate of C0 is near 100 and for K near 0.5. The next iterations further refine the estimates.
Final Iteration: C0 range evaluated was 109.0–109.4 by 0.1; K was 0.556–0.560 by 0.001
The least-squares estimates, at the desired levels of precision, are C0 = 109.2 and K = 0.558.

This next example uses Excel's built-in two-factor ANOVA, without replication, to evaluate the tablet dissolution data given in Table 8.9.

Commands in Analyses

Columns A, B, C, D	Enter dissolution values from Table 8.9.
Main Menu	Tools → Data Analysis → Anova: Two-Factor without Replication
Dialog Box	
Input Range:	Highlight or enter A1:D9
Labels:	Click on this option
Alpha:	Enter 0.05
Output Range	Click on or enter A11
OK	Click to perform calculations

Cell F3	= ABS(D23-D25)/SQRT(2*D32/8)	Calculate pair-wise t-test
Cell F4	= ABS(D24-D25)/SQRT(2*D32/8)	
Cell G3	= TDIST(F3,C32,2)	Determine pair-wise p-value
Cell G4	= TDIST(F4,C32,2)	

This next example uses Excel's built-in two-factor ANOVA, with replication, to evaluate the replicate tablet dissolution data given in Table 8.12.

Commands in Analyses

Columns A,B,C,D	Enter dissolution values from Table 8.12.
Main Menu	Tools → Data Analysis → Anova: Two-Factor with Replication
Dialog Box	
Input Range:	Highlight or enter A1:D17
Rows per sample:	Enter 2
Alpha:	Enter 0.05
New Worksheet Ply:	Click on this option
OK	Click to perform calculations

Workbook 8.9 Two-Way Analysis of Variance of Tablet Dissolution Results

	A	B	C	D	E	F	G
1	**LAB**	**Generic A**	**Generic B**	**Standard**			
2	1	89	83	94		**t-value**	**p-value**
3	2	93	75	78	**A vs Std**	0.09	0.927
4	3	87	75	89	**B vs Std**	2.23	0.043
5	4	80	76	85			
6	5	80	77	84			
7	6	87	73	84			
8	7	82	80	75			
9	8	68	77	75			
10							
11	Anova: Two-Factor Without Replication						
12							
13	*SUMMARY*	*Count*	*Sum*	*Average*	*Variance*		
14	1	3	266	88.6666667	30.33333		
15	2	3	246	82	93		
16	3	3	251	83.6666667	57.33333		
17	4	3	241	80.3333333	20.33333		
18	5	3	241	80.3333333	12.33333		
19	6	3	244	81.3333333	54.33333		
20	7	3	237	79	13		
21	8	3	220	73.3333333	22.33333		
22							
23	A		8	666	83.25	58.78571	
24	B		8	616	77	10	
25	STANDARD	8	664	83	45.14286		
26							
27							
28	ANOVA						
29	*Source of Variation*	*SS*	*df*	*MS*	*F*	*P-value*	*F crit*
30	Rows	391.8333	7	55.9761905	1.931799	0.139436	2.764196
31	Columns	200.3333	2	100.166667	3.456861	0.060239	3.73889
32	Error	405.6667	14	28.9761905			
33							
34	Total	997.8333	23				

Workbook 8.12   Two-Way ANOVA of Replicated Dissolution Results (worksheet 1)

	A	B	C	D
		Generic A	**Generic B**	**Standard**
1	**Lab**	**Generic A**	**Generic B**	**Standard**
2	1	87	81	93
3		91	85	95
4	2	90	74	74
5		96	76	82
6	3	84	72	84
7		90	78	94
8	4	75	73	81
9		85	79	89
10	5	77	76	80
11		83	78	88
12	6	85	70	80
13		89	76	88
14	7	79	74	71
15		85	86	79
16	8	65	73	70
17		71	81	80

Workbook 8.12   Two-Way ANOVA of Replicated Dissolution Results (continued)

	A	B	C	D	E	F	G
58							
59	ANOVA						
60	*Source of Variation*	*SS*	*df*	*MS*	*F*	*P-value*	*F crit*
61	Sample	783.6667	7	111.9524	4.569485	0.00231	2.422631
62	Columns	400.6667	2	200.3333	8.176871	0.001959	3.402832
63	Interaction	811.3333	14	57.95238	2.365403	0.030779	2.129795
64	Within	588	24	24.5			
65							
66	Total	2583.667	47				
67							
68	Drugs	400.6667	2	200.3333	3.456861	0.060239	3.73889

(New Worksheet Ply)

	A	B	C	D	E
1	Anova: Two-Factor With Replication				
2					
3	SUMMARY	Generic A	Generic B	Standard	Total
4	*1*				
5	Count	2	2	2	6
6	Sum	178	166	188	532
7	Average	89	83	94	88.66667
8	Variance	8	8	2	27.86667
9–45	*(Rows not shown)*	*xxxxxxxx*	*xxxxxxxx*	*xxxxxxxx*	*xxxxxxxx)*
46	*8*				
47	Count	2	2	2	6
48	Sum	136	154	150	440
49	Average	68	77	75	73.33333
50	Variance	18	32	50	37.86667
51					
52	*Total*				
53	Count	16	16	16	
54	Sum	1332	1232	1328	
55	Average	83.25	77	83	
56	Variance	65.26667	20.66667	59.6	

Commands in Analyses

Cell A68	Enter "Drugs"	Drugs effect is that for columns in the ANOVA table
Cell B68	= B62	Drugs SS
Cell C68	= C62	Drugs degrees of freedom
Cell D68	= D62	Drugs MS
Cell E68	= D62/D63	F-ratio = Drugs MS/Interaction MS
Cell F68	= FINV(E68,2,14)	p-value for Drugs F-ratio with 2 & 14 degrees of freedom
Cell G68	= FDIST(0.95,2,14)	Critical F-distribution value with 2 & 14 degrees of freedom

Notes on Interpretation

The analysis for Drugs in row 68 is based on the assumption that Drugs is a fixed effect and Laboratories (Rows) is a random effect. The analysis in row 62 for the Column (Drugs) effect assumes that both Drugs and Laboratories are fixed effects. If the laboratories are a random sample of all the available laboratories and the results are to be generalized to all laboratories, then use the row 68 results. If the eight laboratories are the only ones of interest, then the results in row 62 should be used.

The next workbook shows how to perform an Analysis of Covariance using the data from Table 8.18. In this example, two different manufacturing methods (I and II) were used to produce four lots of products whose potency and raw material potency are shown.

Workbook 8.18 Analysis of Covariance to Compare Two Methods (worksheet 1)

	A	B	C	D	E	F
1	**Method**	**MI**	**MII**	**Meth2**	**Material**	**Product**
2	I	98.4	0	0	98.4	98.0
3	I	98.6	0	0	98.6	97.8
4	I	98.6	0	0	98.6	98.5
5	I	99.2	0	0	99.2	97.4
6	II	0	98.7	1	98.7	97.6
7	II	0	99	1	99.0	95.4
8	II	0	99.3	1	99.3	96.1
9	II	0	98.4	1	98.4	96.1
10						
11				**Mean**	98.775	
12	**F-parallel**	**p-value**		**Adj Mean**	**I**	97.8639
13	0.010	0.925			**II**	96.3611
14					**Diff (II-I)**	1.50278
15					**p-value**	0.036637
16	**Slope**	**Intercpt I**	**Intercept II**			
17	−0.81481	178.3472	176.8444			

Commands in Analyses

Columns A, E and F	Enter Method, Material and Product values from Table 8.18.
Column B	Copy Method I values into rows 2–5, enter 0 elsewhere.
Column C	Copy Method II values into rows 6–9, enter 0 elsewhere.
Column D	Enter 0 for Method I row and 1 for Method II row.
Cell E11	= AVERAGE(E2:E9) Mean for Material values.
Main Menu	Tools → Data Analysis → Regression (ANOVA for separate lines)
Dialog Box	
Input Y Range:	Highlight or enter F1:F9
Input X Range:	Highlight or enter B1:D9
Labels:	Click on this option
New Worksheet Ply:	Click on this box
OK	Click to perform calculations
Main Menu	Tools → Data Analysis → Regression (ANOVA for parallel lines)
Dialog Box	
Input Y Range:	Highlight or enter F1:F9
Input X Range:	Highlight or enter D1:E9
Labels:	Click on this option
New Worksheet Ply:	Click on this box
OK	Click to perform calculations
Cell A17	Copy slope (Material coefficient) from parallel lines Worksheet
Cell B17	Copy Intercept coefficient from same Worksheet
Cell C17	= B17 + coefficient for Meth2 from parallel lines Worksheet
Cell F12	= B17 + E11*A17
Cell F13	= C17 + E11*A17
Cell F14	= F12-F13 Difference between adjusted Method means
Cell F15	p-value for difference from Meth2 in parallel lines Worksheet
Cell A13	= (SS resid. parallel lines – SS resid. separate lines)/(SS resid separate/4)
Cell B13	= FDIST(A13,1,4)

Workbook 8.18 Analysis of Covariance to Compare Two Methods (section of worksheet ply for separate lines)

	A	B	C	D	E
10	ANOVA				
11		*df*	*SS*	*MS*	*F*
12	Regression	3	5.82575	1.941916667	2.916885718
13	Residual	4	**2.663**	0.66575	
14	Total	7	8.48875		
15					
16		*Coefficients*	*Standard Error*	*t Stat*	*P-value*
17	Intercept	188.4	134.2219351	1.403645386	0.233093906
18	MI	−0.916666667	1.359891744	−0.674073264	0.537213
19	MII	−0.733333333	1.216324153	−0.602909456	0.579083754
20	Meth2	−19.61	180.1993982	−0.108823893	0.918582825

Notes on Analyses (separate lines)

Cell C13 contains the residual SS for separate lines (2.663) to be used in the test for parallelism (Cell A13 in Worksheet 1). The Intercept (188.4 in B17) is the intercept for the Method I line. The slope for the Method I line is the coefficient for MI (− 0.917 in B18). The intercept for Method II is the addition of the coefficient for Meth2 (B20) to the Method I intercept (B17), which is 188.4– 19.6 = 168.8. The slope for the Method II line is the coefficient for MII (-0.733 in B19).

(section of worksheet ply for parallel lines)

	A	B	C	D	E
10	ANOVA				
11		*df*	*SS*	*MS*	*F*
12	Regression	2	5.819028	2.909514	5.449095
13	Residual	5	**2.669722**	0.533944	
14	Total	7	8.48875		
15					
16		*Coefficients*	*Standard Error*	*t Stat*	*P-value*
17	Intercept	178.3472	80.13591	2.225559	0.076591
18	Meth2	−1.50278	0.530852	−2.83088	**0.036637**
19	Material	−0.81481	0.811906	−1.00358	0.361646

Notes on Analyses (parallel lines)

Cell C13 contains the residual SS for parallel lines (2.67) to be used in the test for parallelism (Cell A13 in Worksheet 1). The coefficient for the Intercept (178.3 in B17) is the intercept for the Method I line. The coefficient for the intercept of Meth2 is the difference between the intercepts for Methods I and II (value −1.50 in B18) which, because the two lines are parallel, is also the difference between the two methods. We estimate that Method II is 1.50 units lower than Method I with the p-value for this difference (0.0366 in E18) being statistically significant at the 0.05 level. The common slope for the parallel lines for the two methods is given by the coefficient for Material (−0.815 in B19).

The next example is taken from Table 9.2. Here we analyze the results from a 2^3 factorial experiment to determine the effect of three components upon the thickness of a tablet.

Workbook 9.2 Evaluation of Results from a 2^3 Factorial Experiment (worksheet 1)

	A	B	C	D	E	F	G	H
	Stearate (A)	**Drug (B)**	**Starch (C)**	**AB**	**AC**	**BC**	**ABC**	**Response**
1								
2	0	0	0	0	0	0	0	475
3	1	0	0	0	0	0	0	487
4	0	1	0	0	0	0	0	421
5	1	1	0	2	0	0	0	426
6	0	0	1	0	0	0	0	525
7	1	0	1	0	2	0	0	546
8	0	1	1	0	0	2	0	472
9	1	1	1	2	2	2	4	522

Commands in Analyses

Column H	Enter response values from Table 9.2.
Columns A, B, C	Enter a 0 where Table 9.2 has a " − " and a 1 where there is a "+"
Cell D2	= 2*A2*B2 Design entry for Stearate-Drug interaction
Cells D3-D9	Copy formula from D2
Cell E2	= 2*A2*C2 Design entry for Stearate-Starch interaction
Cells E3-E9	Copy formula from E2
Cell F2	= 2*B2*C2 Design entry for Drug-Starch interaction
Cells F3-F9	Copy formula from F2
Cell G2	= 4*A2*B2*C2 Design entry for 3-way interaction
Cells G3-G9	Copy formula from G2
Main Menu	Tools → Data Analysis → Regression (Estimate Main Effects)
Dialog Box	
Input Y Range:	Highlight or enter H1:H9

Workbook 9.2 Evaluation of Results from a 2^3 Factorial Experiment (main effects worksheet)

	A	B	C	D	E	F
10	ANOVA					
11		*Df*	*SS*	*MS*	*F*	*Significance F*
12	Regression	3	13768	4589.333	23.91835	0.005135
13	Residual	4	767.5	191.875		
14	Total	7	14535.5			
15						
16		*Coefficients*	*Standard Error*	*t Stat*	*P-value*	*Lower 95%*
17	Intercept	465.25	9.794769	47.49984	1.18E-06	438.0553
18	Stearate (A)	22	9.794769	2.246097	0.088025	−5.19469
19	Drug (B)	−48	9.794769	−4.90057	0.008041	−75.1947
20	Starch (C)	64	9.794769	6.5341	0.002834	36.80531
21		**MS**				
22	A	968				
23	B	4608				
24	C	8192				

Input X Range:	Highlight or enter A1:C9
Labels:	Click on this option
New Worksheet	
Ply:	Click on this box
OK	Click to perform calculations

Rename New Worksheet "Main Effects"

Main Menu	Tools → Data Analysis → Regression (Estimate 2-Factor Interactions)
Dialog Box	
Input Y Range:	Highlight or enter H1:H9
Input X Range:	Highlight or enter A1:F9
Labels:	Click on this option
New Worksheet	
Ply:	Click on this box
OK	Click to perform calculations

Rename New Worksheet "Interaction"

Repeat Regression Analysis with Input "X" Range as A1:G9 to obtain estimate for A*B*C interaction

Commands in Analyses (Main Effects Worksheet)

Cell B22	= D18*D18*D13
Cell B23	= D19*D19*D13
Cell B24	= D20*D20*D13

(2-factor interactions worksheet)

	A	B	C	D	E	F
10	ANOVA					
11		*df*	*SS*	*MS*	*F*	*Significance F*
12	Regression	6	605.5	100.9167	0.622942	0.7478876
13	Residual	1	162	**162**		
14	Total	7	767.5			
15						
16		*Coefficients*	*Standard Error*	*t Stat*	*P-value*	*Lower 95%*
17	Intercept	14.25	11.90588	1.196887	0.443097	−137.02791
18	Stearate (A)	−19	15.58846	−1.21885	0.437411	−217.06928
19	Drug (B)	−15	15.58846	−0.96225	0.512246	−213.06928
20	Starch (C)	−23	15.58846	−1.47545	0.379198	−221.06928
21	AB	**5.5**	9	0.611111	0.650783	−108.85535
22	AC	**13.5**	9	1.5	0.374334	−100.85535
23	BC	**9.5**	9	1.055556	0.482798	−104.85535
24		**MS**				
25	AB	**60.5**				
26	AC	**364.5**				
27	BC	**180.5**				

Commands in Analyses (2 factor interactions worksheet)

Cell B25	= D21*D21*D13
Cell B26	= D22*D22*D13
Cell B27	= D23*D23*D13

(worksheet 1 continued)

	A	B	C	D	E	F	G
11	Effect	Estimate	Df	SS	MS	F	p-value
12	A	22	1	968	968	7.2	0.0748
13	B	−48	1	4608	4608	34.3	0.0099
14	C	64	1	8192	8192	61.0	0.0044
15	AB	5.5	1	60.5	60.5		
16	AC	13.5	1	364.5	364.5	2.7	0.1981
17	BC	9.5	1	180.5	180.5		
18	ABC	9	1	162	162		
19	Error		3	403	134.3333		

Commands in Analyses (ANOVA similar to Table 9.5)

Column A	Enter Effect Names
Column B	Values are coefficients from Main Effects & Interactions Worksheets
	Coefficient for ABC is from regression including all effects (Wrksht not shown).
Column C	Enter 1 for all effects except Error. Enter 3 for Error.
Cells E12-E17	Enter values from Main Effects & 2-Factor Interaction Worksheets
Cell E18	Enter value for Residual MS from Cell D13 of 2-Factor Interaction Worksheet
Cell D12-D18	Enter same values that are in Cells E12-E18
Cell D19	= SUM(D15,D17,D18) Error term is chosen to be sum of AB, BC & ABC terms
Cell E19	= D19/C19 MS = SS/df
Cell F12	= E12/E$19 F = Effect MS/Error MS
Cells F13, F14, F16	Copy formula from F12
Cell G12	= FDIST(F12, 1,3) p-value for Effect from F-distribution
Cell G13,G14,G16	Copy formula from G12

Section 11.5 presents how to perform repeated measures Analysis of Variance. The methods used in the analysis are illustrated using the results of a comparison of two antihypertensive drugs. One group of patients received the standard drug and a second group the new drug. Diastolic blood pressure was recorded for each patient prior to treatment (baseline) and then at 2, 4, 6, and 8 weeks after treatment. The results, presented in Table 11.22, are analyzed in the following workbook.

Commands in Analyses

Cells A3-F10	Enter patient numbers and diastolic blood pressures from Table 11.22
Cells A15-A22	Copy patient numbers from Cells A3:A10
Cell B15	= C3-$B3 Calculates change from baseline
Cells B16-B22	Copy formula from Cell B15
Cells C15-E22	Copy formula from B15 through B22
Cell B25	= Sum(B15:B22) Sum of changes at Week 2
Cells C25-E25	Copy formula from Cell B25
Cell F25	= Sum(B25:E25) Sum of changes for all weeks

Worksheet 11.22   Comparison of Two Antihypertensive Drugs (worksheet 1)

	A	B	C	D	E	F
1		Standard Drug				
2	**Patient**	**Baseline**	**Wk 2**	**Wk 4**	**Wk 6**	**Wk 8**
3	1	102	106	97	86	93
4	2	105	103	102	99	101
5	5	99	95	96	88	88
6	9	105	102	102	98	98
7	13	108	108	101	91	102
8	15	104	101	97	99	97
9	17	106	103	100	97	101
10	18	100	97	96	99	93
14	**Patient**	**Wk 2**	**Wk 4**	**Wk 6**	**Wk 8**	
15	1	4	−5	−16	−9	
16	2	−2	−3	−6	−4	
17	5	−4	−3	−11	−11	
18	9	−3	−3	−7	−7	
19	13	0	−7	−17	−6	
20	15	−3	−7	−5	−7	
21	17	−3	−6	−9	−5	
22	18	−3	−4	−1	−7	
23						
24						Standard
25	Sum	−14	−38	−72	−56	−180
26						

Section for New Drug (not shown):

Cells H2:M2	Enter or Copy the headings in cells A2-F2	
Cell K1	Enter heading "New" for New Drug	
Cells H3-M11	Enter New Drug patient numbers and diastolic readings	
Cell I15	= J3-$I3	Calculate changes from baseline
Cells I16–I23	Copy formula from Cell I15	
Cells J15-L23	Copy formulas from I15 through I23	
Cell I25	= Sum(I15:I23)	New drug sum of changes Week 2
Cells J25-L25	Copy formula from Cell I15	
Cell M25	= Sum(I25:L25)	New drug sum of changes all weeks

(section of analyses shown in Tables 11.24 and 11.25)

	A	B	C	D	E	F	G
27	ANOVA	**Standard**			ANOVA	**New**	
28	*Source*	*SS*	*df*		*Source*	*SS*	*df*
29	Rows	57.5	7		Rows	114.2222	8
30	Columns	232.5	3		Columns	486.9722	3
31	Error	255.5	21		Error	407.7778	24
32							
33	Total	545.5	31		Total	1008.972	35
34							
35	**CT**	**Source**	**df**	**SS**	**MS**	**F**	**p-value**
36	3750.368	Patients	15	171.72	11.45		
37		Weeks	3	669.69	223.23		
38		Drugs	1	196.16	196.16	17.13	0.0009
39		WK × Drug	3	49.78	16.59	1.13	0.3487
40		Error	45	663.28	14.74		
41		Total	67	1750.63			

Commands in Analyses

Main Menu	Tools → Data Analysis → Anova: Two-Factor Without Replication
Dialog Box	
Input Range:	Highlight or enter B15:E22
Alpha Level	Enter or accept default value of 0.05
New Worksheet Ply:	Click on this box
OK	Click to perform calculations
	(Copy ANOVA values from new worksheet to main Worksheet 1)
Cells B27-G33	Copy from cells A19-C25 of new worksheet to get Source, SS & df
Main Menu	Tools → Data Analysis → Anova: Two-Factor Without Replication
Dialog Box	
Input Range:	Highlight or enter I15:L23 (New Drug data not shown)
Alpha Level	Enter or accept default value of 0.05
New Worksheet Ply:	Click on this box
OK	Click to perform calculations
	(Copy ANOVA values from new worksheet to main worksheet 1)
Cells E27-G33	Copy from Cells A19-C25 of new worksheet to get Source, SS & df
Cells A35-G35	Enter Headings CT, Source, df, SS, MS, F and p-value
Cells B36-B41	Enter Source names
Cell A36	= POWER(F25 + M25,2)/68 Correction Term
Cell C36	15 (Combined row df for Standard and New Drugs)
Cell C37	3 (number of weeks – 1)
Cell C38	1 (number of drugs – 1)
Cell C39	= 3*1 (Product of Week df and Drugs df)
Cell C41	= 4*17–1 (#Weeks *#Patients – 1)
Cell C40	= 67 – 15 – 3 – 1 – 3 (error df = Total-Patients-Drugs-WeeksxDrugs)
Cell D36	= B29 + F29 (Combined Row SS for Standard and New Drugs)
Cell D37	= (SUMSQ((B25 + I25),(C25 + J25),(D25 + K25),(E25 + L25))/17)-A36
Cell D38	= F25*F25/32 + M25*M25/36 – A36
Cell D39	= B30 + F30-D37
Cell D40	= B31 + F31 (Combined Error SS for Standard and New Drugs)
Cell D41	= SUM(D36:D40) (Total SS = Sum of all other SS)

Cell E36	= D36/C36 (MS = SS/df)
Cell E37-E40	Copy formula from Cell E36
Cell F38	= E38/E36 (F = MSeffect/MSerror Drugs uses MS Patients as error term)
Cell F39	= E39/E40 (F value for Weeks x Drugs using ANOVA error term)
Cell G38	= FDIST(F38,1,15) (p-value for F with 1 df & 15 df)
Cell G39	= FDIST(F39,3,45)

Table 12.2 shows the average weights of 50 tablets from 30 batches of a tablet product.

In the next example, Excel is used to calculate the three-batch moving average for the weights. These results are then used to construct a control plot of the moving averages along with their upper and lower control limits.

Workbook 12.2 Average Weight of 50 Tablets from 30 Batches of a Product

	A	B	C	D	E	F	G
1	**Batch**	**Average**	**Move Ave**	**Range**	**Mean**	**Low**	**High**
2	0				400.0	397.603	402.397
3	1	398.4	N/A				
4	2	399.5	N/A				
5	3	398.8	398.9	1.1			
6	4	397.4	398.6	2.1			
7	5	402.7	399.6	5.3			
			Rows 8–26 not shown				
27	25	398.4	399.5	3.1			
28	26	398.8	398.6	0.4			
29	27	399.9	399.0	1.5			
30	28	400.9	399.9	2.1			
31	29	399.9	400.2	1.0			
32	30	399.5	400.1	1.4			
33	31				400.0	397.603	402.397
34	Mean	400.0		2.35			

Commands in Analyses

Data Entry:	Enter Batch numbers and averages from Table 12.2 into columns A and B, adding a Batch 0 and 31 for graphing purposes.	
Cell C5	= Average(B3:B5)	Average of first 3 batches
Cell C6-C32	Copy formula from Cell C5	
Cell D5	= MAX(B3:B5)-MIN(B3: B5)	Range (Max-Min) of first 3 batches
Cell D6-D32	Copy formula from Cell D5	
Cell B34	= Average(B3:B32)	Average of the 30 batches
Cell D34	Copy formula from Cell B34	Average of moving ranges
Cell E33	= B34	
Cell F33	= $E$33 – 1.02*$D$34	Lower Limit using factor (1.02) from Table IV.10
Cell G33	= $E$33 + 1.02*$D$34	Upper Limit using factor (1.02) from Table IV.10
Cell E2	= E33	
Cell F2	= F33	
Cell G2	= G33	

Click on Chart Wizard and choose to create a XY scatter plot.
Click Next and then click on Series Tab, then on Add.
Click on worksheet icon for X-values.
Choose cells A2 through A33, click icon to accept this range.
For Y-values, click worksheet icon, choose cells C2 through C33.
Click Add for Series 2. X-values are A2 through A33. Y-values F2 through F33.
Click Add for Series 3. X-values are A2 through A33. Y-values are G2 through G33.
Click Add for Series 3. X-values are A2 through A33. Y-values are E2 through E33.
Click Next and add chart title, X and Y axes labels.
Click Legend tab and remove check mark on Show Legend (by clicking it).
Click tab for Gridlines and make sure all choices are blank.
Click Next and choose the name Plot for the New Worksheet for the chart.
Click on Plot Area and choose None for fill effects.
On Main Menu click Tools, Options & Chart.
Choose to plot empty cells as Interpolated.
Click on Lower & Upper limit points and set symbol to None and line to a dashed, black, custom
    line.
Click on Mean point and set symbol to None and line to a solid, black, custom line.
Click on an X-axis number and then on the Scale tab.
Set Minimum $= 0$, Maximum $= 31$, Major Unit $= 1$.

Three-Batch Moving Average Plot for Table 12.2 Results

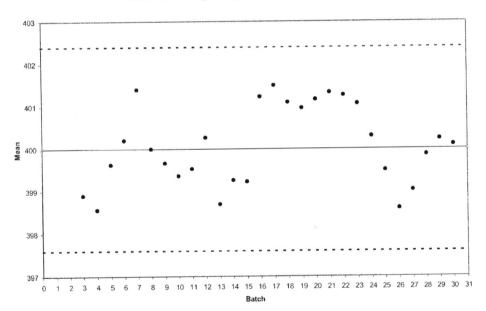

In the next example, the assay results for a sample of theee tablets from four different batches of a product, as shown in Table 12.9, are used to demonstrate how to calculate the variance components. The experiment was a nested design in which the total variance can be divided into its components of between batches, between tablets within batch, and between assays within tablets.

Workbook 12.9 Determination of Variance Components in a Nested Design

	A	B	C	D	E	F	G
1	Batch	Tablet	Assay1	Assay 2	Assay 3	SSQ	df
2	A	1	50.6	50.5	50.8	0.046667	2
3		2	49.1	48.9	48.5	0.186667	2
4		3	51.1	51.1	51.4	0.06	2
5	B	1	50.1	49.0	49.4	0.62	2
6		2	51.0	50.9	51.6	0.286667	2
7		3	50.2	50.0	49.8	0.08	2
8	C	1	51.4	51.7	51.8	0.086667	2
9		2	52.1	52.0	51.4	0.286667	2
10		3	51.1	51.9	51.6	0.326667	2
11	D	1	49.0	49.0	48.5	0.166667	2
12		2	47.2	47.6	47.6	0.106667	2
13		3	48.9	48.5	49.2	0.246667	2
14					Total =	2.5	24
15					MS =	0.104167	
16							
17	A	B	C	D			
18	50.63	49.50	51.63	48.83			
19	48.83	51.17	51.83	47.47			
20	51.20	50.00	51.53	48.87			

Commands in Analyses

Column A,B,C,D,E	Enter values from Table 12.9 into rows 1 through 13	
Cell G2	Enter 2	Assay degrees of freedom for Tablet
Cells G3-G13	Copy G2 value	
Cell F2	= G2*VARA(C2:E2)	Assay SS for Tablet
Cells F3:F13	Copy formula from F2	
Cell F14	= Sum(F2:F13)	Pooled within-tablet assay SS
Cells G14	Copy formula from F14	Pooled degrees of freedom for assay
Cell F15	= F14/G14	MS = SS/df
Cell A18	= Average (C2:E2)	Tablet 1, Batch A average
Cells A19:A20)	Copy formula from A18	Tablets 2 & 3 averages, Batch A
Cell B18	= Average(C5:E5)	Tablet 1, Batch B average
Cells B19:B20)	Copy formula from B18	

Cell C18	= Average(C8:E8)	Tablet 1, Batch C average
Cells C19:C20)	Copy formula from C18	
Cell D18	= Average(C11:E11)	Tablet 1, Batch D average
Cells D19:D20)	Copy formula from D18	

Workbook 12.9   Determination of Variance Components in a Nested Design (continuation of worksheet)

	A	B	C	D	E	F	G
32	ANOVA						
33	*Source of Variation*	*SS*	*df*	*MS*	*F*	*P-value*	*F crit*
34	Between Groups	16.22917	3	5.409722	7.410578	0.01071	4.06618
35	Within Groups	5.84	8	0.73			
36							
37	Total	22.06917	11				
38						**Correct**	**Correct**
39	S^2_w	0.104167				**SS**	**MS**
40	S^2_t	0.695278			**Between**	48.6875	16.22917
41	S^2_b	1.559907			**Within**	17.52	2.190

Commands in Analyses

Main Menu	Tools → Data Analysis → Anova: Single Factor	
Dialog Box		
Input Range:	Highlight or enter A17:D20	
Labels:	Click on this option	
Output Range:	Highlight or enter A32	
OK	Click to perform calculations	
Cell F40	= 3*B34	SS individual = 3 * SS of means
Cell F41	Copy formula from F40	
Cell G40	= F40/C34	MS Between Batches
Cell G41	Copy formula from G40	MS Between Tablets (within batch)
Cell B39	= F15	Between-Assay (within tablet) Variance
Cell B40	= (1/3)*(G41-B39)	Between-Tablet (within batch) Variance
Cell B41	= (1/9)*(G40-G41)	Between-Batch Variance

In the next example, the Day 1 calibration curve results (Peak Area vs. Concentration) from Table 13.8 are used to demonstrate how to obtain the weighted linear regression analysis shown in Table 13.10.

Workbook 13.10 Weighted Linear Regression Analysis

	A	B	C	D	E	F	G	H
1	**X**	**Y**	**wt**	**wt*X**	**wt*X*X**	**wt*Y**	**wt*X*Y**	**wt(Y-Ym)**2**
2	0.05	0.003	400	20	1	1.2	0.06	0.000936
3	0.05	0.004	400	20	1	1.6	0.08	0.000112
4	0.20	0.016	25	5	1	0.4	0.08	0.003289
5	0.20	0.018	25	5	1	0.45	0.09	0.004536
6	1.00	0.088	1	1	1	0.088	0.088	0.006967
7	1.00	0.094	1	1	1	0.094	0.094	0.008005
8	10.00	0.920	0.01	0.1	1	0.0092	0.092	0.008381
9	10.00	0.901	0.01	0.1	1	0.00901	0.0901	0.008037
10	20.00	1.859	0.0025	0.05	1	0.0046475	0.09295	0.008598
11	20.00	1.827	0.0025	0.05	1	0.0045675	0.09135	0.008303
12								
13		**Sum**	852.025	52.3	10	3.859425	0.8584	0.057164
14					**Ym =**	0.0045297		
15	**Slope**	0.09154						
16	**Intercept**	−0.00109						
17								

Commands in Analyses

Columns A and B	Enter Day 1 values from Table 13.8 (X = Conc, Y = Area)	
Cell C2	= 1/(A2^2)	Weight is 1/(X*X)
Cells C3-C11	Copy formula from C2	
Cell D2	= C2*A2	Weight*X = 1/X
Cells D3:D11	Copy formula from D2	
Cell E2	= D2*A2	Weight*X*X = 1
Cells E3:E11	Copy formula from E2	
Cell F2	= C2*B2	Weight*Y = Y/X
Cells F3:F11	Copy formula from F2	
Cell G2	= D2*B2	Weight*X*Y
Cells G3:G11	Copy formula from G2	
Cell C13	= SUM(C2:C11)	Σwt
Cell D13	= SUM(D2:D11)	Σ(wt*X)
Cell E13	= SUM(E2:E11)	Σ(wt*X^2)
Cell F13	= SUM(F2:F11)	Σ(wt*Y)
Cell G13	= SUM(G2:G11)	Σ(wt*X*Y)
Cell F14	= (SUM(F2:F12))/C13	Weighted mean for Y (Ym)
Cell H2	= C2*(B2-$F$14)^2	wt*(Y-Ym)^2
Cells H3:H11	Copy formula from H2	
Cell H13	= SUM(H2:H11)	Σ(wt*(Y-Ym)^2)
Cell B15	= (G13-((D13*F13)/C13))/(E13-((D13*D13)/C13))	
Cell B16	= (F13-(B15*D13))/C13	

(continuation of worksheet)

	A	B	C	D	E	F	G
18	X	Y	Yp	wt(Y-Yp)**2	Yav	wt(Yav-Yp)**2	wt(Y-Yav)**2
19	0.05	0.003	0.003	0.000095	0.003500	0.0000001	0.000100
20	0.05	0.004	0.003	0.000105	0.003500	0.0000001	0.000100
21	0.20	0.016	0.017	0.000037	0.017000	0.0000012	0.000025
22	0.20	0.018	0.017	0.000015	0.017000	0.0000012	0.000025
23	1.00	0.088	0.090	0.000006	0.091000	0.0000003	0.000009
24	1.00	0.094	0.090	0.000013	0.091000	0.0000003	0.000009
25	10.00	0.920	0.914	0.000000	0.910500	0.0000001	0.000001
26	10.00	0.901	0.914	0.000002	0.910500	0.0000001	0.000001
27	20.00	1.859	1.830	0.000002	1.843000	0.0000004	0.000001
28	20.00	1.827	1.830	0.000000	1.843000	0.0000004	0.000001
29							
30			SUM	0.0002754		0.0000043	0.0002711
31							

Commands in Analyses

Columns A and B	Copy values from rows 2–11.	
Cell C19	$= \$B\$16 + \$A19*\$B\$15$	Predicted Y value (Yp)
Cell C20-C28	Copy formula from C19	
Cell D19	$= (1/(A19*A19))*(B19-C19)\hat{\ }2$	$wt^*(Y-Yp)^2$
Cells D20:D28	Copy formula from D19	
Cells E19 and E20	$= (B\$19 + B\$20)/2$	Average Y value:	X = 0.05 (Yav)
Cells E21 and E22	$= (B\$21 + B\$22)/2$		X = 0.20
Cells E23 and E24	$= (B\$23 + B\$24)/2$		X = 1.00
Cells E25 and E26	$= (B\$25 + B\$26)/2$		X = 10.0
Cells E27 and E28	$= (B\$27 + B\$28)/2$		X = 20.0
Cell F19	$= (1/(A19*A19))*(E19-C19)\hat{\ }2$	$wt^*(Yav-Yp)^2$
Cells F20-F28	Copy Formula from F19	
Cell G19	$= (1/(A19*A19))*(B19-E19)\hat{\ }2$	$wt^*(Y-Yav)^2$
Cells G20-G28	Copy Formula from G19	
Cell D30	$= SUM(D19:D28)$	$\$SMwt^*(Y-Yp)^2$
Cell F30	$= SUM(F19:F28)$	$\$SMwt^*(Yav-Yp)^2$
Cell G30	$= SUM(G19:G28)$	$\$SMwt^*(Y-Yav)^2$

Workbook 13.10 Creation of ANOVA Table 13.10

	A	B	C	D	E	F	G
31							
32		**Source**	**df**	**SS**	**MS**	**F**	
33		**Slope**	1	0.056889	0.0568891	1652.7	
34		**Error**	8	0.000275	0.0000344		
35		**Dev Reg**	3	0.000004	0.0000014	0.03	
36		**Within**	5	0.000271	0.0000542		
37		**Total**	9	0.057164			
38							

Commands in Analyses

Cell C37	= 10–1	Number of (x,y) pairs – 1
Cell C33	Enter 1	Slope has a single degree of freedom
Cell C34	= C37-C33	Total df – Slope df
Cell C36	Enter 5	5 concentrations that have duplicate values
Cell C35	= C34-C36	Error df – Within df
Cell D37	= H13	$\Sigma(wt^*(Y\text{-}Ym)^2)$
Cell D34	= D30	$\Sigma(wt^*(Y\text{-}Yp)^2)$
Cell D33	= D37-D34	Total SS – Error SS
Cell D35	= F30	$\Sigma(wt^*(Yav\text{-}Yp)^2)$
Cell D36	= G30	$\Sigma(wt^*(Y\text{-}Yav)^2)$
Cell E33	= D33/C33	SS/df
Cells E34-E36	Copy formula from E33	
Cell F33	= E33/E34	
Cell F35	= E35/E36	

The next set of programs are from Chapter 15, Nonparametric Methods. These programs use only the basic mathematical and sorting functions of Excel.

The first of these examples uses the paired time to peak concentration results from a comparative bioavailability study in 12 subjects. The analysis of the data, shown in Table 15.3, is based on the differences between the results for two oral formulations of a drug, A and B. The program implements the Wilcoxon Signed Rank Test shown in Table 15.4.

Commands in Analyses

Columns A, B and C	Enter values from Table 15.3.	
Cell D2	= C2-B2	Calculates B-A difference
Cells D3-D13	Copy D2	
Cell E2	= ABS(D2)	Absolute value of difference
Cells E3-E13	Copy E2	
Cell F2	= E2/D2	+ 1 if difference >0; – 1 if difference <0
Cells F3-F12	Copy F2	

Workbook 15.4   Wilcoxon Signed Rank Test Analysis of Table 15.3 Data

	A	B	C	D	E	F
1	**Subject**	**A**	**B**	**B-A**	**Abs(B-A)**	**Sign(B-A)**
2	1	2.5	3.5	1	1	1
3	2	3	4	1	1	1
4	3	1.25	2.5	1.25	1.25	1
5	4	1.75	2	0.25	0.25	1
6	5	3.5	3.5	0	0	#DIV/0!
7	6	2.5	4	1.5	1.5	1
8	7	1.75	1.5	−0.25	0.25	−1
9	8	2.25	2.5	0.25	0.25	1
10	9	3.5	3	−0.5	0.5	−1
11	10	2.5	3	0.5	0.5	1
12	11	2	3.5	1.5	1.5	1
13	12	3.5	4	0.5	0.5	1

(worksheet contined)

	G	H	I	J	K	L	M
1	**Index**	**SortVal**	**SortSign**	**Rank**	**SignRank**	**Positive**	**Negative**
2	1	0.25	1	2	2	2	
3	2	0.25	−1	2	−2		2
4	3	0.25	1	2	2	2	
5	4	0.5	−1	5	−5		5
6	5	0.5	1	5	5	5	
7	6	0.5	1	5	5	5	
8	7	1	1	7.5	7.5	7.5	
9	8	1	1	7.5	7.5	7.5	
10	9	1.25	1	9	9	9	
11	10	1.5	1	10.5	10.5	10.5	
12	11	1.5	1	10.5	10.5	10.5	
13		**N=**	11		**Sum**	59	7
14					**Z=**	2.312	
15					**p-value**	0.021	

Commands in Analyses

Column G	Enter the count of the non-zero differences.
Cells H2-H12	Copy nonzero values from Cells E2-E13 using Paste Special, Values option.
Cells 12-I12	Copy corresponding values from cells F2-F13 (Paste Special, Values).
Cells H2-I12	Highlight this Range of cells and under Data choose to sort this selection by column H.
Cell J2-J12	Enter number in column G unless the number in column H is tied with another in column H. Use the average G number for the ties. For example, Cells J2, J3 and J4 get the number 2 because their H value, 0.25, is a three-way tie for index numbers 1, 2 and 3.
Cell K2	$= I2*J2$ Signed Rank
Cells K3-K12	Copy formula from K2
Cell L2	$= IF(K2 > 0, J2, " ")$ Enters rank if sign is positive
Cells L3-L12	Copy formula from L2
Cell M2	$= IF(K2 < 0, J2, " ")$ Enters rank if sign is negative
Cells M3-M12	Copy formula from M2
Cell I13	$= COUNT(I2:I12)$ Determines N, the number of signed ranks
Cell L13	$= SUM(L2:L12)$ Sum of ranks with positive signs
Cell M13	$= SUM(M2:M12)$ Sum of ranks with negative signs
Cell L14	$= ABS(L13-I13*(I13 + 1)/4)/SQRT(I13*(I13 + 0.5)*(I13 + 1)/12)$
Cell L15	$= 2*(1-NORMSDIST(L14))$

Using the Peak Concentration (Cmax) results from a two-way, crossover Bioequivalence study, a method for calculating a nonparametric confidence interval on the mean treatment ratio is shown in the following example.

Commands in Analyses

Columns A, B, and C	Enter values from Table 15.6 into rows 2–13.
Cell D2	$= C2/B2$ Calculates B/A Ratio
Cells D3-D13	Copy formula from D2
Cell D16	$= 1/12$ Power for Geometric Mean
Cell D15	$= Product(D2:D13)$ Product of Ratios
Cell D17	$= Power(D15,D16)$ Product to 1/12th power is Geom. Mean Ratio
Cells E1-L1	Enter Column Labels.
Cells J2, J3	Enter 95% and 90%. Level of Confidence Interval for row
Column E	Start in row 2 (Subject) and enter number 1 twelve times, 2 eleven times, 3 ten times, 4 nine times, etc., until 12 is entered into row 79. These numbers represent the first Subject for each pair.
Column F	Starting in row 2, enter Subject numbers 1–12, next numbers 2–12, next 3–12, next 4–12, etc., until 12 is entered into row 79. These represent the second Subject for each pair.

Workbook 15.6 Nonparametric Confidence Interval for C_{max}

	A	B	C	D
1	**Subject**	**A**	**B**	**B/A**
2	1	135	102	0.755556
3	2	179	147	0.821229
4	3	101	385	3.811881
5	4	109	106	0.972477
6	5	138	189	1.369565
7	6	135	105	0.777778
8	7	158	130	0.822785
9	8	156	125	0.801282
10	9	174	144	0.827586
11	10	147	133	0.904762
12	11	145	114	0.786207
13	12	147	167	1.136054
14				
15			**Product =**	1.080296
16			**1/12 =**	0.083333
17		**Geometric**	**Mean =**	1.006457

Cell G2	= POWER($D$2*D2,0.5)	Geometric mean of Subject 1 ratio paired with itself
Cells G3-G13	Copy G2 formula	Geometric mean ratio of Subject 1 with all others
Cell G14	= POWER($D$3*D3,0.5)	Geometric mean of Subject 2 ratio paired with itself
Cells G15-G24	Copy G14 formula	Geometric mean of Subject 2 with Subjects 3–12
Cell G25	= POWER($D$4*D4,0.5)	Geometric mean of Subject3 ratio paired with itself
Cells G26-G34	Copy G25 formula	Geometric mean of Subject 3 with Subjects 4–12

Cells G35-G79   Continue as above for remaining paired subject ratios.

Cells H2-H3   Enter index numbers 1 & 2 The number for the geometric mean (after sorting)

Cells H4-H79   Highlight Cells H2-H3 and drag copy to obtain index numbers 3–78

Column I   Highlight Cells G2-G79

Choose Copy under Edit on Main Menu toolbar.

Place cursor in Cell I2 and then choose Paste Special under Edit on Main Menu

(worksheet continued)

	E	F	G	H	I	J	K	L
1	1st Subj	2nd Subj	Geomean	Index	Sorted	Confidence	Low	High
2	1	1	0.755556	1	0.755556	95%	0.800	1.247
3	1	2	0.787708	2	0.766586	90%	0.804	1.065
4	1	3	1.697082	3	0.770729			
5	1	4	0.857182	4	0.777778			
6	1	5	1.017243	5	0.778083			
7	1	6	0.766586	6	0.781981			
8	1	7	0.788454	7	0.786207			
9	1	8	0.778083	8	0.787708			
10	1	9	0.790751	9	0.788454			
11	1	10	0.8268	10	0.789442			
12	1	11	0.770729	11	0.790751			
13	1	12	0.926473	12	0.793709			
14	2	2	0.821229	13	0.799208			
15	2	3	1.769301	14	0.799965			
	Rows 16–74 not shown							
75	10	11	0.843404	74	1.857107			
76	10	12	1.013834	75	1.925349			
77	11	11	0.786207	76	2.080986			
78	11	12	0.945079	77	2.284868			
79	12	12	1.136054	78	3.811881			

In Paste Special dialog box, choose to paste Values and then click OK
Next Highlight all entries in Column I
Choose Sort under Data on Main Menu.
Choose to stay with the current selection when prompted about
    expanding.
Choose Sort, Ascending for the column labeled "Sorted."
Click OK.

Use Table 15.5 to obtain the ranking numbers for the upper and lower confidence interval limits

Cell K2	= I15	Lower 95% CI limit is 14th ranked geometric mean ratio
Cell L2	= I66	Upper 95% CI limit is 65th ranked geometric mean ratio
Cell K3	= I19	Lower 90% CI limit is 18th ranked geometric mean ratio
Cell L3	= I62	Upper 90% CI limit is 61 st ranked geometric mean ratio

Workbook 15.8 Wilcoxon Rank Sum Test for Differences Between Two Independent Groups

	A	B	C	D	E	F	G	H
1	**Apparatus**	**Dissolved**	**Index**	**App**	**Sorted**	**Rank**	**O Rank**	**M Rank**
2	O	53	1	O	50	1	1	
3	O	61	2	O	52	2	2	
4	O	57	3	O	53	3	3	
5	O	50	4	O	54	4	4	
6	O	63	5	M	55	5.5		5.5
7	O	62	6	M	55	5.5		5.5
8	O	54	7	M	56	7		7
9	O	52	8	O	57	9	9	
10	O	59	9	O	57	9	9	
11	O	57	10	M	57	9		9
12	O	64	11	M	58	11		11
13	M	58	12	O	59	12.5	12.5	
14	M	55	13	M	59	12.5		12.5
15	M	67	14	O	61	14	14	
16	M	62	15	O	62	15.5	15.5	
17	M	55	16	M	62	15.5		15.5
18	M	64	17	O	63	17	17	
19	M	66	18	O	64	18.5	18.5	
20	M	59	19	M	64	18.5		18.5
21	M	68	20	M	66	20		20
22	M	57	21	M	67	21		21
23	M	69	22	M	68	22		22
24	M	56	23	M	69	23		23
25						**N =**	11	12
26						**Sum =**	105.5	170.5
27						**Z =**	1.631	
28						**p-value =**	0.103	

The next example demonstrates how to perform the Wilcoxon Rank Sum Test for comparing the differences between two independent groups. In this example, Excel is used to perform the necessary calculations on the tablet dissolution results given in Table 15.8. The results from a modified dissolution apparatus are compared with those obtained from the original apparatus to see if they are statistically different from each other.

Commands in Analysis

Columns A & B	Enter apparatus and dissolution results from Table 15.8.
Column C	Enter the index numbers 1 through 23.
Column D & E	Copy values from Column A & B.
	Highlight D2 through E24.
	From Main Menu Toolbar, choose Data and then Sort.
	Sort by column "Sorter", in ascending order, indicating
	there is a Header Row.
Cell F2	= C2 Rank for E2, a unique number in col.
Cell Fx	Copy F2 formula for each row, x, for each Ex that is unique.
Cells F6 & F7	= AVERAGE(C6:C7) Rank for tied E values (2).
Cell Fx & Fy	Copy F6 formula to consecutive Ex & Ey ties of size 2.
Cells F9,F10,F11	= AVERAGE(C9:C11) Rank for tied E values (3).

Commands in Analyses (continued

Cell G2	= IF(D2 = "O", F2, "")	Enters rank for original apparatus O.
Cells G3:G24	Copy G2 formula	
Cell H2	= IF(D2 = "M", F2, "")	Enters rank for modified apparatus.
Cells H3:H24	Copy H2 formula	
Cell G25	= COUNT(G2:G24)	# of original apparatus values.
Cell H25	= COUNT(H2:H24)	# of modified apparatus values.
Cell G26	= SUM(G2:G24)	Original apparatus Rank Sum.
Cell H26	= SUM(H2:H24)	Modified apparatus Rank Sum.
Cell G27	= (ABS(G26-(G25*(G25 + H25 + 1))/2))/(SQRT(G25*H25* (G25 + H25 + 1)/12))	
Cell G28	= 2*(1-NORMSDIST(G27))	2-sided p-value for G27 Z-val

Next we analyze the time-to-sleep values (Table 15.10) from one group of rats given a low dose (L) of an experimental drug, a second group a high dose (H), and a third a dose of a control, sedative (C).

Commands in Analyses

Columns A & B	Enter compound id & time-to-sleep values from Table 15.10	
Column C	Enter the index numbers 1 through 29	
Cells D2-D30	Copy values from A2-A30	
Cells E2-E30	Copy values from B2-B30.	
	Highlight D1 through E30.	
	From Main Menu Toolbar, choose Data and then Sort.	
	Sort by column "Sorter", in ascending order, indicating there is a	
	Header Row.	
Cells F2-F30	= Cn n = 1–30;	If En is a unique value (e. g. F13 = C13)
	= AVERAGE(Cx:Cy),	for the Ex to Ey equal values (ties) e.g. F2-F7 = AVERAGE(C2: C7).
Cells G2-G30	In first cell for a group of tied ranks in F, put # of tied values.	
Cell E32	= COUNT(E2:E30) number of values.	
Cell Hn	= Gn*(Gn*Gn-1)/($E$32*($E$32*$E$32–1)) for each n, where there is	
	an entry in cell Gn.	
	This is the correction factor for the group Gn of ties.	

Workbook 15.10 Kruskal Wallis Test (One-Way Anova) for Differences Between Independent Groups (>2)

	A	B	C	D	E	F	G	H	I	J	K
1	**ID**	**Time**	**Indx**	**Compnd**	**Sorted**	**Rank**	**Tie Size**	**Tie Corr**	**Control**	**Low**	**High**
2	C	8	1	C	1	3.5	6	0.009	3.5		
3	C	1	2	C	1	3.5			3.5		
4	C	9	3	L	1	3.5				3.5	
5	C	9	4	H	1	3.5					3.5
6	C	6	5	H	1	3.5					3.5
7	C	3	6	H	1	3.5					3.5
8	C	15	7	H	2	7.5	2	0.000			7.5
9	C	1	8	H	2	7.5					7.5
10	C	7	9	C	3	10	3	0.001	10		
11	L	10	10	H	3	10					10
12	L	5	11	H	3	10					10
13	L	8	12	H	4	12					12
14	L	6	13	L	5	13				13	
15	L	7	14	C	6	15	3	0.001	15		
16	L	7	15	L	6	15				15	
17	L	15	16	H	6	15					15
18	L	1	17	C	7	18.5	4	0.002	18.5		
19	L	15	18	L	7	18.5				18.5	
20	L	7	19	L	7	18.5				18.5	
21	H	3	20	L	7	18.5				18.5	
22	H	4	21	C	8	22	3	0.001	22		
23	H	8	22	L	8	22				22	
24	H	1	23	H	8	22					22
25	H	1	24	C	9	24.5	2	0.000	24.5		
26	H	3	25	C	9	24.5			24.5		
27	H	1	26	L	10	26				26	
28	H	6	27	C	15	28	3	0.001	28		
29	H	2	28	L	15	28				28	
30	H	2	29	L	15	28				28	
31											
32				**Count**	29		**Sum**	0.016	149.5	191.0	94.5
33							**Correctn**	0.984			
34								**n**	9	10	10
35								**R*R/n**	2483.4	3648.1	893.0
36											
37								**Chi-Sq**	6.89		
38								**p-value**	0.032		
39								**Chi-Sq(c)**	7.00		
40								**p-value**	0.030		

Cells I2–I30	= IF(Dn = "C",Fn," ")	n = 1 to 30; Rank for Control rows.
Cells J2-J30	= IF(Dn = "L",Fn," ")	n = 1 to 30; Rank for Low Dose rows.
Cell K2-K30	= IF(Dn = "H",Fn," ")	n = 1 to 30; Rank for High Dose rows.
Cell H32	= SUM(H2:H30)	Sum of correction factor for ties.
Cell H33	= 1-H32	Correction for ties.
Cell I32	= SUM(I2:I30)	Rank Sum for Control.
Cell J32	= SUM(J2:J30)	Rank Sum for Low Dose.
Cell K32	= SUM(K2:K30)	Rank Sum for High Dose.
Cell I34	= COUNT(I2:I30)	Number of Control Values.
Cell J34	= COUNT(J2:J30)	Number of Low Dose Values.
Cell K34	= COUNT(K2:K30)	Number of High Dose Values.
Cell I35	= I32*I32/I34	(Control Rank Sum Squared)/n.
Cell J35	= J32*J32/J34	(Low Dose Rank Sum Squared)/n.
Cell K35	= K32*K32/K34	(High Dose Rank Sum Squared)/n.
Cell I37	= (12/(E32*(E32 + 1))*(SUM(I35:K35))-3*(E32 + 1))	Chi-Square Statistic
Cell I38	= 2*(1-NORMSDIST(I37))	P-value for I37 Chi-Square Statistic corrected for ties.
Cell I39	= I37/H33	
Cell I40	= 2*(1-NORMSDIST(I39))	P-value for I39 statistic

In the next Workbook, the tablet hardness results in Table 15.11 from five tablet formulations (1–5) produced on four different tablet presses (A-D) are examined by nonparametric, two-way ANOVA to validate that all presses have statistically equivalent performance.

Commands in Analyses

Columns A & B	Enter tablet press and hardness values from Table 15.11 in order shown
Cell B23	Enter 5, the number of tablet formulations
Cell B24	Enter 4, the number of tablet presses
Column C	Enter 5 groups of the index numbers 1–4 (one for each tablet formulation)
Column D	Enter the tablet formulation number for each value in column C
Cell E2	= 10*B2*D2 value is proportional to hardness
Cells E3-E21	Copy formula from Cell E2
Column F	Copy Column A values
Column G	Copy Column E, using the Paste Special, values, option under Edit Highlight Columns F and G, rows 1 through 21.
	From Main Menu Toolbar, choose Data and then Sort.
	Sort by column "SortMod", in ascending order, indicating there is a Header Row.
Cells H2-H21	= IF(Fn = "A",Cn," ")	n = 1 to 21; Enters ranks for Press A values.

Workbook 15.11 Friedman and Modified Friedman Tests (Two-Way Anova)

	A	B	C	D	E	F	G	H	I	J	K
1	Press	Value	Index	Tab	ModVal	Press	SortMod	A Rank	B Rank	C Rank	D Rank
2	A	7.5	**1**	1	75	B	69		1		
3	B	6.9	**2**	1	69	D	70				2
4	C	7.3	**3**	1	73	C	73			3	
5	D	7.0	**4**	1	70	A	75	4			
6	A	8.2	**1**	2	164	D	158				1
7	B	8.0	**2**	2	160	B	160		2		
8	C	8.5	**3**	2	170	A	164	3			
9	D	7.9	**4**	2	158	C	170			4	
10	A	7.3	**1**	3	219	A	219	1			
11	B	7.9	**2**	3	237	D	228				2
12	C	8.0	**3**	3	240	B	237		3		
13	D	7.6	**4**	3	228	C	240			4	
14	A	6.6	**1**	4	264	D	256				1
15	B	6.5	**2**	4	260	B	260		2		
16	C	7.1	**3**	4	284	A	264	3			
17	D	6.4	**4**	4	256	C	284			4	
18	A	7.5	**1**	5	375	D	335				1
19	B	6.8	**2**	5	340	B	340		2		
20	C	7.6	**3**	5	380	A	375	3			
21	D	6.7	**4**	5	335	C	380			4	
22											
23	r =	5					Sum =	14	10	19	7
24	c =	4					SumR*R	706			
25							Chi-Sqr	9.72			
26							p-value	0.0211			
27							A2	150			
28							B2	141.2		CritDiff	5.90
29							T2	7.364			
30							p-value	0.0047			

Cells I2-I21	$= IF(Fn = \text{"B"},Cn,\text{" "})$	$n = 1$ to 21; Enters ranks for Press B values.
Cells J2-J21	$= IF(Fn = \text{"C"},Fn,\text{" "})$	$n = 1$ to 21; Enters ranks for Press C values.
Cells K2-K21	$= IF(Fn = \text{"D"},Fn,\text{" "})$	$n = 1$ to 21; Enters ranks for Press D values.
Cell H23	$= SUM(H2:H21)$	Rank Sum for Press A.
Cell I23	$= SUM(I2:I21)$	Rank Sum for Press B.
Cell J23	$= SUM(J2:J21)$	Rank Sum for Press C.
Cell K23	$= SUM(K2:K21)$	Rank Sum for Press D.
Cell H24	$= SUMSQ(H23:K23)$	Sum of Squared Rank Sums
Cell H25	$= ((12{*}H24)/(B23{*}B24{*}(B24 + 1)))\text{-}3{*}B23{*}(B24 + 1)$	Friedman X^2
Cell H26	$= CHIDIST(H25,B24\text{–}1)$	p-value for Friedman's test
Cell H27	$= SUMSQ(H2:K21)$	A2 = Sum of squares for the 29 individual ranks
Cell H28	$= H24/B23$	B2 = Average Squared Rank Sum
Cell H29	$= ((B23 - 1){*}(H28\text{-}(B23{*}B24{*}(B24 + 1){*}(B24 + 1))/4)/(H27\text{-}H28)$	Modified X^2
Cell H30	$= FDIST(H29, B24\text{–}1,(B23\text{–}1){*}(B24\text{–}1))$	p-value for modified Friedman test
Cell K28	$= TINV(0.05,(B23\text{–}1){*}(B24\text{–}1)){*}SQRT((2{*}B23{*} (H27\text{-}H28))/ ((B23\text{–}1){*}(B24\text{–}1)))$	

Minimum difference between any two Rank Sums that is significant ($p < 0.05$)

The tablet harness values are used again to demonstrate how to perform the Quade Test for randomized block designs as shown in Table 15.12.

Workbook 15.12 Quade Test on Table 15.11 Tablet Hardness Values

Commands in Analyses

Columns A, B, & C	Enter press, formulation and hardness values from Table 15.11 in rows 2–21	
Cell A24	Enter 4, the number of tablet presses (columns)	
Cell A27	Enter 5, the number of tablet formulations (rows)	
Cell D2	$= MAX(B2:B5)\text{-}MIN(B2:B5)$	Determines range of tablet 1 hardness
Cells D6, D10, D14, D18	$= MAX(Bx:By)\text{-}MIN(Bx:By)$	for D6 x, y $= 6, 9$ for D10 x, y $= 10, 13$ for D14 x, y $= 14, 17$ for D18 x, y $= 18, 21$
Cells B23-B27	Enter tablet formulation numbers 1–5	
Cells C23-C27	Copy ranges for each formulation from cells D2, D6, D10, D14 & D18	
Cells D23-D27	Rank the ranges using the average rank for ties (e. g., tied ranks 3 and 4 $= 3.5$)	
Cells E2-E21	Enter 5 groups of index numbers 1–4 (one group per tablet formulation)	
Cell F2	$= 10{*}B2{*}D2$ modifies hardness value to obtain correct sorting within press	

Workbook 15.12 Quade Test on Table 15.11 Tablet Hardness Values

	A	B	C	D	E	F	G	H	I	J	K	L	M
1	Press	Value	Tab	Q	Index	Mod	Press	q	SortMod	A	B	C	D
2	A	7.5	1	0.6	1	75	B	1.5	69		−2.25		
3	B	6.9	1		2	69	D	1.5	70				−0.75
4	C	7.3	1		3	73	C	1.5	73			0.75	
5	D	7.0	1		4	70	A	1.5	75	2.25			
6	A	8.2	2	0.6	1	164	D	1.5	158				−2.25
7	B	8.0	2		2	160	B	1.5	160		−0.75		
8	C	8.5	2		3	170	A	1.5	164	0.75			
9	D	7.9	2		4	158	C	1.5	170			2.25	
10	A	7.3	3	0.7	1	219	A	3.5	219	−5.25			
11	B	7.9	3		2	237	D	3.5	228				−1.75
12	C	8.0	3		3	240	B	3.5	237		1.75		
13	D	7.6	3		4	228	C	3.5	240			5.25	
14	A	6.6	4	0.7	1	264	D	3.5	256				−5.25
15	B	6.5	4		2	260	B	3.5	260		−1.75		
16	C	7.1	4		3	284	A	3.5	264	1.75			
17	D	6.4	4		4	256	C	3.5	284			5.25	
18	A	7.5	5	0.9	1	375	D	5	335				−7.50
19	B	6.8	5		2	340	B	5	340		−2.50		
20	C	7.6	5		3	380	A	5	375	2.50			
21	D	6.7	5		4	335	C	5	380			7.50	
22		TAB	Rng	Rnk									
23	k=c	1	0.6	1.5					Sum =	2.00	−5.50	21.00	−17.50
24	4	2	0.6	1.5									
25		3	0.7	3.5					A	270		CritDiff	
26	r	4	0.7	3.5					B	156.3		21.2	
27	5	5	0.9	5					T	5.499			
28									p-value	0.0131			
29													

Cells F3-F21	Copy F2 formula
Cells G2-G21	Copy cells A2-A21 press values
Cells H2-H21	Copy the D23-D27 tablet ranks for formulations in column C
Cells I2–I21	Copy F2-F21 values using the Paste Special option under Edit.

Highlight rows 2–21 of Columns G, H and I.
From Main Menu Toolbar, choose Data and then Sort.
Sort G2:I21 selection by column ``SortMod'', in ascending order.

Cells J2-J21	$= IF(\$Gn = "A",\$Hn*(\$En-(\$A\$24 + 1)/2),"")$	n = 2 to 21; Sij for Press A.
Cells K2-K21	$= IF(\$Gn = "B",\$Hn*(\$En-(\$A\$24 + 1)/2)," ")$	n = 2 to 21; Sij for Press B.
Cells L2-L21	$= IF(\$Gn = "C",\$Hn*(\$En-(\$A\$24 + 1)/2)," ")$	n = 2 to 21; Sij for Press C.
Cells M2-M21	$= IF(\$Gn = "D",\$Hn*(\$En-(\$A\$24 + 1)/2),"")$	n = 2 to 21; Sij for Press D.
Cell J23	$= SUM(J2:J21)$	Rank Sum for Press A.
Cell K23	$= SUM(K2:K21)$	Rank Sum for Press B.
Cell L23	$= SUM(L2:L21)$	Rank Sum for Press C.
Cell M23	$= SUM(M2:M21)$	Rank Sum for Press D.
Cell J25	$= SUMSQ(J2:M21)$	$A = \Sigma Sij^2$
Cell J26	$= (SUMSQ(J23:M23))/A27$	$B = \Sigma(\Sigma Sij)^2/r)$
Cell J27	$= ((A27–1)*J26)/(J25-J26)$	Quade test statistic $T = (r-1)B/(A-B)$
Cell J28	$= FDIST(J27,A24–1,(A27–1)*(A24–1))$	p-value from $F_{3,12}$ distribution
Cell I26	$= TINV(0.05,(A27–1)*(A24–1))*SQRT((2*A27*(J25-J26))/((A27–1)*(A24–1)))$	

Difference between any two Rank Sums which is significant at p = 0.05.

In the next Workbook, a product made from four lots of raw material each with a different potency (X) is assayed for its potency (Y) after being manufactured using two different methods (I and II). The results, shown in Table 15.13, are used to demonstrate the Quade Nonparametic Covariance Analysis.

Commands in Analysis

Columns A, C & D	Enter method, Assay (Y) and Material (X) values into rows 2–9	
Column B	Enter observation numbers 1–8 into rows 2–9	
Cells A12-A19	Enter Index numbers 1–8 which will be used as a guide when ranking values	
Cells B12-B19	Copy B2-B9 Observation numbers	
Cells C12-C19	Copy D2-D9 X values	
Cells F12-F19	Copy B2-B9 Observation numbers	
Cells G12-G19	Copy C2-C9 Y values	
Cell B21	$= (A19 + 1)/2$	Mean rank, 4.5, for 8 observations
Cells B12-C19	Highlight this section and sort in ascending order (indicate a header row)	
Cells F12-G19	Highlight this section and sort in ascending order (indicate a header row)	
Cells D12-D19	Rank sorted cells C12-C19, using the average for tied ranks	
Cells H12-H19	Rank sorted cells G12-G19 using the average for tied ranks	
Cell E12	$= D12-\$B\$21$	Center X rank by subtracting the mean rank
Cells E13-E19	Copy formula from Cell E12	Center remaining X ranks
Cells I12-I19	Copy formula from Cell E12	Center Y ranks by subtracting the mean rank
Cells E2:E9	Enter centered Y rank, matching sorted Obs number with Obs number in Col B	
Cells F2:F9	Enter centered X rank, matching sorted Obs number with Obs number in Col B	

Workbook 15.13   Quade Nonparametric Covariance Analysis (ANCOVA)
(main worksheet ply)

	A	B	C	D	E	F	G	H	I
1	Method	Obs	Y	X	Adj Ry	Adj Rx			
2	I	1	98.0	98.4	2.5	−3			
3	I	2	97.8	98.6	1.5	−1			
4	I	3	98.5	98.6	3.5	−1			
5	I	4	97.4	99.2	−0.5	2.5			
6	II	5	97.6	98.7	0.5	0.5			
7	II	6	95.4	99.0	−3.5	1.5			
8	II	7	96.1	99.3	−2	3.5			
9	II	8	96.1	98.4	−2	−3			
10									
11	Index	Sort Obs	Sort X	Rank X	Adj Rx	Sort Obs	Sort Y	Rank Y	Adj Ry
12	1	1	98.4	1.5	−3	6	95.4	1	−3.5
13	2	8	98.4	1.5	−3	7	96.1	2.5	−2
14	3	2	98.6	3.5	−1	8	96.1	2.5	−2
15	4	3	98.6	3.5	−1	4	97.4	4	−0.5
16	5	5	98.7	5	0.5	5	97.6	5	0.5
17	6	6	99.0	6	1.5	2	97.8	6	1.5
18	7	4	99.2	7	2.5	1	98.0	7	2.5
19	8	7	99.3	8	3.5	3	98.5	8	3.5
23									
20		(N+1)/2							
21		4.5							

Commands in Analysis (continued)
Main Menu   Tools → Data Analysis → Regression
Dialog Box
   Input Y Range: Highlight or enter E1:E9
   Input X Range: Highlight or enter F1:F9
   Labels: Click on this option
   New Worksheet Ply: Enter "Regression"
   Residuals   Click on this option
   OK   Click to perform calculations

(regression worksheet ply)

	A	B	C
20			
21			
22	RESIDUAL OUTPUT		
23			
24	*Observation*	*Predicted Rank Y*	*Residuals*
25	1	1.445122	1.054878
26	2	0.481707	1.018293
27	3	0.481707	3.018293
28	4	−1.20427	0.704268
29	5	−0.24085	0.740854
30	6	−0.72256	−2.77744
31	7	−1.68598	−0.31402
32	8	1.445122	−3.44512
33			

Commands in Analyses (continued)
Main Worksheet Ply:

| Cells G2:G9 | Copy Predicted values from Cells B25-B32 of Regression Worksheet Ply |
| Cells H2:H9 | Copy Residual values from Cells C25-C32 of Regression Worksheet Ply |

(main worksheet ply)

	G	H	I	J
1	**Predicted**	**Residual**	**Method I**	**Method II**
2	1.4451	1.0549	1.0549	0.7409
3	0.4817	1.0183	1.0183	−2.7774
4	0.4817	3.0183	3.0183	−0.3140
5	−1.2043	0.7043	0.7043	−3.4451
6	−0.2409	0.7409		
7	−0.7226	−2.7774		
8	−1.6860	−0.3140		
9	1.4451	−3.4451		
10				

Commands in Analyses (continued)

Cells I2–I5	Copy Residual values for Method I from Cells H2-H5
Cells J2-J5	Copy Residual values for Method II from Cells H6-H9
Main Menu	Tools → Data Analysis → Anova: Single Factor
Dialog Box	
Input Range:	Highlight or enter $I$1:$J$5
Labels:	Click on this option
New Worksheet Ply:	Enter word "ANOVA"
OK	Click to perform calculations

(ANOVA worksheet ply)

	A	B	C	D	E	F	G
1	Anova: Single Factor						
2							
3	SUMMARY						
4	*Groups*	*Count*	*Sum*	*Average*	*Variance*		
5	Method I	4	5.795732	1.448933	1.119382		
6	Method II	4	−5.79573	−1.44893	3.944294		
7							
8							
9	ANOVA						
10	*Source of Variation*	*SS*	*df*	*MS*	*F*	*P-value*	*F crit*
11	Between Groups	16.79525301	1	16.79525	6.633621	0.042018	5.987374
12	Within Groups	15.19102748	6	2.531838			
13							
14	Total	31.98628049	7				
15							

Note: The ANOVA Worksheet contains the results of the Analysis of Covariance.

The next Workbook shows how to perform an evaluation for comparability of baseline disease severity (mild, moderate, or very severe) for patients randomized to one of two treatment groups (A or B) in a clinical trial. The data are taken from Table 15.16 and the analysis follows that shown in Table 15.17.

Workbook 15.16 Chi-Square Evaluation of a 2×3 Contingency Table

Commands in Analysis

Cells C4-E5	Enter the patient counts from Table 15.16
Cell A8	Enter the number of rows in Table
Cell A11	Enter the number of columns in Table
Cell C6	= SUM(C4:C5)
Cells D6-E6	Copy formula from Cell C6
Cell F4	= SUM(C4:E4)
Cells F5-F6	Copy formula from Cell F4
Cell C12	= (C$6*$F4)/$F$6
Cells C13 & D12-E13	Copy formula from Cell C12
Cells C14-E14	Copy formula from Cells C6-E6
Cells F12-F14	Copy formula from Cells F4-F6
Cell C20	= (C4-C12)*(C4-C12)/C12
Cells C21 & D20-E21	Copy formula from Cell C19
Cell D22	= SUM(C20:E21)
Cell D23	= CHIDIST(D22,(A8–1)*(A10–1))

(patients categorized by disease severity and treatment)

	A	B	C	D	E	F
1				**Observed**		
2						
3		**Severity:**	Very	Moderate	Mild	Total
4	**Treatment:**	A	13	24	18	55
5		B	19	20	12	51
6		Total	32	44	30	106
7	**Rows:**					
8	2					
9	**Cols:**			**Expected**		
10	3					
11		**Severity:**	Very	Moderate	Mild	Total
12	**Treatment:**	A	16.60	22.83	15.57	55
13		B	15.40	21.17	14.43	51
14		Total	32	44	30	106
15						
16						
17				$(0\text{-}E)^2/E$		
18						
19		**Severity:**	Very	Moderate	Mild	
20	**Treatment:**	A	0.782	0.060	0.381	
21		B	0.844	0.065	0.410	
22			$X^2 =$	2.541		
23			$P =$	0.281		

The final Excel Workbook uses the results shown in Table 15.21 on the Incidence of Carcinoma in Drug- and Placebo-Treated Animals to demonstrate the method of calculating exact confidence intervals for a 2×2 contingency table.

Workbook 15.21 Fisher's Exact Test for Carcinoma Results in Drug- and Placebo-Treated Animals

	A	B	C	D	E	F	G
1						A values	p-values
2			Carcinomas			0	0.03043
3			Present	Absent		1	
4		Placebo	0	12	12	2	
5		Drug	5	9	14	3	
6			5	21	26	4	
7			p-value	0.03043		5	0.01204
8							
9			Carcinomas			Fisher's	p-value
10			Present	Absent			**0.04247**
11		Placebo	5	7	12		
12		Drug	0	14	14		
13			5	21	26		
14			p-value	0.01204			
15							

Commands in Analyses

Cells C6,D6,E4,E5	Enter marginal totals from Table 15.21	$(A + B), (C + D), (A + C), (B + D)$
Cell E6	$= \text{SUM}(E4\text{:}E5)$	$N = A + B + C + D$
Cell C4	Enter Placebo-Present count	A
Cell C5	$= C6\text{-}C4$	$B = (A + B)\text{-}A$
Cell D4	$= E4\text{-}C4$	$C = (A + C)\text{-}A$
Cell D5	$= E6\text{-}D4\text{-}C5\text{-}C4$	$D = \text{Total-B-C-A}$
Cell D7	$= (\text{FACT}(C6)^*\text{FACT}(D6)^*\text{FACT}(E4)^*\text{FACT}(E5))/$ $(\text{FACT}(E6)^*\text{FACT}(C4)^*\text{FACT}(C5)^*\text{FACT}(D4)^* \text{FACT}(D5))$ Note: The function FACT(x) returns the factorial of the number x or the number in that cell if x is a cell reference (e. g. $x = C6$).	
Column F	Enter all possible values for A (Placebo-Present count) This is obtained by going from a count of 0 and increasing to a count of $A + B$ (cell C5) or $A + C$ (cell D4), whichever is smaller.	
Cells B9-E14	Highlight and Copy Cells B2:E7 creates a working table	

Set the value for A (Cell C11) to 0 in the working table. If the *p*-value in Cell D14 $\leq$ T Cell D7 then copy that value (use Paste Special, value) to column G beside the appropriate A value in column F. Continue through all the possible values for A shown in column F.

Cell G10	$= \text{SUM}(G2\text{:}G8)$	*p*-value for Fisher's Exact Test

SAS Programs

The following programs written for the SAS System perform the same analyses as those presented in the Excel Workbooks section of this appendix. As such, no commentary is provided for these programs other than that needed to interpret the results of the SAS output. It is assumed that the reader has a basic understanding of the SAS System and knows how to operate SAS in his/her computer environment. The SAS programs utilize only the basic mathematical and statistical functions and standard procedures available in SAS/Base and SAS/STAT. The programs have been kept as simple as possible in hopes that the reader will easily be able to follow each program's logic. All data are contained within the program itself (Cards Statement). The reader should be able to easily modify the program code to input data from an external file.

```
/*    SAS Program to Analyze Table 1.1 */

Options ls=90 ps=60 nodate pagno=1;
Title1 'Serum Cholesterol Changes (mg %) After Administration of a
Cholesterol Lowering Drug';

Data A;
input change;      /* read in serum cholesterol changes */
cards;             /* (only a portion of the 156 data points are shown) */
17
-12
25
-37
-29
-39
-22
0
-22
*
*
*
14
17
-13
-22
-3
-17
1
;

Proc Univariate plot; /* Procedure to obtain descriptive analyses */
var change;           /* identifies variable to be analyzed */

run;
quit;
```

SAS OUTPUT FOR ANALYSES OF TABLE 1.1:

Serum Cholesterol Changes (mg %) After Administration of a Cholesterol Lowering Drug   1

Univariate Procedure

Variable=CHANGE

```
              Moments                              Quantiles(Def=5)

N                 156  Sum Wgts        156      100% Max       55      99%      46
Mean         -10.6218  Sum           -1657       75% Q3        10      95%      34
Std Dev      27.68191  Variance    766.2883      50% Med     -9.5      90%      23
Skewness     -0.28357  Kurtosis    -0.16183      25% Q1     -28.5      10%     -49
USS            136375  CSS         118774.7        0% Min     -97       5%     -60
CV           -260.614  Std Mean    2.216327                            1%     -71
T:Mean=0     -4.79252  Pr>|T|        0.0001      Range          152
Num ^= 0          152  Num > 0           58      Q3-Q1         38.5
M(Sign)           -18  Pr>=|M|       0.0044      Mode           -27
Sgn Rank        -2296  Pr>=|S|       0.0001
```

                               Extremes

```
                  Lowest    Obs     Highest   Obs
                   -97(    107)       38(     109)
                   -71(    142)       39(     105)
                   -69(    127)       40(      60)
                   -66(     48)       46(      91)
                   -64(     13)       55(     125)
```

```
          Stem Leaf                            #       Boxplot
            5 5                                 1          |
            4 06                                2          |
            3 3445589                           7          |
            2 0011344567                       10          |
            1 00112234444666777799            20       +-----+
            0 0000111222455556668999          22       |     |
           -0 9988876665433321                16       |     |
           -1 99988777655433222111100         23       *--+--*
           -2 98777765543222210               17       +-----+
           -3 999877444221110                 15          |
           -4 99977441                         8          |
           -5 8844300                          7          |
           -6 964320                           6          |
           -7 1                                1          |
           -8
           -9 7                                1          0
              ----+----+----+----+---
          Multiply Stem.Leaf by 10**+1
```

```
/* Sas Program to Analyze Table 1.4  */

Options ls=90 ps=60 nodate pagno=1;
Title1 'Tablet Potencies';
Data A;
input @@ wi potency;
cards;      /* data entered as frequency & value */
1 90   0 91   2 92   1 93   5 94   1 95   2 96   7 97   10 98   8 99   13 100   17 101   13 102
9 103   0 104   0 105   5 106   4 107   0 108 0 109 2 110
;
Data B;                    /* sets up file of 27 desired percentiles */
cnt=1; p=1; output;
do p=2 to 16 by 2;
 cnt=cnt+1;  output;
end;
do p=20 to 95 by 5;
 cnt=cnt+1; output;
end;
do p=97 to 100 by 3;
 cnt=cnt+1; output;
end;

Proc sort;
by cnt;

Proc univariate plot pctldef=1 data=a;
freq wi;                              /* desired percentiles */
output out=c pctlpre=P pctlpts=1 2 4 6 8 10 12 14 16 20 25 30 35 40
                               45 50 55 60 65 70 75 80 85 90 95 97 100;
Data d;
set c;                  /* Captures Potencies at desired percentiles */
keep cnt value;
array v[27] P1 -- P100;
do cnt=1 to 27;
 value=v[cnt]+0.5; /* calculates upper end of class intervals for the */
 output;               /* potency values at desired percentiles */
end;                   /* file contains 27 records of class intervals */

Proc sort;
by cnt;

Data e;           /* Merges the 27 desired percentile values with their */
merge b d;        /* corresponding upper limit of the class intervals */
by cnt;
keep value p;
label p = Cumulative Percent;
label value = Potency;

Proc print label data=e;
id value;
var p;

Proc Plot nolegend data=e;
Plot p*value='*'/ vaxis = by 10;

run;
quit;
```

SAS OUTPUT FOR ANALYSES OF TABLE 1.4:

Tablet Potencies   1

                              Univariate Procedure

Variable=POTENCY

	Moments				Quantiles(Def=1)		
N	100	Sum Wgts	100	100% Max	110	99%	110
Mean	100.23	Sum	10023	75% Q3	102	95%	107
Std Dev	3.686873	Variance	13.59303	50% Med	100	90%	106
Skewness	-0.01867	Kurtosis	0.719475	25% Q1	98	10%	95
USS	1005951	CSS	1345.71	0% Min	90	5%	94
CV	3.678412	Std Mean	0.368687			1%	90
T:Mean=0	271.8564	Pr> |T|	0.0001	Range	20		
Num ^= 0	100	Num > 0	100	Q3-Q1	4		
M(Sign)	50	Pr>=|M|	0.0001	Mode	101		
Sgn Rank	2525	Pr>=|S|	0.0001				

                                  Extremes

Lowest	Obs	Highest	Obs
90(	1)	102(	13)
92(	3)	103(	14)
93(	4)	106(	17)
94(	5)	107(	18)
95(	6)	110(	21)

```
Stem Leaf                                 #          Boxplot
 110 00                                   2             0
 109
 108
 107 0000                                 4             |
 106 00000                                5             |
 105                                                    |
 104                                                    |
 103 000000000                            9             |
 102 0000000000000                        13         +-----+
 101 00000000000000000                    17         |     |
 100 0000000000000                        13         *--+--*
  99 00000000                             8          |     |
  98 0000000000                           10         +-----+
  97 0000000                              7             |
  96 00                                   2             |
  95 0                                    1             |
  94 00000                                5             |
  93 0                                    1             |
  92 00                                   2             |
  91
  90 0                                    1             0
    ----+----+----+----+
```

Tablet Potencies

Univariate Procedure

Variable=POTENCY

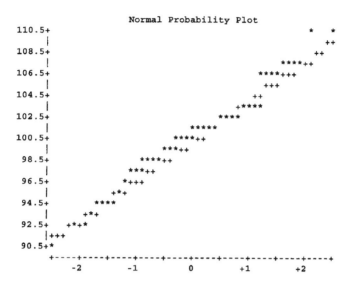

Normal Probability Plot

Tablet Potencies

Potency	Cumulative Percent
90.5	1
92.5	2
93.5	4
94.5	6
94.5	8
95.5	10
96.5	12
97.5	14
97.5	16
98.5	20
98.5	25
99.5	30
99.5	35
100.5	40
100.5	45
100.5	50
101.5	55
101.5	60
101.5	65
102.5	70
102.5	75
102.5	80
103.5	85
106.5	90
107.5	95
107.5	97
110.5	100

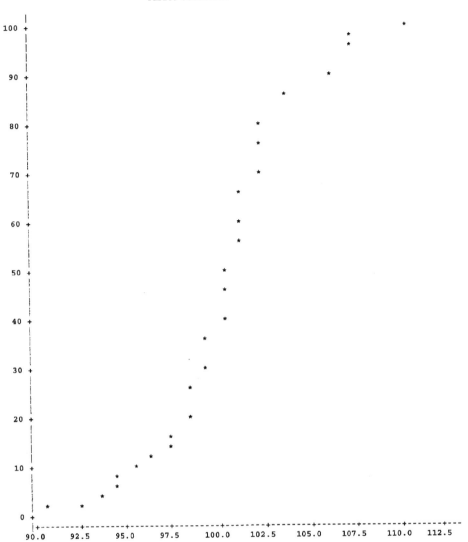

Tablet Potencies

```
/* Sas Program to Analyze Table 5.1  */

Options ls=90 ps=60 nodate pagno=1;
Title1 'Tablet Potencies';
Title3 'Calculation of 95% Confidence Interval on the Mean';

Data A;
input @@ potency;
cards;      /* data */
101.8 102.6 99.8 104.9 103.8 104.5 100.7 106.3 100.6 105.0
;

Proc Means noprint;      /* Calculates mean, N and standard deviation */
Var potency;
Output out=mn N=N Mean=Mean Std=Std;            /* makes results available in
data set mn */

Data A;
set mn;                            /* uses results from Proc Means */
df=N-1;                            /* degrees of freedom */
alpha = 0.05;                      /* interval will have 1-alpha = 95%
coverage */
tval = tinv (1-alpha/2,df);
Ci_low = mean - tval*std/sqrt(N);
Ci_hi = mean + tval*std/sqrt(N);
label tval=critical t-value;
label std=standard deviation;
label Ci_low= Lower Confidence Limit;
label Ci_hi= Upper Confidence Limit;


Proc print label data = a;
Id mean;
var std N df alpha tval std Ci_low Ci_hi;
format tval std F4.2 Ci_low Ci_hi F7.2;

run;
quit;
```

SAS OUTPUT FOR ANALYSES OF TABLE 5.1:

Tablet Potencies   1

Calculation of 95% Confidence Interval on the Mean

MEAN	standard deviation	N	DF	ALPHA	critical t-value	standard deviation	Lower Confidence Limit	Upper Confidence Limit
103	2.22	10	9	0.05	2.26	2.22	101.41	104.59

```
/* Sas Program to Analyze Table 5.9  */

Options ls=90 ps=60 nodate pagno=1;

Title1 'Tablet Percent Dissolution After 15 Minutes';
Title3 'Two-Sided, Independent Group t-Test';

Data A;
input Form_A Form_B;

cards;       /* data */
68        74
84        71
81        79
85        63
75        80
69        61
80        69
76        72
79        80
74        65
;

Proc Print;
id Form_A;
Var Form_B;

Data B;        /* Formats values for t-test */
set A;
keep form percent;
Form='A'; percent=Form_A; output;
Form='B'; percent=Form_B; output;

Proc TTest;
Class Form;
Var percent;

run;
quit;
```

SAS OUTPUT FOR ANALYSES OF TABLE 5.9:

Tablet Percent Dissolution After 15 Minutes   1

Two-Sided, Independent Group t-Test

FORM_A	FORM_B
68	74
84	71
81	79
85	63
75	80
69	61
80	69
76	72
79	80
74	65

Tablet Percent Dissolution After 15 Minutes   2

Two-Sided, Independent Group t-Test

TTEST PROCEDURE

Variable: PERCENT

FORM	N	Mean	Std Dev	Std Error	Minimum	Maximum
A	10	77.10000000	5.78215646	1.82847842	68.00000000	85.00000000
B	10	71.40000000	6.97933458	2.20705938	61.00000000	80.00000000

Variances	T	DF	Prob>|T|
Unequal	1.9888	17.4	0.0627
Equal	1.9888	18.0	0.0621

For H0: Variances are equal, F' = 1.46   DF = (9,9)   Prob>F' = 0.5840

```
/* Sas Program to Analyze Table 5.11  */

Options ls=90 ps=60 nodate pagno=1;

Title1 'Analysis of Variance for Area Under the Curve from a Bioavailability
Study';

Data A;
input Animal Form_A Form_B;
ratio=Form_A/Form_B;

cards;      /* data */
1       136        166
2       168        184
3       160        193
4       94         105
5       200        198
6       174        197
;

Proc Print;
id Animal ;
Var Form_A Form_B Ratio;

Data B;        /* Formats values for test */
set A;
keep animal form type area;

Type='Diffrnce'; Form='A';          area=Form_A;    output;
Type='Diffrnce'; Form='B';          area=Form_B;    output;
Type='Ratio';    Form='A/B';        area=ratio;     output;
Type='Ratio';    Form='Expected';   area=1;         output;

Proc Sort;
By Type

Proc Anova Data=B;
Class Animal Form;
Model Area=Animal Form;
Means Form;
By Type;

run;
quit;
```

SAS OUTPUT FOR ANALYSES OF TABLE 5.11:

Analysis of Variance for Area Under the Curve from a Bioavailability Study   1

ANIMAL	FORM_A	FORM_B	RATIO
1	136	166	0.81928
2	168	184	0.91304
3	160	193	0.82902
4	94	105	0.89524
5	200	198	1.01010
6	174	197	0.88325

Analysis of Variance for Area Under the Curve from a Bioavailability Study   2

TYPE=Diffrnce

Analysis of Variance Procedure
Class Level Information

Class	Levels	Values
ANIMAL	6	1 2 3 4 5 6
FORM	2	A B

Number of observations in data set = 12

Analysis of Variance for Area Under the Curve from a Bioavailability Study   3

TYPE=Diffrnce

Analysis of Variance Procedure

Dependent Variable: AREA

Source	DF	Sum of Squares	Mean Square	F Value	Pr > F
Model	6	13656.16666667	2276.02777778	26.92	0.0012
Error	5	422.75000000	84.55000000		
Corrected Total	11	14078.91666667			

R-Square	C.V.	Root MSE	AREA Mean
0.969973	5.586901	9.19510739	164.58333333

Source	DF	Anova SS	Mean Square	F Value	Pr > F
ANIMAL	5	12629.41666667	2525.88333333	29.87	0.0010
FORM	1	1026.75000000	1026.75000000	12.14	0.0176

Analysis of Variance for Area Under the Curve from a Bioavailability Study   4

```
                           TYPE=Diffrnce

                  Analysis of Variance Procedure

            Level of      ------------AREA------------
            FORM      N      Mean              SD

            A         6    155.333333     36.5002283
            B         6    173.833333     35.7514568
```

```
   Analysis of Variance for Area Under the Curve from a Bioavailability Study      5

                              TYPE=Ratio

                   Analysis of Variance Procedure
                     Class Level Information

              Class     Levels     Values

              ANIMAL       6      1 2 3 4 5 6

              FORM         2      A E

           Number of observations in by group = 12
```

```
   Analysis of Variance for Area Under the Curve from a Bioavailability Study      6

                              TYPE=Ratio

                   Analysis of Variance Procedure
```

Dependent Variable: AREA

Source	DF	Sum of Squares	Mean Square	F Value	Pr > F
Model	6	0.04708292	0.00784715	3.31	0.1052
Error	5	0.01186634	0.00237327		
Corrected Total	11	0.05894926			

	R-Square	C.V.	Root MSE	AREA Mean
	0.798702	5.150647	0.04871621	0.94582700

Source	DF	Anova SS	Mean Square	F Value	Pr > F
ANIMAL	5	0.01186634	0.00237327	1.00	0.5000
FORM	1	0.03521657	0.03521657	14.84	0.0120

```
   Analysis of Variance for Area Under the Curve from a Bioavailability Study      7
```

                              TYPE=Ratio

                    Analysis of Variance Procedure

```
        Level of       -------------AREA------------
        FORM      N       Mean              SD

        A         6     0.89165399       0.06889512
        E         6     1.00000000       0.00000000
```

```
/* Sas Program to Analyze Table 7.5  */

Options ls=90 ps=60 nodate pagno=1;
Title1 'Tablet Stability Based on Assay';

Data A;
input @@ Month Assay;
cards;     /* data */
0 51  0 51  0 53   3 51  3 50   3 52   6 50    6 52   6 48
9 49  9 51  9 51  12 49  12 48  12 47  18 47  18 45  18 49
;

Proc Sort Data=A;
By Month;

Proc Print;
id Month;
var Assay;

Proc Means N Mean Var noprint; /* Calculates mean of Months in assay data */
var month;
output out=mndat N=Nx Mean=MNx Var=Varx ;

Proc Reg outest=est Data=A;                   /*Regression Analysis*/
Model Assay = Month/clm;

Data B;        /* Months 1-60 with mean for months in assay data */
set mndat;
Keep Month N MNx SSQx;
SSQx=17*Varx;     /*sum of squares of deviations from mean for months */
N=Nx;
do Month=1 to 60;
output;
end;

Proc Sort Data=B;
```

```
By Month;

Data C;                        /* Months 1-60 with slope & intercept values */
set est;
slope=Month;
do i=1 to 60;
Month=i;
output;
end;

Proc sort;
by Month;

Data D;
merge B C;
by Month;
T=TINV(0.975,N-2);             /* t-value for 95% confidence inteval */
Predict=Intercep + Slope*Month;
CI95_Lo=predict-t*_RMSE_*sqrt((1/N) + ((month-MNx)**2/SSQx));
CI95_Hi=predict+t*_RMSE_*sqrt((1/N) + ((month-MNx)**2/SSQx));

Proc Print;
id Month;
var Predict CI95_Lo CI95_Hi;

Data E;
set D A;
Proc Plot data=E;
Plot Assay*Month='*' Predict*Month='.' CI95_Hi*Month='-'
   CI95_Lo*Month='-' / overlay vaxis=30 32 34 36 38 40 42 44 46 48 50 52 54;

run;
quit;
```

SAS OUTPUT FOR ANALYSES OF TABLE 7.5 (Partial Printout):

First page of output is a listing of the Month and Assay values in Table 7.5

Tablet Stability Based on Assay   2

Model: MODEL1
Dependent Variable: ASSAY

Analysis of Variance

Source	DF	Sum of Squares	Mean Square	F Value	Prob>F
Model	1	44.80000	44.80000	24.548	0.0001
Error	16	29.20000	1.82500		
C Total	17	74.00000			

Root MSE	1.35093	R-square	0.6054	
Dep Mean	49.66667	Adj R-sq	0.5807	
C.V.	2.71998			

Parameter Estimates

Variable	DF	Parameter Estimate	Standard Error	T for H0: Parameter=0	Prob > |T|
INTERCEP	1	51.800000	0.53552378	96.728	0.0001
MONTH	1	-0.266667	0.05382216	-4.955	0.0001

Obs	Dep Var ASSAY	Predict Value	Std Err Predict	Lower95% Mean	Upper95% Mean	Residual
1	51.0000	51.8000	0.536	50.6647	52.9353	-0.8000
2	51.0000	51.8000	0.536	50.6647	52.9353	-0.8000
3	53.0000	51.8000	0.536	50.6647	52.9353	1.2000
4	51.0000	51.0000	0.417	50.1162	51.8838	2.83E-15
5	50.0000	51.0000	0.417	50.1162	51.8838	-1.0000
6	52.0000	51.0000	0.417	50.1162	51.8838	1.0000
7	50.0000	50.2000	0.336	49.4875	50.9125	-0.2000
8	52.0000	50.2000	0.336	49.4875	50.9125	1.8000
9	48.0000	50.2000	0.336	49.4875	50.9125	-2.2000
10	49.0000	49.4000	0.323	48.7154	50.0846	-0.4000
11	51.0000	49.4000	0.323	48.7154	50.0846	1.6000
12	51.0000	49.4000	0.323	48.7154	50.0846	1.6000
13	49.0000	48.6000	0.384	47.7852	49.4148	0.4000
14	48.0000	48.6000	0.384	47.7852	49.4148	-0.6000
15	47.0000	48.6000	0.384	47.7852	49.4148	-1.6000
16	47.0000	47.0000	0.625	45.6743	48.3257	2.78E-15
17	45.0000	47.0000	0.625	45.6743	48.3257	-2.0000
18	49.0000	47.0000	0.625	45.6743	48.3257	2.0000

Sum of Residuals   0
Sum of Squared Residuals   29.2000
Predicted Resid SS (Press)   38.0436

Tablet Stability Based on Assay   3

MONTH	PREDICT	CI95_LO	CI95_HI
1	51.5333	50.4876	52.5791
2	51.2667	50.3053	52.2281
3	51.0000	50.1162	51.8838
4	50.7333	49.9185	51.5482
5	50.4667	49.7098	51.2235
6	50.2000	49.4875	50.9125
7	49.9333	49.2487	50.6179
8	49.6667	48.9917	50.3417
9	49.4000	48.7154	50.0846
10	49.1333	48.4208	49.8459
11	48.8667	48.1098	49.6235
12	48.6000	47.7852	49.4148
13	48.3333	47.4495	49.2171
14	48.0667	47.1053	49.0281
15	47.8000	46.7543	48.8457
16	47.5333	46.3981	48.6686
17	47.2667	46.0378	48.4955
18	47.0000	45.6743	48.3257
19	46.7333	45.3083	48.1584
20	46.4667	44.9401	47.9932
21	46.2000	44.5704	47.8296
22	45.9333	44.1992	47.6675
23	45.6667	43.8269	47.5064
24	45.4000	43.4536	47.3464
25	45.1333	43.0796	47.1871
26	44.8667	42.7048	47.0285
27	44.6000	42.3295	46.8705
28	44.3333	41.9536	46.7130
29	44.0667	41.5773	46.5560
30	43.8000	41.2007	46.3993
31	43.5333	40.8237	46.2430
32	43.2667	40.4463	46.0870
33	43.0000	40.0688	45.9312
34	42.7333	39.6910	45.7757
35	42.4667	39.3129	45.6204
36	42.2000	38.9347	45.4653
37	41.9333	38.5563	45.3103
38	41.6667	38.1778	45.1555
39	41.4000	37.7991	45.0009
40	41.1333	37.4203	44.8463
41	40.8667	37.0414	44.6919
42	40.6000	36.6624	44.5376
43	40.3333	36.2833	44.3834
44	40.0667	35.9040	44.2293
45	39.8000	35.5248	44.0752
46	39.5333	35.1454	43.9213
47	39.2667	34.7659	43.7674
48	39.0000	34.3864	43.6136
49	38.7333	34.0069	43.4598
50	38.4667	33.6272	43.3061
51	38.2000	33.2476	43.1524
52	37.9333	32.8678	42.9988
53	37.6667	32.4881	42.8453
.	.	.	.
.	.	.	.
.	.	.	.
60	35.8000	29.8286	41.7714

Last page of output is a low-resolution graph of Assay values, Predicted values
and the upper and lower confidence intervals vs. Month.

```
/* Sas Program to Analyze Table 7.6  */

Options ls=90 ps=60 nodate pagno=1;
Title1 'Weighted (1/X**2) Linear Regression of Spectrophotometric Data';

Data A;
input @@ ConcX ODy;
w=(1/ConcX)**2;

cards;      /* data */
5 0.105 5 0.098 10 0.201 10 0.194 25 0.495 25 0.508 50 0.983 50
1.009 100 1.964 100 2.013;

Proc Print;
id ConcX;
var ODy w;

Proc Reg Data=A;                    /*Regression Analysis*/
Model ODy = ConcX;
weight w;

run;
quit;
```

SAS OUTPUT FOR ANALYSES OF TABLE 7.6:

Weighted (1/X**2) Linear Regression of Spectrophotometric Data   1

CONCX	ODY	W
5	0.105	0.0400
5	0.098	0.0400
10	0.201	0.0100
10	0.194	0.0100
25	0.495	0.0016
25	0.508	0.0016
50	0.983	0.0004
50	1.009	0.0004
100	1.964	0.0001
100	2.013	0.0001

Weighted (1/X**2) Linear Regression of Spectrophotometric Data   2

Model: MODEL1
Dependent Variable: ODY

Analysis of Variance

Source	DF	Sum of Squares	Mean Square	F Value	Prob>F
Model	1	0.00187	0.00187	8188.538	0.0001
Error	8	1.8284045E-6	2.2855057E-7		
C Total	9	0.00187			

Root MSE	0.00048	R-square	0.9990
Dep Mean	0.14270	Adj R-sq	0.9989
C.V.	0.33502		

Parameter Estimates

Variable	DF	Parameter Estimate	Standard Error	T for H0: Parameter=0	Prob > |T|
INTERCEP	1	0.001656	0.00215007	0.770	0.4633
CONCX	1	0.019860	0.00021948	90.491	0.0001

```
/* Sas Program to Analyze Table 7.8  */

Options ls=90 ps=60 nodate pagno=1;
Title1 'Nonlinear Regresssion of Stability Data';

Data A;
Input Time Conc;
Cards;            /* Data */
1 63
2 34
3 22
;

Proc NLIN Data=A;
parameters C0=100 60 K=0.1 1;    /* Possible starting values for C0 and K */
model Conc=C0*exp(-K*Time);

Run;
Quit;
```

SAS OUTPUT OF ANALYSES OF TABLE 7.8:

```
            Nonlinear Regresssion of Stability Data                    1

   Non-Linear Least Squares Grid Search     Dependent Variable CONC
                    C0              K           Sum of Squares
               60.000000       0.100000          808.552311
              100.000000       0.100000         5759.703593
               60.000000       1.000000         2706.292436
              100.000000       1.000000         1395.672757


Non-Linear Least Squares Iterative Phase   Dependent Variable CONC   Method: Gauss-Newton
            Iter        C0              K           Sum of Squares
              0      60.000000       0.100000          808.552311
              1      87.927439       0.441778           49.229548
              2     106.998026       0.555363            7.246645
              3     109.207951       0.558231            5.717857
              4     109.221748       0.558272            5.717825
              5     109.221925       0.558273            5.717825
NOTE: Convergence criterion met.


       Non-Linear Least Squares Summary Statistics     Dependent Variable CONC

        Source              DF Sum of Squares      Mean Square

        Regression           2   5603.2821747      2801.6410874
        Residual             1      5.7178253         5.7178253
        Uncorrected Total    3   5609.0000000

        (Corrected Total)    2    888.6666667


        Parameter    Estimate    Asymptotic            Asymptotic 95 %
                                  Std. Error         Confidence Interval
                                                     Lower         Upper
        C0        109.2219253  8.5009601325  1.2087258208  217.23512471
        K           0.5582728  0.0515179098 -0.0963137380    1.21285939


                     Asymptotic Correlation Matrix

            Corr                 C0                K
            ----------------------------------------
            C0                    1        0.9118934069
            K           0.9118934069                  1
```

```
/* Sas Program to Analyze Table 8.9  */

Options ls=90 ps=60 nodate pagno=1;
Title1 'Two-Way ANOVA of Tablet Dissolution Results';

Data A;
Input Lab GenericA GenericB Standard;
Keep Lab Tablet Percent;              /* creates 3 records for each lab */

Tablet='Std'; Percent=Standard;
Output;                               /* record with Standard results */

Tablet='A'; Percent=GenericA;
Output;                               /* record with GenericA results */

Tablet='B'; Percent=GenericB;
Output;                               /* record with GenericB results */

Cards;                                /* Data */
1       89      83      94
2       93      75      78
3       87      75      89
4       80      76      85
5       80      77      84
6       87      73      84
7       82      80      75
8       68      77      75
;

Proc Print;            /* Data reformatted for ANOVA */
Id Lab;
Var Tablet Percent;

Proc ANOVA;
Class Lab Tablet;              /* Lab & Tablet are categorical variables */
Model Percent = Lab Tablet;
Means Tablet/ Tukey T;     /* Tukey & t-test comparisons of Tablet means*/

Run;
Quit;
```

SAS OUTPUT FOR ANALYSES OF TABLE 8.9:

Two-Way ANOVA of Tablet Dissolution Results   1

LAB	TABLET	PERCENT
1	GenA	89
1	GenB	83
1	Std	94
2	GenA	93
2	GenB	75
2	Std	78
3	GenA	87
3	GenB	75
3	Std	89
4	GenA	80
4	GenB	76
4	Std	85
5	GenA	80
5	GenB	77
5	Std	84
6	GenA	87
6	GenB	73
6	Std	84
7	GenA	82
7	GenB	80
7	Std	75
8	GenA	68
8	GenB	77
8	Std	75

Two-Way ANOVA of Tablet Dissolution Results   2

Analysis of Variance Procedure
Class Level Information

Class	Levels	Values
LAB	8	1 2 3 4 5 6 7 8
TABLET	3	GenA GenB Std

Number of observations in data set = 24

Two-Way ANOVA of Tablet Dissolution Results   3

Analysis of Variance Procedure

Dependent Variable: PERCENT

Source	DF	Sum of Squares	Mean Square	F Value	Pr > F
Model	9	592.16666667	65.79629630	2.27	0.0818
Error	14	405.66666667	28.97619048		
Corrected Total	23	997.83333333			

R-Square	C.V.	Root MSE	PERCENT Mean
0.593452	6.638792	5.38295369	81.08333333

Source	DF	Anova SS	Mean Square	F Value	Pr > F
LAB	7	391.83333333	55.97619048	1.93	0.1394
TABLET	2	200.33333333	100.16666667	3.46	0.0602

Two-Way ANOVA of Tablet Dissolution Results   4

Analysis of Variance Procedure

T tests (LSD) for variable: PERCENT

NOTE: This test controls the type I comparisonwise error rate not the
      experimentwise error rate.

Alpha= 0.05 df= 14 MSE= 28.97619
Critical Value of T= 2.14
Least Significant Difference= 5.7726

Means with the same letter are not significantly different.

T Grouping	Mean	N	TABLET
A	83.250	8	GenA
A			
A	83.000	8	Std
B	77.000	8	GenB

Two-Way ANOVA of Tablet Dissolution Results   5

Analysis of Variance Procedure

Tukey's Studentized Range (HSD) Test for variable: PERCENT

NOTE: This test controls the type I experimentwise error rate, but
      generally has a higher type II error rate than REGWQ.

Alpha= 0.05 df= 14 MSE= 28.97619
Critical Value of Studentized Range= 3.701
Minimum Significant Difference= 7.0444

Means with the same letter are not significantly different.

Tukey Grouping	Mean	N	TABLET
A	83.250	8	GenA
A			
A	83.000	8	Std
A			
A	77.000	8	GenB

```
/* Sas Program to Analyze Table 8.12  */

Options ls=85 ps=60 nodate pagno=1;
Title1 'Two-Way ANOVA of Replicated Dissolution Results';

Data A;
Input @@ Lab GenericA GenericB Standard;
Keep Lab Drug Percent;        /* creates 3 records for each record in data */
Drug='GenA';
Percent=GenericA;
Output;                       /* record with Generic A results */
Drug='GenB';
Percent=GenericB;
 Output;                      /* record with Generic B results */
Drug='Std ';
Percent=Standard;
Output;                       /* record with Standard results */

Cards;                        /* Data */
1       87      81      93 1     91      85      95
2       90      74      74 2     96      76      82
3       84      72      84 3     90      78      94
4       75      73      81 4     85      79      89
5       77      76      80 5     83      78      88
6       85      70      80 6     89      76      88
7       79      74      71 7     85      86      79
8       65      73      70 8     71      81      80
;

Proc Sort;
By lab Drug;

Proc Print;            /* Data reformatted for ANOVA */
Id Lab;
Var Drug Percent;

Proc ANOVA;
Class Lab Drug;
Model Percent = Lab Drug Lab*Drug;   /* Lab*Drug is interaction term */
Test H=Drug E=Lab*Drug;              /* Constructs F-test for Drug */

Run;
Quit;
```

SAS OUTPUT FOR ANALYSES OF TABLE 8.12:

(Page 1 Printout of Data not Shown)

Two-Way ANOVA of Replicated Dissolution Results   2

Analysis of Variance Procedure

Class Level Information

Class	Levels	Values
LAB	8	1 2 3 4 5 6 7 8
DRUG	3	GenA GenB Std

Number of observations in data set = 48

Two-Way ANOVA of Replicated Dissolution Results   3

Analysis of Variance Procedure

Dependent Variable: PERCENT

Source	DF	Sum of Squares	Mean Square	F Value	Pr > F
Model	23	1995.66666667	86.76811594	3.54	0.0016
Error	24	588.00000000	24.50000000		
Corrected Total	47	2583.66666667			

R-Square	C.V.	Root MSE	PERCENT Mean
0.772416	6.104519	4.94974747	81.08333333

Source	DF	Anova SS	Mean Square	F Value	Pr > F
LAB	7	783.66666667	111.95238095	4.57	0.0023
DRUG	2	400.66666667	200.33333333	8.18	0.0020
LAB*DRUG	14	811.33333333	57.95238095	2.37	0.0308

Tests of Hypotheses using the Anova MS for LAB*DRUG as an error term

Source	DF	Anova SS	Mean Square	F Value	Pr > F
DRUG	2	400.66666667	200.33333333	3.46	0.0602

```
/* Sas Program to Analyze Table 8.18  */


Options ls=85 ps=60 nodate pagno=1;
Title1 'Analysis of Covariance to Compare Two Methods';

Data A;
Input @@ Method Material Product;
Cards;             /* Data */
1 98.4 98.0   1 98.6 97.8   1 98.6 98.5   1 99.2 97.4
2 98.7 97.6   2 99.0 95.4   2 99.3 96.1   2 98.4 96.1
;

Proc Print;
ID Method;
Var Material Product;

Proc GLM data=A;
Title3 'Separate Lines';
Class Method;
Model Product = Material Method Material*Method;
Estimate 'Intercept I'  Intercept 1 Method 1 0 Material*Method 0 0;
Estimate 'Intercept II' Intercept 1 Method 0 1 Material*Method 0 0;
Estimate 'Slope I' Intercept 0 Method 0 0 Material 1 Material*Method 1 0;
Estimate 'Slope II' Intercept 0 Method 0 0 Material 1 Material*Method 0 1;

Proc GLM data=A;
Title3 'Parallel Lines';
Class Method;
Model Product = Material Method;
Estimate 'Intercept I'  Intercept 1 Method 1 0;
Estimate 'Intercept II' Intercept 1 Method 0 1;
Estimate 'Common Slope' Intercept 0 Method 0 0 Material 1;
Lsmeans Method/stderr pdiff;

Run;
Quit;
```

SAS OUTPUT FOR ANALYSES OF TABLE 8.18:

```
            Analysis of Covariance to Compare Two Methods               1
                METHOD    MATERIAL    PRODUCT

                  1         98.4        98.0
                  1         98.6        97.8
                  1         98.6        98.5
                  1         99.2        97.4
                  2         98.7        97.6
                  2         99.0        95.4
                  2         99.3        96.1
                  2         98.4        96.1
```

Separate Lines

General Linear Models Procedure
Class Level Information

Class   Levels   Values

METHOD   2   1 2

Number of observations in data set = 8

Analysis of Covariance to Compare Two Methods   3

Separate Lines

General Linear Models Procedure

Dependent Variable: PRODUCT

Source	DF	Sum of Squares	Mean Square	F Value	Pr > F
Model	3	5.82575000	1.94191667	2.92	0.1639
Error	4	2.66300000	0.66575000		
Corrected Total	7	8.48875000			

R-Square	C.V.	Root MSE	PRODUCT Mean
0.686291	0.840196	0.81593505	97.11250000

Source	DF	Type I SS	Mean Square	F Value	Pr > F
MATERIAL	1	1.54006579	1.54006579	2.31	0.2029
METHOD	1	4.27896199	4.27896199	6.43	0.0643
MATERIAL*METHOD	1	0.00672222	0.00672222	0.01	0.9248

Source	DF	Type III SS	Mean Square	F Value	Pr > F
MATERIAL	1	0.54450000	0.54450000	0.82	0.4169
METHOD	1	0.00788424	0.00788424	0.01	0.9186
MATERIAL*METHOD	1	0.00672222	0.00672222	0.01	0.9248

Parameter	Estimate	T for H0: Parameter=0	Pr > |T|	Std Error of Estimate
Intercept I	188.400000	1.40	0.2331	134.221935
Intercept II	168.790000	1.40	0.2330	120.234335
Slope I	-0.916667	-0.67	0.5372	1.359892
Slope II	-0.733333	-0.60	0.5791	1.216324

```
              Analysis of Covariance to Compare Two Methods            4

                            Parallel Lines

                    General Linear Models Procedure
                       Class Level Information

                    Class    Levels    Values

                    METHOD      2       1 2

          Number of observations in data set = 8
```

```
              Analysis of Covariance to Compare Two Methods            5

                            Parallel Lines

                    General Linear Models Procedure
```

Dependent Variable: PRODUCT

Source	DF	Sum of Squares	Mean Square	F Value	Pr > F
Model	2	5.81902778	2.90951389	5.45	0.0555
Error	5	2.66972222	0.53394444		
Corrected Total	7	8.48875000			

	R-Square	C.V.	Root MSE	PRODUCT Mean
	0.685499	0.752442	0.73071502	97.11250000

Source	DF	Type I SS	Mean Square	F Value	Pr > F
MATERIAL	1	1.54006579	1.54006579	2.88	0.1502
METHOD	1	4.27896199	4.27896199	8.01	0.0366

Source	DF	Type III SS	Mean Square	F Value	Pr > F
MATERIAL	1	0.53777778	0.53777778	1.01	0.3616
METHOD	1	4.27896199	4.27896199	8.01	0.0366

Parameter	Estimate	T for H0: Parameter=0	Pr > |T|	Std Error of Estimate
Intercept I	178.347222	2.23	0.0766	80.1359137
Intercept II	176.844444	2.20	0.0788	80.2576983
Common Slope	-0.814815	-1.00	0.3616	0.8119056

```
              Analysis of Covariance to Compare Two Methods            6

                            Parallel Lines

                    General Linear Models Procedure
                         Least Squares Means
```

METHOD	PRODUCT LSMEAN	Std Err LSMEAN	Pr > |T| H0:LSMEAN=0	Pr > |T| H0: LSMEAN1=LSMEAN2
1	97.8638889	0.3703972	0.0001	0.0366
2	96.3611111	0.3703972	0.0001	

```
/* Sas Program to Analyze Table 9.2  */

Options ls=90 ps=60 nodate pagno=1;
Title1 'Factorial Experiment on Tablet Thickness';
Title3 'Factors:  A = Stearate    B = Drug      C = Starch';

Data A;
input A B C RESPONSE;
Cards;
0.5 60   30 475
1.5 60   30 487
0.5 120 30 421
1.5 120 30 426
0.5 60   50 525
1.5 60   50 546
0.5 120 50 472
1.5 120 50 522 ;

Proc GLM;
Class A B C;
model RESPONSE = A B C A*B A*C B*C A*B*C;
Estimate 'A effect' A -1 1;
Estimate 'B effect' B -1 1;
Estimate 'C effect' C -1 1;
Estimate 'AB effect' A*B 1 -1 -1 1 /divisor=2;
Estimate 'AC effect' A*C 1 -1 -1 1 /divisor=2;
Estimate 'BC effect' B*C 1 -1 -1 1 /divisor=2;
Estimate 'ABC effect' A*B*C -1 1 1 -1 1 -1 -1 1 /divisor=4;

Proc GLM;                     /* Error = AB + BC + ABC */
Class A B C;
model RESPONSE = A B C A*C;
Estimate 'A effect' A -1 1;
Estimate 'B effect' B -1 1;
Estimate 'C effect' C -1 1;
Estimate 'AC effect' A*C 1 -1 -1 1 /divisor=2;
run;
quit;
```

SAS OUTPUT FOR ANALYSIS OF TABLE 9.2

Factorial Experiment on Tablet Thickness   1

Factors: A = Stearate   B = Drug   C = Starch

General Linear Models Procedure
Class Level Information

Class	Levels	Values
A	2	0.5 1.5
B	2	60 120
C	2	30 50

Number of observations in data set = 8

Factorial Experiment on Tablet Thickness 2

Factors: A = Stearate   B = Drug   C = Starch

General Linear Models Procedure

Dependent Variable: RESPONSE

Source	DF	Sum of Squares	Mean Square	F Value	Pr > F
Model	7	14535.50000000	2076.50000000	.	.
Error	0	.	.		
Corrected Total	7	14535.50000000			

R-Square	C.V.	Root MSE	RESPONSE Mean
1.000000	0	0	484.25000000

Source	DF	Type I SS	Mean Square	F Value	Pr > F
A	1	968.00000000	968.00000000	.	.
B	1	4608.00000000	4608.00000000	.	.
C	1	8192.00000000	8192.00000000	.	.
A*B	1	60.50000000	60.50000000	.	.
A*C	1	364.50000000	364.50000000	.	.
B*C	1	180.50000000	180.50000000	.	.
A*B*C	1	162.00000000	162.00000000	.	.

Source	DF	Type III SS	Mean Square	F Value	Pr > F
A	1	968.00000000	968.00000000	.	.
B	1	4608.00000000	4608.00000000	.	.
C	1	8192.00000000	8192.00000000	.	.
A*B	1	60.50000000	60.50000000	.	.
A*C	1	364.50000000	364.50000000	.	.
B*C	1	180.50000000	180.50000000	.	.
A*B*C	1	162.00000000	162.00000000	.	.

Parameter	Estimate	T for H0: Parameter=0	Pr > |T|	Std Error of Estimate
A effect	22.0000000	99999.99	0.0	0
B effect	-48.0000000	99999.99	0.0	0
C effect	64.0000000	99999.99	0.0	0
AB effect	5.5000000	99999.99	0.0	0
AC effect	13.5000000	99999.99	0.0	0
BC effect	9.5000000	99999.99	0.0	0
ABC effect	9.0000000	99999.99	0.0	0

Factors: A = Stearate   B = Drug   C = Starch

General Linear Models Procedure
Class Level Information

Class	Levels	Values
A	2	0.5 1.5
B	2	60 120
C	2	30 50

Number of observations in data set = 8

Factors: A = Stearate   B = Drug   C = Starch

General Linear Models Procedure

Dependent Variable: RESPONSE

Source	DF	Sum of Squares	Mean Square	F Value	Pr > F
Model	4	14132.50000000	3533.12500000	26.30	0.0113
Error	3	403.00000000	134.33333333		
Corrected Total	7	14535.50000000			

R-Square	C.V.	Root MSE	RESPONSE Mean
0.972275	2.393438	11.59022577	484.25000000

Source	DF	Type I SS	Mean Square	F Value	Pr > F
A	1	968.00000000	968.00000000	7.21	0.0748
B	1	4608.00000000	4608.00000000	34.30	0.0099
C	1	8192.00000000	8192.00000000	60.98	0.0044
A*C	1	364.50000000	364.50000000	2.71	0.1981

Source	DF	Type III SS	Mean Square	F Value	Pr > F
A	1	968.00000000	968.00000000	7.21	0.0748
B	1	4608.00000000	4608.00000000	34.30	0.0099
C	1	8192.00000000	8192.00000000	60.98	0.0044
A*C	1	364.50000000	364.50000000	2.71	0.1981

Parameter	Estimate	T for H0: Parameter=0	Pr > |T|	Std Error of Estimate
A effect	22.0000000	2.68	0.0748	8.19552724
B effect	-48.0000000	-5.86	0.0099	8.19552724
C effect	64.0000000	7.81	0.0044	8.19552724
AC effect	13.5000000	1.65	0.1981	8.19552724

```
/* Sas Program: Repeated Measures (Split-Plot) ANOVA - Table 11.22*/

   Options ls=95 ps=58 nodate pagno=1;
   Title1 'Split-Plot ANOVA of Table 11.22 Data';

   Data A;
   Input Drug $ Patient Baseline WK2 WK4 WK6 WK8;
   Array R[4] WK2 WK4 WK6 WK8;
   Do i=1 to 4;
   R[i]=R[i]-Baseline;
   End;
   Cards;
   Standard 1      102     106     97      86      93
   Standard 2      105     103     102     99      101
   Standard 5      99      95      96      88      88
   Standard 9      105     102     102     98      98
   Standard 13     108     108     101     91      102
   Standard 15     104     101     97      99      97
   Standard 17     106     103     100     97      101
   Standard 18     100     97      96      99      93
   New      3      98      96      97      82      91
   New      4      106     100     98      96      93
   New      6      102     99      95      93      93
   New      8      102     94      97      98      85
   New      10     98      93      84      87      83
   New      11     108     110     95      92      88
   New      12     103     96      99      88      86
   New      14     101     96      96      93      89
   New      16     107     107     96      93      97    ;

   Proc Print;
   Title3 'Change From Baseline';
   Id Drug Patient;
   Var WK2 WK4 WK6 WK8;

   Data B;              /* Format to : Drug Patient Week Change */
   Set A;
   T1=2; T2=4; T3=6; T4=8;
   Array W[4] T1-T4;
   Array C[4] WK2 WK4 WK6 WK8;
   Do I=1 to 4;
   Week=W[I];    Change=C[I];
   Output;
   End;

   Proc GLM Data=B;
   Class Week Drug Patient ;
   Model change = Week Drug Patient(Drug) Week*Drug/SS2 SS3;
   Test H=Drug E=Patient(Drug) ;

   Run;
   Quit;
```

SAS OUTPUT FOR ANALYSES OF TABLE 11.22 DATA:

Split-Plot ANOVA of Table 11.15 Data   1

Change From Baseline

DRUG	PATIENT	WK2	WK4	WK6	WK8
Standard	1	4	-5	-16	-9
Standard	2	-2	-3	-6	-4
Standard	5	-4	-3	-11	-11
Standard	9	-3	-3	-7	-7
Standard	13	0	-7	-17	-6
Standard	15	-3	-7	-5	-7
Standard	17	-3	-6	-9	-5
Standard	18	-3	-4	-1	-7
New	3	-2	-1	-16	-7
New	4	-6	-8	-10	-13
New	6	-3	-7	-9	-9
New	8	-8	-5	-4	-17
New	10	-5	-14	-11	-15
New	11	2	-13	-16	-20
New	12	-7	-4	-15	-17
New	14	-5	-5	-8	-12
New	16	0	-11	-14	-10

Split-Plot ANOVA of Table 11.22 Data   2

Change From Baseline

General Linear Models Procedure
Class Level Information

Class	Levels	Values
WEEK	4	2 4 6 8
DRUG	2	New Standard
PATIENT	17	1 2 3 4 5 6 8 9 10 11 12 13 14 15 16 17 18

Number of observations in data set = 68

Change From Baseline

General Linear Models Procedure

Dependent Variable: CHANGE

Source	DF	Sum of Squares	Mean Square	F Value	Pr > F
Model	22	1087.35457516	49.42520796	3.35	0.0003
Error	45	663.27777778	14.73950617		
Corrected Total	67	1750.63235294			

	R-Square	C.V.	Root MSE	CHANGE Mean
	0.621121	-51.69625	3.83920645	-7.42647059

Source	DF	Type II SS	Mean Square	F Value	Pr > F
WEEK	3	669.69117647	223.23039216	15.15	0.0001
DRUG	1	196.16013072	196.16013072	13.31	0.0007
PATIENT(DRUG)	15	171.72222222	11.44814815	0.78	0.6951
WEEK*DRUG	3	49.78104575	16.59368192	1.13	0.3487

Source	DF	Type III SS	Mean Square	F Value	Pr > F
WEEK	3	654.72222222	218.24074074	14.81	0.0001
DRUG	1	196.16013072	196.16013072	13.31	0.0007
PATIENT(DRUG)	15	171.72222222	11.44814815	0.78	0.6951
WEEK*DRUG	3	49.78104575	16.59368192	1.13	0.3487

Tests of Hypotheses using the Type III MS for PATIENT(DRUG) as an error term

Source	DF	Type III SS	Mean Square	F Value	Pr > F
DRUG	1	196.16013072	196.16013072	17.13	0.0009

Note: Type II SS corresponds to results in Table 11.25 However Type III SS is technically more appropriate with the unequal numbers of patients in each Drug group. The slight difference between the Type II and Type III results is only in the Sum of Squares (SS) and, therefore, the Mean Square for Week. The evaluations of interest, Drug and Week*Drug, are equivalent for both Type II and Type III results.

```
/* SAS PROGRAM TO ANALYZE TABLE 12.9 DATA */


Options LS=95 PS=58 Nodate Pageno=1;

Title1 'Determination of Variance Components for Table 12.9 Results';

Data A;
Input Batch $ Tablet Assay1 Assay2 Assay3;

Cards;
A       1       50.6        50.5        50.8
A       2       49.1        48.9        48.5
A       3       51.1        51.1        51.4
B       1       50.1        49.0        49.4
B       2       51.0        50.9        51.6
B       3       50.2        50.0        49.8
C       1       51.4        51.7        51.8
C       2       52.1        52.0        51.4
C       3       51.1        51.9        51.6
D       1       49.0        49.0        48.5
D       2       47.2        47.6        47.6
D       3       48.9        48.5        49.2
;

Proc Print;
Id Batch;
Var Tablet Assay1 Assay2 Assay3;

Data B;                         /* Format Data for Analyses */
Set A;
Array A[3] Assay1 Assay2 Assay3;
Do I=1 to 3;
Assay=A[I];
Output;
End;

Proc Varcomp;
Class Batch Tablet;
Model Assay=Batch Tablet(Batch);

Run;
Quit;
```

SAS OUTPUT FOR ANALYSES OF TABLE 12.9 RESULTS:

Determination of Variance Components for Table 12.9 Results   1

BATCH	TABLET	ASSAY1	ASSAY2	ASSAY3
A	1	50.6	50.5	50.8
A	2	49.1	48.9	48.5
A	3	51.1	51.1	51.4
B	1	50.1	49.0	49.4
B	2	51.0	50.9	51.6
B	3	50.2	50.0	49.8
C	1	51.4	51.7	51.8
C	2	52.1	52.0	51.4
C	3	51.1	51.9	51.6
D	1	49.0	49.0	48.5
D	2	47.2	47.6	47.6
D	3	48.9	48.5	49.2

Determination of Variance Components for Table 12.9 Results   2

Variance Components Estimation Procedure
Class Level Information

Class	Levels	Values
BATCH	4	A B C D
TABLET	3	1 2 3

Number of observations in data set = 36

Determination of Variance Components for Table 12.9 Results   3

MIVQUE(0) Variance Component Estimation Procedure

SSQ Matrix

Source	BATCH	TABLET(BATCH)	Error	ASSAY
BATCH	243.00000000	81.00000000	27.00000000	438.18750000
TABLET(BATCH)	81.00000000	99.00000000	33.00000000	198.62250000
Error	27.00000000	33.00000000	35.00000000	68.70750000

Variance Component	Estimate ASSAY
Var(BATCH)	1.55990741
Var(TABLET(BATCH))	0.69527778
Var(Error)	0.10416667

Note: Variance due to error is the between-assay variance component

```
/* SAS PROGRAM TO ANALYZE TABLE 13.8, DAY 1 DATA TO CONSTRUCT TABLE 13.10 */

Options ls=95 ps=58 nodate pageno=1;
Title1 'Regression Analysis of Table 13.8 Day 1 Results';

Data A;
input X Y1 Y2;   /* Concentration is X, Peak Areas are Y1 & Y2 */
Keep X Y Wt;
Wt = (1/X)**2;   /* Weight by inverse of concentration squared */
Y=Y1;
Output;
Y=Y2;
Output;
Label X = Conc        (X);
Label Y = Peak Area   (Y);
Label Wt = Weight     [1/(X*X)];

Cards;           /* data in order: Conc(X), Rep1(Y1), Rep2(Y2) */
0.05  0.003 0.004
0.20  0.016 0.018
1.00  0.088 0.094
10.00 0.920 0.901
20.00 1.859 1.827
;

Proc Print label;
ID X;
Var Y Wt;

Proc Reg; /* Fit Calibration Line to (X,Y) pairs */
model Y = X;
Weight Wt;

Proc GLM;  /* Analysis to obtain Within Duplicates SS */
Class X;
Model Y=X;
Weight Wt;

Run;
Quit;
```

SAS OUTPUT FOR ANALYSES OF TABLE 13.8, DAY 1 RESULTS:

Regression Analysis of Table 13.8 Day 1 Results   1

Conc (X)	Peak Area (Y)	Weight [1/(X*X)]
0.05	0.003	400.000
0.05	0.004	400.000
0.20	0.016	25.000
0.20	0.018	25.000
1.00	0.088	1.000
1.00	0.094	1.000
10.00	0.920	0.010
10.00	0.901	0.010
20.00	1.859	0.003
20.00	1.827	0.003

Regression Analysis of Table 13.8 Day 1 Results   2

The REG Procedure
Model: MODEL1
Dependent Variable: Y Peak Area   (Y)

Weight: Wt Weight   [1/(X*X)]

Analysis of Variance

Source	DF	Sum of Squares	Mean Square	F Value	Pr > F
Model	1	0.05689	0.05689	1652.71	<.0001
Error	8	0.00027537	0.00003442		
Corrected Total	9	0.05716			

Root MSE	0.00587	R-Square	0.9952	
Dependent Mean	0.00453	Adj R-Sq	0.9946	
Coeff Var	129.52287			

Parameter Estimates

Variable	Label	DF	Parameter Estimate	Standard Error	t Value	Pr > |t|
Intercept	Intercept	1	-0.00109	0.00024393	-4.46	0.0021
X	Conc (X)	1	0.09154	0.00225	40.65	<.0001

Regression Analysis of Table 13.8 Day 1 Results   3

The GLM Procedure

Class Level Information

Class   Levels   Values

X   5   0.05 0.2 1 10 20

Number of observations   10

Regression Analysis of Table 13.8 Day 1 Results   4

The GLM Procedure

Dependent Variable: Y   Peak Area   (Y)

Weight: Wt   Weight   [1/(X*X)]

Source	DF	Sum of Squares	Mean Square	F Value	Pr > F
Model	4	0.05689338	0.01422334	262.34	<.0001
Error	5	0.00027109	0.00005422		
Corrected Total	9	0.05716446			

R-Square	Coeff Var	Root MSE	Y Mean
0.995258	162.5539	0.007363	0.004530

Source	DF	Type I SS	Mean Square	F Value	Pr > F
X	4	0.05689338	0.01422334	262.34	<.0001

Source	DF	Type III SS	Mean Square	F Value	Pr > F
X	4	0.05689338	0.01422334	262.34	<.0001

Notes on How to Construct Table 13.10 from SAS Output Pages:

- Intercept and Slope are given on output page 2 under Parameter Estimates.
- Slope DF, SS, MS & F are those listed for Model at top of page 2.
- Error and Total DF, SS & MS (error) are listed at top of page 2.
- Within (duplicates) DF, SS and MS are those listed for Error on page 4.
- Deviations from Regression DF = DF Error (page 2) – DF Error (page 4).
- Deviations from Regression SS = SS Error (page 2) – SS Error (page 4).

```
/* Sas Program to Analyze Table 15.3 Data as Shown in Table 15.4  */

Options ls=90 ps=60 nodate pageno=1;
Title1 'Wilcoxon Signed Rank Test of Paired Tmax Results';

Data A;
Input Subject TmaxA TmaxB;
Diff=TmaxB-TmaxA;              /* calculate the B-A difference */
AbsDiff=Abs(Diff);            /* obtain absolute value of difference */
Dsign=Sign(Diff);             /* determines sign of the difference */
If Diff=0 Then Do;            /* exclude zero differences from analysis */
    AbsDiff=.;
    Dsign=.;
  end;
Label Diff=Difference (B-A);
Label AbsDiff=Absolute Value of Difference;
Label Dsign=Sign;

Cards;
1   2.5   3.5
2   3.0   4.0
3   1.25  2.5
4   1.75  2.0
5   3.5   3.5
6   2.5   4.0
7   1.75  1.5
8   2.25  2.5
9   3.5   3.0
10  2.5   3.0
11  2.0   3.5
12  3.5   4.0
;

Proc Print Label;
ID Subject;
Var TmaxA TmaxB Diff AbsDiff Dsign;

Proc Rank;                    /* rank results by absolute value of difference */
Var AbsDiff;                  /* variable AbsDiff will now contain the rank */

Proc Means N Sum;            /* obtain ranksums by sign */
Title3 'Rank Sums for Using Exact Tables';
Var AbsDiff;
Class DSign;
output out=mnsum Sum=;

Data B;
Set mnsum;
If _Type_=0;
N=_Freq_;                     /* get number in analysis */
```

```
Data C;
Set mnsum;
If _Type_=1 and DSign>0;    /* get Ranksum for positive sign */
R=AbsDiff;                  /*AbsDiff has Rank Sum  */

Data D;                     /* calculate large sample approximation */
Title3 'Normal Approximation for Large Samples';
Merge B C;
Z=(Abs(R-N*(N+1)/4))/Sqrt(N*(N+0.5)*(N+1)/12);
pval=2*(1-Probnorm(Z));
Label pval=p-value;

Proc Print Label;
ID N;
Var R Z pval;

Run;
Quit;
```

SAS Output for Analyses Shown in Table 15.4:

Wilcoxon Signed Rank Test of Paired Tmax Results   1

Subject	Tmax A	Tmax B	Difference (B-A)	Absolute Value of Difference	Sign
1	2.50	3.5	1.00	1.00	1
2	3.00	4.0	1.00	1.00	1
3	1.25	2.5	1.25	1.25	1
4	1.75	2.0	0.25	0.25	1
5	3.50	3.5	0.00	.	.
6	2.50	4.0	1.50	1.50	1
7	1.75	1.5	-0.25	0.25	-1
8	2.25	2.5	0.25	0.25	1
9	3.50	3.0	-0.50	0.50	-1
10	2.50	3.0	0.50	0.50	1
11	2.00	3.5	1.50	1.50	1
12	3.50	4.0	0.50	0.50	1

Wilcoxon Signed Rank Test of Paired Tmax Results   2

Rank Sums for Using Exact Tables

The MEANS Procedure

Analysis Variable : AbsDiff Values of AbsDiff Were Replaced by Ranks

Sign	N Obs	N	Sum
-1	2	2	7.0000000
1	9	9	59.0000000

```
                 Wilcoxon Signed Rank Test of Paired Tmax Results              3

                    Normal Approximation for Large Samples

                       N      R      Z        p-value

                      11     59    2.31168    0.020795



------------------------------------------------------------------------------------



/* Sas Program to Produce Table 15.7 and Confidence Intervals  */


Options ls=90 ps=58 nodate pageno=1;
Title1 'Nonparametric Confidence Intervals for Table 15.6 Cmax Ratios';

Data A;
input Subject CmaxA CmaxB;
Ratio=CmaxB/CmaxA;  /* Calculate Cmax Ratio B/A */

Cards;          /* data */
1      135    102
2      179    147
3      101    385
4      109    106
5      138    189
6      135    105
7      158    130
8      156    125
9      174    144
10     147    133
11     145    114
12     147    167
;

Proc Print;
ID Subject;
Var CmaxA CmaxB ratio;

Data B;      /* Creates a one-record file of the ratios for each Subject */
Keep Sub1-Sub12 Geomean;     /* Variables Sub1-Sub12 contain the ratios */
Array Rat[12] Sub1-Sub12;
Product=1;
Do I=1 to 12;
Set A;
Product=Product*ratio;    /* multiplies Product by ratio for Subject I */
Rat[I]=ratio;             /* captures ratio for Subject I */
end;
Geomean=Product**(1/12);    /* computes geometric mean ratio for subjects */
```

```
Data C;            /* Pairing of the ratios */
Set B;
Keep Pair Geomean Geomn;
Array First[12] Sub1-Sub12;    /* array for first ratio in pair */
Array Second[12] Sec1-Sec12;   /* array for second ratio in pair */
Do I=1 to 12;
 Second[I]=First[I];                /* copy ratios to second array */
end;
Do fst=1 to 12;            /* pair ratio of each subject with itself and with*/
  Do sec=fst to 12;    /* that of each subject number higher than it*/
  Geomn=Sqrt(First[fst]*Second[sec]); /* geometric mean for paired subjs */
  Pair=compbl(fst||','||sec);
  Output;
  end;
end;
Label Geomn=Sorted Geometric Mean Ratio;
Label Pair=Subjects;

Proc Sort;            /* Sorts paired Geometric Mean Ratios */
By Geomn;

Proc Print Label;
Var Pair Geomn;

Data D;
Keep Geomean CI95Lo CI90Lo CI90Hi CI95Hi;
Array CI[4] CI95Lo CI90Lo CI90Hi CI95Hi;
Z=1;
Do I=1 to 78;
Set C;
If I=14 or I=18 or I=61 or I=65 then
    do;                      /* capture geomn vals for CI limits */
    CI[Z]=geomn;
    Z=Z+1;
  end;
end;
Label Geomean=Geometric Mean Ratio;
Label CI90Lo = 90% CI Lower Limit;
Label CI90Hi = 90% CI Upper Limit;
Label CI95Lo = 95% CI Lower Limit;
Label CI95Hi = 95% CI Upper Limit;

Proc Print Label;
ID Geomean;
Var CI95Lo CI90Lo CI90Hi CI95Hi;
Attrib Geomean CI95Lo CI90Hi CI90Lo CI95Hi Format=F6.3;

Run;
Quit;
```

SAS Output for Analyses Shown in Table 15.7 (Table 15.6 data):

Nonparametric Confidence Intervals for Table 15.6 Cmax Ratios   1

Subject	Cmax A	Cmax B	Ratio
1	135	102	0.75556
2	179	147	0.82123
3	101	385	3.81188
4	109	106	0.97248
5	138	189	1.36957
6	135	105	0.77778
7	158	130	0.82278
8	156	125	0.80128
9	174	144	0.82759
10	147	133	0.90476
11	145	114	0.78621
12	147	167	1.13605

Nonparametric Confidence Intervals for Table 15.6 Cmax Ratios   2

[Partial Page Shown]

Obs	Subjects	Sorted Geometric Mean Ratio
1	1, 1	0.75556
2	1, 6	0.76659
3	1, 11	0.77073
4	6, 6	0.77778
5	1, 8	0.77808
6	6, 11	0.78198
7	11, 11	0.78621
8	1, 2	0.78771
9	1, 7	0.78845
10	6, 8	0.78944
11	1, 9	0.79075
12	8, 11	0.79371
13	2, 6	0.79921
14	6, 7	0.79996
15	8, 8	0.80128
16	6, 9	0.80230
17	2, 11	0.80353
18	7, 11	0.80429
19	9, 11	0.80663

Nonparametric Confidence Intervals for Table 15.6 Cmax Ratios   3

Obs	Subjects	Sorted Geometric Mean Ratio
53	10, 12	1.01383
54	1, 5	1.01724
55	5, 6	1.03209
56	5, 11	1.03767
57	5, 8	1.04757
58	4, 12	1.05109
59	2, 5	1.06053
60	5, 7	1.06154
61	5, 9	1.06463
62	5, 10	1.11316
63	12, 12	1.13605
64	4, 5	1.15407
65	5, 12	1.24736
66	5, 5	1.36957
67	1, 3	1.69708
68	3, 6	1.72186
69	3, 11	1.73116
70	3, 8	1.74768
71	2, 3	1.76930
72	3, 7	1.77098
73	3, 9	1.77614
74	3, 10	1.85711
75	3, 4	1.92535
76	3, 12	2.08099
77	3, 5	2.28487
78	3, 3	3.81188

Nonparametric Confidence Intervals for Table 15.6 Cmax Ratios   4

Geometric Mean Ratio	95% CI Lower Limit	90% CI Lower Limit	90% CI Upper Limit	95% CI Upper Limit
1.006	0.800	0.804	1.065	1.247

```
/* Sas Program to Analyze Table 15.8  */

Options ls=90 ps=60 nodate pagno=1;
Title1 'Dissolution Results at 30 Minutes';
Title3 'Wilcoxon Rank Sum Test for Difference Between Two Independent Groups';

Data A;
input App $ Diss;
label App=Apparatus;
label Diss=Amount Dissolved;
cards;        /* Data from Table 15.8*/
O      53
.
.
M      56 ;

Proc Rank;               /* Rank Dissolution Data */
Var Diss;
Ranks Rank;

Proc Print Label;      /* Displays Ranks for Data */
ID App;
Var Diss Rank;

Proc Means Sum;        /* Obtain Rank Sums by apparatus */
Var Rank;
Class App;
Output out=mn Sum=Sum;

Data B;        /* Captures Rank Sums and numbers of data values */
Array Num[2] n1 n2;
Array T[2] t1 t2;
Do I= 1 to 3;
Set mn;
If _Type_=1 then
      do;
         Num[I-1]=_Freq_;
         T[I-1]=Sum;
      end;
end;
I=1;
if n2<n1 then I=2;
Z=Abs(T[I]-Num[I]*(n1+n2+1)/2)/Sqrt(n1*n2*(n1+n2+1)/12);   /* Z-statistic */
pval=2*(1-probnorm(z));                     /* p-value for normal approximation */
label pval=p-value;

Proc Print label;
ID n1;
var n2 t1 t2 z pval;

Run;
Quit;
```

SAS OUTPUT FOR TABLE 15.8:

Dissolution Results at 30 Minutes   1

Wilcoxon Rank Sum Test for Difference Between Two Independent Groups

Apparatus	Amount Dissolved	Rank for Variable Diss
O	53	3.0
O	61	14.0
O	57	9.0
O	50	1.0
O	63	17.0
O	62	15.5
O	54	4.0
O	52	2.0
O	59	12.5
O	57	9.0
O	64	18.5
M	58	11.0
M	55	5.5
M	67	21.0
M	62	15.5
M	55	5.5
M	64	18.5
M	66	20.0
M	59	12.5
M	68	22.0
M	57	9.0
M	69	23.0

Dissolution Results at 30 Minutes   2

Wilcoxon Rank Sum Test for Difference Between Two Independent Groups

The MEANS Procedure

Analysis Variable : Rank Rank for Variable Diss

Apparatus	N Obs	Sum
M	12	170.5000000
O	11	105.5000000

Dissolution Results at 30 Minutes   3

Wilcoxon Rank Sum Test for Difference Between Two Independent Groups

n1	n2	t1	t2	Z	p-value
12	11	170.5	105.5	1.63096	0.10290

```
/* Sas Program to Analyze Table 15.10  */

Options ls=90 ps=60 nodate pagno=1;
Title1 'Kruskal-Wallis Test (One-Way ANOVA)';

Data A;
input @@ t1-t29;
keep cmpnd Time;
array v[29] t1-t29;
do i=1 to 29;
cmpnd='H';                           /* compound = High Dose */
if i<10 then cmpnd='C';       /* compound = Control, first 9 elements */
if i>9 and i<20 then cmpnd='L';   /* compound = Low dose, elements 10-20 */
time=v[i];
output;                   /* creates 29 records of two variables: cmpnd time */
end;
cards;      /* data */
8 1 9 9 6 3 15 1 7
10 5 8 6 7 7 15 1 15 7
3  4 8 1 1 3  6 2  2
;

proc npar1way wilcoxon;
class cmpnd;
var time;    /* if sample sizes are small, add statement: Exact Wilcoxon */
             /* to obtain exact tests for the treatment comparison */
run;
quit;
```

SAS OUTPUT:

```
                      Kruskal-Wallis Test (One-Way ANOVA)                     1

                        N P A R 1 W A Y   P R O C E D U R E

                 Wilcoxon Scores (Rank Sums) for Variable TIME
                        Classified by Variable CMPND
```

CMPND	N	Sum of Scores	Expected Under H0	Std Dev Under H0	Mean Score
C	9	149.500000	135.0	21.0479748	16.6111111
L	10	191.000000	150.0	21.6247384	19.1000000
H	10	94.500000	150.0	21.6247384	9.4500000

```
                    Average Scores Were Used for Ties


            Kruskal-Wallis Test (Chi-Square Approximation)

            CHISQ = 6.9981       DF = 2       Prob > CHISQ = 0.0302
```

```
/* Sas Program to Analyze Table 15.11  */

Options ls=90 ps=60 nodate pagno=1;
Title1 'Friedman Test (Two-Way ANOVA) on Tablet Hardness Data';

Data A;
input @@ t1-t20;
keep press form hardness;
array v[20] t1-t20;
do i=1 to 20;
if i<6 then press='A';            /* press is A for first 5 hardness values */
if i>5 and i<11 then press='B';   /* press is B for values 6 through 10 */
if i>10 and i<16 then press='C';  /* press is C for values 11 through 15 */
if i>15 then press='D';           /* press is C for values 15 through 20 */
form=i-5*Int(i/5.1);              /* determines formulation from counter i */
hardness=v[i];
output;           /* creates 20 records of two variables: press hardness */
end;
cards;              /* data */
7.5 8.2 7.3 6.6 7.5
6.9 8.0 7.9 6.5 6.8
7.3 8.5 8.0 7.1 7.6
7.0 7.9 7.6 6.4 6.7
;

proc print;
id form;
var press hardness;

proc freq;                 /* Friedman Test */
tables form*press*hardness/ noprint cmh2 scores=rank;
footnote1 'Friedman Chi-Square is given in Statistic 2';

proc sort data=a;
by form;
footnote1 "   ";

proc rank;          /* Ranks press results (hardness) within each
formulation */
var hardness;
by form;

proc Anova;               /* Modified Test */
Title1 'Modified Friedman Test (Two-Way ANOVA) on Tablet Hardness Data';
class form press;
model hardness=form press;
means press/LSD;     /* Performs multiple pairwise comparisons of Press
means */

run;
quit;
```

SAS OUTPUT TABLE 15.11:

Friedman Test (Two-Way ANOVA) on Tablet Hardness Data   1

FORM	PRESS	HARDNESS
1	A	7.5
2	A	8.2
3	A	7.3
4	A	6.6
5	A	7.5
1	B	6.9
2	B	8.0
3	B	7.9
4	B	6.5
5	B	6.8
1	C	7.3
2	C	8.5
3	C	8.0
4	C	7.1
5	C	7.6
1	D	7.0
2	D	7.9
3	D	7.6
4	D	6.4
5	D	6.7

Friedman Test (Two-Way ANOVA) on Tablet Hardness Data   2

SUMMARY STATISTICS FOR PRESS BY HARDNESS
CONTROLLING FOR FORM

Cochran-Mantel-Haenszel Statistics (Based on Rank Scores)

Statistic	Alternative Hypothesis	DF	Value	Prob
1	Nonzero Correlation	1	0.864	0.353
2	Row Mean Scores Differ	3	9.720	0.021

Total Sample Size = 20

Friedman Chi-Square is given in Statistic 2

Modified Friedman Test (Two-Way ANOVA) on Tablet Hardness Data   3

Analysis of Variance Procedure
Class Level Information

Class	Levels	Values
FORM	5	1 2 3 4 5
PRESS	4	A B C D

Number of observations in data set = 20

Modified Friedman Test (Two-Way ANOVA) on Tablet Hardness Data   4

Analysis of Variance Procedure

Dependent Variable: HARDNESS   VALUE OF HARDNESS REPLACED BY RANK

Source	DF	Sum of Squares	Mean Square	F Value	Pr > F
Model	7	16.20000000	2.31428571	3.16	0.0389
Error	12	8.80000000	0.73333333		
Corrected Total	19	25.00000000			

	R-Square	C.V.	Root MSE	HARDNESS Mean
	0.648000	34.25395	0.85634884	2.50000000

Source	DF	Anova SS	Mean Square	F Value	Pr > F
FORM	4	0.00000000	0.00000000	0.00	1.0000
PRESS	3	16.20000000	5.40000000	7.36	0.0047

Modified Friedman Test (Two-Way ANOVA) on Tablet Hardness Data   5

Analysis of Variance Procedure

T tests (LSD) for variable: HARDNESS

NOTE: This test controls the type I comparisonwise error rate not the
      experimentwise error rate.

Alpha= 0.05 df= 12 MSE= 0.733333
Critical Value of T= 2.18
Least Significant Difference= 1.1801

Means with the same letter are not significantly different.

T Grouping		Mean	N	PRESS
	A	3.8000	5	C
	A			
B	A	2.8000	5	A
B				
B	C	2.0000	5	B
	C			
	C	1.4000	5	D

Notes on Modified Friedman Test Output:

Page 4   Press row is comparison across presses. F-value is Friedman Test.
Page 5   Least Significant Difference is for mean; for Rank Sum: 5 * 1.18 = 5.90.
           Press mean C is significantly greater than means for B & D.
           Press mean A is significantly greater than mean for D.

```
/* Sas Program to Produce Quade's Test Analysis shown in Table 15.12   */

Options ls=90 ps=60 nodate pagno=1;
Title1 'Quade Test for Randomized Block Design on Data of Table 15.11';

Data A;
input @@ t1-t20;
keep press form rn hardness;
array v[20] t1-t20;
do i=1 to 20;
if i<6 then press='A';          /* press is A for first 5 hardness values */
if i>5 and i<11 then press='B';   /* press is B for values 6 through 10 */
if i>10 and i<16 then press='C';  /* press is C for values 11 through 15 */
if i>15 then press='D';          /* press is D for values 15 through 20 */
form=i-5*Int(i/5.1);     /* determines tablet formulation from counter i */
hardness=v[i];
z=form;
rn=range(v[z],v[z+5],v[z+10],v[z+15]); /* determines range for each form */
rn=round(rn,0.01);
output;
end;
cards;                  /* data */
7.5 8.2 7.3 6.6 7.5
6.9 8.0 7.9 6.5 6.8
7.3 8.5 8.0 7.1 7.6
7.0 7.9 7.6 6.4 6.7
;

proc sort data=A;
by form press;

Data B1;              /* creates one record for each formulation containing
range */
set A;
keep form rn;
if press="A";

proc rank data=B1 out=B2;    /* rank the tablet formulations by range */
var rn;
ranks q;                 /* store ranks for each formulation in variable q */

data C;
set B2;              /* creates records of formulation ranks for each press */
press="A"; output;
press="B"; output;
press="C"; output;
press="D"; output;

proc sort data=C;
by form press;
```

```
proc rank data=A out=A2;   /* rank hardness of each form within press */
var hardness;                 /* replace hardness value by its rank */
by form;

data d;
merge A2 C;                 /* Merge hardness and formulation ranks */
by form press;
keep press form rn q hardness;
hardness=q*(hardness-5/2);      /* Quade's test using formulation ranks q */
label rn=range;
label q=rank;

proc print label;           /* Produces Table 15.12 */
id form;
var press hardness rn q;

proc Anova;                 /* Quade's test using modified hardness ranks */
class form press;
model hardness= form press;
means press/LSD;            /* multiple paired comparisons of press means */

run;
quit;
```

SAS OUTPUT:

Quade Test for Randomized Block Design Applied Data of Table 15.11   1

FORM	PRESS	VALUE OF HARDNESS REPLACED BY RANK	range	rank
1	A	2.25	0.6	1.5
1	B	-2.25	0.6	1.5
1	C	0.75	0.6	1.5
1	D	-0.75	0.6	1.5
2	A	0.75	0.6	1.5
2	B	-0.75	0.6	1.5
2	C	2.25	0.6	1.5
2	D	-2.25	0.6	1.5
3	A	-5.25	0.7	3.5
3	B	1.75	0.7	3.5
3	C	5.25	0.7	3.5
3	D	-1.75	0.7	3.5
4	A	1.75	0.7	3.5
4	B	-1.75	0.7	3.5
4	C	5.25	0.7	3.5
4	D	-5.25	0.7	3.5
5	A	2.50	0.9	5.0
5	B	-2.50	0.9	5.0
5	C	7.50	0.9	5.0
5	D	-7.50	0.9	5.0

Quade Test for Randomized Block Design Applied Data of Table 15.11   2

Analysis of Variance Procedure
Class Level Information

Class	Levels	Values
FORM	5	1 2 3 4 5
PRESS	4	A B C D

Number of observations in data set = 20

Quade Test for Randomized Block Design Applied Data of Table 15.11   3

Analysis of Variance Procedure

Dependent Variable: HARDNESS   VALUE OF HARDNESS REPLACED BY RANK

Source	DF	Sum of Squares	Mean Square	F Value	Pr > F
Model	7	156.30000000	22.32857143	2.36	0.0919
Error	12	113.70000000	9.47500000		
Corrected Total	19	270.00000000			

R-Square	C.V.	Root MSE	HARDNESS Mean
0.578889	9999.99	3.07814879	0

Source	DF	Anova SS	Mean Square	F Value	Pr > F
FORM	4	0.00000000	0.00000000	0.00	1.0000
PRESS	3	156.30000000	52.10000000	5.50	0.0131

Quade Test for Randomized Block Design Applied Data of Table 15.11   4

Analysis of Variance Procedure

T tests (LSD) for variable: HARDNESS

NOTE: This test controls the type I comparisonwise error rate not the
      experimentwise error rate.

Alpha= 0.05   df= 12   MSE= 9.475
Critical Value of T= 2.18
Least Significant Difference= 4.2417

Means with the same letter are not significantly different.

T Grouping		Mean	N	PRESS
	A	4.200	5	C
	A			
B	A	0.400	5	A
B				
B		-1.100	5	B
B				
B		-3.500	5	D

Note on Analysis: Press Mean C does not share a grouping letter with presses B and
                     D, indicating C is significantly different than these two
                     presses at the 5% level of significance.

```
/* Sas Program to Perform Quade Nonparametric ANCOVA on Table 15.13 Data */

Options ls=90 ps=60 nodate pagno=1;
Title1 'Quade Nonparametric Covariance Analysis on Table 15.13 Data';

Data A;
input Method$ Assay Material;

Cards;
I       98.0        98.4
I       97.8        98.6
I       98.5        98.6
I       97.4        99.2
II      97.6        98.7
II      95.4        99.0
II      96.1        99.3
II      96.1        98.4
;
```

```
proc rank data=A out=A2;        /* rank assay and material */
var Material Assay;
ranks Rx Ry;                    /* material ranks in Rx, Assay ranks in Ry */

Proc Standard mean=0 out=B;     /* Centers Ranks about 0; puts results in B*/
Var Rx Ry;

Proc Reg noprint data=B;        /* Regression of Assay ranks on Material ranks */
model Ry = Rx;
Output out=C r=resid;           /* Residuals in C */

Data D;                 /* Gets Method, Assay, Material, Ry, Rx & Residuals */
merge A B C;

Proc Print;
Var Assay Material Ry Rx Resid;
Sum Resid;
By Method;

proc Anova;                     /* ANOVA on regression residuals */
class Method;
model resid= Method;

run;
quit;
```

SAS OUTPUT:

```
          Quade Nonparametric Covariance Analysis on Table 15.13 Data          1
-------------------------------------- METHOD=I --------------------------------------

         OBS    ASSAY    MATERIAL    RY      RX      RESID

          1      98.0     98.4      2.5     -3.0    1.05488
          2      97.8     98.6      1.5     -1.0    1.01829
          3      98.5     98.6      3.5     -1.0    3.01829
          4      97.4     99.2     -0.5      2.5    0.70427
                                                   -------
         METHOD                                    5.79573


-------------------------------------- METHOD=II --------------------------------------

         OBS    ASSAY    MATERIAL    RY      RX      RESID

          5      97.6     98.7      0.5      0.5    0.74085
          6      95.4     99.0     -3.5      1.5   -2.77744
          7      96.1     99.3     -2.0      3.5   -0.31402
          8      96.1     98.4     -2.0     -3.0   -3.44512
                                                   --------
         METHOD                                   -5.79573
                                                   ========
                                                    0.00000
```

Analysis of Variance Procedure
Class Level Information

Class	Levels	Values
METHOD	2	I II

Number of observations in data set = 8

Quade Nonparametric Covariance Analysis on Table 15.13 Data   3

Analysis of Variance Procedure

Dependent Variable: RESID   Residual

Source	DF	Sum of Squares	Mean Square	F Value	Pr > F
Model	1	16.79525301	16.79525301	6.63	0.0420
Error	6	15.19102748	2.53183791		
Corrected Total	7	31.98628049			

	R-Square	C.V.	Root MSE	RESID Mean
	0.525077	9999.99	1.59117501	0.00000000

Source	DF	Anova SS	Mean Square	F Value	Pr > F
METHOD	1	16.79525301	16.79525301	6.63	0.0420

```
/* Sas Program to Analyze Table 15.16   */

Options ls=90 ps=60 nodate pagno=1;
Title1 'Patients Categorized by Disease Severity and Treatment';

Data A;
input Trtment$ Severe$ Count;

cards;      /* data */
A Very      13
B Very      19
A Moderate 24
B Moderate 20
A Mildly   18
B Mildly   12
;

Proc Freq order=data;          /* order of Severity should be as entered */
Tables Trtment*Severe/nofreq nopercent norow nocol nocum expected;
                                      /*expected values */
```

```
                                        /*frequency table & stats*/
Tables Trtment*Severe/ chisq nopercent norow nocol nocum;
Weight Count;                           /*data are counts per cell*/

run;
quit;
```

SAS Output:

Patients Categorized by Disease Severity and Treatment　　　1

TABLE OF TRTMENT BY SEVERE

TRTMENT　　　SEVERE

Expected	Very	Moderate	Mildly	Total
A	16.604	22.83	15.566	
B	15.396	21.17	14.434	
Total	32	44	30	106

TABLE OF TRTMENT BY SEVERE

TRTMENT　　　SEVERE

Expected	Very	Moderate	Mildly	Total
A	13	24	18	55
B	19	20	12	51
Total	32	44	30	106

STATISTICS FOR TABLE OF TRTMENT BY SEVERE

Statistic	DF	Value	Prob
Chi-Square	2	2.541	0.281
Likelihood Ratio Chi-Square	2	2.553	0.279
Mantel-Haenszel Chi-Square	1	2.334	0.127
Phi Coefficient		0.155	
Contingency Coefficient		0.153	
Cramer's V		0.155	

Sample Size = 106

```
/* Sas Program to Analyze Table 15.21  */

Options ls=90 ps=60 nodate pagno=1;
Titlel "Fisher's Exact Test for Carcinoma in Drug- and Placebo-Treated Animals";


Data A;
input Trtment$ Carcnoma$ Count;
cards;                       /* data */
Placebo Present 0
Placebo Absent  12
Drug    Present 5
Drug    Absent  9
;

Proc Freq order=data;      /* Trtment order in table should be as entered */
Tables Trtment*Carcnoma/nopercent norow nocol nocum exact; /*table & stats*/
Weight Count;              /* data are counts for each table cell */

run;
quit;
```


SAS OUTPUT:

```
          Fisher's Exact Test for Carcinoma in Drug- and Placebo-Treated Animals       1

                          TABLE OF TRTMENT BY CARCNOMA

                              TRTMENT    CARCNOMA
```

Frequency	Absent	Present	Total
Placebo	12	0	12
Drug	9	5	14
Total	21	5	26

```
                   STATISTICS FOR TABLE OF TRTMENT BY CARCNOMA
```

Statistic	DF	Value	Prob
Chi-Square	1	5.306	0.021
Likelihood Ratio Chi-Square	1	7.208	0.007
Continuity Adj. Chi-Square	1	3.256	0.071
Mantel-Haenszel Chi-Square	1	5.102	0.024
Fisher's Exact Test (Left)			1.000
(Right)			0.030
(2-Tail)			0.042
Phi Coefficient		0.452	
Contingency Coefficient		0.412	
Cramer's V		0.452	

```
          Sample Size = 26
          WARNING: 50% of the cells have expected counts less
                   than 5. Chi-Square may not be a valid test.
```

Note: Fisher's Exact Test (2-Tail) is correct test.

REFERENCES

1. Halvorson M, Young M. Running Microsoft Office 2000 Professional, Part III Microsoft Excel. Washington: Microsoft Press, Redmond, 1999.
2. SAS Institute Inc. SAS® Language Reference, Version 6, 1st ed., Cary, NC: SAS Institute Inc., 1990.
3. SAS Institute Inc. SAS/STAT® User's Guide, Version 6, 4th ed., Volumes 1 and 2. Cary, NC: SAS Institute Inc., 1990.

Appendix IX
An Alternative Solution to the Distribution of the Individual Bioequivalence Metric*

The Office of Generic Drugs (OGD) of the Federal Drug Administration (FDA) has recently published statistical guidelines for determination of bioequivalence [1], see above. Included in that publication is a statistical approach to determining individual bioequivalence (IB), as recommended by Hyslop et al. [1]. Herewith, is a description of an alternative approach. The probability density function (PDF) of the IB metric is determined and used to construct a decision rule for acceptance. The acceptance criterion is based on an upper 95% confidence interval for the metric, defined as 2.4948. Here is shown the derivation here for the reference-scaled metric. However, with minor modifications, this approach is also applicable to the constant denominator metric and to population bioequivalence described in the FDA guidance [1]. The following has been described in chapter 11, but is repeated here for the sake of continuity.

The reference-scaled metric is defined as

$$\phi = [(\mu_t - \mu_r)^2 - \sigma_d^2 + \sigma_t^2 - \sigma_r^2]/\sigma_r^2, \tag{IX.1}$$

or, equivalently as

$$\phi = [(\mu_t - \mu_r)^2 + \sigma_d^2 + \sigma_t^2]/\sigma_r^2 - 1. \tag{IX.2}$$

Here, μ_t is the mean of the parameter for the test product, μ_r is the mean of the parameter for the reference product, σ_d^2 = subject-product interaction variance, σ_t^2 = within-subject test variance, σ_r^2 = within-subject reference variance.

For a four-period replicate design as described by Hyslop and in the FDA guidance [1,2], we can also define [3]

$$\sigma_i^2 = \sigma_d^2 + 0.5\sigma_t^2 + 0.5\sigma_r^2, \tag{IX.3}$$

where σ_i^2 is the variance of $(\mu_t - \mu_r)$. Combining equations (IX.2) and (IX.3),

$$\phi = [(\mu_t - \mu_t)^2 + \sigma_i^2 + 0.5\sigma_i^2]/\sigma_r^2 - 1.5. \tag{IX.4}$$

The parameter estimates, $\overline{X}_t \overline{X}_r$, S_i^2, S_t^2 and S_r^2, are computed using a mixed-effects linear model as described in the FDA guidance [1].

The analysis in the recent guidance is approximate, has reasonably good properties [1,2], and is relatively simple to calculate. It appears to agree well with the results of the previously used bootstrap simulation approach.

The following derivation results in a more direct approach to estimating the upper confidence interval. The idea is to derive the PDF of the metric. Once the PDF is known, the cumulative probability distribution function (CDF), the 95% confidence interval, as well as other parameters of interest can be easily determined.

*Abstracted from a paper submitted to the Journal, *Drug Development and Industrial Pharmacy*, Marcel Dekker.

IX.1   DERIVATION AND RESULTS

In principle, the PDF of ϕ can be determined if the joint distributions of the random variables $\overline{X}_t, \overline{X}_r, S_i^2, S_t^2$ and S_r^2 are known. In general, this would be a formidable task. However, under the usual assumption of statistical independence of these variables [2], it is quite feasible to compute the PDF of ϕ. Further assumptions include [1] that the random variables $\overline{X}_t$ and $\overline{X}_r$ are Gaussian after the usual logarithmic transformation, and [2] that the variances are distributed as $\sigma_i^2 \chi^2/\text{d.f}$. With these assumptions, which are similar to those made by Hyslop [2], the PDF of ϕ can be derived as shown below. In the derivation, we have used the formulae for computing the PDF of the sum of two independent variables and the PDF of the ratio of two independent variables. These may be found in Ref.[4].

For ease of notation, define the following random variables:

$$Y = (\overline{X}_t - \overline{X}_r)^2$$
$$Z = S_i^2$$
$$U = 0.5S_t^2$$
$$V = S_r^2$$

In terms of these, define further the intermediate variables,

$$W = Y + Z$$
$$G = W + U$$

The metric may then be expressed as

$$\phi = \frac{G}{V} - 1.5.$$

Since $\overline{X}_t$ and $\overline{X}_r$ are both Gaussian, their difference is also Gaussian. Let the mean and standard deviation of $(\overline{X}_t - \overline{X}_r)$ be μ and σ, respectively. Then the PDF of Y, $p(y)$ is given by

$$p(y) = \frac{1}{\sigma\sqrt{2\pi}} \exp\left(-\frac{y + \mu^2}{2\sigma^2}\right) \frac{1}{\sqrt{y}} \cosh\left(\frac{\mu\sqrt{y}}{\sigma^2}\right) \quad y \geq 0$$

Let $q(z)$ be the PDF of Z. Since Y and Z are independent, the PDF of W, $r(w)$, is given by the convolution of $p(y)$ and $q(z)$. Thus

$$r(w) = \int_0^w p(y)q(w - y)dy$$

Similarly, if $s(u)$ is the PDF of U, then the PDF of the variable G, $f(g)$, is given by

$$f(g) = \int_0^g r(w)s(g - w)dw$$

Finally, let a(m) be the PDF of ϕ. If t(v) is the PDF of V, then

$$a(m) = \int_0^w vt(v)f[(m + 1.5)v]dv$$

A program was written in MATLAB [5] to evaluate $a(m)$ using numerical integration to compute the various integrals in the above derivation. If the parameters defining the distributions of X_t, X_r, etc. were known, this would be an exact solution. In the absence of such

Table IX.1 Comparison of Results of Convolution Method to Hyslop Method for the Parameter Values Shown

N	Mean	Difference	S^2_i	S^2_t	S^2_r	Hyslop[a]	Convolution[b]
122[c]		0	0.02	0.02	0.0125	−0.028	2.185
		0	0.02	0.02	0.01	−0.001	2.46
		0	0.02	0.03	0.01	+0.005	3.065
		0.2	0.12	0.12	0.065	+0.023	3.175
26[d]		0.05	0.12	0.1	0.085	−0.008	2.43
		0.05	0.198	0.02	0.1075	+0.0004	2.50
		0.05	0.08	0.049	0.05	+0.005	2.68
		0.2	0.12	0.12	0.095	+0.0205	2.96
16		0.05	0.05	0.05	0.05	−0.0085	2.24
		0.05	0.02	0.02	0.02	−0.0014	2.41
		0.05	0.05	0.1	0.05	+0.0296	3.395
		0.05	0.03	0.02	0.02	+0.0623	3.725
12		0.05	0.02	0.02	0.01	−0.0014	2.79
		0.05	0.02	0.022	0.03375	−0.0118	2.46
		0	0.05	0.04	0.0475	+0.0144	3.56
		0.07	0.05	0.04	0.0475	+0.0222	3.175

[a] Hyslop method passes for negative values.
[b] Convolution passes for values less than 2.498.
[c] Sequence sizes are 30,30,30,32.
[d] Sequence sizes are 6,6,6,8.

knowledge, an approximate solution is obtained by using the observed values of the means and variances as the parameter values. Clearly, this solution would approach the exact solution with large sample sizes. With the sample sizes usually used in BE studies, we expect that the solution should be reasonably good. A preliminary spot check of the results and decisions comparing this new approach to that of Hyslop is shown in Table IX.1. Examples are shown where the decisions are borderline.

REFERENCES

1. Guidance FOR Industry. Statistical Approaches to Establishing Bioequivalence. New York: Food and Drug Administration, CDER, 2001
2. Hyslop T. Hsuan, F. Holder, DJ. A small sample confidence interval approach to assess individual bioequivalence. Stat Med 2000; 19:1885–2897.
3. Ekbohm G. Melander, H. The subject-by-formulation interaction as a criterion for inter-changeability of drugs. Biometrics 1989; 45:1249–1254.
4. Rice JA. Mathematical Statistics and Data Analysis. Pacific Grove, CA: Wadsworth and Brooks/Cole, 1994.
5. Matlab, The Mathworks, Inc., Natick, MA.

Matlab Program to Compute CDF of Metric

EXAMPLE

Data from Replicate Design with 8 subjects

A. mean of xt-xr is the estimated difference between the means of Test
   and Reference = -0.1715

B. std err of xt-xr is the standard error of (A) = 0.295

C. si^2 is the estimated interaction variance = 0.0871

D. sr^2 is the estimated within subject variance of the reference =
   0.06605

E. st^2 is the estimated within subject variance of the test = 0.0729

Matlab Program

```
clear
global degfree
global mean
%These variables are used to compute chisq/df distributions
%using chidf.m
%
mtr=-.01715              % mean of xt-xr
str=0.295               % std err of xt-xr
msi=0.0871              % si^2
msr=0.06605             % sr^2
mst=0.0729              % st^2
%mit=msi+.5*mst         % mean of si^2 + .5 st^2
%sit=sqrt(ssi^2+ (.5*sst)^2)
```

```
%
%Note: si^2, sr^2 and st^2 are distributed as sig^2 chi^2/d.f
%         where the degrees of freedom are 6.
%         Therefore the mean = sig^2
%         sig^2 is, of course, unknown. We approximate it by the sample
%         estimate
%         we need the distribution of
%              [(xt-xr)^2 + si^2 + .5st^2 ] / sr^2 -1.5
%prob density of (xr-xt)^2 is ptr2
%prob density of si^2 is psi (deg of free =6)
%prob density of sr^2 is psr
%prob density of st^2 is pst
%prob density of si^2+.5st^2 is pit
%prob density of denominator is pden (= prob density of sr^2)
%prob density of numerator is pnum
%prob density of metric is pmetric
%
xtr2max=0.15
xchimax=0.4
deltr=.00005
ntr2=1+round(xtr2max/deltr)
nchi=1+round(xchimax/deltr)
xtr=linspace(0.,xtr2max,ntr2);
xchi=linspace(0.,xchimax,nchi);
ctr=(1/(str*sqrt(2*pi)))*exp(-mtr^2/(2*str^2));
ptr2(1)=2*ctr/sqrt(deltr);


for i=2:ntr2
    xdel=xtr(i);
```

```
    xsqrt=sqrt(xdel);

    ptr2(i)=ctr*exp(-xdel/(2*str^2))*(1/xsqrt)*cosh(mtr*xsqrt/str^2);

end

sumtr2=deltr*ones(1,ntr2)*ptr2'

ptr2=ptr2/sumtr2;

plot (xtr,ptr2)

title('prob density of (xt - xr)^2')

text(.03,250,'xt-xr is gaussian')

text(.03,225,['mu= ' num2str(mtr), ',   sigma= ', num2str(.129)])

iplot=0


if(iplot==1)

    print -dwin

end

%

mean=msi

degfree=6

psi=chidf(xchi);

sumpsi=deltr*ones(1,nchi)*psi'

plot (xchi,psi)

title('prob density of si^2')

text(.3,5,['sig^2= ' num2str(msi)])

text(.3,4,['df =    ' int2str(degfree)])


if(iplot==1)

    print -dwin

end

mean=msr

degfree=6
```

```
psr=chidf(xchi);

sumpsr=deltr*ones(1,nchi)*psr'

plot(xchi,psr)

title('prob density of sr^2')

text(.3,8,['sig^2= ' num2str(msr)])

text(.3,7,['df =   ' int2str(degfree)])


if(iplot==1)

    print -dwin

end

mean=mst*.5

degfree=6

pst=chidf(xchi);

sumpst=deltr*ones(1,nchi)*pst'

plot(xchi,pst)

title('prob density of .5*st^2')

text(.3,8,['sig^2= ' num2str(mst)])

text(.3,7,['df =   ' int2str(degfree)])


if(iplot==1)

    print -dwin

end

%

psist= deltr*conv(psi,pst);

[junk,npl]=size(psist)

xpl=linspace(0.,2.*xchimax,npl);

plot(xpl,psist)

title('prob density of si^2+.5st^2')

sumsist=deltr* ones(1,npl)*psist'
```

```
if(iplot==1)

    print -dwin

end

pnum=deltr*conv(psist,ptr2);

[junk,nplnum]=size(pnum)

xnum=linspace(0.,2.*xchimax+xtr2max,nplnum);

for i=1:nplnum

    cumnum(i)=deltr*ones(1,i)*pnum(1:i)';

end


plot(xnum,pnum)

sumnum=deltr*ones(1,nplnum)*pnum'

title('prob density of numerator')


if(iplot==1)

    print -dwin

end

maxplot=15.

nplot=301

xmetric=linspace(-1.5,maxplot,nplot);

for i=1:nplot

    factor=xmetric(i)+1.5;

    for k=1:nchi

        index=round((factor*xnum(k)/deltr)+1);

        if(index > nplnum)

            index=nplnum;

        end

        integrand(k)=xnum(k)*psr(k)*pnum(index);
```

```
           cumint(k)=psr(k)*cumnum(index);

     end

     pmetric(i)=deltr*ones(1,nchi)*integrand';

     cumplot(i)=deltr*ones(1,nchi)*cumint';

end

plot(xmetric,pmetric)

title('prob density of metric')

summetric=cumplot(nplot)

%summetric=(maxplot/(nplot-1))*ones(1,nplot)*pmetric'


if(iplot==1)

     print -dwin

end

nstop95=1

nstop90=1

for i=1:nplot

%        cumplot(i)= (maxplot/(nplot-1))*ones(1,i)*pmetric(1:i)';

         if cumplot(i)>.90 & nstop90 > 0

            cum90=xmetric(i);

            nstop90=0

         end

         if cumplot(i)>.95 & nstop95 > 0

            cum95=xmetric(i);

            nstop95=0

         end

end

iplot=1

plot(xmetric,cumplot)

title('cumulative distribution of ratio')
```

```
text(10,.5,['90% is at   ' num2str(cum90)])

text(10,.4,['95% is at   ' num2str(cum95)])


if(iplot==1)

    print -dwin

end

%distribution of chi^2/d.f. when mean is not unity

function y=chidf(x)

global degfree

global mean

n=degfree;

n2=n/2;

n2m=n2-1;

u=degfree*x/mean;

y=(n/mean)* (1/((2^n2)*gamma(n2)))* u.^n2m.*exp(-u/2);

end
```

Appendix X
Some Statistical Considerations and Alternate Designs and Considerations for Bioequivalence

X.1 PARALLEL DESIGN IN BIOEQUIVALENCE

The great majority of bioequivalence studies measure drug in body fluids, such that products can be compared within an individual using crossover designs. In some rare circumstances, this approach is either not possible or impractical. For example, drugs with long half-lives may not be amenable to a crossover design or studies where a clinical endpoint is required in patients because of insufficient blood concentrations. In these cases a parallel design may be used.

In parallel designs comparative products are not given to the same patient. Patients are randomly assigned to one of the test products. In this discussion, we will use examples where two products are to be compared, a test and reference product. Typically, a random device is used to assign product to patients as they enter the study, with an aim of having equal numbers of patients in each product group. For a bioequivalence study, it would be expected that patients would all be entered together, each patient assigned a number. If more patients are needed that can be accommodated at one site, a multicenter study may be necessary. Randomization schemes for parallel studies have been described in the literature [1]. Note that for these designs, the number of observations in each group needs not be identical; dropouts do not invalidate any of the remaining data.

Endpoints in clinical studies can be "continuous" data or discrete. For example, the endpoint could be treadmill time to angina, or a local treatment for ulcers, where the endpoint is dichotomous, that is, success or failure. We will discuss the analysis of both kinds of studies.

Another problem with parallel studies is how to construct a test comparing products. For numerical data, one should consider whether or not to transform the data. The usual bioequivalence study uses a log transform of the pharmacokinetic parameters. In clinical studies, it is not obvious if the clinical result should be transformed. In general, a transformation is not necessary, but may depend on the nature of the resulting data. For dichotomous data, we have a different problem when comparing outcomes.

The analysis will be illustrated using the following hypothetical data. The study is for a drug taken orally that is absorbed, but is in such low concentrations in the blood that an acceptable analysis is not available. The study looks for a clinical endpoint that can be measured objectively. The drug is given once daily for seven days. The endpoint is the average time it takes for patients to fall asleep. A parallel study is used because of the potential for carryover of a physiological or psychological nature. At first, the data are considered to be approximately normal, and no transformation is needed. The study design is single blind, with the evaluator being blinded, as is typical for the usual bioequivalence crossover studies. The results of the study are as follows:

Product	N	Average	Variance
Test	24	0.980	0.228
Reference	26	0.949	0.213

Without a (log) transformation, the confidence interval computation is more complicated than that for the usual crossover design with a log transformation. The ratio of test/reference is not normally distributed. Before the log transformation requirement was initiated, an approximate confidence interval was computed as described by FDA and the literature [1]. However, presently, the FDA is recommending use of Fieller's method for computing confidence intervals. We will calculate the confidence interval using both of these methods for the sake of illustrating the methods and comparing the results.

X.1.1 Old FDA Method

$$\text{Confidence interval(1)} = \frac{[(\text{Average test} - \text{average reference}) \pm t(\text{d.f.0.1}) * \text{sqrt}(\text{variance} * (1/N1 + 1/N2))]}{\text{Average test}}$$

Where the t value is from the t distribution with appropriate degrees of freedom at the (one-sided) 5% level. The variance, in this case would be the pooled variance from the two groups. The computations for the numerator are the same as that computed for a 90% confidence interval in a two independent group t test.

In this example, the point estimate (Test/Reference) is 103.3% with a lower and upper 90% confidence interval equal to 92.3% and 114.3%, respectively (see Table X.1 for raw data and calculations).

One could also use a log transformation if appropriate. Of course, there should be some documentation of the rationale for a transformation. Using a log transform the results are 103.1 with a lower and upper 90% confidence interval equal to 91.8% and 115.8%, respectively

Table X.1 Data for Parallel Design Study (Clinical Endpoint)

Subject	Test	Subject	Reference
1	0.82	1	0.83
2	0.54	2	1.22
3	1.01	3	1.14
4	1.4	4	0.88
5	0.89	5	0.95
6	1	6	1.4
7	0.76	7	1.1
8	1.23	8	0.84
9	0.87	9	0.99
10	0.99	10	0.61
11	1.1	11	0.68
12	1.15	12	1.03
13	0.76	13	0.79
14	0.65	14	1.09
15	1.25	15	0.91
16	1.11	16	1.22
17	0.77	17	1.1
18	0.63	18	0.89
19	0.98	19	1.17
20	1.32	20	0.58
21	1.26	21	1.11
22	0.94	22	0.75
23	0.99	23	0.95
24	1.11	24	1.03
		25	0.88
		26	0.54

	Test	Reference
Average	0.9804167	0.949231
Standard deviation	0.2281967	0.213353
Variance	0.0520737	0.045519
Point estimate =		1.032853863
t =		1.677224191
Pooled variance =		0.048660009
Upper level		114.3170257
Lower level		92.2537469

Table X.2 Data For Parallel Design Study Transformed to Logarithms

Subject	Test	Log	Subject	Ref	Log
1	0.82	−0.19845	1	0.83	−0.186329578
2	0.54	−0.61619	2	1.22	0.198850859
3	1.01	0.00995	3	1.14	0.131028262
4	1.4	0.336472	4	0.88	−0.127833372
5	0.89	−0.11653	5	0.95	−0.051293294
6	1	0	6	1.4	0.336472237
7	0.76	−0.27444	7	1.1	0.09531018
8	1.23	0.207014	8	0.84	−0.174353387
9	0.87	−0.13926	9	0.99	−0.010050336
10	0.99	−0.01005	10	0.61	−0.494296322
11	1.1	0.09531	11	0.68	−0.385662481
12	1.15	0.139762	12	1.03	0.029558802
13	0.76	−0.27444	13	0.79	−0.235722334
14	0.65	−0.43078	14	1.09	0.086177696
15	1.25	0.223144	15	0.91	−0.094310679
16	1.11	0.10436	16	1.22	0.198850859
17	0.77	−0.26136	17	1.1	0.09531018
18	0.63	−0.46204	18	0.89	−0.116533816
19	0.98	−0.0202	19	1.17	0.157003749
20	1.32	0.277632	20	0.58	−0.544727175
21	1.26	0.231112	21	1.11	0.104360015
22	0.94	−0.06188	22	0.75	−0.287682072
23	0.99	−0.01005	23	0.95	−0.051293294
24	1.11	0.10436	24	1.03	0.029558802
			25	0.88	−0.127833372
			26	0.54	−0.616186139

	Test		Reference	
Point estimate =	1.032854		1.03123	
t =	1.677224			
Pooled variance =	0.05942			
Upper level	115.775	(log) 0.146479		
Lower level	91.8533	(log) −0.08498		

(see Table X.2 for raw data and calculations). This result is similar to that for the untransformed data, a result of the relatively low coefficient of variation.

X.1.2 Fieller's Method

Fieller's method can be used to compute confidence intervals for the ratio of two normally distributed variables. There are assumptions when using Fieller's method that include the assumption of normality. Also the value of the denominator in Fieller's equation must show the reference product average to be "statistically significant" when compared to zero. In most cases, the results of this approach should give similar conclusions as the old FDA method above.

The method is described in an FDA document [2], which is duplicated below.

X.1.2.1 Fieller's Calculation for Crossover data (Correlated Values)

For an example of this calculation, see Ref. [2].

$$\frac{[(\text{Average test}/\text{average reference}) - G(\sigma - RT/\sigma - RR) \pm (1/\text{average reference}) \times \text{Sqrt}(K * \sigma - RR/n]}{(1 - G)}$$

$$G = \frac{t^2 * \sigma - RR}{n * \text{average reference}^2}$$

$$K = \left(\frac{\text{Average test}}{\text{average reference}}\right)^2 + (1 - G)(\sigma - TT) + \left(\frac{\sigma - RT}{\sigma - RR}\right)$$

$$* \left(\frac{G * \sigma - RT}{\sigma - RR - 2} * \frac{\text{Average test}}{\text{Average reference}}\right)$$

$\sigma - TT = $ Variance test

$\sigma - RR = $ Variance reference

$$\sigma - RT = \sum \frac{(\text{test} - \text{average test})(\text{reference} - \text{average reference})}{n - 1}$$

X.1.2.2 *Fieller's Calculation for Independent Data*

If the two groups are independent as in the above example, the term that relates to the correlation of the data for the two groups, $\sigma - RT$, is considered to be zero, and is not included in the equation. Applying the data in Table X.1 without a transformation, the calculations are as follows:

Interval $= [(\text{Average test/average reference}) \pm (1/\text{average reference}) \times \text{Sqrt}(K * \sigma - RR/n)]/(1-G)$

$G = t^2 * \sigma - RR/(n * \text{average reference}^2)$

$K = (\text{Average test/average reference}^2 + (1-G)(\sigma - TT/\sigma - RR) - (2 * \text{Average test/average reference})$

	Test	Reference
Average	0.9804167	0.949231
Standard	$\sigma - TT = $	$\sigma - RR = $
deviation	0.2281967	0.213353
Pooled	0.04866	
variance		

G = 0.0054659

K = 2.204524409

Upper interval = 1.09866168

Lower interval = 0.967046045

X.2 OUTLIERS

An outlier is an observation far removed from the bulk of the observations. A more detailed discussion and statistical detection of outliers, as well as their treatment can be found in a number of references [1].

For crossover studies and parallel studies, the detection of an outlier using common statistical methods is straightforward. Using an appropriate statistical model, a single statistical outlier can be identified. Although this alone may be sufficient to suspect an anomaly, usually it would be more definitive if other evidence is available to verify that the suspected datum is indeed "mistaken." A more creative approach is possible in the case of replicate designs (see below). In these situations, we have estimates of within-subject variability that can be used to identify outliers. For example, if the within-subject variance for a given treatment (omitting the subject with the suspected outlier) is 0.04, and the two values for the log-transformed parameter for the suspected data are 3.8 and 4.9 (corrected for period effects if necessary and meaningful), we may perform an F test comparing variances for the suspect data and the remaining data. The F ratio is

$$\frac{0.61}{0.04} = 15.3.$$

If the degrees of freedom for the denominator ($N - 1$, where N is the number of subjects including the outlier) is 25, an F value of 15.3 is highly significant ($P < 0.01$). One may wish to correct the significance level, although there is no precedent for this approach. An alternative analysis could be an ANOVA with and without the suspected outlier. An F test with 1 d.f. in the numerator and appropriate d.f. in the denominator would be

$$\frac{[\text{SS (all data)} - \text{SS (without outlier data)}]}{1}.$$

Another approach that has been used is to compare results for periods 1 and 2 versus periods 3 and 4 in a 4 period fully replicated design.

Of course, if there are is an obvious cause for the outlier, a statistical justification is not necessary. However, further evidence, even if only suspicious, is helpful.

If an outlier is detected, as noted above, the most conservative approach is to find a reason for the outlying observation, such as a transcription error, or an analytical error, or a subject who violated the protocol, and so on. In these cases, the data may be reanalyzed with the corrected data, or without the outlying data if due to analytical or protocol violation, for example.

If an obvious reason for the outlier is not forthcoming, one may wish to perform a new small study, replicating the original study, including the outlying subject along with a number of other subjects (at least 5 or 6) from the original study. The results from the new study can be examined to determine if the data for the outlier from the original study is anomalous. The procedure here is not fixed, but should be reasonable, and makes sense. One can compare the test to reference ratios for the outlying subject in the two studies, and demonstrate that the data from the new study show the outlying subject is congruent with the other subjects in the new study, for example.

X.3 DICHOTOMOUS OUTCOME

Studies with a dichotomous outcome (e.g., cured or not cured) are, typically, clinical studies on patients. They may be parallel or crossover studies. An example of a crossover study with a dichotomous outcome would be an application of a patch or topical product studying sensitivity or evidence of a pharmacodynamic response. It would be difficult to compare products based on a ratio for crossover designs with a dichotomous outcome. Statistical tests for such designs would fall in the category of a McNemar test, where only those results that are different for the two products are considered in the analysis. Thus, the results that are "positive" for both products, or "negative" for both products would not be considered in the analysis. Thus far, no regulatory requirements have been issued for bioequivalence for such designs.

Parallel designs for bioequivalence using dichotomous outcomes are not uncommon. These studies usually use patients with the "disease." The results are analyzed using either the binomial distribution or the normal approximation to the binomial, where the outcome may be cured or not cured. The FDA guidances suggest that the confidence interval for the difference of the proportion of "successes" (or "failures") between the products be within ±20% for equivalence. Some criteria may be based on a one-sided 95% confidence interval in the case of noninferiority studies. Proposals have been made to modify the ±20% window for equivalence depending on the observed proportion [3].

For example, consider the following example:

Test product	160/200 successes = 80%
Reference product	170/200 successes = 85%

The confidence interval for the difference in proportion of successes is calculated as

$$(85-80) \pm \mathrm{sqrt}\left(P0 * Q0 * \left(\frac{1}{N1} + \frac{i}{N2}\right)\right) = 5 \pm 1.96 * \mathrm{sqrt}\left(0.825 * 0.175 * \left(\frac{2}{200}\right)\right) = 5 \pm 7.4.$$

This result would pass the $\pm 20\%$ requirements. The interval is -2.4% to 7.4%.

X.4   STEADY STATE STUDIES

Steady state (SS) studies have been used to study bioequivalence for some drug products, for example, controlled release products and highly variable products. SS is approximately attained after about 5 drug half-lives. For example, if the half-life is 8 hours, the drug should be administered for about 40 hours; for example, five single doses given at 8-hour intervals. At SS, theoretically, C_{max}, C_{min}, and the AUC during a dosing interval remain constant. In particular, the relative amount of drug absorbed is measured by the AUC over the dosing interval at SS. SS studies are now discouraged by the FDA. One reason given for this proposal is that the variability is reduced in SS studies, resulting in a less sensitive test for showing differences. This lowering of the variability, however, could be useful from a practical point of view to compare highly variable drug products. Thus, there is some controversy about the use and utility of SS studies.

The design of SS studies are typically crossover studies with multiple dosing. Two groups of patients are entered into the study similar to the usual two-treatment, two-period design. However in the SS design, multiple dosing is administered, using the usual dosing schedule, for a sufficient period of time to attain SS. One would estimate the total number of doses needed based on a package insert, literature or available experimental results.

SS is achieved if the PK parameters remain constant with a given multiple dosing regimen. Typically, dosing should be administered for at least three or more consecutive days. Appropriate dosage administration and sampling should be carried out to document SS. The trough concentration data should be analyzed statistically to verify that SS was achieved prior to Period 1 and Period 2 pharmacokinetic sampling.

According to the FDA Guidance [4,5], the following parameters should be measured:

a.   Individual and mean blood drug concentration levels.
b.   Individual and mean trough levels ($C_{min\ ss}$).
c.   Individual and mean peak levels ($C_{max\ ss}$).
d.   Calculation of individual and mean steady state $AUC_{interdose}$ ($AUC_{interdose}$ is AUC during a dosing interval at steady state).
e.   Individual and mean percent fluctuation.

$$\left[= 100 * \frac{C_{max\ ss} - C_{min\ ss}}{C_{average\ ss}}\right]$$

f.   Individual and mean time to peak concentration.

The log-transformed AUC and C_{max} data during the final dosing interval should be analyzed statistically using analysis of variance. The 90% confidence interval for the ratio of the geometric means of the pharmacokinetic parameters (AUC and C_{max}) should be within 80% to 125%. Fluctuation for the test product should be evaluated for comparability with the fluctuation of the reference product.

X.5 BIOEQUIVALENCE STUDIES PERFORMED IN GROUPS

Bioequivalence studies are usually performed at a single site, where all subjects are recruited and studied as a single group. On occasion, more than one group is required to complete a study. For example, if a large number of subjects are to be recruited, the study site may not be large enough to accommodate the subjects. In these situations, the study subjects are divided into two cohorts. Each cohort is used to assess the comparative products individually, as might be done in two separate studies. Typically, the two cohorts are of approximately equal size. Another example of a study that is performed in groups is the so–called "Add-on" study. In Canada, if a study fails because it was not sized sufficiently, an additional number of subjects may be studied so that the combined, total number of subjects would be sufficient to pass the study based on results of the initial failing study. This reduces the cost to the pharmaceutical company, which, otherwise, would have to repeat the entire study with a larger number of subjects.

It is not a requirement that each group separately pass the confidence interval requirement. The final assessment is based on a combination of both groups. The totality of data is analyzed with a new term in the analysis of variance (ANOVA), a Treatment × Group interaction term. This is a measure (on a log scale) of how the ratios of test to reference differ in the groups. For example, if the ratios are very much the same in each group, the interaction would be small or negligible. If interaction is large, as tested in the ANOVA, then the groups cannot be combined. However, if at least one of the groups individually passes the confidence interval criteria, then the test product would be acceptable. If interaction is not statistically significant ($P > 0.10$), then the confidence interval based on the pooled analysis will determine acceptability. It is an advantage to pool the data, as the larger number of subjects results in increased power and a greater probability of passing the bioequivalence confidence interval, if the products are truly bioequivalent.

In Canada, a second statistical test (in addition to the test for interaction) is required when an Add-on group is studied. Each group is analyzed separately in the usual manner. The residual variances from the two separate groups are compared using an F test. If the variances are significantly different, the groups cannot be pooled and the product will probably fail. Note that the second group is studied only if the original study failed because of lack of size. It is possible that the Add-on study could pass on its own, and in this case, the test product would be acceptable. This second test comparing variances seems rather onerous, because an analysis is possible for the combined groups with unequal variance. However, it may be the intention of the Canadian HPB to trade the benefit of the add-on design for unnecessarily more stringent regulatory requirements. An intensive study of the appropriateness and properties of add-on designs is being investigated by FDA and industry personnel in the United States at the time of this writing. A final finding is forthcoming.

An interesting question arises if more than two groups are included in a bioequivalence study. As before, if there is no interaction, the data should be pooled. If interaction is evident, at least one group is different from the others. Usually, it will be obvious which group is divergent from a visual inspection of the treatment differences in each group. The remaining groups may then be tested for interaction. Again, as before, if there is no interaction, the data should be pooled. If there is interaction, the aberrant group may be omitted, and the remaining groups tested, and so on. In rare cases, it may not be obvious which group or groups are responsible for the interaction. In that case, more statistical treatment may be necessary, and a statistician should be consulted. In any event, if any single group or pooled groups (with no interaction) passes the bioequivalence criteria, the test should pass. If a pooled study passes in the presence of interaction, but no single study passes, one may still argue that the product should pass, if there is no apparent reason for the interaction. For example, if the groups are studied at the same location under the identical protocol, and there is overlap in time among the treatments given to the different groups, as occurs often, there may be no obvious reason for a significant interaction. Perhaps, the result was merely due to chance, random variation. One may then present an argument for accepting the pooled results.

The following statistical models have been recommended for analysis of data in groups:

Model 1: GRP SEQ GRP*SEQ SUBJ(GRP*SEQ) PER(GRP) TRT GRP*TRT
If the GRP*TRT term is not significant ($P > 0.10$), then reanalyze the data using Model 2.
Model 2: GRP SEQ GRP*SEQ SUBJ(GRP*SEQ) PER(GRP) TRT

X.6 REPLICATE STUDY DESIGNS

Replicate studies in the present context are studies in which individuals are administered one or both products on more than one occasion. For purposes of bioequivalence, either three or four period designs are recommended. The two treatment four-period design is the one most used. FDA [1] gives sponsors the option of using replicate design studies for all bioequivalence studies. Replicate studies may provide information on within-subject variance of each product separately, as well as potential product × subject interactions, although these analyses are not required by FDA.

The FDA recommends that submissions of studies with replicate designs be analyzed for average bioequivalence. The following (Table X.3) is an example of the analysis of a two treatment four-period replicate design to assess average bioequivalence. The design has each of two products, balanced in 2 sequences, ABAB and BABA, over four periods. Table X.1 shows the results for C_{max} for a replicate study. Eighteen subjects were recruited for the study and 17 completed the study. An analysis using the usual approach for the TTTP design, as discussed above, is not recommended. The FDA [1] recommends use of a mixed model approach as in SAS PROC MIXED (11). The recommended code is

Table X.3 Results of a Four-Period, Two-Sequence, Two-Treatment, Replicate Design (C_{max})

Subject	Product	Sequence	Period	C_{max}	$Ln(C_{max})$
1	Test	1	1	14	2.639
2	Test	1	1	16.7	2.815
3	Test	1	1	12.95	2.561
4	Test	2	2	13.9	2.632
5	Test	1	1	15.6	2.747
6	Test	2	2	12.65	2.538
7	Test	2	2	13.45	2.599
8	Test	2	2	13.85	2.628
9	Test	1	1	13.05	2.569
10	Test	2	2	17.55	2.865
11	Test	1	1	13.25	2.584
12	Test	2	2	19.8	2.986
13	Test	1	1	10.45	2.347
14	Test	2	2	19.55	2.973
15	Test	2	2	22.1	3.096
16	Test	1	1	22.1	3.096
17	Test	2	2	14.15	2.650
1	Test	1	3	14.35	2.664
2	Test	1	3	22.8	3.127
3	Test	1	3	13.25	2.584
4	Test	2	4	14.55	2.678
5	Test	1	3	13.7	2.617
6	Test	2	4	13.9	2.632
7	Test	2	4	13.75	2.621
8	Test	2	4	13.25	2.584
9	Test	1	3	13.95	2.635
10	Test	2	4	15.15	2.718
11	Test	1	3	13.15	2.576
12	Test	2	4	21	3.045
13	Test	1	3	8.75	2.169
14	Test	2	4	17.35	2.854
15	Test	2	4	18.25	2.904
16	Test	1	3	19.05	2.947
17	Test	2	4	15.1	2.715
1	Reference	1	2	13.5	2.603
2	Reference	1	2	15.45	2.738
3	Reference	1	2	11.85	2.472
4	Reference	2	1	13.3	2.588

Table X.3   *Continued*

5	Reference	1	2	13.55	2.606
6	Reference	2	1	14.15	2.650
7	Reference	2	1	10.45	2.347
8	Reference	2	1	11.5	2.442
9	Reference	1	2	13.5	2.603
10	Reference	2	1	15.25	2.725
11	Reference	1	2	11.75	2.464
12	Reference	2	1	23.2	3.144
13	Reference	1	2	7.95	2.073
14	Reference	2	1	17.45	2.859
15	Reference	2	1	15.5	2.741
16	Reference	1	2	20.2	3.006
17	Reference	2	1	12.95	2.561
1	Reference	1	4	13.5	2.603
2	Reference	1	4	15.45	2.738
3	Reference	1	4	11.85	2.472
4	Reference	2	3	13.3	2.588
5	Reference	1	4	13.55	2.606
6	Reference	2	3	14.15	2.650
7	Reference	2	3	10.45	2.347
8	Reference	2	3	11.5	2.442
9	Reference	1	4	13.5	2.603
10	Reference	2	3	15.25	2.725
11	Reference	1	4	11.75	2.464
12	Reference	2	3	23.2	3.144
13	Reference	1	4	7.95	2.073
14	Reference	2	3	17.45	2.859
15	Reference	2	3	15.5	2.741
16	Reference	1	4	20.2	3.006
17	Reference	2	3	12.95	2.561

```
PROC MIXED;
CLASSES SEQ SUBJ PER TRT;
MODEL LNCMAX = SEQ PER TRT/DDFM = SATTERTH;
RANDOM TRT/TYPE = FA0(2) SUB = SUBj G;
REPEATED/GRP = TRT SUB = SUBJ;
LSMEANS TRT;
ESTIMATE 'T VS. R' TRT 1 − 1/CL ALPHA = 0.1;
RUN;
```

The abbreviated output is shown in Table X.4.

Table X.4   Analysis of Data from Table X.1 for Average Bioequivalece

ANALYSIS FOR LN-TRANSFORMED CMAX

The MIXED Procedure

Class Level Information

Class	Concentrations	Values
SEQ	2	1 2
SUBJ	17	1 2 3 4 5 6 7 8 9 10 11 12 13
		14 15 16 17
PER	4	1 2 3 4
TRT	2	1 2

Table X.4   (Continued)

Covariance Parameter Estimates (REML)

Cov Parm	Subject	Group	Estimate
FA(1,1)	SUBJ		0.20078553
FA(2,1)	SUBJ		0.22257742
FA(2,2)	SUBJ		-0.00000000
DIAG	SUBJ	TRT 1	0.00702204
DIAG	SUBJ	TRT 2	0.00982420

Tests of Fixed Effects

Source	NDF	DDF	Type III F	Pr > F
SEQ	1	13.9	1.02	0.3294
PER	3	48.2	0.30	0.8277
TRT	1	51.1	18.12	0.0001

ESTIMATE Statement Results

Parameter T VS. R

Alpha = 0.1	**Estimate**	**Std Error**	**DF**	**t**	**Pr > |t|**
	0.09755781	0.02291789	51.1	4.26	0.0001

	Lower	**0.0592**		**Upper 0.1360**

Least Squares Means

Effect	TRT	LSMEAN	Std Error	DF	t	Pr > |t|
TRT	1	2.71465972	0.05086200	15	53.37	0.0001
TRT	2	2.61710191	0.05669416	15.3	46.16	0.0001

ANALYSIS FOR LN-TRANSFORMED CMAX

REFERENCES

1. Bolton S, Bon C. Pharmaceutical Statistics, 4th ed. Rockville, MD: Marcel Dekker, 2004.
2. Guidance, Topical Dermatologic Corticosteroids: In Vivo Bioequivalence, Issue Date June 2, 1995, FDA.
3. Statistical Considerations for Clinical Trials in Developing Antimicrobial Drugs, Anti-infective Drug Products Advisory Committee, July 29, 1998, Daphne Lin, Ph.D., CDER/OFB/DBIV.
4. Guidance for Industry, Bioavailability and Bioequivalence Studies for Orally Administered Drug Products, Genereal Considerations, CDER, 2003.
5. Guidance for Industry, Clozapine Tablets: In Vivo Bioequivalence and In Vitro Dissolution Testing, Center for Drug Evaluation and Research (CDER), June 2005.

Answers to Exercises

CHAPTER 1

1. (a) Tablet hardness, blood concentration of drug, creatinine in urine
   (b) Number of patients with side effects, bottles with fewer than 100 tablets, white blood cell count
   (c) Any continuous variable, rating scale
   (d) Race, placebo group in clinical study, number of bottles of syrup that are cloudy
2. None (This is a simple linear transformation; the C.V. is unchanged.)
3.

Interval	Frequency
−99.5 to −83.5	1
−83.5 to −67.5	2
−67.5 to −51.5	10
−51.5 to −35.5	16
−35.5 to −19.5	26
−19.5 to −3.5	34
−3.5 to +12.5	33
12.5 to 28.5	24
28.5 to 44.5	8
44.5 to 60.5	2

4. −10.27
5. Approximately 82% between 95 and 105 mg (0.91–0.09); approximately 9% above 105 mg
6. (a) Mean $= -12.65$, $S = 31.68$; (b) $\bar{X} = -7$, $S = 30.48$ (read data in columns). Differences probably not significant. The last set is more precise but the standard deviations are virtually identical (the variability is probably not different in the two sets of data).
7. Median $= -16 = (-13 - 19)/2$; range $= 46$ to $- 64 = 110$
8. (a) Median $= -16$ as in Problem 7; range $= 100$ to $- 64 = 164$
   (b) Mean $= -8.5$, $S = 40.09$, $S^2 = 1607$
10. Probably not unbiased
11. $\sigma = \sqrt{2/3} = 0.816, \tilde{S} = 0.6285$
13. $\sqrt{\Sigma(X - \bar{x})^2/(N - 1)} = \sqrt{(0.0001 + 0 + 0.0001)/2} = 0.01$. The s.d. of 2.19; 2.20, and 2.21 is also 0.01. If a constant is added to each value (the constant added here is 1), the s.d. is unchanged. Standard deviation depends on differences among the values, not the absolute magnitude.
14. (a) 101.875; (b) 4.79; (c) 22.98; (d) $4.79/101.875 = 0.047$; (e) 14; (f) 101.5
15. $\Sigma N_i X_i^2 = 1(90.5)^2 + 6(70.5)^2 + \cdots + 16(29.5)^2 + 3(49.5)^2 = 137,219$
    $\Sigma N_i X_i = 1(- 90.5) + 6(- 70.5) + \cdots + 16(29.5) + 3(49.5) = -1658$
    $\Sigma N_i = 156$
    $S^2 = [137,219 - (- 1658)^2/156]/155 = 771.6$
    $S = 27.79$
16. 16.167, 9.865, 7.009

17. $\bar{X}_w = (2 \times 3 + 5 + 7 + 3 \times 11 + 14 + 3 \times 57)/10 = 17.9$

$$S_w^2 = \frac{7149 - 3204.1}{9} = 438.3$$

CHAPTER 2

1.
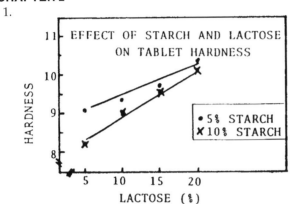

2.
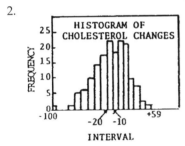

3.
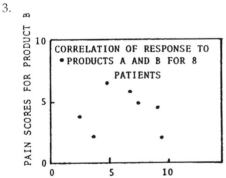

4.

PIE CHART FOR TABLET DEFECTS

5.

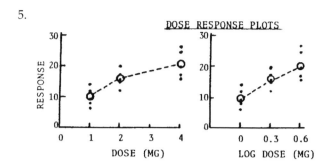

6.

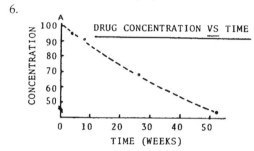

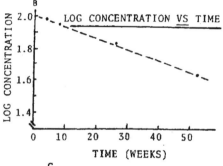

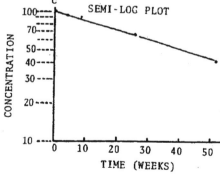

7.

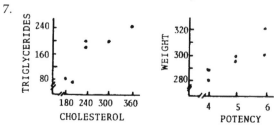

CHAPTER 3

1. Larger sample, more representative, blinded, less bias, etc.
2. All patients with disease who can be treated by antibiotic
3. Preference for new formulation among 24 panelists; number of broken tablets in sample of 100; race of patients in clinical study
4. 50,000 specked, but 20,000 are also chipped. Therefore, 30,000 are only specked. Probability of speck or chip is 0.06 (60,000 tablets have either a speck or a chip).
5. (a) $P(A \text{ and } B) = P(A|B)P(B)$. Let A = high blood pressure and B = diabetic. Then $P(A \text{ and } B) = (0.85)(0.10) = 0.085$.
   (b) If independent, $P(A) = P(A|B)$; $0.25 \neq 0.85$; they are not independent.
6. $(0.75)^2(0.25)^2 = 0.35163 \times 6 = 0.21094$. There are 6 ways of choosing 2 patients out of 4 $\binom{4}{4}$.
7. $(0.6)^3(0.4)^3 = 0.013824 \times 20 = 0.276$. There are 20 ways of choosing 3 patients out of 6 $\binom{6}{4}$.
8. 0.3697
9. (a) Approximately 0.8; (b) 0.2
10. $Z = (170 - 215)/35 = 1.29$; probability = approximately 0.10
11. $Z = (60 - 50)/5 = 2, P(X \leq 60) = 0.977$; $Z = (40 - 50)/5 = -2, P(X \leq 40) = 0.023$; $P(40) \leq X \leq 60) = 0.977 - 0.023 = 0.954$
12. Not necessarily; the patient may have a cholesterol value in the extremes of the normal distribution.
13. $Z = (137 - 140)/2.5 = -1.2$, probability $\leq Z = 0.115$; $Z = (142 - 140)/2.5 = 0.8$, probability $\leq Z = 0.788$; $P(137 \leq Z \leq 142) = 0.788 - 0.115 = 0.673$
14. $Z = (280 - 205)/45 = 1.67$; probability = 0.952; probability $Z > 280 = 1 - 0.952 = 0.048$
15. There are 36 equally likely possibilities, of which one is 2.
16. Yes! The order of heads and tails is not considered in the computation of probability.
17. $P(0 \text{ defects}) = \binom{20}{0}(0.01)^0(0,99)^{20} = 0.818$; $P(1 \text{ defect}) = \binom{20}{1}(0.01)^1(0.99)^{19} = 0.165$; $P(0 \text{ or } 1 \text{ defect}) = 0.818 + 0.165 = 0.983$
18. $\binom{10}{1}(0.5)^1(0.5)^9 = 0.0098$
19. $\binom{4}{2}(0.01)^2(0.99)^2 = 0.00059$. The probability is small; and two of four cures can be considered unlikely. The probability of this event plus equally likely or less likely events (three of four and four of four cures) is close to 0.00059. Thus, we conclude that the new treatment is effective.
20. $\sqrt{(0.01)(0.99)20} = 0.445$; $\sqrt{(0.01)(0.00)/20} = 0.022$ (Problem 17)
    $\sqrt{(0.01)(0.99)4} = 0.199$; $\sqrt{(0.01)(0.99)/4} = 0.497$ (Problem 19)
21. $S = \sqrt{(0.5)(0.5)/20} = 0.112$; $Z = (0.75 - 0.5)/0.112 = 2.24$; $P(Z > 2.24) = 1 - 0.988 = 0.012$

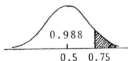

0.988

0.5 0.75

Drug is a promising candidate. The probability of observing such a large response is small if the true proportion of responses is 50%.
22. $P(0 \text{ defects}) = 0.99^{30} = 0.7397$; $P(1 \text{ defect}) = (30)(0.01)(0.99)^{29} = 0.2242$;
    $P(0 \text{ or } 1 \text{ defect}) = 0.7397 + 0.2242 = 0.9639$; $P(\text{more than } 1 \text{ defect}) = 1 - 0.9639 = 0.0361$
23. $85 = 35 + 50 + 50 - 20 - 15 - 25 + P(ABC)$; $P(ABC) = 10\%$

CHAPTER 4

1. Starting at the upper left corner,* going down in Table IV.1. Even numbers to A. Patients assigned to A: 1, 2, 3, 5, 6, 8, 13, 14, 15, 16, 17, and 19.

---

* We started at the upper left and read down for convenience and for the purpose of illustration. Otherwise, the starting point should be random.

2. Start as in Problem 1. If the number is 1 to 3, assign to A; 4 to 6, assign to B; 7 to 9, assign to C; do not count zeros.

Patient	Random number	Treatment
1	4	B
2	8	C
3	2	A
4	5	B
5	8	C
6	4	B
7	9	C
8	2	A
9	1	A
10	5	B
11	5	B
12	5	B
13	4	B
14	6	B (8 B's)
15	8	C
16	3	A
17	9	C
18	3	A
19	8	C
20	8	C
21	9	C (8 C's)
Remaining patients (22, 23, 24) given A		

(May also randomize in groups of three; e.g., the first three patients are B, C, A—random numbers 4 and 8 refer to B and C.)

3. Start as above in Table IV.1. Use two-digit numbers between 1 and 30: 28, 24, 14, 6, 17, 29.
5. Placebo: 1, 2, 4, 5, 7, 8, 9, 10, 12, 18; Drug: 3, 6, 11, 13, 14, 15, 16, 17, 19, 20.
6. Take 20 tablets at a specific time every hour, all at the same time each hour (e.g., on the hour). Take 20 tablets each hour, but randomize the time the 20 are taken; e.g., first hour, take the sample at 5 min past the hour; second hour, take at 25 min past the hour; etc. Take tablets, one every 3 min during each hour. Take tablets at random times during each hour.
7. (see also Problem 3) 44, 8, 28, 55, 88
10. $\bar{X} = 300.7$

CHAPTER 5

1. $Z = (49.8 - 54.7)/2 = -2.45; = 0.0071$
2. $103 \pm 2.58(2.2)/\sqrt{10} = 103 \pm 1.8 = 101.2$ to 104.8
3. (a) $5.95 \pm 2.57(1.16/\sqrt{6} = 5.95 \pm 0.17$
    (b) $0.024 \pm 1.96\sqrt{(0.024)(0.976)/500} = 2.4 \pm 1.34\%$
    (c) $(0.83 - 0.50) \pm 1.06\sqrt{(0.83)(0.17)} >\!sh\!> 60 + (0.50)(0.50)/50 = 0.33 \pm 0.17$
4. (a) $Z = |498 - 502|/(5.3/\sqrt{6} = 1.85$; not significant, $\alpha = 0.05$; two tailed test
    (b) $t = (5.08 - 4.86)/\sqrt{0.095(2/5} = 1.13$; not significant at 5% level
    (c) $T = 4/\sqrt{(15.2)/6} = 2.51; t_5 = 2.57$; just misses significance at 5% level; two-tailed test.
5. (a) 0.098, larger
    (b) 0.350 and 0.261, average s.d. $= 0.305$, pooled s.d. $= 0.308$
6. (a) $\bar{X} = 10.66$, s.d. $= 0.932$
    (b) $\bar{X} = 9.66$, s.d. $= 0.4696$. $t_{18} = 1/(0.738\sqrt{2/10} = 3.03$; difference is significant
    (c) Approximate test: $Z = (0.7 - 0.2)/\sqrt{(0.45)(0.55)(2/10}) = 2.24$; significant. Chi-square test with correction $= 3.23$; not quite significant.
    (d) $0.45 \pm 1.96\sqrt{(0.45)(0.55)(1/20}) = 0.45 \pm 0.22$

7. Paired t test; 3 d.f.; $\alpha = 0.05$; two tailed test
   (a) $t = 0.07\sqrt{0.0039/4} = 2.23$; not significant
   (b) $0.07 \pm 3.18(0.0627)/\sqrt{4} = 0.07 \pm 0.10$
8. (a) Paired t test, 11 d.f.; $t = 0.5/(0.612/\sqrt{12}) = 2.83$; significant at 5% level
   (b) $0.5 \pm 2.2(0.612/\sqrt{12}) = 0.5 \pm 0.39$
9. $9/60$ and $6/65 = 15/125 = 0.12$; $80/1000$ and $57/1000 = 137/2000 = 0.685$
10. $t = (16.7 - 15)/(3.87/\sqrt{10}) = 1.39$; 10% level, one-sided test, this is significant
11. Chi-square $= (3.5)^2(2/12 + 2/88) = 2.32$; not significant
12. $Z = (|0.05 - 0.028| - 1/400)/\sqrt{(0.028)(0.972)/200} = 1.67$;   not   significant. $0.05 \pm$ $1.96\sqrt{(0.95)(0.05)/(200)} = 0.5 \pm 0.03$; $10 \pm 1.96\sqrt{(0.95)(0.05)(200)} = 10 \pm 6$
13. (a) $50 \pm 1.96\sqrt{(0.01)(0.99)(5000)} = 50 \pm 13.79$ in 5000 for 1,000,000 tablets; $10,000 \pm 2758$
    (b) $(0.01 - 0.02)/\sqrt{(0.02)(0.98)/5000} = -5.05$; P #of 0.001;
        very unlikely $1.96\sqrt{(0.01)(0.99)/N} = 0.001$, $N = (1.96)^2(0.99)(0.01/10) = 38{,}032$
14. Chi-square $= (4.5)^2(1/35.45 + 1/24.55 + 1/29.55 + 1.20.45) = 3.07$; not significant at 5% level. $(40/60 - 25/50) \pm 1.96\sqrt{(0.67)(0.33)/60 \pm (0.5(0.5)/50} = 0.167 \pm 0.183$
15. $Z = (|0.75 - 0.5| - 1/80)/\sqrt{(0.5)(0.5)/40} = 3.0$; $P < 0.05$
16. $Z = (|0.45 - 0.2| - 1/40)/\sqrt{(0.8)(0.2)/20} = 2.51$; $P < 0.05$; $0.45 \pm$ $2.58\sqrt{(0.45)(0.55)/20} = 0.45 \pm 0.287$
17. Chi-square $= (3.5)^2(1/13.85 + 1/86.15 + 1/13.15 + 1/81.85) = 2.10$; not significant
18. $(1.8)^2(1/7.2 + 1/7.8 + 1/52.8 + 1/57.2) = 0.98$
19.

80	920
57	943

    $= 2 \times 2$ table

    $\chi^2 = 11^2(1/68.5 + 1/931.5 + 1/68.5 + 1/931.5) = 3.79$; just misses significance at 5% level.
20. $F_{9,9} = 0.869/0.220 = 3.94$, $P < 0.10$ (Table IV.6). This is a two-sided test. A ratio of 3.18 is needed for significance at the 10% level.
21. Correct $\chi^2 = 3.79$; d'Agostino $= 2.04$
22. $\chi^2 = 28.6135 - 20.8591 = 7.75$ ($P < 0.05$)
23. $\sigma^2 = \dfrac{9 \times 5}{0.711} = 63.29$ $\sigma = 7.96$
24. $\sigma^2 = \dfrac{(7.8)^2}{18.49} = 95$ $\sigma = 9.7$

CHAPTER 6

1. $2(5/10)^2(1.96 + 0.84)^2 + 0.25(1.96)^2 =$ approximately 5 per group
2. $2(5/10)^2(1.96 + 0.84)^2 =$ approximately 4 per group
3. $[(0.8 \times 0.2 + 0.9 \times 0.1)/(0.1)^2](1.96 + 1.28)^2 =$ approximately 263 per group
4. $[(0.5 \times 0.5 + 0.5 \times 0.5)/(0.2)^2](1.96 + 1.28)^2 =$ approximately 132 per group
5. $(1.96)^2(0.5 \times 0.5)/(0.15)^2 =$ approximately 43
   $(1.96)^2(0.2)(0.8)/(0.15)^2 =$ approximately 28
6. $(10/10)^2(1.96 + 2.32)^2 + 2 =$ approximately 21 tablets
7. (a) $Z_\beta = (3/5)\sqrt{19/2} - 1.96 = -0.11$; power is approximately 46%
   (b) $Z_\beta = (3/5)\sqrt{49/2} - 1.96 = 1.01$; power $= 84\%$
8. $(10/3)^2(1.96 + 1.28)^2 =$ approximately 117
9. $Z_\beta = (0.2/0.25)\sqrt{10} - 1.96 = 0.57$; power is approximately 71%
10. $2(12/10)^2(1.96 + 1.65)^2 + 0.25Z_\alpha^2 =$ approximately 39
11. $Z_\beta = (15/40)\sqrt{16} - 1.96 = -0.46$; power $=$ approximately 0.32
12. $(1.96)^2(0.90)(0.10)/(0.05)^2 = 138.2 =$ approximately 139
13. $N = 2(5/6)^2(1.96 + 1.28)^2 + 0.25(1.96)^2 = 15.5 =$ approximately 16
14. 23 tablets per formulation

CHAPTER 7

1. (a) $b = 40/10 = 4; a\,12 - (4)(3) = 0$
   (b) $S_{y,x}^2 = (164 - 16.10)/3 = 1.33; S_b^2\ 1.33/10 = 0.133$
   $t = 4/\sqrt{0.133} = 10.95$; significantly different from 0
   (c) $|4 - 5|/\sqrt{0.133} = 2.74$; d.f. $= 3$; not significant, 3.18 needed for significance
   (d) 3 hr;$Y = 4X = 12 \pm 3.18\sqrt{1.33}\sqrt{1/5 + 0.10} = 10.36$ to 13.64.
   5 hr;$Y = 4X = 20 \pm 3.18\sqrt{1.33}\sqrt{1/5 + 4/10} = 17.16$ to 22.84
   (e) $Y = 4(20) = 80 \pm 3.18\sqrt{1 + 1/5 + (20 - 3)^2/10} = 80 \pm 20.1$
   (f) $b = \sum Xy/\sum X^2 = 220/55 = 4$
2. (a) $a = -0.073; b = 0.2159$
   (b) $S_{y,x}^2 = 0.003377; S_a^2 = 0.001848; -1.69$(3 d.f.); not significant; may be due to interfering impurity
   (c) $C = 7.98$; confidence limits are 7.43 to 8.64; see Eq. (7.17)
3. (a) $b = 27/41.2 = 0.655, a = 100 - 0.655(200.4) = -31.3$
   (b) $Y = -31.3 + 0.655)(200) = 99.74$
   (c) $99.74 \pm 3.18\sqrt{0.0102}\sqrt{1/5 + (200 - 200.4^2/41.2} = 99.74 \pm 0.46$
4. (a) 0.9588
   (b) $t_{10} = 10.7$; r is significantly different from 0 at 5% level
5. $r = 0.6519; t_8 = 1.84/0.76 = 2.43$, significant at 5% level
6. $r = -0.93135; t_7 = 6.77$, significant at 5% level
7. $r = 0.2187; F = 6.54/1.067 = 6.135$
   $r_{ds} = (6.135) - 1)/\sqrt{(6.135) + 1)^2 - 4(0.2187^2)6.135} = 0.728$
   $t_8 = 0.728\sqrt{8}/\sqrt{1 - 0.728^2} = 3.00$; $p < 0.05$; drug B is less variable
8. $Y = -3.90082 + 0.99607X$; predicted values: 0.10049 ($X = \ln 5$); 0.20043 ($X = \ln 10$), 0.49928 ($X = \ln 25$), 0.99584 ($X = \ln 50$), 1.98626 ($X = \ln 100$).
9. (a) $C = 2.5482 - 0.01209t$; (b) 24.66 mos; (c) 23.27 mos; (d) 23.55 mos.
10. $a = 0.5055$

CHAPTER 8

1. For significance at the 5% level, t(8 d.f.) ≥ 2.31 (two-sided test) A vs. $B : t = (101.2 - 99.4)/S_p\sqrt{1/5 + 1/5} = 2.84(P < 0.05); S_p = 1.0.$ A vs. $C : t = (101.6 - 101.2)/(1.58\sqrt{1/5 + 1/5}) = 0.40.$ B vs. $C : t = (101.6 - 99.4)/(1.67\sqrt{1/5 + 1/5}) = 2.08$

2.

Source	d.f.	MS	F
Between treatments	2	0.167	0.039
Within treatments	3	4.33	

Treatments are not significantly different.

3. Pooled error term from ANOVA table (Table 8.3) $= 2.10$
   A vs. $B : t = 1.8/\sqrt{2.10(2/5)} = 1.96$
   A vs. $C : t = 0.44$
   B vs. $C : t = 2.40 (P < 0.05)$
   Pooled error results in different values of t. This is appropriate if F is significant and/or tests are proposed a priori (use pooled error, i.e., WMS).

4. (a) $H_0: \mu_1 = \mu_2 = \mu_3 = \mu_4; H_a: \mu_i \neq \mu_j; \alpha = 0.05$
   (b) Fixed
   (c)

Source	d.f.	MS	F
Between analysts	3	2.89	5.78 (<0.05)
Within analysts	8	0.50	

   $LSD = 2.31\sqrt{0.5(2/3)} = 1.33$
   A differs from B, C, and D; B differs from C and D

(d) Tukey test: $4.53\sqrt{0.5(3)} = 1.85$; only analysts A and C differ at 5% level Scheffé test: $\sqrt{0.5(3)4.07(1/3 + 1/3)} = 2.02$; none of the analysts differ at 5% level

5. $H_0: \mu_i = \mu_j; H_a: \mu_i \neq \mu_j; \alpha = 5\%$

(a)

Source	d.f.	MS	F
Between clinics	6	16.425	8.21 ($P < 0.05$)
Within clinics	13	2	

(b) Yes

(c) Fisher's LSD method (for example) at the 5% level
$LSD = 2.16\sqrt{2(1/3 + 1/3)} = 2.49$
Clinic 1 $\neq$ clinics 2, 5, 7; clinic 2 $\neq$ clinics 3, 5, 6; clinic 3 $\neq$ clinics 5, 7; clinic 4 $\neq$ clinic 5; clinic 5 $\neq$ clinics 6, 7; clinic 6 $\neq$ clinic 7 For comparisons to clinic 7,
$LSD = 2.16\sqrt{2(1/3 + 1/2)} = 2.79$

6. (a) Drugs fixed; (b) Machines fixed; (c) formulations fixed; (d) Machines random; (e) Clusters chosen at random

7. $H_0: \mu_1 = \mu_2 = \mu_3; \alpha = 0.05$

Source	d.f.	MS	F
Between batches	2	115.2	10.26 ($P < 0.05$)
Within batches	12	11.24	

t test shows that batch 3 is different from batches 1 and 2; e.g., batch 1 vs. batch 3: $t_{12} = (20.33 - 11.8)/\sqrt{11.24(1/6 + 1/5)}$

8. (a)

Source	d.f.	MS	F
Row	5	1679.0	
Column	2	8.22	0.34 ($P > 0.05$)
Error	10	23.96	

(b)

Source	d.f.	MS	F
Row	5	52.99	
Column	2	26.06	5.37 ($P < 0.05$)
Error	10	4.86	($F_{2,10} = 4.10$ for $\alpha = 0.05$)

(c) Averages of drugs are: placebo $= -0.33$, drug 1 $= -3.67$, and drug 2 $= -4.17$. Tukey test: $3.88\sqrt{4.86/6} = 3.49$; therefore, drug 2 is different from placebo. Newman–Keuls test: Drugs 1 and 2 different from placebo ($P < 0.05$). Dunnett test: Drug 1 and drug 2 different from control ($P < 0.05$).

9. (a) If the six presses comprise all of the presses, the presses are fixed. Hours are fixed (i.e., each hour of the run is represented).

Source	d.f.	MS	F
Hour	4	11.95	6.76 ($P < 0.05$)
Presses	5	2.45	1.38 ($P > 0.05$)
Error	20	1.77	

(b) Presses are not significantly different (5% level)

(c) "Hours" are significantly different.

(d) Assume no interaction

(e) Use Tukey test: $4.23\sqrt{1.77/6} = 2.30$; hour 3 is significantly different from hours 1, 2, and 5.

10.

Source	d.f.	MS	F
Rows	2	7.06	2.05
Columns	2	16.89	4.91 ($P < 0.05$)
Interaction	4	3.03	0.88
Within	9	3.44	

($F_{2,9} = 4.26$ for significance at 5% level.)

"Presses" are significant. "Interaction" is not significant. Interaction means that differences between presses depend on the hour at which tablets are assayed.

11. Average results: $A = 2.90$, $B = 6.50$, $C = 6.07$

If "sites" are random, use CR as error term.

$5.04\sqrt{22.66/24} = 4.90$ (no significant differences).

If "sites fixed," use within error.

$3.4\sqrt{3.215/24} = 1.24$ A is lower than B and C)

12. ANOVA Table:

Source	d.f.	Sum-Squares	Mean Square
C	2	14.29167	7.145834
B	2	9.125	4.5625
Error	3	7.083334	2.361
Total (Adj)	7	30.5	

13. ANOVA Table:

Source	d.f.	Sum-Squares	Mean Square	F-Ratio	Prob > F
A (Method)	1	6.438E-04	7.438E-04	7.15	0.0369
Error	6	5.406E-04	9.010E-05		
Total (Adj)	7	1.184E-03			

Method average
1.9921655
2.974223
$P = 0.0366$ from ANCOVA

CHAPTER 9

1. ANOVA Table:

Source	d.f.	MS	F
Stearate	1	1.56	5.21
Mixing time	1	1.82	6.1
Stearate X mixing time	1	0.72	2.41

Mixing time and stearate are significant at 5% level. Interaction is not significant.

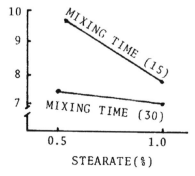

2.
Low starch, low stearate	Low starch, high stearate
0.475	0.487
0.421	0.426
Av. = 0.448	Av. = 0.4565

High starch − low starch = 0.4565 − 0.4480 = 0.0085

3. ANOVA:

Source	d.f.	MS	F
a	1	0.66	14.0*
b	1	0.06	1.3
ab	1	0.03	—
c	1	7.41	158**
ac	1	0.10	—
bc	1	3.25	69**
abc	1	0.01	—

*$P < 0.05$; **$P < 0.01$.

Error = (0.03 + 0.10 + 0.01)/3 = 0.047; d.f. = 3

(a) a, c, bc

(b)

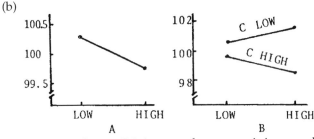

(c) When C is low, as B is increased, recovery is increased.

When C is high, as B is increased, recovery is decreased.

4. Synergism (or antagonism) would be evidenced by a significant AB interaction. If the effects are additive, we would expect an increase of 12 for the AB combination beyond placebo (4 from A and 8 from B). This is close to the observed increase of 14 (35 − 21) for AB. The combination of A and B work better than either one alone, but the evidence for synergism is not strong.

5. Weigh (1), ab, ac, bc: empty, a and b together, a and c together, b and c together.

Source	d.f.	MS	F
A	1	2014	21.3[a]
B	1	356	3.8
AB	1	14	0.2
C	1	45	0.5
AC	1	741	7.9[b]
BC	1	121	1.3
ABC	1	36	—
D	1	5704	60.5[a]
AD	1	114	1.2
BD	1	226	2.4
ABD	1	128	—
CD	1	0.02	0
ACD	1	10	—
BCD	1	10	—
ABCD	1	271	—
Total	15	9806	

Estimate of error = 94.3

[a]$P < 0.01$

[b]$P < 0.05$

AC interaction is significant: at low C, the A effect is 52.2 − 43.3; i.e., changing from low to high level of A has little effect when C is at the low level. At high C, the A effect is 62.4 − 26.4.

CHAPTER 10

1. 1.00, 1.11, 1.60, 1.64, 1.74, 1.80, 2.06, 2.16, 2.30, 2.34, 2.36, 2.57, 2.70, 2.90, 2.90, 2.99, 3.10, 3.12, 3.18, 3,66

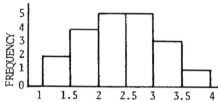

2. $\log Y = -0.127 + 1.068 \log X$
   $\log 47 = -0.127 + 1.068 \log X$
   $\log X = 1.685$
   $X = 48.4 \, \text{mg}$

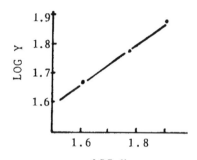

LOG X

3. $\bar{R} = 1.066, S = 0.281; (0.066)/(0.089) = 0.75$ (not significant at 5% level). The t test for log B − log A is identical except for sign as the t test for log A − log B. This example shows the problems of using ratios. The average of A/B is not (in general) the reciprocal of B/A.

4. $(62 − 54)/(62 − 47) = 8/15 = 0.533$. This is an outlier according to the Dixon test. We probably should not omit this value without further verification. The outlier could be due to analytical error and/or the presence of tablets with unusual high potency.

5. Winsorized, 50.7; using all values, 51.4.

6. $t = [2.8 − 0.6]/[1.732\sqrt{1/5 + 1/5}] = 2.01$.
   (Note the difference between the variances of the two groups.)
   Use a square-root transformation:
   Process 1: mean = 1.4363, s.d. = 0.960
   Process 2: mean = 0.6, s.d. = 0.548
   $t = [1.4363 − 0.6]/[0.782\sqrt{1/5 + 1/5}] = 1.69$

CHAPTER 11

1. (b) $$t = \sqrt{\frac{107.2 − (−3.05)}{1983.9(1/20 + 1/20)}} = 7.83 \, (t^2 = F)$$

2.

Source	d.f.	MS	F	
Subjects	11	5.19		
Treatment	1	0.04	0.005	Treatments are not significantly different.
Order	1	2.04	0.25	
Error	10	8.04		

3.

Source	d.f.	MS	F	
Subjects	11	16.41		
Treatments	1	155	13.19	$(P < 0.01)$
Order	1	177	9.96	$(P < 0.05)$
Error	10	11.75		

$$(22.3 - 17.3) \pm 2.23\sqrt{11.75(1/12 + 1/12)} = 5 \pm 3.12$$

Grizzle analysis: Residual effect $= \dfrac{(245)^2 + (230)^2}{12} - \dfrac{(475)^2}{24} = 9.375$;

within MS $= 17.11$; $F_{1,10} = 9.375/17.11 = 0.55$; not significant at 5% level

4. $\overline{A/B} = 1.334, S^2 = 0.238; t = (1.334 - 1.0)/\sqrt{0.238(1/12)} = 2.37; P < 0.05.$

5. $\log \overline{X} = 0.0954265$; antilog $= 1.246$; $S^2 = 0.0309$; $t = 1.88$ (not significant; assume no order effect); $0.0954 \pm 2.20\sqrt{0.031(1/12)} = -0.016$ to 0.207; antilogs: 0.96 to 1.61

6. Two-way ANOVAS:

	Placebo		Active		Combined ANOVA	
Source	d.f.	MS	d.f.	MS	d.f.	MS
Patients	5	2.866	5	2.742	10	2.804
Weeks	3	1.055	3	7.264	3	3.91
Patients × weeks	15	0.956	15	0.897	30	0.926
Drugs	1...1					15.1875
Drugs × weeks	3...3					4.41

For "drugs," $F_{1,10} = 15.1875/2.804 = 5.416(P < 0.05)$; for "drugs × weeks," $F_{3,15} = 4.41/0.926 = 4.76 P < 0.05)$. From the accompanying plot and the F test for interaction, the active effect increases with time while the placebo is relatively constant.

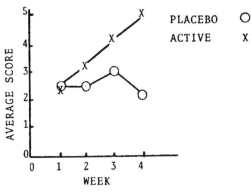

7. $N = 2(55/60)^2(1.96 + 1.28)^2 + 1 \approx 19$
8. $|-4.75 + 7.6|/(3.433\sqrt{1/8 + 1/9}) = 1.71(P > 0.05)$
10. Suppose that we start in column 5 in the Blocks of 6 section of Table 11.1. We can equate numbers 1 and 2 to Treatment A, 3 and 4 to Treatment B, and 5 and 6 to Treatment C. The assignments are as follows:

From Table 11.1

3	2	5	1	5
2	1	2	6	6
1	3	3	5	4
5	5	4	2	3
6	6	1	3	2
4	4	6	4	1

Subject	Treatment	Subject	Treatment	Subject	Treatment
1	B	13	C	25	C
2	A	14	A	26	C
3	A	15	B	27	B
4	C	16	B	28	B
5	C	17	A	29	A
6	B	18	C	30	A
7	A	19	A		
8	A	20	C		
9	B	21	C		
10	C	22	A		
11	C	23	B		
12	B	24	B		

11. $A = 3$, $B = 2$. The effect of A in Period 2 = 3 (Direct effect) + 2 (carryover) + 3 (period) = 8. The effect of B in Period 2 is $2 + 2 + 3 = 7$. $A - B = 8 - 7 = 1$.

12. $N = 2(0.8)(1 - 0.8)\{(1.65 + 1.28)/0.16\}^2 = 108$ per group.

CHAPTER 12

1. $\bar{X} = 9.95$; limits are $9.95 \pm 1.88(0.10) = 9.95 \pm 0.19$
   $\bar{R} = 0.10$; for $N = 2$, limits are 0 to $(3.27)(0.10) = 0.33$

2. $\sigma = \sqrt{0.02(0.98)/1000} = 0.004427$; $3\sigma = 0.0133$; $0.02 \pm 3\sigma = 0.0067$ to 0.0333

3. $\bar{X}$ control chart is centered at 47.6 with limits $47.6 \pm 1.02(1.2) = 47.6 \pm 1.22$. R chart has a target of 1.2 with lower limit of 0 and upper limit of $2.57(1.2) = 3.1$ (see Table IV.10).

4. $P = 1\%$; accept if 0 or 1 rejects. Probability 0 rejects $= 0.99^{100} = 0.366$.
   $P(1 \text{ reject}) = 0.370$; $P(\text{batch rejected}) = 1 - 0.736 = 0.264$.

5. $\bar{X} = 10.02$; limits : $10.02 \pm 0.31(0.38) = 10.02 \pm 0.12$
   $R = 0.38$; limits: lower is $0.22(0.38) = 0.08$; upper is $1.78(0.38) = 0.68$
   Many means are out of limits. Either find cause or, if not possible, use moving average if means are well within official limits.

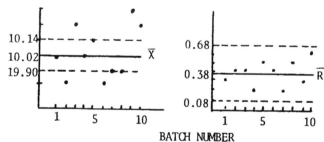

BATCH NUMBER

6. $p = 50/100,000 = 0.005 = $ probability of reject; $q = 0.9995$; therefore, probability of passing batch $= 0.9995^{100} = 0.951$

7.

Source	d.f.	MS
Between	3	483.3
Within	8	87.83

Between-analyst component $= (483.3 - 87.83)/3 = 131.8$; within-analyst component $= 87.83$

Three analysts perform four essays:
$$S^2 = \frac{4(131.8) + 87.83}{12} = 51.3$$

Four analysts perform two assays:
$$S^2 = \frac{2(131.8) + 87.83}{8} = 43.9$$

Cost is $24 for both procedures. The latter procedure (four analysts) is more precise.

8. Limits are $399.6 \pm 1.02(3.48) = 399.6 \pm 3.55$

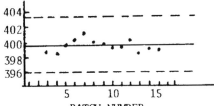

BATCH NUMBER

9. $\bar{X} = 10.21, \bar{R} = 0.24, \bar{S} = \sqrt{0.052} = 0.23$

   Limits for $\bar{X} = 10.21 \pm 1.88(0.24) = 10.21 \pm 0.45$

   Limits for $\bar{R} = 0$ to $3.27(0.24) = 0$ to 0.78

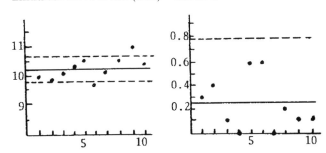

BATCH NUMBER

10. $N = 4$; limit $= 2.28, \bar{R} = 2.28(12.5) = 28.5$ (0 is lower limit)
11. 6.25 vs. 3.8
13. (a) $90 + 1.71(0.3/2 + 0.5 + 4/20)^{1/2} = 91.58$

    $110 - 1.71(0.3/2 + 0.5 + 4/20)^{1/2} = 108.42$

    (b) $90 + 1.71(0.3 + 0.5 + [4/20]/2)^{1/2} = 91.62$

    $110 - 1.71(0.3 + 0.5 + [4/20]/2)^{1/2} = 108.38$

    Consider the advantages and disadvantages of different kinds of replication.

CHAPTER 13
1. $\bar{R} = 3.375$; upper limit $= 3.375 \times 2.57 = 8.7$
2. $\bar{X} = 106.5$; $\bar{R} = 5.4$; limits $= 106.5 \pm 1.02(5.4) = 106.5 \pm 5.5$
3. $S_1^2 = 15.67$; $S_2^2 = 2.83$; $S_3^2 = 3.94$

   $\bar{S}^2 = 7.48$

   $X_2^2 = 72.440 - 61.958 = 10.482 (P < 0.05)$
4. $\bar{R} = 2.38$; upper limit $= 2.38 \times 3.27 = 7.78$
5. $\bar{X} = 102.4$; $\bar{R} = 3.3$; limits $= 102.4 \pm 3(3.3)/1.128 = 102.4 \pm 8.8$

CHAPTER 15
1. $t = 0.583/0.6685\sqrt{1/12} = 3.02$; $P < 0.05$; parametric t test shows significance
2. (a) 9 of 12 comparisons are higher for B: not significant

   (b) (b) $t = 0.5/(0.61\sqrt{1/12}) = 2.83$; $P < 0.05$
3. Σ Ranks for $A = 11$(or 67); Σ ranks for $B = 67$; $N = 12$, $\alpha = 0.05$

   $Z = \dfrac{|67 - 12(13)/4|}{\sqrt{12(12.5)(13)/12}} = 2.20$; $P < 0.05$
4. Use the Wilcoxon signed-rank test. $\sum R = 13.5$ (or 22.5); $P > 0.05$ (not significant).
5. Use the Wilcoxon rank sum test.

   $Z = \dfrac{|74 - 10(10 + 10 + 1)/2|}{\sqrt{10(10(10 + 10 + 1)/12}} = 2.34$; $P < 0.05$

   $t = \dfrac{4.35 - 2.09}{\sqrt{3.816(1/10 + 1/10)}} = 2.59$; $P < 0.05$

6. Use the Kruskal-Wallis test. Sum of ranks = 63.5, 40.5, and 16.

$$\chi_2^2 = \frac{12}{15(16)}(1133.3) - 3(15+1) = 8.67; \; P < 0.05$$

There is a significant difference (batch 3 has lowest dissolution).

7. Sum of ranks = 31, 21.5, and 19.5.

$$\chi_2^2 = \frac{12}{36(3+1)}(31^2 + 21.5^2 + 19.5^2) - 3(12)(4) = 6.29; \; P < 0.05$$

The standard has the highest C_{max} (standard is greater than B, $P < 0.05$; see Ref. 2).

8.

	0	1	2	Total
A	50(38.9)	50(61.1)	75(75)	175
B	20(31.1)	60(48.9)	60(60)	140
Total	70	110	135	315

$X_2^2 = 11.69$; $P < 0.01$. The distribution of scores for A and B is different.

9.

		Capping		
		Yes	No	Total
	Yes	13(1.8)	45(56.2)	58
Specks	No	18(29.2)	924(912.8)	942
	Total	31	969	1000

(a) $S_1^2 = 73.7$ (corrected); $P \ll 0.01$; not independent

(b) $Z = \dfrac{|0.714 - 0.5| - 1/126}{\sqrt{0.5(0.5)/63}} = 3.27; \; P < 0.01$

The difference is significant at the 1% level.

10. The probability of the fourfold table is 0.0304:

$$\frac{12!5!14!21!}{0!12!5!9!26!} = 0.0304$$

The only least likely table has five tumors in the controls and zero tumors in the treated group. This table has a probability of 0.012.4. Therefore, the probability of the given table + more unlikely tables is $0.0304 + 0.01204 = 0.0421$. The χ^2 test (corrected) is equal to 3.98, which is equal to $P = 0.0460$.

11. The median is 303.25. There are nine runs. According to Table IV.14, fewer than 6 or more than 15 runs are needed for significance at the 5% level. Therefore, the sequence is not significantly nonrandom for both one- and two-sided tests.

14. $\chi^2 = 5.44 (P < 0.05)$

15.

Source	d.f.	Sum-Squares	Mean Square
A (Treatment)	1	2.485E-04	2.485E-04
B (Subject)	11	.637813	.057983
Error	11	1.138684	.1035167
Total (Adj)	23	1.776746	

90%C.I. : $(4.9615 - 4.9551) \pm 1.8\sqrt{0.1035167/6} = 0.0064 \pm 0.2364 = 0.795$ to 1.275

16. Sequence 1: $P_1 + T_1 - P_2 - T_2$
Sequence 2: $P_1 + T_2 - P_2 - T_1$
Seq. 1 − Seq. 2 = $2(T_P - T_2)$

17. Sequence: $\dfrac{|73 - 8(8+9+1)/2|}{\sqrt{8 \times 9(8+9+1)/12}} = 0.096 P > 0.5$

Period: $\dfrac{|54 - 8(8+9+1)/2|}{\sqrt{8 \times 9(8+9+1)/12}} = 1.73 P < 0.10$

18. Answer: $p = 0.012$

Source	d.f.	Sum-Squares	Mean Square	F-Ratio	Prob > F
A (Press)	3	.9815	.3271667	5.64	0.012
B (Formula)	4	5.288004	1.322001		
AB	12	.6959966	5.79E-02		
Total (Adj)	19	6.965501			

19.

Source	d.f.	Sum-Squares	Mean Square
Formulation	4	0	0
Press	3	156.3	52.1
Error	12	113.7	9.475
Total (Adj)	19	270	

$$\text{LSD}_{\bar{x}} = 2.18\sqrt{9.475\left(\frac{1}{5} + \frac{1}{5}\right)} = 4.244$$
$$\text{Sum} = 5 \times 4.244 = 21.22$$

CHAPTER 16

1. (a) $'X_1 = 0; 'X_2 = 1; 'X_3 = 0; Y = 10.725 + 2.225 = 12.95$
    (b) $'X_1 = 1; 'X_2 = 1; 'X_3 = 0.6; Y = 15.36$
2. See Eq. (16.4). $'X_1 = (1 - 1)/1 = 0; 'X_2 = (0.5 - 0.5)/0.5 = 0; 'X_3 = (2.5 - 2.5)/2.5 = 0$
3. $Y = (9.7 + 7.2 + 8.4 + 4.1)/4 + (-9.7 + 7.2 - 8.4 + 4.1)X_1/4 + (-9.7 - 7.2 + 8.4 + 4.1)X_2/$
    $4 + (9.7 - 7.2 - 8.4 + 4.1)X_1 X_2/4 = 7.35 - 1.7X_1 - 1.1X_2 - 0.45X_1 X_2$
4. $A' = (8.75 - 7.5)/2.5 = 0.5; B' = (100 - 75)/25 = 1.0; Y = 7.35 - 1.7(0.5) - 1.1(1) - 0.45(0.5) =$
    5.725
5. $Y = 19.75 + 4.25(St) + 3.25(M) - 2.25(M)(St)$. Note: M and St are coded. One possibility is
    $(St) = -0.23$ and $(M) = -1$. This is equivalent to 15 min of mixing and 0.539% stearate, for
    a 15-min dissolution time.
6. $Y + 10A + 15B + 30AB$; let $B = 1 - A.Y = 10A + 15(1 - A) + 30A(1 - A) = -30A^2 + 25A + 15$;
    $dY/dA = -60A + 25 = 0; A = 0.417 = 41.7\%$
7. (a) $Y = 292A + 5.6B + 50.4C - 492.8AB - 186.8AC - 49.6BC + 54.6ABC$
    (b) 100% B is 5.6 min. Combinations between 50 and 100% B and 0 and 50% A may give a
        fast dissolution (e.g., 0.6 of B and 0.4 of A = less than 2 min).
    (c) There are many combinations. For example, 35% of A and 65% of C results in a disso-
        lution of approximately 92 min.

Index